Hydrogen Sulfide

Wiley Series in Drug Discovery and Development

Binghe Wang, Series Editor

Drug Delivery: Principles and Applications
Edited by Binghe Wang, Teruna Siahaan, and Richard A. Soltero

Computer Applications in Pharmaceutical Research and Development
Edited by Sean Ekins

Glycogen Synthase Kinase-3 (GSK-3) and Its Inhibitors: Drug Discovery and Development
Edited by Ana Martinez, Ana Castro, and Miguel Medina

Aminoglycoside Antibiotics: From Chemical Biology to Drug Discovery
Edited by Dev P. Arya

Drug Transporters: Molecular Characterization and Role in Drug Disposition
Edited by Guofeng You and Marilyn E. Morris

Drug-Drug Interactions in Pharmaceutical Development
Edited by Albert P. Li

Dopamine Transporters: Chemistry, Biology, and Pharmacology
Edited by Mark L. Trudell and Sari Izenwasser

Carbohydrate-Based Vaccines and Immunotherapies
Edited by Zhongwu Guo and Geert-Jan Boons

ABC Transporters and Multidrug Resistance
Edited by Ahcène Boumendjel, Jean Boutonnat, and Jacques Robert

Drug Design of Zinc-Enzyme Inhibitors: Functional, Structural, and Disease Applications
Edited by Claudiu T. Supuran and Jean-Yves Winum

Kinase Inhibitor Drugs
Edited by Rongshi Li and Jeffrey A. Stafford

Evaluation of Drug Candidates for Preclinical Development: Pharmacokinetics, Metabolism, Pharmaceutics, and Toxicology
Edited by Chao Han, Charles B. Davis and Binghe Wang

HIV-1 Integrase: Mechanism and Inhibitor Design
Edited by Nouri Neamati

Carbohydrate Recognition: Biological Problems, Methods, and Applications
Edited by Binghe Wang and Geert-Jan Boons

Chemosensors: Principles, Strategies, and Applications
Edited by Binghe Wang and Eric V. Anslyn

Medicinal Chemistry of Nucleic Acids
Edited by Li He Zhang, Zhen Xi, and Jyoti Chattopadhyaya

Plant Bioactives and Drug Discovery: Principles, Practice, and Perspectives
Edited by Valdir Cechinel Filho

Dendrimer-Based Drug Delivery Systems: From Theory to Practice
Edited by Yiyun Cheng

Cyclic-Nucleotide Phosphodiesterases in the Central Nervous System: From Biology to Drug Discovery
Edited by Nicholas J. Brandon and Anthony R. West

Drug Transporters: Molecular Characterization and Role in Drug Disposition, 2nd Edition
Edited by Guofeng You and Marilyn E. Morris

Drug Delivery: Principles and Applications, 2nd Edition
Edited by Binghe Wang, Longqin Hu, and Teruna J. Siahaan

Carbon Monoxide in Drug Discovery: Basics, Pharmacology, and Therapeutic Potential
Edited by Binghe Wang and Leo E. Otterbein

Drug Transporters: Molecular Characterization and Role in Drug Disposition, 3rd Edition
Edited by Guofeng You and Marilyn E. Morris

Hydrogen Sulfide: Chemical Biology Basics, Detection Methods, Therapeutic Applications, and Case Studies, First Edition
Edited by Michael D. Pluth

Hydrogen Sulfide

Chemical Biology Basics, Detection Methods, Therapeutic Applications, and Case Studies

Edited by

Michael D. Pluth
Department of Chemistry & Biochemistry
University of Oregon
Eugene, OR, USA

WILEY

This edition first published 2023

Registered Office
John Wiley & Sons, Inc., 111 River Street, Hoboken, NJ 07030, USA

Editorial Office
111 River Street, Hoboken, NJ 07030, USA

For details of our global editorial offices, customer services, and more information about Wiley products visit us at www.wiley.com.

Wiley also publishes its books in a variety of electronic formats and by print-on-demand. Some content that appears in standard print versions of this book may not be available in other formats.

Library of Congress Cataloging-in-Publication Data

Names: Pluth, Michael D., editor.
Title: Hydrogen sulfide : chemical biology basics, detection methods, therapeutic applications, and case studies / edited by Michael D. Pluth.
Description: First edition. | Hoboken, NJ : Wiley, 2022. | Series: Wiley series in drug discovery and development
Identifiers: LCCN 2022018103 (print) | ISBN 9781119799870 (cloth) |
Subjects: LCSH: Hydrogen sulfide–Toxicology. | Hydrogen sulfide–Therapeutic use.
Classification: LCC RA1247.H8 H9325 2022 (print) | LCC RA1247.H8 (ebook) | DDC 615.9/1–dc23/eng/20220609
LC record available at https://lccn.loc.gov/2022018103
LC ebook record available at https://lccn.loc.gov/2022018104

Cover design by: Wiley
Cover image: Courtesy of Michael D. Pluth

Set in 9.5/12.5pt STIXTwoText by Straive, Chennai, India

Contents

Preface

The last two decades have seen significant advances in our understanding of how reactive sulfur species affect biological systems. Of such species, hydrogen sulfide (H_2S) has emerged as an important small molecule that now joins nitric oxide (NO) and carbon monoxide (CO) in a trio of "gasotransmitters" with broad impacts in various aspects of human health and disease. Why has Nature chosen to use these simple, reactive, typically toxic molecules to carry out complex and precise biological actions? The answer may be simple: In complex biological milieu, the heightened reactivity and affinity for specific targets, often manifested as toxicity at higher concentrations, are needed for the signal to rise above the noise. NO, CO, H_2S, and emerging gasotransmitters all fit this paradigm – they are reactive small molecules that were initially recognized as toxic gasses but later found to be produced endogenously and exert biological action.

Growing from early initial observations of the potential role of H_2S in neuromodulation, investigations into H_2S now span diverse disciplines including chemistry, biology, physiology, and other adjacent fields. Although many biological pathways involving H_2S are complex, all are governed by fundamental chemical interactions between reactive sulfur species and other molecular entities. Adding to this complexity, sulfur has a wide array of available redox states (−2 to +6), and H_2S is the only canonical gasotransmitter with ionizable hydrogens. These properties provide simple pathways to modulate charge, lipophilicity, redox potential, and nucleophilicity – all of which contribute to the rich and rapidly expanding chemical biology of H_2S. Understanding these systems and making connections to new, yet undiscovered pathways and mechanisms of action requires knowledge spanning different disciplines. One goal of this book is to provide a foundation for understanding the fundamental chemical biology of H_2S to help bridge this gap between adjacent fields, while also highlighting selected applications and opportunities poised to advance this multidisciplinary area.

This book is divided into four main sections that cover key aspects of H_2S chemical biology. The first section (Chapters 1–5) focuses on general concepts related to

H_2S and interactions with other reactive species in biological systems. These are the fundamentals – the key points needed to understand the complex interactions between H_2S and other reactive biomolecules. The second section (Chapters 6–9) focuses on methods of measuring and detecting H_2S in complex biological environments. These summarize key advances in available tools to ask the "where, when, and how much" type questions. The third section (Chapters 10–15) focuses on methods for delivering H_2S in biological environments. These summarize key advances in available tools to modulate H_2S levels in biological environments. The fourth and final section (Chapters 16–19) includes selected case studies on the role of H_2S and H_2S delivery related to human health. These highlight key opportunities for applications of H_2S action in complex environments.

As the complex field of H_2S chemical biology continues to advance and evolve, I hope that this book will provide a useful resource for students, researchers, and other professionals alike and help to inspire future work in this rapidly growing area.

Michael D. Pluth
Oregon, USA
December, 2021

List of Contributors

Beatriz Alvarez
Laboratorio de Enzimología, Instituto de Química Biológica
Facultad de Ciencias, Universidad de la República
Montevideo
Uruguay

and

Centro de Investigaciones Biomédicas (CEINBIO)
Universidad de la República
Montevideo
Uruguay

Dayana Benchoam
Laboratorio de Enzimología, Instituto de Química Biológica
Facultad de Ciencias, Universidad de la República
Montevideo
Uruguay

and

Centro de Investigaciones Biomédicas (CEINBIO)
Universidad de la República
Montevideo
Uruguay

and

Graduate Program in Chemistry
Facultad de Química, Universidad de la República
Montevideo
Uruguay

Vincenzo Brancaleone
Department of Science
University of Basilicata
Potenza
Italy

Brock Brummett
Department of Chemistry
Brown University
Providence, RI
USA

Mariarosaria Bucci
Department of Pharmacy, School of Medicine
University of Naples Federico II
Naples
Italy

Giuseppe Cirino
Department of Pharmacy, School of Pharmacy
University of Naples Federico II
Naples
Italy

Valentina Citi
Department of Pharmacy
University of Pisa
Pisa
Italy

Miriam M. Cortese-Krott
Myocardial Infarction Research Laboratory, Department of Cardiology, Pneumology, and Angiology, Medical Faculty
Heinrich Heine University
Düsseldorf
Germany

and

Department of Physiology and Pharmacology
Karolinska Institutet
Stockholm
Sweden

Ernesto Cuevasanta
Laboratorio de Enzimología, Instituto de Química Biológica, Facultad de Ciencias
Universidad de la República
Montevideo
Uruguay

and

Centro de Investigaciones Biomédicas (CEINBIO)
Universidad de la República
Montevideo
Uruguay

and

Unidad de Bioquímica Analítica, Centro de Investigaciones Nucleares, Facultad de Ciencias
Universidad de la República
Montevideo
Uruguay

Jon M. Fukuto
Department of Chemistry
Johns Hopkins University
Baltimore, MD
USA

and

Department of Chemistry
Sonoma State University
Rohnert Park, CA
USA

Annie K. Gilbert
Department of Chemistry and Biochemistry
Institute of Molecular Biology
Knight Campus for Accelerating Scientific Impact, Materials Science Institute, University of Oregon, Eugene
Oregon, 97403
USA

Rynne A. Hankins
Department of Chemistry
Wake Forest University
Winston-Salem, NC
USA

Christopher Hine
Department of Cardiovascular and Metabolic Sciences
Cleveland Clinic Lerner Research Institute
Cleveland, OH
USA

Balakuntalam S. Kasinath
Department of Medicine, Center for Renal Precision Medicine
University of Texas Health
San Antonio, TX
USA

and

Barshop Institute for Longevity and Aging Studies
University of Texas Health
San Antonio, TX
USA

and

Geriatric Research, Education & Clinical Center
South Texas Veterans Health Care System
San Antonio, TX
USA

Christopher G. Kevil
Departments of Pathology and Translational Pathobiology
Molecular and Cellular Physiology and Cell Biology and Anatomy, LSU Health –Shreveport
Shreveport, LA
USA

Hideo Kimura
Department of Pharmacology, Faculty of Pharmaceutical Science
Sanyo-Onoda City University
Sanyo-Onoda, Yamaguchi
Japan

Petr Klán
Department of Chemistry, Faculty of Science
Masaryk University
Brno
Czech Republic

and

RECETOX, Faculty of Science
Masaryk University
Brno
Czech Republic

Yannie Lam
Department of Chemistry
Brown University
Providence, RI
USA

Hak Joo Lee
Department of Medicine, Center for Renal Precision Medicine
University of Texas Health
San Antonio, TX
USA

Zhao Li
Department of Chemistry
Virginia Polytechnic Institute and State University
Blacksburg, VA
USA

and

Macromolecules Innovation Institute
Virginia Polytechnic Institute and State University
Blacksburg, VA
USA

Christopher Link
Department of Cardiovascular and Metabolic Sciences
Cleveland Clinic Lerner Research Institute
Cleveland, OH
USA

Natalia Llarena
Department of Cardiovascular and Metabolic Sciences
Cleveland Clinic Lerner Research Institute
Cleveland, OH
USA

and

Department of Reproductive Endocrinology and Infertility
Cleveland Clinic Women's Health Institute
Cleveland, OH
USA

John C. Lukesh III
Department of Chemistry
Wake Forest University
Winston-Salem, NC
USA

Alma Martelli
Department of Pharmacy
University of Pisa
Pisa
Italy

John B. Matson
Department of Chemistry
Virginia Polytechnic Institute and State University
Blacksburg, VA
USA

and

Macromolecules Innovation Institute
Virginia Polytechnic Institute and State University
Blacksburg, VA
USA

Caitlin McCartney
Department of Chemistry
Brown University
Providence, RI
USA

Matías N. Möller
Centro de Investigaciones Biomédicas (CEINBIO)
Universidad de la República
Montevideo
Uruguay

and

Laboratorio de Fisicoquímica Biológica, Instituto de Química Biológica, Facultad de Ciencias
Universidad de la República
Montevideo
Uruguay

Bindu D. Paul
Department of Pharmacology and Molecular Sciences
Johns Hopkins University School of Medicine
Baltimore, MD
USA

and

Department of Psychiatry and Behavioral Sciences
Johns Hopkins University School of Medicine
Baltimore, MD
USA

and

The Solomon H. Snyder Department of Neuroscience
Johns Hopkins University School of Medicine
Baltimore, MD
USA

Claudio Pellecchia
Dipartimento di Chimica e Biologia
Università degli Studi di Salerno
Fisciano, SA
Italy

Eugenia Piragine
Department of Pharmacy
University of Pisa
Pisa
Italy

Michael D. Pluth
Department of Chemistry and Biochemistry
Institute of Molecular Biology
Knight Campus for Accelerating Scientific Impact, Materials Science Institute, University of Oregon, Eugene Oregon, 97403
USA

Geat Ramush
Department of Chemistry
Brown University
Providence, RI
USA

Xinggui Shen
Departments of Pathology and Translational Pathobiology, LSU Health –Shreveport
Shreveport, LA
USA

Tomáš Slanina
Institute of Organic Chemistry and Biochemistry of the Czech Academy of Sciences
Prague
Czech Republic

Ellen H. Speers
LSU Health–Shreveport
Shreveport, LA
USA

Peter Štacko
Department of Chemistry
University of Zurich
Zurich
Switzerland

Maria Strianese
Dipartimento di Chimica e Biologia
Università degli Studi di Salerno
Fisciano, SA
Italy

Sarah N. Swilley-Sanchez
Department of Chemistry
Virginia Polytechnic Institute and State University
Blacksburg, VA
USA

and

Macromolecules Innovation Institute
Virginia Polytechnic Institute and State University
Blacksburg, VA
USA

Zachary J. Tonzetich
Department of Chemistry
University of Texas at San Antonio
San Antonio, TX
USA

Binghe Wang
Department of Chemistry and Center for Diagnostics and Therapeutics
Georgia State University
Atlanta, GA
USA

Yingying Wang
Department of Chemistry
Brown University
Providence, RI
USA

Zhen Xi
Department of Chemical Biology, State Key Laboratory of Elemento-Organic Chemistry
College of Chemistry, National Pesticide Engineering Research Center
Nankai University
Tianjin
China

Ming Xian
Department of Chemistry
Brown University
Providence, RI
USA

Jie Yang
Department of Cardiovascular and Metabolic Sciences
Cleveland Clinic Lerner Research Institute
Cleveland, OH
USA

Long Yi
Beijing University of Chemical Technology (BUCT)
College of Chemical Engineering
Beijing
China

Bingchen Yu
Department of Chemistry and Center for Diagnostics and Therapeutics
Georgia State University
Atlanta, GA
USA

Zhengnan Yuan
Department of Chemistry and Center for Diagnostics and Therapeutics
Georgia State University
Atlanta, GA
USA

Aili Zhang
Department of Cardiovascular and Metabolic Sciences
Cleveland Clinic Lerner Research Institute
Cleveland, OH, USA

1

Fundamental and Biologically Relevant Chemistry of H_2S and Related Species

Jon M. Fukuto

Department of Chemistry, Johns Hopkins University, Baltimore, MD, USA
Department of Chemistry, Sonoma State University, Rohnert Park, CA, USA

List of Abbreviations

RSH	Thiols
Cys-SH	Cysteine
Cys-SSH	cysteine hydropersulfide
Cys-SS-Cys	cystine
GSH	glutathione
GSSH	glutathione hydropersulfide
GSSG	oxidized glutathione
HSAB	hard-soft acid-base
HOMO	highest occupied molecular orbitals
LUMO	lowest unoccupied molecular orbitals
RSSR	disulfides
BDE	bond dissociation energy
RSSH	hydropersulfide
CcO	cytochrome coxidase
HbFeIII	ferric hemoglobin
HbFeII	ferrous hemoglobin
MbFeIII	ferric myoglobin
MbFeII	ferrous myoglobin
RS	alkylthiyl
MPO	myeloperoxidase
RSOH	sulfenic acid

RSNO	S-nitrosothiol
TEMPO	4-hydroxy-2,2,6,6-tetramethylpiperidin-1-yl)oxyl
RSS	perthiyl radical
Sox	sulfur-oxidizing enzyme

1.1 Introduction

It is now evident that several endogenously synthesized di- and tri-atomic molecules are important physiological bioregulators and/or signaling species (for a review of these molecules, see [1–3]). Nitric oxide (NO), carbon monoxide (CO), and hydrogen sulfide (H_2S) are all biosynthesized and reportedly possess important physiological signaling and regulatory functions. Since these molecules all exist as gasses at room temperature and pressure they have been grouped together under the term "gasotransmitters," although they are solutes in biological media and not gases when they are acting physiologically. Dioxygen (O_2), although not necessarily endogenously biosynthesized, can also be included in this group since it too (along with derived species) has biological signaling functions that are distinct from its role in cellular respiration. Importantly, as a group these signaling molecules likely represent an integrated signaling "web" due to the fact that they can share biological targets, can be involved in each other's biosynthesis, and for some even react with each other as either the parent molecule or via derived species. Thus, for a complete understanding of the chemical biology, mechanism of action and physiological functions of these signaling molecules it is important to consider them as an integrated group and not necessarily as individual species (e.g. [2, 4]). Among these molecules (i.e. NO, CO, O_2, and H_2S), H_2S is distinct in that it has chemical properties and general reactivity not found in the others (*vide infra*). Clearly, the chemical biology is best defined and established for NO and, of course, O_2. The chemical biology of H_2S is much less understood and is therefore the focus of this chapter.

1.2 The Chemical Biology of H_2S

As is the case with all biological effector species, the biological activity and/or utility of H_2S are due to its chemical properties and chemical interactions with biological targets. In the context of this chapter, the term "chemical biology" refers to the chemical properties and reactivity of a species that are responsible for its biological actions. Therefore, an exhaustive and comprehensive discussion of all H_2S chemistry is unwarranted and only the chemistry that is thought to be biologically relevant covered herein.

1.2.1 Basic Chemical Properties of H_2S

Unlike the other small-molecule signaling species (NO, CO, and O_2), H_2S is ionizable. The pK_a's of H_2S and HS^- are 6.8 and 14, respectively. This indicates that at physiological pH (7.4), H_2S is approximately 80% ionized to HS^- with negligible amounts of the dianion S^{2-}. H_2S is the simplest of all thiols (RSH) and yet is somewhat distinct from them in that it is inherently more acidic. For example, the pK_a values for most biologically relevant thiols (e.g. cysteine [Cys-SH], glutathione [GSH]), free in solution are approximately 8.2. Thus, while most free thiols exist predominately as the protonated RSH species at physiological pH, H_2S is mostly ionized. To be sure, the pK_a of thiols in or attached to proteins can vary significantly due to effects associated with the protein environment (e.g. [5]). As a neutral molecule, H_2S favorably partitions into nonpolar (e.g. lipid) environments as indicated by reported partition coefficients between H_2O and organic solvents such as octanol or hexane of approximately 2 (indicating a twofold higher concentration of H_2S in a hydrophobic/lipid environment versus a water environment) [6]. Thus, H_2S can readily traverse membranes by simple diffusion. Of course, the charged, anionic HS^- will not partition into nonpolar environments and therefore will not readily traverse membranes by diffusion. Thus, the rate of H_2S diffusion will be highly dependent on the pH with acidic conditions favoring a more rapid diffusion. It is worthwhile to compare H_2S and H_2O with regard to their relative abilities to diffuse across membranes. The dipole moment for H_2S is 0.97, while that for H_2O is significantly greater at 1.84. Thus, H_2S is distinct from H_2O regarding its interaction with lipid, nonpolar environments. The diffusion of H_2O across membranes is greatly dependent on the nature/composition of the membrane and occurs with a permeability coefficient in the range of $2–50 \times 10^{-4}$ cm/s [7]. The permeability coefficient for H_2S was found to be significantly greater at >0.5–3 cm/s [6, 8] indicating a much greater ability for H_2S to diffuse across membranes compared to H_2O. Importantly, unlike H_2O diffusion across membranes, which can be greatly enhanced by aquaporins, H_2S diffusion is relatively rapid and not enhanced by aquaporins [8].

The predominant biological chemistry of H_2S is that of a Lewis base. That is, H_2S (or more likely HS^-) is an electron donor that reacts (forms bonds) with Lewis acids (i.e. electron acceptors). In organic chemistry terms, Lewis bases can be considered as nucleophiles and Lewis acids considered as electrophiles. Thus, HS^- is an electron-donating nucleophile capable of reacting with electron poor/deficient electrophilic centers. Of course, this is the chemical property that biological RSH species in general possess. One thing that sets H_2S apart from all other RSH molecules is that it possesses two dissociable and exchangeable protons and therefore can potentially react twice with Lewis acids. Thus, the generation of complexes with bridging sulfides that connect two, for example, metal centers (i.e. M–S–M, M = metal) is very commonly seen. It is important to note, however, that Lewis bases/nucleophiles are not all equivalent with respect to their reactivity toward Lewis acids/electrophiles. One simple and useful way of distinguishing and rationalizing nucleophilic/electrophilic reactivity preferences is to utilize the Hard–Soft Acid Base (HSAB) principles developed by Pearson [9, 10]. The idea of HSAB relies on the characterization of Lewis bases and Lewis acids (i.e. nucleophiles and electrophiles) as being either soft or hard and the understanding that hard bases prefer to react with hard acids and soft bases prefer to react with soft acids. Of course, there are some reactants that are considered as "borderline" which can react either way. Generally speaking, soft bases/nucleophiles are qualitatively viewed as species with high polarizability, lower electronegativity, easily oxidized, and possessing low-lying empty orbitals. Soft acids/electrophiles have an acceptor atom with low positive charge, large size, are polarizable, and possess easily excited outer electrons. On the other hand, hard bases have a donor atom with low polarizability, high electronegativity with high-energy/inaccessible empty orbitals. Hard acids possess an acceptor atom with high-positive charge, small size, and not polarizable. These reactivity preferences can also be reconciled on the basis of the energetics and nature of the reacting highest occupied molecular orbitals (HOMOs) of the base/nucleophile and the lowest unoccupied molecule orbitals (LUMOs) of the acid/electrophile (i.e. the Frontier orbitals) [11] with hard acid–hard base interactions viewed as primarily ionic in nature and soft acid–soft base interactions viewed as highly covalent in nature (due to favorable HOMO–LUMO overlap). Regardless, it is clear that RSH, RS^-, H_2S, and HS^- are all soft bases/nucleophiles which readily predicts reaction with soft acids/electrophiles. H_2S-reactive and biologically relevant soft electrophiles include heavy metals (e.g. Cd^{2+}, Hg^{2+}), low oxidation state metals (e.g. Cu^+), α–β unsaturated carbonyl compounds (e.g. acrolein, *N*-ethylmaleimide) and disulfides (RSSR), among others. Importantly, the principles of HSAB also correctly reconcile the fact that RSSR electrophiles are not readily cleaved by hard nucleophiles such as H_2O, HO^-, NH_3, or RNH_2.

1.2.2 H_2S Redox Chemistry

Numerous biologically relevant sulfur species with varying oxidation states have been reported and characterized (e.g. [12]) and of all these H_2S (as well as thiols) represent the fully reduced sulfur species (Figure 1.1).

Thiols, including H_2S, can potentially serve as one-electron reductants, generating the radical intermediates ($RS^•$/$HS^•$). However, RS^-/RSH and HS^-/H_2S are not powerful one-electron reductants as indicated by the relatively high reduction potential for the $RS^•$, H^+/RSH, and $HS^•$/HS^- couples of approximately 0.92 V [13, 14]. Indeed, $RS^•$ is known to be a potent oxidant generated at the active site

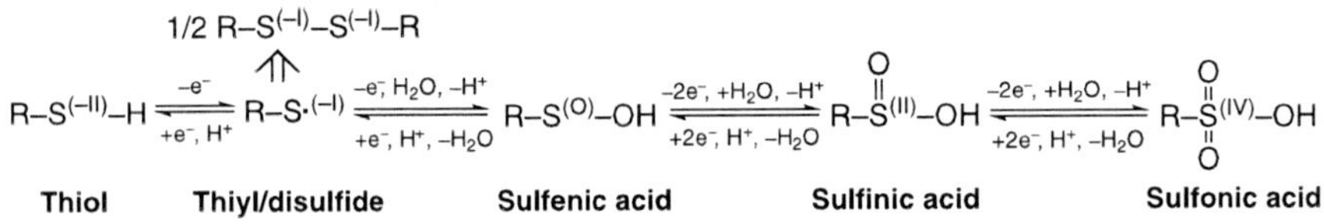

Figure 1.1 Redox relationship of biologically relevant sulfur species (R = alkyl, H). The oxidation states of the sulfur atoms indicated by superscripted numerals.

of the enzyme ribonucleotide reductase capable of abstracting a hydrogen atom from the ribose ring of DNA (e.g. [15]). Thus, the fact that $RS^{\cdot}$ and $HS^{\cdot}$ are relatively strong oxidants generated from RS^-/RSH or HS^-/H_2S oxidation indicates the weak one-electron reducing capabilities of these species. Another indicator of the relatively weak one-electron reducing capabilities of RSH/H_2S are the S—H bond dissociation energies (BDEs) of approximately 87–92 kcal/mol [16, 17]. For comparison, the O—H BDE for tocopherol (e.g. vitamin E) is significantly lower (79 kcal/mol), indicating a higher propensity to serve as a reducing H-atom donor.

1.2.3 Reactions of H_2S with Metals/Metalloproteins

As alluded to above, H_2S is a soft Lewis base and therefore reacts very readily with soft Lewis acid metals. Also, since H_2S has two exchangeable protons, it has the potential to form two bonds to metal species, either to a single metal or bridging two metals. A prime example of sulfide (S^{2-}) bridging can be found in iron–sulfur (FeS) cluster proteins. Although FeS clusters are very important and established biochemical entities involved in numerous functions including electron transfer processes, gene expression, oxygen sensing, regulation of enzymatic function and metal sensing, among many others (e.g. [18]), these will not be discussed further as there are numerous reviews of this topic available (e.g. [19, 20]). It is, however, worth mentioning that FeS biosynthesis involves the intermediacy of a hydropersulfide (RSSH), a functional group that will be discussed in some detail later.

The interactions of H_2S with heme proteins is among the most studied, due in part to the fact that disruption of heme protein function (e.g. cytochrome c oxidase, CcO) is thought to contribute significantly to H_2S-mediated toxicity (e.g. [21]) and that heme proteins can also be involved in physiological H_2S signaling, metabolism, and function [22]. The interactions of H_2S with heme proteins, such as CcO, are potentially complex (e.g. [23]) and highly concentration-dependent. At low levels, H_2S can actually serve as a source of electrons, supporting oxidative phosphorylation [24, 25], an effect supported by a functional model system for CcO whereby low levels of H_2S were capable of reducing both Fe and Cu sites of the model (as well as cytochrome c, which can donate electrons to CcO) [26]. However, at higher levels, H_2S competes with O_2 binding, possibly via binding to both the iron heme a_3 and Cu_B centers of CcO [22] inhibiting respiration.

A particularly interesting interaction between H_2S and a heme protein occurs with the specialized hemoglobin of *Lucina pectinata*, a mollusk that inhabits H_2S-rich environments. The hemoglobin of *L. pectinata* does not carry O_2, but rather carries H_2S as a source of electrons for symbiotic chemoautotrophic bacteria that inhabit its gills [27]. H_2S can bind to and react with hemoglobin/myoglobin from other species (*vide infra*) often leading to oxidized H_2S molecules. However, the stability of H_2S bound to the hemoglobin of *L. pectinata* appears to be the result of a protective, hydrophobic pocket around the bound H_2S that protects it against further chemistry [28–30]. With other globins such as hemoglobin or myoglobin, H_2S binding can result in further reactions. For example, H_2S binding to ferric hemoglobin ($HbFe^{III}$) or ferric myoglobin ($MbFe^{III}$) results in the catalytic oxidation of H_2S to either thiosulfate or polysulfide species (e.g. HS_nH, $n > 1$) via ferrous hemoglobin ($HbFe^{II}$) or ferrous myoglobin ($MbFe^{II}$) intermediates [31–33]. The reduction of Fe^{III} to Fe^{II} by HS^- has also been observed in model "picket-fence" porphyrins as well [34]. Finally, Jensen and Fago [35] have examined the interactions of H_2S with $HbFe^{III}$ and $MbFe^{III}$ at physiological concentrations and postulate that $HbFe^{III}$ can act as an H_2S carrier in blood, whereas $MbFe^{III}$ may be involved in H_2S metabolism. In the reduction of Fe^{III} to Fe^{II} by H_2S, a hydrothiyl ($HS^{\bullet}$) radical is thought to be generated, which can further react with H_2S/HS^- leading to the generation of polysulfide species (e.g. [31]). It is worth noting that the reaction of an alkylthiyl radical ($RS^{\bullet}$) with an alkylthiolate (RS^-) is reported to give a disulfide radical anion [36], which is a potent reducing agent [37] capable of, for example converting O_2 to superoxide (O_2^-). Whether the equivalent processes for $HS^{\bullet}$ and HS^- are involved in the above reactions remain to be determined. Interestingly, another heme protein that is not involved in O_2 transport or storage, myeloperoxidase (MPO), also reacts with H_2S in chemistry that involves coordination of H_2S/HS^- to a ferric heme followed by possible redox chemistry generating ferrous protein and presumably $HS^{\bullet}$ [38, 39]. As with the globins, $HS^{\bullet}$ generation can lead to polysulfur species. Peroxidatic chemistry of MPO also results in the oxidation of H_2S to $HS^{\bullet}$ further contributing to the formation of polysulfurs.

1.2.4 H_2S and Sulfheme Formation

As described above, the interaction of H_2S/HS^- with various hemeproteins results in redox chemistry that could result in $HS^{\bullet}$ intermediacy. Although $HS^{\bullet}$ in these systems can react with other sulfur molecules (giving, for example, thiosulfate or polysulfur species), it has also been proposed to react with the porphyrin ring of the heme protein, generating a modified heme species referred to as sulfheme (e.g. [40]). In sulfheme proteins, a sulfur atom is inserted into a pyrrole of the

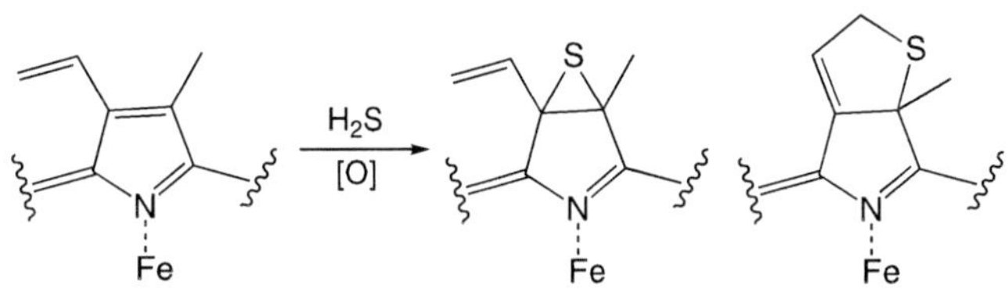

Figure 1.2 Some sulfheme structures generated via H_2S/oxidizing conditions. Source: Arbelo-Lopez et al. [41]/American Chemical Society.

porphyrin ring (pyrrole B) under oxidizing conditions leading to a change in the absorbance (giving an absorbance at approximately 620 nm). Several distinct sulfheme isomers have been characterized (Figure 1.2) (e.g. [41]).

Currently, the exact mechanism of sulfheme formation is not established. However, it appears clear that H_2S-mediated sulfheme formation requires oxidizing conditions, consistent with the idea that H_2S oxidation to $HS^{\cdot}$ may be a critical step. However, it should be noted that H_2S-independent sulfheme formation has been reported, possibly involving sulfenic acid (RSOH) or alkyl thiyl ($RS^{\cdot}$) species (e.g. [42]). Moreover, other oxidized H_2S species such as RSSH and/or H_2S_2 have been implicated in sulfheme formation as well [43]. Regardless, sulfheme formation can alter the function/biochemistry of the heme protein. For example, the O_2 affinities of myoglobin and hemoglobin are dramatically decreased when in the sulfheme form [44, 45] and sulfhemoglobin loses cooperative O_2 binding [46].

1.2.5 H_2S and Heavy Metals

As a soft Lewis base that preferentially reacts with soft Lewis acids, it is expected that H_2S will rapidly react with heavy metal toxins such as Cd^{2+} and mercury species (e.g. methyl mercury, MeHg), forming stable complexes (many of which are known to precipitate). Indeed, H_2S can be used as a reagent to precipitate and rid solutions of a variety of metal ions (e.g. [47]). It has been known for a long time that one strategy microbes utilize to gain resistance against heavy metal toxicity is via H_2S generation [48]. For example bacterial resistance against mercury [49] and cadmium [50] toxicity has been linked to H_2S-biosynthetic capacity. This is also true in mammalian cells where exogenous H_2S administration or increased H_2S biosynthesis resulted in the detoxification of MeHg via the formation of the relatively nontoxic bis-methylmercury sulfide ($(MeHg)_2S$) [51]. In all of these cases, it is thought that H_2S forms complexes with heavy metals (e.g. CdS, NiS, PbS) leading to nonreactive (and often insoluble) metal sulfides, thus protecting other biochemical thiol targets from being disrupted.

1.3 H_2S Reactions with Other Sulfur Species

As mentioned earlier, H_2S is the simplest of all thiols and as such can react in ways typical to other biological thiols. Some of the most prevalent and important reactions of biological thiols are those with other sulfur species. Thus, a discussion of this chemistry will be given.

1.3.1 Sulfane Sulfur

As will be addressed immediately below, when discussing biologically relevant H_2S reactions with other sulfur species, there is the possibility of generating sulfane sulfur-containing species. The term "sulfane sulfur" refers to a sulfur atom that is bonded only to other sulfur atoms [52]. For example, the bolded sulfur atoms of the following molecules are considered sulfane sulfurs since they are bonded only to other sulfur atoms: $^{-}$**S**–S(O)$_2$O^{-} (thiosulfate), RS**S**SR (dialkyl trisulfide), RS**SS**SR (dialkyl tetrasulfide), and RS**S**$^{-}$ (hydropersulfide anion). Elemental sulfur (mostly S_8) contains only sulfane sulfur atoms. In many molecules, the oxidation state of the sulfane sulfur is zero (S^0) (e.g. RS**S**^{0}SR, RS**S**^{0}**S**^{0}SR). However, this is not always the case since the sulfane sulfur atom of RS**S**$^{-}$/RS**S**H has the same formal oxidation state as the sulfur atoms of RSSR and is not in the S^0 state (the formal oxidation state of the sulfane sulfur atoms in these molecules is S^{1-}). Generally speaking, sulfane sulfurs possess reactivity (typically increased lability) that is distinct from other sulfurs and, therefore, this term is useful in drawing attention to this difference in chemical reactivity (e.g. [53]).

1.3.2 Generation of RSSH

Clearly, H_2S itself possesses important chemical properties/reactivity that serve vital biochemical functions. However, there are H_2S-derived species that may also be physiologically important signaling and/or effector molecules. As the simplest of all RSH species, H_2S has the potential to participate in sulfur redox/substitution chemistry. One of the potentially important reactions of RSH is with oxidized thiol species (e.g. RSSR, RSOH, *S*-nitrosothiols [RSNO]) (e.g. [54]). For example, RSH can react with RSOH or RSSR in substitution reactions (Reactions (1.1) and (1.2)).

$$\mathrm{RSH + R'SOH \rightarrow RSSR' + H_2O} \quad (1.1)$$

$$\mathrm{R'SH + RSSR \rightleftharpoons R'SSR + RSH} \quad (1.2)$$

These reactions are important in many biochemical processes. For example, Reaction (1.1) is a key step in the catalytic mechanism of peroxiredoxins (e.g. [55]), and Reaction (1.2) is crucial to the activity of glutaredoxins (e.g. [56]) and thioredoxins (e.g. [57]), just to name a few. Importantly, when the thiol is H_2S (i.e. R = H in Reactions (1.1) and (1.2)), a hydropersulfide (RSSH) is the reaction product (Reactions (1.3) and (1.4), note: reaction products shown as the anions since this will be the prominent species present at physiological pH) [58–61].

$$H_2S + RSOH \rightarrow RSS^- + H_2O + H^+ \tag{1.3}$$

$$H_2S + RSSR \rightleftharpoons RSS^- + RSH + H^+ \tag{1.4}$$

Thus, the presence of H_2S allows for the formation of a novel and potentially important sulfur species, RSSH. Therefore, due to the biological prevalence of oxidized thiol species (e.g. RSSR, RSOH) any discussion of H_2S chemical biology should include, at the very least, a discussion of RSSH. Moreover, a comparison of the chemical properties of RSH versus RSSH may provide insight into the possible functions of RSSH (i.e. the possible reasons, if any, RSH is converted to RSSH). This idea prompted many to begin to examine the prevalence, relevance, and chemical biology of RSSH species, in part, as a way to fully understand potential H_2S biology. Before continuing with the discussion of RSSH chemical biology, it is important to recognize that H_2S and RSSH are intimately linked via the Reaction (1.4) equilibrium. Thus, in a biological system where RSSR and RSH species are likely to exist, RSSH and H_2S will both be present via Reaction (1.4), making it difficult to delineate the individual biological activities associated with these chemically distinct species (*vide infra*). That is, in biological studies where H_2S is administered, there is a strong possibility that RSSH species are also formed and that the biological activities observed are due to these derived species (and vice versa). As will be discussed herein, one way to begin to try to delineate the activities associated with these individual species is to first examine and compare their chemical properties/reactivities and then postulate, based on these chemical differences, which species are most likely responsible for the observed effects. Thus, a brief comparison of the chemical properties of RSSH versus RSH and other RSH-derived species will be given.

1.3.3 RSH Versus RSSH Comparison

First and foremost, it needs to be understood that the conversion of RSH to RSSH represents an oxidation since the sulfur atoms of RSSH are at the same oxidation state as the sulfurs of RSSR (e.g. R = H in Figure 1.1). Thus, the RSH to RSSH conversion may be more prevalent in a biological system under oxidative stress conditions. This is clear by simply examining Reactions (1.3) and (1.4), where RSSH generation would be dependent on an oxidation of RSH to either RSSR or RSOH. Therefore, it is important to remember that RSSH is more likely to be prevalent in a biological system under oxidizing conditions, which are often times associated with cellular stress.

The chemical reactivities of the sulfhydryl (–SH) functional group important to the general functions of biological thiols and thiol proteins are (i) nucleophilicity (or Lewis base activity), (ii) redox activity (e.g. posttranslational redox modifications to RS$^•$, RSOH, RSSR, RSNO), and (iii) metal ligation (which can be related to the nucleophilicity/Lewis base properties) (e.g. [12, 62, 63]). An example of biological thiol nucleophilicity is found in the numerous Cys-SH proteases that utilize the –SH (or –S$^-$) functional group to react with electrophilic carbonyl centers leading to the cleavage of peptide bonds. This same reactivity is also found in cysteine phosphatases, which cleave phosphate groups from phosphorylated sites (e.g. serine, threonine, or tyrosine). In these cases, the reactive species is the more nucleophilic anionic Cys-S$^-$ (as opposed to the protonated and less nucleophilic Cys-SH). The redox activity of thiols is illustrated in part by the prevalence and importance of reversible disulfide formation in protein conformation/folding [64] and regulation [65]. That is, the conversion of protein Cys-SH to an intra- or inter-molecular Cys-SS-Cys is an important reaction for proper protein folding (especially during translation) and in the regulation of thiol protein activity. Thiol protein oxidation to sulfinic and sulfonic acids (Figure 1.1) as well as other sulfur-modifications (e.g. RSNO, sulfenylamide) may also be important to protein activity (for a review, see [5, 66]). One-electron oxidation of a protein Cys-SH to the corresponding thiyl radical, Cys-S$^•$, is also important to the actions of the enzyme ribonucleotide reductase since the Cys-S$^•$ species is vital to the mechanism of deoxyribose generation in DNA biosynthesis [15]. Finally, many metalloproteins bind metals via ligation of protein Cys-S$^-$. For example, the zinc-storage metallothionein proteins bind Zn^{2+} via numerous protein Cys-S$^-$ residues, as do DNA-binding Zn-finger proteins (e.g. [67]).

The diverse biochemical utility of Cys-SH is of course due to the chemical properties of the –SH functional group. As introduced previously, the existence and prevalence of H_2S (the topic of this chapter) in biological systems alludes to a corresponding relevance and prevalence of RSSH species due to Reactions (1.3) and (1.4). Thus, the biochemical utility of the hydropersulfide (–SSH) functional group becomes a matter of significant potential relevance. Importantly, numerous reports indicate that RSSH species, both in small molecules and in proteins, are highly prevalent. For example, Mustafa et al. [1] reported the existence of hydropersulfides in proteins and postulated that H_2S signaling can occur via RSSH generation and Ida et al. [68] discovered that hydropersulfides are also highly prevalent in small molecules such as cysteine- and glutathione-hydropersulfide (Cys-SSH and GSSH) as well as in numerous proteins. A somewhat astounding finding indicates that Cys-SSH can be made by cysteinyl t-RNA synthetase in an H_2S-independent process and incorporated into proteins [69], giving further support to the biological relevance of the –SSH functional group. To be sure, the possible biological relevance of RSSH species has been postulated long ago. For example, almost 60 years ago, it was reported that the enzyme cystathionine β-synthase (from the fungi *Neurospora*) could catalyze the conversion of cystine (Cys-SS-Cys) to Cys-SSH (in an H_2S-independent pathway) [70], a result predicted from earlier chemical studies [71]. Moreover, others postulated the existence of a Cys-SSH in the active forms of the molybdenum enzymes aldehyde oxidase [72] and xanthine oxidase [73] (*vide infra*). Although the origins of the RSSH species in these previous studies was not established (and may not be directly due to prior H_2S generation), it remains an important topic since, again, H_2S and RSSH are intimately linked (*vide infra*) and with regards to their potential biological activity should always be discussed together. Moreover, it is also important to note that some H_2S biosynthetic pathways have RSSH intermediates that are reductively converted to H_2S and the corresponding RSH, further linking RSSH and H_2S biology [74].

The general chemical reactivity profile of the –SSH group is similar to that listed above for the –SH group. Both RS^-/RSH and RSS^-/RSSH are nucleophilic, potentially reducing and capable of coordinating metals (e.g. [75, 76]). However, unlike RSH, RSSH can also be electrophilic (akin to RSSR). Thus, RSS^-/RSSH can be either nucleophilic or electrophilic (primarily a function of the protonation state – the deprotonated RSS^- will be nucleophilic/reducing and the protonated RSSH can be electrophilic/oxidizing), whereas RS^-/RSH are strictly nucleophilic/reducing and RSSR species can be electrophilic/oxidizing. Although RS^- and RSS^- possess similar general biochemical properties, there is a significant difference in the relative degrees of reactivity. In terms of nucleophilicity and reducing capability, RSS^-/RSSH is much more reactive compared to the corresponding RS^-/RSH. For example, Cuevasanta et al. [58] report that a protein-SSH (human serum albumin) is approximately threefold more reactive (pH-independent) than the corresponding protein-SH in reaction with a disulfide. Further examination has shown that RSSH (e.g. GSSH) is significantly more nucleophilic than a pK_a-matched RSH with rate constants that are 3–1670-times higher, depending on the nature of the electrophile [77]. It is also important to note that RSSH is more acidic compared to RSH by about 1.5–4 pK_a units [58, 77, 78]; therefore, there is always a higher percentage of the nucleophilic anionic species with RSSH compared to RSH. The increased nucleophilicity of RSS^- versus RS^- is further supported by theoretical assessments of the energies of the HOMOs (the filled orbital containing the nucleophilic electrons), which are higher in energy for

RSS^- compared to the corresponding RS^- by approximately 7–12 kcal/mol [58, 79]. The increased nucleophilicity of RSS^- versus RS^- can be explained on the basis of an apparent α-effect, which results from having an electron pair on an adjacent atom to the nucleophilic center leading to an increase in reactivity (e.g. [80]).

The reducing capabilities of RSS^-/RSSH are also significantly greater than the corresponding RS^-/RSH species. This can be readily shown by comparison of the one-electron reduction potentials for $RS^•$ and $RSS^•$ [81]. The reduction potentials for the $RS^•$, H^+/RSH, and $RSS^•/RSS^-$ couples (pH 7) are 0.96 and 0.68 V (calculated), respectively, consistent with the idea that RSS^- is a superior reductant compared to RS^-. This is evidenced experimentally as RSS^-/RSSH readily reduces numerous ferric heme proteins to the ferrous species, under conditions where the corresponding RSH does not [61, 82, 83] (Reaction (1.5)). The perthiyl radical ($RSS^•$) formed from these reactions can dimerize giving the corresponding tetrasulfide (RSSSSR) [82] (Reaction (1.6)) without performing any further chemistry (i.e. does not react further as an oxidant).

$$\mathrm{RSSH + Fe^{III}\text{-}heme \rightarrow RSS^{\bullet} + Fe^{II}\text{-}heme + H^+} \tag{1.5}$$

$$\mathrm{RSS^{\bullet} \rightarrow RSSSSR} \tag{1.6}$$

Interestingly, the ability of RSSH to reduce ferric heme proteins was actually reported much earlier in 1971 [84]. In this fascinating report, Massey and coworkers were examining a curious claim made previously by others that the addition of oxidized glutathione (GSSG) facilitated the GSH-mediated reduction of ferric cytochrome c (which is a very slow reaction), a process that was postulated to be involved in oxidative phosphorylation [85]. Since GSSG is an oxidant, it was difficult to surmise how GSSG could increase the reducing ability of GSH (which, as discussed above, is not a very good reductant). However, it was found that GSSG can be contaminated with the trisulfide GSSSG, which could react with GSH forming GSS^- (Reaction (1.7)).

$$\mathrm{GSSSG + GSH \rightarrow GSSG + GSS^- + H^+} \tag{1.7}$$

As indicated above, GSS^- is a far superior reductant compared to GS^-, and it was GSS^- that was the likely reductant in these systems. Moreover, considering the previous reports that the GSSG/GSH system (which actually produces GSS^- via contaminating GSSSG) can be involved in oxidative phosphorylation, this presents the possibility of a role for RSSH in mitochondrial activity (something that has actually been more recently postulated [69]). The reducing capabilities of organic hydropersulfides (e.g. benzylhydropersulfide, benzyl-SSH) have also been demonstrated. For example, benzyl-SSH was capable of reducing benzoquinone to the hydroquinone and the corresponding tetrasulfide, RSSSSR in a reaction that was much more facile than the equivalent process with RSH [86].

Some biological reducing agents can quench otherwise reactive radical oxidants via hydrogen atom donation (e.g. tocopherols [87]) and, therefore, can serve to limit cellular oxidative damage that occurs via radical chain mechanisms. It is important to recognize that RSSH is also a superior hydrogen atom donor compared to RSH (i.e. Reaction (1.8)) and therefore has the potential to act as an antioxidant.

$$X^{\cdot}\text{(radical oxidant)} + RSSH \rightarrow RSS^{\cdot} + X\text{–}H \quad (1.8)$$

The superior H-atom donating ability of RSSH versus RSH is evident by noting that the S—H BDEs for RS–H and RSS–H are 92 and 70 kcal/mol, respectively [76]. The relatively greater reducing capability of RSSH versus RSH is also evidenced by the finding that -hydroxy-2,3,6,6-(tetramethylpiperidin-1-yl)oxyl (TEMPO, a stable nitroxide radical) reacts with RSSH giving the RSS$^{\cdot}$ and the reduced TEMPO species, a reaction proposed to occur via H-atom abstraction [82] (Reaction (1.9)). Expectedly, this react does not occur with RSH.

$$R'SSH + R_2N\text{–}O^{\cdot}\text{(TEMPO)} \rightarrow R'SS^{\cdot} + R_2NOH \quad (1.9)$$

Further, Chauvin et al. [88] reported that RSSH species can act as potent antioxidants via hydrogen-atom transfer reactions, with reactivity equal to or greater than that of tocopherols. Importantly, the perthiyl radical (RSS$^{\cdot}$) generated in this reaction is relatively nonreactive as an oxidant and, as mentioned previously, in chemical systems readily dimerizes to the corresponding tetrasulfide RSSSSR [82, 88].

The increased reducing capabilities of RSS^-/RSSH compared to RS^-/RSH can be explained by examining and comparing the relative stabilities of the oxidized products, RS$^{\cdot}$, and RSS$^{\cdot}$. The RSS$^{\cdot}$ radical is significantly more stable than the corresponding RS$^{\cdot}$ radical (as evidenced by the lower reduction potential and lower S—H BDE for the RSH versus RSSH species, *vide infra*). This increased stability can be explained by the existence of electron delocalization in the RSS$^{\cdot}$ species that does not exist with RS$^{\cdot}$ [82]. Many reactive radical species can be trapped by other paramagnetic species such as nitric oxide (NO). Indeed, RS$^{\cdot}$ easily couples with NO to give directly a stable *S*-nitrosothiol (RS–NO) (Reaction (1.10)). However, RSS$^{\cdot}$ is so stable that it does not even have the tendency to couple with NO and at room temperature RSS–NO is not observed [82]:

$$RS^{\cdot} + {}^{\cdot}NO \rightarrow RS\text{–}NO \quad (1.10)$$

Indeed, RSS^- species may represent novel and efficient effectors for liberating NO from RS–NO compounds. That is, the nucleophilicity of RSS allows facile reaction with the electrophilic RS–NO, leading to a fleeting RSS–NO species, which is unstable and rapidly decomposes via homolysis to RSS$^{\cdot}$ (seen as the coupled RSSSSR) and NO (Reactions (1.11) and (1.12)) [82]:

$$RSS^- + R'S\text{–}NO \rightarrow RSS\text{–}NO + R'S^- \quad (1.11)$$

$$RSS\text{–}NO \rightarrow \tfrac{1}{2}RSSSSR + NO \quad (1.12)$$

1.3.4 RSSH Interactions with Metals/Metalloproteins

Although it is expected that RSS^- should readily coordinate biological metals/metalloproteins (especially "soft" metals, using HSAB terminology) to form stable complexes with chemical properties distinct from RS^- complexes, this has not been extensively reported or established. Regardless, the few current examples may portend future reports as research and interest in persulfide biochemistry continues. Some of the earliest reports of possible RSS^- coordination to a metalloprotein were by the Massey lab who proposed that the molybdenum enzymes xanthine oxidase [73] and aldehyde oxidase [72] contained a Mo-SS-Cys complex that was crucial to activity. This assertion was based primarily on the observation that both enzymes can be treated with cyanide ion (CN^-), resulting in the generation of thiocyanate (SCN^-) and enzyme inhibition. Since CN^- is known to react with the RSS^-/RSSH functional group to generate SCN^- [70, 89] (Reaction (1.13)), it was concluded that both enzymes contained a critical Mo-SSR function:

$$RSSH + CN^- \rightarrow RSH + SCN^- \tag{1.13}$$

To be sure, others contend that these enzymes may not possess a Mo-SS-Cys group but, instead, are more likely molybdenum sulfides (MoS) (e.g. [90]), which can also form SCN^- when reacted with CN^-. Interestingly, with other Mo-containing enzymes a "sulfur shift" mechanism has been proposed to be important to catalysis, providing the possibility that both models are correct and both structures (MoS and Mo-SS-Cys) are relevant to catalysis [91, 92]. Persulfide ligation to a heme protein has also been reported. The sulfur-oxidizing enzyme system (Sox) found in certain bacteria contain SoxAX cytochromes. Crystallographic analysis of SoxAX reveal an iron heme-persulfide complex [93, 94], although the exact role of the persulfide ligand in catalysis remains to be established.

1.3.5 The Electrophilicity of RSSH

As mentioned earlier, RSSH (the protonated species only) can be electrophilic, akin to the electrophilicity of RSSR. Indeed, the chemistry of RSSH can be viewed as directly analogous to that of RSSR. As such, RSSH can serve to either generate H_2S or transfer an –SH functional group to another nucleophile depending on which sulfur is attacked by the nucleophile. Using a thiol nucleophile as an example for this substitution reaction, Reaction (1.14) (which, as written, is simply the reverse of Reaction (1.4)) illustrates how nucleophilic attack by R′SH at the carbon-bound sulfur of RSSH releases H_2S and forms the corresponding disulfide:

$$RSSH + R'SH \rightleftharpoons RSSR' + H_2S \tag{1.14}$$

On the other hand, nucleophilic attack at the terminal sulfur of RSSH transfers the –SH group from one thiol to another (Reaction (1.15), the transferred sulfur is shown in bold):

$$RS\underline{S}H + R'SH \rightleftharpoons RSH + R'S\underline{S}H \quad (1.15)$$

Thus, as with RSSR species, both sulfur atoms of RSSH can be subject to nucleophilic attack. In studies of cyanolysis reactions (cleavage of RSSR by CN^-) of unsymmetrical RSSR′ compounds, it is reported that the preferred site of nucleophilic attack is primarily governed by the quality of the leaving group [95] although steric effects can play a role as well [96]. Thus, it may be expected that nucleophilic attack on the carbon-bound sulfur will be favored based on the quality of the leaving group (i.e. HS^- is a weaker base [H_2S pK_a = 7] than typical aliphatic RS^- groups [RSH pK_a typically 8–9]). However, steric restrictions to attack will be greater for the carbon-bound sulfur atom. Experimental and theoretical examination of the reaction of nucleophiles with RSSH indicates that both sulfurs are susceptible to reaction, with a slight possible preference for reaction at the terminal sulfur atom with soft nucleophiles (Reactions (1.13) and (1.15)) [79, 97] and that steric restriction to attack at the internal sulfur atom can play a factor [97]. In a biological system it is likely that numerous factors (e.g. protein effects, local pH) will be able to promote reaction at one sulfur over the other, and it is unwise to assume any general reaction preference.

1.3.6 Higher-Order Polysulfides

The reaction of H_2S with RSSR to give RSSH and RSH, and vice versa (Reactions (1.4) and (1.14)) as presented is clearly oversimplified. These reactions represent the reaction of possible nucleophiles, RS^-/RSH and HS^-/H_2S with possible electrophiles, RSSR and RSSH, which are all present in solution. What also needs to be considered is that a product of Reaction (1.4), RSS^-/RSSH, is a superior nucleophile to either RS^-/RSH or HS^-/H_2S and therefore should react with electrophiles in solution as well. Moreover, RSSH is likely more acidic than RSH or H_2S and therefore higher levels of the more nucleophilic anion will be present for RSSH. The reaction of RSS^- with either RSSR or RSSH electrophiles generates a trisulfide species, RSSSR (Reactions (1.16) and (1.17)):

$$RSS^- + RSSR + H^+ \rightleftharpoons RSSSR + RSH \quad (1.16)$$

$$RSS^- + RSSH + H^+ \rightleftharpoons RSSSR + H_2S \quad (1.17)$$

Thus, it seems likely that trisulfide species will also be generated under the conditions of Reaction (1.4) or (1.14). Indeed, this has been found to be the case as a reaction mixture containing initially H_2S and RSSR resulted in an equilibrium mixture that also contains the trisulfide RSSSR [98]. Interestingly, higher-order polysulfides (e.g. the tetrasulfide RSSSSR) were not observed in this equilibrium mixture (or were too low in concentration to detect). Clearly, reactions that could conceivably produce an RSSSSR species can be envisioned (for example, Reaction (1.18)):

$$\mathrm{RSS^- + RSSSR + H^+ \rightleftharpoons RSSSSR + RSH} \tag{1.18}$$

Also possible is Reaction (1.19), whereby nucleophilic attack by RSS^- occurs at a carbon-bound sulfur, resulting in an overall nonproductive reaction since the reactants and products are the same (if all R groups are identical):

$$\mathrm{R'SS^- + RSSSR \rightleftharpoons R'SSSR + RSS^-} \tag{1.19}$$

The lack of observable production of RSSSSR (under the conditions of this chemical study, e.g. via for example Reaction (1.18)) and even higher-order species can be rationalized using the arguments above regarding the site of sulfur atom attack on disulfides and the quality of the leaving group (*vide supra*). That is, in Reaction (1.18), RSSSSR generation requires nucleophilic attack at the central sulfur atom with RS^- as the leaving group. Since RS^- would be a poorer leaving group compared to RSS^-, it was proposed that Reaction (1.18) was unfavorable and significant levels of higher-order polysulfides were not observed. To be sure, higher-order polysulfides, including the elemental form of sulfur S_8, can form from these types of reactions over time (e.g. [99]), indicating the potential for Reaction (1.18), and many others (e.g. Reaction (1.6)), to occur under the proper conditions or given enough time.

1.3.7 RSSH Instability

In aqueous and neutral solutions, RSSH is unstable. This is not surprising due to the fact that electrophilic RSSH will be in an acid–base equilibrium with nucleophilic RSS^- and self-reactivity is expected. Thus, at the very least, a solution of RSS^-/RSSH will result in the generation of equilibrium levels of RSSSR and H_2S (Reaction (1.17)) along with many other products associated with other competing reactions (e.g. [100, 101]). Importantly, in chemical reactions where H_2S can volatilize from solution, equilibria can be drawn to the products of Reaction (1.17). The inherent instability of RSSH species is troublesome for those examining RSSH biology since stock aqueous solutions of RSSH would lack long-term stability. This typically necessitates the development and use of donor compounds for studies aimed at determining the biological activity of RSSH. Regardless, once RSSH species (even from donors) are introduced to an aqueous biological system the decomposition reactions discussed herein (e.g. Reactions (1.5), (1.8), (1.14), (1.16), (1.17)) can occur and care must be taken when interpreting the origins/mechanisms of the observed effects.

1.4 The Biochemical Utility of RSSH

Based on the discussion above, it appears that some of the biochemically relevant chemical properties/reactivities of RSSH are distinct from RSH and unique among all biologically relevant sulfur-containing species. Based on these chemical differences, the idea that per- and poly-sulfides have important physiological functions (that may be underappreciated) has been discussed by many. Most notably and early on, Toohey prophetically discussed the likely regulatory role of sulfane sulfur in biology [53] as well as emphasized that H_2S and persulfides may be intimately linked, that experiments using H_2S were also examining other contaminating S^0 species and that H_2S-mediated signaling may in fact be due to persulfide generation [102]. Further reports discussing the possible roles of sulfane sulfur species (e.g. RSSH) in biological signaling have since been published (e.g. [1, 75, 103–105]). Also, the presumed therapeutic benefits of plants in the *Allium* family (e.g. garlic, onions) has been hypothesized to be a result of the high content of sulfane sulfur-containing species, which are capable of RSSH generation (e.g. [106]). To be sure, *Allium*-based per- and poly-sulfides are also potentially toxic since they are oxidized sulfur species [106] that can participate in potentially deleterious redox chemistry [107]. Thus, as with many presumed beneficial, antioxidant species, the levels of the presumed protectant are important considerations.

Since this chapter focuses on the chemistry of H_2S and derived species, it is beyond the scope of this chapter to provide a detailed and comprehensive discussion of the biological effects/actions attributed to H_2S, persulfides, etc. It is, however, important to point out a few aspects of the chemistry discussed above and how this chemistry may be physiologically important/relevant. For example, RSS^-/RSSH are superior nucleophiles and reductants compared to the corresponding RS^-/RSH. The fact that RSSH is oxidized with respect to RSH and therefore RSH conversion to RSSH should occur primarily under oxidizing conditions brings up an interesting and unique situation: The generation of a superior reductant may occur primarily under oxidizing conditions. In a biological system, this may mean that cells exposed to an oxidative stress have a greater ability to generate the potent reductant RSSH that may oppose the stress [108]. Although this is currently speculative, there are reports consistent with this idea. For example, cells with increased intracellular RSSH levels or treated with RSSH donor molecules have been shown to possess an increased resistance against oxidative and/or electrophilic stress (e.g. [68, 98, 109]). To be sure, the fact that RSSH are oxidized with respect to RSH and, therefore, potentially oxidizing species, they also have the potential to be toxic (e.g. [107, 110]).

The increased nucleophilicity of RSSH may also allow some thiol-dependent enzymes to increase activity when converted to a persulfide. Although, in most cases it has been reported that RSH conversion to RSSH decreases normal enzyme activity (e.g. GAPDH [111], papain [61], peroxiredoxin [112]), increased activity has also been reported (e.g. Parkin [113]). Thus, the biological/physiological implications of RSSH generation are likely to be complex and varied and remain to be fully established. However, it would not be at all surprising to find that Nature utilizes the unique chemistry of the –SSH functional group in chemically predictable and important ways and that RSSH species are vital players in sulfur biology/physiology.

1.5 Summary/Conclusion

Hopefully, it is clear that the chemical biology of H_2S is potentially extensive due in part to the numerous distinct and diverse chemical species that can be generated in a biological milieu and to the numerous possible biological targets. Importantly, the presence of per- and poly-sulfides (which are not necessarily the result of initial H_2S formation) are intimately linked to H_2S biochemically and therefore, always need to be considered as possible biological mediators/effectors. Thus, in many studies whereby H_2S is added to an experiment and the resulting biological activity monitored, an old adage important to the study of chemical mechanisms is applicable: "It's not always what you add to solution that's important – it's what's in solution that counts."

References

1 Mustafa, A., Gadalla, M.M., and Synder, S.H. (2009). Signaling by gasotransmitters. *Sci. Signal.* 2: 1–8.

2 Fukuto, J.M., Carrington, S.J., Tantillo, D.J. et al. (2012). Small molecule signaling agents: the integrated chemistry and biochemistry of nitrogen oxides, oxides of carbon, dioxygen, hydrogen sulfide, and their derived species. *Chem. Res. Toxicol.* 25: 769–793.

3 Szabo, C. (2016). Gasotransmitters in cancer: from pathophysiology to experimental therapy. *Nat. Rev. Drug Discov.* 15: 185–203.

4 Fukuto, J.M. and Collins, M.D. (2007). Interactive endogenous small molecule (gaseous) signaling: implications for teratogenesis. *Curr. Pharm. Des.* 13: 2952–2978.

5 Poole, L.B. (2015). The basics of thiols and cysteines in redox biology and chemistry. *Free Radic. Biol. Med.* 80: 148–157.

6 Cuevasanta, E., Denicola, A., Alvarez, B., and Moller, M.N. (2012). Solubility and permeation of hydrogen sulfide in lipid membranes. *PLoS One* 7 (4): e34562.

7 Haines, T.H. (1994). Water transport across biological membranes. *FEBS Lett.* 346: 115–122.

8 Mathai, J.C., Missner, A., Kugler, P. et al. (2009). No facilitator required for membrane transport of hydrogen sulfide. *Proc. Natl. Acad. Sci. U.S.A.* 106: 16633–16638.

9 Pearson, R.G. (1963). Hard and soft acids and bases. *J. Am. Chem. Soc.* 85: 3533–3539.

10 Pearson, R.G. and Songstad, J. (1967). Application of the principle of hard and soft acids and bases to organic chemistry. *J. Am. Chem. Soc.* 89: 1827–1836.

11 Pearson, R.G. (1987). recent advances in the concept of hard and soft acids and bases. *J. Chem. Educ.* 64: 561–567.

12 Jacob, C., Giles, G.I., Giles, N.M., and Sies, H. (2003). Sulfur and selenium: the role of oxidation state in protein structure and function. *Angew. Chem. Int. Ed.* 42: 4742–4758.

13 Buettner, G.R. (1993). The pecking order of free radicals and antioxidants: lipid peroxidation, α-tocopherol, and ascorbate. *Arch. Biochem. Biophys.* 300: 535–543.

14 Das, T.N., Huie, R.E., Neta, P., and Padmaja, S. (1999). Reduction potential of the sulfhydryl radical: pulse radiolysis and laser flash photolysis studies of the formation and reactions of $^{\cdot}SH$ and $HSSH^{\cdot-}$ in aqueous solutions. *J. Phys. Chem. A* 103: 5221–5226.

15 Holmgren, A. and Sengupta, R. (2010). The use of thiols by ribonucleotide reductase. *Free Radic. Biol. Med.* 49: 1617–1628.

16 Benson, S.W. (1978). Thermochemistry and kinetics of sulfur-containing molecules and radicals. *Chem. Rev.* 78: 23–35.

17 Blanksby, S.J. and Ellison, G.B. (2002). Bond dissociation energies of organic molecules. *Acc. Chem. Res.* 36: 255–263.

18 Brzoska, K., Meczynska, S., and Kruszewski, M. (2006). Iron-sulfur cluster proteins: electron transfer and beyond. *Acta Biochim. Pol.* 53: 685–691.

19 Johnson, D.C., Dean, D.R., Smith, A.D., and Johnson, M.K. (2005). Structure, function, and formation of biological iron-sulfur clusters. *Annu. Rev. Biochem.* 74. 247–281.

20 Braymer, J.J. and Lill, R. (2017). Iron-sulfur cluster biogenesis and trafficking in mitochondria. *J. Biol. Chem.* 292: 12754–12763.

21 Guidotti, T.L. (2010). Hydrogen sulfide: advances in understanding human toxicity. *Int. J. Toxicol.* 29 (6): 569–581.

22 Pietri, R., Roman-Morales, E., and Lopez-Garriga, J. (2011). Hydrogen sulfide and hemeproteins: knowledge and mysteries. *Antioxid. Redox Signal.* 15: 393–404.

23 Cooper, C.E. and Brown, G.C. (2008). The inhibition of mitochondrial cytochrome oxidase by the gases carbon monoxide, nitric oxide, hydrogen cyanide and hydrogen sulfide: chemical mechanism and physiological significance. *J. Bioeng. Biomembr.* 40: 533–539.

24 Powell, M.A. and Somero, G.N. (1986). Hydrogen sulfide oxidation is coupled to oxidative phosphorylation in mitochondria of *Solemya reidi*. *Science* 233: 563–566.

25 Yong, R. and Searcy, D.G. (2001). Sulfide oxidation coupled to ATP synthesis in chicken liver mitochondria. *Comp. Biochem. Physiol. B Biochem. Mol. Biol.* 129: 129–137.

26 Collman, J.P., Ghosh, S., Dey, A., and Decreau, R.A. (2009). Using a functional enzyme model to understand the chemistry behind hydrogen sulfide induced hibernation. *Proc. Natl. Acad. Sci. U.S.A.* 106: 22090–22095.

27 Kraus, D.W., Wittenberg, J.B., Jung-Fen, L., and Peisach, J. (1990). Hemoglobins of the *Lucina pectinata*/bacteria symbiosis. *J. Biol. Chem.* 265: 16054–16059.

28 Rizzi, M., Wittenberg, J.B., Coda, A. et al. (1996). Structural bases for sulfide recognition in *Lucina pectinata* hemoglobin I. *J. Mol. Biol.* 258: 1–5.

29 Pietri, R., Lewis, A., Leon, R.G. et al. (2009). Factors controlling the reactivity of hydrogen sulfide with hemeproteins. *Biochemistry* 48: 4881–4894.

30 Nguyen, B., Zhao, X., Vyas, K. et al. (1998). Solution and crystal structures of a sperm whale myoglobin triple mutant that mimics the sulfide-binding hemoglobin from *Lucina pectinata*. *J. Biol. Chem.* 273: 9517–9526.

31 Vitvitsky, V., Yadav, P., Kurthen, A., and Banerjee, R. (2015). Sulfide oxidation by a noncanonical pathway in red blood cells generates thiosulfate and polysulfides. *J. Biol. Chem.* 290: 8310–8320.

32 Bostelaar, T., Vitvitsky, V., Kumutima, J. et al. (2016). Hydrogen sulfide oxidation by myoglobin. *J. Am. Chem. Soc.* 138: 8476–8488.

33 Vitvitsky, V., Yadav, P.K., An, S. et al. (2017). Structural and mechanistic insights into hemoglobin catalyzed hydrogen sulfide oxidation and the fate of polysulfide products. *J. Biol. Chem.* 292: 5584–5592.

34 Hartle, M.D., Prell, J.S., and Pluth, M.D. (2016). Spectroscopic investigations into the binding of hydrogen sulfide to synthetic picket-fence porphyrins. *Dalton Trans.* 45: 4843–4853.

35 Jensen, B. and Fago, A. (2018). Reactions of ferric hemoglobin and myoglobin with hydrogen sulfide under physiological conditions. *J. Inorg. Biochem.* 182: 133–140.

36 Tung, T.-L. and Stone, J.A. (1975). The formation and reactions of disulfide radical anions in aqueous solution. *Can. J. Chem.* 53: 3153–3157.

37 Surdhar, P.S. and Armstrong, D.A. (1987). Reduction potentials and exchange reactions of thiyl radicals and disulfide anion radicals. *J. Phys. Chem.* 91: 6532–6537.

38 Palinkas, Z., Furtmuller, P.G., Nagy, A. et al. (2014). Interactions of hydrogen sulfide with myeloperoxidase. *Br. J. Pharmacol.* 172: 1516–1532.

39 Garai, D., Rios-Gonzalez, B.B., Furtmuller, P.G. et al. (2017). Mechanisms of myeloperoxidase catalyzed oxidation of H_2S by H_2O_2 or O_2 to produce potent protein Cys-polysulfide-inducing species. *Free Radic. Biol. Med.* 113: 551–563.

40 Rios-Gonzalez, B.B., Roman-Morales, E.M., Pietri, R., and Lopez-Garriga, J. (2014). Hydrogen sulfide activation in hemeproteins: the sulfheme scenario. *J. Inorg. Biochem.* 133: 78–86.

41 Arbelo-Lopez, H.D., Simakov, N.A., Smith, J.C. et al. (2016). Homolytic cleavage of both heme-bound hydrogen peroxide and hydrogen sulfide leads to formation of sulfheme. *J. Phys. Chem. B* 120: 7319–7331.

42 Padovani, D., Hessani, A., Castillo, F.T. et al. (2016). Sulfheme formation during homocysteine S-oxygenation by catalase in cancers and neurodegenerative diseases. *Nat. Commun.* 7: 13386.

43 Alvarez, L., Suarez Vega, V., McGinity, C. et al. (2020). The reactions of hydropersulfides (RSSH) with myoglobin. *Arch. Biochem. Biophys.* 687: 108391.

44 Berzofsky, J.A., Peisach, J., and Blumberg, W.E. (1971). Sulfheme proteins. *J. Biol. Chem.* 246: 7366–7372.

45 Carrico, R.J., Blumberg, W.E., and Peisach, J. (1978). The reversible binding of oxygen to sulfhemoglobin. *J. Biol. Chem.* 253: 7212–7215.

46 Park, C.M., Nagel, R.L., Blumberg, W.E. et al. (1986). Sulfhemoglobin, properties of partially sulfurated tetramers. *J. Biol. Chem.* 261: 8805–8810.

47 Alvarez, M.T., Crespo, C., and Mattiasson, B. (2007). Precipitation of Zn(II), Cu(II) and Pb(II) at bench-scale using biogenic hydrogen sulfide from the utilization of volatile fatty acids. *Chemosphere* 66: 1677–1683.

48 Gadd, G.M. and Griffiths, A.J. (1978). Microorganisms and heavy metal toxicity. *Microb. Ecol.* 4: 303–317.

49 Pan-Hou, H.S.K. and Imura, N. (1981). Role of hydrogen sulfide in mercury resistance determined by plasmid of *Clostridium cochlearium* T-2. *Arch. Microbiol.* 129: 49–52.

50 Sharma, P.K., Balkwill, D.L., Frenkel, A., and Vairavamurthy, M.A. (2000). A new *Klebsiella plantacola* strain (Cd-1) grows anaerobically at high cadmium concentration and precipitates cadmium sulfide. *Appl. Environ. Microbiol.* 66: 3083–3087.

51 Yoshida, E., Toyama, T., Shinkai, Y. et al. (2011). Detoxification of methylmercury by hydrogen sulfide-producing enzyme in mammalian cells. *Chem. Res. Toxicol.* 24: 133–1635.

52 Wood, J.L. (1987). Sulfane sulfur. *Methods Enzymol.* 143: 25–29.

53 Toohey, J.I. (1989). Sulphane Sulphur in biological systems: a possible regulatory role. *Biochem. J.* 264: 625–632.

54 Nagy, P. (2013). Kinetics and mechanisms of thiol-disulfide exchange covering direct substitution and thiol oxidation-mediated pathways. *Antioxid. Redox Signal.* 18: 1623–1641.

55 Karplus, P.A. (2015). A primer on peroxiredoxin biochemistry. *Free Radic. Biol. Med.* 80: 183–190.

56 Xiao, Z., La Fontaine, S., Bush, A.I., and Wedd, A.G. (2019). Molecular mechanisms of glutaredoxin enzymes: versatile hubs for thiol-disulfide exchange between protein thiols and glutathione. *J. Mol. Biol.* 431: 158–177.

57 Lu, J. and Holmgren, A. (2014). The thioredoxin antioxidant system. *Free Radic. Biol. Med.* 66: 75–87.

58 Cuevasanta, E., Lange, M., Bonanata, J. et al. (2015). Reaction of hydrogen sulfide with disulfide and sulfenic acid to form the strongly nucleophilic persulfide. *J. Biol. Chem.* 290: 26866–26880.

59 Rao, G.S. and Gorin, G. (1959). Reaction of cystine with sodium sulfide in sodium hydroxide solution. *J. Org. Chem.* 24: 749–753.

60 Cavallini, D., Federici, G., and Barboni, E. (1970). Interactions of proteins with sulfide. *Eur. J. Biochem.* 14: 169–174.

61 Francoleon, N.E., Carrington, S.J., and Fukuto, J.M. (2011). The reaction of H_2S with oxidized thiols: generation of persulfides and implications to H_2S biology. *Arch. Biochem. Biophys.* 516: 146–154.

62 Giles, N.M., Giles, G.I., and Jacob, C. (2003). Multiple roles of cysteine in biocatalysis. *Biochem. Biophys. Res. Commun.* 300: 1–4.

63 Bak, D.W., Bechtel, T.J., Falco, J.A., and Weerapana, E. (2019). Cysteine reactivity across the subcellular universe. *Curr. Opin. Chem. Biol.* 48: 96–105.

64 Narayan, M. (2017). The structure-forming juncture in oxidative protein folding: what happens in the ER? In: *Advances in Experimental Medicine and Biology, 966*, Protein Reviews, vol. 18 (ed. M.Z. Atassi), 163–179. Singapore: Springer.

65 Cook, K.M. and Hogg, P.J. (2013). Post-translational control of protein function by disulfide bond cleavage. *Antioxid. Redox Signal.* 18: 1987–2015.

66 Paulsen, C.E. and Carroll, K.S. (2013). Cysteine-mediated redox signaling: chemistry, biology, and tools for discovery. *Chem. Rev.* 113: 4633–4679.

67 Coleman, J.E. (1992). Zinc proteins: enzymes, storage proteins, transcription factors, and replication proteins. *Annu. Rev. Biochem.* 61: 897–946.

68 Ida, T., Sawa, T., Ihara, H. et al. (2014). Reactive cysteine persulfides and S-polythiolation regulate oxidative stress and redox signaling. *Proc. Natl. Acad. Sci. U.S.A.* 111: 7606–7611.

69 Akaike, T., Ida, T., Wei, F.-W. et al. (2017). Cysteinyl tRNA synthase governs cysteine polysulfidation and mitochondrial bioenergetics. *Nat. Commun.* 8 (1177): 1–15.

70 Flavin, M. (1962). Microbial transsulfuration: the mechanism of an enzymatic disulfide elimination reaction. *J. Biol. Chem.* 237: 768–777.

71 Cavallini, D., de Marco, C., and Mondovi, B. (1960). Cleavage of cystine by a pyridoxal model. *Arch. Biochem. Biophys.* 87: 281–288.

72 Branzoli, U. and Massey, V. (1974). Evidence for an active site persulfide residue in rabbit liver aldehyde oxidase. *J. Biol. Chem.* 249: 4346–4349.

73 Massey, V. and Edmondson, D. (1970). On the mechanism of inactivation of xanthine oxidase by cyanide. *J. Biol. Chem.* 245: 6595–6598.

74 Stipanuk, M.H. and Beck, P.W. (1982). Characterization of the enzymic capacity for cysteine desulphhydration in liver and kidney of rat. *Biochem. J.* 206: 267–277.

75 Ono, K., Akaike, T., Sawa, T. et al. (2014). The redox chemistry and chemical biology of H_2S, hydropersulfides and derived species: implications of their possible biological activity and utility. *Free Radic. Biol. Med.* 77: 82–94.

76 Fukuto, J.M., Lin, J., Khodade, V.S., and Toscano, J.P. (2020). Predicting the possible physiological/biological utility of the hydropersulfide functional group based on its chemistry: similarities between hydropersulfides and selenols. *Antioxid. Redox Signal.* 33: 1295–1307.

77 Benchoam, D., Semelak, J.A., Cuevasanta, E. et al. (2020). Acidity and nucleophilic reactivity of glutathione persulfide. *J. Biol. Chem.* 295: 15466–15481.

78 Everett, S.A. and Wardman, P. (1995). Perthiols as antioxidants: radical-scavenging and prooxidative mechanisms. *Methods Enzymol.* 251: 55–69.

79 Saund, S.S., Sosa, V., Henriquez, S. et al. (2015). The chemical biology of hydropersulfides (RSSH): chemical stability, reactivity and redox roles. *Arch. Biochem. Biophys.* 588: 15–24.

80 Edwards, J.O. and Pearson, R.G. (1961). The factors determining nucleophilic reactivities. *J. Am. Chem. Soc.* 84: 16–24.

81 Koppenol, W.H. and Bounds, P.L. (2017). Signaling by sulfur-containing molecules. Quantitative aspects. *Arch. Biochem. Biophys.* 617: 3–8.

82 Bianco, C.L., Chavez, T.A., Sosa, V. et al. (2016). The chemical biology of the persulfide (RSSH)/perthiyl (RSS•) redox couple and possible role in biological redox signaling. *Free Radic. Biol. Med.* 101: 20–31.

83 Galardon, E., Huguet, F., Herrero, C. et al. (2017). Reactions of persulfides with heme cofactor of oxidized myoglobin and microperoxidase 11: reduction and coordination. *Dalton Trans.* 46: 7939–7946.

84 Massey, V., Williams, C.H. Jr., and Palmer, G. (1971). The presence of S^0-containing impurities in commercial samples of oxidized glutathione

and their catalytic effect on the reduction of cytochrome c. *Biochem. Biophys. Res. Commun.* 42: 730–738.

85 Froede, H.C. and Hunter, F.E. Jr., (1970). Catalytic effect of GSSG on reduction of cytochrome c by GSH – possible model for facilitation of electron transfer and energy conservation by sulfonium ion formation. *Biochem. Biophys. Res. Commun.* 38: 954–961.

86 Nakabayashi, T. and Tsurugi, J. (1963). Aralkyl hydrodisulfides. I. The reaction with 2,2-diphenyl-1-picrylhydrazyl and with benzoquinone. *J. Org. Chem.* 28: 811–813.

87 Jiang, Q. (2014). Natural forms of vitamin E: metabolism, antioxidant, and anti-inflammatory activities and their role in disease prevention and therapy. *Free Radic. Biol. Med.* 72: 76–90.

88 Chauvin, J.-P.R., Griesser, M., and Pratt, D.A. (2017). Hydropersulfides: H-atom transfer agents par excellence. *J. Am. Chem. Soc.* 139: 6484–6493.

89 Cavallini, D., De Marco, C., Mondovi, B., and Mori, B.G. (1960). The cleavage of cystine by cystathionase and the transulfuration of hypotaurine. *Enzymologia* 22: 161–173.

90 Wahl, R.C. and Rajagopalan, K.V. (1982). Evidence for the inorganic nature of the cyanolyzable sulfur of molybdenum hydroxylases. *J. Biol. Chem.* 257: 1354–1359.

91 Cerqueira, N.M.F.S.A., Fernandes, P.A., Gonzalez, P.J. et al. (2013). The sulfur shift: an activation mechanism for periplasmic nitrate reductase and formate dehydrogenase. *Inorg. Chem.* 52: 10766–10772.

92 Cerqueira, N.M.F.S.A., Pakhira, B., and Sarkar, S. (2015). Theoretical studies on the mechanisms of some Mo enzymes. *J. Biol. Inorg. Chem.* 20: 323–335.

93 Bamford, V.A., Bruno, S., Rasmussen, T. et al. (2002). Structural basis for the oxidation of thiosulfate by a sulfur cycle enzyme. *EMBO J.* 21: 5599–5610.

94 Dambe, T., Quentmeier, A., Rother, D. et al. (2005). Structure of the cytochrome complex SoxXA of *Paracoccus pantotrophus*, a heme enzyme initiating chemotrophic sulfur oxidation. *J. Struct. Biol.* 152: 229–234.

95 Hiskey, R.G. and Carroll, F.I. (1961). Chemistry of aliphatic disulfides. II. Cyanide cleavage of unsymmetrical disulfides. *J. Am. Chem. Soc.* 83: 4644–4647.

96 Hiskey, R.G., Bowers, W.H., and Harpp, D.N. (1964). Chemistry of aliphatic disulfides. VI. Effects of α-alkylation on the cyanide cleavage of unsymmetrical disulfides. *J. Am. Chem. Soc.* 86: 2010–2014.

97 Nakabayashi, T., Tsurugi, J., Kawamura, S. et al. (1966). Aralkyl hydrodisulfides. VII. The reaction with trisubstituted phosphites. *J. Org. Chem.* 31: 4174–4178.

98 Bianco, C.L., Akaike, T., Ida, T. et al. (2019). The reaction of hydrogen sulfide with disulfides: formation of a stable trisulfide and implications to biological systems. *Br. J. Pharmacol.* 176: 671–683.

99 Bailey, T.S. and Pluth, M.D. (2015). Reactions of isolated persulfides provide insight into the interplay between H_2S and persulfide reactivity. *Free Radic. Biol. Med.* 89: 662–667.

100 Pickering, T.L., Saunders, K.J., and Tobolsky, A.V. (1967). Disproportionation of organic polysuflides. *J. Am. Chem. Soc.* 89: 2364–2367.

101 Cuevasanta, E., Moller, M.N., and Alvareez, B. (2017). Biological chemistry of hydrogen sulfide and persulfides. *Arch. Biochem. Biophys.* 617: 9–25.

102 Toohey, J.I. (2011). Sulfur signaling: is the agent sulfide or sulfane? *Anal. Biochem.* 413: 1–7.

103 Toohey, J.I. and Cooper, A.J.L. (2014). Thiosulfoxide (sulfane) sulfur: new chemistry and new regulatory roles in biology. *Molecules* 19: 12789–12813.

104 Filipovic, M.R. (2015). Persulfidation (S-sulfyhydration) and H_2S. In: *Chemistry, Biochemistry and Pharmacology of Hydrogen Sulfide, Handbook of Experimental Pharmacology*, vol. 230 (ed. P.K. Moore and M. Whiteman), 29–59. Charm, Switzerland: Springer.

105 Alvarez, L., Bianco, C.L., Toscano, J.P. et al. (2017). Chemical biology of hydropersulfides and related species: possible roles in cellular protection and redox signaling. *Antioxid. Redox Signal.* 27: 622–633.

106 Munchberg, U., Anwar, A., Mecklenburg, S., and Jacob, C. (2007). Polysulfides as biologically active ingredients of garlic. *Org. Biomol. Chem.* 5: 1505–1518.

107 Munday, R., Munday, J.S., and Munday, C.M. (2003). Comparative effects of mono-, di-, tri-, and tetrasulfides derived from plants of the *Allium* family: redox cycling in vitro and hemolytic activity and phase 2 enzyme induction in vivo. *Free Radic. Biol. Med.* 34: 1200–1211.

108 Millikin, R., Bianco, C.L., White, C. et al. (2016). The chemical biology of protein hydropersulfides: studies of a possible protective function of biological hydropersulfide generation. *Free Radic. Biol. Med.* 97: 136–147.

109 Powell, C.R., Dillon, K.M., Wang, Y. et al. (2018). A persulfide donor responsive to reactive oxygen species: insights into reactivity and therapeutic potential. *Angew. Chem. Int. Ed.* 57: 6324–6328.

110 Lin, J., Akiyama, M., Bica, I. et al. (2019). The uptake and release of polysuflur cysteine species by cells: physiological and toxicological implications. *Chem. Res. Toxicol.* 32: 447–455.

111 Jarosz, A.P., Wei, W., Gauld, J.W. et al. (2015). Glyceraldehyde 3-phosphate dehydrogenase (GAPDH) is inactivated by S-sulfuration in vitro. *Free Radic. Biol. Med.* 89: 512–521.

112 Cuevasanta, E., Reyes, A.M., Zeida, A. et al. (2019). Kinetics of formation and reactivity of the persulfide in the one-cysteine peroxiredoxin from *Mycobacterium tuberculosis*. *J. Biol. Chem.* 294: 13593–13605.

113 Vandiver, M.S., Paul, B.D., Xu, R. et al. (2013). Sulfhydration mediates neuroprotective actions of parkin. *Nat. Commun.* 4: 1626. https://doi.org/10.1038/ncomms2623.

2

Signaling by Hydrogen Sulfide (H_2S) and Polysulfides (H_2S_n) and the Interaction with Other Signaling Pathways

Hideo Kimura

Department of Pharmacology, Faculty of Pharmaceutical Science, Sanyo-Onoda City University, Sanyo-Onoda, Yamaguchi, Japan

List of Abbreviations

ATP	adenosine triphosphate
CARS	cysteinyl tRNA synthetase
CAT	cysteine aminotransferase
CBS	cystathionine β-synthase
CysSSH	cysteine persulfide
CO	carbon monoxide
CSE	cystathionine γ-lyase
DAO	D-amino acid oxidase
DATS	diallyl trisulfide
DEA NONOate	diethylamine nonoate
DMTS	dimethyl trisulfide
DTT	dithiothreitol
GAPDH	glyceraldehyde-3-phosphate dehydrogenase
GCL	glutamate cysteine ligase
γGCS	γ-glutamyl cysteine synthetase
GSH	glutathione
GSSH	glutathione persulfide
HIF	hypoxia inducible factor
HNO	nitroxyl
H_2O_2	hydrogen peroxide
HSNO	thionitrous acid
H_2S	hydrogen sulfide
H_2S_n	polysulfides

Keap1	Kelch ECH associating protein 1
LC-MS/MS	liquid chromatograph-mass spectrometry
LTP	long-term potentiation
3MP	3-mercaptopyruvate
3MST	3-mercaptopyruvate sulfurtransferase
NaHS	sodium hydrosulfide
NMDA	*N*-methyl D,L-aspartate
NADH/NAD	nicotinamide adenine dinucleotide
NO	nitric oxide
Nrf2	nuclear factor erythroid 2-related factor 2
O_2	oxygen
PTEN	phosphatase and tensin homolog
SAM	*S*-adenosyl-L-methionine
SOD	super oxide dismutase
SQR	sulfur quinone oxidoreductase
$SSNO^-$	nitrosopersulfide
TRPA1	transient receptor potential ankyrin 1
VEGF	vascular endothelial growth factor

2.1 Introduction

Endogenous H_2S has been identified by Warenycia et al. in rat brain [1]. The levels of H_2S accumulated in the brain were measured when animals were exposed to this toxic gas. Notably, they realized even without the exposure to H_2S, H_2S was detected in a certain amount using the methylene blue method. Other groups also measured H_2S concentrations in the brains of humans and bovine with the same method and obtained 50–160 μM [1–3], although these levels have since been revised to significantly lower levels. Furne et al. reported that the endogenous H_2S concentration in the brain is approximately 14 nM using a gas chromatography [4]. The recent evaluation of H_2S as monobromobimane adduct by LC–MS/MS showed approximately 3.0 μM in this tissue [5].

We found that H_2S facilitates the induction of hippocampal long-term potentiation (LTP), a synaptic model of memory formation, by enhancing the activity of *N*-methyl-D-aspartate (NMDA) receptors [6]. The hinge of the ligand-binding domain located at the external side of the receptors is the redox sensor, which enhances the activity of the receptors when reduced by dithiothreitol (DTT) [7]. Because H_2S is a reducing agent, it was expected to be a mechanism for the enhancement of NMDA receptors and subsequent induction of LTP [6]. However, even at concentrations lower than DTT, H_2S showed a greater effect, suggesting that there is another mechanism.

Synaptic transmission is regulated not only by the postsynaptic receptors but also by signaling from astrocytes that surround synapses [8]. We found that H_2S activates astrocytes to induce Ca^{2+} influx [9], but later identified that H_2S_n activate them much more potently than H_2S [10–12]. H_2S_n activates transient receptor potential ankyrin 1 (TRPA1) channels in astrocytes that make them release the gliotransmitter D-serine, a co-agonist of NMDA receptors, thereby facilitating the induction of LTP [12, 13].

H_2S_n molecules are produced by 3-mercaptopyruvate sulfurtransferase (3MST), which is also known as an H_2S producing enzyme, in concert with thioredoxin [14, 15]. The other enzymes such as sulfur quinone oxidoreductase (SQR) [16], super oxide dismutase (SOD) [17–19], catalase [20], cysteine t-RNA synthetase (CARS) [21], hemoproteins [22], and peroxidases [23] can also produce H_2S_n and other *S*-sulfurated molecules. The endogenous concentration of H_2S_2 was measured to be approximately 2.6 μM, which is almost equivalent to that of H_2S (3.0 μM) measured in the brain [5, 12, 14]. Note that the levels of H_2S and H_2S_2 may vary in other tissues including blood.

The chemical interaction of H_2S with nitric oxide (NO) generates H_2S_n. The synergistic effect of both molecules has been discovered on the relaxation of vascular smooth muscle [24]. In addition to H_2S_n, nitrosothiols, HSNO, HNO, and $SSNO^-$ have been proposed to be produced from the interaction of H_2S and NO [25–30], and they all have the effect on activating TRPA1 channels [12, 26, 30].

The iron–sulfur cluster, which consists of the active center of the enzymes belonging to the respiratory chain such as aconitase, releases H_2S under acidic conditions, and it is designated as acid-labile sulfur [31, 32]. Cells or tissues contain another form of sulfur that releases H_2S under reducing conditions and is designated as bound sulfane sulfur [32–34]. The release rate and amounts of H_2S from bound sulfane sulfur are distinct in different organ types. Tissue homogenates, in turn, have the capacity to absorb H_2S, and their absorption rate and amounts also differ in tissues depending on the organ types [32].

S-Sulfuration (S-sulfhydration) has been proposed as a mode of action of H_2S to modify the activity of the target proteins [35]. H_2S *S*-sulfurates oxidized cysteine residues including *S*-nitrosylated (P-Cys-SNO) and *S*-sulfenylated (P-Cys-SOH) to P-Cys-SSH [36, 37]. H_2S_n and other *S*-sulfurated molecules such as cysteine persulfide (Cys-SSH) and glutathione persulfide (GSSH), which are contained in bound sulfane sulfur, can *S*-sulfurate cysteine residues (P-Cys-SH) to P-Cys-SSH.

2.2 Determination of the Endogenous Concentrations of H_2S

The endogenous H_2S in the brain was identified by Warenycia et al. using the methylene blue method in which *N*,*N*-dimethyl-*p*-phenylendiamine reacts with

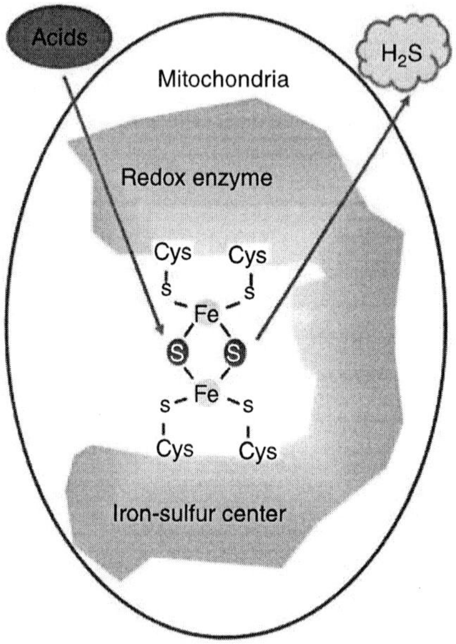

Figure 2.1 Acid-labile sulfur releases H_2S under acidic conditions. Iron–sulfur cluster localized at the active site of enzymes belonging to the respiratory chain in mitochondria release H_2S at pH less than 5.4. Source: Based on Ishigami et al. [32].

H_2S in the presence of Fe^{3+} with strong acids to generate methylene blue that is measured with a spectrophotometer at 670 nm [1]. Two other groups also measured H_2S in the brains of humans and bovine and obtained H_2S levels of 50–160 μM [2, 3]. Because high concentrations of acids are used in the methylene blue method, a large amount of H_2S is released from acid-labile sulfur as an artifact and greatly added to the amount of H_2S quantified (Figure 2.1). This method is, therefore, not appropriate to measure endogenous H_2S in tissues and cells. However, it can be applied for the measurement of H_2S released from producing enzymes *in vitro* where their substrate cysteine is added to the reaction mixture in excess (2–10 mM), and the obtained values can be subtracted from those without cysteine as a control experiment.

The lowest concentration of endogenous H_2S measured in the brain was approximately 14 nM [4]. In these studies, they mixed the tissue with phosphate buffer (pH 5.7) in the syringe and measured H_2S volatilized from homogenates of the tissue to the air space using a gas chromatography. Because the pK_a value of H_2S to dissociate to H^+ and HS^- is 7.04, most of endogenous H_2S will be volatilized at pH 5.7. However, because H_2S is also reabsorbed by tissue homogenates [32], the value must be under estimated.

Recently, the concentration of H_2S was re-evaluated by LC–MS/MS after generating the stable adducts with an alkylating agent. Brains were homogenized in phosphate buffered (pH 7.0) with monobromobimane, the alkylating agent, to react with H_2S to generate a sulfide dibimane adduct, which was analyzed by

LC–MS/MS, to provide a concentration of approximately 3.0 μM [5]. The sulfide dibimane adduct is stable and cannot be reabsorbed by tissue homogenates. However, even with this method, during homogenization the extracellular and intracellular molecules are mixed and can react with each other. For example, extracellular cysteine disulfide can be reduced by intracellular glutathione to generate more enzymatic H_2S. In addition, endogenous *S*-sulfurated molecules may react with each other and some of them release H_2S before generating the sulfide dibimane adducts [38].

2.3 H_2S and H_2S_n as Signaling Molecules

H_2S facilitates the induction of hippocampal LTP by enhancing the activity of NMDA receptors [6]. H_2S reduces cysteine disulfide located at the hinge of the ligand-binding domain of NMDA receptors to enhance their activity. In addition, hydrogen polysulfides (H_2S_n, $n \geq 2$) activate TRPA1 channels in astrocytes, which surround synapses and release D-serine, a gliotransmitter, and co-agonist of NMDA receptors [9–13]. H_2S_n *S*-sulfurates 2 out of 23 cysteine residues located at the amino terminus of TRPA1 channels to activate them [39, 40]. As an alternative mechanism, one of the two cysteine residues is *S*-sulfurated and reacts with another cysteine residue to generate a cysteine disulfide bond that may induce the conformational changes more effectively than just two *S*-sulfurated cysteine residues (Figure 2.2). A tumor suppressor, phosphatase and tensin homolog (PTEN), is regulated in a similar manner [41]. Either of the two cysteine residues Cys 71 and Cys 124 is *S*-sulfurated and then generates a cysteine disulfide

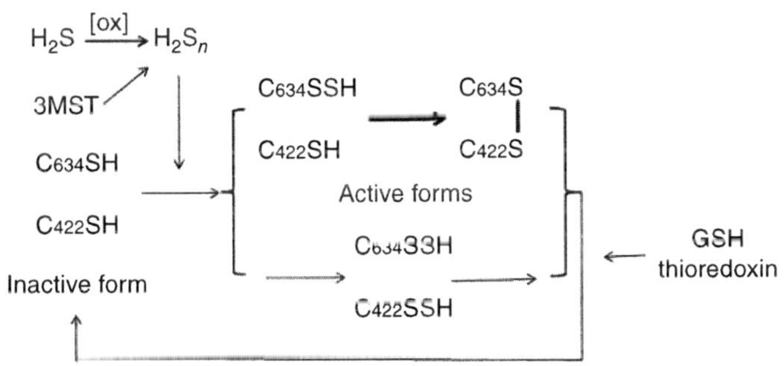

Figure 2.2 TRPA1 channels are activated by S-sulfuration of two cysteine residues located at the amino terminus of the channels. H_2S_n, which are produced by the oxidation of H_2S or by 3MST, *S*-sulfurate two cysteine residues to activate the channels. Alternatively, one of the two cysteine residues is *S*-sulfurated and reacts with the remaining cysteine residue to generate a cysteine disulfide bond. The latter mechanism may more efficiently induce the conformational changes than the former one.

bond resulting in the conformational changes. A monomer protein kinase G1α is inactive, while *S*-sulfurated monomer reacts with another monomer to generate the active homo dimer [42].

H_2S_2 and H_2S_3 exist endogenously in the brain, and the concentration of H_2S_2 is approximately 2.6 μM, which is almost equivalent to the concentration of H_2S (3.0 μM) [5, 12]. H_2S_n was produced from 3-mercaptopyruvate (3MP), which was exogenously applied and incorporated into cells, and observed with the SSP4 sulfane sulfur-specific fluorescence probe developed by Chen et al. [12, 43]. Although this nonreversible probe has a disadvantage of continuing emission of fluorescence even after H_2S_n have disappeared, it has an advantage to clearly show the localization of produced low levels of H_2S_n in cells by accumulating the response until it reaches the sensitivity threshold. H_2S_n and other *S*-sulfurated molecules, such as cysteine persulfide (CysSSH), glutathione persulfide (GSSH), and persulfurated cysteine residues of proteins (P-CysSSH), are produced by 3MST [14, 38]. *S*-Sulfurated molecules are also produced by other enzymes including SQR [16], CARS [21], hemoproteins [22], SOD [17–19], catalase [20], and peroxidases [23, 44].

2.4 Crosstalk Between H_2S and NO

2.4.1 The Chemical Interaction of H_2S and NO Produces H_2S_n

In addition to enzymatic production, H_2S_n are generated by the chemical interaction between H_2S and NO. This finding stems from the discovery that H_2S relaxes vascular smooth muscle in synergy with NO [24]. Whiteman et al. proposed that a nitrosothiol is produced by the interaction of the two molecules [25]. HSNO, HNO, $SSNO^-$, and H_2S_n (mainly H_2S_2 and H_2S_3) have been demonstrated to be generated from H_2S and NO [26–29, 45, 46].

The addition of DEA NONOate, an NO donor, to Na_2S, the sodium salt of H_2S, consumed H_2S to generate H_2S_2 and H_2S_3 [29]. Dibimane adducts of H_2S_n were detected by LC–MS/MS, and H_2S_n localized in cells were detected by the SSiP1 fluorescence probe, which reversibly detects H_2S_n and other *S*-sulfurated molecules. In addition, an increase in the influx of Ca^{2+} by the activation of TRPA1 was detected by the Ca^{2+} sensitive dye Fluo 4, which was in parallel with the increased intracellular levels of H_2S_n generated from H_2S and NO [29].

Eberhardt et al. and Cortese-Krott et al. reported that H_2S_n were produced from H_2S and NO, but that HNO and $SSNO^-$ were proposed to be the major reactive products (Figure 2.3) [26, 27]. The products formed from the reaction of H_2S and NO were degraded by cyanide (HCN) and reducing substances (DTT). Both

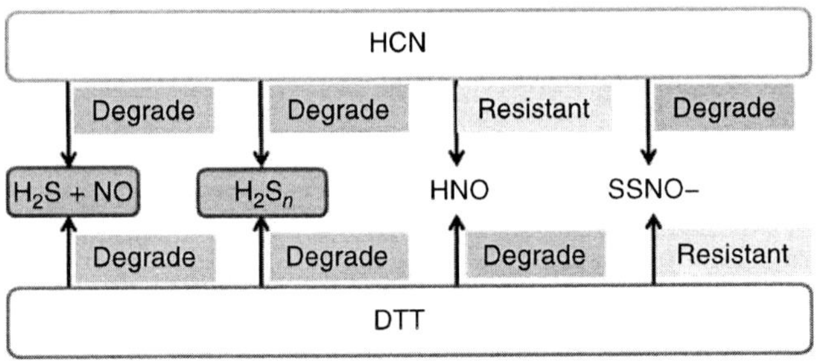

Figure 2.3 Products from the chemical interaction of H_2S and NO. The chemical interaction of H_2S and NO generates H_2S_n, HNO, and $SSNO^-$. Both H_2S_n and the product from H_2S and NO are sensitive to cyanide and DTT. In contrast, HNO is resistant to cyanide, and $SSNO^-$ is resistant to DTT. Based on these observations H_2S_n are the potential products from H_2S and NO to activate TRPA1 channels. Source: Based on Miyamoto et al. [29].

H_2S_n and the product from H_2S and NO are sensitive to cyanide and DTT. In contrast, HNO is resistant to cyanide, and $SSNO^-$ is resistant to DTT. Based on these observations, H_2S_n are likely to be the chemical entities produced from the interaction of the two molecules rather than HNO and $SSNO^-$. H_2S_n and HNO have been proposed to activate TRPA1 channels [12, 26, 29], while $SSNO^-$ has been initially reported as a carrier of NO to release it for relaxing vascular smooth muscle [27]. A recent study by Bogdándi et al. proposed that H_2S_n transiently activate TRPA1 channels at the early phase of the production from H_2S and NO, and that $SSNO^-$, which is more stable than H_2S_n, sustainably activate the channels [30].

2.4.2 Regulation of NO-Producing Enzymes by H_2S and Vice Versa

Another crosstalk between H_2S and NO is the regulation of NO-producing enzymes by H_2S and vice versa. The NO donors nitroprusside and *S*-nitroso-*N*-acetyl-DL-penicillamine increase H_2S production and expression of cystathionine γ-lyase (CSE), respectively [47]. H_2S activates endothelial NO synthetase (eNOS) and also increases the production of NO, which is consistent with the observation that NO metabolites were lower in the blood and heart in CSE-knockout mice compared to the wild type. Via this mechanism, H_2S protects the heart from ischemia–reperfusion injury [48]. H_2S also protects against pressure overloaded-induced heart failure through a similar mechanism [49]. In contrast, H_2S suppresses the activity of neuronal NOS in submucosal- and myenteric-ganglion neurons where endogenous production of NO is significantly greater in CSE-knockout mice than in the wild type [50].

2.5 Cytoprotective Effect of H_2S, H_2S_n, and H_2SO_3

The application of H_2S protects neurons from oxidative stress. It enhances the activity of cystine/glutamate antiporter to increase the intracellular concentrations of cysteine, a substrate for the production of GSH and augments the activity of glutamate cysteine ligase (GCL) a rate-limiting enzyme also known as γ-glutamyl cysteine synthetase for GSH production [51, 52]. H_2S stabilizes the membrane potential to protect neurons from over-excitation by enhancing the activity of adenosine triphosphate (ATP)-dependent K^+ channels and cystic fibrosis transmembrane conductance regulator Cl^- channels [53].

H_2S applied to tissue cultures immediately decreases its level in half by volatilization in around 15 minutes and almost disappears in one hour [53]. H_2S is also oxidized to H_2S_n and further to sulfite (SO_3^{2-}) and thiosulfate ($S_2O_3^{2-}$) [54]. H_2S_n, as well as H_2S, exerts cytoprotective effects by increasing the intracellular concentrations of cysteine and GSH. H_2S_n facilitates S-sulfuration of Keap1 to release Nrf2 from Keap1/Nrf2 complex to the nucleus, leading to the upregulation of redox gene transcription including GCL for the production of GSH [55].

Sulfite also exerts a cytoprotective effect greater than H_2S and H_2S_n by increasing the extracellular concentrations of cysteine, which is efficiently transported into cells and used for the production of GSH [54]. The endogenous sulfite level is approximately 2 μM in the brain [54]. Sulfite reacts with cystine to generate cysteine and *S*-cysteinesulfonate [56], and the transformation efficiency is approximately 50% [54]. To protect neurons from oxidative stress, 1 μM sulfite is enough, while it generates 0.5 μM *S*-cysteinesulfonate, which causes excitotoxicity similarly to glutamate. For excitotoxicity, 2–5 μM glutamate is required [57], and pathophysiological concentration of glutamate for traumatic brain injury is approximately 20 μM [58]. Considering these observations, the toxic effect of 0.5 μM *S*-cysteinesulfonate can be minimal. However, the special attention is required for those who are sensitive to sulfite and those deficient in sulfite oxidase [59, 60].

2.6 Energy Formation in Mitochondria with H_2S

H_2S as a toxic gas inhibits cytochrome c oxidase by suppressing the binding of oxygen and subsequently decreases energy formation [61]. It has been hypothesized that mitochondria originated from sulfide-oxidizing symbionts. Yong and Searcy demonstrated that in the presence of low concentrations of H_2S, chicken liver mitochondria accelerated the consumption of oxygen coupled to ATP generation. They showed that ATP synthesis requires less than 5 μM H_2S, and maximum respiration is induced at 10 μM [62]. The brain contains approximately 3 μM H_2S [5], which may be enough to be involved in the synthesis of ATP.

Low concentrations of H_2S preserve or enhance mitochondrial function, while high concentrations inhibit cytochrome c oxidase [63]. H_2S binds to SQR that transfers two electrons and as many protons to coenzyme Q. At cytochrome c oxidase, a pair of electrons reduces a single oxygen atom [64]. However, in the brain, the expression of SQR is significantly lower [65]; therefore, the hemoprotein neuroglobin is predicted to be involved in the oxidation of H_2S instead of SQR [66].

Tetrazolium salts, including MTT and WST-8, are reduced by cellular activity to form intensely colored formazans, and used for quantitating living cells. As the reduction can be catalyzed by enzymes located in postmitochondrial subcellular fractions, a decrease in MTT reduction has been ascribed to a decrease in the cell viability due to mitochondrial dysfunction [67, 68]. H_2S enhanced WST-8 reducing activity in primary cultures of neurons and a neuronal cell line but not in other cell types [69]. Although H_2S is a reducing molecule, the possibility that H_2S reduced WST-8 was excluded by the fact that a control experiment in the presence of H_2S without cells did not enhance WST-8 reduction. The other reducing agents GSH, DTT, α-tocopherol, and 2-mercaptoethanol did not enhance WST-8 reduction in cells, and the effect was specific to H_2S. H_2S enhanced WST-8 reduction without decreasing ATP production even at 100 μM, while azide, another cytochrome c oxidase inhibitor, did not enhance WST-8 reduction but decreased ATP production [69]. Azide greatly elevated the NADH/NAD ratio, while H_2S increased the ratio less than azide. Tyrosine kinase inhibitors suppressed the effects of H_2S on WST-8 reduction, which may suggest a trophic activity in which the activation of tyrosine kinase is involved.

2.7 *S*-Sulfurated Proteins and Bound Sulfane Sulfur in Cells

Cells have proteins with *S*-sulfurated cysteine residues (P-CysSSH) which occupy 10–20% of the cysteine sites [35]. The *S*-sulfurated molecules including H_2S_n, cysteine persulfide, GSSH, and P-CysSSH release H_2S under reducing conditions and are included in bound sulfane sulfur (Figure 2.4) [32–34]. The amount and identity of the sulfur species of bound sulfane sulfur are distinct among different tissues. For example, liver contains a greater amount of bound sulfane sulfur than brain, and heart contains less than the liver and brain [32]. A release of H_2S from liver homogenates reaches to maximum at 1–2 hours after the addition of DTT and then decreases, while brain homogenates gradually increase the amount of released H_2S for up to 5 hours without significant decreases.

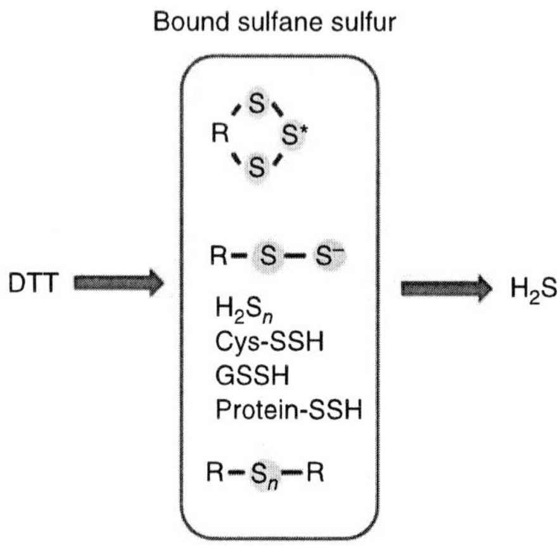

Figure 2.4 Bound sulfane sulfur, which releases H_2S under reducing condition, contains *S*-sulfurated (*S*-sulfhydrated) molecules. Source: Based on Refs. [32–34].

S-Sulfurated molecules including H_2S_n, cysteine persulfide, GSSH, and *S*-sulfurated cysteine residues of proteins in cells release H_2S under reducing conditions.

Tissue homogenates absorb H_2S. For example, liver homogenates absorb the entire applied H_2S immediately after its addition, while heart homogenates immediately absorb approximately 80% of applied H_2S over about 10 minutes. Brain homogenates absorb applied H_2S gradually over the course of about 30 minutes [32]. These observations suggest that few P-CysSSH are contained in heart homogenates, but that a reasonable amount of P-CysSOH and P-CysSNO are present, which can absorb H_2S to generate P-CysSSH.

Some of bound sulfane sulfur or *S*-sulfurated molecules can be reduced to thiols under physiological conditions. In the presence of the endogenous concentrations of GSH together with cysteine or dihydrolipoic acid alone, bound sulfane sulfur in neurons and astrocytes releases H_2S at pH 8.4 [15, 32]. When neurons repetitively excite, the extracellular concentrations of K^+ are increased. To return to the resting potential, the Na^+/HCO_3^- cotransporter is activated, and the entrance of HCO_3^- into cells causes the alkalinization, in which the intracellular pH increase to 8.4 [32, 70]. These observations suggest that the activity of neurons may induce the reduction of *S*-sulfurated cysteine residues of target proteins to thiols to change their activity.

2.8 Regulating the Activity of Target Proteins by H_2S and H_2S_n

Human proteins include approximately 214,000 cysteine residues encoded in the genome. In cells and tissues, 5–12% of total protein cysteine residues are oxidized,

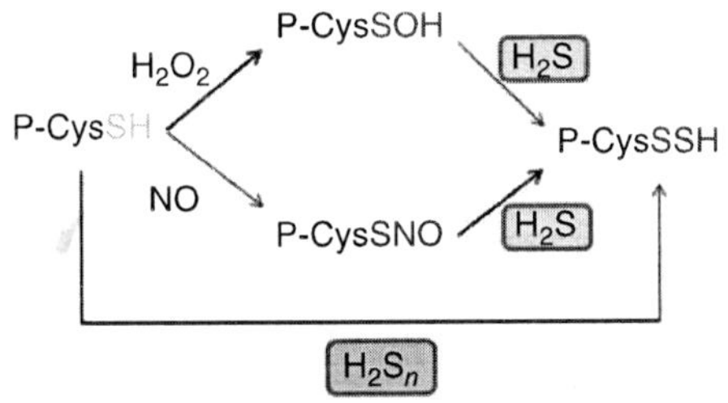

Figure 2.5 S-Sulfuration (S-sulfhydration) of cysteine residues of target proteins modifies their activity by inducing the conformational changes. H_2S *S*-sulfurates oxidized cysteine residues, such as *S*-sulfenylated and *S*-nitrosylated cysteines, while H_2S_n *S*-sulfurates cysteine residues.

and this can be increased to more than 40% under oxidative conditions [71]. The oxidized form includes P-CysSNO and P-CysSOH, which can be *S*-sulfurated by H_2S to generate P-CysSSH, while P-CysSH are *S*-sulfurated by H_2S_n to produce other *S*-sulfurated molecules (Figure 2.5).

2.8.1 S-Sulfuration by H_2S

Specific cysteine residues of parkin, which is an E3 ubiquitin ligase to clear misfolded proteins associated with Parkinson's disease, are *S*-nitrosylated (P-CysSNO) in patients, while they are *S*-sulfurated (P-CysSSH) in normal individuals [72]. Because H_2S can *S*-sulfurate P-CysSNO to P-CysSSH, H_2S may be involved in maintaining parkin in the active form. Hydrogen peroxide (H_2O_2) transmits signals via S-sulfenylation of the target proteins and NO via S-nitrosylation [73, 74], and these oxidized cysteine residues can then be *S*-sulfurated by H_2S.

Most eukaryotic manganese superoxide dismutase (MnSOD), which have at least one cysteine residue, are strongly suppressed by H_2O_2. However, cotreatment with H_2S rescues the enzymatic activity, and Cys193 is *S*-sulfurated [36]. There are two possible mechanisms for the S-sulfuration of MnSOD. The Cys193 is *S*-sulfenylated by H_2O_2, and then *S*-sulfurated by H_2S. Alternatively, the mixture of H_2S and H_2O_2 produces H_2S_n which *S*-sulfurate Cys193. Because the reaction of H_2S with H_2O_2 is slow (0.73 M^{-1} s^{-1}) [75] and because *S*-sulfenylated protein reacts 2 orders of magnitude faster with H_2S than with glutathione [76], the former reaction must be prevalent.

Epidermal growth factor (EGF) activates EGF receptor tyrosine kinase and increases the levels of P-CysSOH in cells at the first 5–15 minutes that returned to the control levels by 30 minutes. P-CysSSH levels are initially decreased but turned into an increase and reached at the maximal at 30 minutes when the expression of H_2S-producing enzymes are increased [36]. The same group reported that a similar result was obtained for vascular endothelial growth factor (VEGF), which activates VEGF receptors, another tyrosine kinase-associated receptors. These observations suggest that the signaling by H_2O_2 through S-sulfenylation of the target proteins is further modified by H_2S to S-sulfuration.

The relation between H_2S and the regulation of kinase activity has been demonstrated in CSE-knockout mice. In these mice, the eNOS activation site serine is less phosphorylated, while eNOS inhibitory site threonine is more phosphorylated than in the wild-type mice. The levels of H_2S and bound sulfane sulfur in the heart and blood are also reduced in CSE-knockout mice [48].

A merit of S-sulfuration is to prevent proteins from over-oxidization such as oxidizing to sulfinic (P-$CysSO_2H$) and sulfonic acids (P-$CysSO_3H$) which deteriorate the function of the proteins. Although *S*-sulfenylated proteins (P-CysSOH) can be reduced by DTT and thioredoxin to P-CysSH, further oxidized proteins P-$CysSO_2H$ and P-$CysSO_3H$ are not able to recover to P-CysSH. In contrast, *S*-sulfurated proteins P-Cys SSO_2H and P-$CysSSO_3H$ can be reduced by thioredoxin to P-CysSH [36, 77]. Thus, S-sulfuration protects proteins from over-oxidation.

2.8.2 S-Sulfuration by H_2S_n

When cysteine residues of target proteins are thiols (P-CysSH), such as TRPA1, PTEN, and protein kinase G1α, H_2S_n and other *S*-sulfurated molecules can *S*-sulfurate them as described previously. These are the examples in which two cysteine residues are involved in the regulation of target proteins.

There are other examples in which S-sulfuration of one cysteine residue is enough to regulate their activity. ATP-activated K^+ (K_{ATP}) channels were initially found to be activated by H_2S [47] through S-sulfuration of Kir subunit Cys43, which is located within the ATP-binding region adjacent to phospholipid, phosphatidylinositol (4,5)-bisphosphate (PIP2) binding site, that enhances the binding to PIP2 while inhibiting the binding to ATP [78]. However, a recent study shows that H_2S_n have a greater potency than H_2S to activate K_{ATP} [79], suggesting that Cys43 may be a thiol rather than an oxidized cysteine residue.

GAPDH is activated through S-sulfuration at the active site Cys150 by H_2S [35], but it was later reported to be suppressed through S-sulfuration at non-active site Cys156 by H_2S_n [80]. It is possible that Cys150 is oxidized like P-CysSNO or P-CysSOH and that H_2S *S*-sulfurates these residues, while Cys156 is a thiol and *S*-sulfurated by H_2S_n. About 5–12% of cysteine residues of protein are oxidized, and these levels are increased up to 40% under oxidative stress [71]. Depending on the redox conditions of cysteine residues, either H_2S or H_2S_n modify the activity of target proteins.

2.9 Perspectives

H_2S has been recognized as a signaling molecule, and then H_2S_n was identified to be more potent than H_2S depending on the target proteins [6, 10–12, 14, 24, 41]. H_2S has been proposed to *S*-sulfurate (*S*-sulfhydrate) cysteine residues of target proteins to regulate their activity [35]. However, the oxidation state of sulfur in both H_2S and cysteine is −2, and atoms in the same oxidation state do not react with each other. Therefore, H_2S is not able to *S*-sulfurate cysteine residues. In contrast, the oxidation state of sulfur in H_2S_n is 0 or −1 and H_2S_n *S*-sulfurate cysteine residues. TRPA1 channels, PTEN, and protein kinase G1α were identified as target proteins for H_2S_n [12, 41, 42].

H_2S_n and other *S*-sulfurated molecules, such as cysteine persulfide, can *S*-sulfurate target proteins. However, 5–12% cysteine residues of total proteins are oxidized in cells and tissues, and the ratio can be increased more than 40% under oxidative stress [71]. Because the oxidation states of sulfur atoms of oxidized cysteine residues such as P-Cys-SOH and P-Cys-SNO are more positive than −2, they can be *S*-sulfurated by H_2S [81]. MnSOD loses its activity by S-sulfenylation, and this activity can be recovered by S-sulfuration by H_2S [36]. S-Sulfuration of Cys150 and Cys156 showed the opposite effects on GAPDH activity. Cys150 is reported to be *S*-sulfurated by H_2S, while Cys156 by H_2S_n [35, 80]. It is necessary to investigate the redox state of the target cysteine residues, namely thiols or oxidized ones, and determine either H_2S or H_2S_n or other *S*-sulfurated molecules transmit the signals.

S-Sulfuration is often compared with phosphorylation. S-Sulfuration of cysteine increases the molecular mass of 32 and −1 charge by CysSSH (pK_a 4.34) dissociating to $CysSS^-$ and H^+, while CysSH (pK_a 8.29) is not dissociated under physiological conditions [76]. Phosphorylation of serine, threonine, and tyrosine increases the molecular mass of 95 with −2 charges under physiological condition. Phosphorylation is more effective than S-sulfuration to cause the conformational changes of the target proteins. However, as described previously, *S*-sulfurated cysteine residues can react with cysteine thiols to make a cysteine disulfide bond, which may be efficient to cause the conformational changes [37].

Lefer and coworkers showed examples in which phosphorylation and S-sulfuration interact with each other. In CSE-knockout mice, the active site Ser1177 of eNOS is less phosphorylated, while the inhibitory site Thr495 is more greatly phosphorylated than in the wild-type mice [48]. The administration of H_2S and polysulfide DATS (diallyl trisulfide) enhanced the phosphorylation of the active site Ser1177 and increase the cardiac nitrite and the levels of nitrosylated proteins [48]. The same group also showed that SG-1002, a H_2S (and H_2S_n) donor, activates VEGF and phosphorylates Akt. The activation of Akt induces phosphorylation of eNOS [49]. Because both DATS and SG-1002 are polysulfides, it is possible that S-sulfuration of cysteine residues of VEGF and its receptors as well as Akt may be involved in the enhancement of eNOS phosphorylation. H_2S may *S*-sulfurate the oxidized cysteine residues of these proteins or may *S*-sulfurate thiols after being oxidized to H_2S_n. It is interesting to know the redox conditions of cysteine residues of VEGF, VEGF receptor, Akt, and eNOS.

Zivanovic et al. showed that EGF, VEGF, and insulin increased the levels of *S*-sulfurated proteins in HeLa cells, human umbilical endothelial vein cells, and SH-SY5Y cells, respectively [36]. Cysteine sulfenylation of EGFR enhances phosphorylation of tyrosine1068 and activates the receptors, while they were suppressed by H_2S. The inhibition of H_2S production increased the half-life of EGFR sulfanylation and its activity. They suggested that H_2O_2 *S*-sulfinates target proteins, which are subsequently *S*-sulfurated by H_2S and that H_2S has a role as a redox switch for controlling cellular signaling initiated by H_2O_2 [36]. Thus, S-sulfuration and phosphorylation interact reciprocally.

In the presence of physiological concentrations of cysteine and GSH, H_2S_2, and H_2S_3 are immediately consumed to *S*-sulfurate cysteine and GSH to generate cysteine persulfide and GSSH [38]. Approximately 3 μM H_2S, 2.6 μM H_2S_2, and less H_2S_3 exist in the brain [5]. H_2S_n, cysteine persulfide, and GSSH as well as H_2S, cysteine, and GSH mostly exist inside the cells, while cystine and GSSG are in the extracellular milieu, which is under more oxidizing conditions than inside the cells. The level of each *S*-sulfurated molecule must be well balanced by interacting with each other. It is interesting to examine how the balance is changed when cells are stimulated to transmit signals, and how it is disturbed under pathophysiological conditions.

H_2S was initially thought to pass through the plasma membrane [82], but later HS^- channels were discovered in bacteria [83] and AE1 channels was subsequently identified to let HS^- pass through in mammals [84]. Although H_2S_n can pass though the plasma membrane, the channels or transporters have yet been identified. Cysteine and GSH remain in the intracellular environment, while H_2S and H_2S_n can not only transduce signals inside the cell but also transmit between cells. It is necessary to identify the channels and transporters for these *S*-sulfurated molecules including H_2S_n.

In order to investigate the real-time localization of these molecules in cells, fluorescence probes specific to them with high sensitivity need to be developed. HSip1 is selective to H_2S and SSP4 is sensitive to *S*-sulfurated molecules, but neither are reversible [43, 85]. SSip1 reversibly detect *S*-sulfurated molecules, but not sensitive enough to detect the endogenous molecules [86]. Fluorescence probes developed so far are not sensitive enough to detect the endogenous H_2S, H_2S_n and other *S*-sulfurated molecules in real-time. Development of such probes is awaited.

Acknowledgments

This work was supported by a grant from Strategic Research Program for Brain Sciences from the Japan Agency for Medical Research and Development, AMED, Grant/Award Number: JP20dm0107085; KAKENHI, Grant/Award Number: 17K08331 to H.K.

Author Disclosure Statement

There is not conflict of interest.

References

1 Warenycia, M.W., Goodwin, L.R., Benishin, C.G. et al. (1989). Acute hydrogen sulfide poisoning. Demonstration of selective uptake of sulfide by the brainstem by measurement of brain sulfide levels. *Biochem. Pharmacol.* 38: 973–981.

2 Goodwin, L.R., Francom, D., Dieken, F.P. et al. (1989). Determination of sulfide in brain tissue by gas dialysis/ion chromatography: postmortem studies and two case reports. *J. Anal. Toxicol.* 13: 105–109.

3 Savage, J.C. and Gould, D.H. (1990). Determination of sulfide in brain tissue and rumen fluid by ion-interaction reversed-phase high-performance liquid chromatography. *J. Chromatogr.* 526: 540–545.

4 Furne, J., Saeed, A., and Levitt, M.D. (2008). Whole tissue hydrogen sulfide concentrations are orders of magnitude lower than presently accepted values. *Am. J. Physiol. Regul. Integr. Comp. Physiol.* 295: R1479–R1498.

5 Koike, S., Kawamura, K., Kimura, Y. et al. (2017). Analysis of endogenous H_2S and H_2S_n in mouse brain by high performance liquid chromatography with fluorescence and tandem mass spectrometric detection. *Free Radic. Biol. Med.* 113: 355–362.

6 Abe, K. and Kimura, H. (1996). The possible role of hydrogen sulfide as an endogenous neuromodulator. *J. Neurosci.* 16: 1066–1071.

7 Aizenman, E., Lipton, D.A., and Loring, R.H. (1989). Selective modulation of NMDA responses by reduction and oxidation. *Neuron* 2: 1257–1263.

8 Dani, J.W., Chernjavsky, A., and Smith, S.J. (1992). Neuronal activity triggers calcium waves in hippocampal astrocyte networks. *Neuron* 8: 429–440.

9 Nagai, Y., Tsugane, M., Oka, J., and Kimura, H. (2004). Hydrogen sulfide induces calcium waves in astrocytes. *FASEB J.* 18: 557–559.

10 Nagai, Y., Tsugane, M., Oka, J.-I., and Kimura, H. (2006). Polysulfides induce calcium waves in rat hippocampal astrocytes. *J. Pharmacol. Sci.* 100: 200.

11 Oosumi, K., Tsugane, M., Ishigami, M. et al. (2010). Polysulfide activates TRP channels and increases intracellular Ca^{2+} in astrocytes. *Bull. Jpn. Soc. Neurochem.* 49: 517.

12 Kimura, Y., Mikami, Y., Osumi, K. et al. (2013, 2013). Polysulfides are possible H_2S-derived signaling molecules in rat brain. *FASEB J.* 27: 2451–2457.

13 Shigetomi, E., Jackson-Weaver, O., Huckstepp, R.T. et al. (2013). TRPA1 channels are regulators of astrocyte basal calcium levels and long-term potentiation via constitutive D-serine release. *J. Neurosci.* 33: 10143–10153.

14 Kimura, Y., Toyofuku, Y., Koike, S. et al. (2015). Identification of H_2S_3 and H_2S produced by 3-mercaptopyruvate sulfurtransferase in the brain. *Sci. Rep.* 5: 14774.

15 Mikami, Y., Shibuya, N., Kimura, Y. et al. (2011). Thioredoxin and dihydrolipoic acid are required for 3-mercaptopyruvate sulfurtransferase to produce hydrogen sulfide. *Biochem. J.* 439: 479–485.

16 Landry, A.P., Ballou, D.P., and Banerjee, R. (2017). H_2S oxidation by nanodisc-embedded human sulfide quinone oxidoreductase. *J. Biol. Chem.* 292: 11641–11649.

17 Searcy, D.G., Whitehead, J.P., and Maroney, M.J. (1995). Interaction of Cu, Zn superoxide dismutase with hydrogen sulfide. *Arch. Biochem. Biophys.* 318: 251–263.

18 Searcy, D.G. (1996). HS^-:O_2 oxidoreductase activity of Cu, Zn superoxide dismutase. *Arch. Biochem. Biophys.* 334: 50–58.

19 Olson, K.R., Gao, Y., Arif, F. et al. (2018). Metabolism of hydrogen sulfide (H_2S) and production of reactive sulfur species (RSS) by superoxide dismutase. *Redox Biol.* 15: 74–85.

20 Olson, K.R., Gao, Y., DeLeon, E.R. et al. (2017). Catalase as a sulfide–sulfur oxido-reductase: an ancient (and modern?) regulator of reactive sulfur species (RSS). *Redox Biol.* 12: 325–339.

21 Akaike, T., Ida, T., Wei, F.-Y. et al. (2017). Cysteinyl-tRNA synthetase governs cysteine polysulfidation and mitochondrial bioenergetics. *Nat. Commun.* 8: 1177.

22 Vitvitsky, V., Yadav, P.K., Kurthen, A., and Banerjee, R. (2015). Sulfide oxidation by a noncanonical pathway in red blood cells generates thiosulfate and polysulfides. *J. Biol. Chem.* 290: 8310–8320.

23 Garai, D., Ríos-González, B.B., Furtmüller, P.G. et al. (2017). Mechanisms of myeloperoxidase catalyzed oxidation of H_2S by H_2O_2 or O_2 to produce potent protein Cys–polysulfide-inducing species. *Free Radic. Biol. Med.* 113: 551–563.

24 Hosoki, R., Matsuki, N., and Kimura, H. (1997). The possible role of hydrogen sulfide as an endogenous smooth muscle relaxant in synergy with nitric oxide. *Biochem. Biophys. Res. Commun.* 237: 527–531.

25 Whiteman, M., Li, L., Kostetski, I. et al. (2006). Evidence for the formation of a novel nitrosothiol from the gaseous mediators nitric oxide and hydrogen sulphide. *Biochem. Biophys. Res. Commun.* 343: 303–310.

26 Eberhardt, M., Dux, M., Namer, B. et al. (2014). H_2S and NO cooperatively regulate vascular tone by activating a neuroendocrine HNO–TRPA1–CGRP signaling pathway. *Nat. Commun.* 5: 4381.

27 Cortese-Krott, M.M., Kuhnle, G.G.C., Dyson, A. et al. (2015). Key bioactive reaction products of the NO/H_2S interaction are S/N-hybrid species, polysulfides, and nitroxyl. *Proc. Natl. Acad. Sci. U.S.A.* 112: E4651–E4660.

28 Moustafa, A. and Habara, Y. (2016). Cross talk between polysulfide and nitric oxide in rat peritoneal mast cells. *Am. J. Physiol. Cell Physiol.* 310: C894–C902.

29 Miyamoto, R., Koike, S., Takano, Y. et al. (2017). Polysulfides (H_2S_n) produced from the interaction of hydrogen sulfide (H_2S) and nitric oxide (NO) activate TRPA1 channels. *Sci. Rep.* 7: 45995.

30 Bogdándi, V., Ditrói, T., Bátai, I.Z. et al. (2020). Nitrosopersulfide ($SSNO^-$) is a unique cysteine polysulfidating agent with reduction-resistant bioactivity. *Antioxid. Redox Signal.* 33: 1277–1294.

31 Ogasawara, Y., Isoda, S., and Tanabe, S. (1994). Tissue and subcellular distribution of bound and acid-labile sulfur, and the enzymic capacity for sulfide production in the rat. *Biol. Pharm. Bull.* 17: 1535–1542.

32 Ishigami, M., Hiraki, K., Umemura, K. et al. (2009). A source of hydrogen sulfide and a mechanism of its release in the brain. *Antioxid. Redox Signal.* 11: 205–214.

33 Warenycia, M.W., Goodwin, L.R., Francom, D.M. et al. (1990). Dithiothreitol liberates non-acid labile sulfide from brain tissue of H_2S-poisoned animals. *Arch. Toxicol.* 64: 650–655.

34 Ogasawara, Y., Ishii, K., Togawa, T., and Tanabe, S. (1993). Determination of bound sulfur in serum by gas dialysis/high-performance liquid chromatography. *Anal. Biochem.* 215: 73–81.

35 Mustafa, A.K., Gadalla, M.M., Sen, N. et al. (2009). H_2S signals through protein S-sulfhydration. *Sci. Signal.* 2: ra72.

36 Zivanovic, J., Kouroussis, E., Kohl, J.B. et al. (2019). Selective persulfide detection reveals evolutionarily conserved antiaging effects of S-sulfuration. *Cell Metab.* 30: 1152–1170.e13.

37 Kimura, H. (2020). Hydrogen sulfide signaling in the central nervous system – comparison with nitric oxide. *Br. J. Pharmacol.* 177: 5031–5045.

38 Kimura, Y., Koike, S., Shibuya, N. et al. (2017). 3 Mercaptopyruvate sulfurtransferase produces potential redox regulators cysteine- and glutathione-persulfide (Cys-SSH and GSSH) together with signaling molecules H_2S_2, H_2S_3 and H_2S. *Sci. Rep.* 7: 10459.

39 Wang, L., Cvetkov, T.L., Chance, M.R., and Moiseenkova-Bell, V.Y. (2012). Identification of in vivo disulfide conformation of TRPA1 ion channel. *J. Biol. Chem.* 287: 6169–6176.

40 Hatakeyama, Y., Takahashi, K., Tominaga, M. et al. (2015). Polysulfide evokes acute pain through the activation of nociceptive TRPA1 in mouse sensory neurons. *Mol. Pain* 11: 24.

41 Greiner, R., Palinkas, Z., Basell, K. et al. (2013). Polysulfides link H_2S to protein thiol oxidation. *Antioxid. Redox Signal.* 19: 1749–1765.

42 Stubbert, D., Prysyazhna, O., Rudyk, O. et al. (2014). Protein kinase G Iα oxidation paradoxically underlies blood pressure lowering by the reductant hydrogen sulfide. *Hypertension* 64: 1344–1351.

43 Chen, W. (2013). New fluorescent probes for sulfane sulfurs and the application in bioimaging. *Chem. Sci.* 4: 2892–2896.

44 Nakamura, S., Nakamura, M., Yamazaki, I., and Morrison, M. (1984). Reactions of ferryl lactoperoxidase (compound II) with sulfide and sulfhydryl compounds. *J. Biol. Chem.* 259: 7080–7085.

45 Filipovic, M.R., Miljkovic, J., Allgauer, A. et al. (2012). Biochemical insight into physiological effects of H_2S: reaction with peroxynitrite and formation of a new nitric oxide donor, sulfinyl nitrite. *Biochem. J.* 441: 609–621.

46 Cortese-Krott, M.M., Fernandez, B.O., Santos, J.L.T. et al. (2014). Nitrosopersulfide ($SSNO^-$) accounts for sustained NO bioactivity of *S*-nitrosothiols following reaction with sulfide. *Redox Biol.* 2: 234–244.

47 Zhao, W., Zhang, J., Lu, Y., and Wang, R. (2001). The vasorelaxant effect of H_2S as a novel endogenous gaseous K_{ATP} channel opener. *EMBO J.* 20: 6008–6016.

48 King, A.L., Polhemus, D., Bhushan, S. et al. (2014). Hydrogen sulfide cytoprotective signaling is endothelial nitric oxide synthase-nitric oxide dependent. *Proc. Natl. Acad. Sci. U.S.A.* 111: 3182–3187.

49 Kondo, K., Bhushan, S., King, A.L. et al. (2013). H_2S protects against pressure overload-induced heart failure via upregulation of endothelial nitric oxide synthase. *Circulation* 127: 1116–1127.

50 Sha, L., Linden, D.R., Farrugia, G., and Szurszewski, J.H. (2014). Effect of endogenous hydrogen sulfide on the transwall gradient of the mouse colon circular smooth muscle. *J. Physiol.* 592: 1077–1089.

51 Kimura, Y. and Kimura, H. (2004). Hydrogen sulfide protects neurons from oxidative stress. *FASEB J.* 18: 1165–1167.

52 Kimura, Y., Goto, Y.-I., and Kimura, H. (2010). Hydrogen sulfide increases glutathione production and suppresses oxidative stress in mitochondria. *Antioxid. Redox Signal.* 12: 1–13.

53 Kimura, Y., Dargusch, R., Schubert, D., and Kimura, H. (2006). Hydrogen sulfide protects HT22 neuronal cells from oxidative stress. *Antioxid. Redox Signal.* 8: 661–670.

54 Kimura, Y., Shibuya, N., and Kimura, H. (2019). Sulfite protects neurons from oxidative stress. *Br. J. Pharmacol.* 176: 571–582.

55 Koike, S., Ogasawara, Y., Shibuya, N. et al. (2013). Polysulfide exerts a protective effect against cytotoxicity caused by *t*-buthylhydroperoxide through Nrf2 signaling in neuroblastoma cells. *FEBS Lett.* 587: 3548–3555.

56 Clarke, H.T. (1932). The action of sulfite upon cystine. *J. Biol. Chem.* 97: 235–248.

57 Mark, L.P., Prost, R.W., Ulmer, J.L. et al. (2001). Pictorial review of glutamate excitotoxicity: fundamental concepts for neuroimaging. *Am. J. Neuroradiol.* 22: 1813–1824.

58 Bullock, R., Zauner, A., Woodward, J.J. et al. (1998). Factors affecting excitatory amino acid release following severe human head injury. *J. Neurosurg.* 89: 507–518.

59 Vally, H. and Misso, N.L. (2012). Adverse reactions to the sulphite additives. *Gastroenterol. Hepatol. Bed. Bench.* 5: 16–23.

60 Kumar, A., Dejanovic, B., Hetsch, F. et al. (2017). *S*-Sulfocysteine/NMDA receptor-dependent signaling underlies neurodegeneration in molybdenum cofactor deficiency. *J. Clin. Invest.* 127: 4365–4378.

61 Hill, B.C., Woon, T.C., Nicholls, P. et al. (1984). Interactions of sulphide and other ligands with cytochrome c oxidase. An electron-paramagnetic-resonance study. *Biochem. J.* 224: 591–600.

62 Yong, R. and Searcy, D.G. (2001). Sulfide oxidation coupled to ATP synthesis in chicken liver mitochondria. *Comp. Biochem. Physiol. B* 129: 129–137.

63 Abou-Hamdan, A., Ransy, C., Roger, T. et al. (2016). Positive feedback during sulfide oxidation fine-tunes cellular affinity for oxygen. *Biochim. Biophys. Acta* 1857: 1464–1472.

64 Goubern, M., Andriamihaja, M., Nübel, T. et al. (2007). Sulfide, the first inorganic substrate for human cells. *FASEB J.* 21: 1699–1706.

65 Linden, D.R., Furne, J., Stoltz, G.J. et al. (2012). Sulphide quinone reductase contributes to hydrogen sulphide metabolism in murine peripheral tissues but not in the CNS. *Br. J. Pharmacol.* 165: 2178–2190.

66 Ruetz, M., Kumutima, J., Lewis, B.E. et al. (2017). A distal ligand mutes the interaction of hydrogen sulfide with human neuroglobin. *J. Biol. Chem.* 292: 6512–6528.

67 Berridge, M.V. and Tan, A.S. (1993). Characterization of the cellular reduction of 3-(4,5-dimethylthiazol-2-yl)-2,5-diphenyltetrazolium bromide (MTT): subcellular localization, substrate dependence, and involvement of mitochondrial electron transport in MTT reduction. *Arch. Biochem. Biophys.* 303: 474–482.

68 Liu, Y., Peterson, D.A., Kimura, H., and Schubert, D. (1997). Mechanism of cellular 3-(4,5-dimethylthiazol-2-yl)-2,5-diphenyltetrazolium bromide (MTT) reduction. *J. Neurochem.* 69: 581–593.

69 Umemura, K. and Kimura, H. (2007). Hydrogen sulfide enhances reducing activity in neurons: neurotrophic role of H_2S in the brain? *Antioxid. Redox Signal.* 9: 2035–2041.

70 Brookes, N. and Turner, R.J. (1994). K^+-Induced alkalinization in mouse cerebral astrocytes mediated by reversal of electrogenic Na^+-$HCO_3{}_-$ cotransport. *Am. J. Physiol.* 267: C1633–C1640.

71 Go, Y.M. and Jones, D.P. (2013). The redox proteome. *J. Biol. Chem.* 288: 26512–26520.

72 Vandiver, M.S., Paul, B.D., Xu, R. et al. (2013). Sulfhydration mediates neuroprotective actions of parkin. *Nat. Commun.* 4: 1626.

73 Paulsen, C.E. and Carroll, K.S. (2013). Cysteine-mediated redox signaling: chemistry, biology, and tools for discovery. *Chem. Rev.* 113: 4633–4679.

74 Stomberski, C.T., Hess, D.T., and Stamler, J.S. (2019). Protein *S*-nitrosylation: determinants of specificity and enzymatic regulation of *S*-nitrosothiol-based signaling. *Antioxid. Redox Signal.* 30: 1331–1351.

75 Carballal, S., Trujillo, M., Cuevasanta, E. et al. (2011). Reactivity of hydrogen sulfide with peroxynitrite and other oxidants of biological interest. *Free Radic. Biol. Med.* 50: 196–205.

76 Cuevasanta, E., Lange, M., Bonanata, J. et al. (2015). Reaction of hydrogen sulfide with disulfide and sulfenic acid to form the strongly nucleophilic persulfide. *J. Biol. Chem.* 290: 26866–26880.

77 Nagahara, N., Nirasawa, T., Yoshii, T., and Niimura, Y. (2012). Is novel signal transducer sulfur oxide involved in the redox cycle of persulfide at the catalytic site cysteine in a stable reaction intermediate of mercaptopyruvate sulfurtransferase? *Antioxid. Redox Signal.* 16: 747–753.

78 Mustafa, A.K., Sikka, G., Gazi, S.K. et al. (2011). Hydrogen sulfide as endothelium-derived hyperpolarizing factor sulfhydrates potassium channels. *Circ. Res.* 109: 1259–1268.

79 Shoji, T., Hayashi, M., Sumi, C. et al. (2019). Pharmacological polysulfide suppresses glucose-stimulated insulin secretion in an ATP-sensitive potassium channel-dependent manner. *Sci. Rep.* 9: 19377.

80 Jarosz, A.P., Wei, W., Gauld, J.W. et al. (2015). Glyceraldehyde 3-phosphate dehydrogenase (GAPDH) is inactivated by *S*-sulfuration in vitro. *Free Radic. Biol. Med.* 89: 512–521.

81 Kimura, H. (2018). MPST produces redox regulators Cys-SSH and GSSH as well as signaling molecules H_2S and H_2S_n. *Atlas of Science* (12 July 2018).

82 Mathai, J.C., Missner, A., Kügler, P. et al. (2009). No facilitator required for membrane transport of hydrogen sulfide. *Proc. Natl. Acad. Sci. U.S.A.* 106: 16633–16638.

83 Czyzewski, B.K. and Wang, D.-N. (2012). Identification and characterization of a bacterial hydrosulphide ion channel. *Nature* 483: 494–497.

84 Jennings, M.L. (2013). Transport of H_2S and HS(−) across the human red blood cell membrane: rapid H_2S diffusion and AE1-mediated Cl(−)/HS(−) exchange. *Am. J. Physiol. Cell Physiol.* 305: C941–C950.

85 Sasakura, K., Hanaoka, K., Shibuya, N. et al. (2011). Development of a highly selective fluorescence probe for hydrogen sulfide. *J. Am. Chem. Soc.* 133: 18003–18005.

86 Takano, Y., Hanaoka, K., Shjimamoto, K. et al. (2017). Development of a reversible fluorescent probe for reactive sulfur species, sulfane sulfur, and its biological application. *Chem. Commun.* 53: 1064–1067.

3

Persulfides and Their Reactions in Biological Contexts

Dayana Benchoam[1,2,3], Ernesto Cuevasanta[1,2,4], Matías N. Möller[2,5], and Beatriz Alvarez[1,2]

[1] *Laboratorio de Enzimología, Instituto de Química Biológica, Facultad de Ciencias, Universidad de la República, Montevideo, Uruguay*
[2] *Centro de Investigaciones Biomédicas (CEINBIO), Universidad de la República, Montevideo, Uruguay*
[3] *Graduate Program in Chemistry, Facultad de Química, Universidad de la República, Montevideo, Uruguay*
[4] *Unidad de Bioquímica Analítica, Centro de Investigaciones Nucleares, Facultad de Ciencias, Universidad de la República, Montevideo, Uruguay*
[5] *Laboratorio de Fisicoquímica Biológica, Instituto de Química Biológica, Facultad de Ciencias, Universidad de la República, Montevideo, Uruguay*

List of Abbreviations

FCSD	flavocytochrome c sulfide dehydrogenase
GSH	glutathione
LMW	low molecular weight
mBrB	monobromobimane
MPST	3-mercaptopyruvate sulfurtransferase
SQOR	sulfide quinone oxidoreductase

3.1 Persulfides Are Key Intermediates in Sulfur Metabolism and Signaling

Persulfides ($RSSH/RSS^-$)[1] are formed in biological systems, where they participate in sulfur trafficking, biosynthesis, and catabolism. They are intermediates in the

1 The term "persulfide" is used herein for the mixture of hydropersulfide (RSSH) and persulfide anion (RSS^-) in equilibrium. For RSSH, IUPAC recommended names are hydridodisulfide, disulfanyl, and dithiohydroperoxide.

pathway for H_2S oxidation in the mitochondrion [1], and persulfides formed in specific proteins participate in the biosynthesis of iron–sulfur clusters and other cofactors [2]. Different biochemical processes can lead to the formation of persulfides at the expense of thiols (RSH) or their oxidized derivatives. Some of these processes require hydrogen sulfide (H_2S/HS^-). H_2S[2] is able to exert potential health benefits in mammals, although it is toxic at high concentrations. Persulfides have been proposed to have roles as transducers of H_2S signaling; the posttranslational modification of critical protein thiols to persulfides would relay at least some of the effects of H_2S [1]. Undoubtedly, the biochemistry of persulfides is intimately linked to the biochemistry of thiols and H_2S.

Persulfides contain a sulfane sulfur, i.e. a sulfur atom covalently bound to two sulfurs or to a sulfur and an ionizable hydrogen. Some other sulfane sulfur compounds are thiosulfate ($S_2O_3^{2-}$ or $^-S\text{-}SO_3^-$), and inorganic or organic polysulfides, as well as their conjugated acids (HSS_nSH, RSS_nSR, and RSS_nH). Persulfides are acidic and tend to be in the anionic form at physiological pH. Their chemistry is complex, and it includes both nucleophilic and electrophilic characters (Figure 3.1). Because of this dual reactivity, isolated persulfides decay in aqueous solution, ultimately producing elemental sulfur, polysulfides, H_2S, and thiols. Thus, working with persulfides is challenging. The increasingly recognized relevance of persulfides in biosynthetic and catabolic pathways, as well as in signaling processes, is guiding efforts to achieve a better understanding of the biological chemistry of persulfides. Herein we describe the properties of persulfides in biological contexts.

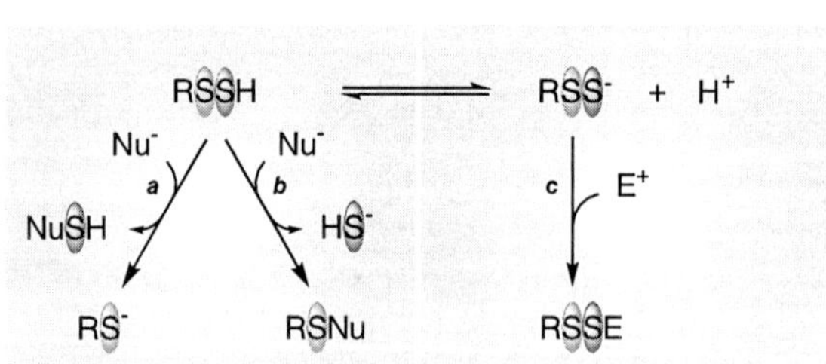

Figure 3.1 Acidity and reactivity of persulfides: Hydropersulfides (RSSH) deprotonate to give the anionic species (RSS^-). Both sulfur atoms in RSSH are electrophilic and can be attacked by nucleophiles producing thiols and sulfur derivatives of the nucleophile when the outer sulfur is attacked (*a*), or releasing H_2S and a sulfur-bridged derivative of the nucleophile when the inner sulfur is attacked (*b*). Persulfide anions are nucleophilic and can react with electrophiles forming disulfide derivatives (*c*).

---

2 The term "H_2S" is used herein for the mixture of dihydrosulfide (H_2S) and hydrosulfide anion (HS^-) in equilibrium.

3.2 Persulfides Are Formed in Biological Systems through Different Pathways

Numerous reactions have been described for the formation and decay of biological persulfides. The active flux of metabolites in sulfur pathways results in significant levels of persulfides in tissues. According to recent measurements based on the reaction of persulfides with a phosphine using an isotope-labeled internal standard and mass spectrometry, the amount of persulfides in bovine aortic endothelial cells was 9 pmol persulfides/mg protein (~2 μM) [3]. In pronase-digested protein from mouse liver tissue, 175 pmol persulfides/mg protein (~37 μM) were quantified by LC-MS after alkylation with an iodoacetamide derivative using an isotope-labeled standard [4]. It has also been estimated with the ProPerDP method (which is based on biotin-labeled alkylation, capture with streptavidin, and selective release with reductants) that ~0.15% of proteins in human embryonic kidney 293 cells or ~1% of proteins in mouse liver tissue bear a persulfide [5]. These numbers are congruent with quantifications of total bound sulfur (persulfides and other sulfane sulfur species) performed decades ago in rat tissues by measuring the H_2S released with dithiothreitol through gas dialysis, derivatization, and HPLC, which yielded 19–324 nmol sulfur/g of tissue (~28–480 μM [6]).

The transformation of thiols to persulfides has been denominated *persulfidation* in the literature and considers diverse processes related to the metabolism of sulfur amino acids. The processes involve the rupture and formation of covalent bonds with sulfur atoms and changes in their oxidation state. In some reactions, the sulfur is provided by H_2S that reacts with thiol oxidation products; others require the transfer of sulfur from partially oxidized sulfur species, or even directly from free cysteine to thiolates. The main known pathways are depicted in Figure 3.2 and described below. Illustrative examples are also provided.

3.2.1 Disulfides Form Persulfides in the Presence of H_2S

H_2S, more precisely the hydrosulfide anion (HS^-), is a nucleophile that can react with diverse biological electrophiles. When it reacts with disulfides, the products formed are a leaving thiol and a persulfide (Eq. (3.1)) [7–11]:

$$HS^- + RSSR' \rightleftharpoons RSSH + R'S^- \rightleftharpoons RSS^- + R'SH \quad (3.1)$$

In mammals, H_2S is mainly produced by three enzymatic systems, cystathionine β-synthase, cystathionine γ-lyase, and the pair cysteine aminotransferase/3-mercaptopyruvate sulfurtransferase (MPST) [1]. The endogenous H_2S can either freely diffuse out of cells [12, 13] or react with various intracellular targets [14]. Among other competing destinations, its consumption by disulfides constitutes a source of persulfides.

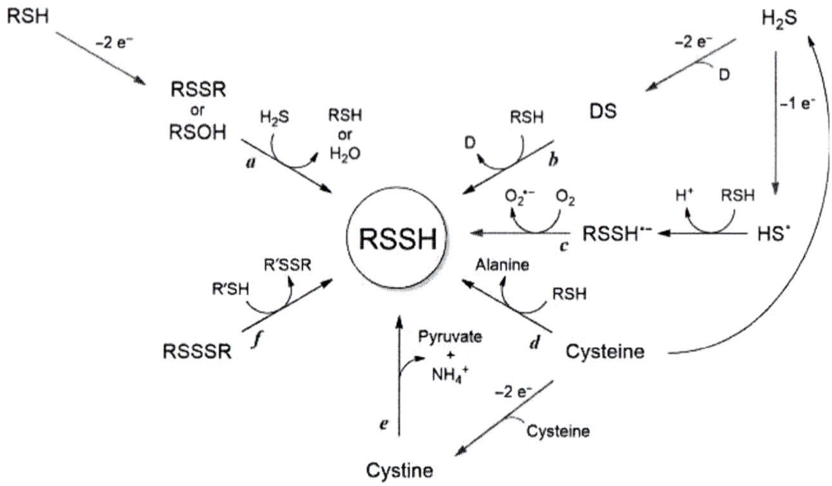

Figure 3.2 Biochemical pathways of persulfide formation: H_2S can react with thiol oxidation products like disulfides or sulfenic acids to form persulfides (path *a*). Thiols can react with oxidation products of H_2S (DS: polysulfides, thiosulfate, persulfides, among others) forming persulfides (path *b*). One-electron oxidation processes can also yield persulfides (path *c*). Cysteine desulfurases can form persulfides at the expense of cysteine (path *d*). β-Elimination in disulfides leads to persulfides and unsaturated products (path *e*). Finally, partial reduction of trisulfides by thiols form persulfides (path *f*).

The rate of the reactions of H_2S with disulfides depends on both the concentration and the nature of the disulfides. At pH 7.4, the rate constants with typical low molecular weight (LMW) disulfides are low, increase with pH as the availability of HS^- increases, and correlate inversely with the pK_a of the leaving thiol. For instance, the values determined for reactions with cystine and glutathione disulfide (GSSG) are 0.6 and 0.23 $M^{-1}\ s^{-1}$ (pH 7.4, 25 °C), respectively [10, 11]. Disulfides are scarce in the cytosol, and the formation of persulfides through this mechanism is expected to occur in disulfide-rich milieus: extracellular environments, intermembrane space of mitochondria, or the endoplasmic reticulum [15–17].

A significant exception to the slowness of this reaction has been identified in enzymes in charge of H_2S catabolism: sulfide quinone oxidoreductases (SQORs) and flavocytochrome *c* sulfide dehydrogenases (FCSDs) [18–21]. Both groups of enzymes oxidize H_2S and deliver the electrons to the transport chain [22, 23]. The mechanism of bacterial FCSD involves a first step that consists of a fast attack of H_2S on an internal disulfide leading to a persulfide and a thiol in the protein [20, 24]. In the human variant of SQOR, a similar mechanism was originally postulated [25–27], but increasing evidence supports that the initial electrophilic species in the enzyme is actually a trisulfide formed between Cys201 and Cys379 that reacts with H_2S forming a leaving group persulfide instead of thiolate (Eq. (3.2)) [28, 29]. The rate constant of the initial reaction is $3 \times 10^5\ M^{-1}\ s^{-1}$ [26], five to six orders of magnitude higher than observed for LMW disulfides:

$$HS^- + Cys201\text{-}SSS\text{-}Cys379 \rightleftharpoons Cys379\text{-}SSH + Cys201\text{-}SS^- \quad (3.2)$$

The persulfide, rapidly formed in SQOR Cys379, is solvent accessible. The sulfur is then transferred to a LMW acceptor to complete the catalytic cycle [25, 28, 30]. If the acceptor is a thiol, e.g. GSH, the product is a new LMW persulfide (Eq. (3.3)):

$$Cys379\text{-}SSH + GS^- \rightarrow Cys379\text{-}S^- + GSSH \quad (3.3)$$

3.2.2 Sulfenic Acids Can Also Form Persulfides by Reaction with H_2S

The endogenous generation of oxidants can lead to the oxidative modification of protein cysteines. When thiolates react with hydroperoxides (ROOH), peroxynitrite ($ONOOH/ONOO^-$) or hypohalous acid (e.g. HOCl), sulfenic acids (RSOH) are formed [31]. Sulfenic acids have electrophilic character and are able to react with H_2S [10, 15, 32, 33]. This reaction forms persulfides and water (Eq. (3.4)):

$$HS^- + RSOH \rightarrow RSSH + OH^- \rightleftharpoons RSS^- + H_2O \quad (3.4)$$

The sulfenic acids formed in serum albumin and in the monothiolic peroxiredoxin alkyl hydroperoxide reductase E were able to form persulfides after incubation with H_2S [10, 15, 33]. Kinetic studies revealed rate constants of 2.7×10^2 and $1.4 \times 10^3\ M^{-1}\ s^{-1}$, respectively [10, 33]. Thiols are likely to outcompete H_2S in the reactions with sulfenic acids in the cytosol, since thiols are present in millimolar concentrations, while H_2S is in the nanomolar range [34], and the rate constants of thiols are comparable in value to those of H_2S [10, 33, 35, 36]. Consequently, the formation of persulfides from the reaction of sulfenic acids with H_2S is predicted to be restricted to sterically protected sulfenic acids or to those located in particularly favorable surroundings in proteins. Nevertheless, experiments performed in cell cultures showed that the induction of sulfenic acid formation by treatment with hydrogen peroxide (H_2O_2) promotes the formation of persulfides to a great extent if compared to cultures enriched in disulfides by treatment with diamide [3, 10]. Moreover, the importance of sulfenic acids in persulfidation is highlighted by the recent observation that, after treatment of different cell lines with H_2O_2, sulfenylation is followed by persulfidation, giving rise to consecutive waves [37].

3.2.3 Other Persulfide Formation Pathways Involve Oxidation Products of H_2S

Some products of H_2S oxidation that contain sulfane sulfur can also provide the outer sulfur of a persulfide. The donor transfers the sulfur atom to nucleophilic thiolates (Eq. (3.5)):

$$RS^- + DS \rightleftharpoons RSS^- + D \tag{3.5}$$

Several routes for H_2S oxidation have been reported. The mitochondrial H_2S oxidation pathway involving SQOR, sulfur dioxygenase, rhodanese, and sulfite oxidase is the central detoxifying route for H_2S and is considered the primary source of sulfane sulfur compounds in cells [19, 22, 38, 39]. This particularly efficient pathway produces thiosulfate and persulfides as intermediates. Additionally, metallic centers can oxidize H_2S [40–50]. For example hemoglobin has been proposed to be an important sink for H_2S in red blood cells where mitochondria are absent, forming thiosulfate and metal-bound polysulfides [51]. Finally, and less probably due to competing targets, the direct reaction of H_2S with one- and two-electron oxidants (e.g. free radicals, hydroperoxides, peroxynitrite, hypohalous acids) generates products containing sulfur in intermediate oxidation states [52–58], including sulfiyl radical ($HS^•$, also called sulfanyl radical), thioperoxide (HSOH), polysulfides, and thiosulfate.

Several of these H_2S derivatives can act as sulfur donors for acceptor thiolates, forming persulfides. The transfer of the sulfane sulfur from thiosulfate to a cysteine has long been known for thiosulfate sulfurtransferases like rhodanese [59]. Thiosulfate generates an internal persulfide in the enzyme that can transfer its sulfur to LMW acceptors [22, 30, 38]. Besides, it has been observed that inorganic polysulfides are able to form persulfides upon simple incubation with protein and LMW thiols [5, 51, 60]. Inorganic polysulfides have also been applied to cell cultures as sulfane donors [61].

The product of the one-electron oxidation of H_2S, the sulfiyl radical ($HS^•$), can induce the formation of persulfides. $HS^•$ can be formed after reaction of H_2S with $O_2^{•-}$, $HO^•$, $NO_2^•$, $CO_3^{•-}$, or heme proteins [48, 53–57, 62, 63], and can form persulfides in a two-step mechanism (Eqs. (3.6) and (3.7)):

$$RS^- + HS^• \rightleftharpoons RSSH^{•-} \tag{3.6}$$

$$RSSH^{•-} + O_2 \rightleftharpoons RSSH + O_2^{•-} \tag{3.7}$$

Alternatively, $RSSH^{•-}$ can be formed after the reaction of $RS^•$ with H_2S. Despite the high rate constants involved (10^8–$10^9\ M^{-1}\ s^{-1}$ [62]), the relative relevance of this free radical pathway will depend on the concentration of radicals that can be reached *in vivo*. Nevertheless, the one-electron oxidation of H_2S can result in oxygen-dependent free radical chain reactions that can amplify oxidative processes.

3.2.4 Some Sulfur Atoms for Persulfides Are Donated by Free Cysteine

Some biological persulfide-forming pathways do not involve H_2S. Several enzymes catalyze the abstraction of sulfur from free cysteine or derivatives such as 3-mercaptopyruvate, and its transfer to an active site thiol. Cysteine desulfurases are PLP-dependent enzymes that produce a persulfide in the catalytic site and alanine as a secondary product (Eq. (3.8)) [2, 64]:

$$\text{Cys} + \text{Enz-S}^- \rightarrow \text{Enz-SS}^- + \text{Ala} \tag{3.8}$$

The sulfur of the persulfide is transferred to other proteins to accomplish biosynthetic roles. For example, the mitochondrial desulfurase Nsf1, involved in the synthesis of iron-sulfur clusters, removes the sulfur from free cysteine to produce a persulfide that is then transferred to the scaffold protein Isu1 [65]. Cysteinyl-tRNA synthetases have been postulated to transfer the sulfur between cysteines in a PLP-dependent process to release free cysteine persulfide [66].

Another pathway involves 3-mercaptopyruvate. The consumption of cysteine by cysteine aminotransferase leads to the formation of 3-mercaptopyruvate (Eq. (3.9)) [67–69]. MPSTs take the sulfur from 3-mercaptopyruvate to form an active site persulfide and pyruvate (Eq. (3.10)) [70–73]:

$$\text{Cys} + \text{2-oxoglutarate} \rightleftharpoons \text{3-mercaptopyruvate} + \text{glutamate} \tag{3.9}$$

$$\text{3-mercaptopyruvate} + \text{MPST-S}^- \rightarrow \text{MPST-SS}^- + \text{pyruvate} \tag{3.10}$$

The persulfide formed in the cysteine of the catalytic site is accessible to various acceptors that can receive the transferred sulfur, mainly thioredoxin and cysteine [71, 73].

Cystathionine β-synthase and cystathionine γ-lyase are cytosolic PLP-dependent enzymes able to catalyze eliminations in sulfur amino acids. Based on the characteristic substrate promiscuity of these enzymes, they can use cystine to form cysteine persulfide, pyruvate, and ammonia (Eq. (3.11)) (or homocysteine persulfide, α-ketobutyrate, and ammonia if homocystine is used, exclusively by the lyase) [74–76]. The cytosolic levels of LMW disulfides are submicromolar, but significant increases could be produced after oxidative stimulus or by upregulation of cystine uptake [76]:

$$\text{Cys-SS-Cys} \rightarrow \text{Cys-SS}^- + \text{pyruvate} + \text{NH}_4^+ \tag{3.11}$$

3.2.5 Trisulfides Are Also a Source of Persulfides

Finally, the nucleophilic substitution of trisulfides by thiols forms persulfides and disulfides (Eq. (3.12)):

$$\text{RSSSR} + \text{R}'\text{S}^- \rightleftharpoons \text{RSS}^- + \text{R}'\text{SSR} \tag{3.12}$$

As mentioned previously, an analogous reaction with HS^- instead of RS^- has been proposed to occur in human SQOR, which can contain a trisulfide in its active site (Eq. (3.2)) [28, 29]. Besides, reducing enzymes, like glutathione reductase, have been proposed to process diglutathionyl trisulfide (GSSSG), forming persulfide and thiol [75, 77]. Due to their relative stability, trisulfides and related compounds have been used as persulfide donors. Cysteine trisulfide and other dialkyl trisulfides have been applied to cell cultures [61, 78, 79]. Various polysulfides and dialkyl polysulfides are found in natural products [80, 81]. Some can be easily synthesized [77, 82] and are even commercially available. The β,γ-unsaturated disulfides, like diallyl disulfides, can also generate persulfides through thiolate attack to electrophilic carbons [81, 83]. The reaction of diallyl trisulfide with GSH generates allylpersulfide as intermediate [61, 81, 83].

3.2.6 Persulfides Can Be Prepared in the Lab

Isolated persulfides in aqueous solution show low stability, with half-lives of minutes, depending on pH [84]. However, persulfides can be obtained in solution in equilibrium with other species. Persulfide stocks have been prepared by incubating disulfides with H_2S [8, 85–87]. The yield was assumed to be complete or was estimated by cyanolysis. Recently, mixtures of GSSG and H_2S were characterized, allowing the quantitative use of this source of glutathione persulfide (GSSH). When H_2S is mixed with excess GSSG, the concentration of GSSH in equilibrium can be estimated considering that initial H_2S equals the sum of H_2S, GSSH and GSSSG in equilibrium, and that the GSSSG formed is approximately half of the GSH formed in equilibrium. For instance, when 10 mM GSSG was mixed with 0.5 mM H_2S, left to equilibrate for two hours, and analyzed by HPLC, GSH was 0.49 mM, GSSH was 0.21 mM, and H_2S was 0.07 mM [11]. Diverse persulfide donors have been designed and synthesized with the aim of releasing persulfides in solution through different strategies [77–79, 88–93].

3.3 Persulfides Are More Acidic Than Thiols

The acidity of persulfides has long been considered to be higher than that of thiols, mainly in view of experimental evidence collected using synthetic thiols and their persulfides [1]. When challenged with tritylthiolate in organic solution, the persulfides tritylpersulfide, adamantylpersulfide, and benzylpersulfide readily transferred the proton to the thiolate, implying that the persulfides are more acidic than the thiol [91]. Following the oxidation of 2-(3-aminopropyl-amino)ethane persulfide by an alkyl radical, it was determined that the pK_a of the persulfide was 6.2, while that of the thiol was 7.6 [94]. Based on its reactivity with peroxyl radicals, the pK_a of cumylpersulfide was calculated to be 7 [95], whereas the pK_a of the cumyl thiol should be around 10 [96].

Using a computational approach, the pK_a of cysteine persulfide was estimated to be 4.34, whereas that of cysteine is 8.29 [10, 97]. A detailed computational study was done recently, aimed at understanding the molecular basis of the increased acidity of persulfides versus thiols; the different steps conceptually involved in thiol and persulfide deprotonation were dissected, and water was included in the system [11]. Using CH_3–SH and CH_3–SSH as simple models, the main difference was found at the free energy of the homolytic S–H bond cleavage, which was ~20% lower for the persulfide than the thiol. This is consistent with S–H bond dissociation energies reported for alkyl thiols and persulfides, which differed by 24% [98], and with IR spectra, which showed shifted S–H stretching frequencies [91, 99, 100]. Thus, the main explanation for the lower pK_a is the weaker S–H bond in the persulfide compared with the thiol.

The pK_a of the biologically relevant persulfide GSSH was only recently determined experimentally [11]. The determination was done by assessing the kinetics of the reaction between GSSH (actually, GSS^-) and monobromobimane (mBrB) at different pHs. The rate of the reaction increased with pH in a sigmoidal fashion, from which a pK_a of 5.45 ± 0.03 was determined. A pK_a of 6.9 for GSSH was also reported elsewhere, but there were several flaws in the experimental design that make this value unreliable [101, 102]. For comparison, the pK_a of GSH is 8.94 [97]; therefore, at pH 7.4, while a small fraction of GSH will be deprotonated (2.8%), nearly all GSSH will be deprotonated (99%) (Figure 3.3).

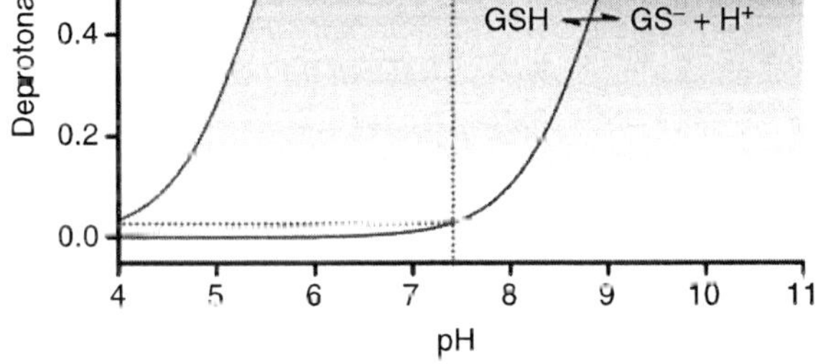

Figure 3.3 Fraction of deprotonated species versus pH for GSSH and GSH: The values of dissociation fraction were calculated using the acid dissociation equilibrium equation, and pK_as of 5.45 for GSSH (light gray trace) [11] and 8.94 for GSH (dark gray trace) [97]. The dashed lines indicate the fraction of deprotonated GSSH (0.99) and GSH (0.028) at the physiological pH of 7.4. Source: Based on Benchoam et al. [11] and Portillo-Ledesma et al. [97].

The higher acidity of persulfides versus thiols is important for the comparison of the electrophilic reactivity of disulfides versus trisulfides. This has implications for SQOR, proposed to contain an active site trisulfide instead of the expected disulfide [28, 29]. This also has implications for the rationalization of the biological reactivity of trisulfides, particularly since they can be administered to cultured cells [61, 78, 79].

The reactions of a disulfide or a trisulfide with a thiolate (R′S⁻) or with HS⁻ as nucleophiles (Nu⁻) are represented by Eqs. (3.13) and (3.14):

$$\text{RSSR} + \text{Nu}^- \rightarrow \text{RSNu} + \text{RS}^- \tag{3.13}$$

$$\text{RSSSR} + \text{Nu}^- \rightarrow \text{RSNu} + \text{RSS}^- \tag{3.14}$$

These reactions occur through S_N2 mechanisms [10, 29, 103]. The rate constants with disulfides and, by extrapolation, with trisulfides, are dependent on the pK_as of the reacting species according to Eq. (3.15) [10, 103]. Typical values for the Brønsted coefficients for the nucleophilic atom, central sulfur, and leaving group sulfur are: β_{nuc}, +0.5; β_c, −0.3; and β_{lg}, −0.5 to −0.7, respectively [10, 103]; C is a constant:

$$\log k = C + \beta_{\text{nuc}} \times \text{p}K_{\text{a,nuc}} + \beta_{\text{c}} \times \text{p}K_{\text{a,c}} + \beta_{\text{lg}} \times \text{p}K_{\text{a,lg}} \tag{3.15}$$

Unless under steric constraints, the alternative reaction of RSSSR in which the nucleophile attacks the internal sulfur atom to produce RSSNu and RS⁻ would not be favored compared to Eq. (3.14) because of the tendency to produce the better leaving group (the one with the lower pK_a, Eq. (3.15)).

Assuming pK_as of persulfides and thiols of 5.45 and 8.94, respectively, as in the case of the glutathione derivatives [11], it can be calculated from Eq. (3.15) that the reaction with the trisulfide would be 10^2–10^3 times faster than the reaction with the analogous disulfide, to account for the better central and leaving group sulfurs. Of course, in the case of protein disulfides and trisulfides, the protein environment is likely to modulate the reactivity beyond these acidity considerations. In fact, in the case of SQOR, computational simulations suggest a rate increase of ~10^5 for reaction with the trisulfide versus the disulfide [29].

3.4 Persulfides Are Stronger Nucleophiles Than Thiols

The nucleophilicity of the anionic species of persulfides (RSS⁻) is manifested by the attack on electrophiles to form disulfide derivatives (Eq. (3.16)):

$$\text{RSS}^- + \text{E}^+ \rightarrow \text{RSSE} \tag{3.16}$$

This nucleophilic reactivity is evidenced when persulfides are derivatized with alkylating agents such as mBrB, iodoacetamide, and *N*-ethylmaleimide

[8, 86, 104]. The reaction of persulfides with electrophiles also occurs *in vivo*, e.g. in the sulfide oxidation pathway. It was recently postulated that, in the catalytic mechanism of human SQOR, H_2S reacts with the active site trisulfide forming two persulfides. One of the persulfides reacts nucleophilically with the FAD cofactor; it forms a charge-transfer complex which likely evolves to a covalent intermediate (a C4a adduct between the persulfide and FAD) [21, 28]. In addition, the mechanism proposed for persulfide dioxygenase involves the attack of GSS^- on the electrophilic metal center of the enzyme [105].

When the electrophiles are disulfides, trisulfides are generated. This occurs in the reactions of persulfides with the synthetic disulfides 5,5′-dithiobis (2-nitrobenzoic acid) or 4,4′-dithiodipyridine [10, 33, 86]. Furthermore, a relatively slow rate constant of 1.6–2.3 M^{-1} s^{-1} (pH 7.4, 25 °C) was recently estimated for the reaction between GSS^- and GSSG, which yields GSSSG and GS^- [11]. Also, dialkyl trisulfides were formed when disulfides were incubated with a persulfide donor or NaHS, which produced persulfide under the working conditions [78, 104].

An in-depth comparison between the nucleophilic reactivity of persulfides and thiolates has been drawn, with mBrB, H_2O_2, and peroxynitrite as electrophiles [11]. Both persulfides and thiolates reacted with these electrophiles through S_N2 mechanisms, although differences in reactivity existed and were related to the electronic structures of the species rather than to solvent effects [11]. In this regard, persulfides had enhanced intrinsic nucleophilicity compared to thiolates with similar basicity [11]. This can be explained by the alpha effect, the increased reactivity caused by the presence of high electron density in the atom adjacent to the nucleophilic atom [106, 107]. Of note, the comparisons should be drawn between persulfide anions and thiolates that react by similar mechanisms [108] and that have equal basicity, not between structural analogs (i.e. $CysSS^-$ with $CysS^-$). Thus, the alpha effect is manifested in Brønsted plots (log k versus nucleophile pK_a) by positive deviations from the expected trend according to the basicity [109]. The magnitude of the alpha effect varies with the reaction. For example, GSS^- reacted 1670 times faster with mBrB than expected for a thiolate with the same pK_a (Figure 3.4) [11].

Although the alpha nucleophile is not necessarily more reactive than the structural analog [11, 110], it often is. Arylpersulfides reacted faster than arylthiolates with alkyl halides in organic solvent [111]. In addition, the persulfides formed in albumin and in the monothiolic peroxiredoxin alkyl hydroperoxide reductase E reacted faster than the corresponding protein thiols with 4,4′-dithiodipyridine [10, 33].

Importantly, at physiological pH, persulfides have higher availability of the anionic species than thiols due to their lower pK_a. Hence, the enhanced nucleophilicity of persulfides compared to thiols at physiological pH is given by a

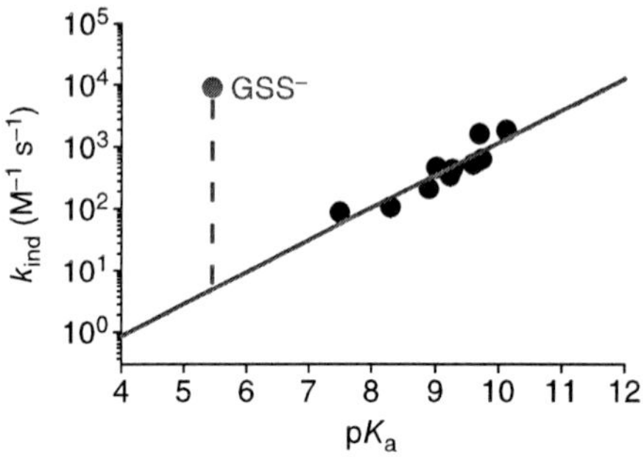

Figure 3.4 Alpha effect of persulfides: Brønsted plot for the reactions of several LMW thiolates (black) or glutathione persulfide anion (GSS^-) (gray) toward mBrB. The magnitude of the alpha effect is given by the positive deviation from the plot (dashed line). Data values were taken from [11, 97]. Source: Adapted from Benchoam et al. [11].

combination of two effects, both the predominance of the deprotonated species and the increased intrinsic nucleophilicity. This is evidenced in the reactions of GSSH and GSH with mBrB. The 3.49 units pK_a difference between GSSH and GSH results in an availability of deprotonated species of 99% for GSSH and 2.8% for GSH at pH 7.4 (Figure 3.3). Therefore, considering both factors, while the intrinsic reactivity of GSS^- is 44 times higher than that of GS^-, at pH 7.4 GSS^-/GSSH reacts 1200 times faster than GS^-/GSH [11].

3.5 Persulfidation Protects Against Irreversible Oxidation

The increased nucleophilicity of persulfides is also manifested in the reactions with typical two-electron oxidants such as H_2O_2 and peroxynitrite. A rate constant of 7.5 M^{-1} s^{-1} (pH 7.02, 25 °C) was determined for the reaction of H_2O_2 with GSSH, which is 22-fold higher than GSH at the same pH. In the case of peroxynitrite, a rate constant of 1.25×10^5 M^{-1} s^{-1} (pH 7.23, 37 °C) was measured for the reaction with GSSH, which is 97 times faster than the reaction with GSH under the same conditions [11]. Furthermore, the rate constant of the reaction of peroxynitrite with albumin persulfide, 1.2×10^4 M^{-1} s^{-1} (pH 7.4, 20 °C), was 4.4 times faster than that of the thiol [10]. Nevertheless, the decays of H_2O_2 and peroxynitrite in the presence of the persulfide formed in the monothiolic peroxiredoxin alkyl hydroperoxide reductase E were several orders of magnitude slower than that with the thiolate. This outcome is related to the geometry of the catalytic site, which favors catalysis with the specific substrates only for the thiolate [33]. Thus, although there is evidence that persulfides have lower pK_as and increased nucleophilicity compared to thiols, this is accurate for LMW compounds. Proteins and enzymes should be independently studied since the local environment could cause different effects.

The two-electron oxidation of persulfides is expected to produce perthiosulfenic acids (RSSOH) (Eq. (3.17)), analogous to the reaction with thiols where sulfenic acids (RSOH) are formed. Computational calculations confirmed that the reactions of methylpersulfide (MeSSH) [11] or ethylpersulfide (EtSSH) [112] with H_2O_2 or peroxynitrite yielded the corresponding perthiosulfenic acids. In addition, the oxidation of cysteine persulfide residues to perthiosulfenic acids was observed in epidermal growth factor receptor (EGFR) [112] and peroxiredoxin 2 [4].

Excess oxidant would lead to subsequent oxidations of perthiosulfenic acids to perthiosulfinic and perthiosulfonic acids ($RSSO_2H$ and $RSSO_3H$) (Eq. (3.18)) [1, 112]. In mouse liver samples, protein levels of $RSSO_3H$ were comparable to those of RSSH [4]. Indeed, these oxidative modifications have been found in papain, albumin, glutathione peroxidase, peroxiredoxin 2, and protein tyrosine phosphatase 1B [4, 10, 15, 86]. Remarkably, unlike sulfinic and sulfonic acids (RSO_2H and RSO_3H) that are considered irreversible modifications, $RSSO_2H$ and $RSSO_3H$ can be reduced to thiols by LMW thiols (Eqs. (3.19) and (3.20)) or cellular reductants belonging to the thioredoxin and glutaredoxin systems. This means that protein thiols can be restored, and thus, protein persulfidation has been proposed to constitute a protective mechanism against irreversible damage in oxidative stress contexts [3, 4, 37, 113–116]. In addition, a complementary mechanism employed by cells for the protective effects of persulfidation was proposed. It consists of already sulfenylated proteins (RSOH) to react with H_2S to form protein persulfides. This way, further irreversible oxidative modifications are avoided and the protein thiols can be rescued as previously explained [37]:

$$RSS^- + [O] \rightarrow RSSOH \quad (3.17)$$

$$RSSOH + [O] \rightarrow RSSO_2H/RSSO_3H \quad (3.18)$$

$$RSSO_2H/RSSO_3H + R'SH \rightarrow HSO_2^-/HSO_3^- + RSSR' \quad (3.19)$$

$$RSSR' + R'SH \rightarrow RSH + R'SSR' \quad (3.20)$$

Note that in Eqs (3.17) to (3.20) protons were sometimes omitted for simplicity.

3.6 Persulfides Interact with Metals and Metalloproteins

Similar to thiols, persulfides can form metal complexes. Persulfide complexes with copper, iridium, platinum, titanium, tungsten, iron, ruthenium, and zinc ions have been characterized [116, 117]. These interactions could provide protection from heavy metal toxicity, as in the case of toxic mercury and cadmium compounds where the presence of persulfides led to a decrease in the toxicity [118, 119].

Although complexes with metalloproteins are known to a lesser extent than LMW complexes, the metal centers of several proteins have been shown to interact with persulfides by coordination and/or reduction [116, 117, 120]. Considering that both sulfur atoms of persulfides are electrophilic, *vide infra*, the complex formed between the metal and the persulfide can be attacked by nucleophiles in three sites, the metal and both sulfurs, while a thiol-complex can only be attacked in the metal center [121], diversifying possible mechanisms of reaction.

Persulfides have been observed to react with heme- and nonheme iron-containing proteins. Some examples are the following: (i) ferric cytochrome *c*: GSSH reduces it to the ferrous state, whereas GSH does it only slowly [8]; (ii) persulfide dioxygenase: GSSH binds directly to the ferrous iron in the active site displacing a water molecule [105, 122]; (iii) N-acetylated microperoxidase 11: CysSSH or *N*-methoxycarbonyl-penicillamine persulfide coordinate its ferric form [120]; and (iv) ferric myoglobin: CysSSH, GSSH, or *N*-methoxycarbonyl-penicillamine persulfide reduces it yielding oxy-ferrous or deoxy-ferrous myoglobin depending on the presence or absence of oxygen [120]. Several other iron-containing proteins, including bacterial, have been shown to interact with persulfides [123–125].

Metalloproteins containing other metal ions besides iron have also been reported to interact with persulfides [116, 117]. For instance, molybdenum-containing proteins that possess a $CysSS^-$ ligand include the bacterial nitrate reductase NapAB [126] and the molybdenum cofactor sulfurase ABA3 from *Arabidopsis thaliana* [127].

3.7 Persulfides Have Electrophilic Character in Both Sulfur Atoms

Persulfides are electrophilic and can be the subject of attack by a nucleophile. Since typical biological thiols are not electrophilic, the modification of a thiol to a persulfide brings about a property that was initially absent in the parent thiol. Electrophilicity should be ascribed to the protonated hydropersulfide (RSSH) instead of the anionic deprotonated persulfide (RSS^-) [128]. Nevertheless, in the case of protein persulfides, the protein environment may facilitate electrophilicity of the deprotonated persulfide [129]. Both sulfur atoms in the persulfide possess electrophilic character. A computational analysis using methylpersulfide revealed that attack by a nucleophile is feasible at either sulfur atom, with a slight preference for attack at the terminal one [128].

Possible nucleophilic partners for electrophilic persulfides are thiols (actually, thiolates), cyanide, sulfite, amines, hydroxide, and tertiary phosphines [86, 91, 130–134]. Reactions are favored with soft nucleophiles (cyanide, thiols, phosphines) rather than hard ones (azide, hydrazine, hydroxylamine) [128]. The reaction with cyanide is exploited in an often used technique for the analysis of persulfides and other sulfane sulfur compounds, in which thiocyanate is formed and reacts with ferric ions to give a red color [135]. In biological settings, protein and nonprotein thiols are likely nucleophilic partners for persulfide electrophilic reactions. Indeed, the glutathione and thioredoxin systems can reduce persulfides to thiols [3–5].

The reaction of a nucleophile with a persulfide can lead to different products depending on the site of attack. If the attack occurs in the inner sulfur, HS^- is released and a sulfur-bridged derivative is formed, e.g. a mixed disulfide if the nucleophile is a thiolate (Eq. (3.21) and Figure 3.1). For example, Keap1 persulfide/Nrf2 was proposed to be attacked by a protein thiolate releasing H_2S and producing a mixed disulfide between the proteins [136]:

$$R\underline{S}SH + Nu^- \rightarrow R\underline{S}Nu + HS^- \tag{3.21}$$

In contrast, if the attack occurs in the outer sulfur, a thiolate is released and a sulfur derivative of the nucleophile is formed. When the nucleophile is a thiolate, a transpersulfidation reaction occurs (Eq. (3.22) and Figure 3.1):

$$RS\underline{S}H + R'S^- \rightarrow RS^- + R'S\underline{S}H \tag{3.22}$$

Attack in the outer sulfur has been proposed for the persulfide formed in MPST. This enzyme transfers a sulfur atom from 3-mercaptopyruvate to a thiol acceptor via a protein persulfide intermediate. The preferential thiol acceptors of the persulfide outer sulfur are thioredoxin and cysteine [71–73]. The secondary persulfides are finally reduced by thiols releasing H_2S. Other examples are the attack of glutathione or coenzyme A on the persulfide formed in SQOR [27, 28, 30] and the attack of active site thiols in thiosulfate sulfurtransferases (e.g. rhodanese) on LMW persulfides, to give the persulfidated protein [38, 39].

The site of attack appears to depend critically on the nature of the leaving group to be released. Usually, the better leaving group is the one with the more acidic conjugate acid, as illustrated by Eq. (3.15). In typical LMW persulfides such as cysteine persulfide or GSSH, the attack of the nucleophile is favored in the inner sulfur and HS^- is released because the pK_a of H_2S is 6.98 [7], much lower than that of cysteine (8.29) or glutathione (8.94) [97]. Indeed, the reaction of GSH with GSSH released H_2S with a rate constant of 1.11 $M^{-1}\ s^{-1}$ (pH 7.4, 25 °C) [11]. However, the situation may be different in protein persulfides in which the parent thiol has unusually low pK_a values, like the peroxiredoxins [33]; transsulfuration would be favored instead of H_2S release. The site of attack also depends on the presence of steric impediments. In persulfides with voluminous groups, attack is favored in the outer sulfur [91].

3.8 Persulfides Are Efficient One-Electron Reductants

Persulfides can be oxidized by one-electron oxidants (e.g. metal ions, oxygen-derived free radicals) forming perthiyl radical ($RSS^\bullet$), and are in fact better reductants than thiols [89, 95, 137] due to their lower reduction potential ($E^{\circ\prime}(RSS^{\cdot}/RSS^{-}) = 0.68$ V versus $E^{\circ\prime}(RS^{\cdot}, H^{+}/RSH) = 0.96$ V) [138] and their lower S–H bond dissociation energy (70 versus 92 kcal/mol for persulfides and thiols, respectively) [95], as well as their faster reaction kinetics [89, 95].

The perthiyl radical product $RSS^{\cdot}$ is stabilized by delocalization of the unpaired electron along the two sulfur atoms [89, 94] and is relatively unreactive. In fact, it does not appear to react with dioxygen or nitric oxide [95, 137, 139]. The main pathway of $RSS^{\cdot}$ decay is dimerization to the tetrasulfide RSSSSR with a rate constant of $1–6 \times 10^{9}\ M^{-1}\ s^{-1}$ [89, 139]. The relatively high stability of persulfide radicals is exploited in a procedure for the synthesis of unsymmetric disulfides through radical substitution of aryl or alkyl radicals on tetrasulfides. The perthiyl radicals formed as intermediates undergo few side reactions, mainly the recombination to the starting tetrasulfide [140].

3.9 Concluding Remarks

The appreciation that persulfides participate in critical biosynthetic and catabolic processes, that they are potential transducers of H_2S actions, and that they can protect against irreversible thiol oxidation has led to increased interest in these elusive species. Efforts directed at better understanding persulfides are revealing a very rich chemistry and are providing a framework to rationalize the molecular basis of their biological properties, summarized as follows: (i) persulfides are more acidic than thiols; (ii) they are better nucleophiles than thiols; (iii) they present electrophilic reactivity in both sulfur atoms, a property absent in thiols; (iv) they are good one- and two-electron reductants. Several questions remain open, and progress in finding answers is inextricable from advances in methodologies to detect and quantify persulfides. It is likely that research in the near future will provide dynamic pictures about the enzymes that use persulfides either as substrates, products, or intermediates, clarify precisely in which pathways persulfides act as mediators of H_2S signaling, and assess the cellular relevance of the reactions of persulfides with oxidants. Hopefully, this will open the way to pharmacological applications.

References

1 Filipovic, M.R., Zivanovic, J., Alvarez, B., and Banerjee, R. (2018). Chemical biology of H_2S signaling through persulfidation. *Chem. Rev.* 118: 1253–1337. https://doi.org/10.1021/acs.chemrev.7b00205.

2 Mueller, E.G. (2006). Trafficking in persulfides: delivering sulfur in biosynthetic pathways. *Nat. Chem. Biol.* 2: 185–194. https://doi.org/10.1038/nchembio779.

3 Wedmann, R., Onderka, C., Wei, S. et al. (2016). Improved tag-switch method reveals that thioredoxin acts as depersulfidase and controls the intracellular levels of protein persulfidation. *Chem. Sci.* 7: 3414–3426. https://doi.org/10.1039/C5SC04818D.

4 Dóka, É., Ida, T., Dagnell, M. et al. (2020). Control of protein function through oxidation and reduction of persulfidated states. *Sci. Adv.* 6: eaax8358. https://doi.org/10.1126/sciadv.aax8358.

5 Dóka, É., Pader, I., Bíró, A. et al. (2016). A novel persulfide detection method reveals protein persulfide- and polysulfide-reducing functions of thioredoxin and glutathione systems. *Sci. Adv.* 2: e1500968. https://doi.org/10.1126/sciadv.1500968.

6 Ogasawara, Y., Isoda, S., and Tanabe, S. (1994). Tissue and subcellular distribution of bound and acid-labile sulfur, and the enzymic capacity for sulfide production in the rat. *Biol. Pharm. Bull.* 17: 1535–1542. https://doi.org/10.1248/bpb.17.1535.

7 Hughes, M.N., Centelles, M.N., and Moore, K.P. (2009). Making and working with hydrogen sulfide: The chemistry and generation of hydrogen sulfide in vitro and its measurement in vivo: a review. *Free Radical Biol. Med.* 47: 1346–1353. https://doi.org/10.1016/j.freeradbiomed.2009.09.018.

8 Francoleon, N.E., Carrington, S.J., and Fukuto, J.M. (2011). The reaction of H_2S with oxidized thiols: generation of persulfides and implications to H_2S biology. *Arch. Biochem. Biophys.* 516: 146–153. https://doi.org/10.1016/j.abb.2011.09.015.

9 Vasas, A., Dóka, É., Fábián, I., and Nagy, P. (2015). Kinetic and thermodynamic studies on the disulfide-bond reducing potential of hydrogen sulfide. *Nitric Oxide* 46: 93–101. https://doi.org/10.1016/j.niox.2014.12.003.

10 Cuevasanta, E., Lange, M., Bonanata, J. et al. (2015). Reaction of hydrogen sulfide with disulfide and sulfenic acid to form the strongly nucleophilic persulfide. *J. Biol. Chem.* 290: 26866–26880. https://doi.org/10.1074/jbc.M115.672816.

11 Benchoam, D., Semelak, J.A., Cuevasanta, E. et al. (2020). Acidity and nucleophilic reactivity of glutathione persulfide. *J. Biol. Chem.* 295: 15466–15481. https://doi.org/10.1074/jbc.RA120.014728.

12 Mathai, J.C., Missner, A., Kügler, P. et al. (2009). No facilitator required for membrane transport of hydrogen sulfide. *Proc. Natl. Acad. Sci. U.S.A.* 106: 16633–16638. https://doi.org/10.1073/pnas.0902952106.

13 Cuevasanta, E., Denicola, A., Alvarez, B., and Möller, M.N. (2012). Solubility and permeation of hydrogen sulfide in lipid membranes. *PLoS One* 7: e34562. https://doi.org/10.1371/journal.pone.0034562.

14 Cuevasanta, E., Möller, M.N., and Alvarez, B. (2017). Biological chemistry of hydrogen sulfide and persulfides. *Arch. Biochem. Biophys.* https://doi.org/10.1016/j.abb.2016.09.018.

15 Zhang, D., Macinkovic, I., Devarie-Baez, N.O. et al. (2014). Detection of protein S-sulfhydration by a tag-switch technique. *Angew. Chem. Int. Ed.* 53: 575–581. https://doi.org/10.1002/anie.201305876.

16 Herrmann, J.M. and Köhl, R. (2007). Catch me if you can! Oxidative protein trapping in the intermembrane space of mitochondria. *J. Cell Biol.* 176: 559–563. https://doi.org/10.1083/jcb.200611060.

17 Hwang, C., Sinskey, A.J., and Lodish, H.F. (1992). Oxidized redox state of glutathione in the endoplasmic reticulum. *Science* 257: 1496–1502. https://doi.org/10.1126/science.1523409.

18 Bouillaud, F. and Blachier, F. (2011). Mitochondria and sulfide: a very old story of poisoning, feeding, and signaling? *Antioxid. Redox Signal.* 15: 379–391. https://doi.org/10.1089/ars.2010.3678.

19 Mishanina, T.V., Libiad, M., and Banerjee, R. (2015). Biogenesis of reactive sulfur species for signaling by hydrogen sulfide oxidation pathways. *Nat. Chem. Biol.* 11: 457–464. https://doi.org/10.1038/nchembio.1834.

20 Sousa, F.M., Pereira, J.G., Marreiros, B.C., and Pereira, M.M. (1859). Taxonomic distribution, structure/function relationship and metabolic context of the two families of sulfide dehydrogenases: SQR and FCSD. *Biochim. Biophys. Acta Bioenerg.* 2018: 742–753. https://doi.org/10.1016/j.bbabio.2018.04.004.

21 Landry, A.P., Ballou, D.P., and Banerjee, R. (2020). Hydrogen sulfide oxidation by sulfide quinone oxidoreductase. *Chembiochem Eur. J. Chem. Biol.* https://doi.org/10.1002/cbic.202000661.

22 Hildebrandt, T.M. and Grieshaber, M.K. (2008). Three enzymatic activities catalyze the oxidation of sulfide to thiosulfate in mammalian and invertebrate mitochondria. *FEBS J.* 275: 3352–3361. https://doi.org/10.1111/j.1742-4658.2008.06482.x.

23 Cherney, M.M., Zhang, Y., James, M.N.G., and Weiner, J.H. (2012). Structure-activity characterization of sulfide:quinone oxidoreductase variants. *J. Struct. Biol.* 178: 319–328. https://doi.org/10.1016/j.jsb.2012.04.007.

24 Chen, Z.W., Koh, M., Van Driessche, G. et al. (1994). The structure of flavocytochrome c sulfide dehydrogenase from a purple phototrophic bacterium. *Science* 266: 430–432. https://doi.org/10.1126/science.7939681.

25 Jackson, M.R., Melideo, S.L., and Jorns, M.S. (2012). Human sulfide:quinone oxidoreductase catalyzes the first step in hydrogen sulfide metabolism and produces a sulfane sulfur metabolite. *Biochemistry* 51: 6804–6815. https://doi.org/10.1021/bi300778t.

26 Mishanina, T.V., Yadav, P.K., Ballou, D.P., and Banerjee, R. (2015). Transient kinetic analysis of hydrogen sulfide oxidation catalyzed by human sulfide quinone oxidoreductase. *J. Biol. Chem.* 290: 25072–25080. https://doi.org/10.1074/jbc.M115.682369.

27 Jackson, M.R., Loll, P.J., and Jorns, M.S. (2019). X-Ray structure of human sulfide:quinone oxidoreductase: Insights into the mechanism of mitochondrial hydrogen sulfide oxidation. *Struct. Lond. Engl.* 1993 (27): 794–805.e4. https://doi.org/10.1016/j.str.2019.03.002.

28 Landry, A.P., Moon, S., Kim, H. et al. (2019). A catalytic trisulfide in human sulfide quinone oxidoreductase catalyzes coenzyme A persulfide synthesis and inhibits butyrate oxidation. *Cell. Chem. Biol.* 26: 1515–1525.e4. https://doi.org/10.1016/j.chembiol.2019.09.010.

29 Landry, A.P., Moon, S., Bonanata, J. et al. (2020). Dismantling and rebuilding the trisulfide cofactor demonstrates its essential role in human sulfide quinone oxidoreductase. *J. Am. Chem. Soc.* https://doi.org/10.1021/jacs.0c06066.

30 Libiad, M., Yadav, P.K., Vitvitsky, V. et al. (2014). Organization of the human mitochondrial hydrogen sulfide oxidation pathway. *J. Biol. Chem.* 289: 30901–30910. https://doi.org/10.1074/jbc.M114.602664.

31 Gupta, V. and Carroll, K.S. (1840). Sulfenic acid chemistry, detection and cellular lifetime. *Biochim. Biophys. Acta BBA – Gen. Subj.* 2014: 847–875. https://doi.org/10.1016/j.bbagen.2013.05.040.

32 Peshenko, I.V. and Shichi, H. (2001). Oxidation of active center cysteine of bovine 1-Cys peroxiredoxin to the cysteine sulfenic acid form by peroxide and peroxynitrite. *Free Radical Biol. Med.* 31: 292–303.

33 Cuevasanta, E., Reyes, A.M., Zeida, A. et al. (2019). Kinetics of formation and reactivity of the persulfide in the one-cysteine peroxiredoxin from *Mycobacterium tuberculosis*. *J. Biol. Chem.* https://doi.org/10.1074/jbc.RA119.008883.

34 Furne, J., Saeed, A., and Levitt, M.D. (2008). Whole tissue hydrogen sulfide concentrations are orders of magnitude lower than presently accepted values. *Am. J. Physiol. Regul. Integr. Comp. Physiol.* 295: R1479–R1485. https://doi.org/10.1152/ajpregu.90566.2008.

35 Nagy, P. and Ashby, M.T. (2007). Reactive sulfur species: kinetics and mechanisms of the oxidation of cysteine by hypohalous acid to give cysteine sulfenic acid. *J. Am. Chem. Soc.* 129: 14082–14091. https://doi.org/10.1021/ja0737218.

36 Turell, L., Botti, H., Carballal, S. et al. (2008). Reactivity of sulfenic acid in human serum albumin. *Biochemistry* 47: 358–367. https://doi.org/10.1021/bi701520y.

37 Zivanovic, J., Kouroussis, E., Kohl, J.B. et al. (2019). Selective persulfide detection reveals evolutionarily conserved antiaging effects of S-sulfhydration. *Cell Metab.* 30: 1152–1170.e13. https://doi.org/10.1016/j.cmet.2019.10.007.

38 Melideo, S.L., Jackson, M.R., and Jorns, M.S. (2014). Biosynthesis of a central intermediate in hydrogen sulfide metabolism by a novel human sulfurtransferase and its yeast ortholog. *Biochemistry* 53: 4739–4753. https://doi.org/10.1021/bi500650h.

39 Libiad, M., Motl, N., Akey, D.L. et al. (2018). Thiosulfate sulfurtransferase like domain containing 1 protein interacts with thioredoxin. *J. Biol. Chem.* https://doi.org/10.1074/jbc.RA117.000826.

40 Sorbo, B. (1958). On the formation of thiosulfate from inorganic sulfide by liver tissue and heme compounds. *Biochim. Biophys. Acta.* 27: 324–329.

41 Baxter, C.F. and Van Reen, R. (1958). The oxidation of sulfide to thiosulfate by metalloprotein complexes and by ferritin. *Biochim. Biophys. Acta.* 28: 573–578.

42 Nakamura, S., Nakamura, M., Yamazaki, I., and Morrison, M. (1984). Reactions of ferryl lactoperoxidase (compound II) with sulfide and sulfhydryl compounds. *J. Biol. Chem.* 259: 7080–7085.

43 Nicholls, P. and Kim, J.K. (1981). Oxidation of sulphide by cytochrome aa3. *Biochim. Biophys. Acta* 637: 312–320.

44 Collman, J.P., Ghosh, S., Dey, A., and Decréau, R.A. (2009). Using a functional enzyme model to understand the chemistry behind hydrogen sulfide induced hibernation. *Proc. Natl. Acad. Sci. U.S.A.* 106: 22090–22095. https://doi.org/10.1073/pnas.0904082106.

45 Pálinkás, Z., Furtmüller, P.G., Nagy, A. et al. (2015). Interactions of hydrogen sulfide with myeloperoxidase. *Br. J. Pharmacol.* 172: 1516–1532. https://doi.org/10.1111/bph.12769.

46 Searcy, D.G., Whitehead, J.P., and Maroney, M.J. (1995). Interaction of Cu,Zn superoxide dismutase with hydrogen sulfide. *Arch. Biochem. Biophys.* 318: 251–263. https://doi.org/10.1006/abbi.1995.1228.

47 Pietri, R., Lewis, A., León, R.G. et al. (2009). Factors controlling the reactivity of hydrogen sulfide with hemeproteins. *Biochemistry* 48: 4881–4894. https://doi.org/10.1021/bi801738j.

48 Bostelaar, T., Vitvitsky, V., Kumutima, J. et al. (2016). Hydrogen sulfide oxidation by myoglobin. *J. Am. Chem. Soc.* 138: 8476–8488. https://doi.org/10.1021/jacs.6b03456.

49 Ruetz, M., Kumutima, J., Lewis, B.E. et al. (2017). A distal ligand mutes the interaction of hydrogen sulfide with human neuroglobin. *J. Biol. Chem.* 292: 6512–6528. https://doi.org/10.1074/jbc.M116.770370.

50 Vitvitsky, V., Miljkovic, J.L., Bostelaar, T. et al. (2018). Cytochrome c reduction by H_2S potentiates sulfide signaling. *ACS Chem. Biol.* 13: 2300–2307. https://doi.org/10.1021/acschembio.8b00463.

51 Vitvitsky, V., Yadav, P.K., An, S. et al. (2017). Structural and mechanistic insights into hemoglobin-catalyzed hydrogen sulfide oxidation and the fate of polysulfide products. *J. Biol. Chem.* 292: 5584–5592. https://doi.org/10.1074/jbc.M117.774943.

52 Nagy, P. and Winterbourn, C.C. (2010). Rapid reaction of hydrogen sulfide with the neutrophil oxidant hypochlorous acid to generate polysulfides. *Chem. Res. Toxicol.* 23: 1541–1543. https://doi.org/10.1021/tx100266a.

53 Carballal, S., Trujillo, M., Cuevasanta, E. et al. (2011). Reactivity of hydrogen sulfide with peroxynitrite and other oxidants of biological interest. *Free Radical Biol. Med.* 50: 196–205. https://doi.org/10.1016/j.freeradbiomed.2010.10.705.

54 Wedmann, R., Bertlein, S., Macinkovic, I. et al. (2014). Working with "H_2S": facts and apparent artifacts, Nitric Oxide Biol. *Chem. Off. J. Nitric Oxide Soc.* 41: 85–96. https://doi.org/10.1016/j.niox.2014.06.003.

55 Vitvitsky, V., Yadav, P.K., Kurthen, A., and Banerjee, R. (2015). Sulfide oxidation by a noncanonical pathway in red blood cells generates thiosulfate and polysulfides. *J. Biol. Chem.* 290: 8310–8320. https://doi.org/10.1074/jbc.M115.639831.

56 Olson, K.R., Gao, Y., DeLeon, E.R. et al. (2017). Catalase as a sulfide-sulfur oxido-reductase: An ancient (and modern?) regulator of reactive sulfur species (RSS). *Redox Biol.* 12: 325–339. https://doi.org/10.1016/j.redox.2017.02.021.

57 Garai, D., Ríos-González, B.B., Furtmüller, P.G. et al. (2017). Mechanisms of myeloperoxidase catalyzed oxidation of H_2S by H_2O_2 or O_2 to produce potent protein Cys-polysulfide-inducing species. *Free Radical Biol. Med.* 113: 551–563. https://doi.org/10.1016/j.freeradbiomed.2017.10.384.

58 Benchoam, D., Cuevasanta, E., Möller, M.N., and Alvarez, B. (2019). Hydrogen sulfide and persulfides oxidation by biologically relevant oxidizing species. *Antioxid. Basel Switz.* 8: https://doi.org/10.3390/antiox8020048.

59 Bordo, D. and Bork, P. (2002). The rhodanese/Cdc25 phosphatase superfamily. Sequence-structure-function relations. *EMBO Rep.* 3: 741–746. https://doi.org/10.1093/embo-reports/kvf150.

60 Greiner, R. and Dick, T.P. (2015). Real-time assays for monitoring the influence of sulfide and sulfane sulfur species on protein thiol redox states. *Methods Enzymol.* 555: 57–77. https://doi.org/10.1016/bs.mie.2014.11.020.

61 Ezeriņa, D., Takano, Y., Hanaoka, K. et al. (2018). N-acetyl cysteine functions as a fast-acting antioxidant by triggering intracellular H_2S and sulfane sulfur production. *Cell. Chem. Biol.* 25: 447–459.e4. https://doi.org/10.1016/j.chembiol.2018.01.011.

62 Das, T.N., Huie, R.E., Neta, P., and Padmaja, S. (1999). Reduction potential of the sulfhydryl radical: pulse radiolysis and laser flash photolysis studies of the formation and reactions of ·SH and HSSH·− in aqueous solutions. *J. Phys. Chem. A.* 103: 5221–5226. https://doi.org/10.1021/jp9907544.

63 Karmann, W., Meissner, G., and Henglein, A. (1967). Pulsradiolyse des schwefelwasserstoffs in wäßriger lösung. *Z. Naturforsch. B* 22: 273–282.

64 Kessler, D. (2006). Enzymatic activation of sulfur for incorporation into biomolecules in prokaryotes. *FEMS Microbiol. Rev.* 30: 825–840. https://doi.org/10.1111/j.1574-6976.2006.00036.x.

65 Braymer, J.J. and Lill, R. (2017). Iron-sulfur cluster biogenesis and trafficking in mitochondria. *J. Biol. Chem.* 292: 12754–12763. https://doi.org/10.1074/jbc.R117.787101.

66 Akaike, T., Ida, T., Wei, F.-Y. et al. (2017). Cysteinyl-tRNA synthetase governs cysteine polysulfidation and mitochondrial bioenergetics. *Nat. Commun.* 8: 1177. https://doi.org/10.1038/s41467-017-01311-y.

67 Ubuka, T., Ohta, J., Akagi, R. et al. (1992). Metabolism of L-cysteine via transamination pathway (3-mercaptopyruvate pathway). *Amino Acids* 3: 243–252. https://doi.org/10.1007/BF00805999.

68 Shibuya, N., Tanaka, M., Yoshida, M. et al. (2009). 3-Mercaptopyruvate sulfurtransferase produces hydrogen sulfide and bound sulfane sulfur in the brain. *Antioxid. Redox Signal.* 11: 703–714. https://doi.org/10.1089/ars.2008.2253.

69 Shibuya, N., Koike, S., Tanaka, M. et al. (2013). A novel pathway for the production of hydrogen sulfide from D-cysteine in mammalian cells. *Nat. Commun.* 4: 1366. https://doi.org/10.1038/ncomms2371.

70 Hylin, J.W. and Wood, J.L. (1959). Enzymatic formation of polysulfides from mercaptopyruvate. *J. Biol. Chem.* 234: 2141–2144.

71 Yadav, P.K., Yamada, K., Chiku, T. et al. (2013). Structure and kinetic analysis of H_2S production by human mercaptopyruvate sulfurtransferase. *J. Biol. Chem.* 288: 20002–20013. https://doi.org/10.1074/jbc.M113.466177.

72 Lec, J.-C., Boutserin, S., Mazon, H. et al. (2018). Unraveling the mechanism of cysteine persulfide formation catalyzed by 3-mercaptopyruvate sulfurtransferases. *ACS Catal.* 8: 2049–2059. https://doi.org/10.1021/acscatal.7b02432.

73 Yadav, P.K., Vitvitsky, V., Carballal, S. et al. (2020). Thioredoxin regulates human mercaptopyruvate sulfurtransferase at physiologically-relevant concentrations. *J. Biol. Chem.* 295: 6299–6311. https://doi.org/10.1074/jbc.RA120.012616.

74 Yamanishi, T. and Tuboi, S. (1981). The mechanism of the L-cystine cleavage reaction catalyzed by rat liver gamma-cystathionase. *J. Biochem. (Tokyo)* 89: 1913–1921.

75 Ida, T., Sawa, T., Ihara, H. et al. (2014). Reactive cysteine persulfides and S-polythiolation regulate oxidative stress and redox signaling. *Proc. Natl. Acad. Sci. U.S.A.* 111: 7606–7611. https://doi.org/10.1073/pnas.1321232111.

76 Yadav, P.K., Martinov, M., Vitvitsky, V. et al. (2016). Biosynthesis and reactivity of cysteine persulfides in signaling. *J. Am. Chem. Soc.* 138: 289–299. https://doi.org/10.1021/jacs.5b10494.

77 Moutiez, M., Aumercier, M., Teissier, E. et al. (1994). Reduction of a trisulfide derivative of glutathione by glutathione reductase. *Biochem. Biophys. Res. Commun.* 202: 1380–1386. https://doi.org/10.1006/bbrc.1994.2083.

78 Bianco, C.L., Akaike, T., Ida, T. et al. (2019). The reaction of hydrogen sulfide with disulfides: formation of a stable trisulfide and implications for biological systems. *Br. J. Pharmacol.* 176: 671–683. https://doi.org/10.1111/bph.14372.

79 Bolton, S.G., Cerda, M.M., Gilbert, A.K., and Pluth, M.D. (2019). Effects of sulfane sulfur content in benzyl polysulfides on thiol-triggered H_2S release and cell proliferation. *Free Radical Biol. Med.* 131: 393–398. https://doi.org/10.1016/j.freeradbiomed.2018.12.025.

80 Münchberg, U., Anwar, A., Mecklenburg, S., and Jacob, C. (2007). Polysulfides as biologically active ingredients of garlic. *Org. Biomol. Chem.* 5: 1505–1518. https://doi.org/10.1039/b703832a.

81 Benavides, G.A., Squadrito, G.L., Mills, R.W. et al. (2007). Hydrogen sulfide mediates the vasoactivity of garlic. *Proc. Natl. Acad. Sci. U.S.A.* 104: 17977–17982. https://doi.org/10.1073/pnas.0705710104.

82 Pilkington, L.I., Deed, R.C., Parish-Virtue, K. et al. (2019). Iterative synthetic strategies and gene deletant experiments enable the first identification of polysulfides in *Saccharomyces cerevisiae. Chem. Commun. Camb. Engl.* 55: 8868–8871. https://doi.org/10.1039/c9cc03020d.

83 Liang, D., Wu, H., Wong, M.W., and Huang, D. (2015). Diallyl trisulfide is a fast H_2S donor, but diallyl disulfide is a slow one: the reaction pathways and intermediates of glutathione with polysulfides. *Org. Lett.* 17: 4196–4199.

84 Smith, D.J. and Venkatraghavan, V. (1985). Synthesis and stability of thiocysteine. *Synth. Commun.* 15: 945–950.

85 Kabil, O. and Banerjee, R. (2012). Characterization of patient mutations in human persulfide dioxygenase (ETHE1) involved in H_2S catabolism. *J. Biol. Chem.* 287: 44561–44567. https://doi.org/10.1074/jbc.M112.407411.

86 Pan, J. and Carroll, K.S. (2013). Persulfide reactivity in the detection of protein S-sulfhydration. *ACS Chem. Biol.* 8: 1110–1116. https://doi.org/10.1021/cb4001052.

87 Capdevila, D.A., Walsh, B.J.C., Zhang, Y. et al. (2021). Structural determinants of persulfide-sensing specificity in a dithiol-based transcriptional regulator. *Nat. Chem. Biol.* 17: 65–70. https://doi.org/10.1038/s41589-020-00671-9.

88 Tsurugi, J. and Nakabayashi, T. (1959). Synthesis of dibenzhydryl and dibenzyl penta- and hexasulfides. *J. Org. Chem.* 24: 807–810. https://doi.org/10.1021/jo01088a019.

89 Everett, S.A. and Wardman, P. (1995). Perthiols as antioxidants: radical-scavenging and prooxidative mechanisms. *Methods Enzymol.* 251: 55–69.

90 Artaud, I. and Galardon, E. (2014). A persulfide analogue of the nitrosothiol SNAP: formation, characterization and reactivity. *Chembiochem Eur. J. Chem. Biol.* 15: 2361–2364. https://doi.org/10.1002/cbic.201402312.

91 Bailey, T.S. and Pluth, M.D. (2015). Reactions of isolated persulfides provide insights into the interplay between H_2S and persulfide reactivity. *Free Radical Biol. Med.* 89: 662–667. https://doi.org/10.1016/j.freeradbiomed.2015.08.017.

92 Yuan, Z., Zheng, Y., Yu, B. et al. (2018). Esterase-sensitive glutathione persulfide donor. *Org. Lett.* 20: 6364–6367. https://doi.org/10.1021/acs.orglett.8b02611.

93 Powell, C.R., Dillon, K.M., Wang, Y. et al. (2018). A persulfide donor responsive to reactive oxygen species: Insights into reactivity and therapeutic potential. *Angew. Chem. Int. Ed.* 57: 6324–6328. https://doi.org/10.1002/anie.201803087.

94 Everett, S.A., Folkes, L.K., Wardman, P., and Asmus, K.D. (1994). Free-radical repair by a novel perthiol: reversible hydrogen transfer and perthiyl radical formation. *Free Radical Res.* 20: 387–400.

95 Chauvin, J.-P.R., Griesser, M., and Pratt, D.A. (2017). Hydropersulfides: H-atom transfer agents par excellence. *J. Am. Chem. Soc.* 139: 6484–6493. https://doi.org/10.1021/jacs.7b02571.

96 Serjeant, E.P. and Dempsey, B. (1979). *Ionisation Constants of Organic Acids in Aqueous Solution*. Oxford; New York: Pergamon Press.

97 Portillo-Ledesma, S., Sardi, F., Manta, B. et al. (2014). Deconstructing the catalytic efficiency of peroxiredoxin-5 peroxidatic cysteine. *Biochemistry* 53: 6113–6125. https://doi.org/10.1021/bi500389m.

98 Benson, S.W. (1978). Thermochemistry and kinetics of sulfur-containing molecules and radicals. *Chem. Rev.* 78: 23–35. https://doi.org/10.1021/cr60311a003.

99 Bailey, T.S., Zakharov, L.N., and Pluth, M.D. (2014). Understanding hydrogen sulfide storage: probing conditions for sulfide release from hydrodisulfides. *J. Am. Chem. Soc.* 136: 10573–10576. https://doi.org/10.1021/ja505371z.

100 Kawamura, S., Horii, T., and Tsurugi, J. (1971). Aryl hydrodisulfides. *J. Org. Chem.* 36: 3677–3680. https://doi.org/10.1021/jo00823a004.

101 Li, H., Liu, H., Chen, Z. et al. (2019). Using resonance synchronous spectroscopy to characterize the reactivity and electrophilicity of biologically relevant sulfane sulfur. *Redox Biol.* 24: 101179. https://doi.org/10.1016/j.redox.2019.101179.

102 Cuevasanta, E., Benchoam, D., Ferrer-Sueta, G. et al. (2019). Commentary on "Using resonance synchronous spectroscopy to characterize the reactivity and electrophilicity of biologically relevant sulfane sulfur" Evidence that the methodology is inadequate because it only measures unspecific light scattering. *Redox Biol.* 26: 101281. https://doi.org/10.1016/j.redox.2019.101281.

103 Szajewski, R.P. and Whitesides, G.M. (1980). Rate constants and equilibrium constants for thiol-disulfide interchange reactions involving oxidized glutathione. *J. Am. Chem. Soc.* 102: 2011–2026. https://doi.org/10.1021/ja00526a042.

104 Bogdándi, V., Ida, T., Sutton, T.R. et al. (2019). Speciation of reactive sulfur species and their reactions with alkylating agents: do we have any clue about what is present inside the cell? *Br. J. Pharmacol.* 176: 646–670. https://doi.org/10.1111/bph.14394.

105 Kabil, O., Motl, N., Strack, M. et al. (2018). Mechanism-based inhibition of human persulfide dioxygenase by γ-glutamyl-homocysteinyl-glycine. *J. Biol. Chem.* 293: 12429–12439. https://doi.org/10.1074/jbc.RA118.004096.

106 Jencks, W.P. and Carriuolo, J. (1960). Reactivity of nucleophilic reagents toward esters. *J. Am. Chem. Soc.* 82: 1778–1786. https://doi.org/10.1021/ja01492a058.

107 Edwards, J.O. and Pearson, R.G. (1962). The factors determining nucleophilic reactivities. *J. Am. Chem. Soc.* 84: 16–24. https://doi.org/10.1021/ja00860a005.

108 Um, I.-H., Im, L.-R., and Buncel, E. (2010). Pitfalls in assessing the α-effect: reactions of substituted phenyl methanesulfonates with HOO(-), OH(-), and substituted phenoxides in H(2)O. *J. Org. Chem.* 75: 8571–8577. https://doi.org/10.1021/jo101978x.

109 Hoz, S. (1982). The α effect: on the origin of transition-state stabilization. *J. Org. Chem.* 47: 3545–3547. https://doi.org/10.1021/jo00139a033.

110 Garver, J.M., Gronert, S., and Bierbaum, V.M. (2011). Experimental validation of the α-effect in the gas phase. *J. Am. Chem. Soc.* 133: 13894–13897. https://doi.org/10.1021/ja205741m.

111 Benaïchouche, M., Bosser, G., Paris, J., and Plichon, V. (1990). Relative nucleophilicities of aryldisulphide and thiolate ions in dimethylacetamide estimated from their reaction rates with alkyl halides. *J. Chem. Soc. Perkin Trans.* 2 (8): 1421–1424. https://doi.org/1421-1424.

112 Heppner, D.E., Hristova, M., Ida, T. et al. (2018). Cysteine perthiosulfenic acid (Cys-SSOH): a novel intermediate in thiol-based redox signaling? *Redox Biol.* 14: 379–385. https://doi.org/10.1016/j.redox.2017.10.006.

113 Ono, K., Akaike, T., Sawa, T. et al. (2014). Redox chemistry and chemical biology of H_2S, hydropersulfides, and derived species: implications of their possible biological activity and utility. *Free Radical Biol. Med.* 77: 82–94. https://doi.org/10.1016/j.freeradbiomed.2014.09.007.

114 Greiner, R., Pálinkás, Z., Bäsell, K. et al. (2013). Polysulfides link H_2S to protein thiol oxidation. *Antioxid. Redox Signal.* 19: 1749–1765. https://doi.org/10.1089/ars.2012.5041.

115 Millikin, R., Bianco, C.L., White, C. et al. (2016). The chemical biology of protein hydropersulfides: studies of a possible protective function of biological hydropersulfide generation. *Free Radical Biol. Med.* 97: 136–147. https://doi.org/10.1016/j.freeradbiomed.2016.05.013.

116 Fukuto, J.M., Lin, J., Khodade, V.S., and Toscano, J.P. (2020). Predicting the possible physiological/biological utility of the hydropersulfide functional group based on its chemistry: similarities between hydropersulfides and selenols. *Antioxid. Redox Signal.* 33: 1295–1307. https://doi.org/10.1089/ars.2020.8079.

117 Fukuto, J.M., Vega, V.S., Works, C., and Lin, J. (2020). The chemical biology of hydrogen sulfide and related hydropersulfides: interactions with biologically relevant metals and metalloproteins. *Curr. Opin. Chem. Biol.* 55: 52–58. https://doi.org/10.1016/j.cbpa.2019.11.013.

118 Abiko, Y., Yoshida, E., Ishii, I. et al. (2015). Involvement of reactive persulfides in biological bismethylmercury sulfide formation. *Chem. Res. Toxicol.* 28: 1301–1306. https://doi.org/10.1021/acs.chemrestox.5b00101.

119 Akiyama, M., Shinkai, Y., Unoki, T. et al. (2017). The capture of cadmium by reactive polysulfides attenuates cadmium-induced adaptive responses and hepatotoxicity. *Chem. Res. Toxicol.* 30: 2209–2217. https://doi.org/10.1021/acs.chemrestox.7b00278.

120 Galardon, E., Huguet, F., Herrero, C. et al. (2017). Reactions of persulfides with the heme cofactor of oxidized myoglobin and microperoxidase 11: reduction or coordination. *Dalton Trans.* 46: 7939–7946. https://doi.org/10.1039/C7DT01638G.

121 Galardon, E., Tomas, A., Selkti, M. et al. (2009). Synthesis, characterization, and reactivity of alkyldisulfanido zinc complexes. *Inorg. Chem.* 48: 5921–5927. https://doi.org/10.1021/ic900238v.

122 Goudarzi, S., Babicz, J.T., Kabil, O. et al. (2018). Spectroscopic and electronic structure study of ETHE1: elucidating the factors influencing sulfur oxidation and oxygenation in mononuclear nonheme iron enzymes. *J. Am. Chem. Soc.* 140: 14887–14902. https://doi.org/10.1021/jacs.8b09022.

123 Álvarez, L., Suarez Vega, V., McGinity, C. et al. (2020). The reactions of hydropersulfides (RSSH) with myoglobin. *Arch. Biochem. Biophys.* 687: 108391. https://doi.org/10.1016/j.abb.2020.108391.

124 Kim, E.J., Feng, J., Bramlett, M.R., and Lindahl, P.A. (2004). Evidence for a proton transfer network and a required persulfide-bond-forming cysteine residue in Ni-containing carbon monoxide dehydrogenases. *Biochemistry* 43: 5728–5734. https://doi.org/10.1021/bi036062u.

125 Kappler, U. and Maher, M.J. (2013). The bacterial SoxAX cytochromes. *Cell. Mol. Life Sci.* 70: 977–992. https://doi.org/10.1007/s00018-012-1098-y.

126 Coelho, C., González, P.J., Moura, J.G. et al. (2011). The crystal structure of *Cupriavidus necator* nitrate reductase in oxidized and partially reduced states. *J. Mol. Biol.* 408: 932–948. https://doi.org/10.1016/j.jmb.2011.03.016.

127 Lehrke, M., Rump, S., Heidenreich, T. et al. (2012). Identification of persulfide-binding and disulfide-forming cysteine residues in the NifS-like domain of the molybdenum cofactor sulfurase ABA3 by cysteine-scanning mutagenesis. *Biochem. J.* 441: 823–832. https://doi.org/10.1042/BJ20111170.

128 Saund, S.S., Sosa, V., Henriquez, S. et al. (2015). The chemical biology of hydropersulfides (RSSH): chemical stability, reactivity and redox roles. *Arch. Biochem. Biophys.* 588: 15–24. https://doi.org/10.1016/j.abb.2015.10.016.

129 Huang, G.-T. and Yu, J.-S.K. (2016). Enzyme catalysis that paves the way for S-sulfhydration via sulfur atom transfer. *J. Phys. Chem. B.* 120: 4608–4615. https://doi.org/10.1021/acs.jpcb.6b03387.

130 Kawamura, S., Otsuji, Y., Nakabayashi, T. et al. (1965). Aralkyl hydrodisulfides. IV. The Reaction of benzyl hydrodisulfide with several nucleophiles. *J. Org. Chem.* 30: 2711–2714. https://doi.org/10.1021/jo01019a048.

131 Kawamura, S., Nakabayashi, T., Kitao, T., and Tsurugi, J. (1966). Aralkyl hydrodisulfides. VI. The reaction of benzhydryl hydrosulfide with several nucleophiles. *J. Org. Chem.* 31: 1985–1987. https://doi.org/10.1021/jo01344a515.

132 Tsurugi, J., Abe, Y., Nakabayashi, T. et al. Aralkyl hydrodisulfides. XI. Reaction with amines. *J. Org. Chem.* 35 (1970): 3263–3266. https://doi.org/10.1021/jo00835a019.

133 Kawamura, S., Kitao, T., Nakabayashi, T. et al. (1968). Aralkyl hydrodisulfides. VIII. Alkaline decomposition and its competition with nucleophiles. *J. Org. Chem.* 33: 1179–1181. https://doi.org/10.1021/jo01267a053.

134 Tsurugi, J., Nakabayashi, T., and Ishihara, T. (1965). Aralkyl hydrodisulfides. III. The reaction with tertiary phosphines. *J. Org. Chem.* 30: 2707–2710. https://doi.org/10.1021/jo01019a047.

135 Wood, J.L. (1987). Sulfane sulfur. *Methods Enzymol.* 143: 25–29.

136 Koike, S., Nishimoto, S., and Ogasawara, Y. (2017). Cysteine persulfides and polysulfides produced by exchange reactions with H_2S protect SH-SY5Y cells from methylglyoxal-induced toxicity through Nrf2 activation. *Redox Biol.* 12: 530–539. https://doi.org/10.1016/j.redox.2017.03.020.

137 Bianco, C.L., Chavez, T.A., Sosa, V. et al. (2016). The chemical biology of the persulfide (RSSH)/perthiyl (RSS·) redox couple and possible role in biological redox signaling. *Free Radical Biol. Med.* 101: 20–31. https://doi.org/10.1016/j.freeradbiomed.2016.09.020.

138 Koppenol, W.H. and Bounds, P.L. (2017). Signaling by sulfur-containing molecules. Quantitative aspects. *Arch. Biochem. Biophys.* https://doi.org/10.1016/j.abb.2016.09.012.

139 Chauvin, J.-P.R., Haidasz, E.A., Griesser, M., and Pratt, D.A. (2016). Polysulfide-1-oxides react with peroxyl radicals as quickly as hindered phenolic antioxidants and do so by a surprising concerted homolytic substitution. *Chem. Sci.* 7: 6347–6356. https://doi.org/10.1039/c6sc01434h.

140 Wu, Z. and Pratt, D.A. (2020). Radical substitution provides a unique route to disulfides. *J. Am. Chem. Soc.* 142: 10284–10290. https://doi.org/10.1021/jacs.0c03626.

4

Hydrogen Sulfide, Reactive Nitrogen Species, and "The Joy of the Experimental Play*"

Miriam M. Cortese-Krott

Myocardial Infarction Research Laboratory, Department of Cardiology, Pneumology, and Angiology, Medical Faculty, Heinrich Heine University, Düsseldorf, Germany
Department of Physiology and Pharmacology, Karolinska Institutet, Stockholm, Sweden

4.1 Introduction

Sulfur and nitrogen, together with carbon and oxygen, are considered as the fundamental building blocks of DNA and proteins, which characterize all living organisms. While carbon constitutes mainly the backbones of these molecules, nitrogen and sulfur define their chemical reactivity and biochemical roles. The chemical properties of both sulfur and nitrogen become very peculiar and multifaceted when the two elements are bound together. The study of the fundamental and quite puzzling properties of S–N compounds has been of interest for centuries, and their application span the production of gunpowder and fireworks, batteries, wastewater disposal, geochemistry, biochemistry, and even studies about the origin of life on earth [1]. in fact, sulfur and nitrogen were both present into the primordial broth.

A systematic analysis of the reactions leading to the formation of S–N compounds and their chemical properties was initiated in Germany more than a century ago [2], followed up until the 1980s [3–13], and put recently into the biological context [1, 14–20]. These reports established that the reaction of sulfur-bearing compounds such as hydrogen sulfide (H_2S), alkali sulfide (e.g. NaHS, Na_2S) or polysulfide (K_2S_x) salts or colloidal sulfur (S_8) can react with N-bearing compounds such as nitric oxide (NO), nitrite (NO_2^-), or nitrous acid (HNO_2) giving rise to a series of multiple chemical reactions. The final products of these reactions are beautifully colored nitrogen and gaseous nitrogen oxides (NO, N_2, N_2O) and

*"die Freude am experimentellen Spiel" (M. Goehring)

sulfur-bearing radical and/or anionic solutes such as polysulfides [12, 13], and S–N hybrid species, i.e. combined sulfur and nitrogen-containing compounds [16].

Not only the complexity of the chemistry behind these reactions but also their beauty were described in great detail in a book published in 1957 by Margot Goehring entitled "Ergebnisse und Probleme der Chemie der Schwefelstickstoffverbindungen" (Results and concerns pertinent to the chemistry of sulfur–nitrogen compounds) [21]. In a public speech in 1978, she described the passion behind her hard scientific experimental work on these molecules as "die Freude am experimentellen Spiel" [22] (the joy of the experimental play). The same passion transpired from the publications of the other pioneers of the field from then and now, which include Frederick Raschig [2], Friz Seel [7–13], Tristram Chivers [23–27], and Ralf Steudel [28–32].

Fritz Seel et al. described the reaction mechanism between either sulfide or polysulfide and NO or nitrite. They identified and chemically characterized a number of reaction intermediates, including short-chain polysulfide ($S_3^{\bullet-}$ and S_4^{2-}) and the S–N compounds, nitrosopersulfide ($SSNO^-$), and dinitrososulfite ($[ONN(O)\text{-}SO_3]^{2-}$) [13]. In addition, F. Seel hypothesized that the same mechanism could be involved in the chain of reactions taking place during the explosion of gun powder [7]. More recently, these compounds and reactions are described in a very detailed monography about molecules containing nitrogen atoms bound to chalcogens (i.e. the elements of group 16 of the periodic table including oxygen (O), sulfur (S), selenium (Se), tellurium (Te))[23].

Nitrosothiols can be also considered as a subcategory of S–N hybrid compounds. They were shown to be present in relevant concentrations in biological tissues and participate in NO signaling [1, 33–41]. Interestingly, the chemistry of nitrosothiols was only recently linked to the traditional systematic of S–N hybrid compounds with biological activity. This began with a report by Munro and Williams [42, 43], who described that in the reaction of S-nitrosothiols and sulfide, multiple compounds were produced including SNO^- and $SSNO^-$, but they observed that the kinetics of the reaction were highly complex and "too complex to be understood" [42, 43].

Of note, the same S–N intermediates and products were originally described by Goehring, Schenk, and Seel in the reactions of sulfide/polysulfide with NO or nitrite, and by Williams for the reaction of S-nitrosothiols and sulfide, and they are also formed under biologically relevant conditions (pH 7.4, air). Moreover, these compounds display potent bioactivity in cultured cells in vitro and rats in vivo [15, 16, 44].

This chapter provides a synopsis of bioactive S–N compounds derived from the interaction of sulfide and NO and their potential role in biological signaling. Specifically, this chapter starts with a brief overview of the basic chemical properties of NO and sulfide and their common metabolites including nitrosothiols,

$HSNO/SNO^-$, $SSNO^-$, and $[ONN(O)\text{-}SO_3]^{2-}$; review what is currently known about their bioactivity; and discuss how some of those fundamental properties relate to the emerging biology and pharmacology.

4.2 Basic Physicochemical Properties of Nitric Oxide and Its Biological Relevant Metabolites

NO is well known to exert a very bright spectrum of physiological actions [45]. To understand the role of NO in physiology and pathophysiology, it is important to take into consideration its physicochemical properties, its reactivity toward other molecules, its localization, and the biochemical characteristics of the enzymes controlling its synthesis [46].

NO is a small, uncharged molecule that can easily cross cell membranes [46]. Its reactivity toward oxygen, free radicals, metal centers, and thiols allows for NO to modify specific targets and thereby modify the activity of enzymes, regulate downstream signaling, and form stable NO metabolites. NO metabolites include nitrite, nitrate, nitrosothiols, nitrosoamines, and nitro-fatty acids and have different bioactivities [47, 48]. For example, reactions with regulatory thiols on target proteins (thiol redox switches) lead to the S-nitrosation of the thiols and modulation of protein structural/functional properties [49, 50].

NO can be synthesized in cells and tissues by three isoforms of a NO synthases (EC 1.14.13.39) converting L-arginine to citrulline and NO [51]. The three isoforms differ substantially for enzyme kinetics, localization, and regulatory mechanisms (dependency on calcium/calmodulin; phosphorylation sites, other posttranslational modifications [52]). NO can also be synthesized from nitrite by globin proteins under hypoxic conditions [53, 54]. Moreover, NO bioactivity is also dependent on the localization and type of target proteins, of which the most important one is the soluble guanylate cyclase [55]. For more information on the biochemical, biological, and physiological properties of NO, please refer to the monography edited by B. Freeman and L. Ignarro [56].

4.2.1 Nitric Oxide

NO is a poor one-electron oxidant. This property allows NO to have a longer half-life in the biological environment in the presence of millimolar concentrations of thiols. For example in the vessel wall, NO can be produced in endothelial cells and can diffuse to the underlying smooth muscle cells, where it reaches the iron atom in the heme group of soluble guanylate cyclase. Some of the NO produced by NO synthases or by other enzymatic/nonenzymatic activities will also undergo autoxidation. In this way, NO will be oxidized to more reactive species (Eqs. (4.1) and (4.2):

$$2NO + O_2 \rightarrow 2NO_2 \tag{4.1}$$

$$NO_2 + NO \rightleftharpoons N_2O_3 \tag{4.2}$$

Because of the reactivity of the NO_2 radical toward cellular antioxidants, in biological environments, N_2O_3 is probably formed efficiently only in lipid membranes (Eq. (4.3)) [57–59]. N_2O_3 is the anhydride of nitrous acid:

$$N_2O_3 + H_2O \rightleftharpoons 2HNO_2 \rightleftharpoons 2H^+ + NO_2^- \tag{4.3}$$

The K value for the equilibrium formation of N_2O_3 is $3.0 \times 10^{-3}\ M^{-1}$ [43, 60]. Nitrous acid, once formed, ionizes with a pK_a of 3.15, and so there is very little of the undissociated HNO_2 present around physiological pH. This reaction will form nitrite (Eq. 4.3).

4.2.2 Nitrite

Nitrite was once considered as a biologically inert product of spontaneous or metalloenzyme-catalyzed oxidation of NO in solution [61] and plasma [62]. However, nitrite can be converted back to NO by globins [54] and was shown to exert signaling effects on its own [63]. Nitrite was considered "inert" in aqueous environments and at biological pH. In fact nitrite is a weak nucleophile [64], so it will not easily react with electrophiles; moreover, water molecules will provide a solvation barrier further impairing chemical reactions. Instead, at very low pH (<3), nitrite can be used to nitrosate thiols to produce nitrosothiols (RSNOs). Under acidic conditions, protonation of HNO_2 results in the formation of NO^+, a particularly powerful electrophile that is able to rapidly react with thiolate according to Eqs. (4.4)–(4.6):

$$HNO_2 + H_3O^+ \rightleftharpoons [H_2NO_2]^+ + H_2O \tag{4.4}$$

$$[H_2NO_2]^+ \rightleftharpoons NO^+ + H_2O \tag{4.5}$$

$$NO^+ + RS^- \rightarrow RSNO \tag{4.6}$$

The K value for equilibrium Eq. (4.5) is $1.2 \times 10^{-8}\ M^{-1}$ [65]. Although there is little thiolate anion present in an acidic solution of thiols (e.g. cysteine has a $pK_a = 8.3$), thiols are sufficiently strong nucleophiles to react once they encounter NO^+ and thus produce S-nitrosocysteine. It is worth pointing out that the release of NO^+ in its "free" form does not occur in biological environments other than perhaps the stomach. Rather, the $[NO^+]$ moiety is formally transferred from a bound form to another acceptor in a so-called "transnitrosation reaction: (Eq. (4.7)):

$$RSH + R'SNO \rightarrow RSNO + R'SH \tag{4.7}$$

At higher pH values (pH 3–7), the reaction pathway is rather different. Undissociated nitrous acid exists in equilibrium with N_2O_3, which is also a nitrosating species. Water is a weak nucleophile. Therefore, N_2O_3 will likely react preferentially with more powerful nucleophiles like glutathione, instead of forming HNO_2 in a biological environment. In fact, according to the Ritchie's N^+ nucleophilicity scale [66], a thiolate is 10^{11} times more nucleophilic than water.

4.2.3 Nitrosothiols (RSNOs)

In nitrosothiols (RSNOs; monothionitrites) an alkyl or aryl substituent is attached to the sulfur atom of a nitroso group. Naturally occurring nitrosothiols in the biological environment are S-nitrosoglutathione (GSNO), S-nitrosoalbumin, and S-nitrosohemoglobin tissues. Although the pathways of formation and degradation/bioactivation of endogenous RSNOs are far from clear at present, the involvement of specific enzymes (nitrosylases/denitrosylases) has been proposed (see for example literature by Stamler and coworkers [67]).

In addition to these enzymatically regulated processes, direct chemical reactions lead to the cleavage of the S—N bond and the release of the NO group under biologically relevant conditions. The Pauling electronegativities for N, O, and S are 3.04, 3.44, and 2.58, respectively [23]. In binary systems, the N—O bond will be polarized with a partially negative charge on oxygen, whereas the S—N bond will be polarized on the nitrogen. Sulfur–nitrogen π-bonds are weaker than oxygen–nitrogen π-bonds, and there is an increased reluctance for sulfur to form multiple bonds with nitrogen as compared to oxygen. For these reasons, the S—N bond in RSNOs is predicted to be more thermally labile than the O—N bond in RONOs.

However, experimental observations confirmed that in the absence of another reaction partner, RSNOs are stable. One mechanism of the decomposition of RSNOs is homolysis of the S–N bond catalyzed by light or temperature to give a thiyl radical and NO (Eq. (4.8)), followed by radical dimerization (Eq. (4.9)) [68]:

$$2RSNO \rightarrow 2RS^{\cdot} + 2NO \tag{4.8}$$

$$2RS^{\cdot} \rightarrow 2RSSR \tag{4.9}$$

Butler and Williams et al. [69] demonstrated that the homolysis rate of RSNO is very sensitive to the presence of copper ions in solution [70], which may form complexes with the sulfur of the S—N bond in a nitrosothiol. More stable nitrosothiols (like S-nitrosoaminopenicillamine, often used in biological experiments) are characterized by steric crowding around the sulfur atom, which may hinder the complexation with metal and may partly explain its stability.

In cells and tissues, the concentration of thiols is very high. In this context, nitrosothiol may undergo transnitrosation with another thiol following a nucleophilic attack of a thiolate on nitrogen [71, 72]. Transnitrosation reactions are discussed in the literature as a way of transferring "NO bioactivity" from one compartment to the other in the body (Eq. (4.10)):

$$RSNO + R'S^- \rightarrow R'SNO + RS^- \tag{4.10}$$

Furthermore, in biological environment, RSNOs can undergo a nucleophilic attack by a thiolate ($R'S^-$) at the sulfur of the nitrosothiol, leading to release of nitroxyl anion (NO^-) as a leaving group (Eq. (4.10)). In fact thiolates are very powerful nucleophile according to Ritchie's scale [66]. Moreover, this reaction is considered as a possible source of nitroxyl in the cell. The outcome of the reaction will depend on the localization, abundance of reduced thiols, and pK_a values of the sulfhydryl groups involved:

$$RSNO + R'S^- \rightarrow RSSR' + NO^- \tag{4.11}$$

In cell and tissues where the concentration of thiols is very high (millimolar in some cells like RBCs) both the reactions according to Eqs. (4.10) and (4.11) are highly likely to take place and be very fast.

4.3 Basic Physicochemical Properties of H_2S and Its Biological Relevant Metabolites

For a long time, H_2S was considered as a toxic molecule. The study of the biological effects and its metabolite in biological environment had been limited to measuring the toxic effects of sulfide or sulfur metabolites. Relatively recently, it was discovered that H_2S is produced endogenously by three independent enzymes and may participate in signaling of redox regulation in multiple pathways including blood pressure regulation, neurotransmission, and suspended animation.

The physicochemical properties of sulfide and its metabolites including polysulfide, persulfide sulfite, and polythionate are very complex, particularly in an aqueous environment. These have been studied in detail for environmental and synthetic industrial purposes [29] as well as in the context of the microbiology of sulfur oxidizing and reducing bacteria [28], and only recently in mammalian tissues [73]. In particular, there are many groups focusing intensively on discovering the nature of the molecular species responsible for the effects of sulfide, which include H_2S/HS^-, inorganic polysulfide, or persulfides [73, 74]. Moreover, another intense field of study is discovering how enzymes involved in sulfide metabolism are regulated and what are the biochemical targets of the sulfide signals [75, 76].

In analogy to the pool of NO metabolites (nitroso-species, nitrite and nitrate, and nitrated fatty acids), sulfide has also been proposed to exist in the form of pools of "free"/available/exchangeable sulfide and "bound" sulfide. Based on the chemical conditions applied to trigger the release of sulfide and/or the assay used for detection (i.e. acidification, alkalization, or reduction), the pools of sulfide were defined as "acid labile," "alkaline labile," and "persulfide"/polysulfide pools [77–81].

Polysulfides have been proposed by the group led by Hideo Kimura to be the bioactive molecules responsible for "H_2S signaling" in cells [74, 82], and they are believed to act as storage or buffer forms of H_2S [83]. More recently, the group of Taka Akaike proposed that low molecular weight persulfide and polysulfidate proteins are highly abundant in cells and tissues and are likely involved in cellular signaling [79]. They demonstrated that persulfide are produced endogenously by enzymatic activity of the cysteinyl-tRNA synthetases (CARSs) [84]. More research is needed to further define the role of persulfide and of these "sulfide pools" in "H_2S signaling" and their interaction with NO and its metabolites. In particular, it is necessary to carry out a further chemical characterization of these pools in biological environments and to refine the analytical techniques to determine the speciation of, per/polysulfides.

4.3.1 H_2S/HS^-

The definition of H_2S as a "gasotrasmitter" was introduced by R. Wang in 2002 [85]. This definition was supposed to simplify and characterize the still unknown role of this molecule by comparing it to the better-known endogenously produced transmitters NO and CO. This definition certainly attracted the attention of many scientists of the respective fields and significantly contributed to the growing in interest in sulfur pharmacology and signaling. This simplification brought a little confusion to the field too.

In the biological environments, none of NO, CO, or H_2S are actually found in gas form. Instead, they are uncharged solutes. At pH 7.4, sulfide is expected to be found mainly (>70%) as hydrosulfide anion (HS^-) and to a lesser extent (<25%) as *dissolved* neutral dihydrogen sulfide (H_2S), while the concentration of the dianion S^{2-} is negligible [86]. These calculations are based on the dissociation constants of H_2S ($pK_{a1} = 6.8$ and $pK_{a2} > 12$), which were obtained experimentally by analyzing the dissociation equilibria of H_2S at different salt concentrations in water [87].

The reactivity of HS^- is mainly dictated by its reducing and electrophilic properties, as well as the ability of sulfur in general to form complexes with transition metals including zinc, copper, and iron. A one-electron oxidation of sulfide form sulfur (thiyl) radicals ($S^{\cdot-}$). This reaction occurs in two steps (Eqs. (4.12) and (4.13)) via the formation of the highly unstable intermediate $HS^{\cdot}$ and its deprotonation at pH 7.4 to form the reactive $S^{\cdot-}$ [29]:

$$HS^- \rightarrow HS^{\cdot} + 1\,e^- \quad (4.12)$$

$$HS^{\cdot} \rightarrow S^{\cdot-} + H^+ \quad (4.13)$$

A one-electron oxidation was proposed to occur by direct reaction with molecular oxygen, which is probably not very efficient in a biological environment (Eq. (4.14)):

$$HS^- + O_2 \rightarrow S^{\cdot -} + H^+ + O_2^{\cdot -} \tag{4.14}$$

A more efficient 1 electron oxidation of HS^- may be catalyzed by transition metals like copper or iron, which have been shown to participate in chemical reactions even while complexed with metal chelators (Eq. (4.15)) [88]:

$$Cu(II) + HS^- \rightarrow Cu(I) + HS^{\cdot} \tag{4.15}$$

The affinity of sulfur species for divalent metals may be important for the effects of sulfide with metal centers in metalloproteins, such as methemoglobin, cytochrome C, and neuroglobin [75]. In fact, the toxicity of sulfide was attributed to its affinity for metal centers in mitochondria and in particular to the inhibition of cytochrome C oxidase, which is the terminal complex in the mitochondrial electron transport chain [89, 90].

Of particular interest is the reaction of sulfide with methemoglobin occurring in red blood cells. Sulfide reacts with methemoglobin and form a metastable complex, which compositions were shown to form deoxy/oxyhemoglobin, polysulfide, and thiosulfate [90–92]. We proposed that formation of these products may occur by *reductive sulfhydration* (Eq. (4.16) and (4.17)) [90]:

$$Hb(III) + H_2S \rightleftharpoons [Hb(III)(SH_2)] \rightleftharpoons [Hb(III)(SH^-)] + H^+ \tag{4.16}$$

$$[Hb(III)(SH^-)] \leftrightarrow [Hb(II)(SH^{\cdot})] \rightarrow Hb\,(II + SH^{\cdot} \tag{4.17}$$

According to these reactions, sulfide produced endogenously in red blood cells (by 3-mercaptopyruvate sulfotransferase or via other mechanisms) may contribute to keep hemoglobin in its reduced Fe(II) oxygen binding form (beside the very powerful cytochrome B2 reductase). Moreover, considering the high abundance of red blood cells in the circulation (10^9 cell/mL blood) and the abundance of methemoglobin (around 100–300 μM/cell) it is likely that these reactions may govern systemic sulfide metabolism in mammals, thus making red blood cell a "central hub" for sulfide metabolism [90].

With a similar mechanism, sulfide was proposed to be able to reduce the heme of an oxydized Fe(III) soluble guanylate cyclase into NO-binding Fe(II) active soluble guanylate cyclase [93]. The physiological relevance of these observations needs to be studied in more detail. Once formed, $S^{\cdot -}$ is a powerful oxidant and may lead to formation of other reactive sulfur species including HS_2^-, for example by reaction with further HS^- or $SO_2^{\cdot -}$ by reaction with oxygen (Eq. (4.18)–(4.22)):

$$HS^{\cdot} \rightarrow S^{\cdot -} + H^+ \tag{4.18}$$

$$2S^{\cdot -} \rightarrow S_2^{2-} \tag{4.19}$$

$$S^{\bullet-} + HS^- \rightarrow HSS^{\bullet 2-} + H^+ \rightarrow HS_2^- \quad (4.20)$$

$$S^{\bullet-} + HS_2^- \rightarrow HSS^{\bullet 2-} + HS^- \quad (4.21)$$

$$S^{\bullet-} + O_2 \rightarrow [SOO^{\bullet-}] \rightarrow SO_2^{\bullet-} \quad (4.22)$$

$S^{\bullet-}$ can also react with other radicals, including NO. This reaction will lead to formation of S–N hybrid molecules, including SNO^-, $SSNO^-$, and SULFI/NO [16]. Please refer to Sections 4.4.1–4.4.3 for further discussion of these activities. Moreover, sulfide is one of the strongest nucleophile according to Richie's scale [94], and therefore allows for nucleophilic attack on electrophiles. One example of electrophiles are sulfane sulfur atoms, i.e. sulfur atoms that are attached to other sulfur atoms as defined by Toohey [95]. This reaction assumes biological importance following the discovery of abundant reservoirs of persulfides and polysulfides in tissues [84, 96]; these chains of sulfur may conceivably act as storage forms and enzyme-independent sources of sulfide. Please refer to Section 4.3.2 for a greater discussion of these points.

A further relevant reaction of sulfide is a nucleophilic attack on the nitrogen in nitrosothiols as proposed by Williams [42, 43]. This is one of the reactions leading to formation of nitrososulfide and nitrosopersulfide. Please refer to Section 4.4.1.

Accumulating evidences indicate that sulfide or its metabolites may affect the enzymatic activity of numerous other proteins (including but not limited to K^+ channels, protein kinases "phophopdiesterases," and endothelial nitric oxide synthase) and to induce posttranslational modifications of cysteines in regulatory proteins such as Keap1 (a process defined as "sulfhydration" or, chemically more correct, "persulfidation"), but in most cases, the precise molecular mode of action is unclear.

4.3.2 Polysulfides and Persulfide

Polysulfides have been proposed to be the bioactive entities responsible for the pharmacological action of sulfide donors as well as the physiological action of sulfide in cells [73, 74]. Polysulfides were also proposed to be possible storage or buffer molecules of H_2S [83]. The chemical properties of polysulfides are necessarily not fully compatible to these proposals. Polysulfides contain a highly reactive sulfane sulfur (e.g. a sulfur atom bound to another sulfur atom), which makes polysulfides highly reactive toward high concentrations of thiols (e.g. glutathione) that are typically found in biological environments [97]. Indeed, the addition of high concentrations of dithiothreitol and reduction of polysulfide to H_2S is used for assessing their concentrations in solution [78, 80]. However, there is compelling evidence indicating that these polysulfides were endogenously produced by the "H_2S" producing enzymes cystathionine β-synthase and cystathionine γ-lyase, and that H_2S may be formed by the decomposition of enzymatically produced organic persulfides/polysulfides [96, 98–100]. Interestingly, 3-mercaptopyruvate sulfotransferase was also shown to produce inorganic polysulfides [100]. These species may be precursors of other reactive sulfur species, including persulfide and sulfoxy radicals ($SS^{\cdot-}$ and $SO_2^{\cdot-}$), thiosulfate ($S_2O_3^-$), and sulfite (SO_3^-) [75].

Interestingly, formation of any of these radicals near an enzymatic source of NO would allow the formation of $SSNO^-$ and SULFI/NO in vivo [16] (as discussed in Sections 4.4.1–4.4.3 of this chapter) [1]. Vice versa, considering the role of polysulfides as both products and intermediates of $SSNO^-$ formation in the reaction of NO or nitrosothiols with sulfide, the reaction itself may provide a further pathway to their synthesis in cells and tissues.

Organic persulfides (including persulfide derivatives of glutathione and cysteine) are stable under reducing conditions and were recently detected in high concentrations in murine organs/tissues [96] and found to be enzymatically endogenously produced by the CARS [84]. Their roles in biology and physiology remain to be fully understood [73].

4.4 Inorganic Sulfur–Nitrogen Compounds

The term *S–N hybrid molecules* is a nonchemical definition used in biological literature to group compounds belonging to different chemical classes, but containing an S—N bond. These compounds typically originate from the reaction of sulfur-centered nucleophiles with a range of inorganic nitrogen species found in biological environment. Recently, these compounds became relevant for understanding the role of the NO/H_2S crosstalk in biology, pharmacology, physiology, and medicine [1, 76, 93, 101].

Considering the basic chemical properties of dissolved NO (a radical) and H_2S/HS^- (an anion) in an aqueous environment, it is clear that a reaction between these two in biological conditions is difficult to predict. There are early thermodynamic studies about this interaction [102]. These studies predicted that NO and H_2S should react to produce water, sulfur, and nitrogen according to the following equation (Eq. (4.23)):

$$2NO + 2H_2S \rightarrow 2H_2O + 2S + N_2 \tag{4.23}$$

The height of the energy barrier for this reaction determines the likelihood of the reaction. In general, the energy barrier of a reaction of a radical with an anion in a solution is likely to be very high; thus this reaction is unlikely to occur in the absence of some kind of catalyst. Reaction 4.23 was studied at first under anaerobic conditions in (unbuffered) alkaline aqueous solutions or in polar organic solvents. The outcome revealed formation of many different products, including sulfur species, nitrogen species, and species containing both elements. Apparently, the nature and the yield of the obtained products were strongly dependent on reaction conditions (pH, temperature, pO_2, and solvent) and peculiarly on the concentration ratio of the reagents (see [1] for a review). These studies suggested that the mechanisms involved in this chemical interaction are very complex.

In an aqueous environment, the reaction appeared to be even more complex to follow as the stability and reactivity of some intermediates changed. The reaction between sulfide and NO leads to formation of at least three very different classes of bioactive intermediates including not only perthionitrites ($SNO^{\bullet-}/SSNO^{\bullet-}$) but also polysulfides (S_n^{2-}), and SULFI/NO ($[ONN(O)\text{-}SO_3]^{2-}$); a NONOate adduct of two molecules of NO with sulfite) [1, 14–16, 44, 103]. These intermediates were shown to be rather distinct in chemical biology, and as a consequence, very different bioactivities in cells and in vivo [15, 16, 44, 104–107].

4.4.1 HSNO/SNO$^-$

Thionitrous acid (HSNO) or its deprotonated form thionitrite (SNO^-) is a highly reactive molecule, which may be formed in a transnitrosation reaction of nitrosothiols (RSNO) with HS^- [43] or by radical reactions involving NO and $S^{\bullet-}$. Similar to RSNO, HSNO/SNO$^-$ was proposed to undergo transnitrosation reactions with other nitrosothiols (leading to formation of R′SNO) or with sulfide leading to formation of $SSNO^-$ and homolysis of the S–N bond leading to formation of highly reactive sulfur-centered radicals and nitrogen radicals (Figure 4.1). However, differently from RSNO, HSNO/SNO$^-$ carries a *mobile* and *ionisable* proton instead of the alkyl group R. The mobile hydrogen promotes isomerization by intramolecular hydrogen shift (Eqs. (4.24) and (4.25)):

(4.24)

(4.25)

Already in 1952 Margot Goehring proposed that compounds like the yellow SONH or the red SNOH may easily isomerize [108] to form different species with the same molecular composition, which are in equilibrium with each other: HNSO, HOSN, HSNO, and HONS. She also observed that these species form insoluble brown materials at temperatures higher than −60°C [108]. Other studies later on confirmed these observations and ordered them according to an increasing ground state energy, as follows: HNSO > HOSN > HSNO > HONS. As a result, the equilibrium was proposed in the order as depicted in Eq. (4.26). Of

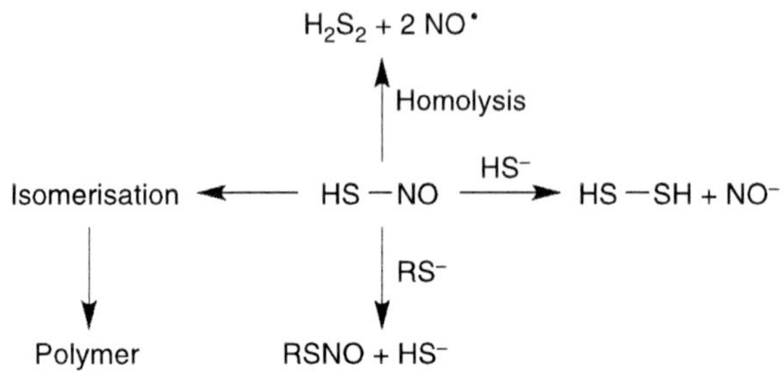

Figure 4.1 Reactivity of HSNO formed by and HS^-: isomerization, homolysis, nucleophilic attack, and transnitrosation. These reactions do not apply with neutral physiological pH, where HSNO is predicted to be present in the ionized form SNO^- or in the form of its isomer NSO^-.

interest, the position of the hydrogen and the O, N, and S atoms will change the chemical reactivity of the molecules:

$$HNSO \rightleftharpoons HOSN \rightleftharpoons HSNO \rightleftharpoons HONS \quad (4.26)$$

The physicochemical properties of HSNO and its isomers have been carefully characterized by Huber et al. at very low temperatures (<–60 °C) and extreme conditions (frozen Ar matrix) [109, 110]. A summary of the reactivity of HSNO is presented in Figure 4.1. Please consider that none of the reactions described in Figure 4.1 apply for SNO^- (the ionized form of HSNO), which is more likely to be found in aqueous environment.

In fact, at physiological pH, HSNO will be ionized and exist mainly as the SNO^- anion (thionitrite), with the exception of some NSO^- (thiazate), formed by isomerization of SNO^- [23]. The pK_a of thionitrous acid has not been experimentally determined but was estimated by Anthony Butler in a recent perspective review [1]. The pK_a of nitrous acid (HONO) is 3.25 and the replacement of oxygen by sulfur in HSNO likely decreases the pK_a of the molecule (as, for example, for cyanic acid and thiocyanic acid). So the pK_a of HSNO it is likely to be less than 3.25. Therefore, under physiological conditions (pH 7.4) is very likely that at any thionitrous acid formed would be largely present in the form of the SNO^- anion rather that undissociated HSNO.

SNO^- undergoes the isomerization into NSO^- (thiazate) [23] (Figure 4.2) as described by T. Chivers [23, 25]. According to Chivers, the middle form is unique and most likely a transition state. Although theoretical calculation [111] suggested a stable ground state structure of the middle form, it has not been isolated yet. Importantly, metal porphyrins in the biological environment may be excellent catalysts of this reaction [112]. Clearly, if SNO^- undergoes the isomerization into NSO^- [23] (Figure 4.2), none of the reaction pathways shown in Figure 4.1 could apply.

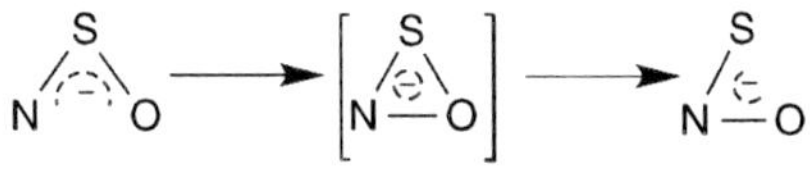

Figure 4.2 Isomerization of SNO^- into NSO^- via formation of a cyclic intermediate as described by Chivers. Source: [25]/Elsevier.

To summarize, there is an overwhelming evidence [1, 14–16, 68, 109, 110, 113, 114] that HSNO in the protonated form will be much less abundant than in protonated form (SNO^-) at biologically relevant conditions, which are characterized by neutral pH and ambient temperatures, and likely be the first product of the reaction of sulfide with a nitrosating agent similar to the reaction of thiols with sulfide.

4.4.2 $SSNO^-$

The nitrosopersulfide anion ($SSNO^-$), also known as perthionitrite, is formed and accumulates in buffered solution at pH 7.4 in the presence of an NO source and excess sulfide. $SSNO^-$ is stable even in the presence of milimolar concentrations of added thiols such as reduced glutathione, cysteine, and dithiothreitol [16]. This unexpected stability is rather atypical for a persulfide and was explained by charge distribution [23] and demonstrated recently by Fukuto et al. [115].

$SSNO^-$ was shown to be formed in the reaction between sulfide or polysulfides with NO in basic deoxygenated aqueous solutions [13] and in polar organic solvents [11]. $SSNO^-$ accumulates in the solution as long as it is in equilibrium with its precursors characterized by homolytic decomposition and formation via radical–radical reaction according to Eqs. (4.27) and (4.28) even after 3 or 7 days (please refer to Figure 4.1b in [13]). Very elegant studies recapitulating, confirming, and broadening these findings were published in the last years [17–20], including one published very recently using NMR [116]:

$$SSNO^- \rightleftharpoons S_2^{\cdot -} + NO^{\cdot} \tag{4.27}$$

$$2SS^{\cdot -} \rightleftharpoons S_4^{2-} \rightleftharpoons S^{\cdot -} + S_3^{\cdot -} \tag{4.28}$$

Indeed, independent groups confirmed that $SSNO^-$ decomposition results in NO release in the headspace [15, 16, 69, 106, 117], even before the nature of $SSNO^-$ as the yellow intermediate responsible for NO donation was known [117]. NO release from $SSNO^-$ is bioactive and forms nitrosylhemoglobin [69], it activates soluble guanylate cyclase in cells [15, 16, 69], it is a potent vasodilator of isolated aortic rings in organ baths [70], and decreases mean arterial pressure in rats [16].

Interestingly, an $SSNO^-$ ligand bound to the iron center of sodium nitroprusside $[Fe(CN)_5N(O)SS]^{4-}$ was recently detected as a product of the reaction between sulfide and SNP using a combination of multinuclear ^{17}O, ^{15}N, and ^{13}C-NMR, as well as UV/visible and IR spectroscopic techniques, accompanied by quantum chemical computational studies [118]. Clearly, the presence of metals and cyanide anions (CN^-) makes the mechanism of the reaction between SNP and sulfide very different from that of NO and sulfide [119]. However, the study of this reaction may be of considerable relevance to understand the complex reactions involving sulfide, NO, and iron centers potentially occurring in cells.

4.4.3 SULFI/NO

SULFI/NO is a compound belonging to the diazeniumdiolates or NONOate class of NO-donor compounds [120]. SULFI/NO is also known as dinitrososulfite [121], "Pelouze's salt" or "*Stickoxid-sulfite*" in the German literature (see [6] and references therein). SULFI/NO is formed as a result of the trapping of two molecules of NO by sulfite. SULFI/NO decomposition generates N_2O in high yield and a small amount of NO and HNO at pH 7.4 [120, 122]. The formation of SULFI/NO may contribute to and explain the NO scavenging effects of sulfide observed by Whiteman and Moore [106], as well as the formation of nitroxyl in the same reaction [15] and lack of it in the $SSNO^-$ decomposition [69].

In an aqueous solution, sulfite may be formed either by reaction of sulfide with O_2 via formation of $SO_2^{\cdot-}$ and $S_2O_4^{2-}$ [29], via formation of $S_2O_3^{2-}$ originating from radical reactions (via $SO_2^{\cdot-}$) or by hydrolysis of HONS (a product of HSNO isomerization), as proposed by Goehring et al. [108]. In cells and tissues, these precursors may be formed by enzymatic oxidation of sulfide or sulfite in the mitochondria, as well as in reactions catalyzed by peroxidases and/or heme proteins [73, 75, 91, 92, 100]. Formation of SULFI/NO in cells and tissue still needs to be confirmed experimentally.

4.5 Putative Biological Relevance of the NO/H_2S Chemical Interaction

The biological significance of the intermediate formed from the observed reaction of sulfide with NO under physiological conditions is still highly debated and unknown. Different aspects were taken into considerations including the pharmacological activity and bioavailability of the molecules. For example how can they be formed in cells and tissues, what are the analytical detection methods capable of determining concentrations, and what are the flux rates of S–N compounds and their precursors in cells and tissues?

4.5.1 Pharmacological Activity

The bioactivities of the sulfur compounds and the *S–N hybrid compounds* were shown in experiments with cells, tissues, or animals, which treated with different concentrations. These experiments demonstrated that S–N compounds may affect cell viability, redox state ad mitochondrial activity. Classical pharmacological tests were carried out on isolated tissues demonstrated the effects of S–N hybrid molecules on vascular function (e.g. on aortic rings) and coronary function and cardiac contractility (e.g. in the Langedorff system); these experiments showed that S–N hybrid molecules are vasodilators. Other tests were carried out by administration of these molecules (or their mixtures) in vivo and found effects on blood pressure and cardiac function. Further investigations are needed to identify the bioactive form(s) in cells and in vivo; considering the complex chemistry and the complex nature of the reactions involved, this is not a simple task and requires specific analytical techniques and genetic models.

After the discovery that the interaction of NO with H_2S leads to accumulation of $SSNO^-$, polysulfide, and SULFI/NO, there was some confusion about whether $SSNO^-$ or SNO^- are signaling molecules. A signaling molecule should be stable enough to survive in the biological environment and deliver the message, but also be reactive enough to hit its target and react with it (once the signaling molecule reached its destination). $SSNO^-$ was found to be stable at pH 7.4 in the presence of millimolar concentrations of cysteine and glutathione and to release bioactive NO as evidenced by activation of NO-dependent cGMP synthesis in cultured cells, vasodilation of aortic rings in organ baths and decrease of blood pressure in rats [16]. SULFI/NO was shown to release low levels of bioactive NO along with HNO in cells [16]. To markedly activate cellular soluble guanylate cyclase, it requires the presence of high concentrations of superoxide dismutase allowing conversion of HNO into NO [16]. If administrated in vivo, SULFI/NO lowers blood pressure in rats and affects cardiac contractility exerting similar effects as other nitroxyl donors [122–125]. Polysulfides appeared to have bioactivity with orders of magnitude greater than that of sulfide itself [99]. Thus, all three classes of compounds formed in the reaction of NO with sulfide are clearly bioactive.

4.5.2 Putative Sources of $SSNO^-$ and SULFI/NO *In Vivo*

An obvious but currently unaddressed question is whether $SSNO^-$ and SULFI/NO can actually be formed in cells and tissues, and if so, by what mechanisms. Concerns have been raised about $SSNO^-$ based on estimates of reaction rates observed in vitro. However, the demonstration that it can be formed from NO and sulfide under physiological pH in a test tube does not mean that this will be its preferred way of production in real life. If there were enzymatically catalyzed reactions giving rise to radical precursors near an NO source, these would react with NO almost instantaneously as soon as they are formed. Interestingly, similar doubts were raised two decades ago when peroxynitrite ($OONO^-$) was discussed as an entity of possible biological relevance. Peroxynitrite is the oxygen analogue of $SSNO^-$ and formed in the reaction of superoxide with NO (Eq. (4.29)):

$$O_2^{\cdot -} + NO^{\cdot} \rightarrow OONO^- \tag{4.29}$$

This is a radical–radical reaction that occurs at a rate close to the diffusion-controlled limit. When this reaction was first proposed to occur in biological environment by Beckman, Freeman et al. [126], some members of the NO community considered $OONO^-$ to be too unstable to be of much biological relevance and, because of the presence of superoxide dismutase whose speed of removing $O_2^{\cdot -}$ was one of the fastest biological reaction known then, its formation in tissues was initially questioned. Since many enzymatic sources of $O_2^{\cdot -}$ (like NADPH oxidases) have been discovered to occur in close proximity to enzymatic NO sources, $OONO^-$ is now an established cellular mediator that contributes to "the dark side" of NO, in particular under inflammatory conditions [127].

According to Eq. (4.30), $SSNO^-$ is in equilibrium with its products of homolysis, $S_2^{\cdot -}$ and $NO^{\cdot}$. We recently put forward the hypothesis that in biological environments $SSNO^-$ may be formed by a radical–radical reaction between $S_2^{\cdot -}$ and NO according to the following equilibrium:

$$S_2^{\cdot -} + NO^{\cdot} \rightarrow SSNO^- \tag{4.30}$$

The disulfide radical ion $S_2^{\cdot -}$ is a well-known species, which can be formed in aqueous environment in nonenzymatic or enzymatic processes [28, 29]. As discussed above, cysteine persulfide (CySSH) and glutathione persulfide (GSSH) have been found in micromolar concentrations in tissues [84, 96] as products of the CARS [84] and previously proposed to be formed from cystine or GSSG by enzymes involved in sulfur amino acid metabolism, including CBS, CSE [79], and MST [100]; in addition, mitochondrial enzymes (SQR) or the hemoglobin-mediated oxidation of sulfur-centered radicals, including $S_2^{\cdot -}$ as well as thiosulfate and sulfite may be involved in their formation [75]. Sulfite may be also enzymatically converted to sulfoxy radicals such as $S_2^{\cdot -}$, $SO_3^{\cdot -}$, $O_3SOO^{\cdot -}$, $SO_4^{\cdot -}$ by peroxidases [75]. Interestingly, all these are precursors of the formation of both $SSNO^-$ and SULFI/NO, or similar intermediates. Since the kinetics of these radical reactions are all very fast, these intermediates are likely to be formed in cells and tissues.

4.5.3 Methods of Detection *In Vivo*

Methods for detection of sulfur compounds in biological environments are essentially based on seperative analytical techniques such as mass spectrometry-based analytical techniques. Recently, chemical probes for detection of SNO^- which probably can also detect $SSNO^-$ were described [128], which will strongly contribute to understanding the biological chemistry of these molecules in cells and tissues [129].

4.6 Summary and Conclusions

The interest for the biological chemistry of the chemical interaction of nitrogen-bearing and sulfur-bearing compounds was recently revived by the discovery that NO and sulfide may react with each other leading to formation of "S–N hybrid species" with highly specific biological and pharmacological activity. Understanding the basic physicochemical properties of these compounds will help identify their characteristics of stability and half-life, their biological targets, specific quantification methods, and their pharmacological and physiological effects. These studies will allow to answer the questions whether and how they are formed in vivo and how they participate to redox physiology hopefully keeping up the same enthusiasm and joy for the experimental play of Margot Goering and the pioneers in this field.

Acknowledgment

I wish to thank Martin Feelisch, Toni Butler, Jon Fukuto, and Taka Akaike and all the scientists who accompanied me in the wonderful land of the sulfur–nitrogen compounds; and Junjie Li for critically reading the chapter.

References

1 Cortese-Krott, M.M., Butler, A.R., Woollins, J.D., and Feelisch, M. (2016). Inorganic sulfur–nitrogen compounds: from gunpowder chemistry to the forefront of biological signaling. *Dalton Trans.* 45 (14): 5908–5919.

2 Raschig, F. (1924). *Schwefel- und Stickstoffstudien*, 310. Leipzig-Berlin: Verlag Chemie.

3 Dunnicliff, H., Mohammad, S., and Kishen, J. (1931). The interaction between nitric oxide and hydrogen sulphide in the presence of water. *J. Phys. Chem.* 35 (6): 1721–1734.

4 Kurtenacker, A. and Löschner, H. (1938). Über die Einwirkung von Stickoxyd auf Thiosulfat und Sulfid. *Z. Anorg. Allg. Chem.* 238 (4): 335–349.

5 Yost, D.M. and Russell, H. (1946). *Systematic Inorganic Chemistry*. Prentice-Hall.

6 Degener, E. and Seel, F. (1956). Zur Kenntnis der Salze der Nitrosohydroxylaminsulfonsäure. I. Chemischer Konstitutionsbeweis der Nitrosohydroxylaminsulfonate. *Z. Anorg. Allg. Chem.* 285 (3-6): 129–133.

7 Seel, F. (1988). Geschichte und Chemie des Schwarzpulvers. Le charbon fait la poudre. *Chem. unserer Zeit* 22 (1): 9–16.

8 Seel, F. and Güttler, H.J. (1973). Polysulfide radical ions. *Angew. Chem. Int. Ed.* 12 (5): 420–421.

9 Seel, F., Kuhn, R., and Simon, G. (1984). Bis (triphenylphosphan) iminium-monothionitrat. *Z. Naturforsch., B: Chem. Sci.* 39 (11): 1622–1623.

10 Seel, F., Kuhn, R., Simon, G. et al. (1985). PNP-Perthionitrit und PNP-Monothionitrit. *Z. Naturforsch., B: Chem. Sci.* 40 (12): 1607–1617.

11 Seel, F., Simon, G., Schuh, J. et al. (1986). Untersuchung der Umsetzung von Schwefel mit Natriumnitrit in DMF, DMSO und HMPT. *Z. Anorg. Allg. Chem.* 538 (7): 177–190.

12 Seel, F. and Wagner, M. (1985). The reaction of polysulfides with nitrogen monoxide in non-acqueous solvents- nitrosodisulfides. *Z. Naturforsch., B: Chem. Sci.* 40 (6): 762.

13 Seel, F. and Wagner, M. (1988). Reaction of sulfides with nitrogen monoxide in aqueous solution. *Z. Anorg. Allg. Chem.* 558: 189–192.

14 Cortese-Krott, M.M., Fernandez, B.O., Kelm, M. et al. (2015). On the chemical biology of the nitrite/sulfide interaction. *Nitric Oxide* 46: 14–24.

15 Cortese-Krott, M.M., Fernandez, B.O., Santos, J.L. et al. (2014). Nitrosopersulfide ($SSNO^-$) accounts for sustained NO bioactivity of S-nitrosothiols following reaction with sulfide. *Redox Biol.* 2: 234–244.

16 Cortese-Krott, M.M., Kuhnle, G.G., Dyson, A. et al. (2015). Key bioactive reaction products of the NO/H_2S interaction are S/N-hybrid species, polysulfides, and nitroxyl. *Proc. Natl. Acad. Sci. U.S.A.* 112 (34): E4651–E4660.

17 Bailey, T.S., Henthorn, H.A., and Pluth, M.D. (2016). The intersection of NO and H_2S: persulfides generate NO from nitrite through polysulfide formation. *Inorg. Chem.* 55 (24): 12618–12625.

18 Marcolongo, J.P., Morzan, U.N., Zeida, A. et al. (2016). Nitrosodisulfide $[S_2NO]^-$(perthionitrite) is a true intermediate during the "cross-talk" of nitrosyl and sulfide. *Phys. Chem. Chem. Phys.* 18 (43): 30047–30052.

19 Marcolongo, J.P., Venâncio, M.F., Rocha, W.R. et al. (2019). NO/H_2S "crosstalk" reactions. The role of thionitrites (SNO^-) and perthionitrites ($SSNO^-$). *Inorg. Chem.* 58 (22): 14981–14997.

20 Marcolongo, J.P., Zeida, A., Slep, L.D., and Olabe, J.A. (2017). Thionitrous acid/thionitrite and perthionitrite intermediates in the "Crosstalk" of NO and H_2S. *Adv. Inorg. Chem.* 70: 277–309.

21 Goehring, M. (1957). *Ergebnisse und Probleme der Chemie der Schwefelstickstoffverbindungen.* Berlin: Akademie-Verl.

22 Becke-Goehring, M. (1978). *Jahrbuch der Heidelberger Akademie der Wissenschaften.* Heidelberg: Heidelberger Akademie der Wissenschaften.

23 Chivers, T. (2005). *A Guide to Chalcogen-Nitrogen Chemistry.* Singapore: World Scientific.

24 Chivers, T. (1974). Ubiquitous trisulphur radical ion S3-[middot]. *Nature* 252 (5478): 32–33.

25 Chivers, T., Da Silva, A.B., Treu, O., and Trsic, M. (1987). Electronic structures and electronic spectra of the linkage isomers NSO^- and SNO^-. *J. Mol. Struct.* 162 (3): 351–357.

26 Chivers, T. and Drummond, I. (1972). Characterization of the trisulfur radical anion S_3^- in blue solutions of alkali polysulfides in hexamethylphosphoramide. *Inorg. Chem.* 11 (10): 2525–2527.

27 Chivers, T. and Elder, P.J. (2013). Ubiquitous trisulfur radical anion: fundamentals and applications in materials science, electrochemistry, analytical chemistry and geochemistry. *Chem. Soc. Rev.* 42 (14): 5996–6005.

28 Dahl, C., Prange, A., and Steudel, R. (2005). Metabolism of natural polymeric sulfur compounds. *Biopolymers* 9: https://doi.org/10.1002/3527600035.bpol9002.

29 Steudel, R. (1996). Mechanism for the formation of elemental sulfur from aqueous sulfide in chemical and microbiological desulfurization processes. *Ind. Eng. Chem. Res.* 35 (4): 1417–1423.

30 Steudel, R. (2003). Inorganic polysulfanes H_2Sn with n > 1. In: *Elemental Sulfur und Sulfur-Rich Compounds II*, 99–126. Springer.

31 Steudel, R. (2003). Inorganic polysulfides Sn2– and radical anions Sn–. In: *Elemental Sulfur und Sulfur-Rich Compounds II*, 127–152. Springer.

32 Steudel, R. and Chivers, T. (2019). The role of polysulfide dianions and radical anions in the chemical, physical and biological sciences, including sulfur-based batteries. *Chem. Soc. Rev.* 48 (12): 3279–3319.

33 Arnelle, D.R. and Stamler, J.S. (1995). NO^+, NO·, and NO^- donation by S-nitrosothiols: implications for regulation of physiological functions by S-nitrosylation and acceleration of disulfide formation. *Arch. Biochem. Biophys.* 318 (2): 279–285.

34 Bartberger, M.D., Mannion, J.D., Powell, S.C. et al. (2001). SN dissociation energies of S-nitrosothiols: on the origins of nitrosothiol decomposition rates. *J. Am. Chem. Soc.* 123 (36): 8868–8869.

35 Pawloski, J.R., Hess, D.T., and Stamler, J.S. (2001). Export by red blood cells of nitric oxide bioactivity. *Nature* 409 (6820): 622–626.

36 Premont, R.T., Reynolds, J.D., Zhang, R., and Stamler, J.S. (2020). Role of nitric oxide carried by hemoglobin in cardiovascular physiology: developments on a three-gas respiratory cycle. *Circ. Res.* 126 (1): 129–158.

37 Stamler, J. and Feelisch, M. (1996). Preparation and detection of S-nitrosothiols. In: *Methods in Nitric Oxide Research*, 521–539.

38 Stamler, J.S., Jaraki, O., Osborne, J. et al. (1992). Nitric oxide circulates in mammalian plasma primarily as an S-nitroso adduct of serum albumin. *Proc. Natl. Acad. Sci. U.S.A.* 89 (16): 7674–7677.

39 Broniowska, K.A. and Hogg, N. (2012). The chemical biology of S-nitrosothiols. *Antioxid. Redox Signaling* 17 (7): 969–980.

40 Bainbrigge, N., Butler, A.R., and Görbitz, C. (1997). The thermal stability of S-nitrosothiols: experimental studies and ab initio calculations on model compounds. *J. Chem. Soc., Perkin Trans.* 1 (2): 351–354.

41 McAninly, J., Williams, D.L.H., Askew, S.C. et al. (1993). Metal ion catalysis in nitrosothiol (RSNO) decomposition. *J. Chem. Soc., Chem. Commun.* (23): 1758.

42 Munro, A.P. and Williams, D.L.H. (2000). Reactivity of sulfur nucleophiles towards S-nitrosothiols. *J. Chem. Soc. Perkin Trans.* 2 (9): 1794–1797.

43 Williams, D.L.H. (2004). *Nitrosation Reactions and the Chemistry of Nitric Oxide*. Amsterdam, The Nederlands: Elsevier.

44 Cortese-Krott, M.M., Pullmann, D., and Feelisch, M. (2016). Nitrosopersulfide ($SSNO^{-}$) targets the Keap-1/Nrf2 redox system. *Pharmacol. Res.* 113: 490–499.

45 Hill, B.G., Dranka, B.P., Bailey, S.M. et al. (2010). What part of NO don't you understand? Some answers to the cardinal questions in nitric oxide biology. *J. Biol. Chem.* 285 (26): 19699–19704.

46 Fukuto, J.M., Carrington, S.J., Tantillo, D.J. et al. (2012). Small molecule signaling agents: the integrated chemistry and biochemistry of nitrogen oxides, oxides of carbon, dioxygen, hydrogen sulfide, and their derived species. *Chem. Res. Toxicol.* 25 (4): 769–793.

47 Bryan, N.S., Rassaf, T., Maloney, R.E. et al. (2004). Cellular targets and mechanisms of nitros(yl)ation: an insight into their nature and kinetics in vivo. *Proc. Natl. Acad. Sci. U.S.A.* 101 (12): 4308–4313.

48 Freeman, B.A., O'Donnell, V.B., and Schopfer, F.J. (2018). The discovery of nitro-fatty acids as products of metabolic and inflammatory reactions and mediators of adaptive cell signaling. *Nitric Oxide* 77: 106–111.

49 Antelmann, H. and Helmann, J.D. (2011). Thiol-based redox switches and gene regulation. *Antioxid. Redox Signaling* 14 (6): 1049–1063.

50 Groitl, B. and Jakob, U. (2014). Thiol-based redox switches. *Biochim. Biophys. Acta (BBA)* 1844 (8): 1335–1343.

51 Bredt, D.S., Hwang, P.M., and Snyder, S.H. (1990). Localization of nitric oxide synthase indicating a neural role for nitric oxide. *Nature* 347 (6295): 768–770.

52 Sessa, W.C. (1994). The nitric oxide synthase family of proteins. *J. Vasc. Res.* 31 (3): 131–143.

53 Cosby, K., Partovi, K.S., Crawford, J.H. et al. (2003). Nitrite reduction to nitric oxide by deoxyhemoglobin vasodilates the human circulation. *Nat. Med.* 9 (12): 1498–1505.

54 Gladwin, M.T. and Kim-Shapiro, D.B. (2008). The functional nitrite reductase activity of the heme-globins. *Blood* 112 (7): 2636–2647.

55 Derbyshire, E.R. and Marletta, M.A. (2012). Structure and regulation of soluble guanylate cyclase. *Annu. Rev. Biochem.* 81: 533–559.

56 Freeman, L.J.I.B.A. (2017). Introduction and overview. In: *Nitric Oxide*, 3e (ed. L.J. Ignarro and B.A. Freeman), xvii–xxii. Academic Press.

57 Wardman, P. (2007). Fluorescent and luminescent probes for measurement of oxidative and nitrosative species in cells and tissues: progress, pitfalls, and prospects. *Free Radical Biol. Med.* 43 (7): 995–1022.

58 Liu, X., Miller, M.J., Joshi, M.S. et al. (1998). Accelerated reaction of nitric oxide with O_2 within the hydrophobic interior of biological membranes. *Proc. Nat. Acad. Sci. U.S.A.* 95 (5): 2175–2179.

59 Möller, M.N., Li, Q., Vitturi, D.A. et al. (2007). Membrane "Lens" effect: focusing the formation of reactive nitrogen oxides from the $^{\bullet}NO/O_2$ reaction. *Chem. Res. Toxicol.* 20 (4): 709–714.

60 Markovits, G.Y., Schwartz, S.E., and Newman, L. (1981). Hydrolysis equilibrium of dinitrogen trioxide in dilute acid solution. *Inorg. Chem.* 20 (2): 445–450.

61 Awad, H.H. and Stanbury, D.M. (1993). Autoxidation of NO in aqueous solution. *Int. J. Chem. Kinet.* 25 (5): 375–381.

62 Shiva, S., Wang, X., Ringwood, L.A. et al. (2006). Ceruloplasmin is a NO oxidase and nitrite synthase that determines endocrine NO homeostasis. *Nat. Chem. Biol.* 2 (9): 486–493.

63 Lundberg, J.O., Weitzberg, E., and Gladwin, M.T. (2008). The nitrate-nitrite nitric oxide pathway in physiology and therapeutics. *Nat. Rev. Drug Discovery* 7 (2): 156–167.

64 Pearson, R.G., Sobel, H.R., and Songstad, J. (1968). Nucleophilic reactivity constants toward methyl iodide and trans-dichlorodi(pyridine)platinum(II). *J. Am. Chem. Soc.* 90 (2): 319–326.

65 Becker, K.H., Kleffmann, J., Kurtenbach, R., and Wiesen, P. (1996). Solubility of nitrous acid (HONO) in sulfuric acid solutions. *J. Phys. Chem.* 100 (36): 14984–14990.

66 Ritchie, C.D. (1972). Nucleophilic reactivities toward cations. *Acc. Chem. Res.* 5 (10): 348–354.

67 Benhar, M., Forrester, M.T., and Stamler, J.S. (2009). Protein denitrosylation: enzymatic mechanisms and cellular functions. *Nat. Rev. Mol. Cell Biol.* 10 (10): 721–732.

68 Schenk, P.W. (1942). Über das Thionylimid SONH. *Ber. dtsch. Chem. Ges. A/B* 75 (1): 94–99.

69 Bolden, C., King, S.B., and Kim-Shapiro, D.B. (2016). Reactions between nitrosopersulfide and heme proteins. *Free Radical Biol. Med.* 99: 418–425.

70 Berenyiova, A., Grman, M., Mijuskovic, A. et al. (2015). The reaction products of sulfide and S-nitrosoglutathione are potent vasorelaxants. *Nitric Oxide* 46: 123–130.

71 Barnett, D.J., McAninly, J., and Williams, D.L.H. (1994). Transnitrosation between nitrosothiols and thiols. *J. Chem. Soc. Perkin. Trans.* 2 (6): 1131–1133.

72 Barnett, D.J., Rios, A., and Williams, D.L.H. (1995). NO-Group transfer (transnitrosation) between S-nitrosothiols and thiols. Part 2. *J. Chem. Soc. Perkin. Trans.* 2 (7): 1279–1282.

73 Fukuto, J.M., Ignarro, L.J., Nagy, P. et al. (2018). Biological hydropersulfides and related polysulfides–a new concept and perspective in redox biology. *FEBS Lett.* 592 (12): 2140–2152.

74 Kimura H (2017) Hydrogen Sulfide and Polysulfide Signaling. (Mary Ann Liebert, Inc. 140 Huguenot Street, 3rd Floor New Rochelle, NY 10801 USA).

75 Mishanina, T.V., Libiad, M., and Banerjee, R. (2015). Biogenesis of reactive sulfur species for signaling by hydrogen sulfide oxidation pathways. *Nat. Chem. Biol.* 11 (7): 457–464.

76 Kevil, C., Cortese-Krott, M.M., Nagy, P. et al. (2017). Cooperative interactions between NO and H2S: chemistry, biology, physiology, pathophysiology. *Nitric Oxide*, (Elsevier) 57–83.

77 Shen, X., Peter, E.A., Bir, S. et al. (2012). Analytical measurement of discrete hydrogen sulfide pools in biological specimens. *Free Radical Biol. Med.* 52 (11-12): 2276–2283.

78 Nagy, P., Palinkas, Z., Nagy, A. et al. (1840). (2014) Chemical aspects of hydrogen sulfide measurements in physiological samples. *Biochim. Biophys. Acta* 2: 876–891.

79 Ono, K., Akaike, T., Sawa, T. et al. (2014). Redox chemistry and chemical biology of HS, hydropersulfides, and derived species: Implications of their possible biological activity and utility. *Free Radical Biol. Med.* 77C: 82–94.

80 Kolluru, G.K., Shen, X., Bir, S.C., and Kevil, C.G. (2013). Hydrogen sulfide chemical biology: pathophysiological roles and detection. *Nitric Oxide* 35: 5–20.

81 Wintner, E.A., Deckwerth, T.L., Langston, W. et al. (2010). A monobromobimane-based assay to measure the pharmacokinetic profile of reactive sulphide species in blood. *Br. J. Pharmacol.* 160 (4): 941–957.

82 Abe, K. and Kimura, H. (1996). The possible role of hydrogen sulfide as an endogenous neuromodulator. *J. Neurosci.* 16 (3): 1066–1071.

83 Bailey, T.S., Zakharov, L.N., and Pluth, M.D. (2014). Understanding hydrogen sulfide storage: probing conditions for sulfide release from hydrodisulfides. *J. Am. Chem. Soc.* 136 (30): 10573–10576.

84 Akaike, T., Ida, T., Wei, F.-Y. et al. (2017). Cysteinyl-tRNA synthetase governs cysteine polysulfidation and mitochondrial bioenergetics. *Nat. Commun.* 8 (1): 1177.

85 Wang, R. (2002). Two's company, three's a crowd: can H_2S be the third endogenous gaseous transmitter? *FASEB J.* 16 (13): 1792–1798.

86 Li, Q. and Lancaster, J.R. Jr., (2013). Chemical foundations of hydrogen sulfide biology. *Nitric Oxide* 35: 21–34.

87 Morse, J.W., Millero, F.J., Cornwell, J.C., and Rickard, D. (2002). The chemistry of the hydrogen sulfide and iron sulfide systems in natural waters. *Earth Sci. Rev.* 24 (1): 1–42.

88 Pawelec, M., Stochel, G., and van Eldik, R. (2004). Mechanistic information on the copper-catalysed autoxidation of mercaptosuccinic acid in aqueous solution. *Dalton Trans.* 2: 292–298.

89 Butler, A.R. and Feelisch, M. (2008). Therapeutic uses of inorganic nitrite and nitrate: from the past to the future. *Circulation* 117 (16): 2151–2159.

90 Cortese-Krott, M.M. (2020). Red blood cells as a "central hub" for sulfide bioactivity: scavenging, metabolism, transport, and cross-talk with nitric oxide. *Antioxid. Redox Signaling* 33 (18): 1332–1349.

91 Bianco, C.L., Savitsky, A., Feelisch, M., and Cortese-Krott, M.M. (2018). Investigations on the role of hemoglobin in sulfide metabolism by intact human red blood cells. *Biochem. Pharmacol.* 149: 163–173.

92 Vitvitsky, V., Yadav, P.K., Kurthen, A., and Banerjee, R. (2015). Sulfide oxidation by a noncanonical pathway in red blood cells generates thiosulfate and polysulfides. *J. Biol. Chem.* 290 (13): 8310–8320.

93 Zhou, Z., Martin, E., Sharina, I. et al. (2016). Regulation of soluble guanylyl cyclase redox state by hydrogen sulfide. *Pharmacol. Res.* 111: 556–562.

94 Ritchie, C.D. and Virtanen, P. (1972). Cation-anion combination reactions. IX. Remarkable correlation of nucleophilic reactions with cations. *J. Am. Chem. Soc.* 94 (14): 4966–4971.

95 Toohey, J.I. (1989). Sulphane sulphur in biological systems: a possible regulatory role. *Biochem. J* 264 (3): 625–632.

96 Ida, T., Sawa, T., Ihara, H. et al. (2014). Reactive cysteine persulfides and S-polythiolation regulate oxidative stress and redox signaling. *Proc. Natl. Acad. Sci. U.S.A.* 111 (21): 7606–7611.

97 Jones, D.P. and Sies, H. (2015). The redox code. *Antioxid. Redox Signaling* 23 (9): 734–746.

98 Kimura, H. (2014). Signaling of hydrogen sulfide and polysulfides. *Antioxid. Redox Signaling* 22 (5): 347–349.

99 Kimura, Y., Mikami, Y., Osumi, K. et al. (2013). Polysulfides are possible H_2S-derived signaling molecules in rat brain. *FASEB J.* 27 (6): 2451–2457.

100 Kimura, Y., Toyofuku, Y., Koike, S. et al. (2015). Identification of $H_2S_3^-$ and H_2S produced by 3-mercaptopyruvate sulfurtransferase in the brain. *Sci. Rep.* 5: 14774.

101 Kolluru, G.K., Yuan, S., Shen, X., and Kevil, C.G. (2015). H_2S regulation of nitric oxide metabolism. *Methods Enzymol.* 554: 271–297.

102 Pierce, J.A. (1928). A study of the reaction between nitric oxide and hydrogen sulphide. *J. Phys. Chem.* 33 (1): 22–36.

103 Filipovic, M.R., Miljkovic, J., Nauser, T. et al. (2012). Chemical characterization of the smallest S-nitrosothiol, HSNO; cellular cross-talk of H_2S and S-nitrosothiols. *J. Am. Chem. Soc.* 134 (29): 12016–12027.

104 Ali, M.Y., Ping, C.Y., Mok, Y.Y. et al. (2006). Regulation of vascular nitric oxide in vitro and in vivo; a new role for endogenous hydrogen sulphide? *Br. J. Pharmacol.* 149 (6): 625–634.

105 Lo Faro, M.L., Fox, B., Whatmore, J.L. et al. (2014). Hydrogen sulfide and nitric oxide interactions in inflammation. *Nitric Oxide* 41: 38–47.

106 Whiteman, M., Li, L., Kostetski, I. et al. (2006). Evidence for the formation of a novel nitrosothiol from the gaseous mediators nitric oxide and hydrogen sulphide. *Biochem. Biophys. Res. Commun.* 343 (1): 303–310.

107 Yong, Q.C., Cheong, J.L., Hua, F. et al. (2011). Regulation of heart function by endogenous gaseous mediators-crosstalk between nitric oxide and hydrogen sulfide. *Antioxid. Redox Signaling* 14 (11): 2081–2091.

108 Goehring, M. and Messner, J. (1952). Zur Kenntnis der schwefligen Säure. III Das Sulfinimid und seine Isomeren. *Z. Anorg. Allg. Chem.* 268 (1-2): 47–56.

109 Müller, R.P., Nonella, M., and Huber, J.R. (1984). Spectroscopic investigation of HSNO in a low temperature matrix. UV, VIS, and IR-induced isomerisations. *Chem. Phys.* 87: 351–361.

110 Nonella, M., Huber, J.R., and Ha, T.K. (1987). Photolytic preparation and isomerization of thionyl imide, thiocyanic acid, thionitrous acid, and nitrogen hydroxide sulfide in an argon matrix: an experimental and theoretical study. *J. Phys. Chem.* 91 (20): 5203–5209.

111 Bohle, D.S., Hung, C.-H., and Smith, B.D. (2005). Theoretical, thermal, and coordination chemistry of the amphoteric thiazate (NSO) 1 ion. *Can. J. Chem.* 83 (12): 2021–2031.

112 Bohle, D.S., Hung, C.-H., Powell, A.K. et al. (1997). Synthesis and characterization of isostructural metalloporphyrin chalconitrosyl complexes Ru (TTP)(NE) Cl (E=O, S) and a remarkable thionitrosyl/nitrite→ nitrosyl/thiazate transformation. *Inorg. Chem.* 36 (10): 1992–1993.

113 Tchir, P.O. and Spratley, R.D. (1975). The photolysis of matrix isolated cis-thionylimide. 1. The identification and infrared spectra of cis-HOSN, HSNO, and SNO. *Can. J. Chem.* 53 (15): 2318–2330.

114 Timerghazin, Q.K., Peslherbe, G.H., and English, A.M. (2008). Structure and stability of HSNO, the simplest S-nitrosothiol. *Phys. Chem. Chem. Phys.* 10 (11): 1532–1539.

115 Bianco, C.L., Chavez, T.A., Sosa, V. et al. (2016). The chemical biology of the persulfide (RSSH)/perthiyl (RSS·) redox couple and possible role in biological redox signaling. *Free Radical Biol. Med.* 101: 20–31.

116 Zhu, X. and Gao, Y. (2020). 17O NMR spectroscopy-assisted in vitro bioactivity studies of the intermediates formed via Na_2S and RSNO cross-linking reactions. *RSC Adv.* 10 (65): 39617–39626.

117 Teng, X., Scott Isbell, T., Crawford, J.H. et al. (2008). Novel method for measuring S-nitrosothiols using hydrogen sulfide. *Methods Enzymol.* 441: 161–172.

118 Gao, Y., Toubaei, A., Kong, X., and Wu, G. (2015). Solving the 170-year-old mystery about red-violet and blue transient intermediates in the Gmelin reaction. *Chem. Eur. J.* 21 (48): 17172–17177.

119 King, B.S. (2013). Potential biological chemistry of hydrogen sulfide (H_2S) with the nitrogen oxides. *Free Radical Biol. Med.* 55: 1–7.

120 Hrabie, J.A. and Keefer, L.K. (2002). Chemistry of the nitric oxide-releasing diazeniumdiolate ("nitrosohydroxylamine") functional group and its oxygen-substituted derivatives. *Chem. Rev.* 102 (4): 1135–1154.

121 Cox, E., Jeffrey, G., and Stadler, H. (1948). Structure of dinitrososulphite ion. *Nature* 162: 770.

122 Keefer, L.K., Nims, R.W., Davies, K.M., and Wink, D.A. (1996). "NONOates" (1-substituted diazen-1-ium-1,2-diolates) as nitric oxide donors: convenient nitric oxide dosage forms. *Methods Enzymol.* 268: 281–293.

123 De Witt, B.J., Marrone, J.R., Kaye, A.D. et al. (2001). Comparison of responses to novel nitric oxide donors in the feline pulmonary vascular bed. *Eur. J. Pharmacol.* 430 (2-3): 311–315.

124 Miranda, K.M., Paolocci, N., Katori, T. et al. (2003). A biochemical rationale for the discrete behavior of nitroxyl and nitric oxide in the cardiovascular system. *Proc. Natl. Acad. Sci. U.S.A.* 100 (16): 9196–9201.

125 Paolocci, N., Katori, T., Champion, H.C. et al. (2003). Positive inotropic and lusitropic effects of HNO/NO- in failing hearts: independence from beta-adrenergic signaling. *Proc. Natl. Acad. Sci. U.S.A.* 100 (9): 5537–5542.

126 Beckman, J.S., Beckman, T.W., Chen, J. et al. (1990). Apparent hydroxyl radical production by peroxynitrite: implications for endothelial injury from nitric oxide and superoxide. *Proc. Natl. Acad. Sci. U.S.A.* 87 (4): 1620–1624.

127 Radi, R. (2013). Peroxynitrite, a stealthy biological oxidant. *J. Biol. Chem.* 288 (37): 26464–26472.

128 Chen, W., Matsunaga, T., Neill, D.L. et al. (2019). Rational design of a dual-reactivity-based fluorescent probe for visualizing intracellular HSNO. *Angew. Chem. Int. Ed.* 58 (45): 16067–16070.

129 Wang, Y., Xu, S., and Xian, M. (2020). Specific reactions of RSNO, HSNO, and HNO and their applications in the design of fluorescent probes. *Chem. Eur. J.* 26 (51): 11673–11683.

5

H_2S and Bioinorganic Metal Complexes

Zachary J. Tonzetich

Department of Chemistry, University of Texas at San Antonio, San Antonio, TX, USA

List of Abbreviations

CcO	cytochrome *c* oxidase
DhHP	deuterohemin hexapeptide
DMF	*N,N*-dimethylformamide
DNIC	dinitrosyl iron complex
FRET	fluorescence resonance energy transfer
Hb	hemoglobin
Mb	myoglobin
Me_2Im	dimethylimidazole
mnt	maleonitriledithiolate
MP11	microperoxidase-11
Pc	phthalocyanine
PDF	peptide deformylase
PDI	pyridinediimine
PPIX	protoporphyrin-IX
RBS	Roussin's black salt
RRE	Roussin's red ester
RRS	Roussin's red salt
SAM	*S*-adenosyl-L-methionine
SCE	saturated calomel electrode
THPP	*tetrakis*-(*p*-hydroxyphenyl)tetraphenylporphyrin
TmPyP	*tetrakis*-(*N*-methylpyridyl)porphyrin
Tp	*tris*-pyrazolylborate
TPP	tetraphenylporphyrinate
XOR	xanthine oxidioreductases

5.1 Introduction

The diverse biological chemistry of hydrogen sulfide involves numerous cellular targets of both physiological and pathological relevance. Among these targets, transition metal ions represent logical sites of reactivity given their affinity for the Lewis-basic sulfur atom of hydrogen sulfide and their proclivity toward redox chemistry. Moreover, much of the biological chemistry of the related small molecules carbon monoxide (CO) and nitric oxide (NO) is known to involve interactions with metal centers. Given that H_2S is often grouped with these species in the context of their behavior as gasotransmitters [1], drawing parallels to their bioinorganic chemistry is logical.

Both H_2S and its conjugate base, the hydrosulfide ion (HS^-), are capable of binding to transition metal ions, although their coordination chemistry is much less developed than that of CO or NO. The relative scarcity of studies concerning metal complexes of H_2S/HS^- stems in large part from the difficulties associated with handling hydrogen sulfide. These difficulties involve practical considerations, such as the toxicity of H_2S gas, as well as aspects inherent to the chemistry of the molecule including its acidity and propensity to engage in redox reactions. Coordination to metal centers can also dramatically alter the fundamental properties of the hydrogen sulfide molecule leading to reactivity distinct from that of free H_2S. Despite these challenges, the bioinorganic chemistry of hydrogen sulfide has a lengthy history and remains an active area of study. Much of the recent focus on H_2S in the context of metal-based chemistry has been spurred by studies demonstrating signaling pathways involving metalloproteins [2–5]. These findings, coupled with the expanding body of evidence implicating H_2S/HS^- in a growing number of biological processes, have ushered in a renaissance of sorts in the coordination chemistry hydrogen sulfide. Both biochemical and synthetic studies have produced new examples of metal complexes featuring H_2S/HS^- as a ligand, and strategies to chemically detect hydrogen sulfide have increasingly targeted metal ion coordination. This chapter will discuss transition metal complexes of H_2S/HS^- from a bioinorganic perspective. Particular attention will be paid to those compounds which have been characterized in detail such that aspects of their bioinorganic chemistry (binding affinity, hydrogen-bonding, spectroscopy, etc.) can be clearly illuminated. Related species such as metal thiolates and sulfides will not be treated, nor will synthetic compounds of H_2S/HS^- unrelated to biology. For readers requiring a more comprehensive treatment of H_2S/HS^- coordination chemistry, exhaustive reviews on the subject that cover abiological systems are available for further reading [6–8].

5.2 Basic Ligative Properties of H_2S/HS^-

Sulfur plays a prominent role in inorganic chemistry, serving as a ligand to metal ions in several forms. The reduced state of sulfur, 2−, is the most Lewis basic and accounts for nearly all the coordination chemistry of the element. In this oxidation state, sulfur finds use as a ligand in the form of thiols, sulfides, thioethers, and as H_2S/HS^-. Due to its moderate electronegativity and polarizability, especially when compared with its lighter congener oxygen, sulfur-based ligands are often quite versatile and can bind to metals across the transition series. This fact is exemplified by HS^- itself, which has been reported as a ligand to nearly all d-block elements.

In considering the coordination of H_2S/HS^- to metal centers, the protonation state of the sulfur atom is often difficult to discern. The pK_a of H_2S at 37 °C is 6.76 resulting in hydrosulfide ion existing as the predominant species (c. 80%) under physiological pH [9]. This fact, paired with the greater Lewis basicity of the ion versus H_2S explains why HS^- accounts for the overwhelming majority of coordination complexes in synthetic and biological systems. Examples of *bona fide* H_2S coordination are much rarer and typically relegated to soft, nonbiologically relevant metal ions such as Ru^{2+} [10, 11]. Nonetheless, the rapid nature of proton transfer equilibria in aqueous solution ensures that both H_2S and HS^- can serve as potential actors in biological chemistry at transition metal sites.

The HS^- ligand can coordinate to metal ions in both a terminal and bridging fashion, much like related thiolates. As a result, complexes of HS^- can encompass both mononuclear species as well as bimetallic compounds and higher order clusters. By contrast, coordination of intact H_2S is limited to the terminal binding mode. The thermodynamics of M—SH bond formation have not been examined in detail, but the strength of related M—SR bonds provides a reasonable baseline. Transition metal-sulfur bond strengths in metal thiolates cover a relatively large range (c. 20–85 kcal/mol) depending upon the identity and oxidation state of the metal involved [12–14]. Extrapolating to hydrosulfide complexes, these numbers indicate that the strength of certain M—S bonds approaches that of the S—H bond of hydrogen sulfide. Accordingly, many complexes of H_2S/HS^- are subject to an additional thermodynamic driving force for formation of insoluble metal-sulfides. Much like the aforementioned acidity, this fact adds another dimension to the challenge of identifying complexes of hydrogen sulfide and its congeners. As a result, great care must be taken when characterizing complexes of these ligands in order to distinguish between H_2S, HS^-, and even sulfide (S^{2-}). Both structural and spectroscopic methods can be valuable in this vein but use of complimentary means of characterization is typically the best way to unambiguously establish the nature of the sulfur ligand.

5.3 H_2S and Heme Iron

Given that iron is the most prevalent transition metal in biology, it is unsurprising that the majority of known bioinorganic H_2S chemistry involves this element. Within the context of iron biochemistry, heme sites have received the greatest attention as targets for H_2S [15]. In many respects, hemes are ideally primed for H_2S/HS^- coordination because the porphyrin cofactor kinetically stabilizes the iron center against metal ion loss by protonation, reduction, and aggregation via sulfide formation, all of which are typical pitfalls of H_2S metal chemistry. Indeed, heme proteins were one of the first systems examined for reactivity with H_2S. Experiments by Hoppe-Seyler over 150 years ago demonstrated formation of green products upon treatment of oxy-myoglobin or oxy-hemoglobin with sulfide [16]. These green products, termed sulfhemes, result from alteration of the porphyrin ring by formation of new carbon–sulfur bonds (Figure 5.1) [17]. Contemporary work has identified three types of porphyrin modification (A, B, and C), all of which result in decreased O_2 affinity for the heme protein [18, 19]. The mechanism of sulfheme formation remains speculative, but the O_2-dependent reactivity suggests a role for oxoferryl species ($Fe^{IV}{=}O$ or $Fe^{IV}{=}O\ Por^{\cdot +}$), which may generate hydrosulfide radicals for further chemistry at the porphyrin ring. The formation of sulfhemes also extends beyond globins and has been observed in catalase and lactoperoxidase enzymes as well [20, 21].

Since the initial studies on sulfhemes, the chemistry of H_2S with heme proteins has been expanded to include a host of different interactions. Conceptually, the simplest of these involves reversible association/dissociation of H_2S/HS^- to the iron center. A physiological role for such chemistry has been identified in the invertebrate *Lucina pectinata* [22]. This mollusk species contains symbiotic bacteria in its gills that generate energy from the metabolism of H_2S. The bacteria produce several hemoglobin proteins for this purpose, one of which, hemoglobin I (HbI), is utilized as an H_2S transporter. In this role, HbI shuttles H_2S from the marine environment to the site of oxidative metabolism within the organism. Among the most notable features of HbI is the fact that it will bind H_2S even in the presence of oxygen to generate ferric hemoglobin sulfide. Studies on the rate of sulfide binding to ferric HbI demonstrate an acceleration upon lowering of the pH consistent with a preference for association of H_2S over HS^-. This result contrasts the majority of studies employing synthetic ferric porphyrinates (*vide infra*) demonstrating the specific role HbI has evolved to perform.

Structural studies of HbI from *L. pectinata* as both the ferric aqua- and sulfide-bound forms have provided further insight into the conditions for optimal H_2S-heme binding [23, 24]. The structure of the sulfide-bound form of HbI displayed in Figure 5.2 reveals a hydrophobic pocket near the distal site of the heme center comprised of four phenylalanine residues. This so-called "Phe-cage"

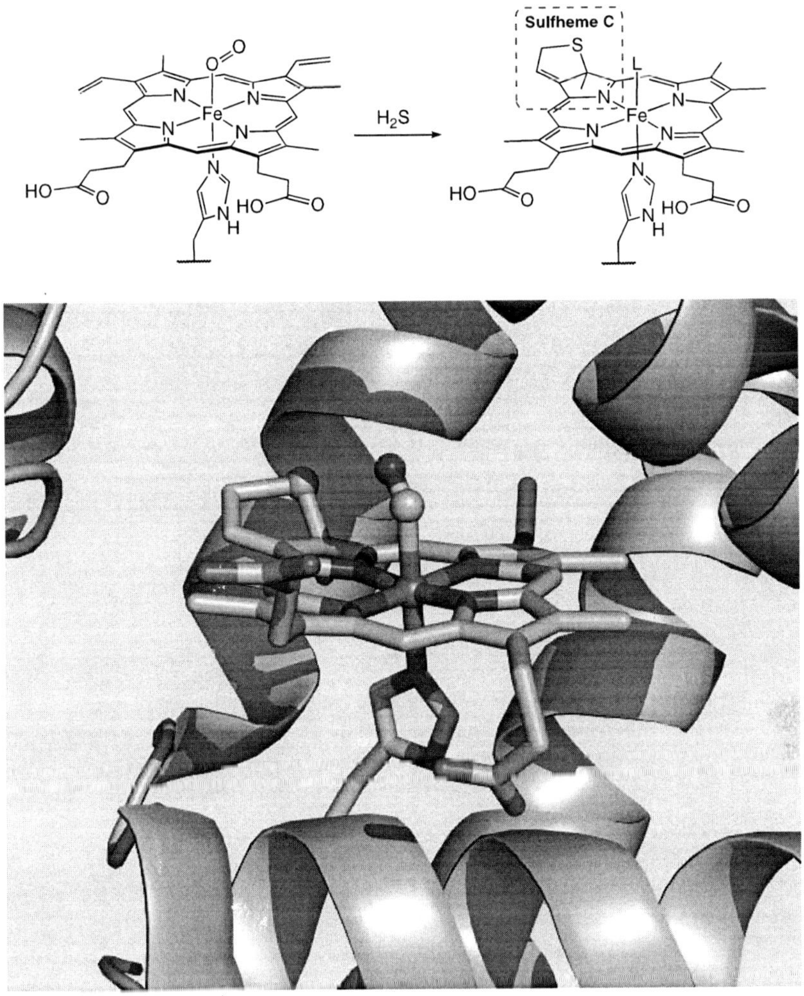

Figure 5.1 Top: Formation of a sulfheme from an oxyheme and H_2S highlighting one of three possible covalent modifications (sulfheme C) to the porphyrin periphery. Bottom: Structure of horse heart cyanomet-sulfmyoglobin C (PDB code: 1YMC)

is unique among heme proteins, and likely engenders a more favorable lipophilic environment for the H_2S molecule near the iron center. The active site of HbI also possesses a distal glutamine residue (Gln-64) capable of hydrogen-bonding with coordinated H_2S, thereby providing a further driving force for sulfide association [25]. Indeed, photodissociation studies on both ferrous- and ferric-sulfide forms of HbI support a role for Gln-64 in stabilizing H_2S adducts [26].

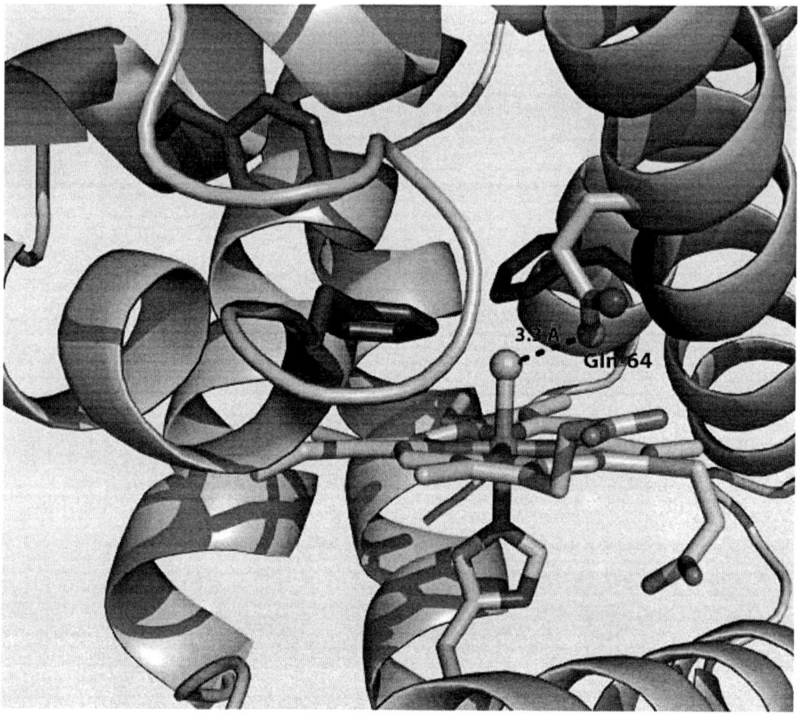

Figure 5.2 Structure of the active site of HbI-sulfide from *L. pectinata* (PDB code: 1MOH). Select Phe residues in the distal pocket of the heme are colored in brown to show the lipophilic pocket.

Binding of sulfide to heme proteins is not limited to HbI and has been observed for a variety of systems upon treatment with H_2S [27–30]. In general, the resulting sulfide adducts feature low-spin ($S = \frac{1}{2}$) iron(III) and demonstrate spectroscopic features comparable to those of the ferric-sulfide from HbI. Unlike HbI however, typical hemes have not evolved to bind H_2S and therefore display several notable differences. For example, the structure of the ferric-sulfide adduct from human hemoglobin is displayed in Figure 5.3 [31]. The Fe—S bond length of 2.2 Å found for the sulfide adduct in both the α- and β-subunits is significantly contracted from that expected for H_2S coordination (c. 2.50 Å). This distance is instead consistent with binding of HS^- to the iron center. Much like HbI, a hydrogen-bonding interaction is present in the distal pocket, but it involves a histine residue (His-58) as opposed to glutamine.

The differences between hydrogen sulfide interactions with HbI versus other heme proteins are not solely of a structural nature. Several heme and related systems have been found to effect the conversion of H_2S to polysulfides and/or

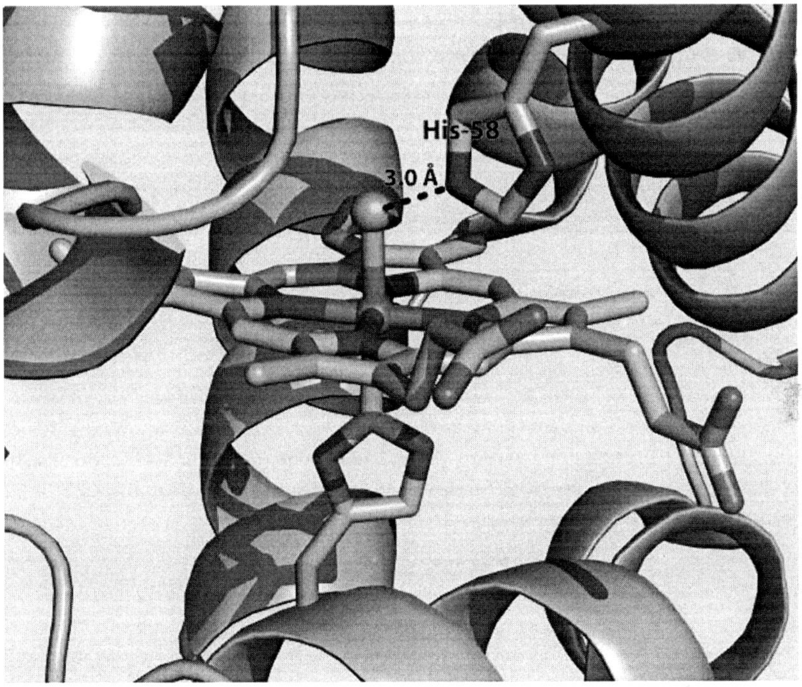

Figure 5.3 Structure of human Hb-sulfide (PDB code: 5UCU).

persulfides [31–33]. This chemistry reflects the fact that prototypical heme proteins are not optimized for H_2S binding and transport and are therefore more apt to undergo redox processes such as sulfide-mediated reduction of the iron(III) center [34]. These reduction events can lead to generation of reactive sulfur species and subsequent downstream chemistry. Examples of such processes have been observed for both ferric hemoglobin and myoglobin, which can oxidize sulfide to thiosulfate through the intermediacy of hydro(poly)sulfides.

Heme sites other than globins have also been identified as targets for H_2S in higher organisms. Cytochrome *c* oxidase (CcO) has been shown to be both stimulated and reversibly inhibited by the presence of H_2S [35–37]. This contrasting behavior arises from concentration-dependent reactivity at the enzyme, which functions as the terminal acceptor in the mitochondrial respiration chain. CcO contains two heme sites (a and a_3) and two Cu centers (Cu_A and Cu_B). Of these metal sites, both the heme a_3 and Cu_B centers can interact with H_2S. At low levels, H_2S can stimulate respiration by reducing the ferric heme a_3 site promoting increased binding of O_2. At high levels of H_2S, however, the same ferric heme a_3 site binds HS^- resulting in conformational changes to the enzyme that lead to inhibition. In the intermediate concentration regime, H_2S can also interact with the

Cu_B site contributing to enzyme inhibition. During H_2S-inhibition, the organism is observed to enter a state of suspended animation due to decreased CcO function. Treatment with O_2 restores normal enzymatic activity reversing the process [38]. CcO therefore demonstrates that in certain cases both redox chemistry and metal binding can occur for H_2S at the same heme site, ultimately resulting in very different organismal responses.

Outside of proteins, H_2S reactivity has been scrutinized for a variety of synthetic and semisynthetic biomimetic heme systems. A major advantage of these systems is that they often permit more detailed characterization of the sulfide-bound product, especially as it concerns the protonation state of sulfur. Additional factors pertaining to the primary and secondary coordination sphere of the heme active site can also be varied with greater ease than with intact proteins. In one such example, the heme peptide microperoxidase MP11 was examined as a platform for H_2S binding [39]. The MP11 peptide is derived from cyctochrome *c*, but lacks residues in the distal region of the heme such as methionine, capable of interacting with the iron center or a bound ligand. Treatment of the ferric form of MP11 with H_2S gives rise to a stable product formulated as the hydrosulfide adduct (Scheme 5.1). Vibrational analysis via resonance Raman spectroscopy identified a band at 366 cm^{-1}, assigned to the Fe-SH unit of a low-spin ferric heme center. The observed binding of hydrosulfide to ferric-MP11 is notable in that the parent protein from which it is derived, cytochrome *c*, instead undergoes reduction to Fe^{II} in the presence of H_2S. Such results underscore the subtle factors at play which select for binding versus redox chemistry at heme centers. Moreover, these factors need not only involve the distal pocket where H_2S binds to iron. In a related example featuring a semisynthetic deuterohemin-His-peptide (DhHP) system, the nature of the proximal heme environment was shown to influence the rate of sulfide binding [40]. Analogs of the deuterohemin system shown in Scheme 5.1 which

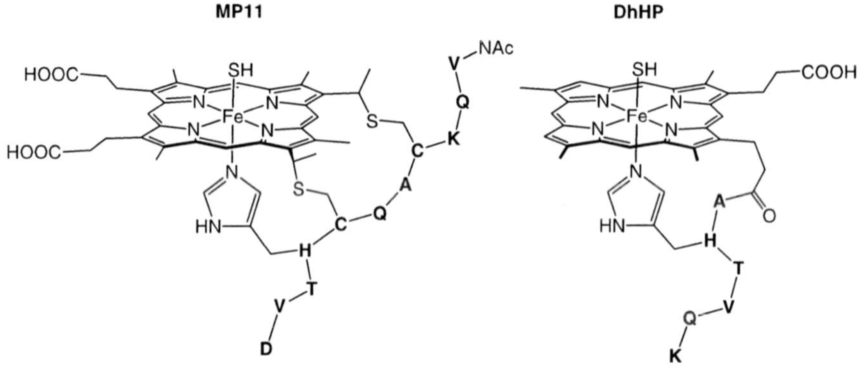

Scheme 5.1 Ferric hydrosulfide complexes of semisynthetic heme complexes.

contain a peptide chain possessing a glutamine residue display significantly faster rates of sulfide association than those containing a truncated unit. Much like HbI from *L. pectinata*, the Glu residue was implicated as a key factor in facilitating binding of H_2S to the iron center. In the case of the DhHP, however, this influence must be exerted via the proximal side of the heme, likely through creation of a hydrogen-bonding network. As these examples demonstrate, the totality of the local environment (proximal and distal) can help direct the chemistry of hydrogen sulfide at heme centers.

In addition to semisynthetic or peptide-derived systems, synthetic porphyrinates have also provided insights into H_2S binding and reactivity, particularly with ferrous iron. The major difference with purely synthetic systems is that only iron(II) is observed to form stable sulfide adducts, all of which feature HS^- as opposed to H_2S. Solution studies with iron(III) porphyrinates have provided evidence for formation of hydrosulfide adducts at low temperatures [41], but these species undergo rapid reduction to iron(II) upon warming and early reported examples of stable ferric hydrosulfide complexes of synthetic porphyrinates have been reconsidered in light of more recent studies. The closest structural analog to a synthetic sulfide-bound iron(III) porphyrinate is the gallium(III) congener, [Ga(TPP)(SH)] (TPP = tetraphenylporphyrinate) [42]. Similar to several of the ferric heme systems described above, gallium(III) was only found to coordinate HS^- and not H_2S. Structural studies of the complex produced a Ga—S bond length of 2.26 Å, which compares well to the Fe–S distance found for the ferric-sulfide adduct of human hemoglobin.

The chemistry of hydrosulfide ion with synthetic iron(II) porphyrinates has been the subject of several studies, which have succeeded in characterizing the binding constants, spectroscopic features, and structural parameters for this class of sulfide adducts. In most instances, the hydrosulfide complexes are generated by direct treatment of the iron(II) porphyrinate with hydrosulfide ion in the form of NaSH or Bu_4NSH (Figure 5.4). The resultant complexes demonstrate similar structural and spectroscopic features consistent with five coordinate, high-spin ($S = 2$) iron(II) (Figure 5.4) [43–46]. The Fe—S bond distances fall in the range of 2.31–2.39 Å, considerably longer than those found for ferric heme sulfides, consistent with the larger ionic radius of high-spin iron(II). Cyclic voltammetry measurements on one such complex, (Bu_4N)[Fe(TPP)(SH)], demonstrated a reversible oxidation event at −0.83 V versus ferrocene/ferrocenium assigned to the Fe(II)/Fe(III) couple [44]. The value of this redox couple is more positive than that found for related thiolate-bound iron(II) porphyrinates, indicating a greater thermodynamic driving force for reduction of Fe(III)-SH over Fe(III)-SR. Formation constants (K_f) for the hydrosulfide adducts depend upon the nature of the porphyrin ligand employed but generally fall in the range of 40 000–200 000, which is comparable to that observed for sulfide binding to ferric-HbI. The ability

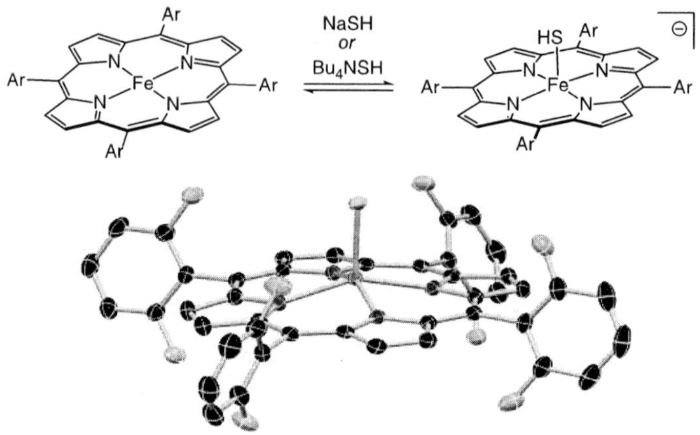

Figure 5.4 Top: Representative reaction for the formation of iron(II) hydrosulfide adducts with HS^- salts. Bottom: Crystal structure of $[Fe(SH)(F_8TPP)]^-$.

to isolate select synthetic hydrosulfide complexes has also permitted studies of their reactivity with other biologically relevant small molecules. In the case of $[Fe(TPP)(SH)]^-$, treatment with 1,2-dimethylimidazole or nitric oxide gas resulted in equilibrium displacement of the hydrosulfide ligand to afford the corresponding ferrous adducts, $[Fe(TPP)(1,2\text{-}Me_2Im)]$ and [Fe(TPP)(NO)], respectively.

5.4 H_2S and Nonheme Iron

In addition to heme systems, the bioinorganic chemistry of H_2S has also been the subject of studies with nonheme iron. Outside of the porphyrin environment, iron can be found coordinated to an array of amino acid residues as well as small molecules. One of the most important forms of nonheme iron occurs in iron-sulfur proteins. These proteins constitute a diverse class that can be found across all kingdoms of life and carry out essential functions from electron transfer to catalysis [47]. The iron-sulfide clusters, $[Fe_nS_m]$, that comprise the key cofactors in these proteins are typically attached to the peptide backbone by cysteinate residues and can exist in a variety of geometries and nuclearities ranging from two iron atoms up to 16. Heterometallic versions of iron–sulfur clusters are also essential components of other enzymes such as nitrogenases.

Production of hydrogen sulfide from iron–sulfur proteins has long been associated with acid treatment, and detection of H_2S from such experiments provided some of the first confirmation for the composition of $[Fe_nS_m]$ cofactors [48]. More recently, structural and biochemical data have demonstrated evidence for hydrosulfide coordination in several iron–sulfur proteins that contain a free

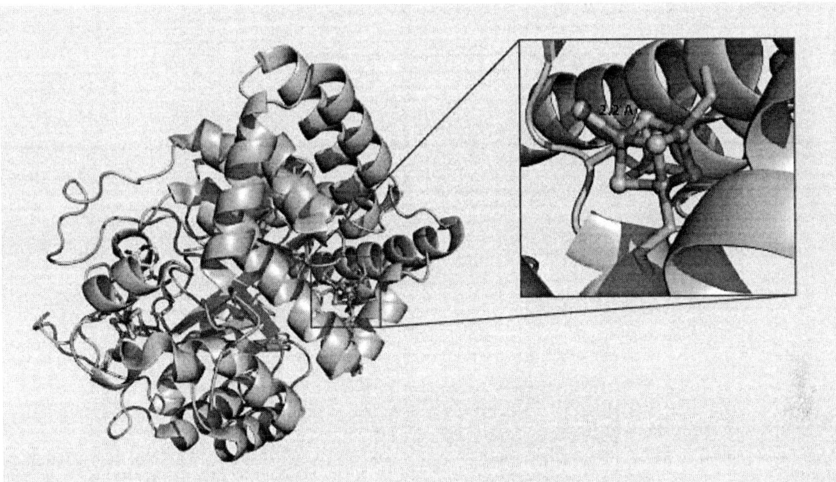

Figure 5.5 Structure of HydG monomer B from *Thermoanaerobacter italicus* (PDB code: 4WCX). Expansion displays the $[Fe_4S_4]$ cluster with terminal hydrosulfide coordination.

coordination site [49]. Crystallographically characterized examples include the $[Fe_4S_4]$ cluster found in the β-subunit of (R)-2-hydroxyisocaproyl-CoA dehydratase [50], as well as the auxiliary $[Fe_4S_4]$ cluster of the *S*-adenosyl-L-methionine (SAM) enzyme HydG [51], which is involved in the H-cluster assembly of [FeFe]-hydrogenase (Figure 5.5). In both these cases, it is uncertain whether the HS^- ligands directly participate in the activity of the enzymes, although in the case of HydG, the hydrosulfide-bound $[Fe_4S_4]$ unit may represent an important intermediate resulting from loss of iron from a novel $[Fe_5S_5]$ cluster.

Studies with the SAM-dependent methylthiotransferase, RimO, implicate a more direct role for hydrosulfide ligation [52]. The resting state of RimO contains two adjacent $[Fe_4S_4]$ clusters linked by a pentasulfide group. Upon reduction, one of the two clusters retains a HS^- ligand. This hydrosulfide moiety undergoes methylation by SAM to produce a MeS^- group, which is ultimately transferred to the enzyme's substrate (Scheme 5.2). Hydrosulfide ligands also play a key role in the function of nitrogenase enzymes where "belt" sulfide ions of the heterometallic cluster cofactors serve as sites of protonation during N_2 activation [53].

Biomimetic work with synthetic iron–sulfur cluster compounds actually predates many of the biochemical studies discussed above. Work in this arena has produced several complexes whose structures closely mimic those found in protein environments. The simplest iron–sulfur clusters possessing hydrosulfide coordination are the $[Fe_4S_4]$ mimic, $[Fe_4S_4(SH)_4]^{2-}$, and the $[Fe_2S_2]$ mimic $[Fe_2S_2(SH)_4]^{2-}$. The latter of these was prepared in the context of studies involving nitric oxide chemistry and will be discussed later. The initial report of

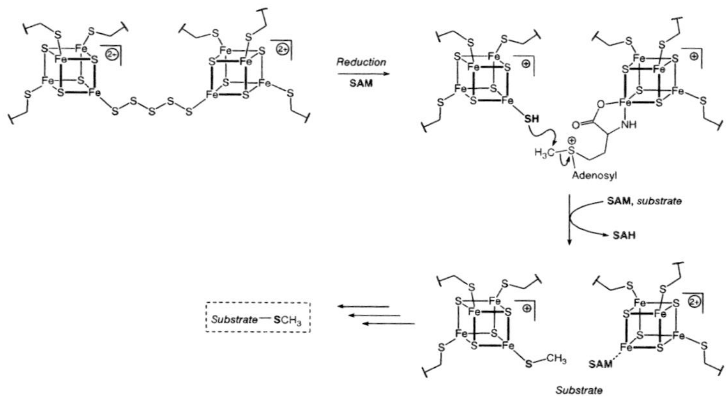

Scheme 5.2 Hydrosulfide-bound [Fe_4S_4] cluster of RimO.

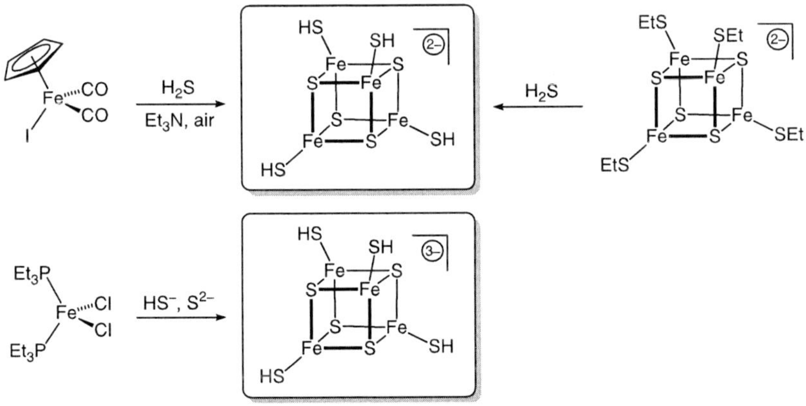

Scheme 5.3 Synthesis of biomimetic $[Fe_4S_4]$ clusters with terminal HS^- ligands.

$[Fe_4S_4(SH)_4]^{2-}$ involves a somewhat unorthodox synthesis involving treatment of organometallic iron species with H_2S in the presence of oxygen (Scheme 5.3) [54]. A higher yielding preparation of $[Fe_4S_4(SH)_4]^{2-}$ can also be accomplished by direct protonlysis of $[Fe_4S_4(SEt)_4]^{2-}$ with H_2S [55]. The structure of $[Fe_4S_4(SH)_4]^{2-}$ is similar to other synthetic thiolate-bound clusters containing the $\{Fe_4S_4\}^{2+}$ cores [56], but features four terminal hydrosulfide ligands. The mean Fe—SH bond length in the cluster is 2.27 Å, which agrees well with that found for the auxiliary cluster of HydG. Moreover, this distance falls in between that of the iron(II) and iron(III) heme hydrosulfide adducts (*vide supra*), consistent with the $\{Fe_4S_4\}^{2+}$ core containing a mixture of formally iron(II) and iron(III). The one-electron reduced analog, $[Fe_4S_4(SH)_4]^{3-}$, can be generated by a reaction of the iron(II) complex, $[Fe(PEt_3)_2Cl_2]$, with a mixture of Et_4NSH and Na_2S [57]. As expected, this more highly reduced cluster displays an elongated Fe—SH bond length of 2.32 Å. Much like the all-thiolate analogs, electrochemical interconversion between $[Fe_4S_4(SH)_4]^{2-}$ and $[Fe_4S_4(SH)_4]^{3-}$ is reversible occurring at a potential −1.08 V (versus SCE) in acetonitrile.

Beyond simple $[Fe_4S_4]$ species, hydrosulfide ligands have been used in the construction of more complex iron–sulfur clusters. In a similar fashion to the synthesis of $[Fe_4S_4(SH)_4]^{2-}$, treatment of the cluster $[Fe_4S_4(LS_3)(SEt)]^{2-}$, which contains the site-differentiating LS_3 ligand, with H_2S generates the hydrosulfide species $[Fe_4S_4(LS_3)(SH)]^{2-}$ (Scheme 5.4) [58]. This complex serves as a higher fidelity model of the 3Cys-HS coordination observed for the enzymes discussed above. Displacement of the ethanethiol ligand by H_2S occurs in an equilibrium fashion, but favors the hydrosulfide species at room temperature ($K_{eq} = 120 \pm 20\,M^{-1}$). Spectroscopic and electrochemical properties of

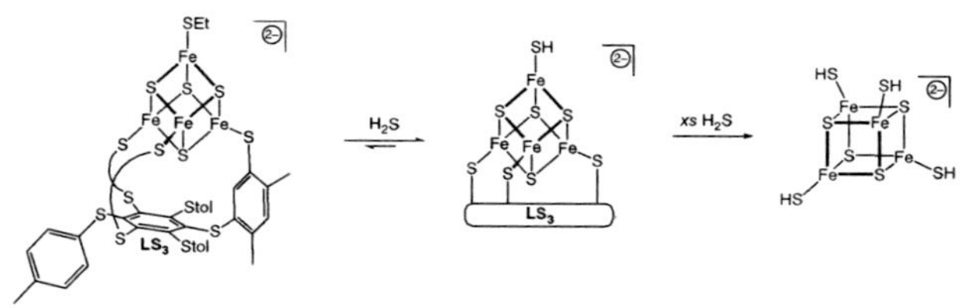

Scheme 5.4 Synthetic site-differentiated $[Fe_4S_4]$ clusters featuring hydrosulfide ligation.

$[Fe_4S_4(LS_3)(SH)]^{2-}$ are akin to those of $[Fe_4S_4(SH)_4]^{2-}$, further consistent with the parallel nature of thiolate and hydrosulfide coordination in Fe–S clusters. Dissolution of $[Fe_4S_4(LS_3)(SH)]^{2-}$ results in an equilibrium mixture containing the sulfide-bridged double cubane, $[\{Fe_4S_4(LS_3)\}_2S]^{4-}$, and free hydrogen sulfide. Reversion to $[Fe_4S_4(LS_3)(SH)]^{2-}$ by treatment of $[\{Fe_4S_4(LS_3)\}_2S]^{4-}$ with excess H_2S is not observed, but instead leads to $[Fe_4S_4(SH)_4]^{2-}$ by complete thiol exchange. Closely related site-differentiated clusters containing modified tridentate ligands can also be prepared in an analogous fashion to $[Fe_4S_4(LS_3)(SH)]^{2-}$ [59]. Crystallographic analyses of several analogs demonstrate Fe—SH bond lengths of 2.26–2.27 Å, similar to $[Fe_4S_4(SH)_4]^{2-}$.

More elaborate heterometallic iron–sulfur clusters containing hydrosulfide ligands have also been subjected to modeling studies. Complexes such as those displayed in Scheme 5.5, can be prepared by HS^- exchange with mono- and dicubane $[MFe_3S_4]$ precursors [60–64]. These clusters bare a structural homology to the P^N and FeMoco clusters of nitrogenase, and feature Mo, W, and V atoms in the capping positions. Notably, the HS^- ligands of these compounds bind to the central Fe atoms of the cluster, which is the same site that sulfide ions become protonated during turnover of nitrogenase [53]. Structural studies of the heterobimetallic species in Scheme 5.5 evince Fe—SH bond distances in a range consistent with that found for the simpler $[Fe_4S_4]$ clusters.

The chemistry of hydrogen sulfide with iron–sulfur proteins also extends to the products of their breakdown. The disassembly of iron–sulfur cofactors in the presence of nitric oxide (NO), another established gasotransmitter, is a well-established feature of NO pathology [65]. The role of H_2S in this process as a by-product or potential reaction partner for NO is gaining increasing attention. The iron-containing species that result from cluster degradation are often low molecular weight complexes featuring $Fe(NO)_2$ units termed dinitrosyl iron compounds (DNICs). DNICs can also take the form of small iron–sulfur clusters of formula $[Fe_2S_2(NO)_4]^{2-}$ and $[Fe_4S_3(NO)_7]^-$, termed Roussin's Red (RRS), and Black salts (RBS), respectively. Due to the relatively simple nature of DNICs, many of the studies in this area have focused on biomimetic compounds that can be prepared in a straightforward fashion from synthetic iron–sulfur clusters or iron nitrosyl precursors.

Reactions of synthetic iron–sulfur clusters with NO produce a mixture of DNICs, RREs, and RBS depending upon the reaction conditions employed [66, 67]. In the presence of excess thiol, however, H_2S can be detected as a product of cluster nitrosylation. Work with both $[Fe_2S_2(SPh)_4]^{2-}$ and $[Fe_4S_4(SPh)_4]^{2-}$, has demonstrated a pathway for H_2S formation in the presence of NO and excess $PhSH/PhS^-$ (Scheme 5.6) [68, 69]. This reactivity suggests that iron–sulfur proteins can serve as a source of H_2S during NO-mediated disassembly of iron–sulfur clusters. Generation of H_2S in the locale of elevated NO concentrations, as are

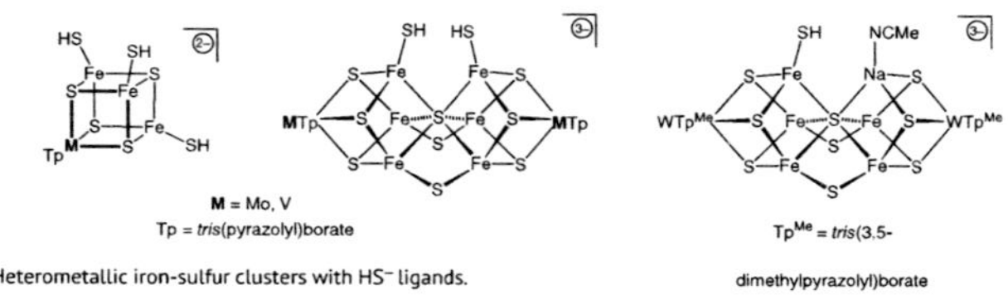

Scheme 5.5 Heterometallic iron-sulfur clusters with HS^- ligands.

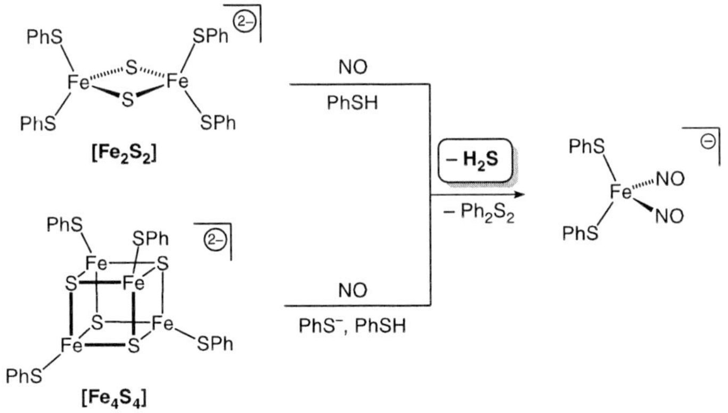

Scheme 5.6 Generation of H_2S during iron-sulfur cluster disassembly.

present during cluster breakdown, might therefore provide a conduit for NO and H_2S crosstalk [70].

Modeling studies have also targeted hydrosulfide-ligated variants of DNICs. The sodium version of Roussin's Red salt, $Na_2[Fe_2S_2(NO)_4]$, undergoes protonation by HCl to generate the corresponding acid, $[Fe_2(\mu\text{-}SH)_2(NO)_4]$, which contains two bridging hydrosulfide ligands [71]. Although not biologically relevant, this species along with RRS serve as a convenient precursors for formation of alkylated variants, $[Fe_2(\mu\text{-}SR)_2(NO)_4]$, known as Roussin's Red esters (RRE). RREs are among the products detected from the breakdown of $[Fe_2S_2]$ cluster-containing proteins. Further protonation of $[Fe_2(\mu\text{-}SH)_2(NO)_4]$ with HCl leads to formation of the chloride-bound DNIC, $[FeCl_2(NO)_2]^-$, with release of H_2S.

In chemistry related to that of $[Fe_2(\mu\text{-}SH)_2(NO)_4]$, treatment of aqueous solutions of RRS with H_2S has been shown to generate the hydrosulfide-bound DNIC $[(HS)_2Fe(NO)_2]^-$ [72]. This DNIC is also accessible by thiol displacement from $[(EtS)_2Fe(NO)_2]^-$ with H_2S, a strategy frequently invoked in the preparation of iron-sulfur clusters (*vide supra*). The hydrosulfide-bound mononitrosyl iron compound, $[Fe(SH)_3(NO)]^-$, can be prepared in a similar fashion from $[Fe(SEt)_3(NO)]^-$, although it is unstable and spontaneously loses H_2S to give the partially nitrosylated bimetallic cluster, $[Fe(\mu\text{-}S)_2(SH)_2(NO)_2]^{2-}$ (Figure 5.6). Electronic structure calculations of this novel cluster support the presence of two antiferromagnetically coupled ferric centers and crystallographic studies identified a Fe—SH bond distance of 2.32 Å. The spontaneous extrusion of H_2S observed for $[Fe(\mu\text{-}S)_2(SH)_2(NO)_2]^{2-}$ also occurs for the homoleptic species, $[Fe(SH)_4]^-$, leading to formation of the $[Fe_2S_2]$ cluster $[Fe(\mu\text{-}S)_2(SH)_4]^{2-}$. Notably,

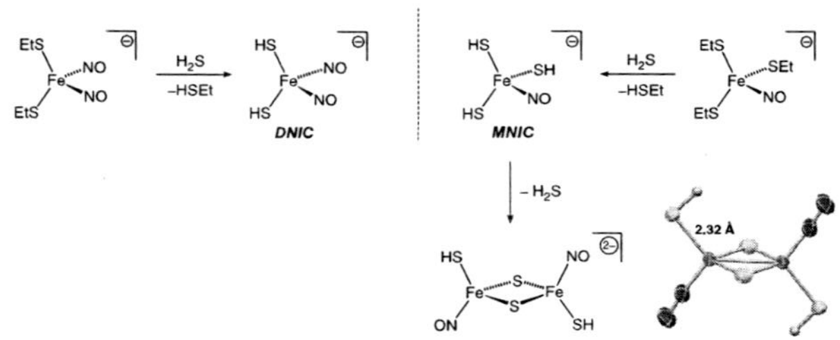

Figure 5.6 Transformations of nitrosyliron species containing hydrosulfide ligands and the crystal structure of $[Fe_2(\mu\text{-}S)_2(SH)_2(NO)_2]^{2-}$.

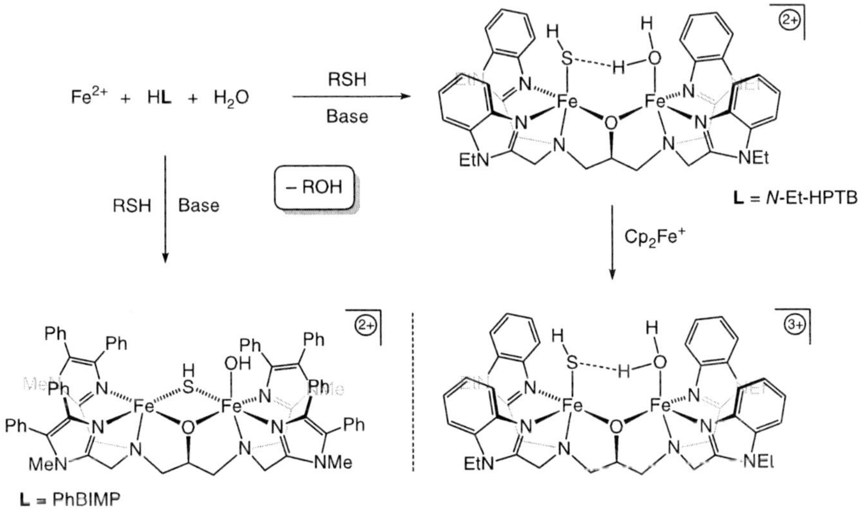

Scheme 5.7 Thiol desulfurization with bimetallic iron(II) species.

the average Fe—SH bond distance of $[Fe_2(\mu\text{-}S)_2(SH)_4]^{2-}$ is 2.32 Å, which is longer than that found in $[Fe_4(\mu\text{-}S)_4(SH)_4]^{2-}$ despite the former containing more formally oxidized iron.

As seen in many of the examples of heme and nonheme species discussed so far, direct addition of H_2S/HS^- to a metal precursor is often the preferred means of synthesizing hydrosulfide complexes. In an alternate approach, biologically relevant hydrosulfide compounds of nonheme iron can additionally be prepared by reactions involving thiol desulfurization [73, 74]. Bimetallic diiron(II) systems possessing both terminal and bridging hydrosulfide ligands have been shown to result from the desulfurization of aromatic and aliphatic thiols in the presence of dinucleating ligand scaffolds (Scheme 5.7). These compounds feature iron(II) and display Fe—SH bond distances in the range of 2.34–2.39 Å. Oxidation of the bimetallic compounds generates mixed-valent iron(II/III) species that retain HS^- ligation. Electronic absorption spectra of one such compound, $[Fe_2(N\text{-Et-HPTB})(SH)(H_2O)(DMF)_2]^{3+}$, demonstrated an intervalence charge transfer band, consistent with a delocalized di-iron core. The mixed-valent complex also displayed a contracted Fe—SH bond distance of 2.18 Å, which falls more in line with the ferric heme hydrosulfides discussed above. In addition to iron, cobalt(II), and nickel(II) have also proven active toward thiol desulfurization in the presence of related dinucleating ligands [75–77]. The products of these reactions are hydrosulfide species of similar structure to the iron analogs in Scheme 5.7.

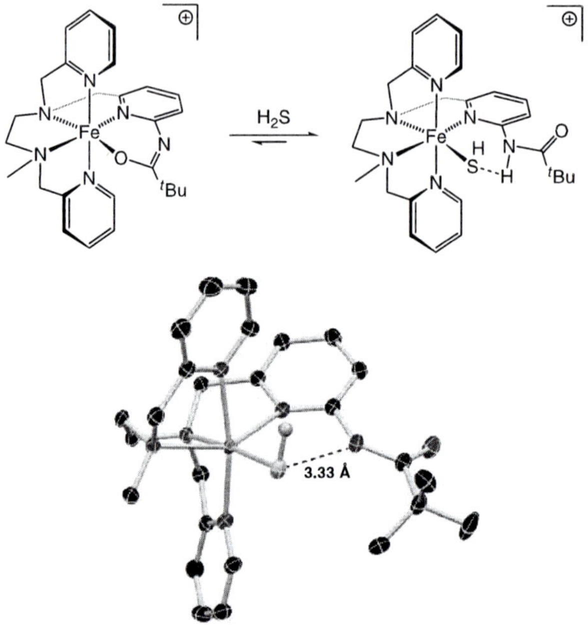

Figure 5.7 Reversible binding of H_2S to a nonheme iron(II) compound and the crystal structure of the corresponding adduct showing the hydrogen bond between the amide group and the hydrosulfide ligand.

Biomimetic nonheme systems have also found success in stabilizing hydrosulfide coordination through reversible interactions with H_2S. The six-coordinate Fe(II) complex shown in Figure 5.7 is capable of forming a hydrosulfide complex by proton transfer to an internal basic site on the chelating ligand [78]. In the absence of H_2S, the carboxamido ligand binds to the iron center in its deprotonated form. Treatment of this precursor with H_2S then leads to protonation of the nitrogen atom and concomitant coordination of HS^- to the metal center. This proton transfer process is reversible, and slow evaporation and/or addition of H_2S acceptors can shift the equilibrium back to the precursor. Within the hydrosulfide complex, the HS^- is stabilized by a hydrogen-bonding interaction with the amide moiety. This H-bonding interaction is maintained in solution and the solid state according to both spectroscopic and structural data, with a Fe—SH bond distance of 2.39 Å.

5.5 H_2S Chemistry with Other Metals

The biological chemistry of hydrogen sulfide with metal ions other than iron is significantly less developed. Nonetheless, H_2S/HS^- coordination has been implicated in several natural systems, and these have in turn provided motivation for a number of biomimetic studies. In the area of zinc chemistry, sulfide is known to inhibit the activity of many Zn^{2+}-containing carbonic anhydrases by binding to the His_3Zn active site [79]. Sulfide coordination to zinc has also been suggested as a means of H_2S transport in the giant tube worm species, *Riftia pachyptila* [80]. Much like *L. pectinata*, *R. pachyptila* inhabits sulfide-rich environments on the ocean floor and has evolved a hemoglobin-based system to deliver H_2S and O_2 to symbiotic bacteria capable of sulfide metabolism. Unlike *L. pectinata*, however, structural studies on the Hb system of *R. pachyptila* have revealed a hollow cavity that binds Zn^{2+} ions through a series of histidine and glutamate residues. The presence of these zinc ions is required for H_2S binding and transport in the organism intimating a possible role for direct sulfide coordination.

Inspired by the prevalence of histidine-rich coordination environments in the biological chemistry of zinc, biomimetic studies of Zn-SH complexes have focused on ligands containing N-donor groups. Among these, tridentate *tris*-pyrazolylborate (Tp) ligands have proven very popular as they reproduce both the threefold symmetry of His_3 sites as well as the N-heterocyclic nature of the imidazole donor. Substituted Tp variants, Tp^R, are readily available and have been employed in the synthesis of several zinc hydrosulfide complexes. Prototypical [Tp^RZn(SH)] species can be prepared by protonolysis of sufficiently basic precursors, such as [Tp^RZnH] or [Tp^RZn(OH)], with H_2S (Scheme 5.8) [81–83]. The resultant hydrosulfide complexes display pseudo-tetrahedral geometries and a ^1H NMR resonance for the sulfhydryl proton upfield of 0 ppm. Bond distances for the Zn-SH unit in these species are found around 2.2 Å, comparable to that of ferric hydrosulfide adducts.

Scheme 5.8 Reactions of Tp^RZn complexes with H_2S.

The reactivity of [$Tp^RZn(SH)$] complexes is most established with acidic functionalities, which result in protonation of the hydrosulfide ligand and loss of H_2S (Scheme 5.8). This chemistry can extend to thiols as well, despite the higher pK_a of RSH versus H_2S. In this instance, loss of H_2S gas is believed to provide the necessary entropic driving force. Alkylation of the Zn-SH unit has also been observed with MeI resulting in formation of MeSH [84]. The process was found to obey a second-order rate law, consistent with direct alkylation of the Zn-SH prior to release of thiol. In chemistry related to the operation of carbonic anhydrases, the zinc hydroxide complex, [($Tp^{Ph,Me}$)Zn(OH)], has been shown to react with carbon disulfide (CS_2) to produce [($Tp^{Ph,Me}$)Zn(SH)] and carbonyl sulfide (COS) [85]. Formed in tandem with COS are small amounts of CO_2, suggesting that COS can undergo further reactivity to generate H_2S, which is the end-product of enzymatic COS hydrolysis by carbonic anhydrase. In addition to *tris*-pyrazolylborates, several other polydentate ligand systems have been used to prepare Zn-SH species. Several such compounds are displayed in Scheme 5.9 [82, 86–88]. Each of these species demonstrates very similar structural and spectroscopic features to the Tp systems discussed above.

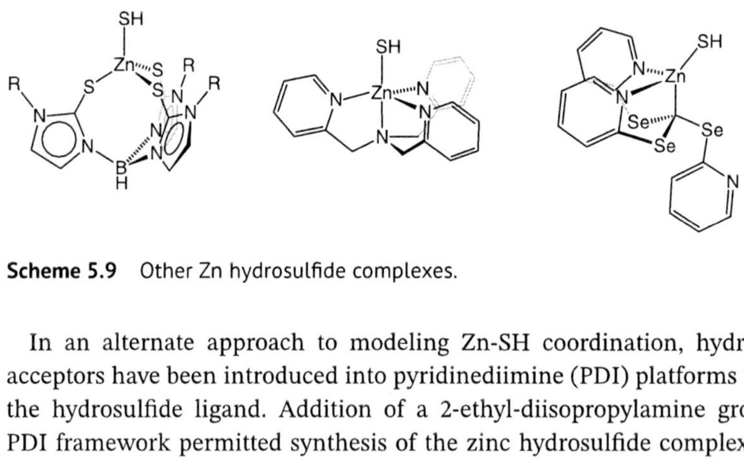

Scheme 5.9 Other Zn hydrosulfide complexes.

In an alternate approach to modeling Zn-SH coordination, hydrogen-bond acceptors have been introduced into pyridinediimine (PDI) platforms to stabilize the hydrosulfide ligand. Addition of a 2-ethyl-diisopropylamine group to the PDI framework permitted synthesis of the zinc hydrosulfide complex shown in Scheme 5.10 [89]. In the absence of this acceptor moiety, treatment of the zinc chloride precursor with HS^- results in metal loss via formation of ZnS. ^1H NMR experiments with the complex identified a resonance for sulfhydryl proton at 11.05 ppm consistent with its involvement in a hydrogen-bonding interaction. Thus, as also seen with iron, H-bonding can provide yet another strategy for stabilization of zinc hydrosulfide complexes.

Another metal ion observed to coordinate sulfide biologically is molybdenum. This element is essential to xanthine oxidoreductases (XORs), which house Mo

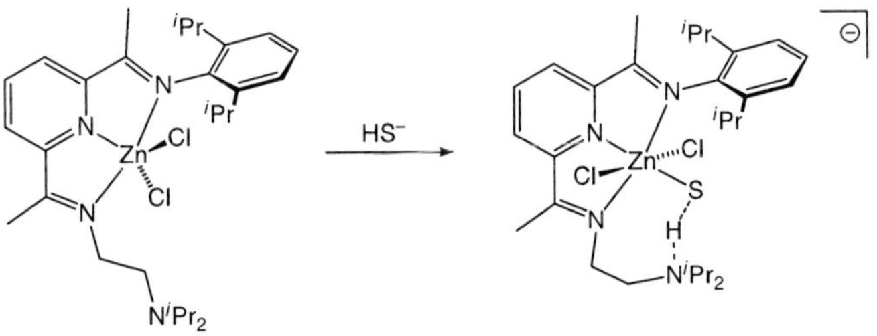

Scheme 5.10 Stabilization of a Zn(II) hydrosulfide complex by hydrogen-bonding.

in their active site in the form of a molybdopterin cofactor [90]. XORs catalyze the oxidation of hypoxanthine to xanthine as well as xanthine to uric acid with concomitant formation of reactive oxygen species. These oxidation reactions are critical steps in the process of purine catabolism in higher organisms including humans. In the resting state of the enzyme, the Mo center contains a sulfido ligand. During catalysis, this sulfide accepts an H atom from purine substrates to produce an intermediate Mo-SH species. Structural studies with the purine-like inhibitor, 4-[5-(pyridin-4-yl)-1H-1,2,4-triazol-3-yl]pyridine-2-carbonitrile (FYX-051), have succeeded in capturing such intermediates and crystallographic data demonstrates a Mo—SH bond length of 2.4 Å, consistent with its formulation as a hydrosulfide as opposed to sulfide complex (Figure 5.8) [91]. Thus, much like RimO discussed above, XO features a hydrosulfide ligand as a crucial intermediate during enzymatic turnover.

Attempts to model the Mo-SH intermediate have produced a few biomimetic complexes that capture many of the essential aspects of the XO active site [92]. The thiolate complexes shown in Scheme 5.11 incorporate a maleonitriledithiolate (mnt) ligand to mimic the dithiolene moiety of the molybdopterin cofactor [93]. These complexes undergo thiol displacement with H_2S to produce the corresponding Mo(IV) hydrosulfides, $[Mo(O)(SH)(mnt)(N^\wedge N)]^-$ (N^N = 2,2′-dipyridine or 1,10 phenanthroline). Monomeric hydrosulfide complexes can be isolated at short reaction times, but further reactivity in the presence of oxygen leads to ligand redistribution and formation of oxo-bridged bimetallic species. Crystallographic studies of $[Mo(O)(SH)(mnt)(N^\wedge N)]^-$ display Mo—SH bond distances of 2.46 Å, very much in line with those found for the XOR intermediate. The simpler Group VI oxo complexes, $[M(O)(mnt)_2]^{2-}$ (M = Mo, W), also react with H_2S to produce a series of sulfide-bridged multimetallic complexes. These compounds are likely formed through a similar process as that in Scheme 5.11 involving the intermediacy of $[M(O)(SH)(mnt)_2]^{2-}$ species.

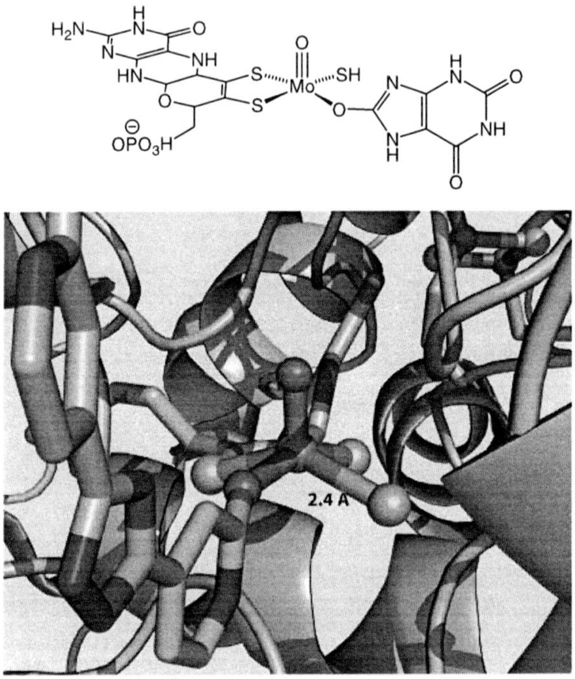

Figure 5.8 Top: Picture of the hydrosulfide-bound XOR intermediate. Bottom: Structure of the active-site of bovine milk XOR with FYX-051 bound showing the hydrosulfide ligand (PDB code: 1V97).

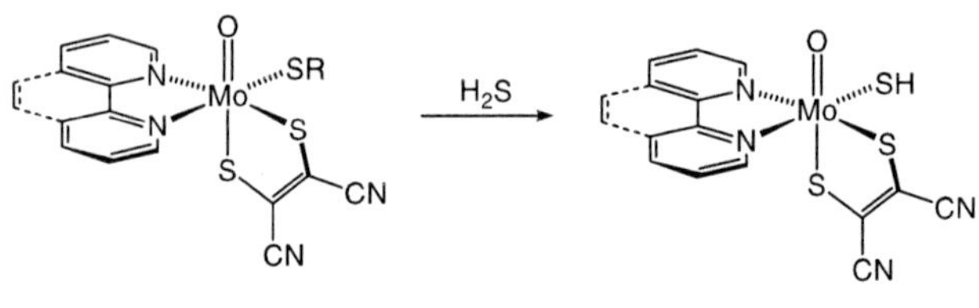

Scheme 5.11 Formation of hydrosulfide complexes in a XOR mimic.

5.6 H_2S Sensing with Transition Metal Complexes

The bioinorganic chemistry discussed in earlier sections of this chapter demonstrates the diversity of reaction pathways available to H_2S in the presence of metal ions. Recognizing that several of these pathways can present an opportunity for H_2S/HS^- detection, systems containing transition metals have been increasingly targeted for sulfide-sensing. These strategies complement a host of other sulfide-sensing mechanisms that have been developed in recent years [94]. In considering the merits of metal-based systems, perhaps their greatest promise lies in the ability to reversibly bind H_2S/HS^-. This section will examine several examples of H_2S detection centered on metal-H_2S interactions. Unsurprisingly, many of these examples draw inspiration from the natural systems described above.

Several research groups have addressed the problem of reversible sulfide detection by focusing on proteins that have already been demonstrated to interact with H_2S. In many cases, changes to the optical absorption spectra of these proteins upon reaction with H_2S have been leveraged to develop new sensors. Fluorescence resonance energy transfer (FRET) between a fluorescent reporter (dye molecule) and an H_2S-responsive protein chromophore provides a convenient conduit for signal transduction in this approach given the commercial availability of many dyes. As a prototypical example, myoglobin can be employed to create a simple FRET-based sensor for H_2S detection [95]. Appending either Cy3 (λ_{max} = 550 nm) or Atto620 (λ_{max} = 620 nm) to the N-terminus of the Mb protein produces a system which demonstrates modulation of dye fluorescence in the presence of H_2S. This modulation arises from FRET quenching of the dye fluorophores due to the similarity of their excitation energies to those of the Q-bands of Mb-sulfide. The sensor is selective to H_2S over other thiols highlighting the crucial role of metal coordination.

In a similar fashion to Mb, dye attachment to recombinant HbI (rHBI) from *L. pectinata* can be used to create a reversible sensor for H_2S [96]. In this case, the coumarin-based dye Pacific Blue is employed as it absorbs at a similar wavelength to the sulfide-free form of rHBI (λ_{max} = 407 nm). In the absence of H_2S, the fluorescence of the dye is partially attenuated due to energy transfer to the heme. Binding of H_2S to rHBI shifts the absorption maxima to 425 nm resulting in less-efficient FRET and an enhanced fluorescence response. Much like the native binding of H_2S to HbI, detection of sulfide by the rHBi-Pacific Blue construct is reversible, which can be demonstrated through successive HS^--treatment and argon-purge cycles.

Nonheme proteins have also been the focus of sensor development through analogous dye-attachment strategies. A cobalt analog of *Escherichia coli* peptide deformylase (PDF) containing the dye Atto620 has been demonstrated to function as a turn-off sensor for H_2S, although the system is not reversible [97]. By substituting Co(II) for Fe(II), the native metal in the enzyme, the modified Co-PDF gives rise to two absorbances at 560 and 660 nm. Interaction of H_2S with the modified enzyme causes a change in the optical spectrum with new absorbances appearing at 625 and 665 nm. The former of these bands occurs at an energy close to that required for excitation of the Atto620 fluorophore. Thus, by enhancing the intensity of the 625 nm absorbance through addition of H_2S, a decrease in fluorophore excitation efficiency is observed leading to a turn-off response. The nature of the H_2S interaction in this system is not completely understood, although crystallographic evidence supports formation of a cobalt-sulfide entity upon treatment of Co-PDF with H_2S (Figure 5.9). The Co—S bond distance of

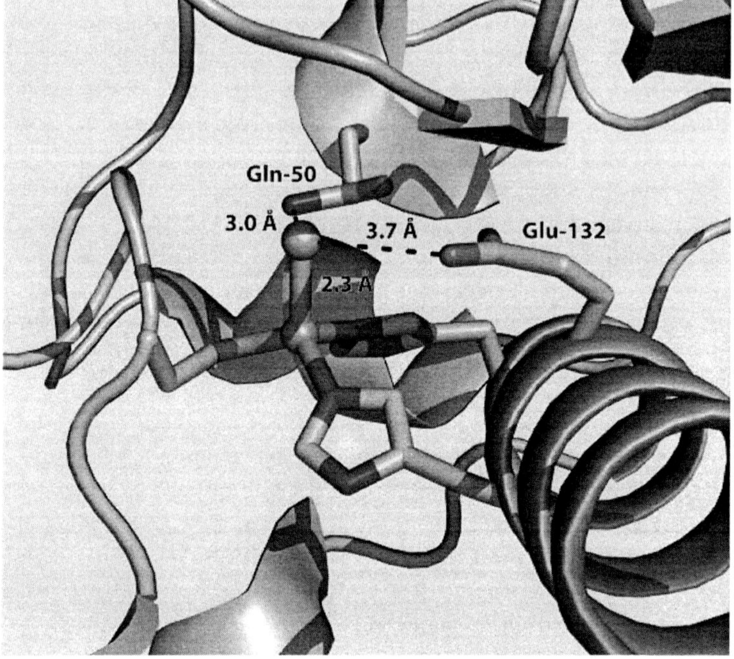

Figure 5.9 Structure of the active site of sulfide-bound Co-PDF from *E. coli* (PDB code: 4AZ4).

2.29 Å is compatible with a Co-SH entity suggesting a possible homology with the heme approaches discussed above. Hydrogen bonding interactions with nearby Gln and Glu residues also appear to play an important role stabilizing the sulfhydryl group.

Related work with Atto620 and azurin protein has demonstrated a metal-based reduction strategy for H_2S detection [98]. Akin to other bacterial blue copper proteins (Type I), azurin functions as a single-electron transfer agent shuttling between Cu(I) and Cu(II) forms. Addition of H_2S to the Cu(II) form reduces the copper center with concomitant bleaching of the charge-transfer absorbances in the 550–660 nm range responsible for the blue color of the oxidized protein. In the dye-labeled conjugate, these charge-transfer transitions quench the fluorescence from Atto620 creating an off-state for the sensor. Reduction to Cu(I) in the presence of H_2S attenuates the quenching resulting in a turn-on fluorescence response. The system can be reset to the off-state by oxidation back to Cu(II) with $K_3Fe(CN)_6$. The reversibility supports the notion that copper is retained in the active site during reaction with H_2S. Interestingly, substitution of the native copper ion for cobalt produces the corresponding Co-azurin which functions as a turn-off sensor for H_2S in conjugation with the dye Alexa350. In a similar fashion to the Co-PDF sensor, the Co-azurin construct displays an increase in absorption upon addition of H_2S that effectively quenches the dye. In this case, the absorption bands occur in the 300–350 nm range accounting for the choice of Alexa350. Much like the copper analog, oxidation of Co-azurin with $K_3Fe(CN)_6$ returns the sensor to its initial state suggesting that it too is amenable to reversible sulfide detection.

Outside of protein-based systems, H_2S sensors based on biomimetic platforms have similarly shown promise in sulfide detection. Many of these examples make use of porphyrins or closely related macrocycles for stabilization of the metal center during reaction with H_2S/HS^-. Protoporphyrin-IX (PPIX), the cofactor found in native hemes has been employed in this regard in its Zn- and Cu-bound forms [99, 100]. Treating solutions of [Zn(PPIX)] with HS^- results in the formation of a species consistent with a hydrosulfide adduct, $[Zn(SH)(PPIX)]^-$ (Scheme 5.12). Notably, this hydrosulfide complex is not observed to form in nonaqueous environments when H_2S is used in place of hydrosulfide. Optical and NMR signatures for $[Zn(SH)(PPIX)]^-$ differ from those of the parent [Zn(PPIX)] complex providing a means of quantifying sulfide concentrations in solution.

In the case of copper, aqueous [Cu(PPIX)] reacts with sulfide to produce a change in the fluorescence spectrum of the molecule with a significant enhancement in the band at 430 nm [101]. Investigations by mass spectrometry implicate the formation of a Cu(II)-hydrosulfide adduct of the type $[Cu(SH)(PPIX)]^-$, although the system does not exhibit the reversibility expected for HS^- coordination. Moreover, similar studies in nonaqueous media demonstrate a lack of reactivity for [Cu(PPIX)] with either H_2S or HS^-.Thus, the processes responsible for the turn-on fluorescence observed for [Cu(PPIX)] in the presence of H_2S remain somewhat speculative with proton transfer events possibility playing an important role.

Related complexes of Zn containing purely synthetic porphyrin derivatives display comparable chemistry to that of [Zn(PPIX)]. Water-soluble zinc complexes of the *tetrakis*-(*p*-hydroxyphenyl)tetraphenyl porphyrin (THPP) and *tetrakis*-(*N*-methylpyridyl)porphyrin (TmPyP) ligands show similar evidence for formation of Zn SH adducts, which can be reversed by addition of acid (Scheme 5.13) [102]. A significant advantage of the synthetic TmPyP ligand

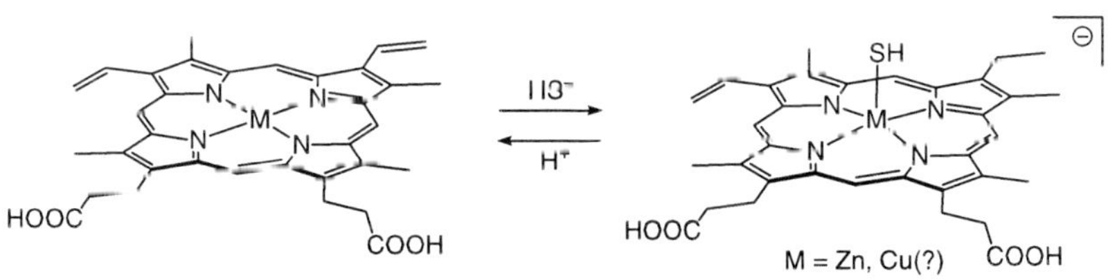

Scheme 5.12 Sulfide detection with metal PPIX complexes.

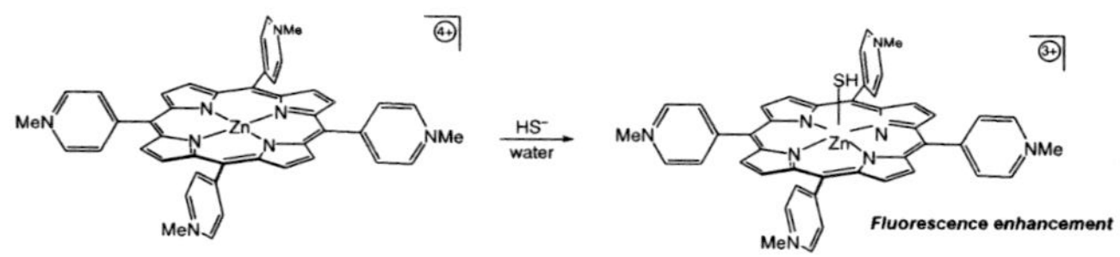

Scheme 5.13 Hydrosulfide detection with $[Zn(TmPyP)]^{4+}$.

over PPIX, however, is the inherent fluorescence of the former. This property allows for real-time monitoring of hydrosulfide binding to $[Zn(TmPyP)]^{4+}$ due to an enhancement in the fluorescence associated with formation of $[Zn(SH)(TmPyP)]^{3+}$.

Yet another synthetic porphyrinogen, phthalocyanine (Pc), has been used in conjunction with Co and Zn for reversible sulfide detection [103]. The parent Zn(II) complex, [Zn(Pc)], reacts with hydrosulfide ion in very a similar fashion to the porphyrin derivatives discussed above to produce the adduct, $[Zn(SH)(Pc)]^-$. Much like [Zn(PPIX)], the reaction of [Zn(Pc)] with HS^- can be followed spectroscopically, demonstrating marked changes by both NMR and UV–Vis. Reversion of $[Zn(SH)(Pc)]^-$ to [Zn(Pc)] occurs upon addition of acid with release of H_2S. In contrast to [Zn(Pc)], the cobalt(II) analog, $[Co^{II}(Pc)]$, displays no sulfide binding but instead undergoes reduction to the monovalent species, $[Co^{I}Pc]^-$ (Scheme 5.14). Reduction from Co(II) to Co(I) is accompanied by large changes in the electronic absorption spectrum of the molecule, which are visible to the naked eye in the form of a color change from blue to yellow. Air oxidation of $[Co^{I}Pc]^-$ regenerates $[Co^{II}(Pc)]$ providing for a means of reversible sulfide detection. Accordingly, successive exposure of the CoPc system to HS^- and air produces a consistent optical response.

Binding of hydrosulfide to Zn(II) has also been used to prepare sensors based on nonporphyrinogen systems [104, 105]. The salen-based Zn(II) compound shown in Scheme 5.15 features two pyridoxal fluorophores and reacts with HS^- in water or buffer to generate the corresponding adduct. As observed for several other Zn systems, no reactivity is observed when H_2S is used in place of hydrosulfide. Attendant with formation of the Zn-SH adduct is a shift in the fluorescence maxima of the molecule from 500 to 380 nm, which can be used to quantify hydrosulfide concentrations. Related salen systems have also shown comparable changes in fluorescence upon hydrosulfide binding.

In an alternate approach, Zn complexes containing fluorophores can detect sulfide via ligand displacement reactions. Treatment of the Tp^RZn complex shown in Scheme 5.16 with buffered solutions of H_2S results in formation of Tp^RZnSH and loss of the of thiocoumarin dye [106]. The free thiocoumarin demonstrates attenuated fluorescence compared with its coordinated form resulting in a turn-off sensor for H_2S. The TpZn system is selective for H_2S against other biologically relevant thiols such as cysteine and glutathione at short reaction and can detect sulfide concentrations as low as 1 μM.

5.7 Summary

As described in this chapter, the bioinorganic chemistry of H_2S involves a variety of different metal ions and coordination environments. In most instances, where

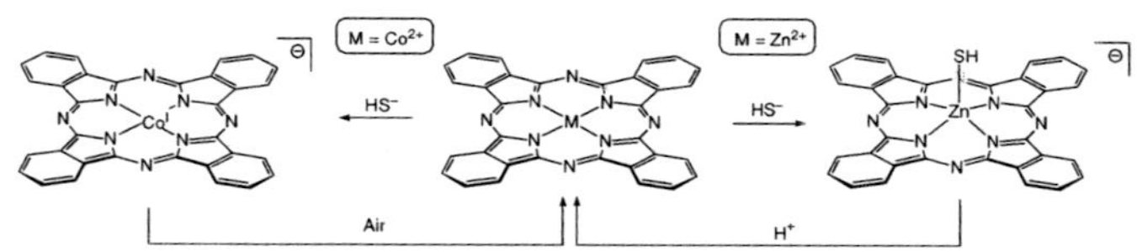

Scheme 5.14 Phthalocyanine-based approaches for sulfide detection.

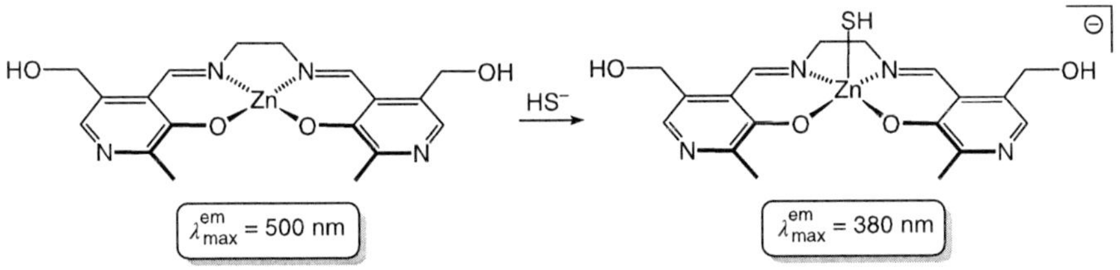

Scheme 5.15 Hydrosulfide detection with a Zn salen complex.

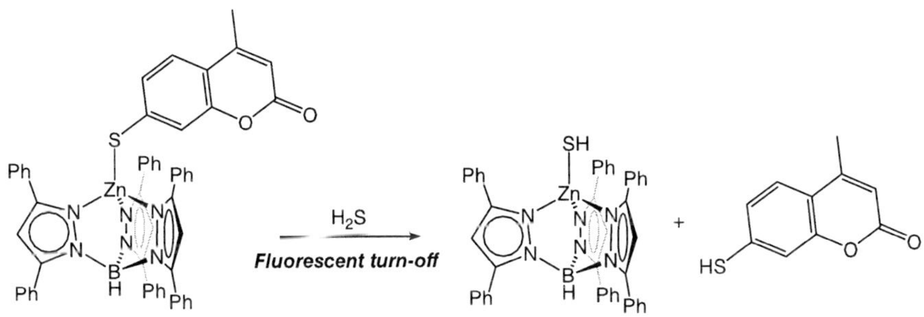

Scheme 5.16 Sulfide detection by fluorophore displacement.

sulfide coordination occurs, it is HS^- ion and not H_2S that dominates. This fact arises from the fundamental chemical differences between the two species, which favor hydrosulfide as a ligand. These differences include charge state, basicity, and the prevalence of HS^- over H_2S in the biological milieu. Regardless of what form hydrogen sulfide takes on, its interactions with metal centers largely encompass two distinct reaction pathways: binding and/or reduction. In the case of iron proteins, the most well-studied targets of H_2S reactivity, both of these processes can be observed owing to the redox activity of the iron(II)/iron(III) couple. The role of the protein environment in selecting for one of these two processes is therefore paramount. A prime example of this phenomenon is exemplified by sulfide binding in natural versus synthetic heme systems. In the former, sulfide coordination takes place to iron(III), whereas in the latter, iron(III) undergoes rapid reduction upon treatment with H_2S/HS^-. In the case of redox inactive metals such as Zn, sulfide binding is more straightforward, although biological roles for such coordination are still not firmly established. In all cases, the field of H_2S bioinorganic chemistry has been greatly assisted by companion studies with biomimetic compounds. The ability to prepare and characterize hydrosulfide complexes of precise composition has aided in establishing structural and spectroscopic benchmarks for the sulfhydryl functional group, and more recently, given rise to a series of H_2S sensors. Moving forward, this symbiosis between biochemical and synthetic chemistry will no doubt continue to help unravel the unique and sometimes perplexing aspects of H_2S biology.

Acknowledgments

Z. J. T. thanks the Welch Foundation (AX-1772) for generously supporting hydrogen sulfide research in his laboratory.

References

1 Wang, R. (2009). Hydrogen sulfide: the third gasotransmitter in biology and medicine. *Antioxid. Redox Signal.* 12 (9): 1061–1064.

2 Filipovic, M.R., Zivanovic, J., Alvarez, B., and Banerjee, R. (2018). Chemical biology of H_2S signaling through persulfidation. *Chem. Rev.* 118 (3): 1253–1337.

3 Kimura, H. (2015). Signaling molecules: hydrogen sulfide and polysulfide. *Antioxid. Redox Signal.* 22 (5): 362–376.

4 Kolluru, G.K., Shen, X., Bir, S.C., and Kevil, C.G. (2013). Hydrogen sulfide chemical biology: pathophysiological roles and detection. *Nitric Oxide* 35: 5–20.

5 Banerjee, R. (2011). Hydrogen sulfide: redox metabolism and signaling. *Antiox. Redox Signal.* 15 (2): 339–341.

6 Kuwata, S. and Hidai, M. (2001). Hydrosulfido complexes of transition metals. *Coord. Chem. Rev.* 213 (1): 211–305.

7 Peruzzini, M., Rios Isaac De, L., and Romerosa, A. (2001). Coordination chemistry of transition metals with hydrogen chalcogenide and hydrochalcogenido ligands. In: *Prog. Inorg. Chem.*, vol. 49 (ed. K.D. Karlin), 169–453. Wiley.

8 Pluth, M.D. and Tonzetich, Z.J. (2020). Hydrosulfide complexes of the transition elements: diverse roles in bioinorganic, cluster, coordination, and organometallic chemistry. *Chem. Soc. Rev.* 49 (12): 4070–4134.

9 May, P.M., Batka, D., Hefter, G. et al. (2018). Goodbye to S^{2-} in aqueous solution. *Chem. Commun.* 54: 1980–1983.

10 Chandrika, D., Ma, E.S., Rettig, S.J. et al. (1997). Synthesis and X-ray structure of an H_2S complex, $RuCl_2$(P-N)(P(*p*-tolyl)$_3$)(SH_2) (P-N = *o*-(diphenylphosphino)-N,N-dimethylaniline). *Inorg. Chem.* 36 (24): 5426–5427.

11 Sellmann, D., Lechner, P., Knoch, F., and Moll, M. (1992). Transition-metal complexes with sulfur ligands. 82. Hydrogen sulfide, disulfur, and carbon disulfide molecules as ligands in sulfur-rich [Ru(PPh_3)('S_4')] complexes (('S_4')$^{2-}$ = 1,2-bis[(2-mercaptophenyl)thio]ethane(2-)). *J. Am. Chem. Soc.* 114 (3): 922–930.

12 Calhorda, M.J., Carrondo, M.A.A.F., and d. C. T., Dias, A. R., Domingos, A. M. T. S., Simoes, J. A. M., Teixeira, C. (1986). Energetics of transition-metal-sulfur and -oxygen bonds in $M(\eta^5\text{-}C_5H_5)_2L_2$ complexes (M = Ti, Mo, W). Molecular structure of $Mo(\eta^5\text{-}C_5H_5)_2(SO_4)$. *Organometallics* 5 (4): 660–667.

13 Calhorda, M.J., de Carrondo, M.A.A.F., C.T., Dias, A.R. et al. (1988). Energetics of metal-sulfur bonds in the complexes $M(\eta^5\text{-}C_5H_5)_2(SC_2H_5)_2$ (M = Ti, W) and $W(\eta^5\text{-}C_5H_5)_2(SC_6H_5)_2$. Molecular structure of $Ti(\eta^5\text{-}C_5H_5)_2(SC_2H_5)_2$. *Inorg. Chem.* 27 (14): 2513–2518.

14 Riordan, C.G. and Halpern, J. (1996). Kinetics, mechanism and thermodynamics of iron carbon bond dissociation in organoiron porphyrin complexes. *Inorg. Chim. Acta* 243 (1): 19–24.

15 Pietri, R., Román-Morales, E., and López-Garriga, J. (2010). Hydrogen sulfide and hemeproteins: knowledge and mysteries. *Antioxid. Redox Signal.* 15 (2): 393–404.

16 Hoppe-Seyler, F. (1866). *Zentbl. Med. Wiss.* 4: 436–438.

17 Evans, S.V., Sishta, B.P., Mauk, A.G., and Brayer, G.D. (1994). Three-dimensional structure of cyanomet-sulfmyoglobin C. *Proc. Natl. Acad. Sci. U.S.A.* 91 (11): 4723.

18 Ríos-González, B.B., Román-Morales, E.M., Pietri, R., and López-Garriga, J. (2014). Hydrogen sulfide activation in hemeproteins: the sulfheme scenario. *J. Inorg. Biochem.* 133: 78–86.

19 Carrico, R.J., Blumberg, W.E., and Peisach, J. (1978). The reversible binding of oxygen to sulfhemoglobin. *J. Biol. Chem.* 253 (20): 7212–7215.

20 Nicholls, P. (1961). The formation and properties of sulphmyoglobin and sulphcatalase. *Biochem. J.* 81 (2): 374–383.

21 Nakamura, S., Nakamura, M., Yamazaki, I., and Morrison, M. (1984). Reactions of ferryl lactoperoxidase (compound II) with sulfide and sulfhydryl compounds. *J. Biol. Chem.* 259 (11): 7080–7085.

22 Kraus, D.W. and Wittenberg, J.B. (1990). Hemoglobins of the *Lucina pectinata*/bacteria symbiosis. I. Molecular properties, kinetics and equilibria of reactions with ligands. *J. Biol. Chem.* 265 (27): 16043–16053.

23 Rizzi, M., Wittenberg, J.B., Coda, A. et al. (1996). Structural bases for sulfide recognition in *Lucina pectinata* hemoglobin I. *J. Mol. Biol.* 258 (1): 1–5.

24 Rizzi, M., Wittenberg, J.B., Coda, A. et al. (1994). Structure of the sulfide-reactive hemoglobin from the clam *Lucina pectinata*: crystallographic analysis at 1.5 Å resolution. *J. Mol. Biol.* 244 (1): 86–99.

25 Fernandez-Alberti, S., Bacelo, D.E., Binning, R.C. et al. (2006). Sulfide-binding hemoglobins: effects of mutations on active-site flexibility. *Biophys. J.* 91 (5): 1698–1709.

26 Pietri, R., Lewis, A., León, R.G. et al. (2009). Factors controlling the reactivity of hydrogen sulfide with hemeproteins. *Biochemistry* 48 (22): 4881–4894.

27 Ruetz, M., Kumutima, J., Lewis, B.E. et al. (2017). A distal ligand mutes the interaction of hydrogen sulfide with human neuroglobin. *J. Biol. Chem.* 292 (16): 6512–6528.

28 Brittain, T., Yosaatmadja, Y., and Henty, K. (2008). The interaction of human neuroglobin with hydrogen sulphide. *IUBMB Life* 60 (2): 135–138.

29 Nicoletti, F.P., Comandini, A., Bonamore, A. et al. (2010). Sulfide binding properties of truncated hemoglobins. *Biochemistry* 49 (10): 2269–2278.

30 Jensen, B. and Fago, A. (2018). Reactions of ferric hemoglobin and myoglobin with hydrogen sulfide under physiological conditions. *J. Inorg. Biochem.* 182: 133–140.

31 Vitvitsky, V., Yadav, P.K., An, S. et al. (2017). Structural and mechanistic insights into hemoglobin-catalyzed hydrogen sulfide oxidation and the fate of polysulfide products. *J. Biol. Chem.* 292 (13): 5584–5592.

32 Bostelaar, T., Vitvitsky, V., Kumutima, J. et al. (2016). Hydrogen sulfide oxidation by myoglobin. *J. Am. Chem. Soc.* 138 (27): 8476–8488.

33 Salnikov, D.S., Kucherenko, P.N., Dereven'kov, I.A. et al. (2014). Kinetics and mechanism of the reaction of hydrogen sulfide with cobalamin in aqueous solution. *Eur. J. Inorg. Chem.* 2014 (5): 852–862.

34 Kabil, O. and Banerjee, R. (2010). Redox biochemistry of hydrogen sulfide. *J. Biol. Chem.* 285 (29): 21903–21907.

35 Hill, B.C., Woon, T.C., Nicholls, P. et al. (1984). Interactions of sulphide and other ligands with cytochrome c oxidase. An electron-paramagnetic-resonance study. *Biochem. J.* 224 (2): 591–600.

36 Nicholls, P. and Kim, J.K. (1982). Sulphide as an inhibitor and electron donor for the cytochrome c oxidase system. *Can. J. Biochem.* 60 (6): 613–623.

37 Blackstone, E., Morrison, M., and Roth, M.B. (2005). H_2S induces a suspended animation-like state in mice. *Science* 308 (5721): 518.

38 Collman, J.P., Ghosh, S., Dey, A., and Decréau, R.A. (2009). Using a functional enzyme model to understand the chemistry behind hydrogen sulfide induced hibernation. *Proc. Natl. Acad. Sci. U.S.A.* 106 (52): 22090.

39 Bieza, S.A., Boubeta, F., Feis, A. et al. (2015). Reactivity of inorganic sulfide species toward a heme protein model. *Inorg. Chem.* 54 (2): 527–533.

40 Zhao, Z., Wang, D., Wang, M. et al. (2016). Proximal environment controlling the reactivity between inorganic sulfide and heme-peptide model. *RSC Adv.* 6 (82): 78858–78864.

41 Mittra, K., Singha, A., and Dey, A. (2017). Mechanism of reduction of ferric porphyrins by sulfide: identification of a low spin Fe^{III}–SH intermediate. *Inorg. Chem.* 56 (7): 3916–3925.

42 Meininger, D.J., Chee-Garza, M., Arman, H.D., and Tonzetich, Z.J. (2016). Gallium(III) tetraphenylporphyrinates containing hydrosulfide and thiolate ligands: structural models for sulfur-bound iron(III) hemes. *Inorg. Chem.* 55 (5): 2421–2426.

43 Pavlik, J.W., Noll, B.C., Oliver, A.G. et al. (2010). Hydrosulfide (HS^-) coordination in iron porphyrinates. *Inorg. Chem.* 49 (3): 1017–1026.

44 Meininger, D.J., Arman, H.D., and Tonzetich, Z.J. (2017). Synthesis, characterization, and binding affinity of hydrosulfide complexes of synthetic iron(II) porphyrinates. *J. Inorg. Biochem.* 167: 142–149.

45 Hartle, M.D., Prell, J.S., and Pluth, M.D. (2016). Spectroscopic investigations into the binding of hydrogen sulfide to synthetic picket-fence porphyrins. *Dalton Trans.* 45 (11): 4843–4853.

46 Dhifet, M., Belkhiria, M.S., Daran, J.-C., and Nasri, H. (2009). (Cryptand-222)potassium(+) (hydrogensulfido)[5,10,15,20-tetrakis(2-pivalamidophenyl)porphyrinato]ferrate(II). *Acta Crystallogr., Sect. E: Struct. Rep.* 65 (8): m967–m968.

47 Beinert, H., Holm, R.H., and Münck, E. (1997). Iron-sulfur clusters: nature's modular, multipurpose structures. *Science* 277 (5326): 653.

48 Lovenberg, W. (1973). *Iron-Sulfur Proteins*. Vol. I and II. New York: Academic Press.

49 Mulliez, E., Duarte, V., Arragain, S. et al. (2017). On the role of additional [4Fe-4S] clusters with a free coordination site in radical-SAM enzymes. *Front. Chem.* 5 (17): 1–5.

50 Knauer, S.H., Buckel, W., and Dobbek, H. (2011). Structural basis for reductive radical formation and electron recycling in (R)-2-hydroxyisocaproyl-CoA dehydratase. *J. Am. Chem. Soc.* 133 (12): 4342–4347.

51 Dinis, P., Suess, D.L.M., Fox, S.J. et al. (2015). X-ray crystallographic and EPR spectroscopic analysis of HydG, a maturase in [FeFe]-hydrogenase H-cluster assembly. *Proc. Natl. Acad. Sci. U.S.A.* 112 (5): 1362.

52 Forouhar, F., Arragain, S., Atta, M. et al. (2013). Two Fe-S clusters catalyze sulfur insertion by radical-SAM methylthiotransferases. *Nat. Chem. Biol.* 9 (5): 333–338.

53 Lukoyanov, D.A., Yang, Z.-Y., Dean, D.R. et al. (2020). Electron redistribution within the nitrogenase active site FeMo-cofactor during reductive elimination of H_2 to achieve N≡N triple-bond activation. *J. Am. Chem. Soc.* 142 (52): 21679–21690.

54 Müller, A., Schladerbeck, N.H., and Bögge, H. (1987). $[Fe_4S_4(SH)_4]^{2-}$, the simplest synthetic analogue for a ferredoxin. *J. Chem. Soc. Chem. Commun.* 1: 35–36.

55 Hoveyda, H.R. and Holm, R.H. (1997). Characterization of the self-condensation equilibrium of $[Fe_4S_4(SH)_4]^{2-}$: spectroscopic identification of a unique sulfido-bridged acyclic tricubane cluster. *Inorg. Chem.* 36 (20): 4571–4578.

56 Venkateswara Rao, P. and Holm, R.H. (2004). Synthetic analogues of the active sites of iron–sulfur proteins. *Chem. Rev.* 104 (2): 527–560.

57 Segal, B.M., Hoveyda, H.R., and Holm, R.H. (1998). Terminal ligand assignments based on trends in metal–ligand bond lengths of cubane-type $[Fe_4S_4]^{2+,+}$ clusters. *Inorg. Chem.* 37 (13): 3440–3443.

58 Cai, L. and Holm, R.H. (1994). Synthesis and electron delocalization of $[Fe_4S_4]$-S-Fe(III) bridged assemblies related to the exchange-coupled catalytic site of sulfite reductases. *J. Am. Chem. Soc.* 116 (16): 7177–7188.

59 Terada, T., Wakimoto, T., Nakamura, T. et al. (2012). Tridentate thiolate ligands: application to the synthesis of the site-differentiated [4Fe-4S] cluster having a hydrosulfide ligand at the unique iron center. *Chem. Asian J.* 7 (5): 920–929.

60 Zhang, Y. and Holm, R.H. (2003). Synthesis of a molecular $Mo_2Fe_6S_9$ cluster with the topology of the PN cluster of nitrogenase by rearrangement of an edge-bridged $Mo_2Fe_6S_8$ double cubane. *J. Am. Chem. Soc.* 125 (13): 3910–3920.

61 Berlinguette, C.P. and Holm, R.H. (2006). Edge-bridged $Mo_2Fe_6S_8$ to PN-type $Mo_2Fe_6S_9$ cluster conversion: structural fate of the attacking sulfide/selenide nucleophile. *J. Am. Chem. Soc.* 128 (36): 11993–12000.

62 Hong, D., Zhang, Y., and Holm, R.H. (2005). Heterometal cubane-type WFe_3S_4 and related clusters trigonally symmetrized with hydrotris(3,5-dimethylpyrazolyl)borate. *Inorg. Chim. Acta* 358 (7): 2303–2311.

63 Zhang, Y., Zuo, J.-L., Zhou, H.-C., and Holm, R.H. (2002). Rearrangement of symmetrical dicubane clusters into topological analogues of the P cluster of nitrogenase: nature's choice? *J. Am. Chem. Soc.* 124 (48): 14292–14293.

64 Zuo, J.-L., Zhou, H.-C., and Holm, R.H. (2003). Vanadium–iron–sulfur clusters containing the cubane-type $[VFe_3S_4]$ core unit: synthesis of a cluster with the topology of the P^N cluster of nitrogenase. *Inorg. Chem.* 42 (15): 4624–4631.

65 Crack, J.C., Green, J., Thomson, A.J., and Brun, N.E.L. (2014). Iron–sulfur clusters as biological sensors: the chemistry of reactions with molecular oxygen and nitric oxide. *Acc. Chem. Res.* 47 (10): 3196–3205.

66 Tonzetich, Z.J., McQuade, L.E., and Lippard, S.J. (2010). Detecting and understanding the roles of nitric oxide in biology. *Inorg. Chem.* 49 (14): 6338–6348.

67 Tsai, M.-L., Tsou, C.-C., and Liaw, W.-F. (2015). Dinitrosyl iron complexes (DNICs): from biomimetic synthesis and spectroscopic characterization toward unveiling the biological and catalytic roles of DNICs. *Acc. Chem. Res.* 48 (4): 1184–1193.

68 Fitzpatrick, J. and Kim, E. (2015). Synthetic modeling chemistry of iron–sulfur clusters in nitric oxide signaling. *Acc. Chem. Res.* 48 (8): 2453–2461.

69 Oakley, K.M., Zhao, Z., Lehane, R.L. et al. (2021). Generation of H_2S from thiol-dependent NO reactivity of model [4Fe-4S] cluster and Roussin's black anion. *Inorg. Chem.* 60 (21): 15910–15917.

70 Pieretti, J.C., Junho, C.V.C., Carneiro-Ramos, M.S., and Seabra, A.B. (2020). H_2S- and NO-releasing gasotransmitter platform: a crosstalk signaling pathway in the treatment of acute kidney injury. *Pharmacol. Res.* 161: 105121.

71 Beck, W., Grenz, R., Götzfried, F., and Vilsmaier, E. (1981). Zur Kenntnis der "Säure" des Roten Roussinschen salzes, $(ON)_2Fe(SH)_2Fe(NO)_2$ und ihrer Derivate. *Chem. Ber.* 114 (9): 3184–3187.

72 Tsou, C.-C., Chiu, W.-C., Ke, C.-H. et al. (2014). Iron(III) bound by hydrosulfide anion ligands: NO-promoted stabilization of the $[Fe^{III}–SH]$ motif. *J. Am. Chem. Soc.* 136 (26): 9424–9433.

73 Ganguly, T., Das, A., and Majumdar, A. (2019). Iron(II) mediated desulfurization of organosulfur substrates produces nonheme diiron(II)-hydrosulfides. *Inorg. Chem.* 58 (15): 9998–10011.

74 Pal, N. and Majumdar, A. (2019). Transfer of hydrosulfide from thiols to iron(II): a convenient synthetic route to nonheme diiron(II)–hydrosulfide complexes. *Dalton Trans.* 48 (18): 5903–5908.

75 Das, A., Ganguly, T., and Majumdar, A. (2021). Thiolate coordination vs C–S bond cleavage of thiolates in dinickel(II) complexes. *Inorg. Chem.* 60 (2): 944–958.

76 Jana, M. and Majumdar, A. (2018). C–S bond cleavage, redox reactions, and dioxygen activation by nonheme dicobalt(II) complexes. *Inorg. Chem.* 57 (2): 617–632.

77 Ganguly, T., Das, A., Jana, M., and Majumdar, A. (2018). Cobalt(II)-mediated desulfurization of thiophenes, sulfides, and thiols. *Inorg. Chem.* 57 (18): 11306–11309.

78 Galardon, E., Roger, T., Deschamps, P. et al. (2012). Synthesis of a $Fe^{II}SH$ complex stabilized by an intramolecular N–H···S hydrogen bond, which acts as a H_2S donor. *Inorg. Chem.* 51 (19): 10068–10070.

79 Krishnamurthy, V.M., Kaufman, G.K., Urbach, A.R. et al. (2008). Carbonic anhydrase as a model for biophysical and physical-organic studies of proteins and protein–ligand binding. *Chem. Rev.* 108 (3): 946–1051.

80 Flores, J.F., Fisher, C.R., Carney, S.L. et al. (2005). Sulfide binding is mediated by zinc ions discovered in the crystal structure of a hydrothermal vent tubeworm hemoglobin. *Proc. Natl. Acad. Sci. U.S.A.* 102 (8): 2713.

81 Looney, A., Han, R., Gorrell, I.B. et al. (1995). Monomeric alkyl and hydride derivatives of zinc supported by poly(pyrazoly)hydroborato ligation: synthetic, structural, and reactivity studies. *Organometallics* 14 (1): 274–288.
82 Galardon, E., Tomas, A., Roussel, P., and Artaud, I. (2011). Synthesis, stability, and reactivity of $[(TPA)Zn(SH)]^+$ in aqueous and organic solutions. *Eur. J. Inorg. Chem.* 2011 (25): 3797–3801.
83 Ruf, M. and Vahrenkamp, H. (1996). Small molecule chemistry of the pyrazolylborate–zinc unit $Tp^{Cum,Me}Zn$. *Inorg. Chem.* 35 (22): 6571–6578.
84 Rombach, M. and Vahrenkamp, H. (2001). Pyrazolylborate–zinc–hydrosulfide complexes and their reactions. *Inorg. Chem.* 40 (24): 6144–6150.
85 Bräuer, M., Anders, E., Sinnecker, S. et al. (2000). The TpZn–OH/CS reaction: theoretical and preparative visualization of an essential bioinorganic reaction path. *Chem. Commun.* 8: 647–648.
86 Rong, Y. and Parkin, G. (2013). The synthesis and structures of tris(2-pyridylseleno)methyl zinc compounds with κ^2-, κ^3-, and κ^4-coordination modes. *Aust. J. Chem.* 66 (10): 1306–1310.
87 Ibrahim, M.M., Seebacher, J., Steinfeld, G., and Vahrenkamp, H. (2005). Tris(thioimidazolyl)borate-zinc-thiolate complexes for the modeling of biological thiolate alkylations. *Inorg. Chem.* 44 (23): 8531–8538.
88 Melnick, J.G., Docrat, A., and Parkin, G. (2004). Methyl, hydrochalcogenido, and phenylchalcogenolate complexes of zinc in a sulfur rich coordination environment: syntheses and structural characterization of the tris(2-mercapto-1-tert-butylimidazolyl)hydroboratozinc complexes $[Tm^{But}]ZnMe$, $[Tm^{But}]ZnEH$ (E = S, Se) and $[Tm^{But}]ZnEPh$ (E = O, S, Se, Te). *Chem. Commun.* 24: 2870–2871.
89 Hartle, M.D., Delgado, M., Gilbertson, J.D., and Pluth, M.D. (2016). Stabilization of a Zn(II) hydrosulfido complex utilizing a hydrogen-bond accepting ligand. *Chem. Commun.* 52 (49): 7680–7682.
90 Hille, R., Hall, J., and Basu, P. (2014). The mononuclear molybdenum enzymes. *Chem. Rev.* 114 (7): 3963–4038.
91 Okamoto, K., Matsumoto, K., Hille, R. et al. (2004). The crystal structure of xanthine oxidoreductase during catalysis: Implications for reaction mechanism and enzyme inhibition. *Proc. Natl. Acad. Sci. U.S.A.* 101 (21): 7931.
92 Majumdar, A., Mitra, J., Pal, K., and Sarkar, S. (2008). Mono-oxo bis(dithiolene) Mo(IV)/W(IV) complexes as building blocks for sulfide bridged bi- and tri-nuclear complexes. *Inorg. Chem.* 47 (12): 5360–5364.
93 Mitra, J. and Sarkar, S. (2013). Modelling the reduced xanthine oxidase in active sulfo and inactive desulfo forms. *Dalton Trans.* 42 (9): 3050–3058.
94 Lin, V.S., Chen, W., Xian, M., and Chang, C.J. (2015). Chemical probes for molecular imaging and detection of hydrogen sulfide and reactive sulfur species in biological systems. *Chem. Soc. Rev.* 44 (14): 4596–4618.

95 Maria, S., Franco De, M., Claudio, P. et al. (2011). Myoglobin as a new fluorescence probe to sense H_2S. *Prot. Pept. Lett.* 18 (3): 282–286.

96 Dulac, M., Melet, A., and Galardon, E. (2018). Reversible detection and quantification of hydrogen sulfide by fluorescence using the hemoglobin I from *Lucina pectinata*. *ACS Sensors* 3 (10): 2138–2144.

97 Strianese, M., Palm, G.J., Milione, S. et al. (2012). A FRET enzyme-based probe for monitoring hydrogen sulfide. *Inorg. Chem.* 51 (21): 11220–11222.

98 Strianese, M., Palm, G.J., Kohlhause, D. et al. (2019). Azurin and HS^-: towards implementation of a sensor for HS^- detection. *Eur. J. Inorg. Chem.* 2019 (6): 885–891.

99 Hartle, M.D., Tillotson, M.R., Prell, J.S., and Pluth, M.D. (2017). Spectroscopic investigation of the reaction of metallo-protoporphyrins with hydrogen sulfide. *J. Inorg. Biochem.* 173: 152–157.

100 Strianese, M., Mirra, S., Lamberti, M., and Pellecchia, C. (2017). Zinc (II) porphyrins as viable scaffolds to stabilize hydrogen sulfide binding at the metal center. *Inorg. Chim. Acta* 466: 426–431.

101 Mirra, S., Milione, S., Strianese, M., and Pellecchia, C. (2015). A copper porphyrin for sensing H_2S in aqueous solution via a "coordinative-based" approach. *Eur. J. Inorg. Chem.* 2015 (13): 2272–2276.

102 Strianese, M., Lamberti, M., and Pellecchia, C. (2017). Chemically reversible binding of H_2S to a zinc porphyrin complex: towards implementation of a reversible sensor via a "coordinative-based approach". *Dalton Trans.* 46 (6): 1872–1877.

103 Hartle, M.D., Sommer, S.K., Dietrich, S.R., and Pluth, M.D. (2014). Chemically reversible reactions of hydrogen sulfide with metal phthalocyanines. *Inorg. Chem.* 53 (15): 7800–7802.

104 Strianese, M., Lamberti, M., and Pellecchia, C. (2018). Interaction of monohydrogensulfide with a family of fluorescent pyridoxal-based Zn(II) receptors. *Dalton Trans.* 47 (48): 17392–17400.

105 Strianese, M., Guarnieri, D., Lamberti, M. et al. (2020). Fluorescent salen-type Zn(II) complexes As probes for detecting hydrogen sulfide and its anion: bioimaging applications. *Inorg. Chem.* 59 (21): 15977–15986.

106 Galardon, E., Tomas, A., Roussel, P., and Artaud, I. (2009). New fluorescent zinc complexes: towards specific sensors for hydrogen sulfide in solution. *Dalton Trans.* 42: 9126–9130.

6

Measurement of Hydrogen Sulfide Metabolites Using the Monobromobimane Method

Xinggui Shen[1], Ellen H. Speers[1], and Christopher G. Kevil[2]

[1] *Departments of Pathology and Translational Pathobiology, LSU Health -Shreveport, 1501 Kings Hwy, Shreveport, LA, USA*
[2] *Departments of Pathology and Translational Pathobiology, Molecular and Cellular Physiology and Cell Biology and Anatomy, LSU Health -Shreveport, 1501 Kings Hwy, Shreveport, LA, USA*

List of Abbreviations

DTPA	diethylenetriamine pentaacetic acid
EDTA	ethylene diamine tetraacetic acid
ESI	electrospray ionization
FA	formic acid
HPLC	high-performance liquid chromatography
LC-MS/MS	liquid chromatography tandem mass spectrometry
MBB	monobromobimane
NEM	*N*-ethylmaleimide
SDB	sulfide-dibimane
SSA	5-sulfosalicylic acid
TCEP	tris(2-carboxyethyl)phosphine
TFA	trifluoroacetic acid

6.1 Introduction

6.1.1 Hydrogen Sulfide: Biological Significance

In the centuries following its discovery by Carl Wilhelm Scheele and Claude Louis Berthollet in the 1770s, hydrogen sulfide (H_2S) was found to be an environmental toxin, causing inflammation, respiratory damage, genotoxic effects, and more [1]. It was not until a trio of papers around 1990 by Warenycia et al., Goodwin et al., and Savage and Gould that H_2S was found to have any positive effect on mammalian biology [2–4]. Hydrogen sulfide's toxicity was the motivation for these studies of its effect on the mammalian central nervous system (CNS); however, H_2S was found in even normal (unexposed to exogenous H_2S) CNS tissues, suggesting that endogenous H_2S has a biological function. Hydrogen sulfide is produced predominantly from cysteine or cystathionine, or 3-mercaptopyruvate by the enzymes cystathionine gamma-lyase, cystathionine beta-synthase, and 3-mercaptosulfurtransferase, respectively [5]. Early studies in the lab of Hideo Kimura demonstrated that endogenous levels of H_2S affects the function of neuronal [6], smooth muscle, and vascular tissues [7]. Since then, there has been a flood of studies demonstrating H_2S as an endogenous signaling molecule, which plays important roles in physiological and pathological processes [8–10].

6.1.2 Hydrogen Sulfide Chemistry

Hydrogen sulfide (H_2S) is a colorless gas with the odor of rotten eggs. The sulfur atom of H_2S has the lowest oxidation state of −2 (sulfur can range from −2 to +6) and can be oxidized to form sulfur dioxide, sulfates, sulfite, and elemental sulfur. H_2S is slightly soluble in water and acts as a weak acid. The pK_a values governing the protonation states of hydrogen sulfide are 7.05 (H_2S:$SH^- + H^+$) and approximately 19 (HS^-:$S^{2-} + H^+$). It can dissociate into H^+ and hydrosulfide anion (HS^-), which in turn may dissociate to H^+ and a sulfide anion (S^{2-}) in the following reaction (Eq. (6.1)):

$$H_2S \leftrightarrow H^+ + HS^- \leftrightarrow 2H^+ + S^{2-} \tag{6.1}$$

Hydrogen sulfide exists in three charge states: H_2S, HS^-, and S^{2-}. At physiological conditions (pH 7.4, 37 °C), hydrogen sulfide is composed of 19% H_2S, 81% HS^-, and very small amounts of S^{2-}. This ratio shifts with temperature and pH. For example, at room temperature (25 °C) and pH 7.4 about 40% of hydrogen sulfide is in the fully protonated form [11, 12]. The relevance of S^{2-} in aqueous solutions has recently been brought into question [13].

Hydrogen sulfide's stability depends on several factors. At low pH, more of it is present as the fully protonated H_2S, and it is less vulnerable to oxidation. However, since H_2S is uncharged and weakly polar, it volatalizes from solution more readily at or below pH 5.4. At higher pH hydrogen sulfide has a larger proportion as HS^- and is less volatile but more vulnerable to oxidation [11, 14].

6.1.3 Bioavailable Sulfide

In vivo, bioavailable sulfide exists in different forms, such as free sulfide (H_2S), acid-labile sulfide, and bound sulfane sulfur (Figure 6.1). Acid-labile sulfide is found as heme and iron–sulfur clusters including [2Fe-2S] and [4Fe-4S]. Bound

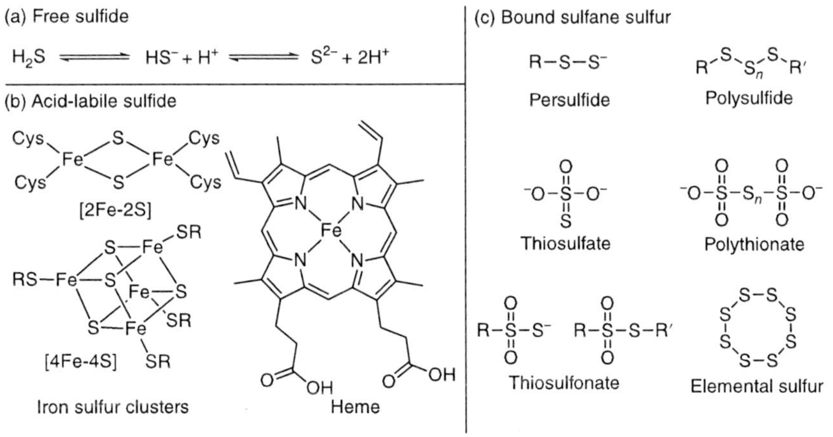

Figure 6.1 Bioavailable sulfide: free sulfide, acid-labile sulfide, and bound sulfane sulfur.

sulfane sulfur is comprised of sulfide bound to other sulfur atom(s), including persulfides, polysulfides, thiosulfate, thiosulfonates, polythionates, and elemental sulfur [12, 15, 16]. The distribution of these bioavailable sulfides varies across different tissues or organs, i.e. bound sulfane sulfur is the largest pool in rat brain [14], while acid-labile sulfide is the largest pool in mouse and human plasma [16].

The homeostasis of bioavailable sulfide involves the transformation of these different forms. In 2009, Ishigami et al. demonstrated that brain homogenates absorbed exogenously applied Na_2S into the bound sulfane sulfur pool, which could be released as free hydrogen sulfide by treatment with the physiological reducing agents glutathione and cysteine. In the brain, the largest sulfide pool was this bound sulfane sulfur [14].

6.2 Monobromobimane: An Optimal Method of Bioavailable Sulfur Detection

Biological samples are complex, and hydrogen sulfide is not stable. Reliable methods of hydrogen sulfide detection are required for accurate characterizations of its biological roles. However, it is now clear that a wide range of values for hydrogen sulfide level have been reported in the same samples [16–20].

The methylene blue method, developed over a century ago, is the most commonly reported method for sulfide measurement [21–24]. Briefly, a solution of *N*,*N*-dimethyl-*p*-phenylenediamine sulfate and ferric chloride ($FeCl_3$) is mixed with the sample under acidic conditions and react with hydrogen sulfide to produce methylene blue, which is detected with a spectrophotometer by absorbance at 670 nm. However, this method can be highly problematic, making it inappropriate for measuring biological levels of hydrogen sulfide [17]. Key problems include (i) interference of other colored substances, (ii) methylene blue dimer and trimer formation, (iii) strong acid chemical pretreatment, and (iv) low sensitivity.

The purpose of this chapter is to provide detailed techniques to perform the MBB derivatization method for detecting bioavailable sulfide. The main methodology used is RP-HPLC (high-performance liquid chromatography) with fluorescence detection with a 2.0 nM limit of detection that is sensitive and reliable enough for use with most biological samples [16].

6.2.1 Monobromobimane Derivatization of Hydrogen Sulfide

Hydrogen sulfide is reacted with monobromobimane (MBB) in basic solution to produce fluorescent sulfide-dibimane (SDB). This derivatization process occurs under alkaline conditions and the product is then stabilized with an acidic pH solution. As shown in Figure 6.2, two molecules of MBB react sequentially with hydrogen sulfide. MBB reacts preferentially with hydrogen sulfide in the HS^- form yielding bimane-SH, and the generated bimane-SH undergoes a second bimolecular reaction with mBB to form the sulfide-dibimane. This later reaction is slower than the initial MBB reaction with HS^- due to decreased nucleophilicity of bimane sulfhydryl groups when compared to HS^-. At pH 9.5, free hydrogen sulfide mainly exist as HS^- [11], and as HS^- is consumed by the MBB reaction the $H_2S{:}HS^-$ equilibrium shifts toward HS^- until all hydrogen sulfide is reacted thus making MBB an effective "sulfide sink" [25]. Additionally, the hydrolysis of the SDB increases with increasing pH values, with only 5% SDB remaining after a 20-hour incubation at pH 9.5. The highest yield of generated SDB was achieved after 30 minutes of incubation with 0.1 mM DTPA (diethylenetriamine pentaacetic acid) and 1% oxygen. The best pH to use for SDB stability is 4.5. In this acidic environment, the generated SDB is stable at 4 °C for up to two weeks, or over one year at −20 °C. Thus, the above facts require that the reaction time of MBB with hydrogen sulfide be

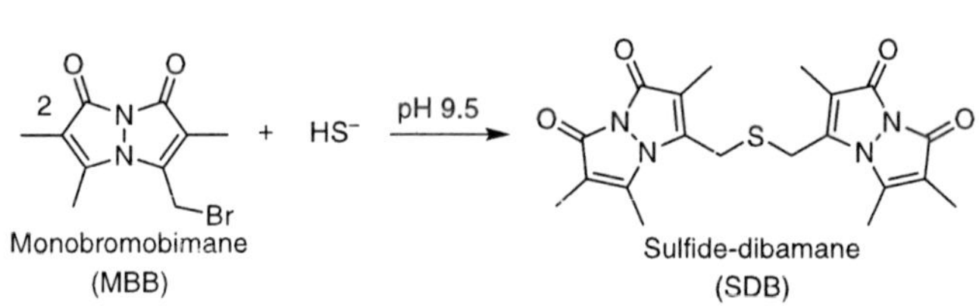

Figure 6.2 Schematic representation of the derivatization of hydrogen sulfide with monobromobimane.

carefully controlled and standardized for all samples. Moreover, unlike dibromobimane (dBB) [26], MBB does not significantly extract sulfur from other sulfhydryl sources such as small molecular weight and protein thiols making it a reliable method to determine bioavailable sulfide levels.

The SDB product is separated from MBB and other species by reverse-phase liquid chromatography and detected by fluorescence ($\lambda_{ex} = 390$ nm, $\lambda_{em} = 475$ nm) or mass spectrometry. SDB is more hydrophobic than MBB and elutes first [17, 25, 27]. MBB has a limit of detection of 2 nM by fluorescence [28] and 0.25 nM by mass spectrometry [27].

6.2.2 History of the Monobromobimane Method

MBB was originally synthesized and applied for fluorescent labeling of human red blood cells by Kosower et al. in 1978–1979 [29, 30]. In 1981, MBB was used to derivatize biological thiols (not for sulfide), and then separation by cation-exchange chromatography [31] or reverse-phase HPLC [32]. In 1987, Kosower et al. summarized applications of MBB labeling of small molecule and protein thiols (though not H_2S). In 1992, Togawa et al. used HPLC with fluorescence detection to measure MBB-derivatized bound sulfide, sulfite, and thiosulfate in human serum [33, 34].

In 2010–2014, two research groups refined use of MBB for hydrogen sulfide measurement. Wintner et al. developed a MBB procedure for blood samples. First, the blood samples were dispensed into the derivatization solution (degassed 25 mM HEPES, 50% water, 50% acetonitrile, 5 mM MBB, pH 8.0) for 10 minutes. Second, ethyl acetate was used to extract the SDB, and the solvent was evaporated. Lastly, the SDB was suspended in acetonitrile, and then analyzed by RP-HPLC [25]. Shen et al. optimized the derivatization reaction between MBB and sulfide, including derivatization buffer, oxygen concentration, temperature, etc. RP-HPLC with a fluorescence detector, stabilization buffer, and stop buffer making sulfide measurement convenient and more reliable [17]. Furthermore, Shen et al. expanded the MBB method to measure different bioavailable sulfide forms including free H_2S, acid-labile, and bound sulfane-sulfur [16]. In 2014, Shen et al. utilized electrospray ionization (ESI)-MS/MS (positive mode) to confirm the identity of the SDB peak (using ^{34}S-labelled SDB as a standard) and validating reaction products and conditions to achieve quantitative and reliable detection. Importantly, use of either mass spectrometry or RP-HPLC methods revealed similar H_2S level values from identical samples. Moreover, the mass spectrometry method is most sensitive by a log order of magnitude, with a lower limit of detection of 0.25 nM [27].

In 2015–2019, several lab-on-a-chip applications of the MBB derivatization method have been developed [35, 36]. According to the MBB method [16], each chip contained a selective silicone layer and different chambers with releasing and trapping buffers. Trapped sulfide is then analyzed after MBB derivatization that is detailed by an issued US patent [36] for this technology.

6.3 Procedures

This section contains procedures for SDB standard preparation (Section 6.3.1), bioavailable sulfide preparation (Section 6.3.2), MBB derivatization of H_2S (Section 6.3.3), HPLC-fluorescence (Section 6.3.4), and mass spectrometric (Section 6.3.5) detection. Caveats and considerations for the protocols are also discussed.

Solutions

- Ultrapure ddH_2O
- Reaction buffer: 100 mM Tris-HCl (pH 9.5), 0.1 mM DTPA
- Stop solution: 200 mM sulfosalicylic acid (SSA) in ddH_2O
- MMB solution: 10 mM in deoxygenated acetonitrile, and protected from light.
- Sulfide standard solution: 0–10 μM of sulfide solution
- Releasing buffer: 100 mM phosphate buffer (pH 2.6), 0.1 mM DTPA
- Releasing buffer with 1 mM TCEP (tris(2-carboxyethyl)phosphine): 100 mM phosphate buffer (pH 2.6), 0.1 mM DTPA, 1 mM TCEP. Make this solution fresh.

6.3.1 Sulfide-Dibimane Standard Synthesis

The following steps are performed in a hypoxic chamber (1% O_2) at room temperature:

(1) Add 10 mL of deoxygenated reaction buffer to a 50 mL tube.
(2) Add 4 mL of 6 mM Na_2S solution.
(3) Add 5 mL of 10 mM MBB solution, and incubate 30 minutes at room temperature.
(4) Add 1 mL of 2-mercaptoethanol to quench excess MBB.
(5) Extract the SDB with 10 mL of ethyl acetate.
(6) Evaporate the solvent with a nitrogen stream.
(7) Re-dissolve them in 5% methanol, and purify the SDB with an Alltech Prevail SPE cartridge a 50% methanol solution (v/v) elution.
(8) Combine the fractions containing pure SDB, and confirm them by HPLC or LC-MS.
(9) Evaporate the solvent, then weigh the solid SDB.
(10) Make standard solutions at appropriate concentrations.

(11) Verify the SDB concentrations with spectrophotometry. SDB has extinction coefficients of 4883 and 4694 $dm^3/mol\ cm$ at 370 nm (methanol solution) and 380 nm (HCl solution), respectively.
(12) Store the standards at −80 °C.

6.3.2 Bioavailable Sulfide Preparation

For free sulfide, skip the below steps and directly continue to Section 6.3.3.

For acid-labile sulfide and bound sulfane sulfur, the following steps are performed in a hypoxic chamber (1% O_2) at room temperature.

Procedure

(1) Degas the releasing buffer with or without 1 mM TCEP.
(2) Transfer 50 μL of sample into two sets of 4 mL BD vacutainer tubes.
(3) Add 450 μL of releasing buffer into one vacutainer tube, and add 450 μL of releasing buffer with 1 mM TCEP into the other vacutainer tubes.
(4) Incubate tubes on a nutator for 30 minutes.
(5) Remove the solution through the cap with a 25 gauge spinal needle or other similar long needle and 1 mL plastic syringe.
(6) Trap the sulfide gas by adding 500 μL of reaction buffer to the BD vacutainer tubes, using a 30 G needle, and incubate for 30 minutes on a nutator.
(7) Proceed with the MBB derivatization (Section 6.3.3).
(8) Calculate acid-labile sulfide and bound sulfane sulfur.
   a. Acid labile sulfide = value obtained by releasing buffer-the value of free H_2S
   b. Bound sulfane sulfur = value obtained by releasing buffer with 1 mM TCEP − the value of acid labile sulfide

6.3.3 Monobromobimane Derivatization

The following steps are performed in a hypoxic chamber (1% O_2) at room temperature:

Procedure

(1) Degas the water and reaction buffer.
(2) Centrifuge the samples.
(3) Add 70 μL of reaction buffer to a PCR tube.
(4) Add 30 μL of sample to the tube.
(5) Add 50 μL of 10 mM MBB solution to the tube in the dark room (MBB is light-sensitive).
(6) 30 minutes later, add 50 μL of stop solution to halt the reaction (in the dark room).
(7) Remove the samples from the hypoxic chamber and store at 4 °C (in the dark) until analysis.

6.3.4 HPLC with Fluorescence Detection

In our lab, sulfide-dibimane is detected using a Shimadzu Prominence HPLC (LC-20) with a fluorescence detector (model RF-10AXL or RF-20AXL). The HPLC column used is an Agilent Eclipse XDB-C18 column (4.6 × 250 mm, 100 Å) with a guard column (4.6 × 12.5 mm, 80 Å). The data processing software is LCsolution (Version 1.23).

Procedure

(1) Centrifuge the derivatized samples and transfer 150 μL supernatant into amber HPLC vials with microinserts. Keep the HPLC vials at 4 °C in the autosampler.
(2) 10 μL of sample is injected into the RP-HPLC for separation on a C18 column and fluorescence detection (excitation/emission 390/475 nm).
(3) The mobile phase for SDB elution is composed of 0.1% TFA (trifluoroacetic acid) in water (Solvent A) and 0.1% TFA in acetonitrile (Solvent B). The flow rate is a 0.6 mL/min, and the gradient is shown below.
(4) The retention times of SDB and MBB are approximately 16.5 and 17.8 minutes, respectively.
HPLC gradient:

Time (min)	% A	% B
0	85	15
5	65	35
16	45	55
23	30	70
24	10	90
26	10	90
28	85	15

6.3.5 Mass Spectrometry Detection

The LC-ESI-MS/MS detection is performed in our laboratory using a Thermo LCQ Deca XP Plus mass spectrometer with a Surveyor HPLC system. The data processing software is Xcalibur and Quan Browser (Version 2.0).

Procedure

(1) Add 3 μL of 1.0 mg/mL ^{34}S-labelled SDB into 30 μL of derivatized sample [37].
(2) Dilute the samples with 10% acetonitrile (0.1% formic acid [FA]).
(3) Inject 5 μL of the samples into the LC-MS.

(4) The separations are performed on Phenomenex C18 column (50 × 2.1 mm, 5 μm). The mobile phase is 0.1% FA in water (Solvent A) and 0.1% FA in acetonitrile (Solvent B). The linear gradient is performed from 10% to 60% of mobile phase B (0.1% FA in acetonitrile) for 15 minutes at a flow rate of 200 μL/min.

(5) MS analyses are carried out on a Thermo LCQ Deca XP with the ESI source in positive mode. Other MS settings include a sheath gas of 10 units, sweep gas of 5 units, spray voltage of 5.0 kV, and capillary temperature of 250 °C. The ion gauge pressure is 2.41×10^{-5} Torr. The max injection time is 400 ms, the isolation width is 1.0, with full scan mass spectra of m/z 150–500. SDB is quantified with selective reaction monitoring; the precursor ion of SDB is m/z 415.3 $[M+H]^+$ (and m/z 417.1 for the ^{34}S-labeled internal standard) and the product ion is m/z 192.9. The MS2 fragmentation uses He at 400 kPa as the collision gas, collision energy 35 V, an isolation width of 1.0, and activation q of 0.25. The automatic gain control is set to 5×10^6. Data analysis is performed with Xcalibur.

6.4 Caveats and Considerations

The MBB method for measuring bioavailable sulfide is a powerful and sensitive quantitative method in biological samples. These procedures here have been highly optimized to ensure bioavailable sulfide measurement is accurate and reliable [16, 17].

The sulfide solution may be readily oxidized by oxygen and metal ions; hence, we preferentially perform the derivatization reaction between sulfide and MBB under hypoxic conditions with a metal chelator [17, 38]. Under 1% O_2 and 100 mM DTPA, we have found the yield of sulfide-dibimane was increased from 50% to nearly 100%. Furthermore, solid Na_2S should be white, but it becomes yellow due to its oxidation. For making sulfide solution, anhydrous Na_2S from Alfa Aesar (Cat. No. 65122) and degassed solvents are recommended. The sulfide solution, once made, should not be degassed or bubbled – this will cause sulfide loss via H_2S volatilization [11, 12].

The pH of the reaction solution is also important for the bioavailable sulfide measurement. First, H_2S is less volatile in alkaline pH than in neutral pH [12, 39], and MBB can effectively react with sulfide at alkaline pH. The yield of SDB in the pH 9.5 of solution is almost 100%, but only 80% at pH 7.0 [17]. Hence, the reaction buffer is chosen as 100 mM Tris-HCl (pH 9.5, 0.1 mM DTPA). Second, acidic solution can halt the S-alkylation reaction. Importantly, SDB is stable for up to two weeks at room temperature and pH 4.5 [16, 17]. This is why 200 mM SSA solution is used as stop solution. Third, TCEP is a strong reductant without sulfur atoms and can reduce disulfide bridges at a wide range of pH (1.5–8.0). In the phosphate buffer (pH 2.6), sulfide gas can be released from the free H_2S and iron–sulfur clusters, and TCEP can efficiently reduce disulfide bonds more rapidly [40]. All of this volatilized sulfide can be trapped with reaction buffer, following MBB derivatization [16].

In conclusion, the MBB method of measuring bioavailable sulfide by RP-HPLC with fluorescence detector is indeed useful and sensitive quantitation method. The degree of sensitivity of the HPLC method meets the requirements of bioavailable sulfide measurement in most biological samples. However, several important facts must be kept in mind, including

(1) MBB is a light-sensitive reagent. Photolysis of MBB results in the loss of the bromine, converting the nonfluorescent MBB into the fluorescent bimane [30, 33]. Hence, H_2S/MBB derivatization should ideally be performed in the dark or with red LED lighting [39].
(2) Binding of H_2S to experimental vessels is minimized by using polypropelene reaction vessels [18].
(3) SDB is stable at 4 °C under acidic solution, but is sensitive to light.
(4) For blood sample, lithium-heparin coated tubes are preferred because ethylene diamine tetraacetic acid (EDTA)-coated tubes cause higher levels of hemolysis [41].

Acknowledgment

This work was supported by an Institutional Development Award (IDeA) from the National Institutes of General Medical Sciences of the National Institutes of Health under grant number P20GM121307 and HL149264 to C.G.K.

Disclosures

C.G.K and X.S. have intellectual property regarding hydrogen sulfide and metabolite measurement chemistry.

References

1 Szabo, C. (2018). A timeline of hydrogen sulfide (H_2S) research: from environmental toxin to biological mediator. *Biochem. Pharmacol.* 149: 5–19.

2 Warenycia, M.W., Goodwin, L.R., Benishin, C.G. et al. (1989). Acute hydrogen sulfide poisoning: demonstration of selective uptake of sulfide by the brainstem by measurement of brain sulfide levels. *Biochem. Pharmacol.* 38 (6): 973–981.

3 Goodwin, L.R., Francom, D., Dieken, F.P. et al. (1989). Determination of sulfide in brain tissue by gas dialysis/ion chromatography: postmortem studies and two case reports. *J. Anal. Toxicol.* 13 (2): 105–109.

4 Savage, J.C. and Gould, D.H. (1990). Determination of sulfides in brain tissue and rumen fluid by ion-interaction reversed-phase high-performance liquid chromatography. *J. Chromatogr. B Biomed. Appl.* 526: 540–545.

5 Stipanuk, M.H. and Beck, P.W. (1982). Characterization of the enzymic capacity for cysteine desulphhydration in liver and kidney of the rat. *Biochem. J* 206 (2): 267–277.

6 Abe, K. and Kimura, H. (1996). The possible role of hydrogen sulfide as an endogenous neuromodulator. *J. Neurosci.* 16 (3): 1066–1071.

7 Hosoki, R., Matsuki, N., and Kimura, H. (1997). The possible role of hydrogen sulfide as an endogenous smooth muscle relaxant in synergy with nitric oxide. *Biochem. Biophys. Res. Commun.* 237 (3): 527–531.

8 Kolluru, G.K., Shen, X., Bir, S.C., and Kevil, C.G. (2013). Hydrogen sulfide chemical biology: pathophysiological roles and detection. *Nitric Oxide* 35: 5–20.

9 Wang, R. (2011). Signaling pathways for the vascular effects of hydrogen sulfide. *Curr. Opin. Nephrol. Hypertens.* 20 (2): 107–112.

10 Kolluru, G.K., Shen, X., and Kevil, C.G. (2013). A tale of two gases: NO and H_2S, foes or friends for life? *Redox Biol.* 1 (1): 313–318.

11 Hughes, M.N., Centelles, M.N., and Moore, K.P. (2009). Making and working with hydrogen sulfide. The chemistry and generation of hydrogen sulfide in vitro and its measurement in vivo: a review. *Free Radical Biol. Med.* 47 (10): 1346–1353.

12 Nagy, P., Pálinkás, Z., Nagy, A. et al. (2014). Chemical aspects of hydrogen sulfide measurements in physiological samples. *Biochim. Biophys. Acta, Gen. Subj.* 1840 (2): 876–891.

13 May, P.M., Batka, D., Hefter, G. et al. (2018). Goodbye to S. *Chem. Commun.* 54 (16): 1980–1983.

14 Ishigami, M., Hiraki, K., Umemura, K. et al. (2009). A source of hydrogen sulfide and a mechanism of its release in the brain. *Antioxid. Redox Signaling* 11 (2): 205–214.

15 Ubuka, T. (2002). Assay methods and biological roles of labile sulfur in animal tissues. *J. Chromatogr. B: Anal. Technol. Biomed. Life Sci.* 781 (1-2): 227–249.

16 Shen, X., Peter, E.A., Bir, S. et al. (2012). Analytical measurement of discrete hydrogen sulfide pools in biological specimens. *Free Radical Biol. Med.* 52 (11-12): 2276–2283.

17 Shen, X., Pattillo, C.B., Pardue, S. et al. (2011). Measurement of plasma hydrogen sulfide in vivo and in vitro. *Free Radical Biol. Med.* 50 (9): 1021–1031.

18 Levitt, M.D., Abdel-Rehim, M.S., and Furne, J. (2011). Free and acid-labile hydrogen sulfide concentrations in mouse tissues: anomalously high free hydrogen sulfide in aortic tissue. *Antioxid. Redox Signaling* 15 (2): 373–378.

19 Nagy, P., Palinkas, Z., Nagy, A. et al. (2014). Chemical aspects of hydrogen sulfide measurements in physiological samples. *Biochim. Biophys. Acta* 1840 (2): 876–891.

20 Zheng, Y., Liao, F., Du, J.B. et al. (2012). Modified methylene blue method for measurement of hydrogen sulfide level in plasma. *Acta Physiol. Sinica* 64 (6): 681–686.

21 Fischer, E. (1883). Bildung von Methylenblau als Reaction auf Schwefelwasserstoff. *Ber. Dtsch. Chem. Ges.* 16 (2): 2234–2236.

22 Mecklenburg, R. (1914). A new method for the colorimetric detection of traces of hydrogen sulphide. *Ztschr. Anorg. Chem.* 86: 143–153.

23 Fogo, J.K. and Popowsky, M. (1949). Spectrophotometric determination of hydrogen sulfide. *Anal. Chem.* 21 (6): 732–734.

24 Cline, J.D. (1969). Spectrophotometric determination of hydrogen sulfide in natural waters. *Limnol. Oceanogr.* 14 (3): 454–458.

25 Wintner, E.A., Deckwerth, T.L., Langston, W. et al. (2010). A monobromobimane-based assay to measure the pharmacokinetic profile of reactive sulphide species in blood. *Br. J. Pharmacol.* 160 (4): 941–957.

26 Montoya, L.A., Shen, X., McDermott, J.J. et al. (2015). Mechanistic investigations reveal that dibromobimane extrudes sulfur from biological sulfhydryl sources other than hydrogen sulfidedagger. *Chem. Sci.* 6 (1): 294–300.

27 Shen, X., Chakraborty, S., Dugas, T.R., and Kevil, C.G. (2014). Hydrogen sulfide measurement using sulfide dibimane: critical evaluation with electrospray ion trap mass spectrometry. *Nitric Oxide* 41: 97–104.

28 Shen, X., Kolluru, G.K., Yuan, S., and Kevil, C.G. (2015). Measurement of H2S in vivo and in vitro by the monobromobimane method. *Methods Enzymol.* 554: 31–45.

29 Kosower, E.M., Pazhenchevsky, B., and Hershkowitz, E. (1978). 1.5-Diazabicyclo[3.3.0]octadienediones (9,10-dioxabimanes). strongly fluorescent syn isomers. *J. Am. Chem. Soc.* 100 (20): 6516–6518.

30 Kosower, N.S., Kosower, E.M., Newton, G.L., and Ranney, H.M. (1979). Bimane fluorescent labels: labeling of normal human red cells under physiological conditions. *Proc. Natl. Acad. Sci. U.S.A.* 76 (7): 3382–3386.

31 Fahey, R.C., Newton, G.L., Dorian, R., and Kosower, E.M. (1981). Analysis of biological thiols: quantitative determination of thiols at the picomole level based upon derivatization with monobromobimanes and separation by cation-exchange chromatography. *Anal. Biochem.* 111 (2): 357–365.

32 Newton, G.L., Dorian, R., and Fahey, R.C. (1981). Analysis of biological thiols: derivatization with monobromobimane and separation by reverse-phase high-performance liquid chromatography. *Anal. Biochem.* 114 (2): 383–387.

33 Kosower, N.S. and Kosower, E.M. (1987). Thiol labeling with bromobimanes. *Methods Enzymol.* 143 (C): 76–84.

34 Togawa, T., Ogawa, M., Nawata, M. et al. (1992). High performance liquid chromatographic determination of bound sulfide and sulfite and thiosulfate at their low levels in human serum by pre-column fluorescence derivatization with monobromobimane. *Chem. Pharm. Bull.* 40 (11): 3000–3004.

35 Baniya, A., Thapa, S., Borquist, E. et al. (2019). Lab-on-a-chip for hydrogen sulphide detection—Part I: Sulphide separation from plasma sample. *Med. Devices Sens.* 2 (3-4): 1–7.

36 Kevil, C.G., Glawe, J.D., Frilot, C.F., Weiss, L. (2019). Hydrogen sulfide detecting apparatus. US 10,254,262 B2.

37 Tan, B., Jin, S., Sun, J. et al. (2017). New method for quantification of gasotransmitter hydrogen sulfide in biological matrices by LC-MS/MS. *Sci. Rep.* 7 (1): 46278.

38 Fahey, R.C. and Newton, G.L. (1987). Determination of low-molecular-weight thiols using monobromobimane fluorescent labeling and high-performance liquid chromatography. *Methods Enzymol.* 143 (C): 85–96.

39 Ditrói, T., Nagy, A., Martinelli, D. et al. (2019). Comprehensive analysis of how experimental parameters affect H_2S measurements by the monobromobimane method. *Free Radical Biol. Med.* 136: 146–158.

40 Han, J.C. and Han, G.Y. (1994). A procedure for quantitative determination of tris(2-carboxyethyl)phosphine, an odorless reducing agent more stable and effective than dithiothreitol. *Anal. Biochem.* 220 (1): 5–10.

41 Unger, J., Filippi, G., and Patsch, W. (2007). Measurements of free hemoglobin and hemolysis index: EDTA- or lithium-heparinate plasma? *Clin. Chem.* 53 (9): 1717–1718.

7

Fluorescent Probes for H_2S Detection: Cyclization-Based Approaches

Yingying Wang, Yannie Lam, Caitlin McCartney, Brock Brummett, Geat Ramush, and Ming Xian

Department of Chemistry, Brown University, Providence, RI, USA

List of Abbreviations

CBS	Cystathionine β-synthase
CSE	Cystathionine γ-lyase
CTAB	Hexadecyltrimethylammonium Bromide
Cys	Cysteine
DMSO	Dimethyl Sulfoxide
ER	Endoplasmic Reticulum
GSH	Glutathione
H2S	Hydrogen Sulfide
Hcy	Homocysteine
HUVEC	Human Umbilical Vein Cells
LPS	Lipopolysaccharide
Na2S	Sodium Sulfide
NADP	Nicotinamide Adenine Dinucleotide Phosphate
NaHS	Sodium Hydrogen Sulfide
NIR	Near Infrared
NO	Nitric Oxide
PBS	Phosphate Buffered Saline
PMA	Phorbol-12-myristate-13-acetate

7.1 Introduction

The physiological and biomedical importance of H_2S has been recognized in the cardiovascular system and other organ systems throughout the body [1–8].

In blood vessels, cystathionine γ-lyase (CSE) is the major H_2S producing enzyme expressed in both smooth muscle and endothelium as well as in periadventitial adipose tissues [9–16]. Regulation of H_2S production from CSE is controlled by a complex integration of transcriptional, posttranscriptional, and posttranslational mechanisms in blood vessels. The production of endogenous H_2S and the exogenous administration of H_2S have been demonstrated to exert protective effects in a number of pathologies [1–16]. For example, H_2S regulates diseases such as hypertension, atherosclerosis, coronary heart disease, and hypoxia-induced pulmonary hypertension. H_2S relaxes vascular smooth muscle to promote vasodilation and reduces blood pressure. Additionally, H_2S acts as a potent antioxidant and, under chronic conditions, upregulates antioxidant defenses. While these results strongly suggest that modulation of H_2S levels could have potential therapeutic values for cardiovascular diseases, the underlying mechanisms of H_2S are still under investigation and H_2S-based therapeutic agents are to still be demonstrated.

To clarify the links between H_2S and cytoprotection, as well as to develop H_2S-based therapies, it is necessary to determine the detailed H_2S profiles under disease states; therefore, the detection of H_2S in biological samples is critical. However, this is still a challenge in the field. As a gaseous molecule, H_2S is quickly degraded in aqueous and biological samples due to volatilization and high reactivity. For example, the $t_{1/2}$ of H_2S loss is about 5 minutes in open solutions because of volatilization [17]; therefore, although the initial H_2S concentrations applied to the test subjects (by using H_2S gas or $NaHS/Na_2S$ salts) are often provided (normally in the range of μM to mM), the actual H_2S concentrations inside cells or tissues which exhibit the optimal protective action in disease models are still unclear. It is expected that the active H_2S concentrations needed for protective effects are much lower and in a tight window (likely in the range of nM to low μM, given the fact that H_2S is toxic at higher concentrations). However, these values have not been determined. This lack of knowledge presents a major burden in the development of H_2S therapy, especially for the use of synthetic H_2S releasing compounds (also known as H_2S donors). To solve this problem, methods that allow accurate and "real-time" detection of H_2S in biological samples, such as live cells and tissues, are needed.

Traditional methods for H_2S detection are known, which include gas chromatography, colorimetric/electrochemical assays, and sulfide precipitation [18–20]. Normally, complicated postmortem processing and/or destruction of samples are needed for these methods. Due to the high reactivity of H_2S under biological environments, these methods can lead to inconsistent results that span 5 orders of magnitude [21–24]. Fluorescence-based assays, however, are convenient to run and allow for accurate and nondestructive measurement in biological samples. In 2011, our group and several others reported the first reaction-based fluorescent probes for H_2S detection [25–29]. These studies inspired many researchers to develop H_2S fluorescent probes in the past several years [30–37]. So far, three types of reactions have been used in the design of H_2S fluorescent probes: (i) H_2S-mediated reductions, often using azide ($-N_3$) containing fluorophores; (ii) H_2S-mediated metal-sulfide formations; and (iii) H_2S-mediated nucleophilic reactions. Each reaction has its pros and cons. In this chapter, we will focus on H_2S nucleophilic reaction-based probes, mainly cyclization approaches. We will discuss the design principles, reaction mechanisms, the properties of representative probes, and their corresponding applications.

7.2 General Design of Nucleophilic Reaction-Cyclization Based Fluorescent Probes

The key to design H_2S fluorescent probes is to find reactions that can differentiate H_2S from other biomolecules, especially thiols. From a reactivity point-of-view, H_2S is a nonsubstituted thiol that can undergo nucleophilic reaction two times via an intramolecular cyclization process, while other thiols like cysteine and GSH are mono-substituted thiols which can only undergo nucleophilic reaction once (with their –SH residue). This property would allow compounds containing dual-electrophilic centers to be useful reagents for H_2S detection (Scheme 7.1a). This approach was used in the design of the first nucleophilic reaction-based fluorescent probes for H_2S (Scheme 7.1b). The idea was to use one reactive electrophile to trap H_2S, forming a –SH containing intermediate. This –SH residue would then undergo an intramolecular reaction with a second electrophilic side (the ester group) to release a –OH containing fluorophore. It is known that many –OH containing fluorophores can be quenched upon –OH acylation; therefore, the probes should be nonfluorescent while the reaction with H_2S would restore their strong fluorescence. The advantage of this strategy is that one can easily modify the structures of all three subunits of the probe template (electrophile, linker, and fluorophore). This approach has the potential to produce excellent fluorescent probes for H_2S.

7.2.1 WSP Probes

WSP series are the first nucleophilic reaction-based probes reported by our lab in 2011 [26, 38]. The fluorescence turn-on mechanism is shown in Scheme 7.2 by using WSP1 as the example [26]. WSP1 possesses low fluorescence due to –OH acylation on fluorescein, which forces the fluorophore to exist in a lactone (ring-closed) form. As a potent nucleophile, H_2S would react with the most electrophilic component of the probe, e.g. 2-pyridiyl disulfide, to form a –SH containing intermediate (persulfide). Then, this persulfide residue would further react

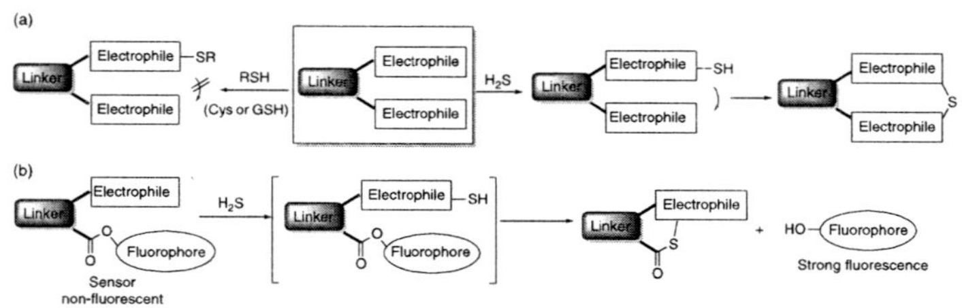

Scheme 7.1 (a) Proposed reactivity difference between H_2S and thiols. (b) General design approach for cyclization-based fluorescent probes that respond to H_2S.

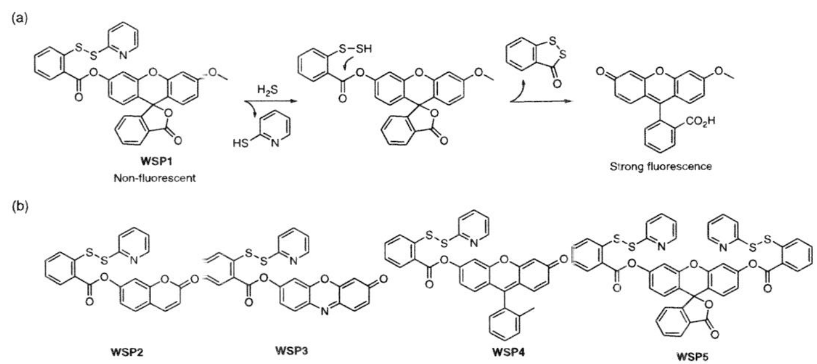

Scheme 7.2 (a) Fluorescence turn-on mechanism of WSP1 as the example. (b) Related probes WSP2–WSP5.

with the nearby ester group to release the free –OH of the fluorescein and restore the strong fluorescence. Several probes (WSP2–5) adopting the same strategy were reported [38]. In general, these probes showed excellent selectivity for H_2S versus biothiols (such as GSH and Cys). They also showed decent sensitivity for H_2S with nanomolar detection limits (47∼266 nM). It should be noted that WSP probes are not without limitations. Since the 2-pyridiyl disulfide site can also react with biothiols (although not turning on fluorescence), high loading of the probes might be needed when WSP are used in H_2S detection with the presence of biothiols. In addition, due to solubility problems in aqueous solutions, a cationic surfactant hexadecyltrimethylammonium bromide (CTAB) is often used at micromolar levels as a solubilizing agent. CTAB facilitates the reaction with sulfide anions by increasing the probe's solubility. However, it should be cautioned as it can cause cell membranes damage when CTAB is used under relatively high concentrations or long incubation times [39, 40]. WSP1 and WSP5 are currently commercially available, for example via Cayman Chemicals (www.caymanchem.com/product/11179/wsp-1; www.caymanchem.com/product/16929/wsp-5). They have been used by many researchers in H_2S studies. For instance, Sen and coworkers reported that H_2S could mediate the IL-1β-induced memory impairment in brain by WSP1 detection [41]. Smith and coworkers used WSP5 to evaluate H_2S levels in retinas of mice [42]. Corpas and coworkers found that H_2S acted as an inhibitor for NADP-dependent isocitrate dehydrogenase by WSP5 analysis [43]. WSPs is also used in evaluating novel H_2S donors. For instance, Martelli and coworkers revealed H_2S releasing ability of erucin by WSP1 [44]. Urara et al. applied WSP1 in human umbilical vein cells (HUVEC) lysate to evaluate H_2S release from amphiphilic block copolymer micelles [45].

Following the report of WSPs, several other probes were also developed using –OH containing fluorophores and WSP's H_2S reaction site (Scheme 7.3). For example, imidazo[1,5-*a*]pyridine and dicyanoisopentanone derivatives were used for the development of fluorescence turn-on probes (NIPY-PBA and EW-H) [46, 47]. NIPY-PBA showed good water solubility which led to a 43 nM detection limit and 1.5 minutes response time. NIPY-PBA was used to monitor concentration changes of exogenous H_2S in live A549 cells. To improve the probes' photophysical properties, a series of red-emitting and near infrared (NIR) fluorophores, such as ketone-Cy [48], dicyanomethylene-4*H*-pyran (DCM) [49], seminaphtharhodafluor (SNARF) [50], and Washington Red 6 (WR6) [51] were employed in the WSP template (Scheme 7.3). The emission wavelengths of these probes ranged from 635 to 680 nm, which enabled deeper tissues penetration and minimized photo damages. For example, in PBS buffers containing 10% DMF, probe **1** emitted red fluorescence at 630 nm when treated with different concentrations (1–40 μM) of H_2S [48]. Its detection limit was calculated to be 36 nm. Increased red fluorescence signals were observed with the addition of various Na_2S (50 nM–50 μM)

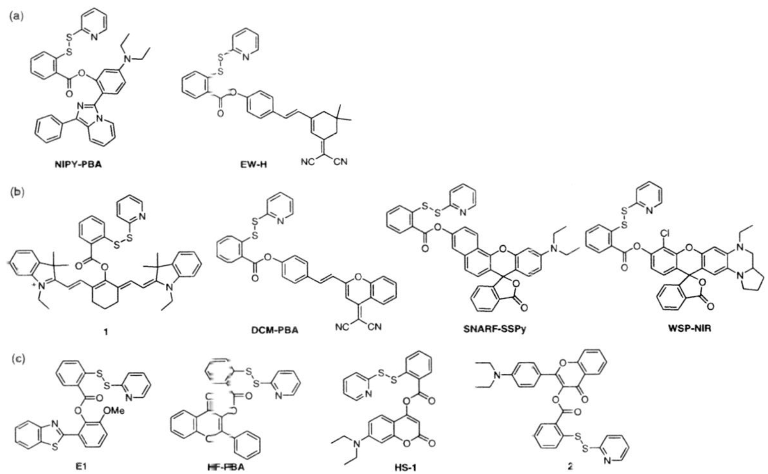

Scheme 7.3 (a) Structures of NIPY-PBA and EW-H. (b) Structures of selected NIR probes. (c) Structures of selected ratiometric probes.

and 100 μM NO donor sodium nitroprusside (an endogenous H_2S activator) to probe **1** (5 μM) pretreated HeLa cells. This indicated the usefulness of probe **1** in detecting exogenous and endogenous H_2S. Probe **1** was further applied in imaging H_2S in a mouse model. After injecting probe **1** (1 mM) and Na_2S solution (1 mM) into the right ankle of mice, a time-dependent fluorescence enhancement was observed around the injected area. While this looks promising, the concentrations of the probe and H_2S used were too high to be biologically friendly. In addition to fluorescence turn-on probes, ratiometric fluorophores, such as 2-(20-hydroxyphenyl)benzothiazole (HMBT) [52], 3-hydroxyflavone (3-HF) [53], 7-diethylamino-4-hydroxy-coumarin [54], and 4′-diethylamino-3-hydroxyflavone [55], have also been used in conjunction with WSP template (Scheme 7.3c). With a specific excitation light, ratiometric fluorophores emit two different emission wavelengths. The ratio of these two emissions depends on the concentration of analytes, regardless of variations in fluorophore's concentration. This property enabled ratiometric probes to achieve high resolution in H_2S detection in complex systems. Using **E1** as an example, under 295 nm excitation, the original emission bands were at 356 and 487 nm. After the treatment of Na_2S, the emission bands were increased at 487 nm while these decreased at 356 nm. This significant shift of emission peaks minimized the band overlap and allowed **E1** to detect exogenous H_2S level changes (100 μM) in HeLa cells.

7.2.2 2,2′-Dithiosalicylic Ester-Based Probes

2,2′-Dithiosalicylic esters are another disulfide-based template that were used in the design of H_2S probes. Similar to the WSP template, it should react with H_2S to break the disulfide bond and form the persulfide intermediate, which then should undergo cyclization and release the fluorophore (Scheme 7.4). Three such probes (YQ1, BHNP-DA, DCN-S) are shown in Scheme 7.4. YQ1 showed good sensitivity to H_2S in PBS buffer containing 30% acetonitrile [56]. In the presence of H_2S (100 μM), it exhibited a 14-fold fluorescence enhancement of the signals, which was significantly higher than signals from Cys and GSH at millimolar levels. Its response to H_2S was fast (10 minutes), and the detection limit was 52 nM. BHNP-DA was reported by Ye and coworkers [57]. This probe could be completely turned on by H_2S within 1 minute with the assistance of CTAB (1 mM). In addition to traditional one-photon microscopy, BHNP-DA was also applied in two-photon microscopy to improve the accuracy of intracellular H_2S detection. Furthermore, BHNP-DA contained the lysosome-targeted moiety morpholine, which allowed it to detect subcellular H_2S in live 4T1 cells. DCN-S was reported by Yin and coworkers [58]. It gave a 5.7-fold fluorescence increase in response to H_2S in 80% DMSO/PBS buffer. Apparently, these 2,2′-dithiosalicylic ester-based probes share the common problem – their poor water solubility.

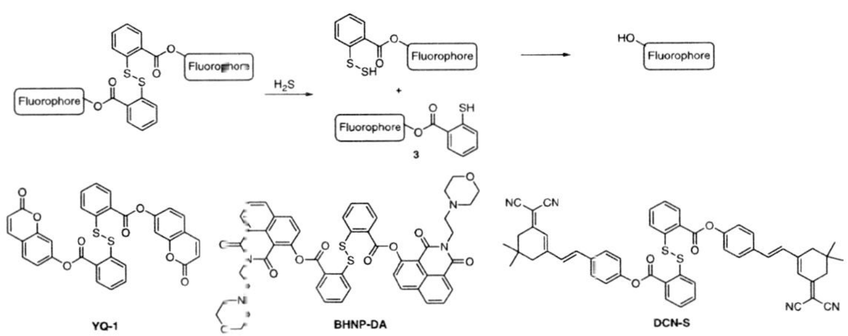

Scheme 7.4 2,2′-Dithiosalicylic ester-based probes for H_2S detection.

As such, high concentrations of surfactant or organic cosolvents are often needed in their tests. Another problem is that 2,2′-dithiosalicylic esters can also react with thiols; therefore, the probe could be consumed by cellular thiols. For example, comparing to a H_2S only system, the fluorescence enhancement of DCN-S toward 1 : 1 cysteine/H_2S was decreased by 14%. Finally, as shown in Scheme 7.4, these probes are dimerized compounds with two fluorophores. Upon reacting with H_2S, it produces a fluorophore-containing by-product (**3** in Scheme 7.4). It is unclear what would happen to this product in the detection process.

7.2.3 Alkyl Halide-Based Probes

Alkyl halides are known to be thiol-labeling agents, and they are also H_2S-reactive functional groups. They can be used to trap H_2S and promote the subsequent cyclization to release ester-masked fluorophores (Scheme 7.5). Moreover, certain alkyl halides may be able to differentiate H_2S from other thiols because H_2S is a stronger nucleophile than biothiols (Cys/Hcy/GSH) under physiological conditions. The pK_a of H_2S (7.0) is significantly lower than those of Cys/Hcy/GSH (8.5). Guo and coworkers explored this strategy and reported a probe **4** [59]. In PBS buffers with 20 mM CTAB, the turn-on fold of probe **4** (5 μM) toward H_2S (100 μM) was about 50. When excess biothiols (500 μM cysteine, 500 μM homocysteine, or 1 mM glutathione) were added with H_2S (100 μM), the same fluorescence enhancement was maintained by probe **4**, which indicated that the interference of cellular thiols was negligible. Probe **5** and probe **6** used 2-bromoethyl-linked carbonate ester and benzyl bromide as H_2S-trapping residues, respectively [60, 61]. Probe **5** (1 μM) gave a 32-fold fluorescence increase to H_2S (80 μM) in 10 minutes. No noticeable change in fluorescence was observed with or without thiol containing amino acids, such as Cys (500 μM) and GSH (500 μM). Probe **5** was applied in a time-dependent detection for exogenous H_2S (20 μM) and

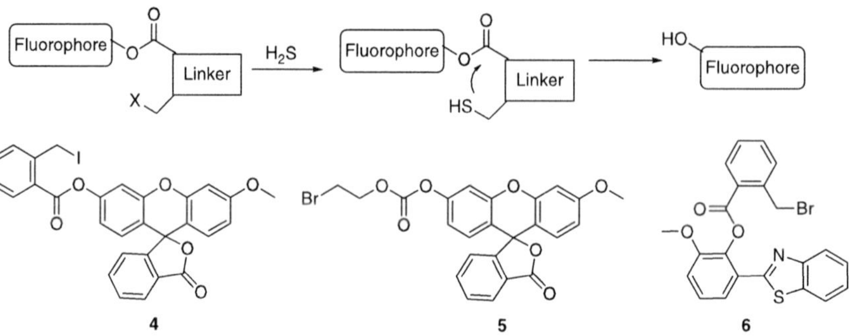

Scheme 7.5 Alkyl halide-based probes for H_2S detection.

lipopolysaccharide (LPS) stimulated H_2S production in normal (BV-2 microglial) and cancer (C6 glioma) cells. In addition, Kumar et al. dissected and incubated rat brain tissues with probe **5** (5 μM) at 37 °C for 1 hour. After treating these hippocampal slices (5 μm) with 20 μM NaHS or 10 mM L-Cys (a CBS stimulator) increased fluorescence intensity was observed in the tissue images. Meanwhile, *in vivo* studies were also carried out with probe **5**. Compared to probe **5** pretreated nematodes alone, a stronger fluorescence was observed with NaHS (300 μM) incubation for 6 hours. These results suggested that probe **5** is a useful tool for visualizing H_2S in living systems.

7.2.4 Diselenide-Based Probes

To solve a major problem of WSP probes, i.e. probe consumption by biothiols, a diselenide-based template was designed (Scheme 7.6). These probes (named as SeP) were expected to react rapidly with H_2S to form a thio-benzeneselenol derivative **7** and a benzeneselenol derivative **8** [62, 63]. The intermediate **7** should undergo a fast and spontaneous cyclization to release the fluorophore. On the other hand, benzeneselenols are known to be unstable and can be easily oxidized in the air to re-form the probe. It is also possible that SeP could react with biothiols (RSH). Such reactions should form intermediates **8** and **9**. Again, **8** should reform the probe and **9** should possess high reactivity toward H_2S to produce **7** and then to release the fluorophore. Overall, the presence of biothiols should not interfere with the response of SeP toward H_2S. As shown in Scheme 7.6, a series of SeP probes were prepared and evaluated. In a 50% acetonitrile/PBS buffer system, SeP1 and SeP2 completed the reaction with H_2S within 2 minutes and showed a 35 and 11 fold fluorescence turn on, respectively [62]. More importantly, the presence of high concentrations of biothiols (Cys or GSH at 1 mM) did not lead to significant fluorescence decrease in the probes' response to H_2S (50 μM). This confirmed that diselenide-based probes solved the probe consumption problem of WSPs. SeP2 was able to sense endogenous H_2S concentration changes caused by CBS activator (*S*-adenosylmethyonine) or inhibitor (aminooxoyacetic acid) using a plate-reader. HBTSeSe was used to detect intracellular H_2S level changes induced by phorbol-12-myristate-13-acetate (PMA) in RAW264.7 macrophages [64]. While promising, these diselenide-based probes still suffer from water solubility problems, presumably due to their dimeric structures. High concentrations of organic solvents (50% of acetonitrile or DMSO) were required in the evaluation of these probes.

7.2.5 Selenenyl Sulfide-Based Probes

To further improve the property of diselenide-based probes, a selenenyl-sulfide (Se-S) template was reported in 2019 [65]. This design took the advantage of the

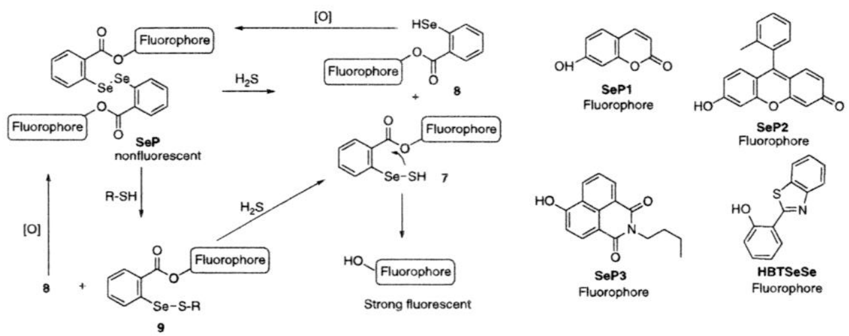

Scheme 7.6 Diselenide-based probes for H_2S detection.

reactive intermediate (**9** in Scheme 7.6) formed between diselenides and thiols. It also simplified the probe's structure as only one fluorophore was employed, which avoided the problem caused by the dimeric structures. The mechanism of these probes, named as SeSP, is shown in Scheme 7.7. The selenenyl sulfide linkage of SeSP should be highly reactive to both H_2S and thiols. With H_2S, it should be converted to a –Se–SH motif and then trigger the intramolecular cyclization to release the fluorophore. With thiols (RSH), the original –Se–S– linkage should be converted into another –Se–S– linkage in an equilibrium process as previous studies revealed that nucleophilic attack by RSH at the Se atom is thermodynamically favorable and kinetically faster [66]. This reaction should not generate a turn on fluorescence response, nor would it consume the probe. This equilibrium will be broken when H_2S is present and thus fluorescence will be turned on. A series of SeSP probes (SeSP1-9) were prepared and tested. The results demonstrated that these probes were highly selective to H_2S, and their responses to H_2S were not significantly affected by Cys or GSH at biologically relevant concentrations. SeSP's solubility appeared to be much improved compared to diselenide probes. SeSP performed well in PBS buffers with 1% DMSO. For example, SeSP1 gave a 1800-fold fluorescent increase to H_2S in 10 minutes. It was also found that *S*-aryl probes worked better than *S*-alkyl probes. Lukesh and coworkers applied this strategy to the development of AL1 and AL2 [67]. Due to the implementation of *gem*-dimethyl groups, ester hydrolysis of AL1/AL2 was prevented. AL1 showed excellent stability and selectivity in PBS buffers. With 50 μM H_2S AL1 (20 μM) exhibited more than 100-fold fluorescence increase in 10 minutes while no fluorescence enhancement was found with thiols such as Cys (100 μM) and GSH (1 mM). Furthermore, the AL template was used to design cancer cell-specific probes. It is known that biotin transporters exist on the surface of some cancer cells. A biotin moiety was conjugated to the AL template to prepare probe AL2. The uptake of AL2 in HeLa cells (biotin transporter positive) was accelerated. On the other hand, the pretreatment of AL2 for the biotin transporter negative noncancerous HEK293 cells was less efficient. In this way, AL2 was used to specifically target cancerous H_2S detection.

7.2.6 Aldehyde Addition-Based Probes

Aldehydes are H_2S-reactive functional groups. Tandem aldehyde addition-intramolecular Michael reaction has been used as a strategy in the design of H_2S probes. In 2011, He and coworkers reported two such probes (SFP-1 and SFP-2) based on benzaldehyde/acrylate containing structures [28]. The proposed mechanism of SFP is shown in Scheme 7.8. The aldehyde group should first react with H_2S to form a hemi-thioacetal intermediate **11**. Then the adjacent α,β-unsaturated acrylate should trap the exposed thiol via an intramolecular Michael addition. The resulting product **12** should result in a turn-on fluorescence due to the

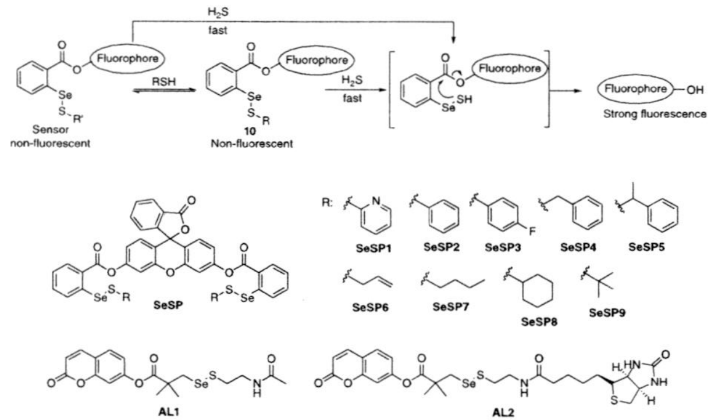

Scheme 7.7 Selenenyl sulfide-based probes for H_2S detection.

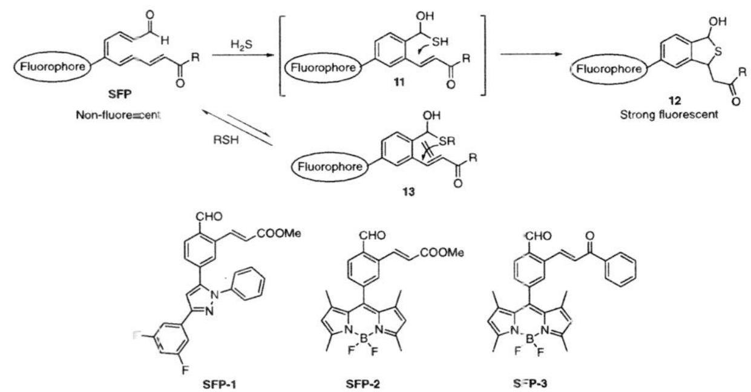

Scheme 7.8 Aldehyde addition-based probes for H_2S using a tandem aldehyde addition-intramolecular Michael reaction.

disappearance of the photoinduced electron transfer effect. The –SH group of thiols could also react with the aldehyde group in a reversible process and free cysteine can react irreversibly with aldehyde through amine-thiol-dual addition. However, such reactions should not trigger the following Michael addition to turn-on fluorescence. As expected, SFP-1 and SFP-2 showed good selectivity toward H_2S in PBS buffers. SFP-2 was applied in visualizing both exogenous and endogenous generated H_2S in HeLa cells. However, the response rate of SFP-2 to H_2S was somewhat slow, and it needed 2 hours to reach a plateau in fluorescence intensity. To improve the response rate, Zhao and coworkers replaced the acrylate ester with a more reactive α,β-unsaturated phenyl ketone and reported SFP-3 [68]. SFP-3 could be turned on by H_2S within 20 minutes. However, the difference in evaluating systems for SFP-3 versus SFP-2 should not be ignored. In SFP-3 studies, 40% acetonitrile was employed, while 1% DMSO was used in SFP-2 evaluations. Thus, the increased reactivity of SFP-3 may also be due to the use of organic solvents.

Ahn et al. applied a similar design and reported a series of two-photon H_2S probes (P1, P3, and P5) [69, 70]. Interestingly, product analysis showed that the probes reacted with H_2S through a different mechanism. Instead of the expected –SH Michael addition product **14**, compound **17** was isolated as the major product (Scheme 7.9). It was suggested that the reaction was started by H_2S addition to aldehyde to form the intermediate **15**. Then the –SH moiety was somehow transferred to enone to form the Michael addition intermediate **16**. Finally, an intramolecular Aldol addition produced the fluorescent product **17**. As one can imagine, **16** could also be generated via the direct Michael addition of H_2S on the enone group. However, a model reaction using compound **18** showed that no product was formed with H_2S; therefore, the authors suggested the proposed mechanism. Probe P1 was found to have high sensitivity for H_2S, but it was significantly interfered by biothiols. To address this problem, two electron-donating groups (methoxy groups) were introduced (P3). The increased steric hindrance and electronic positivity on enone β-carbon prevented interference from cellular thiols and granted P3 "near-complete" selectivity to H_2S. A similar probe, P5, was also developed [70]. With the extended π system from the pendant pyrrolidine ring, P5 showed a significant longer emission wavelength (625 nm) than P3 (501 nm). All three probes were capable of detecting H_2S in HeLa cells.

Another aldehyde addition-based H_2S recognition site, *O*-carboxybenzaldehyde, was employed by Zhang and coworkers [71]. As shown in Scheme 7.10, H_2S would react with the aldehyde group of the probe (M–H_2S) to form the thiobenzyl

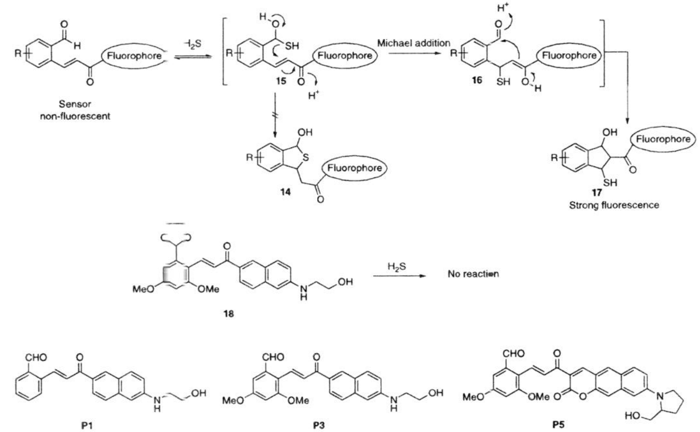

Scheme 7.9 Aldehyde addition-based probes proposed to function through –SH migration to an enone intermediate.

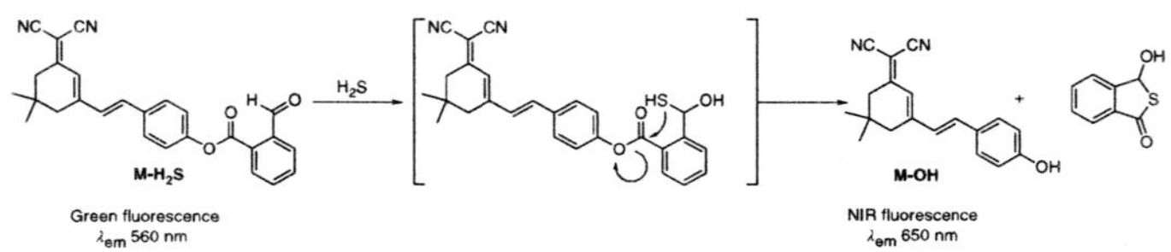

Scheme 7.10 *O*-Carboxybenzaldehyde-based probes for H_2S detection.

alcohol intermediate, which then underwent a nucleophilic cyclization to release the fluorophore M–OH and 3-hydrobenzothiophene ketone. M–H_2S was a near-infrared and ratiometric probe. Owing to the intermolecular charge transfer effect, M–H_2S exhibited a fluorescence emission at 560 nm by itself. After reacting with H_2S, the near-infrared emission of M–OH was recovered at 650 nm. In 10% DMSO/PBS buffer solutions, M–H_2S showed decent sensitivity and selectivity for H_2S. The detection limit was calculated to be 39.1 nM. More importantly, M–H_2S was found to specifically target the endoplasmic reticulum (ER). Based on this property, Zhang et al. stimulated ER stress by tunicamycin (20 μg/mL) and found that ER stress enhanced H_2S levels in cells.

7.2.7 Michael Addition-Cyclization Based Probes

Michael addition is another type of reaction that could occur with thiols/H_2S under physiological conditions and has also been used in the design of H_2S probes. Holmes and coworkers discovered that thiols reacted readily with certain Michael acceptors at physiological pH, but the products could not be obtained [72]. This indicated that Michael addition with these substrates was a rapid-equilibrium process and no stable covalent product was formed. Based on this property, Michael addition-based probes may be able to differentiate H_2S from biothiols. As shown in Scheme 7.11, probe **19** should react with H_2S to form the intermediate **20**, which then undergoes an intramolecular cyclization to release the fluorophore. The probe could also react with thiols, but the reaction should be reversible and the presence of H_2S could compete with thiols and lead to the generation of fluorophore. Based on this idea, probes **22** and **23** were prepared and tested [73]. They showed good selectivity to H_2S. They also gave comparable fluorescence responses to H_2S only solution (100 μM) and H_2S/thiol solution (100 μM/1 mM). 4-Hydroxy-1,8-naphthalimide was also used in this design in probe **24** [74].

Samanta et al. adopted the Michael addition strategy and developed a ratiometric probe for H_2S (Scheme 7.12) [75]. In this work, a 3-indolylacrylate derivative **3-IA** was found to be blue fluorescent (emission at 408 nm). The addition of H_2S resulted in a drastic decrease in the emission intensity at 408 nm and the concurrent appearance of a new emission at 463 nm, corresponding to the formation of **3-IAS**. The probe provided a substantial ratiometric fluorescence response (15.6-fold enhancement) to 12 equiv of H_2S with a detection limit of 1.7 μM.

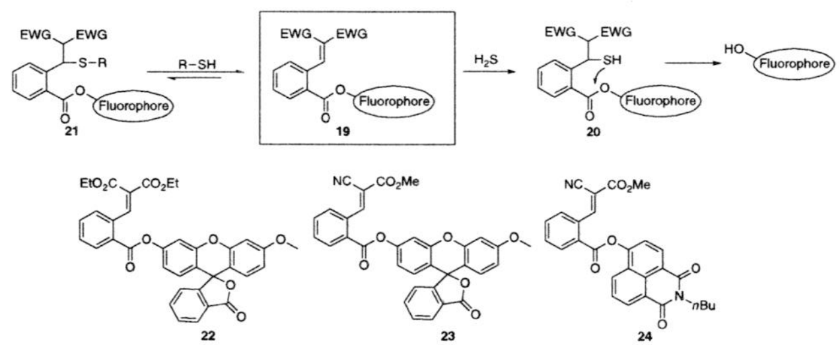

Scheme 7.11 Michael addition cyclization-based probes for H_2S.

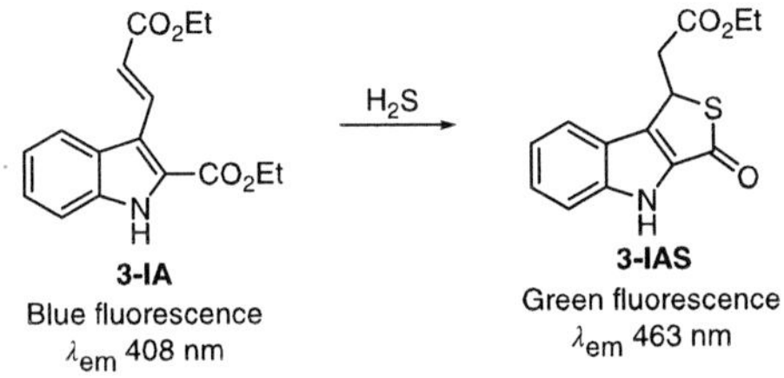

Scheme 7.12 Michael addition cyclization-based probes that generate a ratiometric response for H_2S.

7.3 Conclusions and Perspectives

The development of H_2S-specific fluorescent probes has advanced our understanding of H_2S biology. So far, a large number (>400) of such probes have been reported. Among the three typical reactions used in the design of H_2S probes, nucleophilic reaction-cyclization based strategies have some advantages. For instance, they do not generate redox-sensitive H_2S-derived by-products or cytotoxic metal-complexes. They normally exhibit high selectivity for H_2S over biothiols. However, some problems still exist in this group of probes: (i) Fluorescence turn-on rates of many probes are slow (e.g. need 30 minutes or even longer to reach maximum fluorescence). (ii) Solubility of many probes are unsatisfied. In many cases, ether organic cosolvents or surfactants such as CTAB are required in the detection. (iii) Since biothiols (such as GSH and Cys) are present in much higher concentrations than H_2S, the probe's possible reactions with biothiols could be problematic. Such reactions may not necessarily turn-on fluorescence but could consume the probe and lead to decreased sensitivity. Some recent works have attempted to solve these problems, but more research is still needed. In addition, current probes are all based on irreversible reactions with H_2S, which can only sense the accumulation of H_2S at certain time points. Reversible probes that can detect H_2S concentration changes (both increases and decreases) are still lacking. This will be a challenge waiting to be addressed in coming years.

Acknowledgments

This work was supported by the National Institute of Health (R21DA046386).

References

1 Olson, K.R., Donald, J.A., Dombkowski, R.A., and Perry, S.F. (2012). Evolutionary and comparative aspects of nitric oxide, carbon monoxide and hydrogen sulfide. *Respir. Physiol. Neurobiol.* 184 (2): 117–129.

2 Li, L., Rose, P., and Moore, P.K. (2011). Hydrogen sulfide and cell signaling. *Annu. Rev. Pharmacol. Toxicol.* 51 (1): 169–187.
3 Wang, R. (2012). Physiological implications of hydrogen sulfide: a whiff exploration that blossomed. *Physiol. Rev.* 92 (2): 791–896.
4 Fukuto, J.M., Carrington, S.J., Tantillo, D.J. et al. (2012). Small molecule signaling agents: the integrated chemistry and biochemistry of nitrogen oxides, oxides of carbon, dioxygen, hydrogen sulfide, and their derived species. *Chem. Res. Toxicol.* 25 (4): 769–793.
5 Szabó, C. (2007). Hydrogen sulphide and its therapeutic potential. *Nat. Rev. Drug Discov.* 6 (11): 917–935.
6 Mani, S., Untereiner, A., Wu, L., and Wang, R. (2014). Hydrogen sulfide and the pathogenesis of atherosclerosis. *Antioxid. Redox Signal.* 20 (5): 805–817.
7 Papapetropoulos, A., Whiteman, M., and Cirino, G. (2015). Pharmacological tools for hydrogen sulphide research: a brief, introductory guide for beginners. *Br. J. Pharmacol.* 172 (6): 1633–1637.
8 Wallace, J.L., Blackler, R.W., Chan, M.V. et al. (2015). Anti-inflammatory and cytoprotective actions of hydrogen sulfide: translation to therapeutics. *Antioxid. Redox Signal.* 22 (5): 398–410.
9 Kimura, H. (2011). Hydrogen sulfide: its production and functions. *Exp. Physiol.* 96 (9): 833–835.
10 Kimura, H., Shibuya, N., and Kimura, Y. (2012). Hydrogen sulfide is a signaling molecule and a cytoprotectant. *Antioxid. Redox Signal.* 17 (1): 45–57.
11 Leffler, C.W., Parfenova, H., Jaggar, J.H., and Wang, R. (2006). Carbon monoxide and hydrogen sulfide: gaseous messengers in cerebrovascular circulation. *J. Appl. Physiol.* 100 (3): 1065–1076.
12 Olson, K.R. (2012). A practical look at the chemistry and biology of hydrogen sulfide. *Antioxid. Redox Signal.* 17 (1): 32–44.
13 Qu, K., Lee, S.W., Bian, J.S. et al. (2008). Hydrogen sulfide: neurochemistry and neurobiology. *Neurochem. Int.* 52 (1): 155–165.
14 Predmore, B.L., Lefer, D.J., and Gojon, G. (2017). Hydrogen sulfide in biochemistry and medicine. *Antioxid. Redox Signal.* 17 (1): 119–140.
15 Whiteman, M. and Moore, P.K. (2009). Hydrogen sulfide and the vasculature: a novel vasculoprotective entity and regulator of nitric oxide bioavailability? *J. Cell. Mol. Med.* 13 (3): 488–507.
16 Stipanuk, M.H. and Ueki, I. (2011). Dealing with methionine/homocysteine sulfur: cysteine metabolism to taurine and inorganic sulfur. *J. Inherit. Metab. Dis.* 34 (1): 17–32.
17 DeLeon, E.R., Stoy, G.F., and Olson, K.R. (2012). Passive loss of hydrogen sulfide in biological experiments. *Anal. Biochem.* 421 (1): 203–207.
18 Ubuka, T. (2002). Assay methods and biological roles of labile sulfur in animal tissues. *J. Chromatogr. B* 781 (1): 227–249.

19 Tangerman, A. (2009). Measurement and biological significance of the volatile sulfur compounds hydrogen sulfide, methanethiol and dimethyl sulfide in various biological matrices. *J. Chromatogr. B* 877 (28): 3366–3377.

20 Doeller, J.E., Isbell, T.S., Benavides, G. et al. (2005). Polarographic measurement of hydrogen sulfide production and consumption by mammalian tissues. *Anal. Biochem.* 341 (1): 40–51.

21 Kabil, O. and Banerjee, R. (2010). Redox biochemistry of hydrogen sulfide. *J. Biol. Chem.* 285 (29): 21903–21907.

22 Olson, K.R. (2009). Is hydrogen sulfide a circulating "gasotransmitter" in vertebrate blood? *Biochim. Biophys. Acta* 1787 (7): 856–863.

23 Furne, J., Saeed, A., and Levitt, M.D. (2008). Whole tissue hydrogen sulfide concentrations are orders of magnitude lower than presently accepted values. *Am. J. Physiol. Regul. Integr. Comp. Physiol.* 295 (5): R1479–R1485.

24 Han, Y., Qin, J., Chang, X. et al. (2006). Hydrogen sulfide and carbon monoxide are in synergy with each other in the pathogenesis of recurrent febrile seizures. *Cell. Mol. Neurobiol.* 26 (1): 101–107.

25 Lippert, A.R., New, E.J., and Chang, C.J. (2011). Reaction-based fluorescent probes for selective imaging of hydrogen sulfide in living cells. *J. Am. Chem. Soc.* 133 (26): 10078–10080.

26 Liu, C., Pan, J., Li, S. et al. (2011). Capture and visualization of hydrogen sulfide by a fluorescent probe. *Angew. Chem. Int. Ed.* 50 (44): 10327–10329.

27 Peng, H., Cheng, Y., Dai, C. et al. (2011). A fluorescent probe for fast and quantitative detection of hydrogen sulfide in blood. *Angew. Chem. Int. Ed.* 50 (41): 9672–9675.

28 Qian, Y., Karpus, J., Kabil, O. et al. (2011). Selective fluorescent probes for live-cell monitoring of sulphide. *Nat. Commun.* 2 (1): 495.

29 Sasakura, K., Hanaoka, K., Shibuya, N. et al. (2011). Development of a highly selective fluorescence probe for hydrogen sulfide. *J. Am. Chem. Soc.* 133 (45): 18003–18005.

30 Peng, B. and Xian, M. (2014). Fluorescent probes for hydrogen sulfide detection. *Asian J. Org. Chem.* 3 (9): 914–924.

31 Lin, V.S., Chen, W., Xian, M., and Chang, C.J. (2015). Chemical probes for molecular imaging and detection of hydrogen sulfide and reactive sulfur species in biological systems. *Chem. Soc. Rev.* 44 (14): 4596–4618.

32 Peng, H., Chen, W., Cheng, Y. et al. (2012). Thiol reactive probes and chemosensors. *Sensors* 12 (11): 15907–15946.

33 Lin, V.S. and Chang, C.J. (2012). Fluorescent probes for sensing and imaging biological hydrogen sulfide. *Curr. Opin. Chem. Biol.* 16 (5): 595–601.

34 Lin, V.S., Lippert, A.R., and Chang, C.J. (2013). Cell-trappable fluorescent probes for endogenous hydrogen sulfide signaling and imaging H_2O_2-dependent H_2S production. *Proc. Natl. Acad. Sci. U.S.A.* 110 (18): 7131–7135.

35 Montoya, L.A. and Pluth, M.D. (2012). Selective turn-on fluorescent probes for imaging hydrogen sulfide in living cells. *Chem. Commun.* 48 (39): 4767–4769.

36 Hammers, M.D., Taormina, M.J., Cerda, M.M. et al. (2015). A bright fluorescent probe for H_2S enables analyte-responsive, 3D imaging in live zebrafish using light sheet fluorescence microscopy. *J. Am. Chem. Soc.* 137 (32): 10216–10223.

37 Chen, W., Pacheco, A., Takano, Y. et al. (2016). A single fluorescent probe to visualize hydrogen sulfide and hydrogen polysulfides with different fluorescence signals. *Angew. Chem. Int. Ed.* 55 (34): 9993–9996.

38 Peng, B., Chen, W., Liu, C. et al. (2014). Fluorescent probes based on nucleophilic substitution–cyclization for hydrogen sulfide detection and bioimaging. *Chem. Eur. J.* 20 (4): 1010–1016.

39 Peng, B. and Xian, M. (2015). Hydrogen sulfide detection using nucleophilic substitution–cyclization-based fluorescent probes. Chapter 3. In: *Methods in Enzymology*, vol. 554 (ed. E. Cadenas and L. Packer), 47–62. Academic Press.

40 Cortesi, R., Esposito, E., Menegatti, E. et al. (1996). Effect of cationic liposome composition on in vitro cytotoxicity and protective effect on carried DNA. *Int. J. Pharm.* 139 (1): 69–78.

41 Mir, S., Sen, T., and Sen, N. (2014). Cytokine-induced GAPDH sulfhydration affects PSD95 degradation and memory. *Mol. Cell* 56 (6): 786–795.

42 Cui, X., Navneet, S., Wang, J. et al. (2017). Analysis of MTHFR, CBS, glutathione, taurine, and hydrogen sulfide levels in retinas of hyperhomocysteinemic mice. *Invest. Ophthalmol. Vis. Sci.* 58 (4): 1954–1963.

43 Muñoz-Vargas, M.A., González-Gordo, S., Cañas, A. et al. (2018). Endogenous hydrogen sulfide (H_2S) is up-regulated during sweet pepper (*Capsicum annuum* L.) fruit ripening. In vitro analysis shows that NADP-dependent isocitrate dehydrogenase (ICDH) activity is inhibited by H_2S and NO. *Nitric Oxide* 81: 36–45.

44 Citi, V., Piragine, E., Pagnotta, E. et al. (2019). Anticancer properties of erucin, an H_2S-releasing isothiocyanate, on human pancreatic adenocarcinoma cells (AsPC-1). *Phytother. Res.* 33 (3): 845–855.

45 Chen, J.J.Y., van der Vlies, A.J., and Hasegawa, U. (2020). Hydrogen sulfide-releasing micelles for promoting angiogenesis. *Polym. Chem.* 11 (27): 4454–4463.

46 Ren, Y., Zhang, L., Zhou, Z. et al. (2018). A new fluorescent probe for quick and highly selective detection of hydrogen sulfide and its application in living cells. *New J. Chem.* 42 (16): 13884–13888.

47 Wu, Q., Huo, F., Wang, J., and Yin, C. (2020). Fluorescent probe for detecting hydrogen sulfide based on disulfide nucleophilic substitution–addition. *Spectrochim. Acta A* 238: 118437–118442.

48 Wang, J., Wen, Y., Huo, F., and Yin, C. (2019). Based 'successive' nucleophilic substitution mitochondrial-targeted H_2S red light emissive fluorescent probe and its imaging in mice. *Sens. Actuators, B* 297: 126773–126779.

49 Men, J., Yang, X., Zhang, H., and Zhou, J. (2018). A near-infrared fluorescent probe based on nucleophilic substitution–cyclization for selective detection of hydrogen sulfide and bioimaging. *Dyes Pigm.* 153: 206–212.

50 Zhang, X., Qu, W., Liu, H. et al. (2020). Visualizing hydrogen sulfide in living cells and zebrafish using a red-emitting fluorescent probe via selenium-sulfur exchange reaction. *Anal. Chim. Acta* 1109: 37–43.

51 Chen, W., Xu, S., Day, J.J. et al. (2017). A general strategy for development of near-infrared fluorescent probes for bioimaging. *Angew. Chem. Int. Ed.* 56 (52): 16611–16615.

52 Xu, Z., Xu, L., Zhou, J. et al. (2012). A highly selective fluorescent probe for fast detection of hydrogen sulfide in aqueous solution and living cells. *Chem. Commun.* 48 (88): 10871–10873.

53 Li, D. P., Zhang, J.-F., Cui, J. et al. (2016). A ratiometric fluorescent probe for fast detection of hydrogen sulfide and recognition of biological thiols. *Sens. Actuators, B* 234: 231–238.

54 Kang, J., Huo, F., and Yin, C. (2017). A novel ratiometric fluorescent H_2S probe based on tandem nucleophilic substitution/cyclization reaction and its bioimaging. *Dyes Pigm.* 146: 287–292.

55 Cao, T., Teng, Z., Gong, D. et al. (2019). A ratiometric fluorescent probe for detection of endogenous and exogenous hydrogen sulfide in living cells. *Talanta* 198: 185–192.

56 Zhao, Q., Kang, J., Wen, Y. et al. (2018). "Turn-on" fluorescent probe for detection of H_2S and its applications in bioimaging. *Spectrochim. Acta A* 189: 8–12.

57 Li, G., Ma, S., Tang, J., and Ye, Y. (2019). Lysosome-targeted two-photon fluorescent probes for rapid detection of H_2S in live cells. *New J. Chem.* 43 (3): 1267–1274.

58 Wang, J., Wen, Y., Huo, F., and Yin, C. (2019). A highly sensitive fluorescent probe for hydrogen sulfide based on dicyanoisophorone and its imaging in living cells. *Sens. Actuators, B* 294: 141–147.

59 Zhang, J., Sun, Y.-Q., Liu, J. et al. (2013). A fluorescent probe for the biological signaling molecule H_2S based on a specific H_2S trap group. *Chem. Commun.* 49 (96): 11305–11307.

60 Reja, S.I., Sharma, N., Gupta, M. et al. (2017). A highly selective fluorescent probe for detection of hydrogen sulfide in living systems: in vitro and in vivo applications. *Chem. Eur. J.* 23 (41): 9872–9878.

61 Zheng, H.-R., Niu, L.-Y., Chen, Y.-Z. et al. (2016). Cascade reaction-based fluorescent probe for detection of H_2S with the assistance of CTAB micelles. *Chin. Chem. Lett.* 27 (12): 1793–1796.

62 Peng, B., Zhang, C., Marutani, E. et al. (2015). Trapping hydrogen sulfide (H_2S) with diselenides: the application in the design of fluorescent probes. *Org. Lett.* 17 (6): 1541–1544.

63 Zhang, C., Peng, B., Chen, W. et al. (2015). A ratiometric fluorescent probe for hydrogen sulfide based on the nucleophilic substitution–cyclization of diselenides. *Dyes Pigm.* 121: 299–304.

64 Guan, H., Zhang, A., Li, P. et al. (2019). ESIPT fluorescence probe based on double-switch recognition mechanism for selective and rapid detection of hydrogen sulfide in living cells. *ACS Omega* 4 (5): 9113–9119.

65 Wang, Y., Yang, C.-T., Xu, S. et al. (2019). Hydrogen sulfide mediated tandem reaction of selenenyl sulfides and its application in fluorescent probe development. *Org. Lett.* 21 (18): 7573–7576.

66 Steinmann, D., Nauser, T., and Koppenol, W.H. (2010). Selenium and sulfur in exchange reactions: a comparative study. *J. Org. Chem.* 75 (19): 6696–6699.

67 Suarez, S.I., Ambrose, R., Kalk, M.A., and Lukesh, J.C. III, (2019). Selenosulfides tethered to *gem*-dimethyl esters: a robust and highly versatile framework for H_2S probe development. *Chem. Eur. J.* 25 (69): 15736–15740.

68 Qian, Y., Zhang, L., Ding, S. et al. (2012). A fluorescent probe for rapid detection of hydrogen sulfide in blood plasma and brain tissues in mice. *Chem. Sci.* 3 (10): 2920–2923.

69 Singha, S., Kim, D., Moon, H. et al. (2015). Toward a selective, sensitive, fast-responsive, and biocompatible two-photon probe for hydrogen sulfide in live cells. *Anal. Chem.* 87 (2): 1188–1195.

70 Ryu, H.G., Singha, S., Jun, Y.W. et al. (2018). Two-photon fluorescent probe for hydrogen sulfide based on a red-emitting benzocoumarin dye. *Tetrahedron Lett.* 59 (1): 49–53.

71 Shu, W., Zang, S., Wang, C. et al. (2020). An endoplasmic reticulum-targeted ratiometric fluorescent probe for the sensing of hydrogen sulfide in living cells and zebrafish. *Anal. Chem.* 92 (14): 9982–9988.

72 Pritchard, R.B., Lough, C.E., Currie, D.J., and Holmes, H.L. (1968). Equilibrium reactions of *n*-butanethiol with some conjugated heteroenoid compounds. *Can. J. Chem.* 46 (5): 775–781.

73 Liu, C., Peng, B., Li, S. et al. (2012). Reaction based fluorescent probes for hydrogen sulfide. *Org. Lett.* 14 (8): 2184–2187.

74 Gao, B., Cui, L., Pan, Y. et al. (2017). A highly selective fluorescent probe based on Michael addition for fast detection of hydrogen sulfide. *Spectrochim. Acta A* 173: 457–461.

75 Samanta, S.K., Ali, S.S., Gangopadhyay, A. et al. (2019). A highly selective ratiometric fluorescent probe for H_2S based on new heterocyclic ring formation and detection in live cells. *Supramol. Chem.* 31 (5): 349–360.

8

Fluorescent Probes for H_2S Detection: Electrophile-Based Approaches

Long Yi[1] *and Zhen Xi*[2]

[1] *State Key Laboratory of Organic-Inorganic Composites and Beijing Key Lab of Bioprocess, Beijing University of Chemical Technology (BUCT), Beijing, China*
[2] *Department of Chemical Biology, State Key Laboratory of Elemento-Organic Chemistry, College of Chemistry, National Pesticide Engineering Research Center, Collaborative Innovation Center of Chemical Science and Engineering, Nankai University, Tianjin, China*

8.1 Introduction

H_2S is the third gasotransmitter along with NO and CO and is involved in many physiological and pathological processes [1, 2]. Endogenous H_2S in mammals can be enzymatically produced from Cys and Hcy by cystathionine-γ-lyase (CSE), cystathionine-β-synthase (CBS), and 3-mercaptopyruvate sulfurtransferase (3-MST)/cysteine aminotransferase (CAT). Recent studies indicate that low levels of H_2S show cytoprotective effects [3], whereas high concentrations of H_2S are cytotoxic. Endogenous H_2S or low levels of exogenous H_2S can exhibit pro-cancer effects, whereas exposure to higher levels of H_2S for long periods can lead to cancer cell death [4]. Like mammalian-derived H_2S, bacterial- and plant-derived H_2S also has important biological functions [5, 6]. Due to its wide distribution in different organisms as well as its complex activities in many human diseases, the molecular mechanisms by which H_2S is involved in different biological processes need further thorough investigation. These different aspects of H_2S biology support the need for development and refinement of chemical tools for H_2S detection *in vivo* or in related complex environments.

Fluorescence probes are excellent tools for *in situ* and real-time monitoring of biological molecules. Many different research groups have successfully developed fluorescent probes to sense H_2S via different reaction mechanisms [7–10]. For example, the H_2S-mediated reduction of aryl azides and sulfonyl azides has emerged as the most common strategy for H_2S fluorescent probes since the initial reports from the Chang group [11] and the Wang group [12], respectively. In

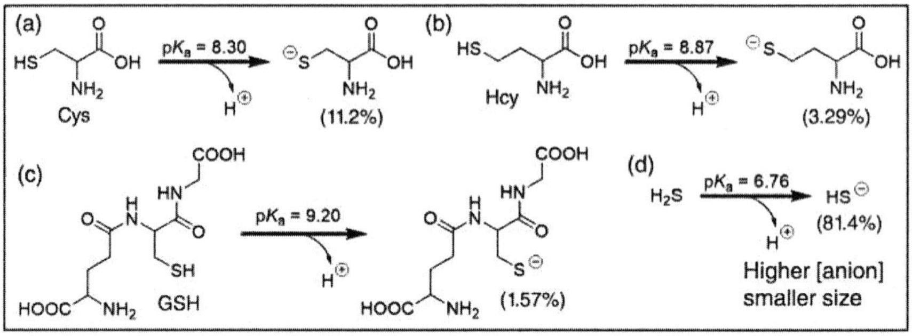

Figure 8.1 (a–d) Dissociation reactions for three biothiols (Cys, Hcy, and GSH) and H_2S in pH 7.4 aqueous buffer at 37 °C. The pK_a as well as the calculated anion concentration is indicated for each reaction.

addition, the reaction mechanism reveals that aryl azides act as electrophiles for the initial addition with HS^- [13]. In this chapter, we focus on the topic of H_2S fluorescent probes based on electrophile-based approaches [11–29].

The dissociation constant (pK_a) of H_2S is 6.76 at 37 °C and therefore, 81.4% of the total H_2S exists as anionic HS^- under physiological conditions (pH 7.4). Considering these pK_a values of biothiols (Figure 8.1) and the smaller size of H_2S when compared to biothiols [1, 2, 30], HS^- should be a stronger nucleophile than biothiols *in vivo*. However, the levels of endogenous glutathione (GSH) (~1–10 mM) and Cys (<200 μM)/Hcy (<15 μM) are much higher than that of H_2S (<1 μM) [1, 31]. Therefore, a major challenge in development of H_2S fluorescent probes is discovering and developing highly selective reactions to selectively detect submicromolar H_2S in the presence of millimolar biothiols under physiological conditions.

Unlike H_2S fluorescent probes, which have only been developed in the past decade, thiol fluorescent probes have been developed since 1960s, and there are many types of organic reactions available for thiol detection [32, 33]. As shown in Figure 8.2, thiolysis of 2,4-dinitrophenylsulfonamide (or sulfonate), thiol-exchange reactions, nucleophilic substitution of electrophilic iodide derivatives, and thiol Michael addition are the commonly used reactions for thiol fluorescent probes for thiol labeling reagents. Inspired by these reaction mechanisms, the rational design of electrophiles with lower reactivity may provide useful receptors and reporters for the development of H_2S-selective probes. For example, Szabo and coworkers reported a method to detect H_2S by using monobromobimane, which was originally used for thiol quantification [34, 35]. In this chapter, we provide early and representative examples of H_2S fluorescent probes in which sulfide attacks an electrophilic appended on a fluorophore to generate a fluorescence response. This chapter contains six sections that are organized by

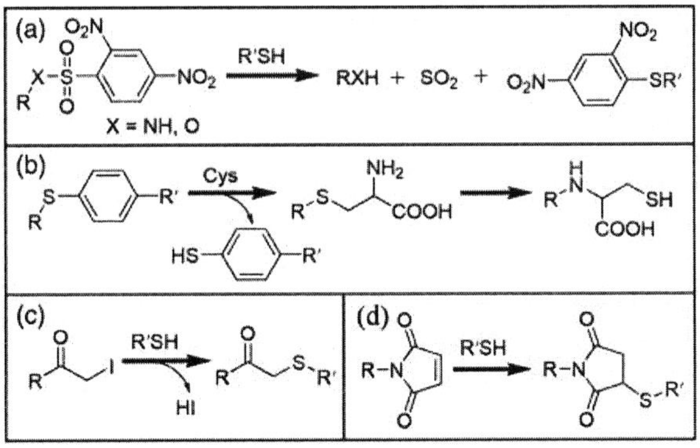

Figure 8.2 (a–d) Four kinds of chemical reactions using for the development of fluorescent probes based on the strong nucleophilic reactivity of thiols. Many of these approaches have been repurposed and refined for electrophile-based detection strategies for H_2S.

the bond changes that occur during fluorophore activation: (i) cleavage of C—O bond [14–17]; (ii) cleavage of C—S bond [18, 19]; (iii) cleavage of C—Cl bond [20–22]; (iv) Michael addition [23–25]; (v) cleavage of C—N bond [26, 27]; and (vi) reduction of aryl azide [11, 28, 29]. We anticipate that this chapter would help readers gain a general understanding about contemporary design strategies, sensing mechanisms, and potential biomedicine applications of the present topic.

8.2 Selected Probes Based on Different Reaction Types

8.2.1 Cleavage of C—O Bond

Thiols can cleave dinitrophenyl groups under basic conditions, whereas the stronger nucleophile H_2S can work under neutral conditions. As an example of this approach, Lin and coworkers reported a near-infrared (NIR) fluorescent H_2S probe **1** in 2012 (Figure 8.3a) [14], which is the first such probe based on thiolysis of dinitrophenyl ether. The probe is highly sensitive for H_2S with a low limit of detection (LOD, 50 nM) and an 18-fold fluorescence enhancement at 708 nm. Notably, small-molecule thiols (e.g. 10 mM GSH or 1 mM Cys) triggered only about 3-fold fluorescence enhancement and had nearly no interference to H_2S detection. Furthermore, probe **1** showed low cytotoxicity and could be used for H_2S imaging in live cells. Subsequently, over 100 H_2S fluorescent probes have been developed based on such thiolysis reaction due to the plethora of oxygen-functionalized fluorophores available for facile modification. Compared with the thiolysis of 2,4-dinitrophenylsulfonate groups in thiol probes, the 2,4-dinitrophenyl ether is less reactive and therefore, the thiolysis reaction with H_2S results in higher selectivity for H_2S over other biothiols.

Building from the success of 2,4-dinitrophenyl ethers as fluorescent probes, nitrobenzodioxzole (NBD) ethers have also been used to develop H_2S fluorescent

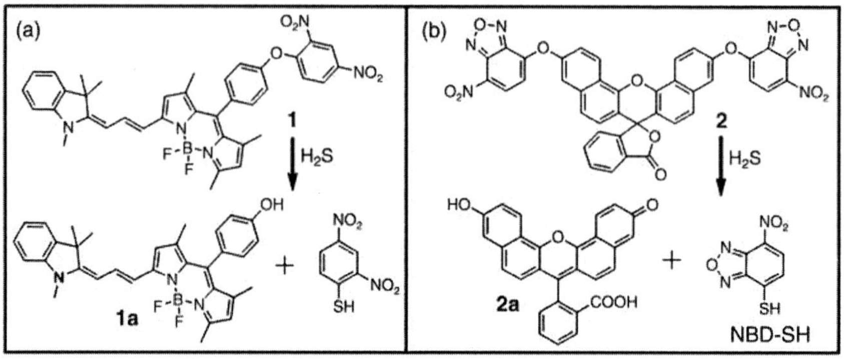

Figure 8.3 (a) Chemical structure of probe **1** and its reaction with H_2S. Source: Cao et al. [14]/Royal Society of Chemistry. (b) Chemical structure of probe **2** and its reaction with H_2S.

probes (Figure 8.3b). In 2014, Yi, Xi, and coworkers developed a NIR H_2S probe **2** based on dual-NBD-capped naphthofluorescein [15, 36]. Upon reaction with H_2S, the fluorescence enhancement of probe **2** is 64-fold at 662 nm, while other tested species (e.g. 1 mM GSH or Cys) are all less than 5.5-fold. Compared with a dual-NBD-capped fluorescein probe, probe **2** exhibits better selectivity toward H_2S over biothiols, but a slower thiolysis rate. Subsequently, the thiolysis of NBD ethers has been used broadly to develop fluorescent probes for sensing biothiols and H_2S [9]. In general, H_2S reacts faster than biothiols with NBD ethers, and therefore, less reactive NBD ethers are often more suitable for the development of selective H_2S probes, whereas more reactive NBD ethers can be used as more general thiol probes.

In 2018, Yuan, Zhang, and coworkers developed a luminescence probe **3** for H_2S detection [16], which is constructed by using an H_2S-responsive masking moiety to a red-emitting Ru^{II} luminophore (Figure 8.4a). The metal-to-ligand charge transfer (MLCT) excited states of Ru^{II} complexes could be interrupted by an intramolecular photoinduced electron transfer (PET). Upon cleavage of the C—O bond of the ester by H_2S, an 86-fold enhancement at 612 nm was observed. Taking advantages of long-lived MLCT emissions ($\tau = 336$ ns for **3a**), the probe was applied for development of a time-gated luminescence assay of H_2S. Probe **3** demonstrates high specificity and sensitivity for H_2S detection, and low cytotoxicity, and was applied for monitoring of lysosomal H_2S in live cells. Moreover, the long-lived phosphorescence enables probe **3** as an efficient tool for monitoring of H_2S in live zebrafish and mice (Figure 8.4b,c).

Photoacoustic (PA) imaging combines diffusive optical and focused ultrasound detection, offering high contrast in deep tissue [37]. On the other

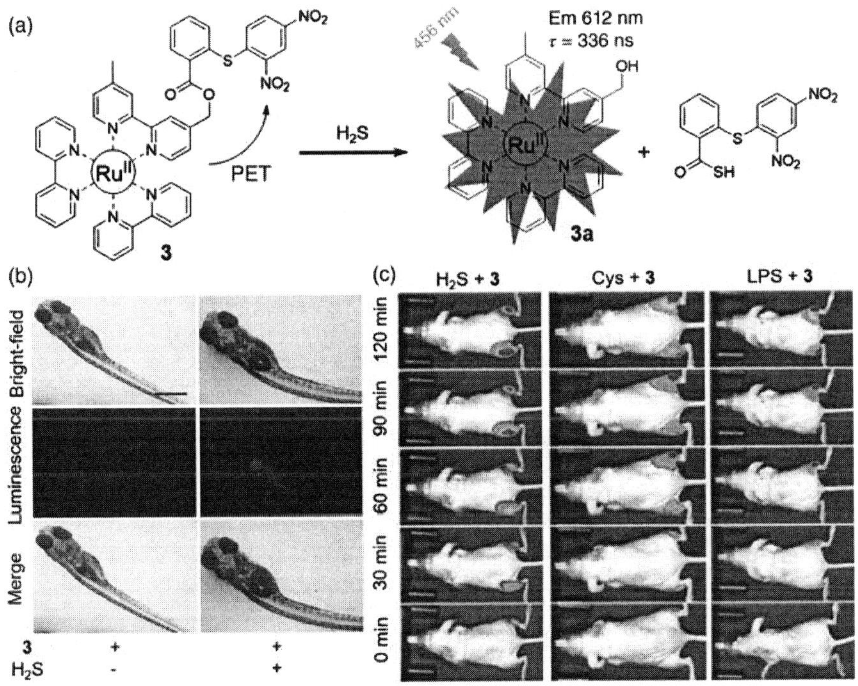

Figure 8.4 (a) A phosphorescence response mechanism of **3** toward H_2S. (b) Visualization of H_2S in zebrafish using probe **3**. (c) Luminescence imaging of H_2S in live mice using probe **3**. Left column: **3** was subcutaneously injected into the left and right legs, followed by the injection of H_2S into the left leg. Middle and right columns: Cys and lipopolysaccharide (LPS) were injected into the right leg, followed by injection of **3** into the left and right legs. Source: (b, c) Du et al. [16]/with permission of John Wiley & Sons, Inc.

hand, ratiometric fluorescent probes, which can cancel out variability caused by microenvironments and uneven distribution of the probes in cells, are highly appealing for bioimaging [38]. To this end, He, Guo, Yang, and coworkers developed an NIR ratiometric optical/PA dual-modality probe **4** for H_2S detection [17]. Upon cleavage of C—O bond of the ester by H_2S, the enolic heptamethine cyanine in **4** undergoes a keto-enol transition to produce **4a** (Figure 8.5a), resulting in fluorescence enhancement at 630 nm and fluorescence decrease at 805 nm as well as relatively good selectivity for H_2S (Figure 8.5b,c). In addition, the absorption spectra of **4** and **4a** are centered at 775 and 850 nm, respectively, with an isosbestic point at 825 nm, suggesting that **4** could be used as a ratiometric PA probe for H_2S. Moreover, the *S*-adenosyl-L-methionine-stimulated H_2S enhancement in the liver of mice was confirmed by using the probe and optical/PA dual-modality imaging, highlighting such strategy for sensing H_2S *in vivo*.

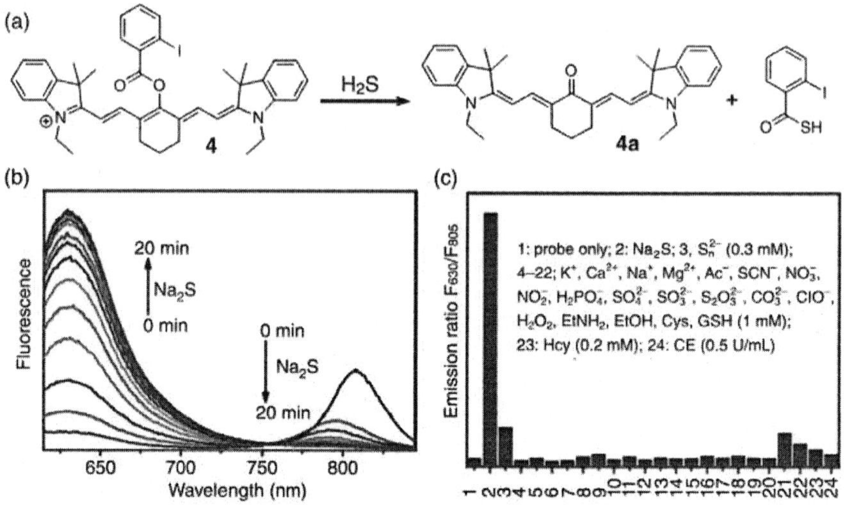

Figure 8.5 (a) Chemical structure of probe **4** and its reaction with H_2S. (b) Time-dependent emission (λ_{ex} = 595 nm) spectra of **4** in buffer (pH 7.4) in the presence of Na_2S. (c) Emission ratio F_{630}/F_{805} of **4** in the presence of different analytes. Source: (b, c) Chen et al. [17]/with permission of American Chemical Society.

8.2.2 Cleavage of C—S Bond

Compared to NIR-I (650–900 nm) imaging, NIR-II (1000–1700 nm) imaging provides a higher signal-to-noise ratio and deeper tissue penetration capability [39]. In 2018, Zhao, Fan, Zhu, and coworkers developed an activatable NIR-II probe **5** (Figure 8.6a) for visualizing endogenous H_2S in colorectal cancers [18]. The probe contains a boron dipyrromethene difluoride (BODIPY) derivative bearing 4-nitrobenzenethioester as the H_2S receptor. To improve water-solubility of **5** and to decrease the response time with H_2S, the probe was incorporated into the hydrophobic interior of water-dispersible silica nanocomposites to produce nanoprobe **5@Si**, which demonstrates H_2S-activated turn-on emission at 900–1300 nm with 780 nm excitation (Figure 8.6b). **5@Si** enables the selective identification of H_2S-rich colon cancer cells and differentiation different types of live cells based on the levels of endogenous H_2S. Importantly, H_2S specifically activates the fluorescence of **5@Si** in NIR-II region, which allows *in vivo* detection and differentiation of cancers at a depth and spatial resolution based on the NIR-II imaging.

Photodynamic therapy (PDT) is emerged as a potent therapeutic strategy for various cancers [40]. In 2019, Zhao, Shi, and coworkers reported a theranostic nanoplatform composed of a H_2S-responsive NIR probe **6** with a light-sensitive moiety for efficient $^1O_2$ generation (Figure 8.6a) [19]. The

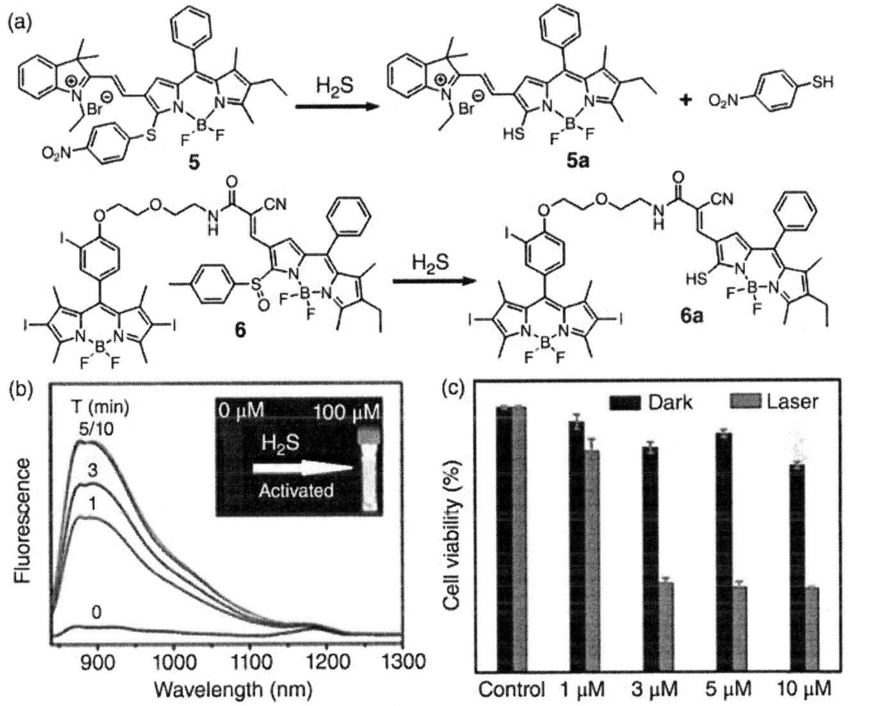

Figure 8.6 (a) Chemical structures of **5** and **6** and their reactions with H_2S. Source: Xu et al. [18]/John Wiley & Sons. (b) Time-dependent NIR-II emission spectra of **5**@Si in the presence of NaSH. Inset: photograph of the H_2S-activated NIR-II emission. Source: Xu et al. [18]/with permission of John Wiley & Sons. (c) The cytotoxicity of **6**@Si toward HCT116 cells under dark or 530 nm light irradiation. Source: Wang et al. [19]/Royal Society of Chemistry.

H_2S-mediated cleavage of C—S bond in **6** results in an about 137-fold emission enhancement at 712 nm. In addition, **6** was employed for fabrication of a nanoprobe **6**@Si with good aqueous solubility and photostability. The 3-(4,5-dimethylthiazol-2-yl)-2,5-diphenyltetrazolium bromide (MTT) assay clearly demonstrated that the **6**@Si showed low cytotoxicity in the absence of light. In contrast, greater cell death was seen in H_2S-rich cancer cells upon simultaneous treatment with **6**@Si and light irradiation at 530 nm (Figure 8.6c). The **6**@Si was successfully employed for *in vivo* PDT, indicating that cancer imaging precisely directed where to implement the light illumination to produce cytotoxic $^1O_2$ for on-demand cancer therapy. Such a strategy could maximize the therapeutic efficiency and minimize side effects, which may facilitate significant advances toward precise diagnosis and treatment of cancers that overexpress endogenous H_2S.

8.2.3 Cleavage of C—Cl Bond

The cleavage of C—I bonds is widely used for labeling of biothiols. Compared with the C—I bond, the cleavage of C—Cl bond should be more difficult for biothiols but might be suitable for the more reactive H_2S (Figure 8.7). In 2017, Zhao and coworkers reported an activatable PA probe **7** for real-time imaging of H_2S in live mice [20]. The probe can generate a 44-fold turn-on absorption at 780 nm in the presence of H_2S, thus providing a strong PA signal output in the NIR region. Furthermore, this probe exhibits extremely fast and highly selective responsiveness, good water-solubility, and excellent biocompatibility. Considering these outstanding features, the probe was successfully applied for the direct PA trapping of endogenous H_2S generation in a HCT116 tumor-bearing mouse model.

In 2019, Zhao, Fan, Zhu, and coworkers developed a theranostic nanoplatform for imaging-guided and photo-controlled drug release [21]. The nanoplatform is fabricated by coencapsulation of **8** (Figure 8.7) and anticancer drug camptothecin-11 (CPT). Probe **8** selectively reacts with H_2S to generate **8a** with a strong absorption at 741 nm, which is useful for efficient photothermal conversion. In addition, the nanoplatform shows H_2S-activatable NIR-II emission under 808 nm excitation and photo-controllable release of CPT. Importantly, the HCT116 tumor in mice was greatly suppressed by administering the nanoplatform

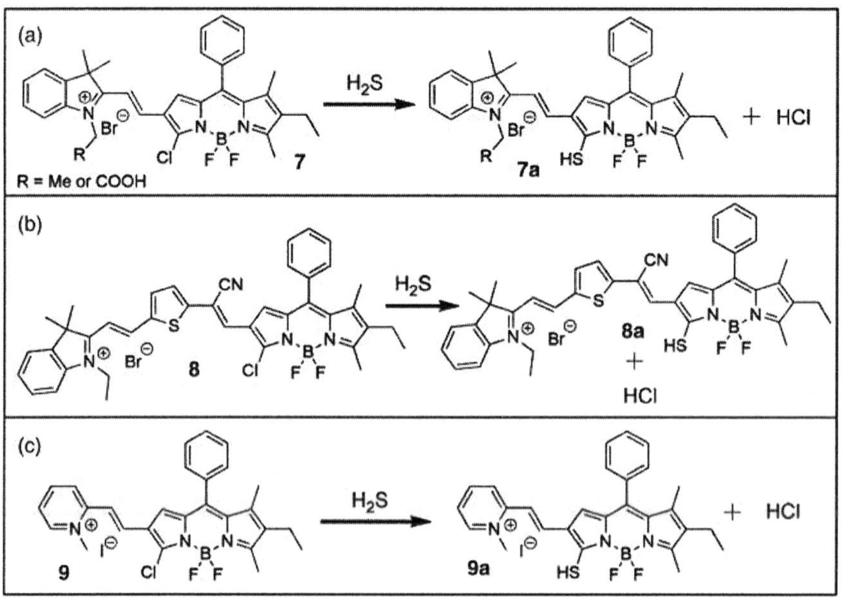

Figure 8.7 (a–c) Chemical structures of probes **7–9** and their reactions with H_2S via cleavage of C—Cl bonds.

with minimal side effects, because of the synergy of the cancer-specific and NIR light-activated therapy. This theranostic nanoplatform sheds light on precision medicine with guidance through NIR-II fluorescent imaging [21].

In 2020, Zhao and coworkers reported a molecular probe **9** with aggregation-enhanced response to H_2S for *in vivo* imaging of H_2S-rich cancers [22]. Probe **9** has the capability of spontaneously self-assembling into nanoprobe under physiological conditions. The self-assembled probe shows significant NIR fluorescence enhancement at 718 nm upon reaction with H_2S (Figure 8.7). The probe is highly selective and biocompatible and was successfully used to differentiate H_2S-rich and H_2S-deficient cancer cells and to visualize H_2S-rich tumor in mice. This work represents the first examples of aggregation-enhanced response to analytes (e.g. H_2S) under physiological conditions.

8.2.4 Michael Addition

The Michael addition of biothiols to maleimide and related derivatives has been widely used for bioconjugation for many years. In order to selectively label H_2S, electrophiles with less reactivity than maleimide should be employed. To this end, a number of research groups have successfully developed fluorescent probes **10–12** for selective detection of H_2S via the Michael addition mechanism (Figure 8.8).

In 2013, Guo, He, and coworkers reported a ratiometric H_2S probe **10** [23], which is constructed by using a combination of coumarin and merocyanine chromophores with a positively charged moiety for mitochondria targeting. The merocyanine moiety acts as the H_2S receptor because of its indolenium C-2 atom as the electrophile. Probe **10** shows two well-separated emission bands centered at 510 and 652 nm with quantum yield (Φ) of 0.03. Upon reaction with H_2S, the emission band at 652 nm of the probe decreases and the band at 510 nm increases simultaneously. The probe is highly selective for H_2S and has a quick response time (less than 30 seconds), thus favoring the real-time H_2S detection. The ratiometric imaging of H_2S by **10** is promising for the practical exploration of H_2S-related processes in mitochondria. In addition, the possibility to recover **10** from **10a** in acidic conditions (pH < 3.0) shows the potential of this reaction in the design of reversible H_2S probes.

In 2015, Peng and coworkers reported a ratiometric fluorescent probe **11** for the highly selective and sensitive detection of H_2S (68.2 nM) with a fast response time (15 seconds) [24]. The probe contains flavylium as the NIR fluorophore and takes advantage of Michael addition of H_2S toward the benzopyrylium moiety, and can be an NIR emission at 707 nm, while upon reaction with H_2S, the emission at 707 nm decreases but a new emission increases at 487 nm. Such large emission shift (220 nm) and 168-fold ratiometric enhancement (F_{487}/F_{707}) make this probe

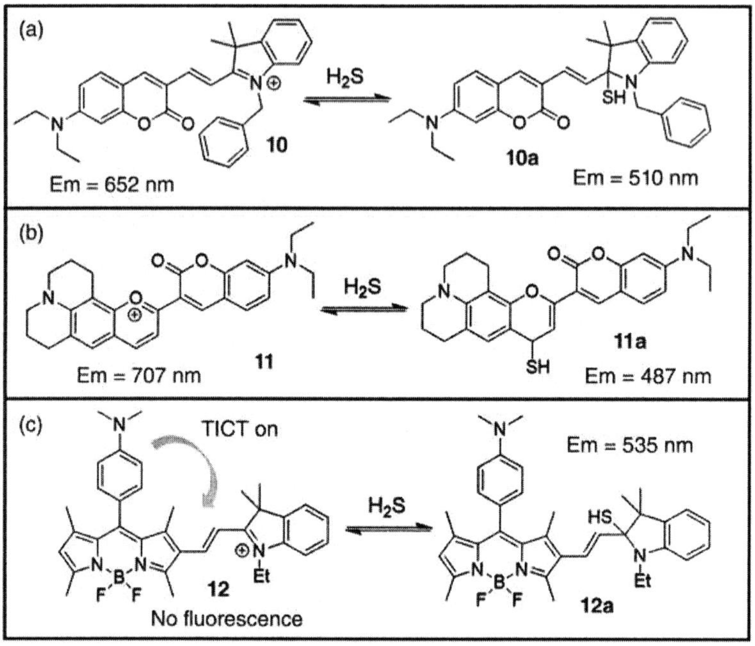

Figure 8.8 (a–c) Chemical structures of probes **10–12** and their reversible reactions with H_2S as well as the fluorescent sensing mechanisms.

favorable for ratiometric bioimaging. In addition, probe **11** localizes in mitochondria because of the positively charged moiety, and can be applied for real-time monitoring of exogenous and endogenous H_2S in live MCF-7 cells and mice.

In 2016, Lin and coworkers reported a fluorescent probe **12** (Figure 8.8) as the first twisted internal charge transfer (TICT)-based H_2S probe [25]. The probe is nonfluorescent ($\Phi = 0.006$) due to the TICT. Michael addition of H_2S toward probe **12** results in a 57-fold fluorescence enhancement at 535 nm ($\Phi = 0.13$) because of the inhibition of TICT. The probe demonstrates favorable properties such as a fast response time (within 2 minutes), high selectivity, and low cytotoxicity. In addition, probe **12** is membrane-permeable and suitable for visualization of endogenous H_2S in live cells.

In total, Michael addition of H_2S toward probes **10–12** demonstrates good selectivity for H_2S over competing analytes and a rapid reaction equilibrium. H_2S disturbs the extended π conjugated structure and then causes a ratiometric response with relatively large Stokes shifts, which is very good for intracellular ratiometric imaging. However, the longer emission wavelength of these probes is a turn-off motif, while the turn-on NIR probes are a preferred approach for imaging in mammals.

8.2.5 Cleavage of C—N Bond

Normally, the C—N bond seems to be stable toward biothiols and H_2S. However, in 2013, Yi, Xi, Pluth, and coworkers discovered that NBD tertiary amines can react with micromolar H_2S and developed colorimetric and fluorescent probes for H_2S detection [9, 36]. Inspired by the success of sulfonyl azides as H_2S scavengers from Xian and coworkers [41], these authors also synthesized NBD-based scavengers **NBD-S1** to **NBD-S6** (Figure 8.9) as part of a structure-function investigation into how the amine identity impacts H_2S reactivity and selectivity [26]. The piperidyl-based NBD compounds react efficiently with micromolar H_2S in buffer (pH 7.4), and the addition of a positively charged group helps significantly increase the thiolysis rates. Crystal structure analysis reveals that the piperazinyl heterocycles in **NBD-S4** and **NBD-S6** are in twist-boat and chair conformations, respectively, where the chair conformation is favorable for the thiolysis reaction [26]. In addition, the positive charge of **NBD-S6** may favor reaction with anionic HS^-. Importantly, **NBD-S6** can effectively scavenge endogenous H_2S in live cells and mice.

Building from the success of NBD amines for H_2S scavenging, this motif has also been widely used to develop H_2S-responsive fluorescent probes [9, 27]. For example, Yi, Zhang, and coworkers developed a NIR fluorescent probe **13** for imaging of endogenous H_2S in live cells and mice (Figure 8.10). The probe demonstrates an 87-fold fluorescence enhancement at 796 nm (with excitation at 730 nm) upon reaction with micromolar H_2S. In general, NBD-based probes are fluorescently quenched by PET from the NBD nitro group, so the distance between the fluorophore and NBD should be relatively short for controlling the quenching efficiency. Probe **13** shows high selectivity and sensitivity toward H_2S and is water-soluble, cell-permeable, minimally cytotoxic, and can be used to monitor endogenous H_2S in live cells. In addition, probe **13** was used to demonstrate that D-Cys induces endogenous H_2S generation to enhance angiogenesis *in vitro*, and intratumoral H_2S in mice can be visualized by NIR fluorescence imaging (Figure 8.10d). These results highlight that fluorescent probes based on cleavage of C—N bond can serve as efficient tools for the highly selective and sensitive detection of H_2S in live animals and even for cancer diagnosis.

8.2.6 Reduction of Aryl Azide

In 2011, Chang and coworkers developed a H_2S probe **14** (Figure 8.11), which is one of the first such probes based on reduction of aryl azide [11]. The probe adopts a closed lactone conformation and exhibits nearly no fluorescence. Upon reaction with H_2S, the probe demonstrates a 9-fold increase in emission and a new absorption in the visible region (λ_{max} = 492 nm, ε = 75,000 M^{-1} cm^{-1};

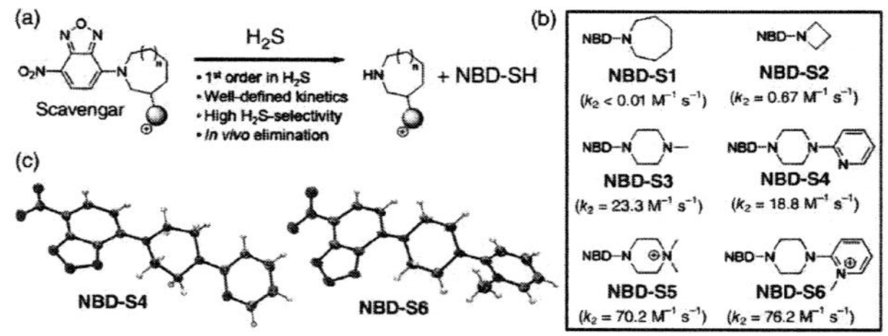

Figure 8.9 (a) Inclusion of cationic charge in the system enables access to efficient H_2S scavengers by thiolysis of NBD cyclic amines. (b) Chemical structures and reaction rates of **NBD-S1** to **NBD-S6**. (c) Oak Ridge thermal ellipsoid plot (ORTEP) diagram for the molecular structures of **NBD-S4** and **NBD-S6**. Source: (a, c) Ismail et al. [26]/Royal Society of Chemistry/CC BY-3.0.

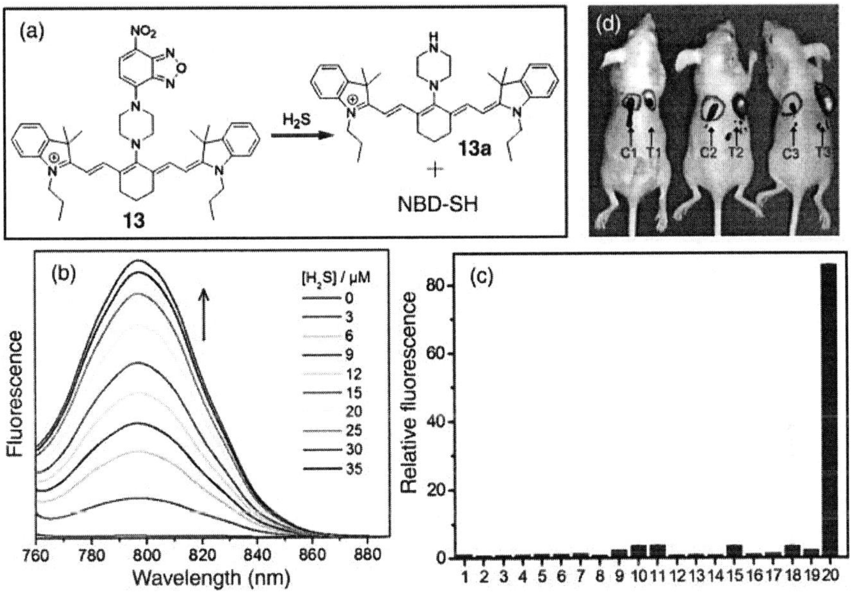

Figure 8.10 (a) Chemical structure of probe **13** and its reaction with H_2S. (b) Emission spectra of **13** in the presence of various concentrations of H_2S. (c) Relative fluorescence intensity at 796 nm of **13** with various species: 1, only probe; 2, Fe^{3+}; 3, Zn^{2+}; 4, H_2O_2; 5, ClO^-; 6, NO_2^-; 7, $S_2O_3^{2-}$; 8, SO_3^{2-}; 9, Hcy; 10, Cys; 11, GSH; 12, 2,2′-dithiodipyridine; 13, 2,2′-dithiodipyridine + H_2S; 14, cystamine; 15, cystamine + H_2S; 16, lysozyme; 17, lysozyme + dithiodipyridine + H_2S; 18, BSA; 19, BSA + dithiodipyridine + H_2S; 20, H_2S. (d) Fluorescence images of mice with skin-pop (s.p.) or intratumoral injection of **13**. The left injection positions (C1, C2, C3) were control, while the right injection positions (T1, T2, T3) were for FHC, HCT116, and HT29 grafted cell positions, respectively. Observed tumors were formed in the HCT116 (T2) and HT29 (T3) xenograft mice. Source: (b–d) Zhang et al. [27]/Royal Society of Chemistry/CC BY-3.0.

λ_{em} = 525 nm, Φ = 0.60). In addition, the probe showed good selectivity for H_2S over GSH (~5-fold), sulfite (~4-fold), and O_2^- (~4-fold), and was applied for imaging of intracellular H_2S [11]. Subsequently, the H_2S-mediated reduction of aryl azides has been widely employed for development of over 300 fluorescent probes for H_2S detection [7, 10, 41]. These probes provide powerful chemical tools for various biological applications such as fluorescent imaging of live animals [28], identification of H_2O_2–H_2S crosstalk using a cell-trappable probe [29], high-throughput enzymatic assay [42], and even azide-based H_2S donors [43]. In addition, the H_2S-mediated reduction of azides could be used for the controllable release of functional groups in live systems [44].

On the other hand, the sensing performances of the azide-based fluorescent probes can be tuned via using different fluorophores. For example, in 2015, Pluth

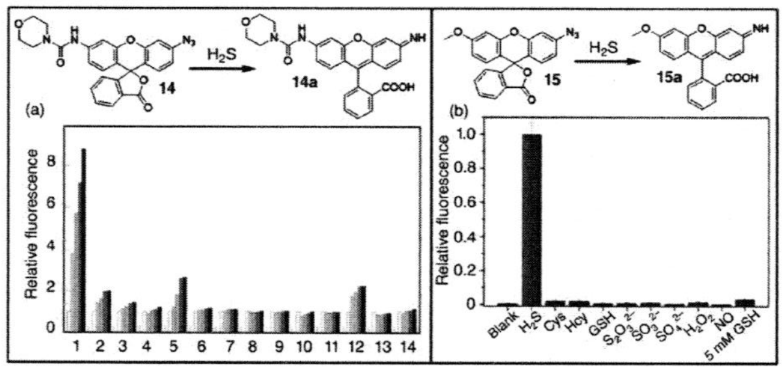

Figure 8.11 (a) Chemical structure of probe **14** and its reaction with H_2S and other species. Relative fluorescence intensity at 525 nm of **14** with various species: 1, H_2S; 2, GSH; 3, Cys; 4, lipoic acid; 5, Na_2SO_3; 6, NaS_2O_3; 7, KSCN; 8, *S*-nitrosoglutathione; 9, $NaNO_2$; 10, NO; 11, H_2O_2; 12, O_2^-; 13, *t*BuOOH; 14, HOCL Source: Lippert et al. [11]/with permission of American Chemical Society. (b) Chemical structure of probe **15** and its reaction with H_2S and other species. Source: Hammers et al. [28]/with permission of American Chemical Society.

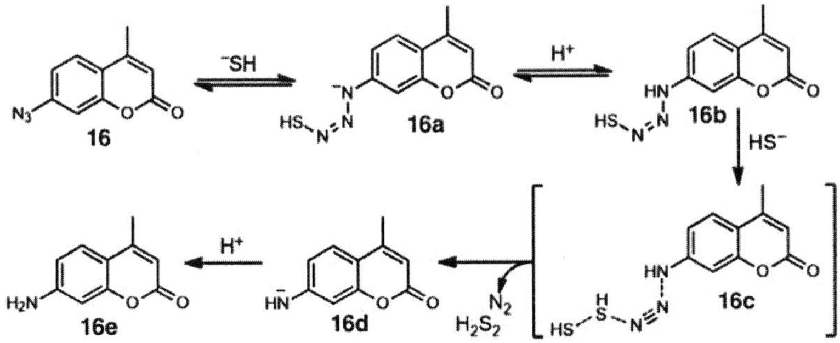

Figure 8.12 The reaction mechanism of H_2S-mediated reduction of aryl azide **16** to produce **16e**. Source: Henthorn and Pluth [13]/American Chemical Society.

and coworkers developed an improved azide-based probe **15** for 3D imaging of H_2S in zebrafish using light sheet fluorescence microscopy (LSFM) [28]. The second-generation probe **15** is nonfluorescence ($\Phi < 0.01$) and gives strong fluorescence ($\Phi = 0.57$) upon reaction with H_2S. The probe is highly sensitive (LOD, 86 nM) and highly selective (Figure 8.11b). Probe **15** exhibits 32-fold preferential reactivity with H_2S relative to 5 mM GSH loading. In addition, probe **15** exhibits significantly less photoactivation than other azide-based probes. Based on these excellent properties, probe **15** was successfully applied to obtain a 3D image of zebrafish intestinal bulb using LSFM, which opens the door for new investigations of whole-organism imaging of H_2S.

Finally, although this is a reduction reaction for these azide-based H_2S probes, mechanistic investigations by Pluth and coworker demonstrated that it occurs by nucleophilic attack of HS^- on the electrophilic azide [13]. As shown in Figure 8.12, the initial nucleophilic attack of HS^- toward **16** generates anionic azidothiol intermediate **16a** and then neutral **16b**, which would be significantly more electrophilic. Subsequent attack on **16b** by a second HS^- generates **16c** and then **16d** with release of N_2 and H_2S_2.

8.3 Conclusion and Future Prospects

The above examples highlight the recently reported H_2S fluorescent probes via electrophile-based approaches. Up to now, there are over 300 probes based on reduction of azides and over 100 probes based on thiolysis of dinitrophenyl ethers, and here we only select some early and representative examples to outline design strategies, sensing mechanisms, and potential bioapplications.

In general, different sensing reactions have their intrinsic advantages and limits. For example, the selectivity for cleavage of C—O bond via H_2S over 5 mM

GSH normally is not very high. Specially, we remind readers again that the high levels of intracellular GSH may result in activation of NBD–OR probes, which may complicate interpretation of results from this class of fluorescent probes for endogenous H_2S (<1 μM) in live organisms. The selectivity for cleavage of C—S bond has not been widely studied, while the probes **5** and **6** are incorporated into silica nanocomposites, and selective reaction of **5**@Si or **6**@Si is achieved possibly due to the smaller size of H_2S than biothiols that cannot enter into the nanocomposites for the reaction.

Using electrophilic halides as examples, the cleavage of C—Cl bond should be difficult for thiols but more readily accessed by H_2S. Therefore, probes **7–9** are selective for H_2S detection. However, the attachment of thiol with the fluorophore after the sensing reaction results in relatively low quantum yield products, which is also the disadvantage for probes **10–12** based on Michael addition. The cleavage of C—N bond of NBD amines is very high selectivity for H_2S. However, such probes normally work in the pH range 5.0–9.0, and the NBD tertiary amine is a relatively large H_2S receptor. In addition, two primary strategies should be considered to improve the fluorescence response, namely: (i) using bright fluorophores, and (ii) reducing the distance between the fluorophore and NBD. Azide-based probes are most widely developed and highly selective, which can work in a wide pH range. However, a potential problem with this type of probes is the inherent photosensitivity of aryl azides. In addition, considering the complexity of *in vivo* biological investigations, it is possible that azides can also participate in other reactions not associated with H_2S sensing (e.g. azides can be potentially reduced by cytochrome P450) [45, 46], and one should be careful for long-term using of the azide-based probes in live mammals.

On the other hand, although the sensing reactions are similar for many of the electrophile-based approaches, the fluorescent probes still have different properties due to the use of different fluorophores, which results in different reactivities, selectivity, and optical properties. For example, the second-generation probe **15** has better photo-stability and larger fluorescent turn-on upon reaction with H_2S than probe **14**. The probe **13** is much better in bioimaging than many other NBD-based probes [9].

Finally, probes based on cleavage of chemical bonds and reduction of aryl azides are irreversible, and such sensing methods perturb the sulfide levels and the redox state of live systems. Michael addition reactions can be reversible below pH 3, but there is still challenge to achieve the reactions at physiological pH. To this end, the idea of development of analyte-replacement fluorescent probes [43] may be preferred in the future.

Overall, fluorescent probes by electrophile-based approaches are the most widely used for H_2S detection. These probes have enabled NIR imaging of intratumoral H_2S in mice as well as LSFM of gut H_2S at whole-organism level. In addition, combination of the fluorescent probes with PA imaging or PDT further provides new opportunities for better imaging or theranostic tools for diverse applications. We expect that electrophile-based probes will become increasingly important tools in revealing different biological mechanisms of endogenous H_2S as well as in the diagnosis and treatment of H_2S-related diseases in the future.

References

1 Filipovic, M.R., Zivanovic, J., Alvarez, B. et al. (2018). Chemical biology of H_2S signaling through persulfidation. *Chem. Rev.* 118 (3): 1253–1337.

2 Wang, R. (2012). Physiological implications of hydrogen sulfide: a whiff exploration that blossomed. *Physiol. Rev.* 92 (2): 791–896.

3 Wallace, J.L. and Wang, R. (2015). Hydrogen sulfide-based therapeutics: exploiting a unique but ubiquitous gasotransmitter. *Nat. Rev. Drug Discov.* 14 (5): 329–345.

4 Szabo, C. and Papapetropoulos, A. (2017). International union of basic and clinical pharmacology. CII: pharmacological modulation of H_2S levels: H_2S donors and H_2S biosynthesis inhibitors. *Pharmacol. Rev.* 69 (4): 497–564.

5 Shatalin, K., Shatalina, E., Mironov, A. et al. (2011). H_2S: a universal defense against antibiotics in bacteria. *Science* 334 (6058): 986–990.

6 Scuffi, D., Nietzel, T., Di Fino, L.M. et al. (2018). Hydrogen sulfide increases production of NADPH oxidase-dependent hydrogen peroxide and phospholipase D-derived phosphatidic acid in guard cell signaling. *Plant Physiol.* 176 (3): 2532–2542.

7 Lin, V.S., Chen, W., Xian, M. et al. (2015). Chemical probes for molecular imaging and detection of hydrogen sulfide and reactive sulfur species in biological systems. *Chem. Soc. Rev.* 44 (14): 4596–4618.

8 Hartle, M.D. and Pluth, M.D. (2016). A practical guide to working with H_2S at the interface of chemistry and biology. *Chem. Soc. Rev.* 45 (22): 6108–6117.

9 Jiang, C.Y., Huang, H.J., Kang, X.Y. et al. (2021). NBD-based synthetic probes for sensing small molecules and proteins: design, sensing mechanisms and biological applications. *Chem. Soc. Rev.* 50 (13): 7436–7495. https://doi.org/10.1039/D0CS01096K

10 Bezner, B.J., Ryan, L.S., and Lippert, A.R. (2020). Reaction-based luminescent probes for reactive sulfur, oxygen, and nitrogen species: analytical techniques and recent progress. *Anal. Chem.* 92 (1): 309–326.

11 Lippert, A.R., New, E.J., and Chang, C.J. (2011). Reaction-based fluorescent probes for selective imaging of hydrogen sulfide in living cells. *J. Am. Chem. Soc.* 133 (26): 10078–10080.

12 Peng, H., Cheng, Y., Dai, C. et al. (2011). A fluorescent probe for fast and quantitative detection of hydrogen sulfide in blood. *Angew. Chem. Int. Ed.* 50 (41): 9672–9675.

13 Henthorn, H.A. and Pluth, M.D. (2015). Mechanistic insights into the H_2S-mediated reduction of aryl azides commonly used in H_2S detection. *J. Am. Chem. Soc.* 137 (48): 15330–15336.

14 Cao, X., Lin, W., Zheng, K. et al. (2012). A near-infrared fluorescent turn-on probe for fluorescence imaging of hydrogen sulfide in live cells based on thiolysis of dinitrophenyl ether. *Chem. Commun.* 48 (85): 10529–10531.

15 Wei, C., Zhu, Q., Liu, W.W. et al. (2014). NBD-based colorimetric and fluorescent turn-on probes for hydrogen sulfide. *Org. Biomol. Chem.* 12 (3): 479–485.

16 Du, Z.B., Song, B., Zhang, W.Z. et al. (2018). Quantitative monitoring and visualization of hydrogen sulfide *in vivo* using a luminescent probe based on a ruthenium(II) complex. *Angew. Chem. Int. Ed.* 57 (15): 3999–4004.

17 Chen, Z.Y., Mu, X.L., Han, Z. et al. (2019). An optical/photoacoustic dual-modality probe: ratiometric *in/ex vivo* imaging for stimulated H_2S upregulation in mice. *J. Am. Chem. Soc.* 141 (45): 17973–17977.

18 Xu, G., Yan, Q.L., Lv, X.G. et al. (2018). Imaging of colorectal cancers using activatable nanoprobes with second near-infrared window emission. *Angew. Chem. Int. Ed.* 57 (14): 3626–3630.

19 Wang, R.C., Dong, K.K., Xu, G. et al. (2019). Activatable near-infrared emission-guided on-demand administration of photodynamic anticancer therapy with a theranostic nanoprobe. *Chem. Sci.* 10 (9): 2785–2790.

20 Shi, B., Gu, X.F., Fei, Q. et al. (2017). Photoacoustic probes for real-time tracking of endogenous H_2S in living mice. *Chem. Sci.* 8 (3): 2150–2155.

21 Shi, B., Ren, N., Gu, L.Y. et al. (2019). Theranostic nanoplatform with hydrogen sulfide activatable NIR responsiveness for imaging-guided on-demand drug release. *Angew. Chem. Int. Ed.* 58 (47): 16826–16830.

22 Wang, R.C., Gu, X.F., Li, Q.Z. et al. (2020). Aggregation enhanced responsiveness of rationally designed probes to hydrogen sulfide for targeted cancer imaging. *J. Am. Chem. Soc.* 142 (35): 15084–15090.

23 Chen, Y.C., Zhu, C.C., Yang, Z.H. et al. (2013). A ratiometric fluorescent probe for rapid detection of hydrogen sulfide in mitochondria. *Angew. Chem. Int. Ed.* 52 (6): 1688–1691.

24 Li, H.D., Yao, Q.C., Jiang, N. et al. (2015). A fluorescent probe for H_2S *in vivo* with fast response and high sensitivity. *Chem. Commun.* 51 (90): 16225–16228.

25 Ren, M.G., Deng, B.B., Kong, X.Q. et al. (2016). A TICT-based fluorescent probe for rapid and specific detection of hydrogen sulfide and its bio-imaging applications. *Chem. Commun.* 52 (38): 6415–6418.

26 Ismail, I., Chen, Z.Y., Sun, L. et al. (2020). Highly efficient H_2S scavengers *via* thiolysis of positively charged NBD amines. *Chem. Sci.* 11 (30): 7823–7828.

27 Zhang, K., Zhang, J., Xi, Z. et al. (2017). A new H_2S-specific near-infrared fluorescence-enhanced probe that can visualize the H_2S level in colorectal cancer cells in mice. *Chem. Sci.* 8 (4): 2776–2781.
28 Hammers, M.D., Taormina, M.J., Cerda, M.M. et al. (2015). A bright fluorescent probe for H_2S enables analyte-responsive, 3D imaging in live zebrafish using light sheet fluorescence microscopy. *J. Am. Chem. Soc.* 137 (32): 10216–10223.
29 Lin, V.S., Lippert, A.R., and Chang, C.J. (2013). Cell-trappable fluorescent probes for endogenous hydrogen sulfide signaling and imaging H_2O_2-dependent H_2S production. *Proc. Natl. Acad. Sci. U.S.A.* 110 (18): 7131–7135.
30 Iciek, M., Chwatko, G., Lorenc-Koci, E. et al. (2004). Plasma levels of total, free and protein bound thiols as well as sulfane sulfur in different age groups of rats. *Acta Biochim. Pol.* 51 (3): 815–824.
31 Kolluru, G.K., Shen, X., Bir, S.C. et al. (2013). Hydrogen sulfide chemical biology: pathophysiological roles and detection. *Nitric Oxide* 35 (30): 5–20.
32 Yue, Y.K., Huo, F.J., and Yin, C.X. (2021). The chronological evolution of small organic molecular fluorescent probes for thiols. *Chem. Sci.* 12 (4): 1220–1226.
33 Liu, Y.N., Yu, Y.N., Zhao, Q.Y. et al. (2021). Fluorescent probes based on nucleophilic aromatic substitution reactions for reactive sulfur and selenium species: recent progress, applications, and design strategies. *Coord. Chem. Rev.* 427 (15): 213601.
34 Shen, X.G., Peter, E.A., Bir, S. et al. (2012). Analytical measurement of discrete hydrogen sulfide pools in biological specimens. *Free Radic. Biol. Med.* 52 (11): 2276–2283.
35 Montoya, L.A., Shen, X., McDermott, J.J. et al. (2015). Mechanistic investigations reveal that dibromobimane extrudes sulfur from biological sulfhydryl sources other than hydrogen sulfide. *Chem. Sci.* 6 (1): 294–300.
36 Montoya, L.A., Pearce, T.F., Hansen, R.J. et al. (2013). Development of selective colorimetric probes for hydrogen sulfide based on nucleophilic aromatic substitution. *J. Org. Chem.* 78 (13): 6550–6557.
37 Cheng, H.B., Li, Y., Tang, B.Z. et al. (2020). Assembly strategies of organic-based imaging agents for fluorescence and photoacoustic bioimaging applications. *Chem. Soc. Rev.* 49 (1): 21–31.
38 Wei, L., Yi, L., Song, F. et al. (2014). FRET ratiometric probes reveal the chiral-sensitive cysteine-dependent H_2S production and regulation in living cells. *Sci. Rep.* 4: 4521.
39 He, S.Q., Song, J., Qu, J.L. et al. (2018). Crucial breakthrough of second near-infrared biological window fluorophores: design and synthesis toward multimodal imaging and theranostics. *Chem. Soc. Rev.* 47 (12): 4258–4278.

40 Zhou, Z., Song, J., Nie, L. et al. (2016). Reactive oxygen species generating systems meeting challenges of photodynamic cancer therapy. *Chem. Soc. Rev.* 45 (23): 6597–6626.

41 Yang, C., Wang, Y., Marutani, E. et al. (2019). Data-driven identification of hydrogen sulfide scavengers. *Angew. Chem. Int. Ed.* 58 (32): 10898–10902.

42 Thorson, M.K., Majtan, T., Kraus, J.P. et al. (2013). Identification of cystathionine β-synthase inhibitors using a hydrogen sulfide selective probe. *Angew. Chem. Int. Ed.* 52 (17): 4641–4644.

43 Steiger, A.K., Pardue, S., Kevil, C.G. et al. (2016). Self-immolative thiocarbamates provide access to triggered H_2S donors and analyte replacement fluorescent probes. *J. Am. Chem. Soc.* 138 (23): 7256–7259.

44 Wang, J., Wang, X., Fan, X. et al. (2021). Unleashing the power of bond cleavage chemistry in living systems. *ACS Cent. Sci.* 7 (6): 929–943.

45 Brandenberg, O.F., Miller, D.C., Markel, U. et al. (2019). Engineering chemoselectivity in hemoprotein-catalyzed indole amidation. *ACS Catal.* 9 (9): 8271–8275.

46 O'Connor, L.J., Mistry, I.N., Collins, S.L. et al. (2017). CYP450 enzymes effect oxygen-dependent reduction of azide-based fluorogenic dyes. *ACS Cent. Sci.* 3 (1): 20–30.

9

Fluorescent Probes for H_2S Detection: Metal-Based Approaches

Maria Strianese and Claudio Pellecchia

Dipartimento di Chimica e Biologia, Università degli Studi di Salerno, Fisciano, SA, Italy

9.1 Introduction

Hydrogen sulfide (H_2S) is a colorless, flammable gas with a characteristic odor of rotten eggs [1]. Over the past two decades, next to its traditional toxic and harmful role, H_2S has gained increasing attention because it was recognized to occur naturally in the human body [2–4]. The concentration of H_2S is reported to be in the submicromolar range for endogenous H_2S with exogenous levels often being higher to generate a physiologic response, whereas the intracellular H_2S levels generated in response to physiological stimuli are still controversial [5, 6]. Innovative and revolutionary studies established that H_2S is a biological molecule that mediates important functions within the human body through its action on bioinorganic targets, joining NO and CO in the family of gasotransmitters [2, 7]. By now, the biological reactivity of NO and CO has been widely clarified thanks to the numerous papers focusing on their coordination chemistry to bioinspired metal complexes [8–11]. H_2S is a weak acid that equilibrates in aqueous solution with its anions HS^- and S^{2-} thus complicating the studies on H_2S reactivity in biological media and the clarification of the specific chemistry associated with the specific protonation state [12]. Selective tracking of this molecule in physiological conditions is particularly relevant to deepen its complex contributions to both healthy and disease states. In this context, the challenge calls for H_2S sensors that are sensitive, rapid in their response, and cost-effective so that they can be widely deployed [13, 14]. The possibility of dynamic monitoring of this gas molecule in biological samples resulted in a surge of interest with the aim of devising systems for its detection. Early approaches for H_2S detection are based on electrochemical and chromatographic techniques [15]. Also methods based on polarography and carbon nanotubes coupled with a Raman and/or a confocal laser have shown promising results [16–18]. While each technique has its own advantages, there remains a need for simple ones to be implemented and cheap H_2S sensors that can be used repeatedly and that do not require sample processing and/or destruction of tissues or cell lysates. With fluorescence-based methods it is possible, in principle, to develop such a sensor. Fluorescence-based methods offer several advantages over other analytical procedures, such as the use of very simple and inexpensive instrumentation. Indeed, fluorescence is by far the method most often applied and is the dominant analytical approach in a large variety of schemes in the fields of medical testing, biotechnology, and drug discovery [19, 20]. In its numerous variations, fluorescence has become the most powerful bioanalytical and diagnostic tool and, in terms of versatility, seems to be second only to NMR spectroscopy. One of the reasons of the tremendous increase in the popularity of fluorescence spectroscopy in the last two decades is probably the fact that in the 1980s the first chemically synthesized fluorescence probes for specific analytes became available [20].

An effective fluorescence-based probe should respond in a selective manner and with high sensitivity to its target analyte. Furthermore, water solubility of the sensing material is a highly favorable characteristic for practical biological measurements. "Turn-on" systems that harness fluorescence enhancement as a result of the recognition event are inherently more sensitive than "turn-off" sensors that exploit fluorescence quenching upon analyte binding.

This chapter focuses on the recent progress in the field of fluorescence-based probes for the detection of H_2S. Existing fluorescent probes for H_2S detection are currently classified considering the mechanism of the reaction by which analyte recognition occurs [21–24]. So far they have been subdivided into six different categories: (*i*) azide to amine reduction [25–32]; (*ii*) nucleophilic addition [33–36]; (*iii*) nitro to amine reduction [37, 38]; (*iv*) metal-displacement [39–45]; (*v*) coordinative-based approach [12, 13, 46–52]; and (*vi*) H_2S-mediated reduction of the metal center [46, 53].

Sensors belonging to categories *i–iii* are based on the use of organic molecules, which change their fluorescence intensity upon interaction with H_2S. Differently, for sensors belonging to categories *iv–vi*, the sensing probe is a metal-containing molecule. However, while for sensors belonging to category *iv*, recognition occurs via the displacement of the metal from the fluorescent ligand with the concomitant H_2S-mediated precipitation of Mt-S to generally produce fluorescence "turn-on" changes [21, 39–41, 54–57], sensors of category *v* exhibit H_2S or HS^- binding to the metal center. As for the sensors of category *vi*, their interaction with H_2S results in the reduction of the metal center.

Usually fluorescent sensors of categories *i–iv* are "one-pot devices" because of the irreversibility of the reaction by which H_2S recognition occurs. With a coordinative-based approach, however, it is possible to remove H_2S from the metal center of the sensor to generate a reversible or cyclable H_2S binding process. This condition is highly desirable for practical sensing applications, which would allow for reusability of the sensing device. Also sensors belonging to category *vi* may be reusable in the case of cyclable systems [46].

More specifically, in this chapter, we focus on the most representative fluorescence-based probes which work via a metal-dependent strategy, thus readers interested in sensors belonging to categories *i–iii* are referred to previous reviews on this topic [22, 24, 35, 58].

9.2 Metal Displacement Approach

9.2.1 Copper-Based Systems

The presence of paramagnetic Cu^{2+} in a molecule usually results in a significant quenching of the fluorescence emission, as largely documented in the literature [39, 40, 42, 43, 59]. Indeed, Cu^{2+} is a transition metal with an incompletely filled d shell, which allows for quenching of the fluorophore fluorescence by energy or electron transfer between the d orbitals of the metal ion and the excited states of the fluorophore [60]. On the other hand, it is well acknowledged that sulfide reacts with copper complexes, resulting in the formation of a very stable CuS species (with a low solubility-product constant, $K_{sp} = 1.27 \times 10^{-36}$) [61, 62]. By exploiting these two favorable conditions, in the past few years, several copper complexes have been implemented as efficient fluorescent H_2S sensors. Their working principle builds on copper-mediated precipitation of sulfide, with the concomitant release of an organic ligand to restore the initial fluorescence intensity of the free ligand [39–44, 55, 62–64]. Under these experimental conditions, the free sulfide anion (S^{2-}) does not exist [65], but this does not negatively impact MtS precipitation.

Most copper complexes proposed as H_2S sensors via the above metal displacement recognition approach are di-(2-picolyl)amine (DPA) based molecules, quinoline-based compounds, cyclen and cyclam derivatives, imidazole and benzimidazole complexes, or Schiff-base complexes.

One of the first examples of H_2S sensing using this approach was reported in 2009 [40] and makes use of a copper complex with a DPA-fluorescein ligand (**1**).

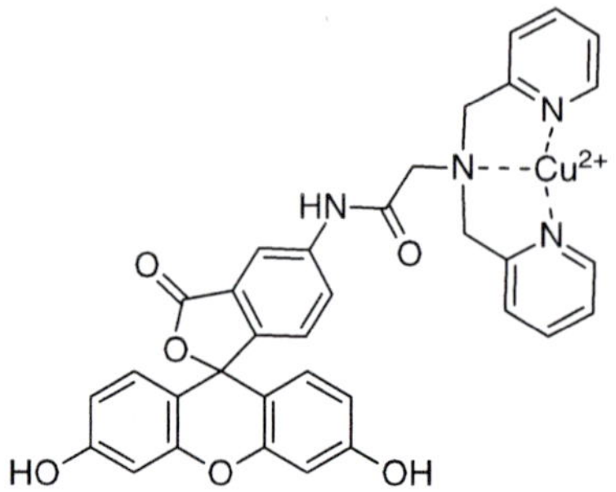

(**1**)

The authors report a selective "turn-on" response of the system to sulfide: an 87-fold increment in the emission intensity. The limit of detection (LOD) was reported as 10 μM of H_2S in aqueous solution at pH 7.4. These features render this system useful for measurements of H_2S in physiological media. By using 7-diethylamino-coumarin as fluorescent unit coordinated to the DPA moiety, Yin and coworkers in 2013 proposed a different copper complex (**2**) for H_2S detection, which works with a similar approach [66]:

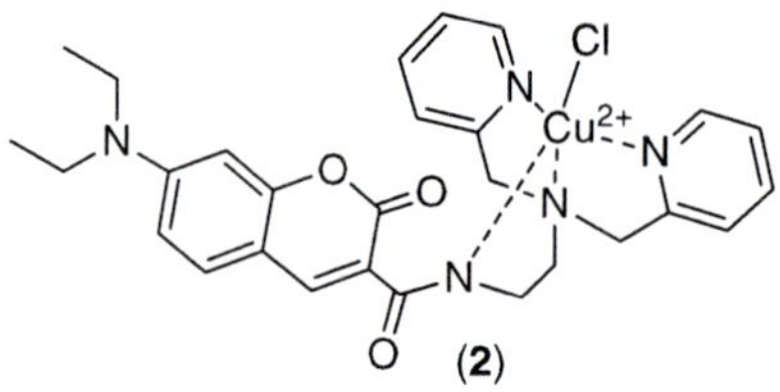

(**2**)

Again, a selective "turn-on" response to sulfide was observed and LOD was quoted to be 14 nM of H_2S. Accurate determination of sulfide in tap water samples by (**2**) established its high potential for real samples measurements. By further changing the fluorescent unit with boron-dipyrromethene (BODIPY) and keeping the DPA-copper chelating unit, another example of a sensor working via the metal extrusion strategy was reported in 2014 (**3**) [67]. The authors claimed that the new ligand framework resulted in enhanced solubility for the resulting system and in a more striking fluorescence response, however, lower sensitivity for **3** was measured (a LOD of 250 nM was quoted in the case of **3** versus 14 nM for **2**). Compound **3** was shown to function also for HeLa cell imaging studies.

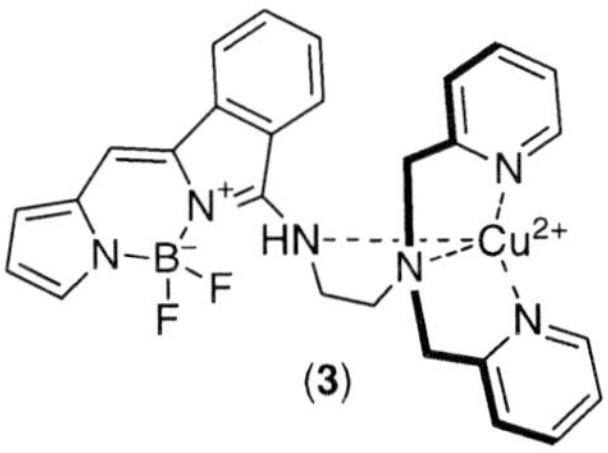

(3)

By introducing a new disubstituted polyacetylene as the fluorescent unit and keeping the DPA motif as a chelating unit for the copper ion, the group of Li and coworkers reported a new and well-performing H_2S sensor via the same copper displacement approach (**4**) [68].

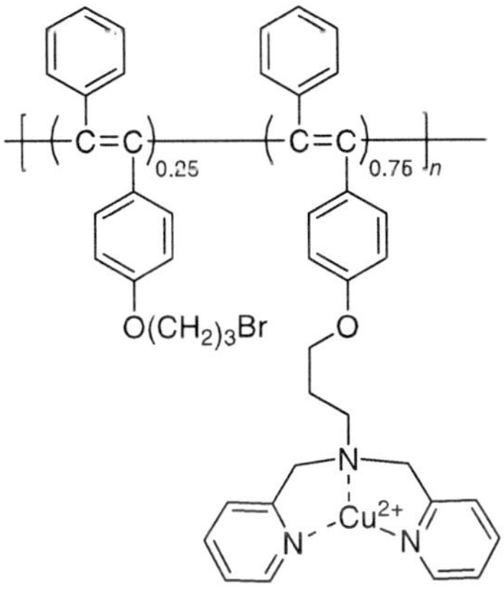

(**4**)

The authors report that the strong green fluorescence of the polyacetylene bearing pyridine moieties in the side chains could be completely quenched by Cu^{2+} even at a very low concentration (2.0×10^{-6} mol/L) and restored upon addition of sulfide anion, with the detection limit down to 5.0×10^{-7} mol/L. High selectivity against a number of possible competing species was observed. Compound **4** represents one of the main examples of polymer-based fluorescent probe for H_2S detection.

In 2011, the group of Zhu and coworkers reported complex **5** [69]. In the case of **5**, the DPA-Cu chelating unit is maintained and an additional 8-aminoquinoline metal coordinating unit and a BODIPY moiety were introduced. The choice of this modification of the ligand framework was driven by the fact that quinoline and DPA had been previously used for the implementation of copper-sensing systems [70–73]. The authors reported a 50 nm red shift of the absorption (from 520 to 569 nm) upon addition of HS^- to the sensing construct and a 34-fold ratiometric enhancement, which indicates that **5** can detect HS^- both via absorption colorimetry and via absorption ratiometry. Furthermore, the authors claim high selectivity for this system toward H_2S against a range of competitive thiols [74] and anions. Probe **5**, however, was reported to be less sensitive than the other DPA-systems with a LOD of 0.167 μM of H_2S, which is possibly due to the strong coordination of copper to the ligand bearing the additional 8-aminoquinoline coordination moiety.

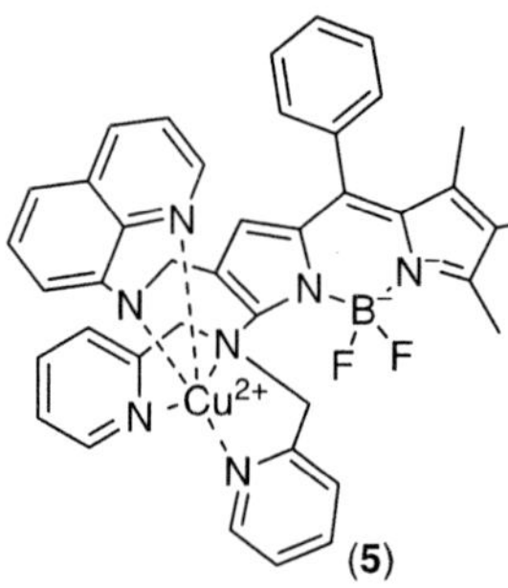

(**5**)

By changing the DPA chelating unit with 8-hydroxyquinoline as copper coordinating group and using fluorescein as signaling unit, Zheng, and coworkers synthesized probe **6** [41]. The authors claim high selectivity for the implemented compound toward H_2S against other common ions and sulfate. In addition, experiments in living cells demonstrated that the sensor has promising biological applications. A clear color variation from yellow to pink in the presence of H_2S constitutes a further positive feature for probe **6**.

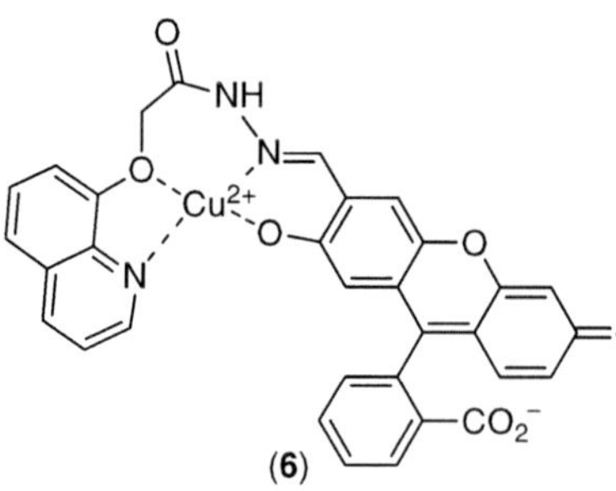

(**6**)

When coupling the fluorescein signaling unit with the azamacrocyclic ligand (cyclen) as Cu^{2+}-chelating moiety, the group of Nagano introduced a new family of very successful Cu-based sensors for H_2S that act via the above metal displacement approach [39]. The working mechanism of probe **7** is displayed in Figure 9.1. Upon addition of 10 μM H_2S, the H_2S-triggered Cu^{2+} removal results in a large

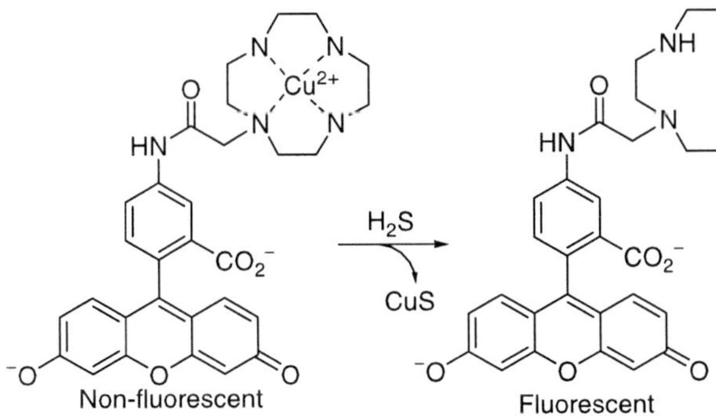

(7)

Figure 9.1 Working mechanism for compound (**7**).

and immediate increment of fluorescence intensity by 50-fold, whereas almost no fluorescence increase is observed upon addition of 10 mM GSH. Furthermore, fluorescence imaging measurements of H_2S in intracellular environments demonstrated the high potential of **7**. In the same paper, the authors use different macrocycles such as triazacyclononane (TACN), tetraazacyclotetradecane (cyclam), and trimethylcyclen (TMCyclen) for the coupling with fluorescein, but the resulting Cu complexes resulted in less-efficient H_2S sensing systems.

Using a related approach, a multifunctional fluorescent Cu(II)-containing sensor based on a cyclen-appended BINOL (1,1′-Bi-2-naphthol) derivative (**8**) was prepared by the group of Yu et al [43]. The LOD for this system was estimated to be 16 μM of H_2S in 100% HEPES buffer (10 mM, pH 7.4). Furthermore, this compound showed high selectivity for H_2S over competitive biothiols, commonly tested anions, and inorganic sulfur compounds.

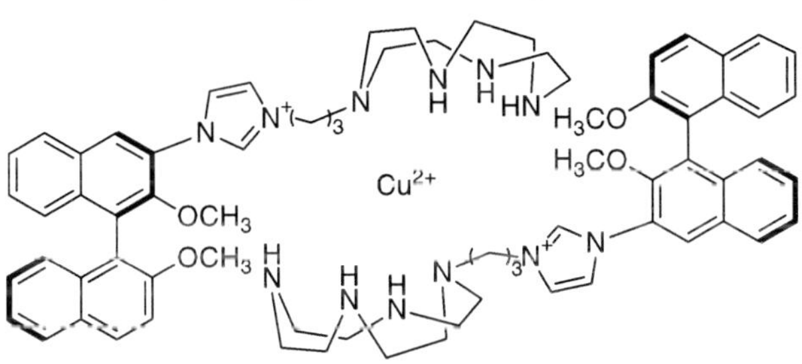

(**8**)

By using a cyclam scaffold as Cu-chelating moiety and an anthracene group as fluorescent unit, the group of Martinez-Manez prepared (**9**) an H_2S sensor, which functions via the same approach [55]. High selectivity toward H_2S over other competing species and a 3.9 μM LOD in HEPES buffer (3 mM, pH 7.5) were reported.

Real-time fluorescence imaging experiments demonstrated that (**9**) can be used to successfully detect HS^- in intracellular environments.

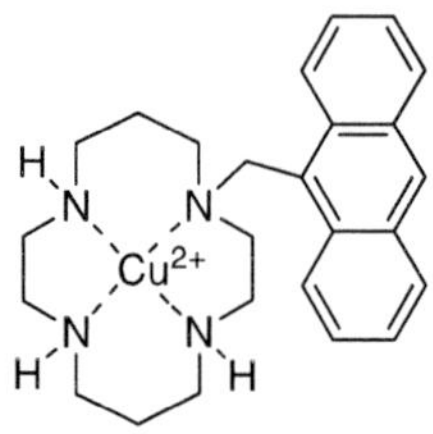

(**9**)

Compounds **10** [75] and **11** [62] represent two other members of the azamacrocyclic-containing Cu^{2+}-based probes for H_2S detection that function via the metal displacement strategy. More specifically, **10** is composed of a macrocyclic ring as the Cu^{2+}-coordinating unit and a naphthalimide moiety as fluorescent signaling unit, whereas **11** uses a tetraazacyclododecane chelated with Cu(II) and a methyl substituted antracenyl group as fluorophore. The ligand surrounding the copper in **11** also exhibits a triazole group. The authors reported high selectivity of probe **10** for H_2S, a 24-fold fluorescence enhancement, and a detection limit of 0.7 μM. The probe also provides a color change from yellow to green in the presence of H_2S that is visible by the naked eye. Probe **11** is completely water soluble, very selective toward H_2S, and has a detection limit of 205 nM. Fluorescence cell imaging and in vivo imaging of H_2S in zebrafish were also reported for **11**. These experiments render probe **11** highly effective when compared to other cyclen-based probes for the detection of sulfide in real sample analysis.

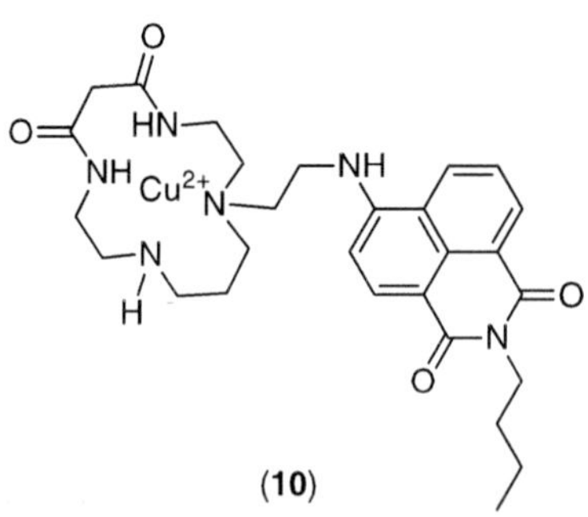

(**10**)

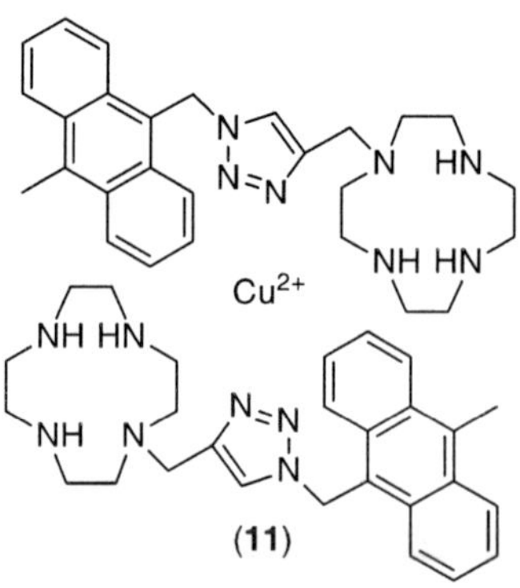

(**11**)

A few years later, our group also proposed a cyclam-based fluorescent copper complex for H_2S detection (**12**) [44]. In this case, a pendant amide-pyrene moiety was included in the ligand skeleton as fluorescent unit. Upon H_2S addition, **12** provides a color change (visible to the naked eye) from blue-green to yellow with a concomitant enhancement of the fluorescence emission. NMR spectroscopy in addition to ESI mass spectrometry provided evidence that **12** operates by the copper displacement recognition mechanism. Selectivity experiments demonstrated that compound **12** was selective for H_2S recognition against a wide range of possible competing species.

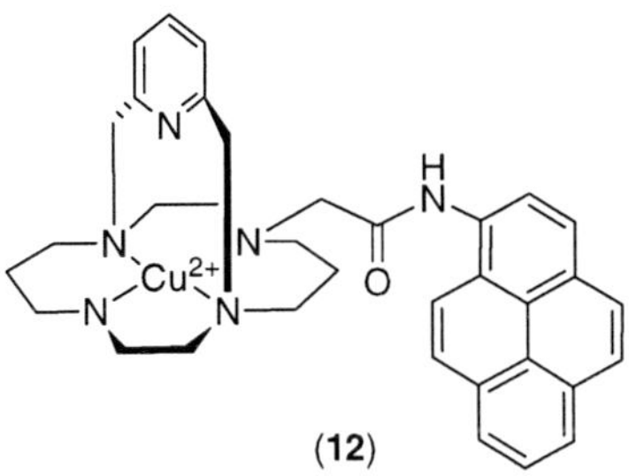

(12)

Imidazole and benzimidazole-based copper complexes have also been largely proposed as effective H_2S fluorescent probes by harnessing the copper displacement sensing approach. Figure 9.2 shows the main examples of probes belonging to this class (**13–18**).

Probe **13** was implemented by the group of Wang [76] and displays a 8-hydroxyquinoline and 1H-phenanthro [9,10-*d*]imidazolo motif as chelating units for Cu^{2+}. In the presence of H_2S, **13** showed a 20-fold fluorescence enhancement with a LOD of 0.18 μM in Tris–HCl/DMF (20 mM, 6/4, v/v). Experiments in live cells reflect the biological potential of **13**. Compound **13** functions over a wide range of pH (3.5–8.5), which is typically a challenge in most of the DPA or macrocyclic-based receptors for the detection of H_2S and highlight its potential.

Compound **14** represents another member of the benzimidazole-based fluorescent sensors for the recognition of sulfide in water [77]. The advantages of **14** are the selectivity for H_2S over biothiols as well as other competitive anions and the LOD of 1.35 μM. Compound **14** was also shown to successfully function as an H_2S indicator in real water samples. Compound **15** represents an example of benzimidazole-based quinoline containing sensor for the selective detection of H_2S [78]. The LOD for **15** was reported to be 0.46 μM. The potential for biological applications of **15** was assessed by successful cell imaging experiments. Probe **16** is composed of the 3-(1H-benzimidazol-2yl)-2-hydroxy-benzoate-sodium ligand chelating the copper center, which renders the sensing molecule water-soluble. For **16**, a LOD of 2.51 μM and a color change from colorless to yellow in the presence of H_2S was reported [79]. Fluorescence lifetime experiments with this construct confirmed that the usual copper displacement approach is operative. Complex **17** exhibits a phenylbenzimidazole unit as a fluorescent molecule. The authors report 100% water solubility for **17** and a two wavelength response for the recognition of sulfide in water [80]. A LOD of 0.912 μM was reported and a good

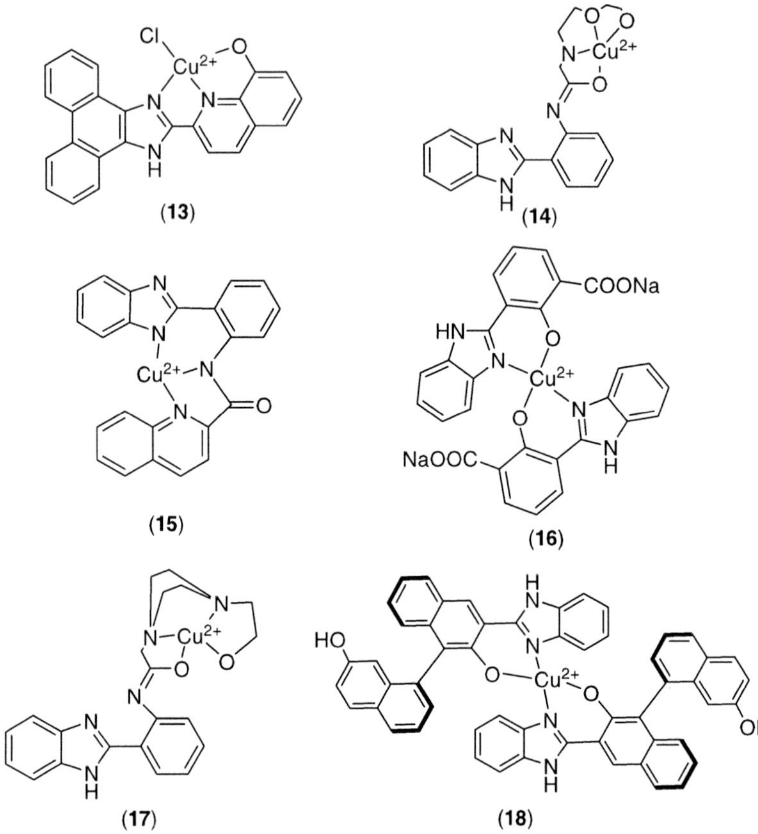

Figure 9.2 Imidazole and benzimidazole-based copper complexes for H_2S sensing.

selectivity was demonstrated; however, no biological experiments were conducted for probe **17**. A binaphthol-benzimidazole ligand is used as the copper chelating motif in probe **18** [81]. For **18**, a LOD of 11 μM was reported, which in addition to the good selectivity for H_2S over different competing thiols and the possibility of functioning for a range of pH values render it a successful sensor. Test strips doped with **17** open the possibility of gaseous H_2S detection in air, which enhances the potential of this sensing construct respect to other benzimidazole-based copper complexes.

Another family of representative copper complexes for use as H_2S sensors functioning via the sulfide-mediated precipitation of copper are those containing the rhodamine moiety as the fluorescent unit. Figure 9.3 displays the main examples belonging to this family for Cu^{2+} based H_2S probes.

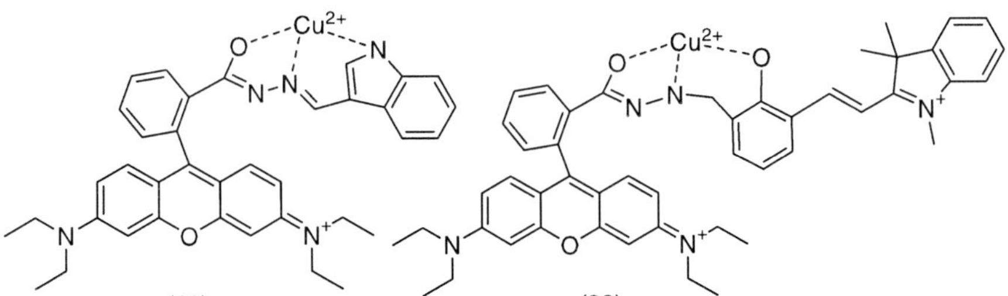
(19) (20)

Figure 9.3 Rhodamine-based systems for Cu^{2+}-mediated H_2S sensing.

The two fluorophores in probe **19** (indole and rhododamine) were chosen since they are a FRET pair [82]. Binding of Cu^{2+} to the indole-functionalized rhodamine results in opening of the spirolactam ring of the rhodamine derivative, whose absorption spectra shows a significant spectral overlap with the emission spectrum of the indole moiety and makes the FRET process possible. An increased absorbance in the near-infrared (NIR) region of the UV-vis spectra was also registered. A concomitant change from colorless to blue was observed when Cu^{2+} binds to the indole-functionalized rhodamine. When adding sulfide to the solution containing the copper complex, the usual precipitation of copper was observed, and the fluorescence of the functionalized rhodamine restored. The LOD for sulfide was reported to by 15 μM, and live cell imaging experiments in HeLa cell lines assessed the biological potential of **19**.

A related approach building on the spirolactam ring reformation, after the displacement of the metal ion, was harnessed by Cai and coworkers in the case of probe **20** [83]. In this example, the FRET pair is comprised of rhodamine and 2,3,3-trimethylindolinium. The resultant probe showed good selectivity toward possible competitive thiols and could be used for bioimaging experiments in live cells. However, the time-consuming multistep synthesis of probe **20** constitutes a drawback over some of the more simple alternatives.

Schiff-base copper complexes working as H_2S sensors via the metal displacement approach have been also proposed over the years. Key examples of these are shown in Figure 9.4 (**21–24**).

More specifically, probe **21** behaves quite similarly to the above copper sensors for H_2S detection, showing a turn-on signal in the presence of sulfide with a LOD of 0.949 μM [84]. Probe **22** ((E)-7-(dimethylamino)-3-((2-(pyridine-2-yl)hydrazono) methyl)naphthalene-2-ol)) was reported as a successful turn-on sensor for H_2S detection in HEPES buffer (20 mM, pH 7.4) [85]. Good selectivity against a series of possible competing thiols was reported. Further, cell-imaging experiments in HeLa cells demonstrated that the system can be used to successfully detect HS^- in intracellular environments.

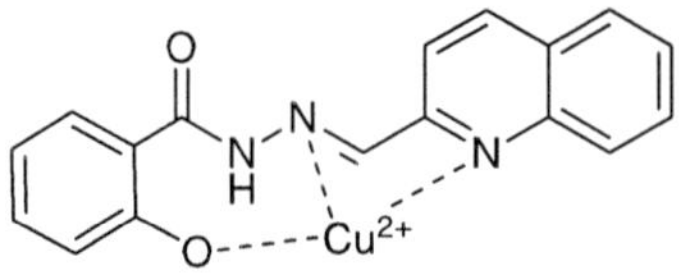

(21)

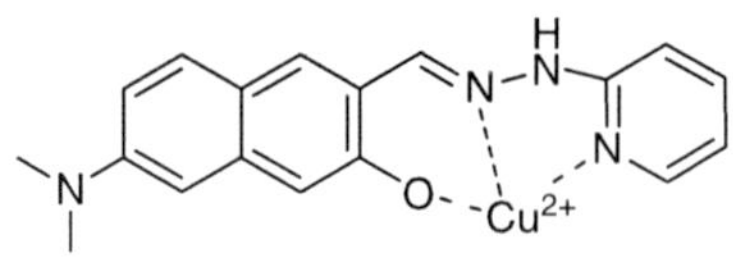

(22)

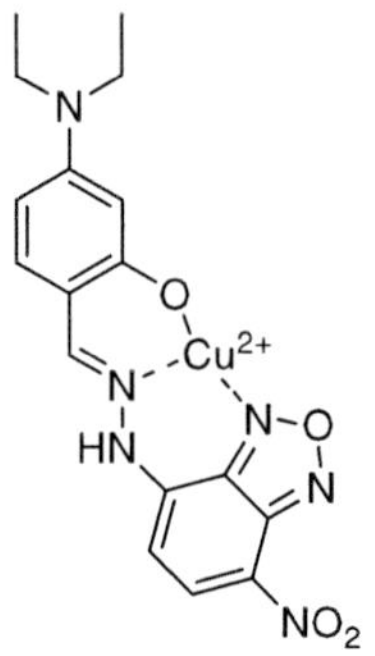

(23)

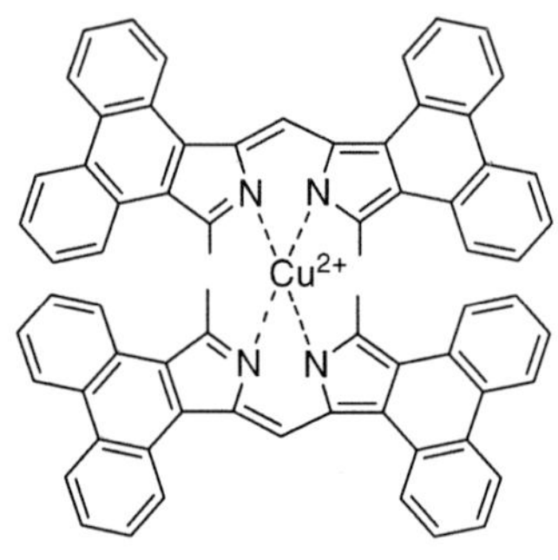

(24)

Figure 9.4 Schiff-base copper-containing systems for H_2S sensing.

Complex **23** contains a 7-nitrobenz-2-oxa-1,3diazole motif as copper chelating ligand and was reported to successfully work for H_2S detection in HEPES buffer (20 mM, pH 7.4) with a LOD of 0.17 μM [86]. Probe **23** showed high selectivity for H_2S and was successful used in live cell imaging experiments using MDA-MB-231 cells.

Probe **24** is a copper-based phenanthrene-fused dipyrromethene complex, which acts as a further example of a well-performing H_2S sensor via the metal displacement approach [87]. The LOD was reported to be 1 μM, and the probe was also shown as highly selective toward H_2S among competitive anions and biothiols. Live cell imaging experiments in HeLa cells assessed its biological potential.

9.2.2 Zinc-Based Systems

After copper, zinc-based complexes have been the most widely exploited to implement H_2S sensors harnessing the metal displacement approach, due to the low solubility product of ZnS (1.6×10^{-24}) [88]. However, unlike the copper systems discussed above, metal ions such as Zn^{2+} are not fluorescent quenchers, thus zinc complexes frequently result in turn-OFF fluorescent sensors for H_2S when working via the metal extrusion strategy. In this context, our group reported one of the first zinc complexes for H_2S harnessing this strategy (**25**) [57]. As stated in the paper, addition of H_2S significantly quenched the fluorescence of **25**. NMR experiments provided evidence that addition of H_2S to **25** results in zinc precipitation with the concomitant progressive hydrolysis of the heteroscorpionate ligand initially coordinated to the zinc center.

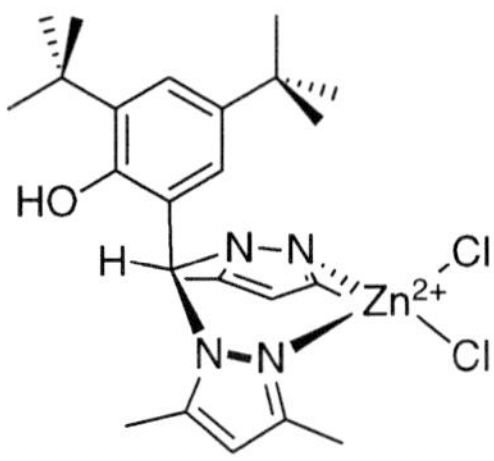

(**25**)

The Alizarin Red S (ARS)-based fluorescent zinc complex **26** represent another member of this class of complexes that was reported for the successful selective detection of H_2S [89]. An advantage of **26** is that it is a commercially available compound and exhibits a very low LOD of 92 nM. In addition, **26** was also shown to work for monitoring H_2S release from an H_2S donor molecule (benzoic(methyl carbonic) dithioperoxyanhydride) and for the detection of H_2S in live cells.

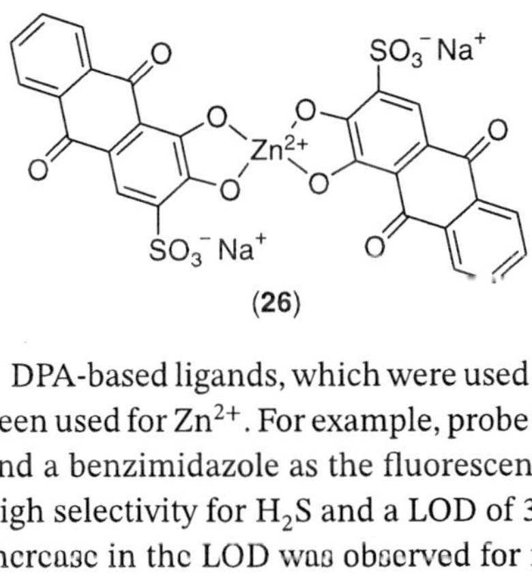

(**26**)

DPA-based ligands, which were used above for Cu^{2+}-based H_2S probes, have also been used for Zn^{2+}. For example, probe **27** exhibits a DPA ligand to chelate the Zn^{2+} and a benzimidazole as the fluorescent signaling unit [90]. The authors reported high selectivity for H_2S and a LOD of 3.46 μM for in vitro measurements. A small increase in the LOD was observed for real water samples (10–50 μM), which renders probe **27** a good candidate for environmental monitoring of H_2S.

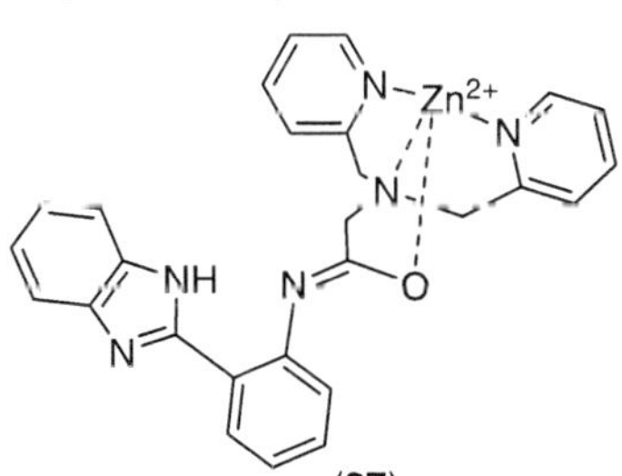

(**27**)

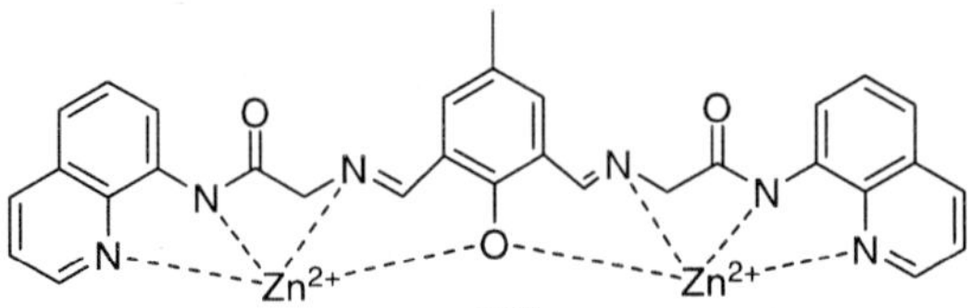

(28)

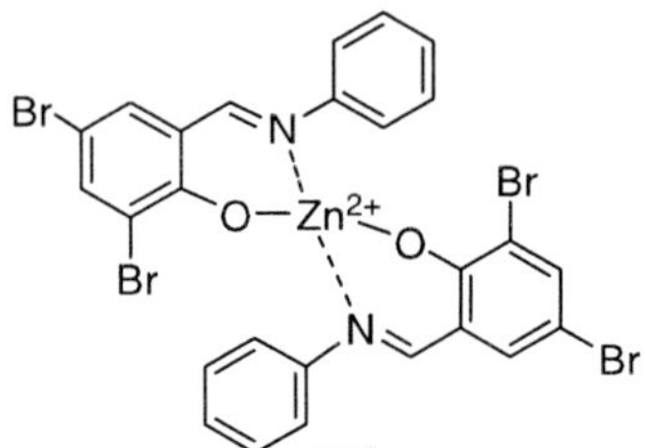

(29)

Figure 9.5 Schiff-base zinc systems for H_2S detection.

Figure 9.5 shows two main zinc-based salen complexes (**28**, **29**) that have been reported as viable H_2S sensors and use the metal extrusion approach. Probe **28** is composed of glycine-*N*-8-quinolylamide and 4-methyl-2,6-diformyl phenol moieties, and showed high selectivity for H_2S among various anions. Probe **28** also has a LOD for H_2S that was found to be in the micromolar range [91]. In 2019, Wang and coworkers reported probe **29** [92] as a colorimetric HS^- sensor that was selective for H_2S with a 192 μM LOD for H_2S. Naked-eye detection of H_2S is also possible with **29** because of the color change from yellow to red that occurs in the presence of H_2S.

9.2.3 Different Metal-Based Systems

The sulfide-mediated extrusion of a metal to develop H_2S sensors is, of course, strictly dependent on the affinity of the sulfide ions to the metal center, which in turn reflects the solubility product of the MtS final product. That is to say, the higher the affinity of sulfide ions to the metal, the lower the K_{sp} of the MtS. Given this, it is easy to understand why early H_2S sensors harnessing the metal displacement approach were based on Hg complexes (for HgS $K_{sp} = 2\times10^{-53}$) [93]. Indeed, probe **30** dates back to 1999, and it is a complex of Hg(II) and 2,2′-pyridylbenzimidazole for which the authors claimed possible applications for the monitoring of sulfides in the oral cavity [94]. A LOD of 4 ppb for H_2S was reported for compound **30**.

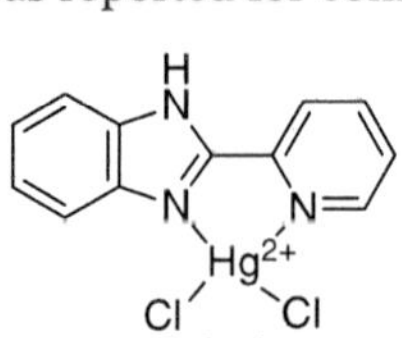

(30)

Much more recently, the mercury displacement strategy has been further leveraged by Chattopadhyay and coworkers who designed rhodamine-based-complex (**31**). This compound harnesses the FRET process, which starts with the opening of the spirolactam ring of rhodamine upon Hg^{2+} coordination. Upon addition of H_2S, Hg^{2+} precipitates as HgS and abolishes the FRET between the two fluorophores [95]. Probe **31** displayed high selectivity for H_2S over various possible competing anions, and the LOD for H_2S was reported as 0.17 μM. Detection of H_2S in HeLa cells was also performed showing the biological potential of **31**.

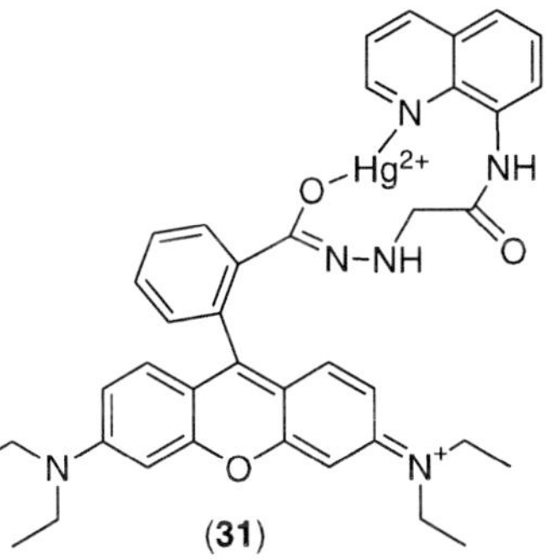

(**31**)

Another example of the mercury-based probes contains a phenothiazine-diaminomalenonitrile ligand coordinating to the mercury ion (**32**) [96]. In addition to responding to H_2S, both cysteine and homocysteine induced an increase in the fluorescence intensity of **32**, so this system is not highly selective. The detection limit was not reported, but successful measurements in HeLa cells confirmed the potential of the probe in a biological context.

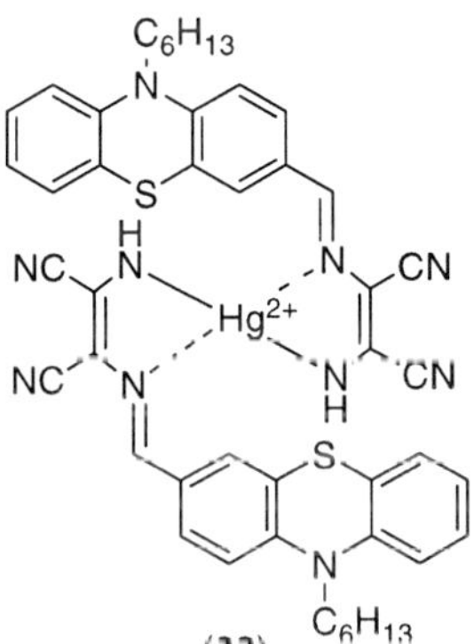

(**32**)

There is also an example of a tin complex as an H_2S sensor via the same metal displacement approach. This compound is a rhodamine-based tin complex (**33**) bearing an iminophenol-azodye [97]. A huge fluorescence quenching (2300 fold) was reported for **33** upon reaction with H_2S, and imaging experiments were also reported in live cells, which highlights its biological potential.

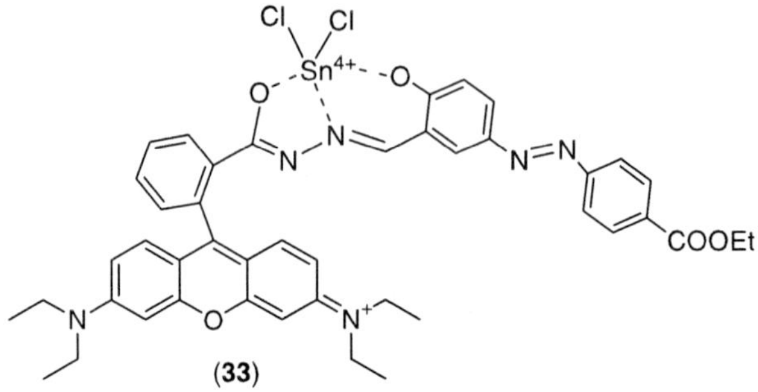

(33)

9.3 Coordinative-Based Approach

For practical measurements, a limitation of fluorescent sensors is the irreversibility of the reaction with the analyte, which means that the probes are not reusable. Seeking to expand the number of reusable sensing devices, we and others have advanced an approach that builds on a coordinative-based approach [13, 46, 98]. With a coordinative-based approach one may, in principle, be able to remove the analyte from the metal center of the sensor and allow for reversible binding. This reversibility is advantageous for practical-sensing applications ensuring reusability of the sensing device.

9.3.1 Metalloporphyrin-Based Systems

Over the last few decades, metalloporphyrin-based sensors have been extensively investigated [99]. Extensive conjugation of 18 electrons in porphyrins gives rise to $\pi \rightarrow \pi^*$ transitions that result into two distinct bands within the visible region of the electromagnetic spectrum: an intense absorption band called B band or Soret band between 350 and 500 nm and a less intense band known as Q-band usually observed between 500 and 700 nm. These optical features render porphyrins promising candidates to develop optical sensors.

On the other hand, the well-known interaction between sulfide and heme iron is very important in organisms that live in sulfide-rich environments and have specialized hemoglobins for binding and delivering H_2S to symbiotic bacteria that extract energy from sulfide oxidation [100]. Thus, heme centers represent the most common targets of metal-based reactivity, and an upsurge of investigations on H_2S reactivity and sensing systems have focused on iron-porphyrin systems [101, 102]. In this section of the chapter, we focus on metalloporphyin-based optical devices for H_2S detection either when the sensing unit is embedded in a natural protein matrix or when it is a synthetic complex.

9.3.1.1 Synthetic Systems

By making use of the abovementioned coordinative-based approach, our research group reported a simple copper protoporphyrin complex (CuPPIX, **34**) as a "turn on" fluorescent probe for the detection of H_2S in aqueous media [98]. We observed a significant enhancement of the S_2 fluorescence emission. We also observed a clear dependence of the fluorescence intensity on KSH concentration. However, even though a selective "turn on" fluorescence in the presence of H_2S was observed against a wide range of possible competing species, the compound required basic aqueous solution to function, which is a limitation of the system.

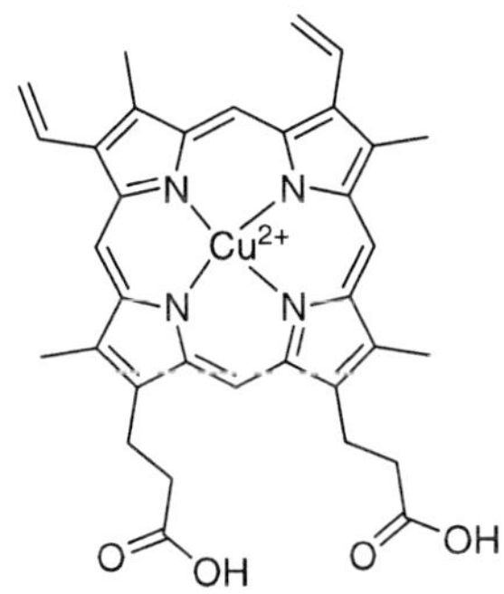

(**34**)

To improve the copper porphyrin complex, we synthesized a series of zinc porphyrins [48, 103]. These complexes are efficient scaffolds to stabilize HS^- binding at the zinc center and thus can successfully act as HS^- sensors via a coordinative-based approach. In this context, our best performing probe was found to be the *tetra*-(*N*-methylpyridyl)porphine zinc complex (TMPyPZn, complex **35**), which works in buffered solutions, at physiological pH and in a reversible manner [48].

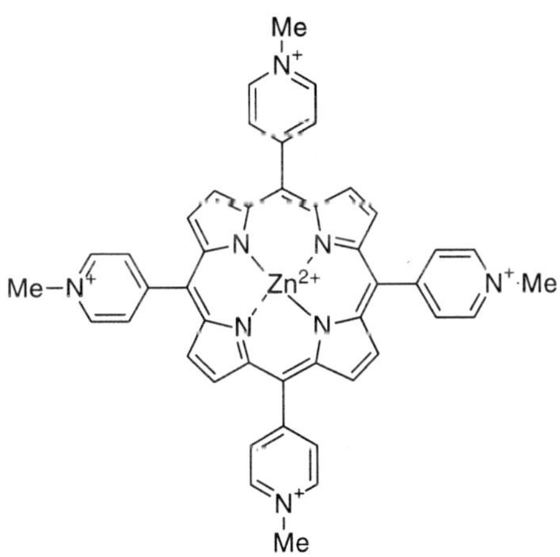

(**35**)

Contemporarily, the Pluth group independently reported the reaction of H_2S and/or NBu_4SH with a group of metal-protoporphyrin-IX (PPIX) complexes [12]. They explored the reactivity of H_2S/HS^- with Mg(II), Cu(II), Zn(II), Sn(IV), Cr(III), Co(II), and Mn(III/II) PPIX complexes to better clarify the fundamental reactivity of each of these metals with sulfide species. In particular, as for the CuPPIX system discussed above (**34**), they reported that in organic solution, CuPPIX does not react with H_2S or HS^-, suggesting that solvent effects and/or the different protonation states of H_2S may play an important role in the observed reactivity of sulfide with metal systems.

9.3.1.2 Natural Systems

The sperm whale Fe(III)-myoglobin (Mb) and human Fe(III)-Hemoglobin (Hb) both possess strong affinity for sulfide [104, 105]. Drawing upon these considerations, we and others have designed fluorescent sensors for H_2S detection which builds on the use of Mb and Hb. We reported a FRET-based sensing device for monitoring H_2S which harnesses the optical characteristics of Mb [47]. We successfully explored the applicability of the system not only in solution but also when applied in a solid sol–gel matrix.

More recently, Galardon and coworkers reported hemoglobin (HbI) from *Lucina pectinata* as efficient H_2S sensor [106]. The system they implemented exploits a detection scheme that had been previously reported by us for the design of sensors for analytes different than H_2S [107]. The detection strategy is based on tuning of the fluorescence spectrum of a coumarin dye by the absorption spectrum of the HbI upon H_2S coordination. The absorption band of the sensor has to overlap with the excitation band of the fluorophore and also has to shift upon reaction with the analyte. The HbI system works in a reversible manner and the detection limit is 200 nM H_2S.

9.3.2 Salen-Based Systems

Among the several applications of salen-based systems spanning from different kinds of catalysis, asymmetric organic synthesis, and molecular sensing [52], our group recently reported that the first zinc salen-based complex for monitoring HS^- (**36**), which exploits the abovementioned coordination-based approach [49]. Probe **36** acts as a "turn-on" fluorescent sensor for HS^- under physiological conditions (MOPS buffer, pH 7.4). Most likely, coordination of the HS^- to the zinc center in **36** is stabilized by hydrogen-bonding interactions with the alcohol groups on the pyridoxal units. Stabilization of a Zn(II) hydrosulfide complex by hydrogen-bond assistance of the ligand was previously reported [108].

When substituting the ethylene bridge between the two coordinating nitrogens with a phenylene moiety, the resulting complex belonged to the salphen ligand family (**37**). These compounds function as efficient "turn-on" fluorescent HS$^-$ sensors working via the coordination-based approach in a reversible manner [50]. Also, this system works in physiological conditions and its fluorescence switching in the presence of HS$^-$ is clearly evident.

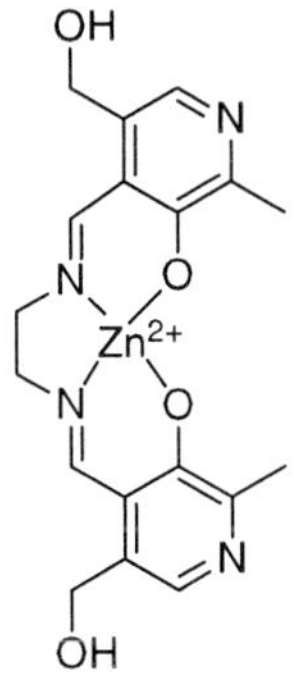

(**36**)

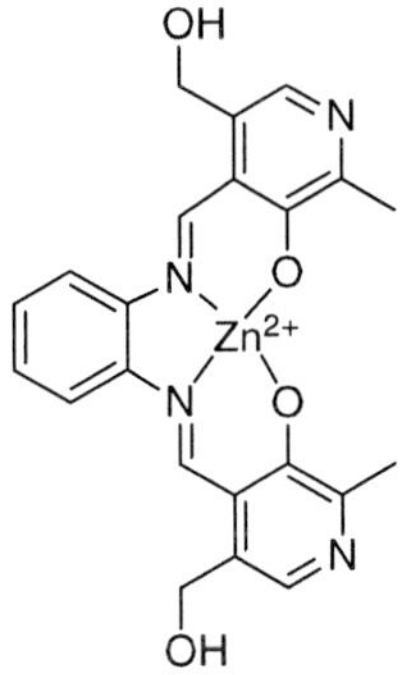

(**37**)

Very recently, we further demonstrated the recognition ability of salen-based zinc complexes by showing that a family of diaminomaleonitrile (DAMN)-salen-zinc complexes (**38**) can successfully detect HS$^-$ with a fluorescence responses [109]. Aqueous solubility, reversibility of the analyte coordination at the zinc center, and clearly visible color variations (visible to the naked eye) represent some of their advantageous characteristics. With this family of complexes, we also reported preliminary biological experiments showing the potential in using this class of compounds as probes for the detection of H_2S in living cells.

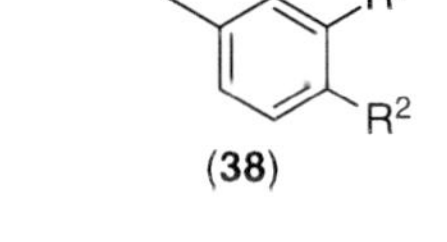

R^1	R^2
H	H
OCH_3	H
H	$N(CH_2CH_3)_2$
H	OH

(**38**)

9.3.3 Systems with Different Organic Ligands

A different organic ligand chelating the zinc center was used by the group of Galardon and coworkers who reported one of the first HS$^-$ sensors harnessing the coordinative-based approach [13]. The system is a thiolato complex of zinc

($Tp^{Ph,Me}Zn(MUS)$) (**39**), where MUS is 7-mercapto-4-methylcoumarin. HS^- recognition occurs due to MUS^- release upon HS^- binding to the zinc center, which results in the concomitant color change of the sample solution from colorless to yellow. The fluorescence intensity of the zinc complex is higher than that of free MUSH, thus the complex acts as "turn-off" H_2S sensor with a LOD of 1 μM.

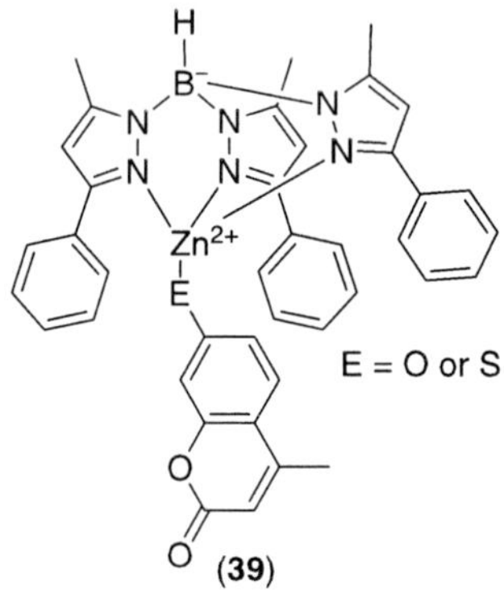

(**39**)

Much more recently, cobaloxime complexes bearing an axial pyridyl-BODIPY ligand (**40**) were proposed as sensitive and selective fluorescent HS^- reporting molecules in aqueous solution. The sensing mechanism involves the substitution of the BODIPY ligand by the HS^- anion at the cobalt center, which results in the releasing of the BODIPY unit together with the concomitant restoration of its strong fluorescence intensity [110]. The finding that when adding KSH to an aqueous solution of the alkyl-cobaloxime, the pyridyl-cobaloxime complexes experience a substitution reaction of the pyridine with the HS^- anion, had been already advanced by us in the past [111], as clearly stated by the authors in the paper.

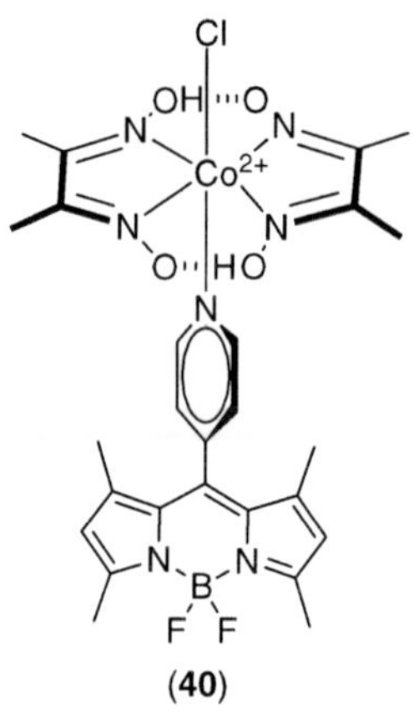

(**40**)

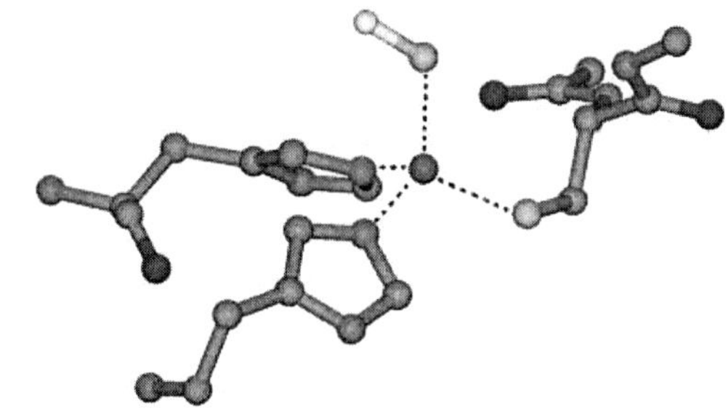

Figure 9.6 Hydrosulfide bound to Co-PDF (hydrosulfide in black and water in dark gray).

As fluorescent sensors that exploit binding of the analyte at the cobalt center, our group reported cobalt containing peptide deformylase (Co-PDF) as a FRET-based H_2S sensor [112]. We also reported that the X-ray structure of the Co-PDF hydrosulfide adduct (Figure 9.6) corroborating the hypothesis that HS^- binds to the cobalt center. Unlike probe **40**, the Co-PDF system operates as a "turn-off" sensor.

Several examples of well-characterized metal hydrosulfides species with more diverse ligands than those described in the above sections for a variety of different applications have been recently and comprehensively reviewed by Pluth and Tonzetich [51].

9.4 H_2S-Mediated Reduction of the Metal Center

As mentioned in the introduction of this chapter, when H_2S interacts with a metal complex, different scenarios are possible, one of which is reduction of the metal center by the H_2S. Over the years, some H_2S fluorescent sensors reported in the literature have been implemented by harnessing this possibility. One of the first sensors that exploits this sensing strategy was reported in 2011 (**41**) [113]. The molecule effecting the HS^- recognition is a mixed-valence diruthenium complex $[Ru_2TIEDCl_4]Cl$ (TIED = tetraiminoethylenedimacrocycle), which exhibits an unpaired electron completely delocalized on the whole structure. The working mechanism by which **41** acts as HS^- sensor is shown in Figure 9.7.

The authors claim that **41** behaves as a selective and chemically reversible HS^- probe with a LOD of 1.35 μM. In 2014, Pluth and coworkers reported that cobalt(II) phthalocyanine (Pc) complexes (**42**) react with HS^- in a chemically reversible manner and can be successfully used as sensing devices for H_2S detection in a reversible manner via subsequent cycles of reduction and oxidation [46]. Figure 9.8 displays the H_2S sensing strategy of **42**.

In 2019, our group reported a natural metalloprotein as an H_2S sensor [53] working via a mechanism similar to **41** and **42**. In our case, the sensing molecule was a natural protein: wild-type azurin, which is a very robust copper protein from

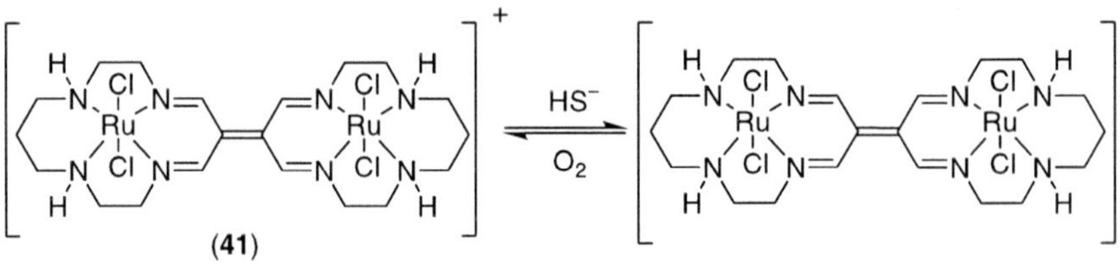

Figure 9.7 Recognition mechanism of H_2S by **41**.

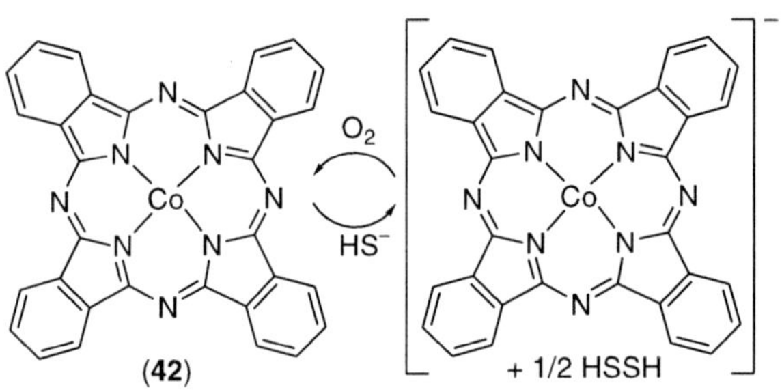

Figure 9.8 Recognition mechanism of H_2S by **42**. Source: Hartle et al. [46]/Royal Society of Chemistry.

Pseudomonas aeruginosa. UV-Vis, fluorescence, and EPR spectroscopy provided strong evidences of the operating mechanism. Furthermore, upon HS^- addition, we observed loss of the blue color (typical of the type I copper center), which is a further confirmation of the reduction of the copper center. The X-ray structure of Na_2S-treated azurin corroborates the copper reduction promoted by H_2S. The azurin-based H_2S sensor works by the means of a FRET mechanism, when covalently attaching a fluorescent label (e.g. Atto 620) to the protein.

9.5 Conclusions and Future Outlooks

Traditional methods for H_2S detection include colorimetry, gas chromatography, electrochemical analysis, and polarography. Although they seem more sensitive in some aspects compared to fluorescence-based devices, fluorescence-sensing strategies are more appealing, and have some other advantages. Traditional methods, such as gas chromatography, are extremely sensitive to H_2S with the reported detection limits below 0.2 pM [114]. However, extreme conditions such as high alkalinity (pH = 8.5) are required for these systems to function. Moreover, traditional methods cannot be used in cell imaging, often require complicated and expensive instruments for quantitative H_2S detection, and cause cell damage. This means that many traditional H_2S detection methods can be used for imaging *in vitro*, but not *in vivo*.

On the other hand, the compactness, portability, high specificity, sensitivity, and a clear optical response (possibly visible at naked eye) represent the main favorable characteristics for a fluorescence-based sensor. Ideally, an effective fluorescence-based probe should respond selectively and with high sensitivity to its target analyte, furthermore, water solubility of the sensing material is a highly desirable characteristic for practical biological measurements.

Among the different strategies devised so far for the implementation of H_2S fluorescence-based sensors, there are three metal-based approaches: metal-displacement, coordinative-based, and H_2S-mediated reduction of the metal center, which are the focus of the present chapter.

For the coordinative-based sensors, it is possible to remove H_2S from the metal center of the sensor to generate a reversible or cyclable H_2S binding process. Also for the sensors functioning via the H_2S-mediated reduction of the metal center, it is usually possible to regenerate the sensor by using O_2 or a proper oxidizing agent (e.g. potassium ferricyanide $\{K_3[Fe(CN)_6]\}$). Cyclable and reusable systems are highly favored both when thinking of applications on an industrial scale (for economic reasons) and when thinking of clinical applications (for accuracy reasons).

In the case of sensors working via the metal displacement, after H_2S binds to the metal center subsequent precipitation of the Mt-S occurs, which results in a chemical modification of the sensing molecule. For several systems belonging to this class, it is reported that by consecutively adding the metal precursor and H_2S to the solution of the sensing system, a switchable change in the fluorescence intensity is observed but often further manipulation of the sensing system connected with elimination of the precipitated MtS is required.

Although the H_2S sensors described here appear promising, it is clear that much work still remains to be done on the implementation of probes tailored for specific models of interest including cells, tissues, and whole organisms, to allow for the elucidation of intricate and dynamic inter- and intracellular events involving H_2S. Further investigation of H_2S biology in a variety of systems will reveal how this reactive sulfur species modulates cellular and physiological conditions, clarifying the mechanisms of its reactivity.

We hope that the design strategies described here may trigger the development of new and improved probes to meet specific research needs and for practical applications not only in the field of H_2S sensing but also for the entire area of molecular recognition.

References

1 Li, L. and Moore, P.K. (2008). Putative biological roles of hydrogen sulfide in health and disease: a breath of not so fresh air? *Trends Pharmacol. Sci.* 29 (2): 84–90.

2 Li, L. and Moore, P.K. (2007). An overview of the biological significance of endogenous gases: new roles for old molecules. *Biochem. Soc. Trans.* 35 (Pt 5): 1138–1141.

3 Wang, R. (2012). Physiological implications of hydrogen sulfide: a whiff exploration that blossomed. *Physiol. Rev.* 92 (2): 791–896.

4 Lin, V.S., Lippert, A.R., and Chang, C.J. (2013). Cell-trappable fluorescent probes for endogenous hydrogen sulfide signaling and imaging H_2O_2-dependent H_2S production. *Proc. Natl. Acad. Sci. U.S.A.* 110 (18): 7131–7135.

5 Yang, G.D., Wu, L.Y., Jiang, B. et al. (2008). H_2S as a physiologic vasorelaxant: Hypertension in mice with deletion of cystathionine gamma-lyase. *Science* 322 (5901): 587–590.

6 Papapetropoulos, A., Pyriochou, A., Altaany, Z. et al. (2009). Hydrogen sulfide is an endogenous stimulator of angiogenesis. *Proc. Natl. Acad. Sci. U.S.A.* 106 (51): 21972–21977.

7 Abe, K. and Kimura, H. (1996). The possible role of hydrogen sulfide as an endogenous neuromodulator. *J. Neurosci.* 16 (3): 1066.

8 Strianese, M., De, M.F., Pavone, V. et al. (2010). A FRET-based biosensor for NO detection. *J. Inorg. Biochem.* 104 (6): 619–624.

9 Hoshino, M., Laverman, L., and Ford, P.C. (1999). Nitric oxide complexes of metalloporphyrins: an overview of some mechanistic studies. *Coord. Chem. Rev.* 187 (1): 75–102.

10 Spiro, T.G. and Wasbotten, I.H. (2005). CO as a vibrational probe of heme protein active sites. *J. Inorg. Biochem.* 99 (1): 34–44.

11 Marìn-Hernàndez, C., Toscani, A., Sancenòn, F. et al. (2016). Chromo-fluorogenic probes for carbon monoxide detection. *Chem. Commun.* 52 (35): 5902–5911.

12 Hartle, M.D., Tillotson, M.R., Prell, J.S., and Pluth, M.D. (2017). Spectroscopic investigation of the reaction of metallo-protoporphyrins with hydrogen sulfide. *J. Inorg. Biochem.* 173: 152–157.

13 Galardon, E., Tomas, A., Roussel, P., and Artaud, I. (2009). New fluorescent zinc complexes: towards specific sensors for hydrogen sulfide in solution. *Dalton Trans.* 9126–9130.

14 Cardoso, A.A., Liu, H., and Dasgupta, P.K. (1997). Fluorometric fiber optic drop sensor for atmospheric hydrogen sulfide. *Talanta* 44 (6): 1099–1106.

15 Choi, M.M. and Hawkins, P. (2003). Development of an optical hydrogen sulphide sensor. *Sens. Actuators, B* 90 (1–3): 211–215.

16 Doeller, J.E., Isbell, T.S., Benavides, G. et al. (2005). Polarographic measurement of hydrogen sulfide production and consumption by mammalian tissues. *Anal. Biochem.* 341 (1): 40–51.

17 Wu, X.C., Sammynaiken, R., Zhang, W.J. et al. (2007). Measurement of low concentration and nano-quantity hydrogen sulfide in aqueous solution: measurement mechanisms and limitations. *Meas. Sci. Technol.* 18: 1315–1320.

18 Borisov, S.M. and Wolfbeis, O.S. (2008). Optical biosensors. *Chem. Rev.* 108 (2): 423–461.

19 Staiano, M., Bazzicalupo, P., Rossi, M., and D'Auria, S. (2005). Glucose biosensors as models for the development of advanced protein-based biosensors. *Mol. Biosyst.* 1 (5–6): 354–362.

20 Wolfbeis, O.S. (2000). Fiber-optic chemical sensors and biosensors. *Anal. Chem.* 72 (12): 81R–89R.

21 Lin, V.S. and Chang, C.J. (2012). Fluorescent probes for sensing and imaging biological hydrogen sulfide. *Curr. Opin. Chem. Biol.* 16 (5-6): 595–601.

22 Kumar, N., Bhalla, V., and Kumar, M. (2013). Recent developments of fluorescent probes for the detection of gasotransmitters (NO, CO and H_2S). *Coord. Chem. Rev.* 257 (15–16): 2335–2347.

23 Strianese, M. and Pellecchia, C. (2016). Metal complexes as fluorescent probes for sensing biologically relevant gas molecules. *Coord. Chem. Rev.* 318: 16–28.

24 Yang, M., Fan, J., Du, J., and Peng, X. (2020). Small-molecule fluorescent probes for imaging gaseous signaling molecules: current progress and future implications. *Chem. Sci.* 11 (20): 5127–5141.

25 Chen, S., Chen, Z.J., Ren, W., and Ai, H.W. (2012). Reaction-based genetically encoded fluorescent hydrogen sulfide sensors. *J. Am. Chem. Soc.* 134 (23): 9589–9592.

26 Lippert, A.R., New, E.J., and Chang, C.J. (2011). Reaction-based fluorescent probes for selective imaging of hydrogen sulfide in living cells. *J. Am. Chem. Soc.* 133 (26): 10078–10080.

27 Peng, H., Cheng, Y., Dai, C. et al. (2011). A fluorescent probe for fast and quantitative detection of hydrogen sulfide in blood. *Angew. Chem Int. Ed.* 50 (41): 9672–9675.

28 Yu, F., Li, P., Song, P. et al. (2012). An ICT-based strategy to a colorimetric and ratiometric fluorescence probe for hydrogen sulfide in living cells. *Chem. Commun.* 48 (23): 2852–2854.

29 Das, S.K., Lim, C.S., Yang, S.Y. et al. (2012). A small molecule two-photon probe for hydrogen sulfide in live tissues. *Chem. Commun.* 48 (67): 8395–8397.

30 Montoya, L.A. and Pluth, M.D. (2012). Selective turn-on fluorescent probes for imaging hydrogen sulfide in living cells. *Chem. Commun.* 48 (39): 4767–4769.

31 Zheng, K., Lin, W., and Tan, L. (2012). A phenanthroimidazole-based fluorescent chemosensor for imaging hydrogen sulfide in living cells. *Org. Biomol. Chem.* 10 (48): 9683–9688.

32 Wu, Z., Li, Z., Yang, L. et al. (2012). Fluorogenic detection of hydrogen sulfide via reductive unmasking of o-azidomethylbenzoyl-coumarin conjugate. *Chem. Commun.* 48 (81): 10120–10122.

33 Qian, Y., Karpus, J., Kabil, O. et al. (2011). Selective fluorescent probes for live-cell monitoring of sulphide. *Nat. Commun.* 2: 495.

34 Liu, C., Pan, J., Li, S. et al. (2011). Capture and visualization of hydrogen sulfide by a fluorescent probe. *Angew. Chem Int. Ed.* 50 (44): 10327–10329.

35 Liu, C., Peng, B., Li, S. et al. (2012). Reaction based fluorescent probes for hydrogen sulfide. *Org. Lett.* 14 (8): 2184–2187.

36 Zhao, Y., Zhu, X., Kan, H. et al. (2012). A highly selective colorimetric chemodosimeter for fast and quantitative detection of hydrogen sulfide. *Analyst* 137 (23): 5576–5580.

37 Wu, M.Y., Li, K., Hou, J.T. et al. (2012). A selective colorimetric and ratiometric fluorescent probe for hydrogen sulfide. *Org. Biomol. Chem.* 10 (41): 8342–8347.

38 Xuan, W., Pan, R., Cao, Y. et al. (2012). A fluorescent probe capable of detecting H2S at submicromolar concentrations in cells. *Chem. Commun.* 48 (86): 10669–10671.

39 Sasakura, K., Hanaoka, K., Shibuya, N. et al. (2011). Development of a highly selective fluorescence probe for hydrogen sulfide. *J. Am. Chem. Soc.* 133 (45): 18003–18005.

40 Choi, M.G., Cha, S., Lee, H. et al. (2009). Sulfide-selective chemosignaling by a Cu2+ complex of dipicolylamine appended fluorescein. *Chem. Commun.* 47: 7390–7392.

41 Hou, F., Huang, L., Xi, P. et al. (2012). A retrievable and highly selective fluorescent probe for monitoring sulfide and imaging in living cells. *Inorg. Chem.* 51 (4): 2454–2460.

42 Cao, X., Lin, W., and He, L. (2011). A near-infrared fluorescence turn-on sensor for sulfide anions. *Org. Lett.* 13 (17): 4716–4719.

43 Wang, M.Q., Li, K., Hou, J.T. et al. (2012). BINOL-based fluorescent sensor for recognition of Cu(II) and sulfide anion in water. *J. Org. Chem* 77 (18): 8350–8354.

44 Mirra, S., Strianese, M., and Pellecchia, C. (2017). A cyclam-based fluorescent ligand as a molecular beacon for Cu^{2+} and H_2S detection. *Eur. J. Inorg. Chem.* 2017 (33): 3900–3907.

45 Kaushik, R., Ghosh, A., and milan Jose, D. (2017). Recent progress in hydrogen sulphide (H_2S) sensors by metal displacement approach. *Coord. Chem. Rev.* 347: 141–157.

46 Hartle, M.D., Sommer, S.K., Dietrich, S.R., and Pluth, M.D. (2014). Chemically reversible reactions of hydrogen sulfide with metal phthalocyanines. *Inorg. Chem.* 53 (15): 7800–7802.

47 Strianese, M., De Martino, F., Pellecchia, C. et al. (2011). Myoglobin as a new fluorescence probe to sense H_2S. *Protein Pept. Lett.* 18 (3): 282–286.

48 Strianese, M., Lamberti, M., and Pellecchia, C. (2017). Chemically reversible binding of H_2S to a zinc porphyrin complex: towards implementation of a reversible sensor via a "coordinative-based approach". *Dalton Trans.* 46 (6): 1872–1877.

49 Strianese, M., Lamberti, M., and Pellecchia, C. (2018). Interaction of monohydrogensulfide with a family of fluorescent pyridoxal-based Zn(II) receptors. *Dalton Trans.* 47 (48): 17392–17400.

50 Strianese, M., Lamberti, M., Persico, A., and Pellecchia, C. (2019). Reactivity of monohydrogensulfide with a suite of pyridoxal-based complexes: A combined NMR, ESI-MS, UV-visible and fluorescence study. *Inorg. Chim. Acta* 119235.

51 Pluth, M.D. and Tonzetich, Z.J. (2020). Hydrosulfide complexes of the transition elements: diverse roles in bioinorganic, cluster, coordination, and organometallic chemistry. *Chem. Soc. Rev.* 49 (12): 4070–4134.

52 Strianese, M., Pappalardo, D., Mazzeo, M. et al. (2020). Salen-type aluminum and zinc complexes as two-faced Janus compounds: contribution to molecular sensing and polymerization catalysis. *Dalton Trans.* 49 (46): 16533–16550.

53 Strianese, M., Palm, G.J., Kohlhause, D. et al. (2019). Azurin and HS^-: towards implementation of a sensor for HS^- detection. *Eur. J. Inorg. Chem.* 2019 (6): 885–891.

54 Zhang, D. and Jin, W. (2012). Highly selective and sensitive colorimetric probe for hydrogen sulfide by a copper (II) complex of azo-dye based on chemosensing ensemble approach. *Spectrochim. Acta A Mol. Biomol. Spectrosc.* 90: 35–39.

55 Santos-Figueroa, L.E., de la Torre, C., El Sayed, S. et al. (2014). Highly selective fluorescence detection of hydrogen sulfide by using an anthracene-functionalized cyclam-Cu-II complex. *Eur. J. Inorg. Chem.* 2014 (1): 41–45.

56 Ye, Z., An, X., Song, B. et al. (2014). A novel dinuclear ruthenium(II)-copper(II) complex-based luminescent probe for hydrogen sulfide. *Dalton Trans.* 43 (34): 13055–13060.

57 Strianese, M., Milione, S., Bertolasi, V. et al. (2011). Heteroscorpionate-based Co^{2+}, Zn^{2+}, and Cu^{2+} complexes: coordination behavior, aerobic oxidation, and hydrogen sulfide detection. *Inorg. Chem.* 50 (3): 900–910.

58 Guo, Z., Chen, G.Q., Zeng, G.M. et al. (2015). Fluorescence chemosensors for hydrogen sulfide detection in biological systems. *Analyst* 140 (6): 1772–1786.

59 Hou, F., Cheng, J., Xi, P. et al. (2012). Recognition of copper and hydrogen sulfide in vitro using a fluorescein derivative indicator. *Dalton Trans.* 41 (19): 5799–5804.

60 Chang, J.H., Choi, M.Y., and Shim, Y.-K. (2001). A significant fluorescence quenching of anthrylaminobenzocrown ethers by paramagnetic metal cations. *Bull. Korean Chem. Soc.* 22 (5): 527–530.

61 Zhu, Y.F., Fan, D.H., and Shen, W.Z. (2008). A general chemical conversion route to synthesize various ZnO-based core/shell structures. *J. Phys. Chem. C* 112 (28): 10402–10406.

62 Palanisamy, S., Lee, L.Y., Wang, Y.L. et al. (2016). A water soluble and fast response fluorescent turn-on copper complex probe for H2S detection in zebra fish. *Talanta* 147: 445–452.

63 Xu, Q.C., Gu, Z.Y., and Xing, G.W. (2017). Design and synthesis of a Cu(II)-complex-based carbazole-hemicyanine hybrid for fluorescent sensing of H_2S in SDS micellar solution. *Tetrahedron* 73 (15): 2123–2130.

64 Lv, K., Chen, J., Wang, H. et al. (2017). One-pot fabrication of FRET-based fluorescent probe for detecting copper ion and sulfide anion in 100% aqueous media. *Spectrochim. Acta A Mol. Biomol. Spectrosc.* 177: 63–68.

65 May, P.M., Batka, D., Hefter, G. et al. (2018). Goodbye to S^{2-} in aqueous solution. *Chem. Commun.* 54 (16): 1980–1983.

66 Wu, X., Li, H., Kan, Y., and Yin, B. (2013). A regeneratable and highly selective fluorescent probe for sulfide detection in aqueous solution. *Dalton Trans.* 42 (46): 16302–16310.

67 Li, X., Gong, Y., Wu, K. et al. (2014). A highly sensitive and water soluble fluorescent probe for rapid detection of hydrogen sulfide in living cells. *RSC Adv.* 4 (68): 36106–36109.

68 Zhang, L., Lou, X., Yu, Y. et al. (2011). A new disubstituted polyacetylene bearing pyridine moieties: convenient synthesis and sensitive chemosensor toward sulfide anion with high selectivity. *Macromolecules* 44 (13): 5186–5193.

69 Gu, X., Liu, C., Zhu, Y.C., and Zhu, Y.Z. (2011). Development of a boron-dipyrromethene-Cu^{2+} ensemble based colorimetric probe toward hydrogen sulfide in aqueous media. *Tetrahedron Lett.* 52 (39): 5000–5003.

70 Ballesteros, E., Moreno, D., Gómez, T. et al. (2009). A new selective chromogenic and turn-on fluorogenic probe for copper(II) in water-acetonitrile 1:1 solution. *Org. Lett.* 11 (6): 1269–1272.

71 Guo, Z., Zhu, W., and Tian, H. (2010). Hydrophilic copolymer bearing dicyanomethylene-4H-pyran moiety as fluorescent film sensor for Cu^{2+} and pyrophosphate anion. *Macromolecules* 43 (2): 739–744.

72 Kim, S.K., Lee, D.H., Hong, J.I., and Yoon, J. (2009). Chemosensors for pyrophosphate. *Acc. Chem. Res.* 42 (1): 23–31.

73 Martìnez-Manez, R. and Sancenòn, F. (2003). Fluorogenic and chromogenic chemosensors and reagents for anions. *Chem. Rev.* 103 (11): 4419–4476.

74 Selectivity of metal-containing probes for HS- is commonly checked against thiols (RSH) since by reaction of metal complexes with thiols form

metal-thiols/thiolates. The fluorescence signals of these products may interfere with the signal of the sensor itself.
75 Lou, X., Mu, H., Gong, R. et al. (2011). Displacement method to develop highly sensitive and selective dual chemosensor towards sulfide anion. *Analyst* 136 (4): 684–687.
76 Wang, J., Long, L., Xie, D., and Zhan, Y. (2013). Highly selective fluorescence turn-on sensor for hydrogen sulfide and imaging in living cells. *J. Lumin.* 139: 40–46.
77 Tang, L., Cai, M., Huang, Z. et al. (2013). Rapid and highly selective relay recognition of Cu(II) and sulfide ions by a simple benzimidazole-based fluorescent sensor in water. *Sensor Actuat. B: Chem.* 185: 188–194.
78 Liu, H., Wu, F., Zhang, B. et al. (2016). A simple quinoline-derived fluorescent sensor for the selective and sequential detection of copper(II) and sulfide ions and its application in living-cell imaging. *RSC Adv.* 6 (81): 77508–77514.
79 Fu, Y., Feng, Q.C., Jiang, X.J. et al. (2014). New fluorescent sensor for Cu^{2+} and S^{2-} in 100% aqueous solution based on displacement approach. *Dalton Trans.* 43 (15): 5815–5822.
80 Tang, L., Dai, X., Cai, M. et al. (2014). Relay recognition of Cu^{2+} and S^{2-} in water by a simple 2-(2'-aminophenyl)benzimidazole derivatized fluorescent sensor through modulating ESIPT. *Spectrochim. Acta A Mol. Biomol. Spectrosc.* 122: 656–660.
81 Sun, M., Yu, H., Li, H. et al. (2015). Fluorescence signaling of hydrogen sulfide in broad pH range using a copper complex based on BINOL-benzimidazole ligands. *Inorg. Chem.* 54 (8): 3766–3772.
82 Kar, C., Adhikari, M.D., Ramesh, A., and Das, G. (2013). NIR- and FRET-based sensing of Cu2+ and S2– in physiological conditions and in live cells. *Inorg. Chem.* 52 (2): 743–752.
83 Zhou, Q., Zhu, Y., Sheng, P. et al. (2014). A highly selective and reversible fluorescent Cu^{2+} and S2GêÆ probe under physiological conditions and in live cells. *RSC Adv.* 4 (87): 46951–46954.
84 Gao, C., Liu, X., Jin, X. et al. (2013). A retrievable and highly selective fluorescent sensor for detecting copper and sulfide. *Sensor Actuat. B: Chem.* 185: 125–131.
85 Qian, Y., Lin, J., Liu, T., and Zhu, H. (2015). Living cells imaging for copper and hydrogen sulfide by a selective "on-off-on" fluorescent probe. *Talanta* 132: 727–732.
86 Meng, Q., Shi, Y., Wang, C. et al. (2015). NBD-based fluorescent chemosensor for the selective quantification of copper and sulfide in an aqueous solution and living cells. *Org. Biomol. Chem.* 13 (10): 2918–2926.
87 Qu, X., Li, C., Chen, H. et al. (2013). A red fluorescent turn-on probe for hydrogen sulfide and its application in living cells. *Chem. Commun.* 49 (68): 7510–7512.

88 Wu, P., Zhang, J., Wang, S. et al. (2014). Sensing during in situ growth of Mn-doped ZnS QDs: a phosphorescent sensor for detection of H_2S in biological samples. *Chem. Eur. J.* 20 (4): 952–956.
89 Kaushik, R., Kumar, P., Ghosh, A. et al. (2015). Alizarin red S-zinc(ii) fluorescent ensemble for selective detection of hydrogen sulphide and assay with an H_2S donor. *RSC Adv.* 5 (97): 79309–79316.
90 Tang, L., Cai, M., Zhou, P. et al. (2013). A highly selective and ratiometric fluorescent sensor for relay recognition of zinc(II) and sulfide ions based on modulation of excited-state intramolecular proton transfer. *RSC Adv.* 3 (37): 16802–16809.
91 Dong, Z., Le, X., Zhou, P. et al. (2014). Sequential recognition of zinc ion and hydrogen sulfide by a new quinoline derivative with logic gate behavior. *RSC Adv.* 4 (35): 18270–18277.
92 Liu, L., Shang, X., He, R. et al. (2019). Anion binding ability and cytotoxicity of a selective colorimetric chemosensor for H_2S based on Zn(II) complex. *Inorg. Chim. Acta* 495: 118994.
93 Kaushik, R., Singh, A., Ghosh, A., and Jose, D.A. (2016). Selective colorimetric sensor for the detection of Hg^{2+} and H_2S in aqueous medium and waste water samples. *Chem. Select* 1 (8): 1533–1540.
94 Rodrìguez-Fernàndez, J., Manuel Costa, J., Pereiro, R., and Sanz-Medel, A. (1999). Simple detector for oral malodour based on spectrofluorimetric measurements of hydrogen sulphide in mouth air. *Anal. Chim. Acta* 398 (1): 23–31.
95 Sen, B., Mukherjee, M., Pal, S. et al. (2014). A water soluble FRET-based ratiometric chemosensor for Hg(II) and S^{2-} applicable in living cell staining. *RSC Adv.* 4 (29): 14919–14927.
96 Vengaian, K.M., Britto, C.D., Sekar, K. et al. (2016). Phenothiazine-diaminomalenonitrile based colorimetric and fluorescence 'Turn-off-on' Sensing of Hg^{2+} and S^{2-}. *Sensor Actuat. B: Chem.* 235: 232–240.
97 Mahapatra, A.K., Manna, S.K., Maiti, K. et al. (2014). Imino-phenolic-azodye appended rhodamine as a primary fluorescence off-on chemosensor for tin (Sn^{4+}) in solution and in RAW cells and the recognition of sulphide by [Ar-Sn]. *RSC Adv.* 4 (69): 36615–36622.
98 Mirra, S., Milione, S., Strianese, M., and Pellecchia, C. (2015). A copper porphyrin for sensing H2S in aqueous solution via a "coordinative-based" approach. *Eur. J. Inorg. Chem.* 13: 2272–2276.
99 Paolesse, R., Nardis, S., Monti, D. et al. (2017). Porphyrinoids for chemical sensor applications. *Chem. Rev.* 117 (4): 2517–2583.
100 Bostelaar, T., Vitvitsky, V., Kumutima, J. et al. (2016). Hydrogen sulfide oxidation by myoglobin. *J. Am. Chem. Soc.* 138 (27): 8476–8488.

101 Meininger, D.J., Caranto, J.D., Arman, H.D., and Tonzetich, Z.J. (2013). Studies of iron(III) porphyrinates containing silanethiolate ligands. *Inorg. Chem.* 52 (21): 12468–12476.

102 Meininger, D.J., Arman, H.D., and Tonzetich, Z.J. (2017). Synthesis, characterization, and binding affinity of hydrosulfide complexes of synthetic iron(II) porphyrinates. *J. Inorg. Biochem.* 167: 142–149.

103 Strianese, M., Mirra, S., Lamberti, M., and Pellecchia, C. (2017). Zinc (II) porphyrins as viable scaffolds to stabilize hydrogen sulfide binding at the metal center. *Inorg. Chim. Acta* 466: 426–431.

104 Kraus, D.W., Wittenberg, J.B., Lu, J.F., and Peisach, J. (1990). Hemoglobins of the Lucina pectinata/bacteria symbiosis. II. An electron paramagnetic resonance and optical spectral study of the ferric proteins. *J. Biol. Chem.* 265 (27): 16054–16059.

105 Pavlik, J.W., Noll, B.C., Oliver, A.G. et al. (2010). Hydrosulfide (HS^-) coordination in iron porphyrinates. *Inorg. Chem.* 49 (3): 1017–1026.

106 Dulac, M., Melet, A., and Galardon, E. (2018). Reversible detection and quantification of hydrogen sulfide by fluorescence using the hemoglobin I from *Lucina pectinata*. *ACS Sens.* 3 (10): 2138–2144.

107 Strianese, M., Varriale, A., Staiano, M. et al. (2011). Absorption into fluorescence. A method to sense biologically relevant gas molecules. *Nanoscale* 3 (1): 298–302.

108 Hartle, M.D., Delgado, M., Gilbertson, J.D., and Pluth, M.D. (2016). Stabilization of a Zn(II) hydrosulfido complex utilizing a hydrogen-bond accepting ligand. *Chem. Commun.* 52 (49): 7680–7682.

109 Strianese, M., Guarnieri, D., Lamberti, M. et al. (2020). Fluorescent salen-type Zn(II) complexes As probes for detecting hydrogen sulfide and its anion: bioimaging applications. *Inorg. Chem.* 59 (21): 15977–15986.

110 Bartelmess, J., Valderrey, V., and Rurack, K. (2019). Development of a turn-on fluorescent probe-based sensing system for hydrogen sulfide in liquid and gas phase. *Front. Chem.* 7: 641.

111 Strianese, M., Mirra, S., Bertolasi, V. et al. (2015). Organometallic sulfur complexes: reactivity of the hydrogen sulfide anion with cobaloximes. *New J. Chem.* 39 (5): 4093–4099.

112 Strianese, M., Palm, G.J., Milione, S. et al. (2012). A FRET enzyme-based probe for monitoring hydrogen sulfide. *Inorg. Chem.* 51 (21): 11220–11222.

113 Quek, Y.L., Tan, C.H., Bian, J., and Huang, D. (2011). Air oxidation of HSGÇô catalyzed by an mixed-valence diruthenium complex, an near-IR probe for HS detection. *Inorg. Chem.* 50 (16). 7379–7381.

114 Furne, J., Saeed, A., and Levitt, M.D. (2008). Whole tissue hydrogen sulfide concentrations are orders of magnitude lower than presently accepted values. *Am. J. Physiol. Regul. Integr. Comp. Physiol.* 295 (5): R1479–R1485.

10

H_2S Release from P=S and Se−S Motifs

Rynne A. Hankins and John C. Lukesh III

Department of Chemistry, Wake Forest University, Wake Downtown Campus, Winston-Salem, NC, USA

List of Abbreviations

3-MST	3-mercaptopyruvate sulfur transferase
CAT	cysteine aminotransferase
CBS	cystathionine β-synthase
CO	carbon monoxide
CSE	cystathionine γ-lyase
DMF	dimethyl formamide
DMSO	dimethyl sulfoxide
DTNB	5,5′-dithiobis-(2-nitrobenzoic acid)
GPx	glutathione peroxidase
H_2S	hydrogen sulfide
LR	Lawesson's reagent
NO	nitric oxide
PLP	pyridoxal-5′-phosphate
RSeSH	thioseleninic acid
RSeSR	selenylsulfide
RSSH	persulfide
TrxR	thioredoxin reductase

10.1 Introduction

Hydrogen sulfide (H_2S) has long been regarded as a highly toxic environmental and industrial pollutant [1]. This limited viewpoint began to shift, however, in 1996 when it was first suggested by Abe and Kimura that H_2S may play a key role in neurotransmission [2]. Since this initial landmark report, numerous studies have been carried out which further highlight the vast physiological and pathological effects of H_2S within mammalian systems, including its ability to lower blood pressure [3], reduce inflammation [4, 5], and promote angiogenesis and wound healing [6, 7]. As such, H_2S is now widely recognized as an important endogenous signaling molecule and, along with nitric oxide (NO) and carbon monoxide (CO), is considered to be the third member of the gasotransmitter family [8, 9].

Endogenous H_2S is primarily produced enzymatically, allowing for strict control over its production. The enzymes principally involved in H_2S biosynthesis are cystathionine β-synthase (CBS), cystathionine γ-lyase (CSE), and 3-mercaptopyruvate sulfur transferase (3-MST) coupled with cysteine aminotransferase (CAT) [10–13]. In each pathway, the ensuing sulfur atom of H_2S is derived from the desulfhydration of either cysteine or homocysteine. Although these enzymes are broadly distributed in mammalian tissues, CBS is regarded as the primary producer of H_2S in the central nervous system while CSE is amply expressed in the cardiovascular and respiratory systems. Both CBS and CSE are pyridoxal-5′-phosphate (PLP)-dependent enzymes that reside in the cytosol. Conversely, 3-MST is a PLP-independent enzyme that is expressed predominantly in mitochondria and is largely responsible for H_2S production within that subcellular space.

To further examine H_2S pathophysiology, along with its primary and secondary roles in cellular signaling, the development of small molecules that slowly liberate the transient gas under biologically relevant conditions have provided much insight [14–19]. Additionally, with a mounting body of evidence that links H_2S deficiency to numerous clinical disorders (especially cardiovascular [20], neurological [21], and other age-related pathologies [22]), it is possible, or perhaps even likely, that some of these donor compounds may prove to be more than useful chemical tools for assisting H_2S research and may, in fact, afford useful therapies for a wide range of diseases by simply replenishing the body's reduced levels of H_2S.

In this chapter, we will focus on two specific classes of donors that release H_2S as a result of either P—S or Se—S bond cleavage. In addition to their biological applications and therapeutic potential, much emphasis will be placed on the chemistry and development of these two donor motifs and, in particular, how structural modifications to both templates impact the rate and efficiency at which H_2S is released.

10.2 H_2S Release from P=S Motifs

Prior to the development of small molecule donors that afforded the controlled and sustained release of H_2S, the majority of scientific studies aimed at unearthing the physiological roles and therapeutic effects of hydrogen sulfide employed simple sulfide salts (e.g. Na_2S and NaHS) as expedient precursors to H_2S. While convenient, the addition of sulfide salts to buffered solutions results in an instantaneous surge in H_2S concentration that often engenders toxicity and poorly mimics the slow and steady enzymatic production of the gasotransmitter. Instead, it has been demonstrated on numerous occasions, both *in vitro* and *in vivo*, that slow releasing donors provide a more favorable therapeutic profile in addition to offering a more accurate depiction of the true physiological functions of H_2S within specific cells and organelles.

Although a sizable number of donors have been developed over the past decade that enable the controlled cellular delivery of H_2S, and at concentrations that are more likely to be biologically relevant, by far the most studied and characterized small molecule donor is GYY4137 – a phosphorodithioate-based donor that provides the slow but prolonged liberation of H_2S via chemical hydrolysis and P—S bond cleavage in aqueous solutions [23].

10.2.1 GYY4137: Synthesis and Characterization of H_2S Release

Lawesson's reagent (LR) is a commercially available sulfurization reagent that has found widespread use in organic synthesis despite its strong and unpleasant odor (Scheme 10.1) [24]. The noxious smell of LR is actually derived from its slow release of H_2S upon contact with moisture in the air. Not surprisingly, LR has been used as a slow-releasing donor in biological studies aimed at extracting the therapeutic effects of H_2S within various disease models [25–27]. Although some promising results have been obtained with LR, its poor solubility in water severely limits its potential as a chemical tool for probing the physiology and pathology of H_2S.

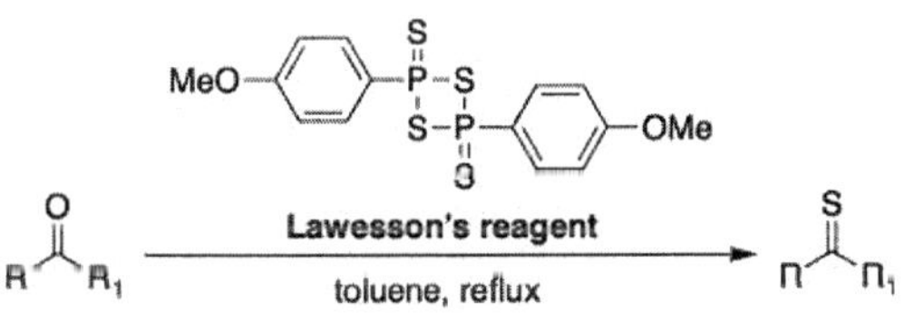

Scheme 10.1 General reaction involving LR as a sulfurization reagent in organic synthesis. Source: Ozturk et al. [24]/American Chemical Society.

To this end, morpholin-4-ium 4-methoxyphenyl(morpholino) phosphinodithioate (GYY4137) is a water-soluble derivative that is easily prepared by treating Lawesson's reagent with morpholine in dichloromethane (Scheme 10.2). GYY4137 is the first organic small molecule with high water solubility to be fully

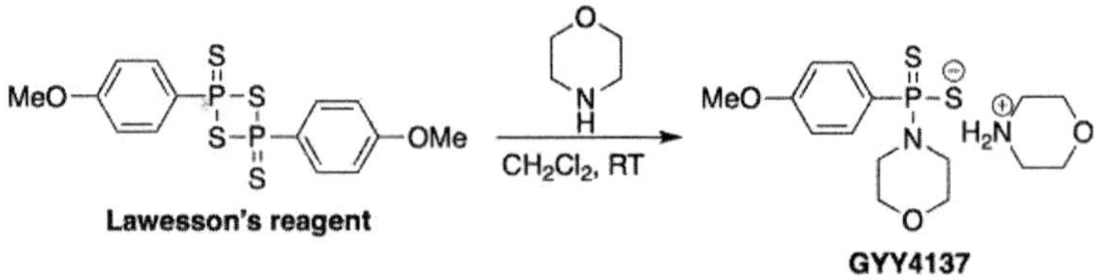

Scheme 10.2 Synthesis of GYY4137.

characterized as a slow-releasing H_2S donor. Although its synthesis had been previously described ~50 years earlier, it wasn't until 2008 when Li and coworkers first reported on its release of H_2S in buffered water [23]. In addition to facilitating its use in biological assays, the high water solubility of GYY4137 (up to 30 mg/mL at pH 7.4) also permits concentrated stock solutions to be prepared with ease and without the use of dimethyl sulfoxide (DMSO), dimethylformamide (DMF), ethanol, or other organic solvents. While certainly an added benefit, the repeated reuse of stock solutions, even if frozen, should still be avoided at all costs given its instability and observed hydrolysis in aqueous solutions [28]. Instead, it is recommended that fresh stock solutions are prepared prior to each experiment to ensure an accurate concentration of donor.

H_2S release from GYY4137 in buffer was initially assessed using 5,5′-dithiobis-(2-nitrobenzoic acid) (DTNB), a common biochemical reagent used to quantify reduced sulfhydryl groups in solution, and via amperometry using an H_2S-selective microelectrode (World Precision Instruments) [23]. The rate and efficiency of H_2S release from GYY4137 was found to be both temperature and pH dependent with larger amounts being liberated at higher temperatures and lower pH. Under physiologically relevant conditions (37 °C, pH 7.4), H_2S production from GYY4137 is extremely sluggish and provides just a 4–5% yield from a starting concentration of 1 mM. However, unlike NaHS (or Na_2S), which results in a rapid increase in H_2S concentration upon its addition to buffer that then swiftly declines to undetectable levels within ~60–90 minutes, GYY4137 has been touted for its ability to provide the continuous and prolonged release of H_2S for up to a week after its addition to aqueous solutions [29]. It is worth mentioning that these contrasting profiles of H_2S release among two of the most commonly employed donors may explain the disparate observations that have been reported in the literature regarding the biological and therapeutic effects of H_2S.

10.2.2 GYY4137: Biological Studies

As stated previously, GYY4137 is far and away the most studied small molecule H_2S donor, especially in terms of its physiological and therapeutic effects (Figure 10.1). In this section, we will briefly summarize some key findings and

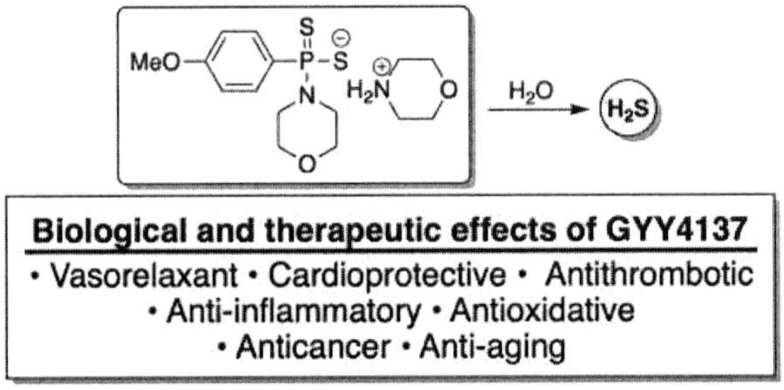

Figure 10.1 A summary of the major physiological and pharmacological effects of GYY4137.

observations. A more in-depth analysis on the biological implications of GYY4137 can be found in several recent reviews [28, 30–32].

Generally speaking, hydrogen sulfide has been shown to exert multiple positive effects on the cardiovascular system and is believed to be a key contributor to overall heart health and protection [33–37]. In fact, the discovery that H_2S could dilate blood vessels, both *in vitro* and *in vivo*, is what sparked significant interest in the field of H_2S biology during the late 1990s and early 2000s [38]. Not surprisingly, similar effects on the cardiovascular system have been reported with GYY4137. In the initial study that first characterized GYY4137 as a slow-releasing H_2S donor, its ability to relax precontracted rat aortic rings was also confirmed [23]. Furthermore, this observed vasodilation was effectively inhibited by the addition of glibenclamide, indicating that, similar to sulfide salts, the vasorelaxant effects of GYY4137 can be ascribed to its activation of vascular smooth muscle K_{ATP} channels [39, 40]. Yet, unlike sulfide salts whose effects were rapid but transient, GYY4137 was found to be a more potent vasorelaxant, presumably due to its sustained release of H_2S which results in its prolonged interaction with aortic rings. This same study also highlighted the *in vivo* antihypertensive effects of GYY4137 in anesthetized rats. Apart from its vasorelaxant properties, GYY4137 has also exhibited cytoprotective effects within various cardiovascular injury models. Perhaps most notably, GYY4137 has been shown to provide superb protection against myocardial reperfusion injury via an array of different mechanisms, highlighting its therapeutic potential [41–44].

Outside of these reports, intriguing studies are beginning to emerge that highlight the anticancer properties of GYY4137 [45–48]. This effect was first demonstrated in 2011 by Lee et al. within a panel of cancer cell lines (MCF-7, MV4-11, HL-60, HeLa, HCT-116, Hep G2, and U2OS) and in comparison to identical concentrations of NaHS (400–800 μM) [29]. Interestingly, it was found that only GYY4137 resulted in a concentration-dependent reduction in cancer cell viability within these cell lines. To rationalize these observations, it was proposed that although NaHS produces large amounts of H_2S upon its addition to water, its concentration falls off quickly suggesting that little, if any, H_2S would be present in cell culture media after only a few hours. Conversely, cells exposed to GYY4137 would be in the presence of small amounts of H_2S for several days and perhaps it is this sustained exposure to low levels of H_2S that is responsible for the observed anticancer activity. The mechanism of action was also probed in this study. It was found that treatment of MCF-7 cells with GYY4137 resulted in G2/M phase cell cycle arrest and the promotion of apoptosis as indicated by the presence of both cleaved caspase-9 and PARP. However, GYY4137 did not reduce cell viability in noncancerous cells (WI-38 and IMR90), suggesting that H_2S may be selectivity toxic towards cancerous cells.

Many studies have also probed the effects of GYY4137 within other physiological and pathophysiological models. While the details of these reports are being omitted from this review, we direct those interested to recent studies highlighting the professed effects of GYY4137 on inflammation [49–53], aging [54, 55], and the reproductive system [56, 57].

10.2.3 GYY4137: Mechanistic Studies

The accelerated release of H_2S under acidic conditions is consistent with a mechanism involving a weak nucleophile, such as water, promoting the requisite sulfur–oxygen exchange reaction with GYY4137. Still, the actual hydrolysis byproducts of the donor were not unequivocally established until a detailed mechanistic study was carried out by Alexander and co-workers using a combination of ^{31}P NMR spectroscopy and mass spectrometry [58]. In light of the fact that slow-releasing donors, such as GYY4137, require high concentrations (>100 μM) to provide workable amounts of H_2S for biological studies, it was recognized that the identification of these byproducts was necessary to provide effective control compounds and to ensure that they themselves were not responsible for any of the observed biological effects.

During the early stages of this study, it was determined that the hydrolysis of GYY4137 in pure water is actually too slow to be observed by NMR. Consequently, the hydrolysis of GYY4137 was monitored in "wet" deuterated chloroform ($CDCl_3$). Under these conditions, it was noted that the complete decomposition of GYY4137 is actually a two-step process with 4-methoxy-phenylphosphonamidothioate **1** being initially generated with an approximate 50% conversion after 13 days (Scheme 10.3). The formation of this hydrolysis byproduct is consistent with H_2S release being acid mediated and initiated by a simple sulfur–oxygen exchange reaction between GYY4137 and water. This study also uncovered that the complete hydrolysis – and the formation of arylphosphonate **2** (Scheme 10.3) – occurs in a second but significantly slower step with only trace amounts of **2** being detected after a 71-day incubation period. Considering the timescale of most experiments involving GYY4137, it was

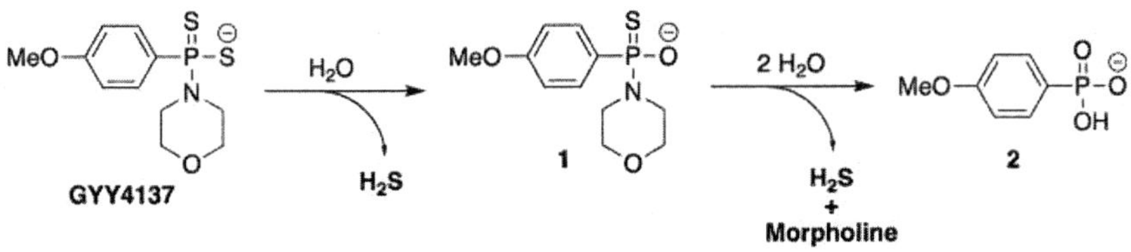

Scheme 10.3 Two-step hydrolysis of GYY4137 as outlined by Alexander and coworkers. The counterion for each species is morpholinium.

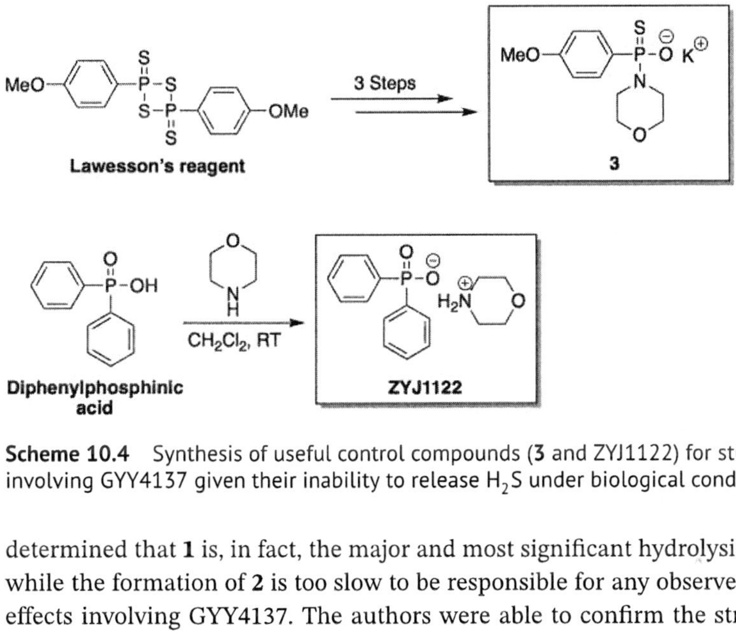

Scheme 10.4 Synthesis of useful control compounds (**3** and ZYJ1122) for studies involving GYY4137 given their inability to release H_2S under biological conditions.

determined that **1** is, in fact, the major and most significant hydrolysis byproduct while the formation of **2** is too slow to be responsible for any observed biological effects involving GYY4137. The authors were able to confirm the structure of **1** from an independent synthesis of its potassium salt (**3**) from Lawesson's reagent (Scheme 10.4). Once in hand, compound **3** was employed as a control compound alongside GYY4137 to certify that the physiological and therapeutic effects of the latter could be attributed to its release of H_2S. To this end, additional studies confirmed that GYY4137, but not **3**, could prevent oxidant-induced cell death and LPS-induced NO and PGE_2 production, suggesting that H_2S liberation from GYY4137 is likely to be responsible for this observed activity.

In addition to hydrolyzed byproducts, it should also be noted that the use of high concentrations of GYY4137 will introduce biological systems to additional species, such as morpholinium counterions, that may confound the interpretation of results. To control for any potential consequences that may arise due to the presence of high amounts of morpholine, additional "time-expired" donors, such as ZYJ1122 [29], have been prepared and used in control experiments alongside

GYY4137 (Scheme 10.4). Moreover, it should also be cited that the initial synthesis and purification of GYY4137, as descried by Li and coworkers, results in its isolation as a dichloromethane complex [23]. Importantly, dichloromethane is known to be metabolized to carbon monoxide (CO) – another distinct "gasotransmitter" [59]. Given that CO is reported to elicit many of the same physiological effects as H_2S [9, 60], it is possible that the observed activity in some of the initial reports involving GYY4137 could also be due, at least in part, to its adventitious release of CO. To further circumvent both of these potential issues, a CH_2Cl_2-free synthesis of GYY4137 as its more pharmaceutically acceptable sodium sulfide salt has also been described [58].

10.2.4 GYY4137: Structural Modifications and Activity of Analogs

The conflicting activities displayed by NaHS and GYY4137 indicate that concentration, along with the rate at which it is delivered, has a profound impact on the observed biological and therapeutic profile of H_2S. While the slow but sustained release of H_2S from GYY4137 appears to afford promising therapeutic potential in response to certain pathological conditions, it is likely that this fixed rate of H_2S production does not provide optimal efficacy, and certainly not for all biological applications. Moreover, the high concentration of GYY4137 that is required to produce biologically relevant amounts of H_2S may complicate the interpretation of results and to what extent the unhydrolyzed donor compound itself may be responsible for any experimental observations. For these reasons, many analogs of GYY4137 have been prepared over the years, primarily in an effort to increase the rate and efficiency of H_2S release.

Structural modifications to the phosphorodithioate core were first explored by Park et al. by replacing the phenyl-phosphorus (C—P) bond of GYY4137 with an aryloxy or alkoxy linkage (**4a–e**, Figure 10.2) [61, 62]. The ensuing rates of H_2S release within this small series of *O*-substituted *N*-phenyl phosphorodithioates was then monitored using 2,6-dansyl azide as an H_2S-selective fluorescent probe [63]. While the authors noted that only trace amounts of H_2S were released from *O*-alkyl substituted donors (**4d** and **4e**), *O*-aryl substituted donors (**4a–c**) provided a comparable release profile to that of GYY4137 and similar protection against H_2O_2-induced oxidative stress within H9c2 cardiomyocytes. Although detailed mechanistic studies have yet to be carried out within this series of donors, it is likely to operate via a similar hydrolytic pathway.

While this initial series of analogs appeared to offer little difference in terms of reactivity, it was theorized that further tuning of the electronic properties on phosphorous may have a more significant impact on the rates of H_2S release from phosphorodithioate-based donors. It was also speculated that ^{31}P NMR would provide a convenient method for monitoring the electronic environment surrounding

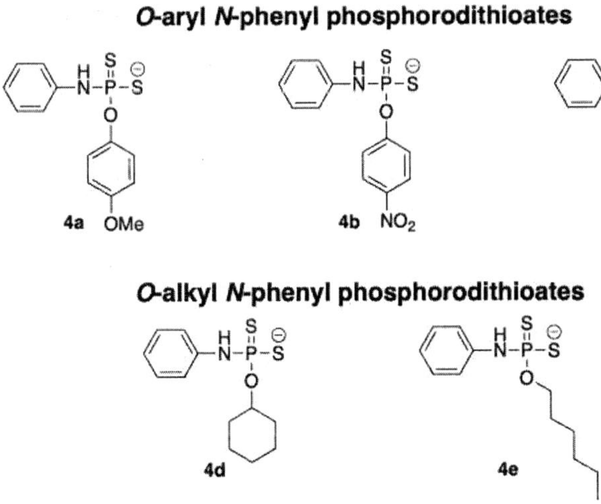

Figure 10.2 Structures of O-substituted N-phenyl phosphorodithioate-based H_2S donors. While little H_2S was liberated from O-alkyl phosphorodithioates, the O-aryl series displayed a similar H_2S release profile to that of GYY4137. Source: Based on Park et al. [61] and Park and Xian [62].

the phosphorous atom based on the ensuing ^{31}P chemical shift of each analog. This concept was explored in depth by Feng and coworkers by replacing the morpholino substituent on GYY4137 with an assortment of alcohols and phenols via their nucleophilic ring opening of Lawesson's reagent (Table 10.1A) [64]. In addition, a series of derivatives with two heteroatoms attached to phosphorous was procured in a single step from P_4S_{10} (Table 10.1B).

Using the same H_2S-selective fluorescent probe (2,6-dansyl azide), H_2S release from each donor (5 mM) was monitored at 22 °C over a four-day period. The results of this study are compiled in Figure 10.3. Within this particular series, analysis of data indeed indicates an apparent correlation between H_2S production and the ^{31}P NMR chemical shift of each donor, with a decreasing chemical shift (greater shielding of the phosphorous nucleus) typically resulting in an increased yield of H_2S. This is especially true considering the fact that compounds **18–21** were tested at one-tenth the concentration (0.5 mM) of the other donors listed. Thus, within this series, phosphorodithioates labeled as slow, medium, and fast H_2S-releasers were obtained, providing a useful set of chemical tools for further explorations into the relationship between H_2S concentration and biological activity. Not surprisingly, when their anticancer effects were investigated, several analogs were identified with improved potency relative to that of GYY4137. Compound **14** (FW1131), in particular, was found to be significantly more potent (by nearly an order of magnitude) in terms of its inhibition of solid tumor cell proliferation in various cancer

Table 10.1 Synthesis and ^{31}P NMR chemical shifts (DMSO-d_6) of phosphorodithioate-based donors with (A): A single heteroatom attached to phosphorous and (B): Two heteroatoms attached to phosphorous.

(A)

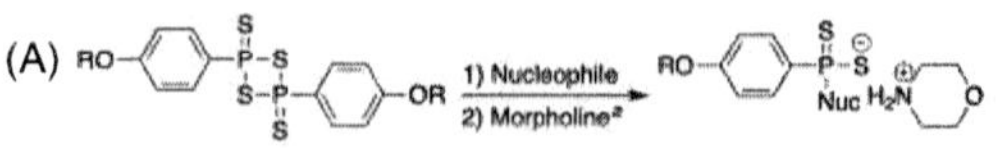

Lawesson's or Belleau's reagent

Compound	R	Nucleophile	^{31}P NMR δ (ppm)
GYY4137	Me	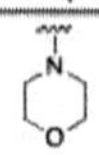	89.00
5	Me	–OMe	107.75
6	Me	–OPh	106.7
7	Me	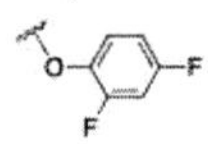	111.90
8	Me	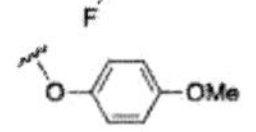	107.39
9	Me	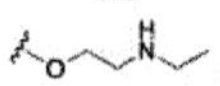	108.49
10	Me		106.95
11	Me	–O^{i}Pr	103.33
12	Me	–O^{n}Bu	104.82
13	Me	–NHCH$_2$Ph	80.31
14 (FW1131)	Ph	–NHCH$_2$Ph	79.71
15	Ph	–OMe	106.83
16	Ph	–O^{i}Pr	102.63
17	Ph	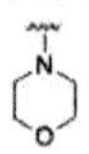	87.17

(B)

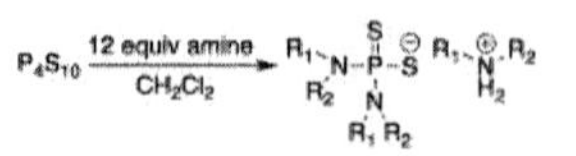

Compound	N(R$_1$)R$_2$	^{31}P NMR δ (ppm)
18	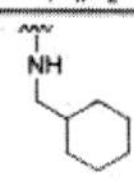	0.19
19	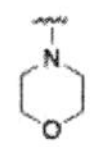	0.32
20	–NHCH$_2$Ph	0.33
21		0.74

a) Each donor was isolated as its morpholinium salt except for compounds **9** and **10** (isolated as a zwitterion) and compounds **13** and **14** (isolated as a benzylamine salt).
Source: Data from Feng et al. [64].

cell lines, including breast (MCF7, MBA-MD231), ovarian (SKOV3), and prostate (PC3) (Figure 10.4).

Intrigued by these results, Feng and coworkers also explored a series of monothiol derivatives by replacing one of the sulfur atoms with either a carbon or another heteroatom [64]. It was discovered that cyclic structures bearing a single sulfur atom could be prepared by treating various amino-phenols with either benzenephosphonothioyl dichloride or Lawesson's reagent (Table 10.2). While a good range of H_2S release profiles were obtained from these cyclized donors, no apparent correlation between ^{31}P NMR chemical shift and efficiency of H_2S release was noted. Nevertheless, the most efficient donor from this series, compound **22** (FW1256), was also found to be the most potent anticancer agent, displaying low micromolar activity against both MCF7 and SKOV3 cells and

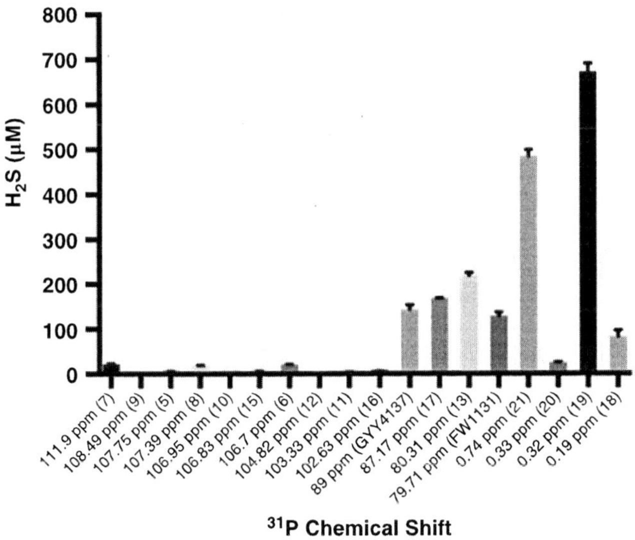

Figure 10.3 Relationship between ^{31}P NMR chemical shift and H_2S production for compounds listed in Table 10.1. Compounds **18–21** were tested at a concentration of 0.5 mM; all other donors were assayed at 5 mM. Reproduced using data from Feng et al. Source: Reproduced using data from Feng et al. [64]/with permission of American Chemical Society.

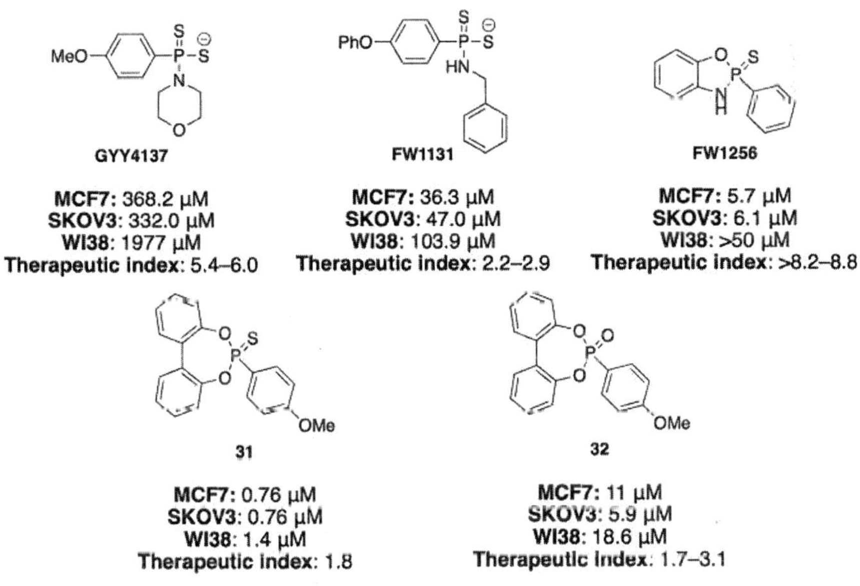

Figure 10.4 Reported IC_{50} (μM) values for cell growth inhibition of compounds in different cancer (MCF7 and SKOV3) and normal (WI38) cell lines. Source: Feng et al. [65]/Elsevier.

Table 10.2 Synthesis and ^{31}P NMR chemical shifts ($CDCl_3$) of cyclized donors and monothio derivatives of GYY4137.

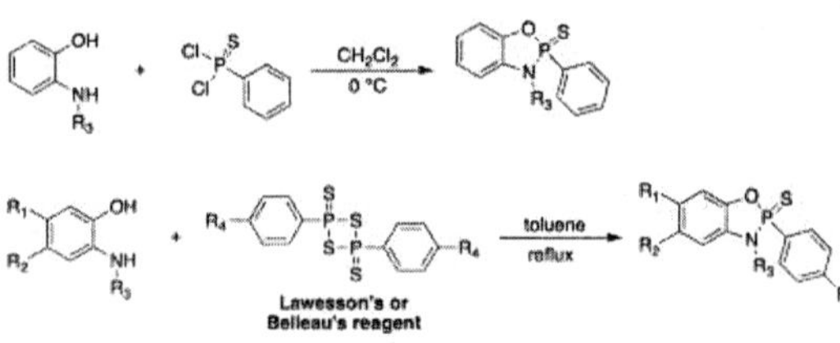

Compound	R_1	R_2	R_3	R_4	^{31}P NMR δ (ppm)
22 (FW1256)	H	H	H	H	93.87
23	H	H	Me	H	97.72
24	H	H	H	OMe	94.34
25	H	H	Me	OMe	98.04
26	Me	H	H	OMe	94.99
27	Me	H	H	OPh	94.15
28	H	OMe	H	OMe	95.30
29	H	F	H	OMe	96.30
30	H	F	H	OPh	95.52

Source: Data from Feng et al. [64].

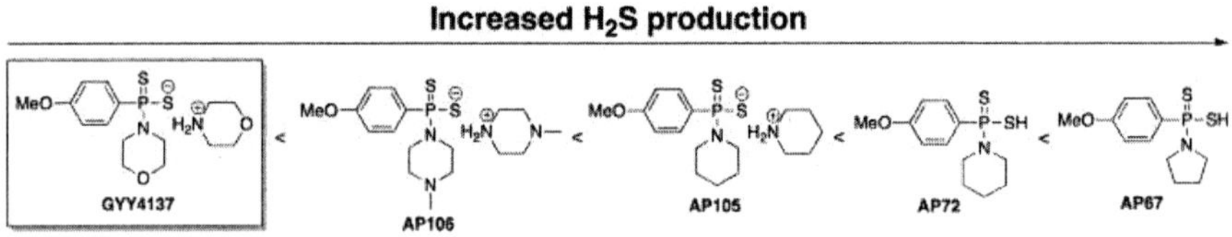

Figure 10.5 Phosphorodithioate-based donors in order of increasing H_2S production. Source: Whiteman et al. [28]/Springer Nature.

with a relatively high therapeutic index (Figure 10.4). Later studies aimed at optimizing these results led investigators to identify another cyclized structure (**31**) with a further 10-fold improvement in potency (Figure 10.4) [65]. However, control experiments with **32**, a non-H_2S-releasing oxygen analog, indicated that the impressive anticancer activity of **31** cannot be entirely attributed to its release of hydrogen sulfide.

A separate but related set of phosphorodithioate-based donors with improved rates of H_2S production, both in buffered water and in human cells (synoviocytes), were reported by Whiteman and coworkers (AP67 > AP72 > AP105 > AP106 > GYY4137, Figure 10.5) [28]. It is interesting to note that the biological activity of these donors, such as their inhibition of LPS-promoted NO and PGE_2 production in RAW264.7 macrophages and human synovial fibroblasts, correlated perfectly with their production of H_2S (AP67 > AP72 > AP105 > AP106 = GYY4137). In other words, donors with a more efficient H_2S release profile were found to provide greater anti-inflammatory and cytoprotective activity.

A very recent report by Carter et al. explored the use of dialkyldithiophosphates as slow-releasing H_2S donors (Scheme 10.5) [66]. The authors of this study were particularly interested in studying the effects of H_2S on plant growth and reasoned that donors of this type would be more biocompatible compared to GYY4137 given that fatty alcohols and phosphoric acid would serve as the sole byproducts

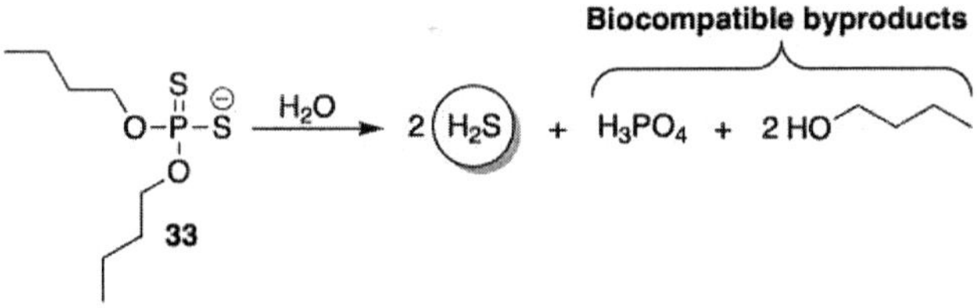

Scheme 10.5 Hydrolysis products of dibutyldithiophosphate. Source: Carter et al. [66]/American Chemical Society.

of H_2S release. Prior work also demonstrated that the rates of hydrolysis from such scaffolds would be highly dependent on the size of the alkyl groups and would therefore provide a convenient method for controlling and tuning the rates of H_2S release. Throughout the course of their studies, the authors noted that dibutyldithiophosphate (**33**) produced H_2S at a similar rate to that of GYY4137 while having a positive impact on the growth of corn plants. Considering the improved biocompatibility of their degradation byproducts, along with the ease at which their physicochemical properties (solubility, lipophilicity, cell permeability, etc.) and H_2S release rates can be modified, dialkyldithiophosphates may offer an important advancement within the field.

While methods for modifying the rates and efficiency of H_2S release from P=S motifs have been extensively explored, the precise spatiotemporal control over the H_2S production from these scaffolds is still limited. This is due to the fact that the release of H_2S from GYY4137 (and related analogs) is immediate upon its introduction to water. However, it was recently demonstrated by Woods and coworkers that H_2S liberation from GYY4137 could be completely halted in aqueous solutions through its coordination to a photolabile ruthenium(II) scaffold $[Ru(tpy)(biq)(L)]^{n+}$ [67]. The ensuing complex (**34**) provided one of the early examples of a red light activated H_2S releasing molecule (Scheme 10.6). Moreover, it also afforded the first ever example of a GYY4137 prodrug that could, in theory, provide the on-demand delivery of this important donor compound within specific cells and tissues. After demonstrating that **34** could increase cellular concentrations of H_2S upon irradiation with red light, the authors then

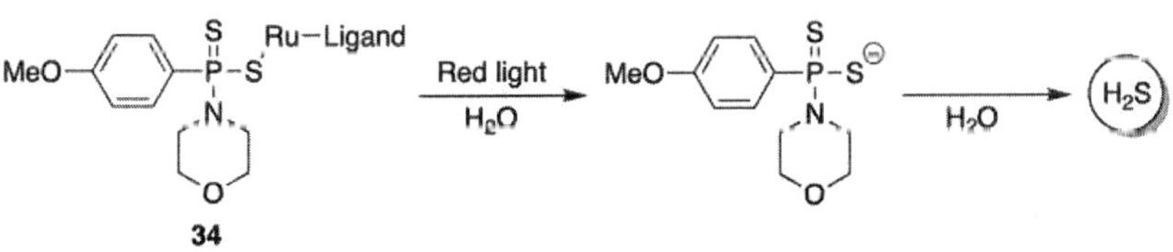

Scheme 10.6 A red light activated H_2S donor. Coordination of GYY4137 to a photolabile Ru(II) complex prohibits its release of H_2S in water.

highlighted its cardioprotective effects in H9c2 cells that were subjected to an ischemia-reperfusion injury model.

10.2.5 JK Donors: Cyclization-Assisted H_2S Release from P=S Motifs

Although the P=S motif has been fully established as an invaluable functional group for the development of useful H_2S donors with high therapeutic potential, the intermolecular hydrolysis reaction that occurs between water and donors of this type is only capable of delivering H_2S at a fixed rate and with little spatiotemporal control over its production. Therefore, alternative activation strategies for this template have been explored in an effort to identify donors that are activated under specific biological conditions and with tunable rates of release. With this in mind, Kang and coworkers were the first to develop a novel series of phosphorothioate-based donors that undergo an intramolecular cyclization reaction to release H_2S (JK-1–JK-5, Scheme 10.7) [68].

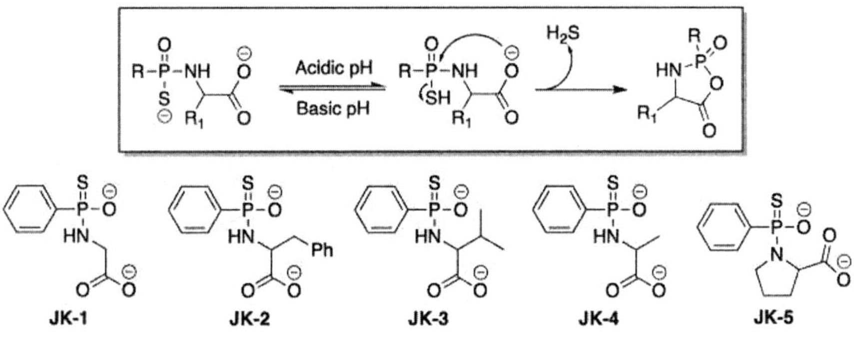

Scheme 10.7 Structures of JK donors along with their putative mechanism for H_2S release under acidic conditions. Source: Kang et al. [68]/American Chemical Society.

Using a modified methylene blue assay for H_2S detection, the authors discovered that the rates of H_2S release from the JK donor series was highly dependent on pH. Typically, it was noted that weakly acidic solutions (pH 5–6) triggered much faster rates of H_2S production, while neutral to slightly basic conditions (pH 7.4–8) provided less efficient release and lower amounts of H_2S. In comparison, the same dependence on pH was not observed for GYY4137 as very little H_2S was detected under the same slightly acidic, neutral, and basic conditions (pH 5–8). The structure of the linked carboxylic acid also appeared to offer another avenue to dramatically alter the reaction kinetics within this series. For example, while JK-1 produced very little H_2S at neutral pH (similar to GYY4137), JK-2 was able to deliver appreciable amounts of hydrogen sulfide under neutral and even slightly basic conditions. This enhanced rate likely stems from the introduction of

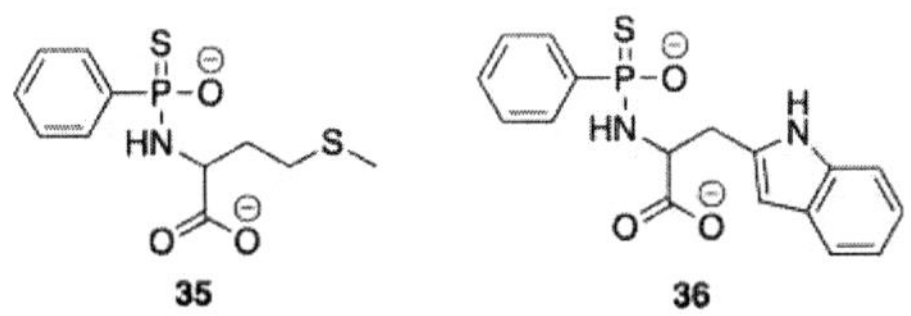

Figure 10.6 Recently reported phosphorothioate-based donors that undergo cyclization-promoted H_2S release under acidic conditions.

the benzyl group at the alpha position, which facilitates the requisite ring closing reaction (Scheme 10.7). In line with this hypothesis, JK-3 and JK-4 displayed similar reaction kinetics to that of JK-2. JK-5, however, failed to afford any detectable levels of H_2S, presumably due to the presence of the rigid proline ring which likely inhibits cyclization.

In light of the fact that numerous diseases and ailments are known to lead to noticeable decreases in pH (inflammation, cancer, and cardiovascular disorders), JK donors – with their efficient release of H_2S under acidic conditions – may prove to be especially useful for providing therapeutically relevant concentrations of H_2S under certain pathological conditions. In their initial study, the authors were able to demonstrate that both JK-1 and JK-2 could provide significant cardioprotection in both cellular and *in vivo* murine models of myocardial ischemia-reperfusion (MI/R) injury. In a later study, JK-1 was shown to prevent aspirin-induced gastric lesions through its selective release of H_2S under acidic conditions, leading to a reduction in inflammation and oxidation [69]. Even more recently, JK-1 was used to facilitate wound healing as a result of its prolonged production of H_2S in wounds exhibiting an acidic pH [70, 71]. The released H_2S is believed to stimulate angiogenesis – a crucial element of the wound healing process.

Since this initial report by Kang and coworkers, additional donors of this type have also been prepared by modifying the substituent at the alpha position. Compound **36** (Figure 10.6), for example, was recently accessed and found to display low toxicity and good pharmacokinetics in addition to its impressive protection of H9c2 cardiomyocytes from MI/R injury [72]. Given the purported drug-like properties of **36**, JK donors (and related analogs) are likely to be good candidates for further preclinical development.

10.3 H_2S Release from Se–S Motifs

Persulfides (RSSH) are a distinct class of reactive sulfur species that play a fundamental role in cellular signaling [73–75]. However, in a similar fashion to H_2S, a complete understanding of their principal functions in redox biology has not been fully elucidated due to their inherent reactivity and instability. Therefore, similar methods for generating reactive persulfides under biologically relevant conditions have been explored with great interest [76]. It has been recently reported that acyl

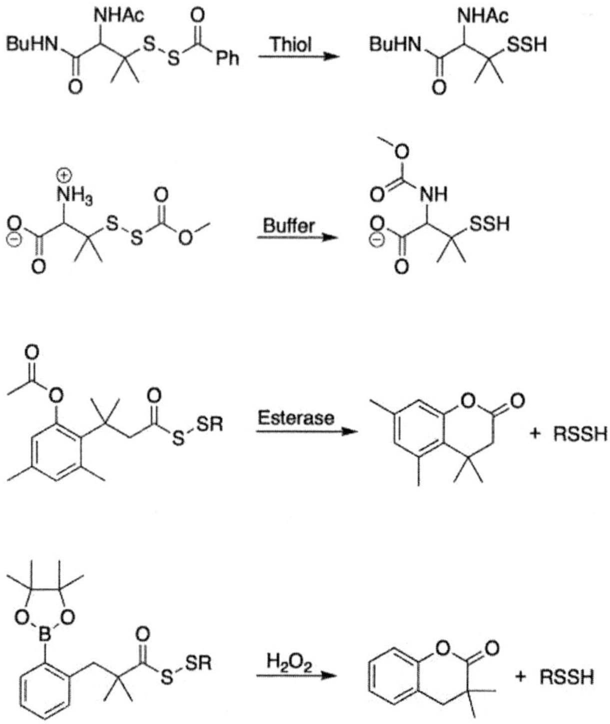

Scheme 10.8 Release of persulfides from acyl disulfide precursors. Source: Based on Refs. [77–80].

disulfides have the potential to serve as useful persulfide pro-moieties given their tendency to undergo chemical hydrolysis or acyl transfer reactions (either intermolecular or intramolecular) to liberate an RSSH functional group (Scheme 10.8) [77–80]. Acyl disulfides have also been touted as advantageous H_2S donors given the ability of the ensuing persulfide to undergo reductive S—S bond cleavage in the presence of cellular thiols [81].

Selenylsulfides (RSeSR) and thioseleninic acids (RSeSH) are the monoselenium equivalent to disulfides and persulfides, respectively [82]. Selenylsulfides, in particular, have garnered significant attention in recent years due to the discovery of their existence within the active sites of key antioxidant enzymes, such as thioredoxin reductase (TrxR) and glutathione peroxidase (GPx) [83]. While sulfur and selenium do share many of the same physical and chemical properties, there are significant distinctions between the two. The majority of these differences are a direct result of selenium being the heavier of the two elements. This increased size leads to increased polarizability and superior nucleophilic and electrophilic character [84]. Therefore, nearly all chemical reactions involving selenium are found to be significantly faster than analogous reactions involving sulfur. For these reasons, acyl selenylsulfides have also been explored for their propensity to liberate H_2S through the resultant Se—S bond cleavage of an initially released thioseleninic acid.

10.3.1 Acyl Selenylsulfides: Synthesis and Characterization of H_2S Release

It was first envisioned by the Xian group that acyl selenylsulfides might provide a useful template for the production of RSeSH under physiological conditions [85]. To access a series of acyl selenylsulfides, the authors first generated phenylselenyl chloride *in situ* upon the treatment of diphenyl diselenide with sulfuryl chloride. The subsequent addition of various thioacids then furnished compounds **37a–37e**. (Scheme 10.9).

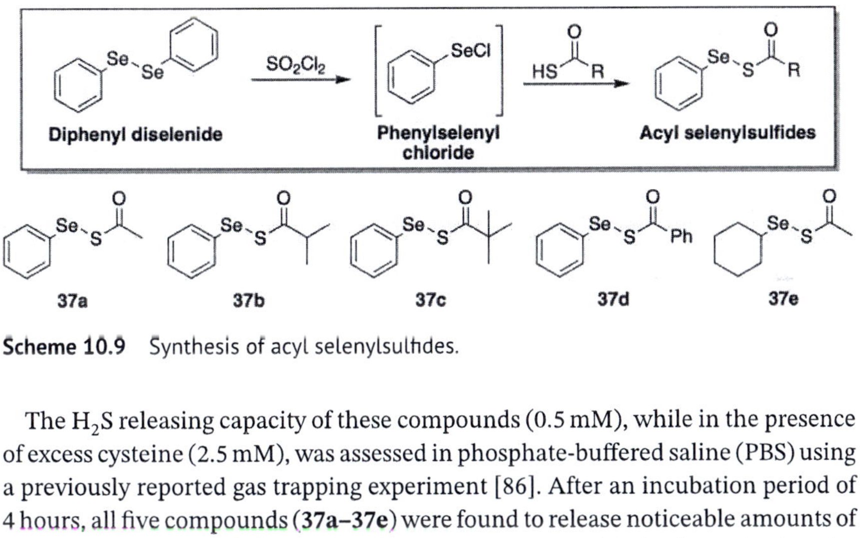

Scheme 10.9 Synthesis of acyl selenylsulfides.

The H_2S releasing capacity of these compounds (0.5 mM), while in the presence of excess cysteine (2.5 mM), was assessed in phosphate-buffered saline (PBS) using a previously reported gas trapping experiment [86]. After an incubation period of 4 hours, all five compounds (**37a–37e**) were found to release noticeable amounts of H_2S. Moreover, the efficiency of H_2S generation appeared to be dependent on the nature of the acyl substituent with **37a** producing the highest amounts of H_2S (79% yield), while more sterically hindered substrates (**37b–d**) afforded a visible reduction in H_2S production (~40%). Cell imaging experiments were also conducted with **37a**, verifying its ability to release H_2S within the live cells.

10.3.2 Acyl Selenylsulfides: Mechanistic Studies

To probe the mechanism of H_2S release from acyl selenylsulfides, the products of the reaction between **37a** and excess cysteine were characterized. Through

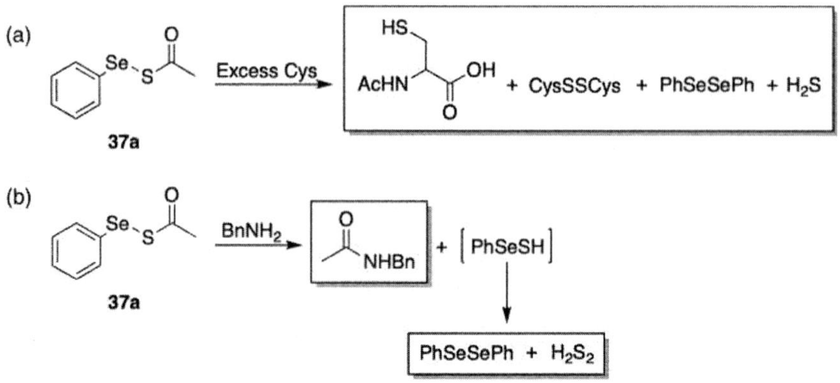

Scheme 10.10 Degradation products of **37a** while exposed to (a) excess cysteine or (b) benzylamine.

this examination it was determined that *N*-acetyl cysteine, cystine, and diphenyl diselenide unequivocally accompany the liberation of H_2S (Scheme 10.10a). Furthermore, when treated with benzylamine ($BnNH_2$), it was discovered that **37a** went on to produce *N*-benzylacetamide. Diphenyl diselenide and H_2S_2 were also observed and proposed to be degradation byproducts of an initial PhSeSH intermediate (Scheme 10.10b).

Bearing in mind these observations – along with the fact that the authors had previously established that thioacids were incapable of releasing H_2S, even in the presence of excess cysteine or glutathione [77] – the following mechanism was put forth (Scheme 10.11). Initially, an *N*-acyl transfer reaction between the acyl selenylsulfide and cysteine takes place, generating a thioseleninic acid (PhSeSH). From this intermediate, two distinct pathways could potentially lead to the eventual release of H_2S. Path a, which entails a sulfide transfer reaction, generating a cysteine persulfide that then reacts with a second equivalent of cysteine to generate cystine and H_2S. Or, path b, which provides the direct release of H_2S from the PhSeSH intermediate via the nucleophilic attack of cysteine at selenium.

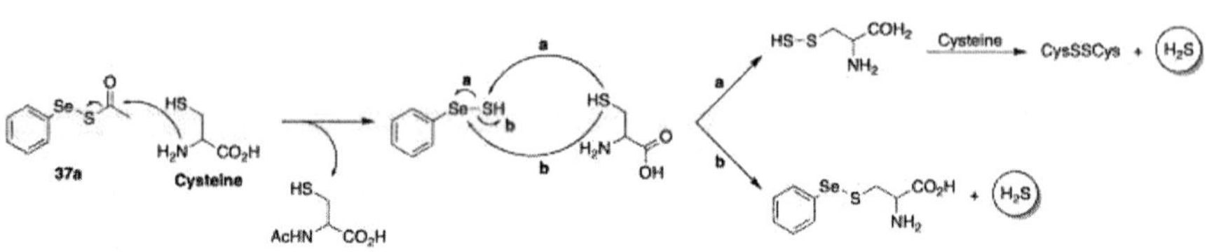

Scheme 10.11 Putative mechanism for H_2S release from acyl selenylsulfides while in the presence of excess cysteine.

10.4 Acyl Selenylsulfides: Structural Modifications and Activity of Analogs

Cyclic acyl selenylsulfides have also been examined by the Xian group as a potential source of H_2S in buffered water [87]. The group initially showed that when compound **38** was treated with *n*-butylamine ($BuNH_2$) in organic solvents, it went on to produce diselenide **40** and diselenylsulfide **41** (Scheme 10.12). Based on this observation, it was suggested that selenylsulfide **39** was generated en route to these final products. Subsequent trapping experiments with thiol blocking reagent MSBT confirmed this hypothesis [88]. Given that cellular thiols – namely cysteine and glutathione – contain both amine and thiol nucleophiles, it was speculated that while in their presence, **38** would exhibit a propensity to liberate H_2S via the subsequent Se—S bond cleavage of **39**. Although **38** was found to be significantly less efficient at producing H_2S relative to acyclic acyl selenylsulfides, this hypothesis was confirmed from gas trapping experiments in PBS buffer.

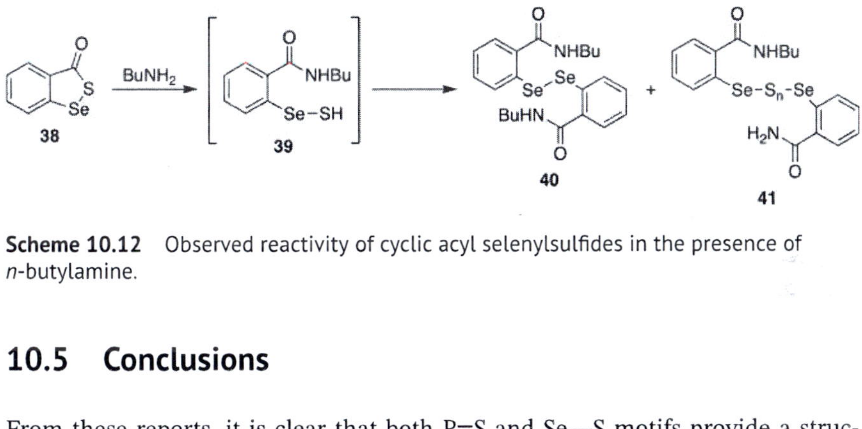

Scheme 10.12 Observed reactivity of cyclic acyl selenylsulfides in the presence of *n*-butylamine.

10.5 Conclusions

From these reports, it is clear that both P=S and Se—S motifs provide a structural basis for the development of useful donors with a predilection for H_2S liberation under biologically relevant conditions. While P=S donor scaffolds, such as GYY4137, have been extensively studied and their effects on inflammation, the cardiovascular system, and cancer cell growth inhibition have been well-established in both *in vitro* and *in vivo* models, H_2S liberation from Se—S motifs has only recently been discovered and the ability of these donors to function within authentic biological systems has yet to be fully explored. However, considering that both templates are amendable to structural modifications that enhance their physicochemical and drug-like properties while further refining their rates and efficiency of H_2S release, the true potential of these scaffolds as useful chemical tools for probing the chemical biology and therapeutic effects of H_2S has likely just begun to scratch the surface.

References

1 Fiedler, N., Kipen, H., Ohman-Strickland, P. et al. (2008). Sensory and cognitive effects of acute exposure to hydrogen sulfide. *Environ. Health Perspect.* 116 (1): 78–85.

2 Abe, K. and Kimura, H. (1996). The possible role of hydrogen sulfide as an endogenous neuromodulator. *J. Neurosci.* 16 (3): 1066–1071.

3 Cacanyiova, S., Berenyiova, A., and Kristek, F. (2016). The role of hydrogen sulphide in blood pressure regulation. *Physiol. Res.* 65 (3): S273–S289.

4 Burguera, E.F., Vela-Anero, Á., Magalhães, J. et al. (2014). Effect of hydrogen sulfide sources on inflammation and catabolic markers on interleukin 1β-stimulated human articular chondrocytes. *Osteoarthr. Cartil.* 22 (7): 1026–1035.

5 Whiteman, M. and Winyard, P.G. (2011). Hydrogen sulfide and inflammation: the good, the bad, the ugly and the promising. *Expert Rev. Clin. Pharmacol.* 4 (1): 13–32.

6 Papapetropoulos, A., Pyriochou, A., Altaany, Z. et al. (2009). Hydrogen sulfide is an endogenous stimulator of angiogenesis. *Proc. Natl. Acad. Sci. U.S.A.* 106 (51): 21972–21977.

7 Liu, F., Chen, D.-D., Sun, X. et al. (2014). Hydrogen sulfide improves wound healing via restoration of endothelial progenitor cell functions and activation of angiopoietin-1 in type 2 Diabetes. *Diabetes* 63 (5): 1763–1778.

8 Gadalla, M.M. and Snyder, S.H. (2010). Hydrogen sulfide as a gasotransmitter. *J. Neurochem.* 113 (1): 14–26.

9 Wang, R. (2012). Shared signaling pathways among gasotransmitters. *Proc. Natl. Acad. Sci. U.S.A.* 109 (23): 8801–8802.

10 Kimura, H. (2011). Hydrogen sulfide: Its production, release and functions. *Amino Acids* 41 (1): 113–121.

11 Miles, E.W. and Kraus, J.P. (2004). Cystathionine β-synthase: structure, function, regulation, and location of homocystinuria-causing mutations. *J. Biol. Chem.* 279 (29): 29871–29874.

12 Pan, L.L., Liu, X.H., Gong, Q.H. et al. (2012). Role of cystathionine γ-lyase/hydrogen sulfide pathway in cardiovascular disease: A novel therapeutic strategy? *Antioxid. Redox Signal.* 17 (1): 106–118.

13 Shibuya, N., Tanaka, M., Yoshida, M. et al. (2009). 3-Mercaptopyruvate sulfurtransferase produces hydrogen sulfide and bound sulfane sulfur in the brain. *Antioxid. Redox Signal.* 11 (4): 703–714.

14 Zheng, Y., Ji, X., Ji, K., and Wang, B. (2015). Hydrogen sulfide prodrugs—a review. *Acta Pharm. Sin. B* 5 (5): 367–377.

15 Hartle, M.D. and Pluth, M.D. (2016). A practical guide to working with H_2S at the interface of chemistry and biology. *Chem. Soc. Rev.* 45 (22): 6108–6117.

16 Powell, C.R., Dillon, K.M., and Matson, J.B. (2018). A review of hydrogen sulfide (H_2S) donors: chemistry and potential therapeutic applications. *Biochem. Pharmacol.* 149: 110–123.

17 Xu, S., Hamsath, A., Neill, D.L. et al. (2019). Strategies for the design of donors and precursors of reactive sulfur species. *Chem. Eur. J.* 25 (16): 4005–4016.

18 Kaur, K., Carrazzone, R.J., and Matson, J.B. (2020). The benefits of macromolecular/supramolecular approaches in hydrogen sulfide delivery: a review of polymeric and self-assembled hydrogen sulfide donors. *Antioxid. Redox Signal.* 32 (2): 79–95.

19 Levinn, C.M., Cerda, M.M., and Pluth, M.D. (2020). Activatable small-molecule hydrogen sulfide donors. *Antioxid. Redox Signal.* 32 (2): 96–109.

20 Li, Z., Polhemus, D.J., and Lefer, D.J. (2018). Evolution of hydrogen sulfide therapeutics to treat cardiovascular disease. *Circ. Res.* 123 (5): 590–600.

21 Cao, X., Cao, L., Ding, L., and Bian, J. (2017). A new hope for a devastating disease: hydrogen sulfide in Parkinson's disease. *Mol. Neurobiol.* 55: 3789–3799.

22 Zhang, Y., Tang, Z.-H., Ren, Z. et al. (2013). Hydrogen sulfide, the next potent preventive and therapeutic agent in aging and age-associated diseases. *Mol. Cell. Biol.* 33 (6): 1104–1113.

23 Li, L., Whiteman, M., Guan, Y.Y. et al. (2008). Characterization of a novel, water-soluble hydrogen sulfide–releasing molecule (GYY4137): new insights into the biology of hydrogen sulfide. *Circulation* 117 (18): 2351–2360.

24 Ozturk, T., Ertas, E., and Mert, O. (2007). Use of Lawesson's reagent in organic syntheses. *Chem. Rev.* 107 (11): 5210–5278.

25 Wallace, J.L., Vong, L., McKnight, W. et al. (2009). Endogenous and exogenous hydrogen sulfide promotes resolution of colitis in rats. *Gastroenterology* 137 (2): 569–578.

26 Medeiros, J.V.R., Bezerra, V.H., Gomes, A.S. et al. (2009). Hydrogen sulfide prevents ethanol-induced gastric damage in mice: Role of ATP-sensitive potassium channels and capsaicin-sensitive primary afferent neurons. *J. Pharmacol. Exp. Ther.* 330 (3): 764–770.

27 Nicolau, L.A.D., Silva, R.O., Damasceno, S.R.B. et al. (2013). The hydrogen sulfide donor, Lawesson's reagent, prevents alendronate-induced gastric damage in rats. *Braz. J. Med. Biol. Res.* 46 (8): 708–714.

28 Whiteman, M., Perry, A., Zhou, Z. et al. (2015). Phosphinodithioate and phosphoramidodithioate hydrogen sulfide donors. In: *Chemistry, Biochemistry and Pharmacology of Hydrogen Sulfide* (ed. P.K. Moore and M. Whiteman), 337–363. Cham: Springer International Publishing.

29 Lee, Z.W., Zhou, J., Chen, C.-S. et al. (2011). The slow-releasing hydrogen sulfide donor, GYY4137, exhibits novel anti-cancer effects in vitro and in vivo. *PLoS One* 6 (6): e21077.

30 Rose, P., Dymock, B.W., and Moore, P.K. (2015). GYY4137, a novel water-soluble, H_2S-releasing molecule. *Methods Enzymol.* 554: 143–167.

31 Kang, J., Neill, D.L., and Xian, M. (2017). Phosphonothioate-based hydrogen sulfide releasing reagents: chemistry and biological applications. *Front. Pharmacol.* 8: 457–466.

32 Nin, D.S., Idres, S.B., Song, Z.J. et al. (2020). Biological effects of morpholin-4-ium 4 methoxyphenyl (morpholino) phosphinodithioate and other phosphorothioate-based hydrogen sulfide donors. *Antioxid. Redox Signal.* 32 (2): 145–158.

33 Liu, Y.-H., Lu, M., Hu, L.-F. et al. (2012). Hydrogen sulfide in the mammalian cardiovascular system. *Antioxid. Redox Signal.* 17 (1): 141–185.

34 Wu, D., Hu, Q., and Zhu, D. (2018). An update on hydrogen sulfide and nitric oxide interactions in the cardiovascular system. *Oxid. Med. Cell. Longev.* 2018: 1–16.

35 Shen, Y., Shen, Z., Luo, S. et al. (2015). The cardioprotective effects of hydrogen sulfide in heart diseases: From molecular mechanisms to therapeutic potential. *Oxid. Med. Cell. Longev.* 2015: 1–13.

36 Szabó, G., Veres, G., Radovits, T. et al. (2011). Cardioprotective effects of hydrogen sulfide. *Nitric Oxide* 25 (2): 201–210.

37 Whiteman, M., Le Trionnaire, S., Chopra, M. et al. (2011). Emerging role of hydrogen sulfide in health and disease: critical appraisal of biomarkers and pharmacological tools. *Clin. Sci.* 121 (11): 459–488.

38 Hosoki, R., Matsuki, N., and Kimura, H. (1997). The possible role of hydrogen sulfide as an endogenous smooth muscle relaxant in synergy with nitric oxide. *Biochem. Biophys. Res. Commun.* 237 (3): 527–531.

39 Zhao, W. (2001). The vasorelaxant effect of H_2S as a novel endogenous gaseous K_{ATP} channel opener. *EMBO J.* 20 (21): 6008–6016.

40 Webb, G.D., Lim, L.H., Oh, V.M.S. et al. (2008). Contractile and vasorelaxant effects of hydrogen sulfide and its biosynthesis in the human internal mammary artery. *J. Pharmacol. Exp. Ther.* 324 (2): 876–882.

41 Karwi, Q.G., Whiteman, M., Wood, M.E. et al. (2016). Pharmacological postconditioning against myocardial infarction with a slow-releasing hydrogen sulfide donor, GYY4137. *Pharmacol. Res.* 111: 442–451.

42 Lilyanna, S., Peh, M.T., Liew, O.W. et al. (2015). GYY4137 attenuates remodeling, preserves cardiac function and modulates the natriuretic peptide response to ischemia. *J. Mol. Cell. Cardiol.* 87: 27–37.

43 Meng, G., Wang, J., Xiao, Y. et al. (2015). GYY4137 protects against myocardial ischemia and reperfusion injury by attenuating oxidative stress and apoptosis in rats. *J. Biomed. Res.* 29 (3): 203–213.

44 Qiu, Y., Wu, Y., Meng, M. et al. (2018). GYY4137 protects against myocardial ischemia/reperfusion injury via activation of the PHLPP-1/Akt/Nrf2 signaling pathway in diabetic mice. *J. Surg. Res.* 225: 29–39.

45 Lee, Z.-W., Teo, X.-Y., Song, Z.J. et al. (2017). Intracellular hyper-acidification potentiated by hydrogen sulfide mediates invasive and therapy resistant cancer cell death. *Front. Pharmacol.* 8: 763–772.

46 Lu, S., Gao, Y., Huang, X., and Wang, X. (2014). GYY4137, a hydrogen sulfide (H_2S) donor, shows potent anti-hepatocellular carcinoma activity through blocking the STAT3 pathway. *Int. J. Oncol.* 44 (4): 1259–1267.

47 Markova, J., Hudecova, S., Soltysova, A. et al. (2014). Sodium/calcium exchanger is upregulated by sulfide signaling, forms complex with the β1 and β3 but not β2 adrenergic receptors, and induces apoptosis. *Pflüg. Arch. Eur. J. Physiol.* 466 (7): 1329–1342.

48 Szadvari, I., Hudecova, S., Chovancova, B. et al. (2019). Sodium/calcium exchanger is involved in apoptosis induced by H_2S in tumor cells through decreased levels of intracellular pH. *Nitric Oxide* 87: 1–9.

49 Li, L., Fox, B., Keeble, J. et al. (2013). The complex effects of the slow-releasing hydrogen sulfide donor GYY4137 in a model of acute joint inflammation and in human cartilage cells. *J. Cell. Mol. Med.* 17 (3): 365–376.

50 Li, L., Salto-Tellez, M., Tan, C.-H. et al. (2009). GYY4137, a novel hydrogen sulfide-releasing molecule, protects against endotoxic shock in the rat. *Free Radical Biol. Med.* 47 (1): 103–113.

51 Rodrigues, L., Ekundi-Valentim, E., Florenzano, J. et al. (2017). Protective effects of exogenous and endogenous hydrogen sulfide in mast cell-mediated pruritus and cutaneous acute inflammation in mice. *Pharmacol. Res.* 115: 255–266.

52 Whiteman, M., Li, L., Rose, P. et al. (2010). The effect of hydrogen sulfide donors on lipopolysaccharide-induced formation of inflammatory mediators in macrophages. *Antioxid. Redox Signal.* 12 (10): 1147–1154.

53 Zhang, H.-X., Liu, S.-J., Tang, X.-L. et al. (2016). H_2S attenuates LPS-induced acute lung injury by reducing oxidative/nitrative stress and inflammation. *Cell. Physiol. Biochem.* 40 (6): 1603–1612.

54 Qabazard, B., Ahmed, S., Li, L. et al. (2013). *C. elegans* aging is modulated by hydrogen sulfide and the sulfhydrylase/cysteine synthase cysl-2. *PLoS One* 8 (11): e80135.

55 Qabazard, B., Li, L., Gruber, J. et al. (2014). Hydrogen sulfide is an endogenous regulator of aging in *Caenorhabditis elegans*. *Antioxid. Redox Signal.* 20 (16): 2621–2630.

56 Ning, N., Zhu, J., Du, Y. et al. (2014). Dysregulation of hydrogen sulphide metabolism impairs oviductal transport of embryos. *Nat. Commun.* 5 (1): 4107–4114.

57 Wang, K., Ahmad, S., Cai, M. et al. (2013). Dysregulation of hydrogen sulfide producing enzyme cystathionine γ-lyase contributes to maternal hypertension and placental abnormalities in preeclampsia. *Circulation* 127 (25): 2514–2522.

58 Alexander, B.E., Coles, S.J., Fox, B.C. et al. (2015). Investigating the generation of hydrogen sulfide from the phosphonamidodithioate slow-release donor GYY4137. *MedChemComm* 6 (9): 1649–1655.

59 Fagin, J., Bradley, J., and Williams, D. (1980). Carbon monoxide poisoning secondary to inhaling methylene chloride. *Br. Med. J.* 281 (6253): 1461–1461.

60 Peers, C., Boyle, J.P., Scragg, J.L. et al. (2015). Diverse mechanisms underlying the regulation of ion channels by carbon monoxide: CO modulation of ion channels. *Br. J. Pharmacol.* 172 (6): 1546–1556.

61 Park, C.-M., Zhao, Y., Zhu, Z. et al. (2013). Synthesis and evaluation of phosphorodithioate-based hydrogen sulfide donors. *Mol. Biosyst.* 9 (10): 2430–2434.

62 Park, C.-M. and Xian, M. (2015). Use of phosphorodithioate-based compounds as hydrogen sulfide donors. *Methods Enzymol.* 554: 127–142.

63 Peng, H., Cheng, Y., Dai, C. et al. (2011). A fluorescent probe for fast and quantitative detection of hydrogen sulfide in blood. *Angew. Chem. Int. Ed.* 50 (41): 9672–9675.

64 Feng, W., Teo, X.-Y., Novera, W. et al. (2015). Discovery of new H_2S releasing phosphordithioates and 2,3-dihydro-2-phenyl-2-sulfanylenebenzo[*d*][1,3,2] oxazaphospholes with improved antiproliferative activity. *J. Med. Chem.* 58 (16): 6456–6480.

65 Feng, W., Novera, W., Peh, K. et al. (2017). Discovery of medium ring thiophosphorus based heterocycles as antiproliferative agents. *Bioorg. Med. Chem. Lett.* 27 (4): 967–972.

66 Carter, J.M., Brown, E.M., Irish, E.E., and Bowden, N.B. (2019). Characterization of dialkyldithiophosphates as slow hydrogen sulfide releasing chemicals and their effect on the growth of maize. *J. Agric. Food Chem.* 67 (43): 11883–11892.

67 Woods, J.J., Cao, J., Lippert, A.R., and Wilson, J.J. (2018). Characterization and biological activity of a hydrogen sulfide-releasing red light-activated ruthenium(II) complex. *J. Am. Chem. Soc.* 140 (39): 12383–12387.

68 Kang, J., Li, Z., Organ, C.L. et al. (2016). pH-Controlled hydrogen sulfide release for myocardial ischemia-reperfusion injury. *J. Am. Chem. Soc.* 138 (20): 6336–6339.

69 Yang, C., Lai, Z., Zheng, Z. et al. (2017). A novel pH-controlled hydrogen sulfide donor protects gastric mucosa from aspirin-induced injury. *J. Cell. Mol. Med.* 21 (10): 2441–2451.

70 Wu, J., Li, Y., He, C. et al. (2016). Novel H_2S releasing nanofibrous coating for in vivo dermal wound regeneration. *ACS Appl. Mater. Interfaces* 8 (41): 27474–27481.

71 Zhao, X., Liu, L., An, T. et al. (2020). A hydrogen sulfide-releasing alginate dressing for effective wound healing. *Acta Biomater.* 104: 85–94.

72 Zhang, J., Zhang, Q., Wang, Y. et al. (2019). Toxicity, bioactivity, release of H_2S in vivo and pharmaco-kinetics of H_2S-donors with thiophosphamide structure. *Eur. J. Med. Chem.* 176: 456–475.

73 Park, C.-M., Weerasinghe, L., Day, J.J. et al. (2015). Persulfides: current knowledge and challenges in chemistry and chemical biology. *Mol. Biosyst.* 11 (7): 1775–1785.

74 Filipovic, M.R., Zivanovic, J., Alvarez, B., and Banerjee, R. (2018). Chemical biology of H_2S signaling through persulfidation. *Chem. Rev.* 118 (3): 1253–1337.

75 Yang, C., Devarie-Baez, N.O., Hamsath, A. et al. (2020). S-persulfidation: chemistry, chemical biology, and significance in health and disease. *Antioxid. Redox Signal.* 33 (15): 1092–1114.

76 Yuan, Z.-N., Zheng, Y.-Q., and Wang, B.-H. (2020). Prodrugs of hydrogen sulfide and related sulfur species: recent development. *Chin. J. Nat. Med.* 18 (4): 296–307.

77 Zhao, Y., Bhushan, S., Yang, C. et al. (2013). Controllable hydrogen sulfide donors and their activity against myocardial ischemia-reperfusion injury. *ACS Chem. Biol.* 8 (6): 1283–1290.

78 Artaud, I. and Galardon, E. (2014). A persulfide analogue of the nitrosothiol SNAP: Formation, characterization and reactivity. *ChemBioChem* 15 (16): 2361–2364.

79 Yuan, Z., Zheng, Y., Yu, B. et al. (2018). Esterase-sensitive glutathione persulfide donor. *Org. Lett.* 20 (20): 6364–6367.

80 Hankins, R.A., Suarez, S.I., Kalk, M.A. et al. (2020). An innovative hydrogen peroxide-sensing scaffold and insight towards its potential as an ROS-activated persulfide donor. *Angew. Chem. Int. Ed.* 59 (49): 22238–22245.

81 Bailey, T.S., Zakharov, L.N., and Pluth, M.D. (2014). Understanding hydrogen sulfide storage: probing conditions for sulfide release from hydrodisulfides. *J. Am. Chem. Soc.* 136 (30): 10573–10576.

82 Hamsath, A. and Xian, M. (2020). Chemistry and chemical biology of selenenyl sulfides and thioseleninic acids. *Antioxid. Redox Signal.* 33 (16): 1143–1157.

83 Steinbrenner, H., Speckmann, B., and Klotz, L.-O. (2016). Selenoproteins: antioxidant selenoenzymes and beyond. *Arch. Biochem. Biophys.* 595: 113–119.

84 Reich, H.J. and Hondal, R.J. (2016). Why nature chose selenium. *ACS Chem. Biol.* 11 (4): 821–841.

85 Hamsath, A., Wang, Y., Yang, C. et al. (2019). Acyl selenyl sulfides as the precursors for reactive sulfur species (hydrogen sulfide, polysulfide, and selenyl sulfide). *Org. Lett.* 21 (14): 5685–5688.

86 Peng, B., Liu, C., Li, Z. et al. (2017). Slow generation of hydrogen sulfide from sulfane sulfurs and NADH models. *Bioorg. Med. Chem. Lett.* 27 (3): 542–545.

87 Kang, J., Ferrell, A.J., Chen, W. et al. (2018). Cyclic acyl disulfides and acyl selenylsulfides as the precursors for persulfides (RSSH), selenylsulfides (RSeSH), and hydrogen sulfide (H_2S). *Org. Lett.* 20 (3): 852–855.

88 Zhang, D., Devarie-Baez, N.O., Li, Q. et al. (2012). Methylsulfonyl benzothiazole (MSBT): a selective protein thiol blocking reagent. *Org. Lett.* 14 (13): 3396–3399.

11

Hydrogen Sulfide: The Hidden Player of Isothiocyanates Pharmacology

Valentina Citi, Eugenia Piragine, Vincenzo Calderone, and Alma Martelli

Department of Pharmacy, University of Pisa, Pisa, Italy

11.1 Organic Isothiocyanates as H_2S-Donors

H_2S emerged in the last few years as a fundamental endogenous gasotransmitter involved in the modulation of a plethora of pathophysiological processes. Indeed, an impaired H_2S production often leads to the onset of a variety of pathologies, i.e. hypertension, atherosclerosis, neurodegeneration, etc. [1]. For this reason, finding molecules capable of donating H_2S in an endogenous-like manner may be considered an innovative strategy for the pharmacological treatment of those pathologies whose onset is related to a reduction of endogenous H_2S production. Na_2S, NaHS, and CaS were the first prototypes of H_2S-donors, but they have been only used for research purpose since they show a massive and fast H_2S generation, which makes them difficult to use in a clinical environment due to the difficulties in the control of the dosage. Indeed, a fast donation of H_2S is characterized by toxic and severe adverse effects related to the exacerbation of hypotensive effect and the inhibition of the mitochondrial respiratory chain [2]. Consequently, the development of novel compounds which could satisfy the need to discover original moieties acting as slow H_2S-donors urged research activities to characterize several innovative and intriguing chemotypes [3].

Beside the drug discovery of new chemical entities, many studies have demonstrated that the pharmacological activity of natural polysulfide compounds used for nutraceutical and phytotherapeutic purpose was related to their ability to release H_2S [4]. For instance, polysulfide compounds widely contained in plants belonging to the Alliaceae family, such as diallyl disulfide (DADS) and diallyl trisulfide (DATS) from garlic (*Allium sativum* L.), have been reported to promote H_2S-dependent vascular relaxation [5]. The research for new chemical entities that are able to donate H_2S led to focus the attention on isothiocyanate (ITC) chemical group which seemed to represent an intriguing candidate: indeed, a small library of aryl ITCs were first investigated in 2014 by Martelli et al. demonstrating that synthetic ITCs release H_2S and are endowed with vasoactive activities [6]. Interestingly, looking at the natural occurring compounds, several edible plants belonging to *Brassicaceae* family contain sulfur compounds. Similarly with the previous synthetic ITCs, in 2014, Citi et al. demonstrated that many natural ITCs exhibited H_2S generating properties, leading the authors to hypothesize that many pharmacological activities ascribed to edible plants rich in ITCs actually involve their ability to release H_2S [7].

Natural ITCs derive from the metabolism of glucosinolates (GLS), which represent the secondary metabolites of several plant species belonging to *Brassicaceae* family (Crucifers), such as broccoli, brussels sprouts, watercress, horseradish, and cabbage. In particular, GLS are transformed in ITCs when the enzyme myrosinase (thioglucosidase) is activated and thus cleaves off the glucose group from the GLS. The resulting aglycone undergoes a spontaneous Lossen-like rearrangement, leading to the formation of the corresponding ITC (Figure 11.1). This reaction occurs when these plants are cut or crunched, and the cells broken so that the GLS and enzyme myrosinase can interact leading to the formation of ITCs [8].

The intuition of evaluating ITCs as H_2S-donors originated by observing an intriguing overlapping between the pharmacological effects of many ITCs, independently by their structure, and the effects described for H_2S: both ITCs and H_2S activate different classes of potassium channels exhibiting vasodilatory effects [9], behave as antiviral, antioxidant, anti-inflammatory agents [10], prevent tumorigenesis, and are neuroprotective agents [11]. Furthermore, both ITCs and H_2S show a particular behavior that defined hormetic profile: low concentrations of ITCs and H_2S show cytoprotective effects and are endowed with antioxidant properties, while increasing concentrations of ITCs and H_2S promote cytotoxic effects widely exploited for studying the potential antitumoral effect of several natural ITCs [12].

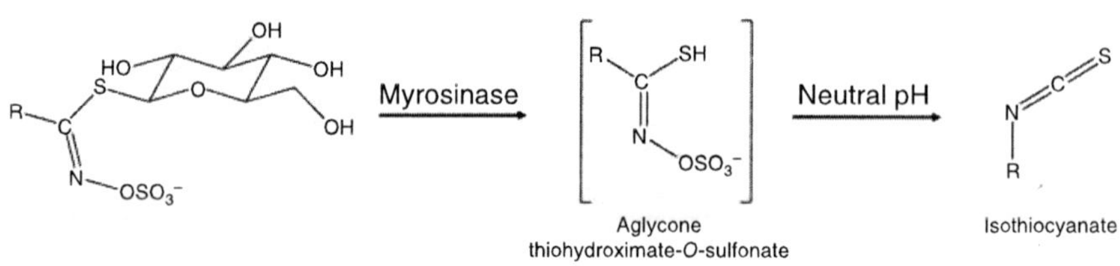

Figure 11.1 The reaction shows the mechanism leading to the formation of the ITCs starting from GLS. First, the β-thioglucoside linkage is cleaved by myrosinase and the resulting aglycone undergoes a spontaneous Lossen-like rearrangement. Finally, at neutral pH, the ITC is formed representing the primary product.

The H_2S-donating activity of some natural ITCs have been first evaluated by Citi and coworkers. In particular, erucin (ERU) contained in several edible plants such as broccoli (*Brassica oleracea* L.) and rocket salad (*Eruca sativa* Mill.), benzyl isothiocyanate (BITC) deriving from garden cress (*Lepidium sativum* L.), as well as allyl isothiocyanate (AITC) present in black mustard (*Brassica nigra* L.) and 4-hydroxy benzyl isothiocyanate (HBITC) contained in white mustard (*Sinapis alba* L.) showed H_2S-releasing properties if incubated in the presence of L-cysteine, an amino acid characterized by the presence of a free thiol group [7]. More recently, in addition to the previously described ITCs as H_2S-donors, also the sulforaphane (SFN), which represents the glucoraphanine aglycone, has been clearly identified as a molecule able to release H_2S in a L-cysteine-dependent manner. Interestingly, SFN promoted the reduction of the oxaliplatin-induced neuropathic pain along with the donation of H_2S. This specific property will be discussed in detail in Section 11.4 [13].

Wang et al. [14] elegantly demonstrated the H_2S-releasing properties of the natural ITC moringin derived from *Moringa oleifera* Lam., a plant belonging to *Moringaceae* family, which is widely used for phytotherapeutic purposes for treating joint and stomach pain, ulcers, and for aiding digestion [15]. In particular, the H_2S-releasing properties of different *Moringa* tissue extracts, characterized by different content of GLS and ITCs, have been evaluated measuring the H_2S formation using the lead acetate test, by exploiting the generation of an evident and visible black precipitate (PbS) [14].

Beside the natural occurring ITCs, also synthetic ITCs have been characterized as H_2S-releasing agents in L-cysteine dependent manner: aryl-ITCs derivatives showed a slow H_2S generation in an aqueous buffer added of L-cysteine and promoted vasoactive effects. In particular, due to their H_2S-releasing properties, they promoted cardio-protection in ischemia/reperfusion injury of rat hearts [16], showed a direct vasodilation on rat aortic rings [6] and inhibited mast cell degranulation [10].

Furthermore, a structure–activity relationship of ITCs for H_2S releasing properties have been recently published by Citi et al. [17]. In particular, a small library of aromatic and aliphatic ITCs characterized by different substituents (electron donating/withdrawing and different steric hindrance groups) have been amperometrically characterized. The H_2S release was measured using an electrode connected to Apollo 4000 Free Radical Analyzer instrument (WPI, Sarasota, FL) able to selectively detect H_2S thanks to a membrane permeable to gaseous molecules and to a specific potential difference. The compounds were incubated at the concentration of 1 mM in phosphate saline buffer, pH 7.4, in the absence or in the presence of 4 mM L-cysteine. The H_2S release was monitored for 30 minutes or until reaching a stable steady state which could vary depending on the H_2S kinetic release. Indeed, sulfur salts, such as NaHS or CaS, reach a stable steady state in few seconds, reflecting their nature of fast H_2S donors, thus quickly running out of the capability of donating H_2S; contrarily, organic compounds, including ITCs, are characterized by a slow increasing H_2S release, generating low, but long-lasting and constant amount of H_2S.

As a general trend, the aromatic compounds were more efficient H_2S-donors with respect to the aliphatic ones. The greater ability to release H_2S is associated with the presence on the aromatic ring of electron-withdrawing groups with a fairly linear trend observable in the series of ortho-, meta-, and para-substituted compounds. ITCs behave as H_2S-donors only in the presence of free thiol such as L-cysteine. The H_2S release implies a nucleophilic attack by the cysteine thiol group on the ITC moiety central carbon leading to an ITC-cysteine adduct. The presence of an electron poor ring (such as pyridine or an electron-withdrawing substituted aromatic ring) facilitates the intramolecular nucleophilic addition of the amino group leading to a 4,5-dihydrothiazole intermediate that produces H_2S, as recently demonstrated (Figure 11.2) [18].

This increasing evidence, which demonstrates the H_2S-relesing properties of the natural and synthetic ITCs, may be the explanation of the pharmacological and nutraceutical properties of all those edible plants characterized by a high amount of sulfur compounds. The structure of the most studied ITCs have been reported in Figure 11.3. Interestingly, the thiol-dependent donation of H_2S from ITCs has been already demonstrated for other sulfur compounds characterized as H_2S-donors (i.e. DADS, DATS, GYY4137) [6, 19] and has been also reported for other synthetic H_2S-donating molecules (i.e. thiazolidinediones, arylthioamides, iminothioethers, mercaptopyruvate, dithioates) [20–23]: even if not fully elucidated, this behavior is probably due to a nucleophilic reaction of the thiol group of L-cysteine with the H_2S-donor moiety similarly to what happens if L-cysteine is incubated with ITCs. This behavior makes these compounds "smart H_2S-donors," a definition indicating that they release H_2S only in biological substrate in which there is high amount of free thiols.

The below mentioned studies have been reported and represented in Figure 11.4 to highlight the perfect overlap between the mechanisms shown by

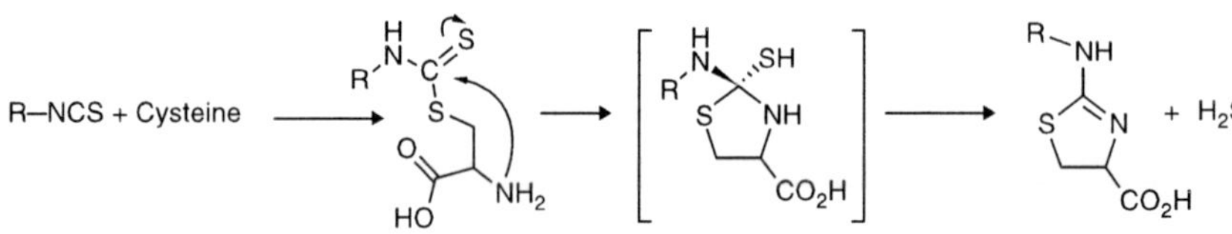

Figure 11.2 The reaction shows the formation of ITC adducts with cysteine that underwent intramolecular cyclization followed by releasing H_2S and 2-carbylamino-4,5-dihydrothiazole-4-carboxylic acids as minor products. Source: Lin et al. [18]/American Chemical Society.

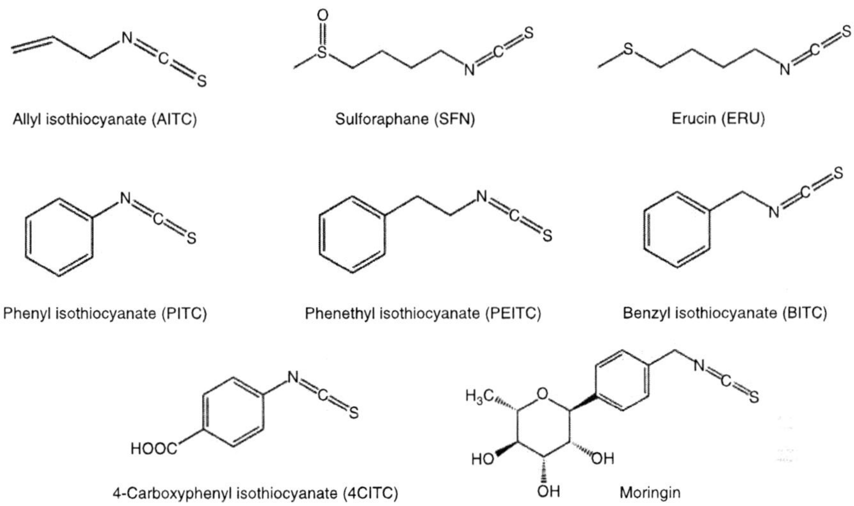

Figure 11.3 Structures of the main isothiocyanates discussed in the chapter.

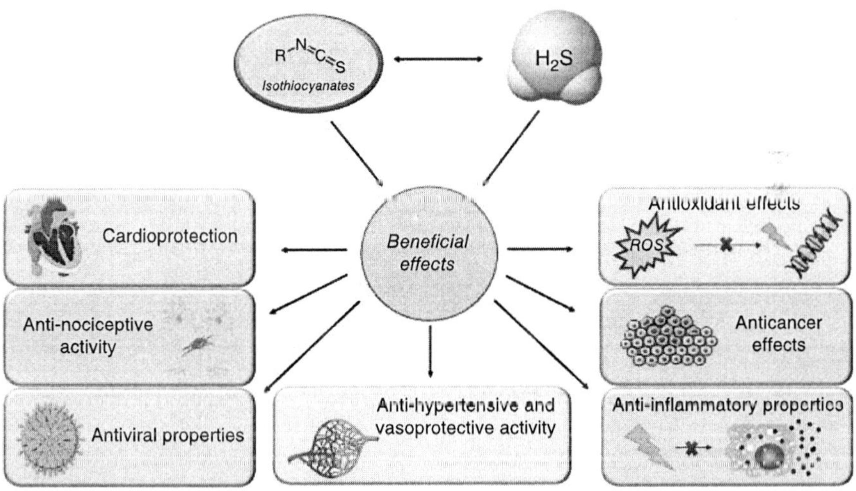

Figure 11.4 Schematic representation of the beneficial effects exhibited by ITCs. Notably, the effects promoted by this heterogeneous class of H_2S-donors are strictly superimposable with those exhibited by H_2S, suggesting a crucial role for the endogenous gasotransmitter in the health promoting activity of both natural and synthetic ITCs.

the ITCs and those exhibited by H_2S, even if some of them have been published before demonstrating the H_2S-generating properties of ITCs. There are now several studies that confirm the hypothesis that the effects observed for ITCs are related to their effective ability in donating H_2S, attributing "a posteriori" the role of H_2S as a "real player" accounting for the pharmacological effects of ITCs. For this reason, ITCs have also been used for the development of hybrid H_2S-donor molecules, for exploiting the well-characterized pharmacological effects of the "native" drug and the beneficial effects of H_2S or with the purpose to limit side effects [24–26].

11.2 Organic ITCs and Cardiovascular System

11.2.1 Effect of ITCs as H_2S Donors in Vascular Inflammation

In the last decades, there have been many *in vitro* and *in vivo* studies aimed at demonstrating the potential protective effects of organic ITCs on the vessel endothelium for being employed as protective agents against vascular inflammation [27, 28]. In this condition, the endothelium and the vascular smooth muscle are continuously exposed to harmful stimuli such as oxidative stress with consequent inflammatory response due to different cardiovascular pathologies including diabetes, hypertension, and atherosclerosis, all of which can lead to vascular impairment [29–31]. The hyperglycemic condition directly promotes the formation of advanced glycated end products (AGEs), which bind the receptor of advanced end glycated products (RAGE). This interaction triggers important intracellular cascades that mediate the increase of intracellular reactive oxygen species (ROS) and the activation of nuclear factor kB (NF-kB), which promotes the expression of several genes involved in atherosclerotic plaque formation and progression, such as intracellular adhesion molecule-1 (ICAM-1), vascular cell adhesion molecule-1 (VCAM-1), monocyte chemoattractant protein-1 (MCP-1), plasminogen activator inhibitor-1 (PAI-1), and vascular endothelial growth factor (VEGF) [32].

SFN has been widely studied as an ITC able to counteract vascular inflammation. Indeed, SFN inhibited AGEs-induced increase in RAGE, MCP-1, ICAM-1, and VCAM-1 gene expression in human umbilical vein endothelial cells (HUVECs), significantly suppressing the AGEs-derived ROS formation. SFN suppressed ICAM-1, VCAM, and aortic RAGE expression also *in vivo* in AGEs-administered rats [33]. These preliminary data and the demonstration of the H_2S-generating properties of SFN may explain the decrease in the AGEs serum levels observed in a Japanese clinical trial carried out on 25 healthy subjects daily receiving 25 g of SFN-precursor containing broccoli sprouts for 2 months

[34]. The same protective role has been also demonstrated for NaHS that reduced AGEs formation in D-galactose treated cells, counteracting ROS formation.

Beside the inhibition of AGE/RAGE pathway, organic ITCs such as SFN, BITC, and phenethyl-isothiocyanate (PEITC) are able to protect vascular cells also against oxidized low-density lipoproteins (oxLDL), which are responsible for the atherogenic plaque formation and progression. Indeed, several *in vitro* studies demonstrated that SFN, BITC, and PEITC are able to inhibit oxLDL-induced ROS formation and prevent the NF-kB translocation into the nucleus, reducing VCAM-1 and ICAM-1 expression counteracting the monocyte adhesion to endothelial cells, exerting a general protective effect against endothelial dysfunction [35]. The vascular protection against atherosclerosis has been also demonstrated in *in vivo* models of hypercholesterolemic rabbits treated with SFN 0.25 mg/kg/day, resulting in an improvement of endothelium integrity associated with antioxidant effect [36].

The antioxidant effect of ITCs and H_2S also relies in the activation of nuclear factor erythroid 2-related factor 2 (Nrf2) and the down-stream antioxidant machinery. Indeed, type I diabetic mice chronically administered with SFN for 3 months showed reduced aortic damages, such as fibrosis, inflammation, and oxidative injury due to up-regulation of Nrf2 and its down-stream antioxidants [37]. Consistently, the attenuation of aortic fibrosis, reduced tumor necrosis factor-α (TNF-α) and VCAM-1 (markers of inflammation) mediated by a significant upregulation of the Nrf2 pathway that also occurred in type II diabetic mice treated with SFN for 4 months [38].

In these studies, the ITCs-induced activation of Nrf2 that was not associated to the H_2S-releasing properties of ITCs, but H_2S and ITCs share the same antioxidant effect by activating Nrf2. Indeed, H_2S modifies the conformation of Nrf2 due to S-sulfhydration on C151 and triggers the consequent dissociation between Keap1 and Nrf2 which is able to translocate into the nucleus promoting the production of antioxidant mediators (Figure 11.5) [39].

ITCs are well-documented Nrf2 activators and their demonstrated H_2S-donor properties may provide a molecular explanation for the antioxidant properties, supporting the use of ITCs in the prevention of vascular alterations in those pathologies (diabetes, atherosclerosis) characterized and sustained by a massive ROS production. ITCs and H_2S also share the same mechanism that leads to vascular anti-inflammatory response: in particular, SFN was reported to inhibit TNF-α production in THP-1 monocytes and in HUVEC, consequently suppressing the expression of important adhesion molecules, such as MCP-1, VCAM-1, and E-Selectin, implied in the atherosclerotic plaque formation and progression [40]. These findings have been also confirmed in mice treated with SFN, which improved the endothelium integrity and functionality and preserved the elastin fibers organization [41]. Other ITCs as 6-methylsulfinylhexyl isothiocyanate

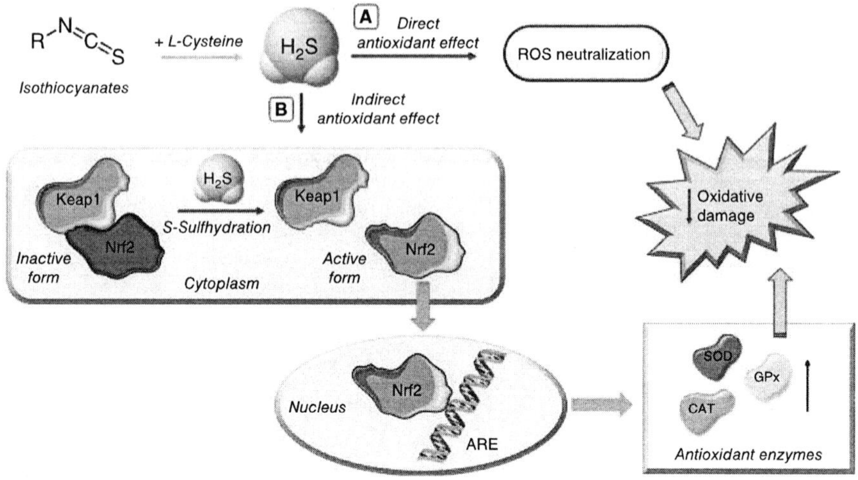

Figure 11.5 Cysteine-dependent H_2S-release is responsible for the antioxidant activity of ITCs. (A) As an electron donor, H_2S directly neutralizes reactive oxygen species (ROS), leading to reduction of oxidative damage in many organs and tissues. (B) Physiologically, the antioxidant transcription factor Nrf2 is sequestered into the cytoplasm by the regulatory protein Keap1. Under oxidative stress conditions, H_2S *S*-sulfhydrates Keap1 at cysteine residues, inducing a rapid dissociation of the Keap1/Nrf2 complex and subsequent Nrf2 activation. Once activated, Nrf2 moves into the nucleus where they bind to the antioxidant response element (ARE), thus inducing the expression of endogenous antioxidant enzymes (such as superoxide dismutase, SOD; glutathione peroxidase, GPx; catalase, CAT). Source: Based on Xie et al. [39].

(6-MSITC), an organic ITC which is mainly present in *Wasabia japonica* Matsum reduced TNFα-induced expression of VCAM-1 and E-selectin in activated endothelial cells [42]. Similarly, H_2S suppressed TNFα-induced expression of ICAM-1, VCAM-1, P-selectin, E-selectin, and heme-oxidase (HO-1) in HUVEC cells, and inhibited monocytes adhesion [43]. The hypothesis that ITCs act by the release of H_2S, which thus represents the "real player" is further strengthened by the previously described overlapping results on suppression of TNFα-induced markers of vascular inflammation, even if the direct link between ITCs and H_2S has not been investigated.

In addition, several clinical trials reported the beneficial effect of the ITC SFN in cardiovascular-related diseases. In particular, SFN 112 or 225 μmol/day for 4 weeks reduced fasting glucose, inflammatory markers, and serum insulin compared to placebo in 81 type II diabetes (T2D) patients. SFN also ameliorated the lipid profile reporting a decrease in serum triglycerides, oxLDL/LDL ratio, and atherogenic index of plasma [44, 45]. Similar results were obtained in another recent randomized double-blind study: the consumption of 400 g of HG-broccoli (i.e. broccoli containing high concentrations of glucoraphanin, which is the GLS converted in SFN by myrosinase) positively influenced the lipid profile in 130 volunteers after 12 weeks [46]. The specific mechanism leading to beneficial effect on metabolic profile associated with the administration of ITCs or the consumption of edible plants characterized by a high content of ITCs still remains unclear, but may be attributed to the ability of ITCs to suppress NF-kB-mediated inflammation [36] and to upregulate Nrf2 signaling against oxidation [35].

11.2.2 Vasorelaxing Effect of ITCs as H_2S Donors

Vasorelaxing properties of ITCs were evaluated for the first time in 2002 by Wilson and coworkers on strips of dog carotid artery. In particular, the authors observed that BITC, selected as active compound of papaya seed extract, promoted direct vasorelaxing effect on phenylephrine (PE) pre-contracted with strips and attenuated the KCl- or PE-induced contraction. At the same time, they recorded also a strong influx of Ca^{2+} into endothelial cells, as observed for AITC, but in this case, they did not investigate the involvement of any type of ion channels [47].

H_2S has been widely characterized as a regulator of the vascular tone, promoting vasorelaxing effects by tuning of vascular tone and one of the main molecular targets involved in this effect is the opening the ATP-sensitive (K_{ATP}) potassium channels. More recently, several other targets have been discovered accounting for the ability of H_2S to induce vasodilatory effect. In particular, among the potassium channels, the voltage gated Kv7.4 was clearly demonstrated and recognized as a pivotal target responsible for the vasorelaxing effect [9]. Moreover, H_2S has been described as responsible for the vasorelaxing effect due to the inhibition of vascular type 5 phosphodiesterase (PDE-5), and the consequent inhibition of cyclic guanosine monophosphate (cGMP) degradation [48, 49]. The vasorelaxing properties of ITCs have been demonstrated and related to their H_2S-releasing ability by using two synthetic "smart" H_2S-donors, such as ITCs (phenyl isothiocyanate [PITC] and 4-carboxyphenyl isothiocyanate [4CITC]). Indeed, the H_2S donation has been recorded when these ITCs were incubated with L-cysteine for mimicking the endogenous presence of free thiols. Furthermore, accordingly to their clear H_2S release, PITC and 4CITC induced vasorelaxation on KCl-precontracted rat aortic rings due to the opening of voltage-operated Kv7 potassium channels a Kv7. The vasodilatory effect was also recorded in the coronary district, since the incubation of PITC and 4CITC caused both a direct vasodilation on the basal coronary flow and reversed the Angiotensin II – induced coronary spasm. The involvement of Kv7 channels was also confirmed using human aortic smooth muscle cells (HASMC) treated with PITC and 4CITC which induced a clear hyperpolarizing effect that was counteracted by incubating the selective Kv7 blocker XE991 [6]. Similar results were also reported in previous work using NaHS, which induces Kv7-dependent HASMC membrane hyperpolarization, rat aortic rings vasorelaxation, and vasodilation on the coronary circle [9]. Another possible mechanism accounting for the vasorelaxing effect and a consequent lowering of blood pressure values of ITCs and H_2S is the histone deacetylase 6 (HDAC6) inhibition [50, 51].

Also, natural ITCs have been investigated as H_2S-donor molecules able to induce vasorelaxing response and consequently able to reduce blood pressure values, again modulating the same H_2S molecular targets. In particular, ERU, an ITC deriving from the seeds of *E. sativa* Mill., has been characterized as a molecule able to release H_2S in a thiol-dependent manner, able to cross cell membrane of HASMC, and donate H_2S into the cytosol. As reported for H_2S and other natural sulfur compounds that are considered H_2S-donors, for example DADS, also ERU promoted a clear membrane hyperpolarization of HASMC, induced a potent vasorelaxing effect on rat aortic rings, and inhibited the vasoconstriction induced by increasing concentrations of norepinephrine (NE).

Finally, ERU acutely administered *in vivo* (10 mg/kg i.p.) induced a marked lowering in the systolic blood pressure values in spontaneously hypertensive rats (SHRs), restoring the values observed in normotensive rats [52]. These data highlight the almost complete overlap between the vasoactive effect of H_2S and those observed for ITCs, strongly suggesting that the ITCs effects may be due to their property to act as H_2S-donors.

Another interesting overlapping mechanism of action among ITCs, sulfur compounds, and H_2S [53] is represented by the activation of Ca^{2+}-permeable nonselective ankyrin transient receptor potential (TRPA1) channels expressed in native endothelial cells, in the proximity to vascular smooth muscle cells (VSMCs). In particular, allicin and AITC activate these channels promoting an increase in intracellular Ca^{2+} levels and consequently triggering vasodilation. The two abovementioned sulfur compounds have been demonstrated to be H_2S donors: the first one is an organosulfur compound from garlic, which spontaneously decomposes to allylic polysulfides such as DADS and DATS [5], while the latter is an organic ITC which exhibits H_2S-donor activity [54].

11.2.3 Organic ITCs and Heart

ITCs and H_2S have been reported to exert benefical effects in primary and secondary cardiomyopathy, which leads to structural and morphological changes usually related to heart failure and ischemia events. Primary cardiomyopathy usually represents idiopathic forms of myocardial disease clinically recognized as hypertrophic cardiomyopathy, while secondary cardiomyopathies are the manifestation of other cardiovascular disease as hypertension, atherosclerosis, or diabetes [55].

The most-studied ITC for preventing the development of primary or secondary cardiomyopathies is SFN, a natural sulfur compound occurring in Brassicaceae family. In particular, SFN has been reported to limit cardiac hypertrophy development in an *in vitro* model of H9c2 treated with AngII, due to the inhibition of the activaton of NF-kB, which represents the main key mediator that sustains the cardiac remodelling process [56]. Furthermore, as widely reported for H_2S, the administration of SFN to Ang-II treated rats activates the Nrf2 antioxidant machinery leading to a massive trascription of antioxidant molecules, which play a crucial role in limiting oxidative stress damage [57]. SFN has been also demonstrated to prevent secondary diabetic cardiomyopathy through the upregulation of Nrf2 whose involvement in counteracting the development of cardiomyopathy has been demonstrated in Nrf2-knockout mice which showed a progressive severe cardiac dysfunction [58] even after the administration of SFN.

Beside primary and secondary cardiomyopathy, iatrogenic cardiomypathy mainly occurs in cancer patients receiving anthracyclines which often cause long-term effects on cardiac structure and function. The mechanism underpinning anthracycline cardiomyopathy is well described in the literature and is mainly based on cell damage due to free radical formation and oxidative stress leading to chronic heart failure. In accordance with the antioxidant effect of ITCs due to the activation of Nrf2 machinery, SFN promoted cardioprotective effect in patients treated with doxorubicin, slowing down the otherwise unavoidable development of myocardium fibrosis [59, 60].

H_2S and ITCs both contribute to reduce cardiac damage after ischema/reperfusion injury by modulating several intracellular pathways [61]. First, as reported for the prevention of cardiomyopathies, ischemia/reperfusion damage is sustained by the massive production of ROS, mainly during the reperfusion phase. SFN limits ROS production by increasing the production of antioxidant molecules by activating Nrf2 in an *in vitro* model of H_2O_2 injury, preserving cell viability [62]. The cardioprotective effect of SFN was also observed in rats receiving SFN 500 μg/kg/day for 3 days before inducing I/R, improving myocardial parameters and limiting the infarct size, by opening mitoK_{ATP} channels [63] which represent the target accounting also for the H_2S-mediated cardio-protection. Indeed, the activation of mitochondrial potassium (mitoK) channels stimulates cardio-protective signaling due to a mild depolarization of the inner mitochondrial membrane, consequently reducing the entry of calcium in the mitochondrial matrix, which avoids the opening of mitochondrial permeability transition pore responsible for the apoptotic death of cardiomyocytes. Moreover, it has been demonstrated that H_2S is effective in protecting the heart submitted to myocardial I/R injury. Interestingly, exogenous administration of H_2S results in a significant reduction of infarct size in several models of I/R injury [64–68]. For instance, Na_2S, administered at a dose of 50 μg/kg during reperfusion significantly reduced the infarct size [64, 69]. Exogenous administration of H_2S reduced the left ventricular remodelling in hearth failure [70]. At the cardiac level, H_2S has been demonstrated to promote "ischemic preconditioning" through the opening of mitoK_{ATP} channels and Kv7 potassium channels in myocardial ischemia/reperfusion damage. Interestingly, impressive overlap in mediating cardioprotective effects emerged in the experimental work by Testai and coworkers. In particular, for the first time, the cardioprotective effect of a synthetic ITC (4CITC) has been related to its ability in donating H_2S. 4CITC promoted a massive reduction in ROS production after the I/R injury in rat hearts, promoted the opening of mitoK_{ATP} whose involvement was demonstrated by administering the mitoK_{ATP} selective blocker 5-Hydroxydecanoic acid (5-HD), which almost completely abolished the beneficial effects of 4CITC. This hypothesis was strengthened by using isolated cardiac mitochondria, where the ITC showed the typical profile of a mitoK channel opener, producing a modest depolarization of the mitochondrial membrane, associated with a reduction of the driving force for the calcium uptake into mitochondria [16].

11.3 Chemopreventive Properties of ITCs

In the last decades, a positive correlation has been observed between regular consumption of edible plants belonging to the *Brassicaceae* botanical family and a reduced risk of developing cancer [71]. Among all phytochemicals, sulfur-containing compounds (i.e. GLSs and their derivatives ITCs) are presumably the main responsible for these beneficial effects [72]. In addition, GLSs and ITCs are receiving great nutraceutical and pharmacological interest also due to their favorable pharmacokinetic profile; indeed, they show good absorption and appreciable bioavailability after oral administration [73].

Emerging evidence suggests that many *Brassicaceae*-derived ITCs promote anticancer effects in various types of cancer, including neuroblastoma [74], breast cancer [75], lung cancer [76], glioma [77], and colon cancer [78]. The chemopreventive effects exhibited by ITCs are the result of a pleiotropic mechanism of action, which mainly involves modulation of several intracellular signaling pathways that control cell proliferation and survival. According to their H_2S-releasing properties, the mechanism of action exhibited by ITCs is strictly superimposable with that shown by H_2S. For instance, both the endogenous gasotransmitter and ITCs prevent cancer development and progression through regulation of the expression/activity of phase I and phase II enzymes, reduction of inflammation, inhibition of cell growth, prevention of metastasis and angiogenesis, and modulation of the epigenetic machinery [79].

Noteworthy, ITCs show hormetic behavior, which depends on both the exposure time and concentration. In particular, ITCs are cytoprotective at low concentrations and cytotoxic at high doses. For instance, low levels of the ITC SFN (1–5 μM) induced tumor cell growth in numerous human cell lines, while higher concentrations (10–20 μM) significantly reduced cell viability [80]. Similarly, the gasotransmitter H_2S also shows such a hormetic effect [81]. Indeed, high concentrations of H_2S (released by the "fast" H_2S-donor salt NaHS) induced both autophagy and early apoptosis of hepatocarcinoma cells (HepG2) via modulation of the PI3K/Akt/mTOR signaling pathway [82]. In addition, the synthetic H_2S-donor GYY4137 (400 μM) promoted cell cycle arrest in the G2/M phase in human breast adenocarcinoma Michigan Cancer Foundation-7 (MCF-7) breast cancer cells [83]. This peculiar dualism in the effects shown by both ITCs and H_2S strengthens the hypothesis that H_2S may be the leading actor in the health-promoting effects of these sulfur compounds. In this regard, the crucial role of H_2S in the anticancer properties of ITCs has been recently demonstrated by Jurkowska and coworkers, who investigated the correlation between the antiproliferative effect of HBITC, from *S. alba* L., with its H_2S-releasing properties [7] on glioblastoma (U87MG) and human neuroblastoma cells (SH-SY5Y) [84]. In particular, authors correlated the increase in intracellular levels of H_2S with mitochondrial depolarization that represents a crucial event during apoptosis. Moreover, the H_2S-donor HBITC (20–80 μM) induced cell cycle arrest in cancer cells through upregulation of the cyclin-dependent kinase p21 and downregulation of p53. Interestingly, this toxic effect was correlated with S-sulfhydration on p53 cysteine 277 residues, further suggesting that H_2S might contribute to the anticancer effects of this natural ITC [84]. Similar chemopreventive properties have been also described for other ITCs, including BITC [85]. In particular, micromolar concentrations of this natural ITC (10–40 μM) significantly reduced cell viability via induction of both apoptosis and oxidative stress in cancer prostate cells. Moreover, SFN (5–10 μM) inhibited the proliferation of human breast cancer cells (SKBR-3 and MDA-MB-231 cells) in a concentration-dependent manner, by inducing apoptosis, oxidative stress, mitochondrial dysfunction, and autophagy [75].

Recently, it has been also shown that the natural ITC ERU may be a promising anticancer agent. ERU derives from the myrosinase-operated hydrolysis of its precursor glucoerucin, a GLS mainly distributed in *E. sativa* Mill. (the common rocket salad). *In vivo*, ERU is also obtained from the enzymatic reduction of SFN. The anticancer effects of ERU have been demonstrated in various tumor cells, including colorectal adenocarcinoma cells (CACO-2; [86]), acute promyelocytic leukemia cells (HL-60; [87]), and breast cancer cells (MCF-7 and MDA-MB-231; [88]). These effects have been mainly associated with cell cycle arrest in the G2/M phase and changes in mitochondrial membrane potential, with subsequent apoptotic cell death [89]. Furthermore, ERU exhibited anticancer effects also *in vivo*. Indeed, a 3-week treatment with ERU (50 mg/kg/day, *per os*) abrogated telomerase in an orthotopic xenograft tumor model of hepatocellular carcinoma [90]. Very recently, Citi and coworkers described the antiproliferative effects of ERU in human pancreatic adenocarcinoma cells (AsPC-1). Once again, the anticancer activity exhibited by ERU has been associated with its H_2S-releasing activity [91]. Indeed, ERU is a "smart" H_2S-donor, that means a molecule that releases the gasotransmitter exclusively inside the cells [4]. Accordingly, ERU (10–100 μM) promoted a slow and concentration-dependent generation of H_2S in AsPC-1 cells [91]. In addition, ERU showed proapoptotic activity in these cancer cells via activation of caspase 3/7 and mitochondrial depolarization, induced cell cycle arrest in the G2 phase and counteracted cell migration, thus preventing the development of metastases. Finally, ERU inhibited the phosphorylation of ERK1/2, which is hyperphosphorylated in AsPC-1 cells. Interestingly, also the gasotransmitter H_2S showed a similar inhibitory effect on phosphorylation of ERK1/2 [92].

Besides natural products, promising results have been also obtained with many synthetic ITCs. For instance, phenylbutyl-ITC and phenylhexyl-ITC caused cell cycle arrest in liver and prostate carcinoma cells *in vitro* [93]. Moreover, sugar molecules (such as glucose and fructose) linked to an ITC functional group are also receiving great interest for the pharmacological treatment of various types of cancer [52]. Indeed, glucose transporters (GLUTs) are overexpressed in cancer cells, as these highly replicating cells require a lot of energy to survive and proliferate [94]. Therefore, novel ITCs glycoconjugates have been recently developed. Briefly, these molecules hijack the GLUT1 transporter, increasing the uptake of the anticancer compound from cancer cells. Once crossed cell membrane, these compounds release H_2S concentration-dependently, thus inducing cell cycle arrest and subsequent cell death [52].

In conclusion, both natural and synthetic "smart" H_2S-donors (including ITCs) seem to represent an effective therapeutic strategy in the context of cancer chemotherapy to reverse, suppress, or prevent the multistage process of carcinogenesis.

11.4 Anti-nociceptive Effects of ITCs

Given the widespread and heterogeneity of pain, the discovery of novel anti-nociceptive agents represents a very timely challenge. Neuropathic pain, which is the most refractory form of pain, commonly occurs in subjects with diabetes, ischemia, spinal cord injury, and virus infections [95–97]. The first-line treatment options currently available (tricyclic antidepressants, calcium-channel-acting modulators, and serotonin-noradrenalin reuptake inhibitors) are still not completely resolutive. Noteworthy, several edible species of the *Brassicaceae* botanical family have been traditionally employed for their analgesic properties, although their mechanism of action remained unclear for years. Recently, the anti-nociceptive effects of broccoli sprouts have been described in an experimental model of abdominal pain in mice. In particular, both broccoli sprouts (100 mg/kg i.p. and 1000 mg/kg orally) and their characteristic ITC SFN (0.1–10 mg/kg orally) produced anti-nociception through activation of μ opioid receptors; accordingly, naltrexone (antagonist of μ opioid receptors) inhibited their analgesic effects [98]. However, the potential synergic effect between ITCs and endogenous opioids is not a fully explanation of the anti-nociceptive activity of these sulfur-containing molecules. In this regard, several ITCs are demonstrated to be effective activators of TRPA1 channels. This channel is strongly involved in the somatosensory perception of pain and also in neurogenic inflammation. Therefore, TRPA1 agonists are actually considered as promising anti-nociceptive and anti-inflammatory molecules [99]. Several compounds are able to desensitize TRPA1, thus promoting analgesia. Among them, polysulfides and ITCs are emerging as novel modulators of these channels. Noteworthy, both classes of sulfur compounds share a peculiar long-lasting and "smart" H_2S-donor profile [5, 7]. Accordingly, a similar TRPA1-activating profile has been also described for H_2S [99], thus suggesting that H_2S may be effectively responsible for the anti-nociceptive effects of sulfur-containing compounds. Together with TRPA1 activation, opening of voltage-dependent calcium channels further contributes to the anti-nociceptive effects promoted by H_2S [100–102].

However, the role of H_2S in nociception is still controversial. In general, physiological levels of H_2S seem to play a relevant role in nociception, as H_2S-syntethysing enzyme cystathionine γ-lyase (CSE) is upregulated in various models of chronic pain [103]. Conversely, some studies reported the pro-nociceptive activity of H_2S in many types of pain. This dualism is presumably due to selected concentrations of H_2S-releasing molecules and nonhomogeneous experimental conditions employed [104]. An evidence of the analgesic properties of H_2S was given by Kida et al. by means of a sciatic nerve constriction injury experimental model. In these conditions, inhalation of H_2S promoted significant neuroprotective effects mainly due to reduction of inflammation in the spinal cord and microglia [105]. In addition, inhalation of H_2S delayed paraplegia in mice after spinal cord ischemia [106]. Very recently, it has been also proposed that it has a crucial role for the antioxidant Nrf2/ARE pathway in the neuroprotective properties of gaseous H_2S after transient spinal cord ischemia [107].

Modulation of TRPA1 channels does not fully explain the anti-nociceptive effects promoted by numerous ITCs, as concentration that is required to desensitize TRPA1 channels is very high and not compatible with clinical use.

Accordingly, there are other mechanisms responsible for analgesic activities of these sulfur-containing molecules. As is well known, activation of inflammatory signaling pathways plays a pivotal role in neuropathic pain. Therefore, anti-inflammatory molecules may represent potential candidates in the pharmacological treatment of this refractory form of pain [108, 109]. In this regard, AITC (1–20 μM) reduced inflammation in microglial cells (BV2) exposed to lipopolysaccharides (LPS) [110]. Moreover, the well-described anti-inflammatory properties of SFN deeply contributed to the reduction of neuropathic pain in diabetic rats and osteoarthritic mice [111, 112]. Similar results have also been obtained in chronic constriction injury-induced neuropathic pain in mice, where treatment with SFN (0.1–100 mg/kg for 7 days, i.p.) dose-dependently reduced plasma levels of many pro-inflammatory mediators (i.e. the cytokines IL-1β, TNF-α, and IL-6) [113]. In this work, pretreatment with the antagonist of opioid receptors naloxone consistently reversed the beneficial effects promoted by SFN on behavioral hypersensitivity without affecting modulation of cytokine levels, furthermore, demonstrating that SFN exhibits independent but synergic anti-inflammatory and anti-nociceptive properties. In addition, SFN alleviated chronic inflammation in mice through modulation of Nrf2 [114]. As ITCs, also H_2S exhibits such anti-inflammatory effects. Inhaled H_2S, for example, prevented endotoxin-induced systemic inflammation in mice [115], while the H_2S-donor GYY4137 counteracted the pro-inflammatory effects of LPS in human cells [116]. Once again, this overlapping in the anti-inflammatory activity of ITCs and H_2S suggests a crucial, hidden role of the gasotransmitter in the health-promoting effects of these sulfur compounds. In this regard, the possible role of H_2S in the analgesic effects of SFN has been demonstrated in a model of chemotherapy-induced neuropathic pain *in vivo*. In particular, the dose-dependent anti-nociceptive activity of SFN was fully abolished following administration of the H_2S-scavenger hemoglobin. Moreover, the analgesic properties of SFN and its precursor glucoraphanin were markedly reduced in the presence of XE991 (a selective Kv7-blocker), suggesting that Kv7 channels play a crucial role in the analgesic effects of H_2S-donors [13]. These channels, which are largely expressed in the central nervous system, represent a suitable pharmacological target for analgesic drugs [117]. Noteworthy, this peculiar mechanism of action is shared with H_2S that is a well-known opener of Kv7 channels [9].

The crucial role of H_2S in the anti-nociceptive effects of ITCs has been also investigated by Di Cesare Mannelli et al., who reported that the analgesic activity of both natural and synthetic ITCs was completely abolished in the presence of hemoglobin [118]. In particular, they investigated the anti-nociceptive properties of structurally heterogeneous H_2S-releasing ITCs (such as AITC, PITC, and 3-carboxyphenyl ITC [6, 7]) in mice treated with the neurotoxic anticancer drugs paclitaxel and oxaliplatin. In these experimental conditions, selected ITCs showed higher values of both potency and efficacy than other analgesic drugs, including pregabalin, duloxetine, and morphine [118]. Once again, the anti-nociceptive activity exhibited by selected ITCs was fully abolished in the presence of XE991, indicating that modulation of Kv7 channels plays a crucial role in the H_2S-mediated analgesic effects of ITCs.

Finally, promising anti-nociceptive effects have been showed by the ITC moringin (from *M. oleifera* Lam.). Indeed, topic administration of moringin significantly reduced neuropathic pain in mice affected by multiple sclerosis through modulation of voltage-gated ion channels [119]. Along with natural compounds, the potential anti-inflammatory and analgesic effects of seven semisynthetic BITCs derived from a BITC present in *M. oleifera* Lam. have been also investigated. These molecules exhibited a dose-dependent anti-nociceptive activity in a temporo-mandibular joint inflammatory hyper-nociception model *in vivo* [120].

To conclude, all these data indicate a promising usefulness of H_2S-releasing molecules in the prevention and treatment of various forms of pain. Therefore, further studies in this very timely research field are strongly encouraged.

11.5 Anti-inflammatory and Antiviral Effects of ITCs

Chronic inflammation is a subclinic pathological condition that progressively leads to deterioration of many organs and tissues, making them more susceptible to damage. A low-grade, sneaky, and persistent inflammatory state is generally associated to aging. This harmful and un-healthy condition, known as "inflamm-aging" [121, 122], enhances the risk of developing several inflammatory-based diseases (i.e. cancer, cardiovascular diseases, and neurodegenerative disorders) in aged subjects. Furthermore, chronic inflammation is a common feature of other pathologies, including asthma, hepatitis, chronic kidney disease (CKD), and inflammatory bowel disease (IBD) [123]. Besides chronic inflammation, also acute inflammation has a relevant role in the onset and progression of several pathological conditions. Usually, it is the primary consequence of a sudden damage, such as mechanical damage or pathogen infections. This acute inflammatory response is generally referred to as "cytokine storm," as it is characterized by an immediate and massive production of pro-inflammatory mediators that leads to substantial, rapid, and almost irreversible tissue damage [124]. Noteworthy, acute inflammation also represents a typical feature of virus infections. For instance, the "cytokine storm" caused by new beta coronavirus SARS-CoV-2 dramatically contributes to progression of COVID-19 disease [125, 126].

In this context, H_2S-releasing molecules play a crucial role, as H_2S has been recognized as an endogenous modulator of inflammation. In particular, H_2S directly inhibits the pro-inflammatory transcription factor NF-κB through S-sulfhydration reactions, thus reducing intracellular production of cytokines and other mediators of inflammation [127, 128]. Noteworthy, such inhibitory effect on NF-κB activity has been also reported for many H_2S-donors, including ITCs. For instance, daily administration of SFN (0.5 mg/kg for 2 weeks, i.p.) significantly prevented multiorgan inflammatory damage in diabetic rats through inhibition of NF-κB [112] (Figure 11.6).

Similarly, micromolar concentrations of AITC (1–20 μM) reduced neuroinflammation in BV2 microglial cells exposed to LPS through inhibition of NF-κB [110]. Moreover, Prawan et al. showed that also several synthetic ITCs (i.e. tetrahydrofurfuryl ITC, 3-morpholinopropyl ITC, and 3,4-methyelendioxybenzyl ITC) exhibit anti-inflammatory activity via modulation of the NF-κB pathway [129].

Along with NF-κB, also mast cells play a crucial role in both acute and chronic inflammation. In particular, activation of mast cells during inflammation is associated with a consistent influx of extracellular calcium ions. This crucial event leads to mast cell degranulation and release of inflammatory mediators, thus directly supporting the inflammatory process in many districts [130].

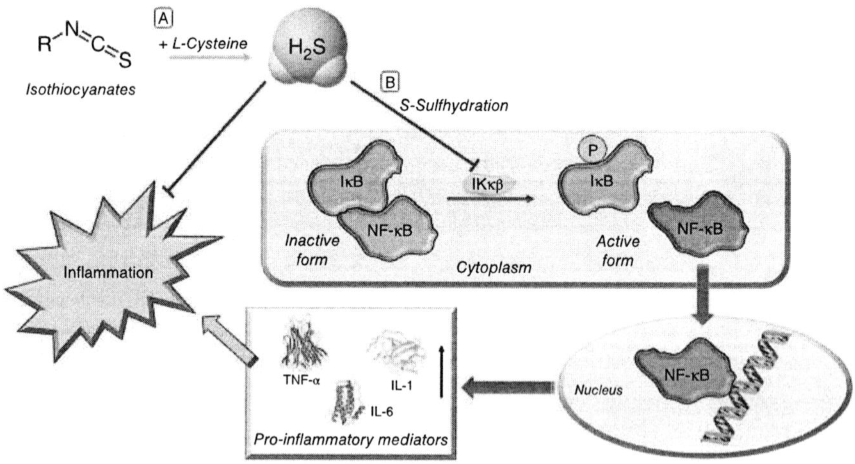

Figure 11.6 H_2S-mediated anti-inflammatory effects of ITCs. (A) ITCs release H_2S in the presence of organic thiols (i.e. L-Cysteine). (B) H_2S promotes anti-inflammatory effects through modulation of the NF-κB signaling pathway. In particular, H_2S blocks the IKκβ kinase (a positive regulator of NF-κB) via S-sulfhydration reactions at cysteine residues. As a consequence, both translocation of the transcription factor NF-κB into the nucleus and subsequent production of several pro-inflammatory mediators (such as tumor necrosis factor-α, TNF-α; interleukin-1, IL-1; interleukin-6, IL-6) are inhibited. Source: Based on Negi et al. [112].

Intriguingly, activation of the H_2S-synthetizing enzyme CSE with its endogenous substrate L-cysteine significantly reduced IgE-induced degranulation of mast cells concentration-dependently *in vitro* [10], indicating a pivotal role for the endogenous gasotransmitter in the antigenic-mediated inflammatory process. In particular, H_2S inhibits mast cells degradation by decreasing calcium availability. In this regard, Marino et al. [10] demonstrated that the synthetic H_2S-donor 4CITC counteracts mast cells degranulation in rat basophilic leukemia cells (RBL-2H3) and bone marrow-derived mast cells (BMMC) via consistent reduction in the intracellular levels of calcium. Moreover, 4CITC significantly decreased phosphorylation of Syk, an intracellular protein phosphorylated after activation of the FcεRI receptor, which occurs when mast cells are exposed to antigens. In particular, aggregation of the FcεRI receptor requires the formation of disulfide bonds [131]. Noteworthy, the main mechanism of action described for H_2S is S-sulfhydration of proteins and enzymes, suggesting that H_2S may interfere with disulfide bond inducing a transient alteration of the function in the FcεRI receptor [10].

In another study, AITC (1–50 μM) caused a deep reduction in intracellular calcium levels in human mast cells (HMC-1). Moreover, AITC reduced the activity of caspase-1, which plays a crucial role in inflammation, as it activates interleukin-18 (IL-18) and interleukin IL-1β under inflammatory conditions [132]. The same results have been also obtained after treatment of HMC-1 cells with the synthetic molecule methallyl isothiocyanate (MAITC, 1–18 μM) [133]. Finally, the anti-inflammatory activity of many wasabi-derived ITCs in RBL-2H3 cells exposed to antigenic stimuli have been also investigated [134]. Noteworthy, the effects exhibited by wasabi ITCs strongly depend on their side chains. In particular, ITCs with three or four carbon chains (i.e. *sec*-butyl ITC, AITC, and 3-butenyl ITC) counteracted the release of histamine but did not show any effect on leukotriene B4 (LTB4) or cysteinyl leukotrienes (CysLTs) release. Conversely, wasabi-ITCs with five carbon chains or more (including 5-hexenyl ITC and 4-pentenyl ITC) only inhibited the release of LTB4. Finally, molecules with a second sulfur atom at the end of a six-carbon chain (such as 6-methylsulfinylhexyl ITC, 6-methylthiohexyl ITC, and 6-methylsulfonylhexyl ITC) inhibited both the release of all the inflammatory mediators and also the increase of intracellular calcium levels. Therefore, natural and synthetic H_2S-donors may be useful in the pharmacological treatment of inflammatory disorders characterized by consistent mast cells degranulation.

As reported above, acute inflammation is a common feature of several pathogen infections, including those induced by viruses. Hence, molecules endowed with both anti-inflammatory and antiviral properties may represent an effective strategy to prevent multi-organ damage associated with virus infections [135]. In this regard, ITCs demonstrated antiviral activity in several studies, presumably due to their H_2S-releasing properties. Micromolar amounts of SFN, for instance, reduced replication of the influenza A/WSN/33 (H1N1) virus *in vitro* [136], while broccoli sprouts counteracted virus replication in humans. Indeed, in a clinical trial conducted on healthy subjects receiving a standard nasal vaccine dose of live attenuated influenza virus, broccoli sprouts increased the production of the pro-apoptotic protease granzyme B in natural killer (NK) cells. Briefly, granzyme B levels are inversely associated with influenza RNA levels, thus suggesting a reliable mechanism of action for broccoli sprouts [137]. Once again, the effects promoted by *Brassicaceae* and their sulfur metabolites are strictly superimposable with that promoted by H_2S. In this regard, the H_2S-donor GYY4137 showed antiviral activity against respiratory syncytial virus (RSV). Interestingly, endogenous generation of H_2S was reduced in airway epithelial cells infected by RSV, as indicated by low levels of CSE expression in these cells [138]. Moreover, CSE knockout mice infected by RSV exhibited both enhanced airway inflammation and viral replication [139]. In accordance with its H_2S-releasing properties, GYY4137 significantly reduced viral replication in several paramyxovirus infections *in vitro*, mainly due to inhibition of viral assembly and release. In addition, GYY4137 prevented virus-induced inflammation via modulation of the transcription factor NF-κB [138]. The antiviral activity of the H_2S-donor GYY4137 against other highly pathogenic RNA viruses (i.e. Ebola virus, influenza A and B, Rift Valley fever, and Crimean-Congo hemorrhagic fever virus) have been also demonstrated [140]. Importantly, other structurally heterogeneous H_2S-donors inhibited viral replication, improved airways dysfunction, and reduced the release of pro-inflammatory mediators in several experimental models. This strongly suggests that the H_2S may effectively be the hidden player in the antiviral and anti-inflammatory properties promoted by structurally different H_2S-donors, including ITCs [141]. Finally, H_2S exhibited antiviral activity against SARS-CoV-2 by interfering with both angiotensin-converting enzyme 2 (ACE2) receptor and transmembrane serine protease 2 (TMPRSS2) [142].

To conclude, the potential anti-inflammatory properties of H_2S pave the way to a more rational use of H_2S-donors in the prevention and treatment of inflammatory-based diseases. Moreover, the emerging antiviral activity of H_2S-donors, including ITCs, makes them possible candidates as antiviral agents. Indeed, their H_2S-releasing properties represent an exciting mechanism of action also in the context of novel virus-induced pathologies, such as COVID-19 [135].

11.6 Conclusion

H_2S has been considered only as a toxic pollutant gas for many years, but in the last decade, its beneficial properties have been widely investigated due to an intense scientific research activity, highlighting the importance of this gaseous molecule in the regulation of several pathophysiological processes. The need to find molecules able to donate H_2S in an endogenous-like manner, to be employed for the treatment of those pathologies related to an impairment in H_2S endogenous production, led to "unmask" the ITC functional group as a good candidate that meets such a purpose.

Indeed, as other natural polysulfide molecules, the ITCs have always been studied for a variety of pharmacological effects, from chemopreventing agents to anti-inflammatory molecules, not knowing that behind the ITC chemical group, H_2S actually represents the "real player." Indeed, the impressive overlapping between both the pharmacological effects and the mechanisms of action exerted by the ITCs and H_2S (for example the ability to *S*-sulfhydrate specific protein cysteine residues) led many research groups to investigate the potential H_2S-donor properties of natural and synthetic ITCs.

The evidence of the effective H_2S-releasing profile of ITCs allows to consider such functional group as an interesting moiety that can be exploited as a novel pharmacological strategy in the treatment of many pathological conditions (i.e. hypertension, vascular inflammation, and cancer) and that may be useful for developing innovative H_2S-donor hybrid drugs for improving the already known properties of "native" drugs or for counteracting possible severe adverse effects.

Acknowledgment

This research was partially funded by the "Bando per il finanziamento di Dimostratori Tecnologici DT (DT grant)," acronym "MEDICA," financed by University of Pisa, MIUR, Italy, and partially by the grant "Consulenza farmacologica per la caratterizzazione cineticochimica del rilascio di acido solfidrico (H_2S) da farmaci e molecole di origine naturale in soluzione, in vitro e in vivo e per la valutazione degli effetti di una modulazione della concentrazione di H_2S in modelli in vitro e in vivo di patologie cardiovascolari e tumorali," (PON COMETA "COLTURE AUTOCTONE MEDITERRANEE E LORO VALORIZZAZIONE CON TECNOLOGIE AVANZATE DI CHIMICA VERDE" (ARS01_00606) PON "R&S" 2014 – 2020 – Azione II – CUP B26G18000200004 – RNA-COR 545910).

References

1 Polhemus, D.J. and Lefer, D.J. (2014). Emergence of hydrogen sulfide as an endogenous gaseous signaling molecule in cardiovascular disease. *Circ. Res.* 114 (4): 730–737.

2 Yoo, D., Jupiter, R.C., Pankey, E.A. et al. (2015). Analysis of cardiovascular responses to the H_2S donors Na_2S and NaHS in the rat. *Am. J. Physiol. Heart Circ. Physiol.* 309 (4): H605–H614.
3 Calderone, V., Martelli, A., Testai, L. et al. (2016). Using hydrogen sulfide to design and develop drugs. *Expert Opin. Drug Discov.* 11 (2): 163–175.
4 Martelli, A., Citi, V., Testai, L. et al. (2020). Organic isothiocyanates as hydrogen sulfide donors. *Antioxid. Redox Signal.* 32 (2): 110–144.
5 Benavides, G.A., Squadrito, G.L., Mills, R.W. et al. (2007). Hydrogen sulfide mediates the vasoactivity of garlic. *Proc. Natl. Acad. Sci. U.S.A.* 104 (46): 17977–17982.
6 Martelli, A., Testai, L., Citi, V. et al. (2014). Pharmacological characterization of the vascular effects of aryl isothiocyanates: is hydrogen sulfide the real player? *Vascul. Pharmacol.* 60 (1): 32–41.
7 Citi, V., Martelli, A., Testai, L. et al. (2014). Hydrogen sulfide releasing capacity of natural isothiocyanates: is it a reliable explanation for the multiple biological effects of Brassicaceae? *Planta Med.* 80 (8, 9): 610–613.
8 Andreasson, E., Bolt Jorgensen, L., Hoglund, A.S. et al. (2001). Different myrosinase and idioblast distribution in Arabidopsis and *Brassica napus*. *Plant Physiol.* 127 (4): 1750–1763.
9 Martelli, A., Testai, L., Breschi, M.C. et al. (2013). Vasorelaxation by hydrogen sulphide involves activation of Kv7 potassium channels. *Pharmacol. Res.* 70 (1): 27–34.
10 Marino, A., Martelli, A., Citi, V. et al. (2016). The novel H_2S donor 4-carboxy-phenyl isothiocyanate inhibits mast cell degranulation and renin release by decreasing intracellular calcium. *Br. J. Pharmacol.* 173 (22): 3222–3234.
11 Higdon, J.V., Delage, B., Williams, D.E. et al. (2007). Cruciferous vegetables and human cancer risk: epidemiologic evidence and mechanistic basis. *Pharmacol. Res.* 55 (3): 224–236.
12 Calabrese, V., Cornelius, C., Dinkova-Kostova, A.T. et al. (2010). Cellular stress responses, the hormesis paradigm, and vitagenes: novel targets for therapeutic intervention in neurodegenerative disorders. *Antioxid. Redox Signal.* 13 (11): 1763–1811.
13 Lucarini, E., Micheli, L., Trallori, E. et al. (2018). Effect of glucoraphanin and sulforaphane against chemotherapy-induced neuropathic pain: Kv7 potassium channels modulation by H_2S release in vivo. *Phytother. Res.* 32 (11): 2226–2234.
14 Wang, X., Liu, Y., Liu, X. et al. (2018). Hydrogen sulfide (H_2S) releasing capacity of isothiocyanates from *Moringa oleifera* Lam. *Molecules* 23 (11): 2809.

15 Popoola, J.O. and Obembe, O.O. (2013). Local knowledge, use pattern and geographical distribution of *Moringa oleifera* Lam. (Moringaceae) in Nigeria. *J. Ethnopharmacol.* 150 (2): 682–691.

16 Testai, L., Marino, A., Piano, I. et al. (2016). The novel H_2S-donor 4-carboxyphenyl isothiocyanate promotes cardioprotective effects against ischemia/reperfusion injury through activation of mitoKATP channels and reduction of oxidative stress. *Pharmacol. Res.* 113 (Pt A): 290–299.

17 Citi, V., Corvino, A., Fiorino, F. et al. (2021). Structure–activity relationships study of isothiocyanates for H_2S releasing properties: 3-pyridyl-isothiocyanate as a new promising cardioprotective agent. *J. Adv. Res.* 27: 41–53.

18 Lin, Y., Yang, X., Lu, Y. et al. (2019). Isothiocyanates as H_2S donors triggered by cysteine: reaction mechanism and structure and activity relationship. *Org. Lett.* 21 (15): 5977–5980.

19 Martelli, A., Testai, L., Citi, V. et al. (2013). Arylthioamides as H_2S donors: L-cysteine-activated releasing properties and vascular effects in vitro and in vivo. *ACS Med. Chem. Lett.* 4 (10): 904–908.

20 Barresi, E., Nesi, G., Citi, V. et al. (2017). Iminothioethers as hydrogen sulfide donors: from the gasotransmitter release to the vascular effects. *J. Med. Chem.* 60 (17): 7512–7523.

21 Severino, B., Corvino, A., Fiorino, F. et al. (2018). 1,2,4-Thiadiazolidin-3,5-diones as novel hydrogen sulfide donors. *Eur. J. Med. Chem.* 143: 1677–1686.

22 Ercolano, G., De Cicco, P., Frecentese, F. et al. (2019). Anti-metastatic properties of naproxen-HBTA in a murine model of cutaneous melanoma. *Front. Pharmacol.* 10: 66.

23 Mitidieri, E., Tramontano, T., Gurgone, D. et al. (2018). Mercaptopyruvate acts as endogenous vasodilator independently of 3-mercaptopyruvate sulfurtransferase activity. *Nitric Oxide* 75: 53–59.

24 Sestito, S., Pruccoli, L., Runfola, M. et al. (2019). Design and synthesis of H_2S-donor hybrids: a new treatment for Alzheimer's disease? *Eur. J. Med. Chem.* 184: 111745.

25 Rapposelli, S., Gambari, L., Digiacomo, M. et al. (2017). A novel H_2S-releasing amino-bisphosphonate which combines bone anti-catabolic and anabolic functions. *Sci. Rep.* 7 (1): 11940.

26 Sestito, S., Daniele, S., Pietrobono, D. et al. (2019). Memantine prodrug as a new agent for Alzheimer's disease. *Sci. Rep.* 9 (1): 4612.

27 Linert, W., Bridge, M.H., Huber, M. et al. (1999). In vitro and in vivo studies investigating possible antioxidant actions of nicotine: relevance to Parkinson's and Alzheimer's diseases. *Biochim. Biophys. Acta* 1454 (2): 143–152.

28 Citi, V., Martelli, A., Gorica, E. et al. (2021). Role of hydrogen sulfide in endothelial dysfunction: pathophysiology and therapeutic approaches. *J. Adv. Res.* 27: 99–113.

29 Petrie, J.R., Guzik, T.J., and Touyz, R.M. (2018). Diabetes, hypertension, and cardiovascular disease: clinical insights and vascular mechanisms. *Can. J. Cardiol.* 34 (5): 575–584.

30 Piragine, E. and Calderone, V. (2020). Pharmacological modulation of the hydrogen sulfide (H_2S) system by dietary H_2S-donors: a novel promising strategy in the prevention and treatment of type 2 diabetes mellitus. *Phytother. Res.* 35 (4): 1817–1846.

31 Martelli, A., Citi, V., and Calderone, V. (2020). Recent efforts in drug discovery on vascular inflammation and consequent atherosclerosis. *Expert Opin. Drug Discov.* 16 (4): 1, 411–17, 427.

32 Nowotny, K., Jung, T., Hohn, A. et al. (2015). Advanced glycation end products and oxidative stress in type 2 diabetes mellitus. *Biomolecules* 5 (1): 194–222.

33 Matsui, T., Nakamura, N., Ojima, A. et al. (2016). Sulforaphane reduces advanced glycation end products (AGEs)-induced inflammation in endothelial cells and rat aorta. *Nutr. Metab. Cardiovasc. Dis.* 26 (9): 797–807.

34 Yamagishi, S. and Matsui, T. (2016). Protective role of sulphoraphane against vascular complications in diabetes. *Pharm. Biol.* 54 (10): 2329–2339.

35 Huang, C.S., Lin, A.H., Liu, C.T. et al. (2013). Isothiocyanates protect against oxidized LDL-induced endothelial dysfunction by upregulating Nrf2-dependent antioxidation and suppressing NFκB activation. *Mol. Nutr. Food Res.* 57 (11): 1918–1930.

36 Shehatou, G.S. and Suddek, G.M. (2016). Sulforaphane attenuates the development of atherosclerosis and improves endothelial dysfunction in hypercholesterolemic rabbits. *Exp. Biol. Med.* 241 (4): 426–436.

37 Miao, X., Bai, Y., Sun, W. et al. (2012). Sulforaphane prevention of diabetes-induced aortic damage was associated with the up-regulation of Nrf2 and its down-stream antioxidants. *Nutr. Metab.* 9 (1): 84.

38 Wang, Y., Zhang, Z., Sun, W. et al. (2014). Sulforaphane attenuation of type 2 diabetes-induced aortic damage was associated with the upregulation of Nrf2 expression and function. *Oxid. Med. Cell Longev.* 2014: 123963.

39 Xie, L., Gu, Y., Wen, M. et al. (2016). Hydrogen sulfide induces Keap1 S-sulfhydration and suppresses diabetes-accelerated atherosclerosis via Nrf2 activation. *Diabetes* 65 (10): 3171–3184.

40 Kim, J.Y., Park, H.J., Um, S.H. et al. (2012). Sulforaphane suppresses vascular adhesion molecule-1 expression in TNF-α-stimulated mouse vascular smooth muscle cells: involvement of the MAPK, NF-κB and AP-1 signaling pathways. *Vascul. Pharmacol.* 56 (3, 4): 131–141.

41 Nallasamy, P., Si, H., Babu, P.V. et al. (2014). Sulforaphane reduces vascular inflammation in mice and prevents TNF-α-induced monocyte adhesion to primary endothelial cells through interfering with the NF-κB pathway. *J. Nutr. Biochem.* 25 (8): 824–833.
42 Okamoto, T., Akita, N., Nagai, M. et al. (2014). 6-Methylsulfinylhexyl isothiocyanate modulates endothelial cell function and suppresses leukocyte adhesion. *J. Nat. Med.* 68 (1): 144–153.
43 Pan, L.L., Liu, X.H., Gong, Q.H. et al. (2011). Hydrogen sulfide attenuated tumor necrosis factor-α-induced inflammatory signaling and dysfunction in vascular endothelial cells. *PLoS One* 6 (5): e19766.
44 Bahadoran, Z., Tohidi, M., Nazeri, P. et al. (2012). Effect of broccoli sprouts on insulin resistance in type 2 diabetic patients: a randomized double-blind clinical trial. *Int. J. Food Sci. Nutr.* 63 (7): 767–771.
45 Bahadoran, Z., Mirmiran, P., Hosseinpanah, F. et al. (2012). Broccoli sprouts powder could improve serum triglyceride and oxidized LDL/LDL-cholesterol ratio in type 2 diabetic patients: a randomized double-blind placebo-controlled clinical trial. *Diabetes Res. Clin. Pract.* 96 (3): 348–354.
46 Armah, C.N., Derdemezis, C., Traka, M.H. et al. (2015). Diet rich in high glucoraphanin broccoli reduces plasma LDL cholesterol: evidence from randomised controlled trials. *Mol. Nutr. Food Res.* 59 (5): 918–926.
47 Wilson, R.K., Kwan, T.K., Kwan, C.Y. et al. (2002). Effects of papaya seed extract and benzyl isothiocyanate on vascular contraction. *Life Sci.* 71 (5): 497–507.
48 Bucci, M., Papapetropoulos, A., Vellecco, V. et al. (2010). Hydrogen sulfide is an endogenous inhibitor of phosphodiesterase activity. *Arterioscler. Thromb. Vasc. Biol.* 30 (10): 1998–2004.
49 Testai, L., D'Antongiovanni, V., Piano, I. et al. (2015). Different patterns of H_2S/NO activity and cross-talk in the control of the coronary vascular bed under normotensive or hypertensive conditions. *Nitric Oxide* 47: 25–33.
50 Leucker, T.M., Nomura, Y., Kim, J.H. et al. (2017). Cystathionine γ-lyase protects vascular endothelium: a role for inhibition of histone deacetylase 6. *Am. J. Physiol. Heart Circ. Physiol.* 312 (4): H711–H720.
51 Howarth, F.P. and Maitland, J.M. (1988). Referral to hospital: can we do better? *BMJ* 297 (6652): 856.
52 Martelli, A., Piragine, E., Citi, V. et al. (2020). Erucin exhibits vasorelaxing effects and antihypertensive activity by H_2S-releasing properties. *Br. J. Pharmacol.* 177 (4): 824–835.
53 Streng, T., Axelsson, H.E., Hedlund, P. et al. (2008). Distribution and function of the hydrogen sulfide-sensitive TRPA1 ion channel in rat urinary bladder. *Eur. Urol.* 53 (2): 391–399.

54 Earley, S., Gonzales, A.L., and Crnich, R. (2009). Endothelium-dependent cerebral artery dilation mediated by TRPA1 and Ca^{2+}-activated K^+ channels. *Circ. Res.* 104 (8): 987–994.

55 Wilson, A.J., Gill, E.K., Abudalo, R.A. et al. (2018). Reactive oxygen species signalling in the diabetic heart: emerging prospect for therapeutic targeting. *Heart* 104 (4): 293–299.

56 Wu, Q.Q., Zong, J., Gao, L. et al. (2014). Sulforaphane protects H9c2 cardiomyocytes from angiotensin II-induced hypertrophy. *Herz* 39 (3): 390–396.

57 Xin, Y., Bai, Y., Jiang, X. et al. (2018). Sulforaphane prevents angiotensin II-induced cardiomyopathy by activation of Nrf2 via stimulating the Akt/GSK-3ss/Fyn pathway. *Redox Biol.* 15: 405–417.

58 de Haan, J.B. (2011). Nrf2 activators as attractive therapeutics for diabetic nephropathy. *Diabetes* 60 (11): 2683–2684.

59 Bai, Y., Chen, Q., Sun, Y.P. et al. (2017). Sulforaphane protection against the development of doxorubicin-induced chronic heart failure is associated with Nrf2 upregulation. *Cardiovasc. Ther.* 35 (5): e12277.

60 Singh, P., Sharma, R., McElhanon, K. et al. (2015). Sulforaphane protects the heart from doxorubicin-induced toxicity. *Free Radic. Biol. Med.* 86: 90–101.

61 Citi, V., Piragine, E., Testai, L. et al. (2018). The role of hydrogen sulfide and H_2S-donors in myocardial protection against ischemia/reperfusion injury. *Curr. Med. Chem.* 25 (34): 4380–4401.

62 Fernandes, R.O., De Castro, A.L., Bonetto, J.H. et al. (2016). Sulforaphane effects on postinfarction cardiac remodeling in rats: modulation of redox-sensitive prosurvival and proapoptotic proteins. *J. Nutr. Biochem.* 34: 106–117.

63 Piao, C.S., Gao, S., Lee, G.H. et al. (2010). Sulforaphane protects ischemic injury of hearts through antioxidant pathway and mitochondrial K(ATP) channels. *Pharmacol. Res.* 61 (4): 342–348.

64 Elrod, J.W., Calvert, J.W., Morrison, J. et al. (2007). Hydrogen sulfide attenuates myocardial ischemia-reperfusion injury by preservation of mitochondrial function. *Proc. Natl. Acad. Sci. U.S.A.* 104 (39): 15560–15565.

65 Sivarajah, A., Collino, M., Yasin, M. et al. (2009). Anti-apoptotic and anti-inflammatory effects of hydrogen sulfide in a rat model of regional myocardial I/R. *Shock* 31 (3): 267–274.

66 Sivarajah, A., McDonald, M.C., and Thiemermann, C. (2006). The production of hydrogen sulfide limits myocardial ischemia and reperfusion injury and contributes to the cardioprotective effects of preconditioning with endotoxin, but not ischemia in the rat. *Shock* 26 (2): 154–161.

67 Zhu, Y.Z., Wang, Z.J., Ho, P. et al. (2007). Hydrogen sulfide and its possible roles in myocardial ischemia in experimental rats. *J. Appl. Physiol.* 102 (1): 261–268.

68 Bucci, M. and Cirino, G. (2011). Hydrogen sulphide in heart and systemic circulation. *Inflamm. Allergy Drug Targets* 10 (2): 103–108.

69 Calvert, J.W., Jha, S., Gundewar, S. et al. (2009). Hydrogen sulfide mediates cardioprotection through Nrf2 signaling. *Circ. Res.* 105 (4): 365–374.

70 Calvert, J.W., Elston, M., Nicholson, C.K. et al. (2010). Genetic and pharmacologic hydrogen sulfide therapy attenuates ischemia-induced heart failure in mice. *Circulation* 122 (1): 11–19.

71 van Poppel, G., Verhoeven, D.T., Verhagen, H. et al. (1999). Brassica vegetables and cancer prevention. Epidemiology and mechanisms. *Adv. Exp. Med. Biol.* 472: 159–168.

72 Mandrich, L. and Caputo, E. (2020). Brassicaceae-derived anticancer agents: towards a green approach to beat cancer. *Nutrients* 12, 12 (3, 3): 868–883.

73 Barba, F.J., Nikmaram, N., Roohinejad, S. et al. (2016). Bioavailability of glucosinolates and their breakdown products: impact of processing. *Front. Nutr.* 3: 24.

74 Jadhav, U., Ezhilarasan, R., Vaughn, S.F. et al. (2007). Iberin induces cell cycle arrest and apoptosis in human neuroblastoma cells. *Int. J. Mol. Med.* 19 (3): 353–361.

75 Pawlik, A., Wala, M., Hac, A. et al. (2017). Sulforaphene, an isothiocyanate present in radish plants, inhibits proliferation of human breast cancer cells. *Phytomedicine* 29: 1–10.

76 Huang, Y.P., Jiang, Y.W., Chen, H.Y. et al. (2018). Benzyl isothiocyanate induces apoptotic cell death through mitochondria-dependent pathway in gefitinib-resistant NCI-H460 human lung cancer cells in vitro. *Anticancer Res.* 38 (9): 5165–5176.

77 Zhu, Y., Zhang, L., Zhang, G.D. et al. (2014). Potential mechanisms of benzyl isothiocyanate suppression of invasion and angiogenesis by the U87MG human glioma cell line. *Asian Pac. J. Cancer Prev.* 15 (19): 8225–8228.

78 Lai, K.C., Hsu, S.C., Kuo, C.L. et al. (2010). Phenethyl isothiocyanate inhibited tumor migration and invasion via suppressing multiple signal transduction pathways in human colon cancer HT29 cells. *J. Agric. Food. Chem.* 58 (20): 11148–11155.

79 Mitsiogianni, M., Koutsidis, G., Mavroudis, N. et al. (2019). The role of isothiocyanates as cancer chemo preventive, chemo-therapeutic and anti-melanoma agents. *Antioxidants* 8 (4): 106–141.

80 Bao, Y., Wang, W., Zhou, Z. et al. (2014). Benefits and risks of the hormetic effects of dietary isothiocyanates on cancer prevention. *PLoS One* 9 (12). e114764.

81 Wu, D., Si, W., Wang, M. et al. (2015). Hydrogen sulfide in cancer: friend or foe? *Nitric Oxide* 50: 38–45.

82 Wang, S.S., Chen, Y.H., Chen, N. et al. (2017). Hydrogen sulfide promotes autophagy of hepatocellular carcinoma cells through the PI3K/Akt/mTOR signaling pathway. *Cell Death Dis.* 8 (3): e2688.

83 Lee, Z.W., Zhou, J., Chen, C.S. et al. (2011). The slow-releasing hydrogen sulfide donor, GYY4137, exhibits novel anti-cancer effects in vitro and in vivo. *PLoS One* 6 (6): e21077.

84 Jurkowska, H., Wrobel, M., Szlezak, D. et al. (2018). New aspects of antiproliferative activity of 4-hydroxybenzyl isothiocyanate, a natural H_2S-donor. *Amino Acids* 50 (6): 699–709.

85 Lin, J.F., Tsai, T.F., Yang, S.C. et al. (2017). Benzyl isothiocyanate induces reactive oxygen species-initiated autophagy and apoptosis in human prostate cancer cells. *Oncotarget* 8 (12): 20220–20234.

86 Jakubikova, J., Sedlak, J., Mithen, R. et al. (2005). Role of PI3K/Akt and MEK/ERK signaling pathways in sulforaphane- and erucin-induced phase II enzymes and MRP2 transcription, G2/M arrest and cell death in Caco-2 cells. *Biochem. Pharmacol.* 69 (11): 1543–1552.

87 Jakubikova, J., Bao, Y., and Sedlak, J. (2005). Isothiocyanates induce cell cycle arrest, apoptosis and mitochondrial potential depolarization in HL-60 and multidrug-resistant cell lines. *Anticancer Res.* 25 (5): 3375–3386.

88 Li, G., Zhou, J., Budhraja, A. et al. (2015). Mitochondrial translocation and interaction of cofilin and Drp1 are required for erucin-induced mitochondrial fission and apoptosis. *Oncotarget* 6 (3): 1834–1849.

89 Azarenko, O., Jordan, M.A., and Wilson, L. (2014). Erucin, the major isothiocyanate in arugula (*Eruca sativa*), inhibits proliferation of MCF7 tumor cells by suppressing microtubule dynamics. *PLoS One* 9 (6): e100599.

90 Herz, C., Hertrampf, A., Zimmermann, S. et al. (2014). The isothiocyanate erucin abrogates telomerase in hepatocellular carcinoma cells in vitro and in an orthotopic xenograft tumour model of HCC. *J. Cell. Mol. Med.* 18 (12): 2393–2403.

91 Citi, V., Piragine, E., Pagnotta, E. et al. (2019). Anticancer properties of erucin, an H_2S-releasing isothiocyanate, on human pancreatic adenocarcinoma cells (AsPC-1). *Phytother. Res.* 33 (3): 845–855.

92 Satyan, K.S., Swamy, N., Dizon, D.S. et al. (2006). Phenethyl isothiocyanate (PEITC) inhibits growth of ovarian cancer cells by inducing apoptosis: role of caspase and MAPK activation. *Gynecol. Oncol.* 103 (1): 261–270.

93 Crowley, E., Rowan, N.J., Faller, D. et al. (2019). Natural and synthetic isothiocyanates possess anticancer potential against liver and prostate cancer in vitro. *Anticancer Res.* 39 (7): 3469–3485.

94 Macheda, M.L., Rogers, S., and Best, J.D. (2005). Molecular and cellular regulation of glucose transporter (GLUT) proteins in cancer. *J. Cell. Physiol.* 202 (3): 654–662.

95 Schlereth, T. and Birklein, F. (2008). The sympathetic nervous system and pain. *Neuromolecular Med.* 10 (3): 141–147.

96 Haanpaa, M., Attal, N., Backonja, M. et al. (2011). NeuPSIG guidelines on neuropathic pain assessment. *Pain* 152 (1): 14–27.

97 Woolf, C.J. and Decosterd, I. (1999). Implications of recent advances in the understanding of pain pathophysiology for the assessment of pain in patients. *Pain* 82 (Suppl 6, supp. 6): S141–S147.

98 Guadarrama-Enriquez, O., Gonzalez-Trujano, M.E., Ventura-Martinez, R. et al. (2018). Broccoli sprouts produce abdominal antinociception but not spasmolytic effects like its bioactive metabolite sulforaphane. *Biomed. Pharmacother.* 107: 1770–1778.

99 Logashina, Y.A., Korolkova, Y.V., Kozlov, S.A. et al. (2019). TRPA1 channel as a regulator of neurogenic inflammation and pain: structure, function, role in pathophysiology, and therapeutic potential of ligands. *Biochemistry* 84 (2): 101–118.

100 Noda, S. and Sato, A. (1990). Effects of infection with *Angiostrongylus cantonensis* on the circulating haemocyte population and the haematopoietic organ of the host snail M-line *Biomphalaria glabrata. J. Helminthol.* 64 (3): 239–247.

101 Roa-Coria, J.E., Pineda-Farias, J.B., Barragan-Iglesias, P. et al. (2019). Possible involvement of peripheral TRP channels in the hydrogen sulfide-induced hyperalgesia in diabetic rats. *BMC Neurosci.* 20 (1): 1.

102 Tsubota-Matsunami, M., Noguchi, Y., Okawa, Y. et al. (2012). Colonic hydrogen sulfide-induced visceral pain and referred hyperalgesia involve activation of both Ca(v)3.2 and TRPA1 channels in mice. *J. Pharmacol. Sci.* 119 (3): 293–296.

103 Syhr, K.M.J., Boosen, M., Hohmann, S.W. et al. (2015). The H_2S-producing enzyme CSE is dispensable for the processing of inflammatory and neuropathic pain. *Brain Res.* 1624: 380–389.

104 Terada, Y. and Kawabata, A. (2015). H_2S and pain: a novel aspect for processing of somatic, visceral and neuropathic pain signals. *Handb. Exp. Pharmacol.* 230: 217–230.

105 Kida, K., Marutani, E., Nguyen, R.K. et al. (2015). Inhaled hydrogen sulfide prevents neuropathic pain after peripheral nerve injury in mice. *Nitric Oxide* 46: 87–92.

106 Kakinohana, M., Marutani, E., Tokuda, K. et al. (2019). Breathing hydrogen sulfide prevents delayed paraplegia in mice. *Free Radic. Biol. Med.* 131: 243–250.

107 Chen, H., Xie, K., Chen, Y. et al. (2019). Nrf2/HO-1 signaling pathway participated in the protection of hydrogen sulfide on neuropathic pain in rats. *Int. Immunopharmacol.* 75: 105746.

108 Hughes, J.P., Chessell, I., Malamut, R. et al. (2012). Understanding chronic inflammatory and neuropathic pain. *Ann. N.Y. Acad. Sci.* 1255: 30–44.

109 Sommer, C., Leinders, M., and Uceyler, N. (2018). Inflammation in the pathophysiology of neuropathic pain. *Pain* 159 (3): 595–602.

110 Subedi, L., Venkatesan, R., and Kim, S.Y. (2017). Neuroprotective and anti-inflammatory activities of allyl isothiocyanate through attenuation of JNK/NF-κB/TNF-α signaling. *Int. J. Mol. Sci.* 18 (7): 1423–1439.

111 Davidson, R.K., Jupp, O., de Ferrars, R. et al. (2013). Sulforaphane represses matrix-degrading proteases and protects cartilage from destruction in vitro and in vivo. *Arthritis Rheum.* 65 (12): 3130–3140.

112 Negi, G., Kumar, A., and Sharma, S.S. (2011). Nrf2 and NF-κB modulation by sulforaphane counteracts multiple manifestations of diabetic neuropathy in rats and high glucose-induced changes. *Curr. Neurovasc. Res.* 8 (4): 294–304.

113 Wang, C. and Wang, C. (2017). Anti-nociceptive and anti-inflammatory actions of sulforaphane in chronic constriction injury-induced neuropathic pain mice. *Inflammopharmacology* 25 (1): 99–106.

114 Redondo, A., Chamorro, P.A.F., Riego, G. et al. (2017). Treatment with sulforaphane produces antinociception and improves morphine effects during inflammatory pain in mice. *J. Pharmacol. Exp. Ther.* 363 (3): 293–302.

115 Tokuda, K., Kida, K., Marutani, E. et al. (2012). Inhaled hydrogen sulfide prevents endotoxin-induced systemic inflammation and improves survival by altering sulfide metabolism in mice. *Antioxid. Redox Signal.* 17 (1): 11–21.

116 Yurinskaya, M.M., Krasnov, G.S., Kulikova, D.A. et al. (2020). H_2S counteracts proinflammatory effects of LPS through modulation of multiple pathways in human cells. *Inflamm. Res.* 69 (5): 481–495.

117 Rivera-Arconada, I., Vicente-Baz, J., and Lopez-Garcia, J.A. (2017). Targeting Kv7 channels in pain pathways. *Oncotarget* 8 (8): 12554–12555.

118 Di Cesare Mannelli, L., Lucarini, E., Micheli, L. et al. (2017). Effects of natural and synthetic isothiocyanate-based H_2S-releasers against chemotherapy-induced neuropathic pain: role of Kv7 potassium channels. *Neuropharmacology* 121: 49–59.

119 Giacoppo, S., Iori, R., Bramanti, P. et al. (2017). Topical moringin-cream relieves neuropathic pain by suppression of inflammatory pathway and voltage-gated ion channels in murine model of multiple sclerosis. *Mol. Pain* 13: 1744806917724318.

120 Dos Santos, A.O., do Val, D.R., da Silveira, F.D. et al. (2018). Antinociceptive, anti-inflammatory and toxicological evaluation of semi-synthetic molecules obtained from a benzyl-isothiocyanate isolated from *Moringa oleifera* Lam. in a temporomandibular joint inflammatory hypernociception model in rats. *Biomed. Pharmacother.* 98: 609–618.

121 Franceschi, C., Bonafe, M., Valensin, S. et al. (2000). Inflamm-aging. An evolutionary perspective on immunosenescence. *Ann. N.Y. Acad. Sci.* 908: 244–254.

122 Testai, L., Citi, V., Martelli, A. et al. (2020). Role of hydrogen sulfide in cardiovascular ageing. *Pharmacol. Res.* 160: 105125.

123 Pawelec, G., Goldeck, D., and Derhovanessian, E. (2014). Inflammation, ageing and chronic disease. *Curr. Opin. Immunol.* 29: 23–28.

124 Tisoncik, J.R., Korth, M.J., Simmons, C.P. et al. (2012). Into the eye of the cytokine storm. *Microbiol. Mol. Biol. Rev.* 76 (1): 16–32.

125 Tang, Y., Liu, J., Zhang, D. et al. (2020). Cytokine storm in COVID-19: the current evidence and treatment strategies. *Front. Immunol.* 11: 1708.

126 Hojyo, S., Uchida, M., Tanaka, K. et al. (2020). How COVID-19 induces cytokine storm with high mortality. *Inflamm. Regen.* 40: 37.

127 Sen, N., Paul, B.D., Gadalla, M.M. et al. (2012). Hydrogen sulfide-linked sulfhydration of NF-κB mediates its antiapoptotic actions. *Mol. Cell* 45 (1): 13–24.

128 Rao, C.Y., Fu, L.Y., Hu, C.L. et al. (2015). H_2S mitigates severe acute pancreatitis through the PI3K/AKT-NF-κB pathway in vivo. *World J. Gastroenterol.* 21 (15): 4555–4563.

129 Prawan, A., Saw, C.L., Khor, T.O. et al. (2009). Anti-NF-κB and anti-inflammatory activities of synthetic isothiocyanates: effect of chemical structures and cellular signaling. *Chem. Biol. Interact.* 179 (2, 3): 202–211.

130 Jeong, H.J., Hong, S.H., Lee, D.J. et al. (2002). Role of Ca(2+) on TNF-α and IL-6 secretion from RBL-2H3 mast cells. *Cell Signal.* 14 (7): 633–639.

131 Siraganian, R.P. (2003). Mast cell signal transduction from the high-affinity IgE receptor. *Curr. Opin. Immunol.* 15 (6): 639–646.

132 Rajamaki, K., Lappalainen, J., Oorni, K. et al. (2010). Cholesterol crystals activate the NLRP3 inflammasome in human macrophages: a novel link between cholesterol metabolism and inflammation. *PLoS One* 5 (7): e11765.

133 Han, N.R., Kim, I.K., Kim, H.M. et al. (2012). Methallyl isothiocyanate inhibits the caspase-1 activity through the inhibition of intracellular calcium levels. *Biochimie* 94 (3): 816–822.

134 Yamada-Kato, T., Nagai, M., Ohnishi, M. et al. (2012). Inhibitory effects of wasabi isothiocyanates on chemical mediator release in RBL-2H3 rat basophilic leukemia cells. *J. Nutr. Sci. Vitaminol.* 58 (4): 303–307.

135 Citi, V., Martelli, A., Brancaleone, V. et al. (2020). Anti-inflammatory and antiviral roles of hydrogen sulfide: rationale for considering H_2S donors in COVID-19 therapy. *Br. J. Pharmacol.* 177 (21): 4931–4941.

136 Zhansheng, L., Yumai, L., Zhiyuan, F. et al. (2019). Natural sulforaphane from broccoli seeds against influenza A virus replication in MDCK cells. *Nat. Prod. Commun.* 14 (6): 1–8.

137 Muller, L., Meyer, M., Bauer, R.N. et al. (2016). Effect of broccoli sprouts and live attenuated influenza virus on peripheral blood natural killer cells: a randomized, double-blind study. *PLoS One* 11 (1): e0147742.

138 Li, H., Ma, Y., Escaffre, O. et al. (2015). Role of hydrogen sulfide in paramyxovirus infections. *J. Virol.* 89 (10): 5557–5568.

139 Ivanciuc, T., Sbrana, E., Ansar, M. et al. (2016). Hydrogen sulfide is an antiviral and antiinflammatory endogenous gasotransmitter in the airways. Role in respiratory syncytial virus infection. *Am. J. Respir. Cell Mol. Biol.* 55 (5): 684–696.

140 Bazhanov, N., Ivanciuc, T., Wu, H. et al. (2018). Thiol-activated hydrogen sulfide donors antiviral and anti-inflammatory activity in respiratory syncytial virus infection. *Viruses* 10 (5): 249–261.

141 Bazhanov, N., Ansar, M., Ivanciuc, T. et al. (2017). Hydrogen sulfide: a novel player in airway development, pathophysiology of respiratory diseases, and antiviral defenses. *Am. J. Respir. Cell Mol. Biol.* 57 (4): 403–410.

142 Yang, G. (2020). H_2S as a potential defense against COVID-19? *Am. J. Physiol. Cell Physiol.* 319 (2): C244–C249.

12

Persulfide Prodrugs

Bingchen Yu, Zhengnan Yuan, and Binghe Wang

Department of Chemistry and Center for Diagnostics and Therapeutics, Georgia State University, Atlanta, GA, USA

List of Abbreviations

GAPDH	glyceraldehyde 3-phosphate dehydrogenase
GMP	guanosine monophosphate
GSH	glutathione
GSSH	glutathione persulfide
H_2S	hydrogen sulfide
H_2S_2	hydrogen persulfide
ROS	reactive oxygen species
TRPA1	transient receptor potential cation channel, subfamily A, member 1

12.1 Introduction

With the firm establishment of the endogenous signaling roles and other functions of hydrogen sulfide (H_2S) comes the question of the exact molecular nature of H_2S-mediated signaling processes. It is generally accepted that H_2S and other sulfur species exert their actions by (i) coordination and/or chelation to metal ions [1, 2], (ii) serving as antioxidants to attenuate oxidative stress [3, 4], (iii) providing sulfur species as structural and functional units such as forming sulfur clusters in cytochrome c [5], (vi) functioning as a nucleophile in detoxifications of both organic species and metal ions [2, 6, 7], and (v) performing protein modifications by processes such as persulfidation [8–10]. Along these lines, there are

Bingchen Yu and Zhengnan Yuan contributed equally.

questions as to exactly what the signaling species are for these various functions. Part of this question arises because of the reversible inter-conversions among various sulfur species including sulfide, persulfide, and to a degree sulfoxide.

As part of the effort to understand the exact chemical nature of sulfur signaling species, research in the past several years has led to some very interesting findings in understanding the roles of persulfides. Persulfides are sulfur species containing a sulfur-sulfur bond. They include hydrogen persulfide (H_2S_2) and R-SS$_n$H ($n \geq 1$) with at least one sulfur atom at the oxidative state of −1 or 0 [11, 12]. Briefly, persulfides can fulfill almost all the functions of H_2S described above. In some cases, persulfides are probably the obligatory pathway to get to the goal, such as in persulfidation. For example, H_2S is known to bind to an open metal-coordination site or displace a weak ligand [1, 2]. Similarly, persulfides have been found to interact with hemoproteins and detoxify heavy metals such as Cd^{2+} and CH_3Hg^+ the same way as H_2S [13–15]. Fukuto et al. observed that persulfide species, such as glutathione persulfide (GSSH), could reduce ferric cytochrome c to its ferrous form fivefold faster than H_2S [16]. Persulfides have been found to show elevated nucleophilicity and electron-donating/reduction ability than H_2S and to react with reactive oxygen species (ROS) faster than H_2S [16–19]. For example, Wang et al. found that a GSSH prodrug was able to provide elevated protective effects compared with Na_2S against oxidative stress induced by H_2O_2 in H9c2 cells [20]. Using a *N*-acetyl cysteine persulfide prodrug, Matson et al. observed that persulfide species was able to rescue cells from oxidative stress to a greater degree than H_2S donors such as Na_2S and GYY4137 [21]. These findings highlight the promising therapeutic potentials of persulfides in oxidative stress-related diseases. H_2S has been found to react with 8-nitro-cGMP to form 8-SH-cGMP and to ameliorate chronic heart failure after myocardial infarction [7, 22, 23]. Not surprisingly, persulfides have been found to convert 8-nitro-cGMP to 8-SH-cGMP with elevated activity compared with H_2S, suggesting the value of persulfide species in regulating signaling processes [7, 9, 19, 24]. Furthermore, persulfide species also exhibit increased potency in inducing protein S-persulfidation, a post-translational protein modification process during which the –SH group of the cysteine residue is converted to a –SSH group [8, 12, 25–29]. Importantly, while persulfide species such as H_2S_2 can directly induce S-persulfidation because of their electrophilicity [30, 31], H_2S fails to do so and can only achieve protein S-persulfidation in the presence of ROS [32, 33].

Due to all these factors, one can argue that in disease models where protein S-persulfidation is the major signaling mechanism involving sulfur species, persulfide species are the signaling species and would have improved potency than H_2S. For example, Kimura et al. demonstrated that H_2S_3 was able to induce Ca^{2+} influx in astrocytes with an EC_{50} of 91 nM by activating the transient receptor potential cation channel, subfamily A, member 1 (TRPA1) channel, showing a more than 300 fold potency increase when compared to Na_2S [34]. Wang et al. demonstrated that H_2S_2 efficiently inactivated glyceraldehyde 3-phosphate dehydrogenase (GAPDH) through protein S-persulfidation with an IC_{50} of 1 μM, while Na_2S showed no inhibitory effect at a similar concentration [30]. Kevil et al. showed that 50 μM Na_2S_3 increased endothelial solute permeability on human umbilical vein endothelial cells (HUVECs) monolayer while Na_2S at same concentration failed to do so [35].

With all the desirable characteristics of persulfides as research tools and potential therapeutics, they also have the problem of instability. Therefore, past few years have seen much effort in developing prodrugs of persulfides to allow for its stable storage as well as controlled release [8, 11, 12, 30]. This chapter examines the progress that has been made so far and remaining challenges in designing such prodrugs.

12.2 Persulfide Prodrugs

The development of persulfide prodrugs to achieve controllable delivery as research tools and for therapeutic purpose can be challenging. Persulfide species are highly reactive species with half-lives on the scale of minutes and are more reactive than H_2S or thiol species such as cysteine and glutathione (GSH) under various circumstances [16, 19]. For example, Filipovic et al. estimated the half-life of 50 μM persulfide in ammonium carbonate buffer (pH = 7.38) to be 2.7 minutes [36], and Wang et al. found the half-live of 100 μM H_2S_2 to be ~5 minutes in PBS (pH = 7.4) [30]. Unlike H_2S, persulfide species can go through disproportion reactions and degrade to a mixture of sulfur species [20]. Persulfide species can act as both a nucleophile and an electrophile under physiological conditions, while H_2S only shows nucleophilicity. This means while H_2S is inert to various free biological thiol species such as cysteine and GSH, persulfide species are sensitive to them. In fact, persulfide intermediates have been used to deliver H_2S in the presence of excess free thiol species as described in Section 12.2.6 [37, 38].

Due to their reactivity and instability, persulfide species need to be caged in a relatively stable prodrug form to achieve controllable delivery. The most common and effective strategy deployed so far is to cage the nucleophilic sulfur atom using a protecting group (Figure 12.1). The protecting group used should be stable enough to allow for the synthesis and storage of the persulfide prodrugs and yet be readily cleaved by various stimuli including enzymes such as esterase and phosphatase, ROS, pH changes, and the presence of free thiol species. Sometimes light is also used as a release trigger for research purposes. Besides triggered release, other characteristics such as tunable release rate are also desired since

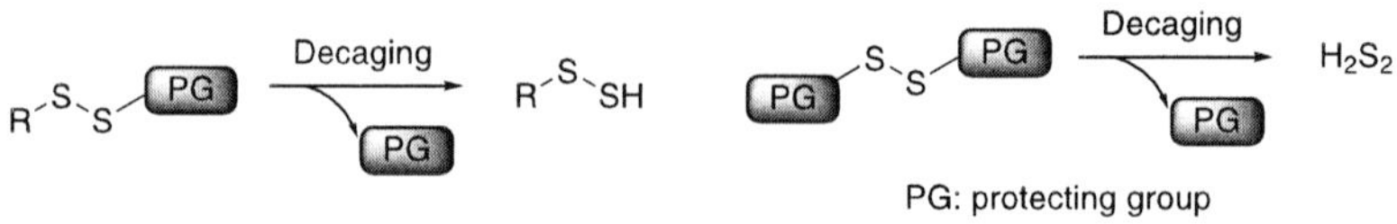

Figure 12.1 Strategies to deliver persulfide species.

the release rate determines the local concentration and thus the effectiveness. Ideally, the by-products after cargo release should be benign with a known toxicity profile. Below, we discuss the various types of prodrugs based on their release mechanisms.

12.2.1 Structural Moieties That Have Been Studied for Their Ability to Cage and Release Persulfide Species

To better understand the basic chemical properties of persulfide such as their stability, pK_a, and nucleophilicity, numerous model reactions to generate persulfide species have been examined. The most straightforward triggers to use for persulfide release are the known chemical reactions including acid-catalyzed hydrolysis and sulfur exchange (Figure 12.2) [16, 41, 42]. Pluth et al. has prepared model persulfides by taking advantage of the instability of thiol ester bond under acidic conditions (Figure 12.2a). Removal of the acetyl protecting group in compounds **1** by acidic ethanolysis would lead to the release of persulfide species **3**. Both bulky persulfide species such as trityl persulfide (TrtSSH) and adamantyl persulfide (AdSSH), as well as primary persulfide species such as benzyl persulfide (BnSSH) can be generated through this reaction. This model reaction was used to study the fundamental chemistry of persulfides such as their stability, degradation product(s), and their roles as nucleophiles and electrophiles [43, 44]. It was found that under a reducing environment, persulfide released H_2S, suggesting the role of persulfide species as a sulfur pool for sulfide storage. 9-Fluorenylmethyl (Fm) disulfide **5** was prepared by Xian et al. as precursors to generate persulfide species under basic conditions (Figure 12.2b) [39]. Compound **5** contains an activated pyridinyl disulfide bond and can react with a thiol group to generate compound **7** and by product **6**. The Fm group in compound **7** is base-labile, leading to the release of persulfide species as the result of degradation. The generated persulfide species was used to study the nucleophilicity of persulfides toward several thiol-blocking reagents. In doing so, bovine serum albumin (BSA) was used in a proof-of-concept study of protein persulfidation. When DBU was used as the base, the cysteine of BSA was converted to cysteine persulfide as a result of persulfidation (Figure 12.2c). Cyclic acyl disulfide **10** can also serve as precursors of persulfide (Figure 12.2d). The acyl group can be cleaved by a primary amine to generate persulfide species **11**, which was found to decompose

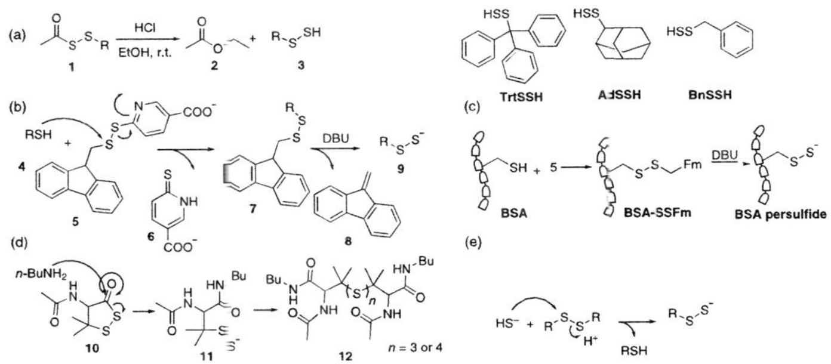

Figure 12.2 Various examples of chemical methods for releasing persulfide species from a protected form: (a) acid-promoted hydrolysis of acylated disulfides; (b and c) the deprotection of 9-fluorenylmethyl (Fm) disulfides using base; (d) the reaction between cyclic acyl disulfide and primary amine; and (e) the reaction between H_2S and a disulfide. Source: (b) Park et al. [39]/American Chemical Society. (e) Based on Francoleon et al. [16] and Pan and Carroll [40].

to form a polysulfide mixture **12** [45]. Another way to generate persulfide species is through the reaction between H_2S and a disulfide (Figure 12.2e) [16, 40]. This method has been used to generate GSSH to study its reactivity with *S*-methyl methanethiosulfonate, an alkylating reagent that has been used to detect protein S-persulfidation [40]. Alvarez et al studied the reactivity of H_2S with various low-molecular-weight disulfide compounds such as GSSG (GSH disulfide), cystine, and Hcy-Hcy (homocysteine disulfide). The second-order rate constants for the reactions between H_2S and these disulfides were determined to be in a range of $1.6 - 6 \times 10^{-1}$ $M^{-1}s^{-1}$ at pH 7.4 and 25 °C [17]. The chemistry described in generating persulfide species lays a strong foundation for the design of prodrugs of persulfide species for application under physiological conditions.

Disulfide or polysulfide salts such as Na_2S_n ($n \geq 2$) can spontaneously produce the corresponding hydrogen persulfide or polysulfide H_2S_n ($n \geq 2$) upon hydrolysis in aqueous solutions. As a result, these salts have been widely used as research tools to study the chemical and biological properties of H_2S_n [30, 34, 35]. However, such salt forms lead to a bolus dose of the persulfide with little control of the release profiles. In the next section, persulfide prodrugs that can generate persulfide species in a controllable manner under near physiological conditions are discussed.

12.2.2 Enzyme-Sensitive Prodrugs

Enzyme-facilitated prodrug activation has been widely used for decades for controllable and biologically compatible drug deliveries [46, 47]. Common enzymes used includes esterase, phosphatases, and proteases [46]. By taking advantage of enzyme-triggered payload release, we developed a series of esterase-sensitive persulfide prodrugs. As shown in Figure 12.3a, we used an ester bond to build the precursor of an unstable "hydroxymethyl disulfide" intermediate [48]. Porcine liver esterase (PLE) was used as a representative esterase. Incubation of the prodrugs (**BW-HP-201** to **206**) with PLE led to the release of persulfide **16** and the corresponding aldehyde byproduct **15** in PBS (pH = 7.4), presumably through esterase-catalyzed ester hydrolysis followed by the spontaneous collapse of the "hemiacetal" intermediate **14**. To examine the release mechanism, the benzaldehyde byproduct from **BW-HP-201** was characterized by HPLC, indirectly indicating the generation of the persulfide. Since the persulfide is not sufficiently stable for quantitative analysis under physiological conditions, dinitrofluorobenzene (DNFB) was used as a trapping agent to react with the persulfide to form a more stable disulfide product **17** for detection purposes [49]. By using such a method, persulfide release yield of 87% was obtained from the prodrug (**BW-HP-201**) in PBS (pH = 7.4) at room temperature. To tune the release rate, the masking ester group (**R_2 in** Figure 12.3a) was modified with substituents

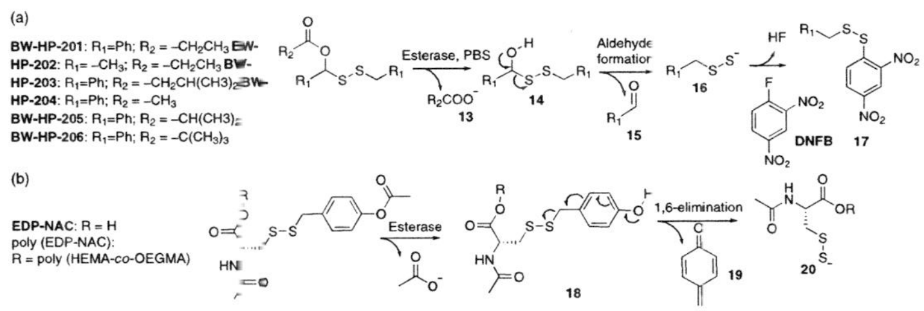

Figure 12.3 Esterase-sensitive persulfide prodrugs. (a) Esterase-promoted persulfide species release via a "hydroxymethyl disulfide" intermediate; (b) esterase-promoted *N*-acetyl cysteine persulfide release via 1,6-elimination.

of various steric hindrance. It was found that increasing the bulkiness of the ester moiety led to an increase in the release $t_{1/2}$ from 12 seconds (**BW-HP-201**) to 145 seconds (**BW-HP-206**) when 0.5 unit/mL PLE was used. To explore the biological activities of these prodrugs, cytotoxicity studies were first carried out on H9c2 cells. No toxicity was observed upon incubation with up to 100 μM of the prodrugs (**BW-HP-201** to **206**) for up to 24 hours. Then, the pharmacological activity of this prodrug was assessed in a well-known heart myocardial infarction reperfusion (MI/R) injury model. The infarct size was measured after 45 minutes of ischemia induced by left coronary artery occlusion followed by 24 hours of reperfusion and treatment with the persulfide prodrug **BW-HP-202**. When compared to the vehicle group, prodrug treatment (50 and 100 μg/kg) was found to effectively decrease the infarct size per area-at-risk (INF/AAR) and the circulating cardiac troponin I level, a marker for acute myocardial infarction [50]. Interestingly, the prodrug at a lower (12.5 μg/kg) or higher (500 μg/kg) doses showed no beneficial effects judged from both the INF/AAR ratio and troponin I level, suggesting persulfide as having a bell-shaped dose-response curve like H_2S [51].

Matson et al. reported a polymeric esterase-sensitive persulfide prodrug based on a self-immolative linker [52]. The release is triggered by esterase-mediated hydrolysis of the phenol ester moiety. The unstable phenyl hydroxy intermediate **18** undergoes 1,6-elimination to release *N*-acetyl cysteine persulfide (Figure 12.3b). DNFB trapping method qualitatively showed the release of persulfide upon incubation of EDP-NAC with PLE. Next, the small molecule persulfide prodrug (EDP-NAC) was conjugated to a block copolymer consisting of 2-hydroxyethyl methacrylate, oligo ethylene glycol methacrylate, and poly (HEMA-*co*-OEGMA) through an ester linkage. By doing so, the release kinetics profile of the polymeric prodrug poly (EDP-NAC) was quite different from that of EDP-NAC. The $t_{1/2}$ of poly (EDP-NAC) increased from 1.6 h (EDP-NAC) to 36 h (in D_2O/DMSO-d_6, monitored by NMR). The prodrug poly (EDP-NAC) exhibited a more potent cytoprotective effect toward 5-fluorouricil (5-FU)-induced toxicity on MCF-7 cells compared to that of EDP-NAC. Co-incubation of 200 μM of the poly (EDP-NAC) with 5-FU (200 μM) and the cells led the viability 100% as opposed to ~50% in the absence of the prodrug. Further, co-treatment with the non-polymeric prodrug EDP-NAC led to a survival rate of about 60%. These results indicated the influence of the release kinetics of the persulfide prodrug on the pharmacological activities.

Previously, we reported several esterase-sensitive H_2S prodrugs by taking advantages of a "trimethyl lock"-accelerated lactonization system [53]. The "trimethyl lock" system utilizes entropic advantages to promote lactonization by introducing three pendant methyl groups (labeled in bold in Figure 12.4), leading to rate enhancement on the order of 5×10^{10} fold [54]. In this case, we envisioned that

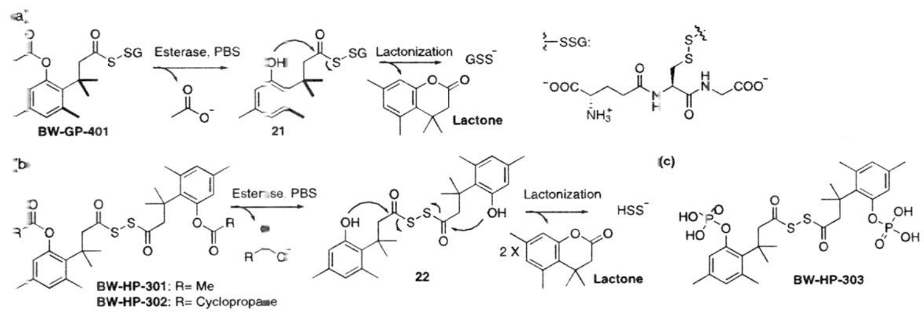

Figure 12.4 Enzyme-sensitive persulfide and H_2S_2 prodrugs utilizing a "trimethyl lock" system. (a) An esterase-sensitive GSSH prodrug;(b) esterase-sensitive H_2S_2 prodrugs; and (c) a phosphatase-sensitive H_2S_2 prodrug. Source: (a) Yuan et al. [20]/American Chemical Society.

this system should also be able to cage and deliver GSSH, which is known to be the most abundant persulfide species in mammalian tissue [55–57]. GSSH plays an important role in biological processes, such as glutathionylation and antioxidation. Therefore, a GSSH prodrug (**BW-GP-401**) was built by caging GSSH in a thiol ester form (Figure 12.4a) [20]. By design, the ester bond can be hydrolyzed by an esterase, revealing the phenol hydroxyl group, which can trigger the lactonization to release GSSH. Although the lactone was detected by HPLC in near quantum yield, suggesting a high yield of GSSH generation, only ~10% of the released GSSH was trapped by DNFB, indicating the highly reactive nature of GSSH. Detailed mass spectrometric studies confirmed this lack of stability at near physiological condition (PBS buffer, pH 7.4, 37 °C), leading to GSSSG and H_2S through disproportionation reactions. GSSH was found to show more potent H_2O_2-scavenging effects as compared with H_2S and GSH. To study the antioxidative effects of the GSSH prodrug, H9c2 cells was pretreated with prodrug prior to exposure with H_2O_2. Cell viability with treatment of 150 μM of the GSSH prodrug was about 60% as compared 40% for the vehicle group, showing the protective effects of the prodrug. In contrast, pretreatment with the lactone, Na_2S or GSH did not significantly alter the H_2O_2-induced toxicity, indicating the improved cytoprotective effects of GSSH as to other sulfide species.

Esterase-sensitive H_2S_2 prodrugs were also prepared using the same "trimethyl lock" system (Figure 12.4b), which was inspired by the previous finding that hydrogen polysulfide showed increased biological activity compared to H_2S [31, 34]. For quantification of H_2S_2 release from the prodrugs, a trapping reagent, monobromobimane (mBB), was incubated with the H_2S_2 prodrug (**BW-HP-301**) in PBS (pH 7.4, 37 °C) in the presence of PLE. The H_2S_2 trapping yield from 100 μM of the prodrug was determined to be ~68% as measured by HPLC. It should be noted that the H_2S_2 trapping yield from 100 μM prodrug was comparable to that from the same concentration of Na_2S_2, indicating excellent release efficiency of the prodrug. The release rate is tunable by modulating the steric bulkiness of the ester moiety ("R" group in Figure 12.4b). With increasing bulkiness, release $t_{1/2}$ also increased from 24 minutes (**BW-HP-301**) to 172 minutes (**BW-HP-302**) in the presence of 2 units/mL PLE. A phosphatase-sensitive H_2S_2 prodrug was also prepared (Figure 12.4c) for improved water solubility. This H_2S_2 delivery system was used to study protein S-persulfidation using GAPDH as a model protein [58]. It was found that the H_2S_2 prodrug (**BW-HP-301**) itself induced protein S-persulfidation and reversible inactivation of GAPDH with an IC_{50} of about 1 μM. In comparison, H_2S itself neither induced protein S-persulfidation nor affected GAPDH activity.

These enzyme-triggered persulfide prodrugs take advantage of environmental stimuli that are ubiquitous in the biological system to achieve controllable delivery of persulfide species. They are easy to prepare and stable for storage (r.t. for several days and −20 °C for several month); and yet readily release the caged persulfide in the presence of the relevant enzyme. The released byproducts such as the lactone are relatively benign. Especially important, their releasing rates can be tuned either by introducing groups with different steric hindrance or conjugating with a block copolymer. Their biological activities have been examined in several *in vitro* and *in vivo* models with promising therapeutic applications. However, using enzymes such as an esterase does not offer specificity and targeted delivery. Protease-sensitive persulfide prodrugs could be one approach to increase specificity. Some prodrugs such as the "trimethyl lock"-based GSSH and H_2S_2 prodrugs cage the persulfide in a "thiol acid" form (Figure 12.4), which is unstable toward free thiols such as cysteine. This means that these prodrugs could decompose in the presence of biological thiols, leading to decreased persulfide release yield. The esterase-sensitive *N*-acetyl cysteine persulfide prodrug utilizing a 1,6-elimination reaction generates a reactive quinonemethide byproduct, which could lead to potential toxicity issues and limit its therapeutic applications. All these are issues that will need to be considered in future studies.

12.2.3 ROS-Sensitive Persulfide Prodrugs

It is well known that many diseases including cancer and inflammation are associated with elevated level of ROS [59, 60]. In developing persulfide-based therapeutic agents, several groups also reported ROS-sensitive persulfide prodrugs. In the first example, a self-immolative system was introduced by Matson et al to deliver *N*-acetyl cysteine (NAC) persulfide [21]. For the ROS-sensing moiety, a widely used aryl boronic ester was installed at the *para*-position to the disulfide substitution on the prodrug BDP-NAC (Figure 12.5a). In the presence of H_2O_2, conversion of the aryl boronic ester to a hydroxyl group was designed to initiate a 1,6-elimination reaction to release NAC persulfide **25**. The formation of NAC persulfide was characterized by LC-MS. However, the release yield was not quantified. The cytoprotective effect of the ROS-sensitive NAC persulfide prodrug was assessed in H_2O_2-treated H9c2 cells. It was found that, in the presence of 200 μM BDP-NAC, cell viability remained at 100% when challenged with H_2O_2 (100 μM). In contrast, in the absence of BDP-NAC, the same treatment by H_2O_2 decreased cell viability to 30%. Further, H_2S and NAC did not present significant protective effect. Under the same conditions, the control compound with the aryl boronic ester moiety only that had no ability to deliver persulfide also exhibited cytoprotective effects and the cell viability was determined to be 70%. Such results suggest both the boronic ester moiety and the released persulfide contribute to the observed cytoprotective effects against H_2O_2. It is possible that the latter effect was due to the ability for the boronate moiety to consume peroxide.

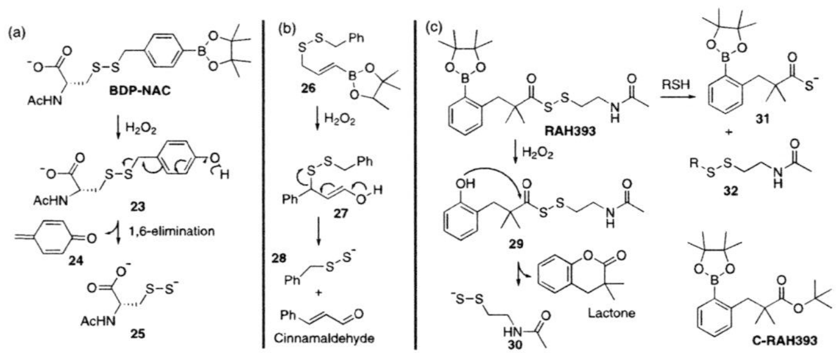

Figure 12.5 ROS-sensitive persulfide prodrugs. (a) A ROS-sensitive NAC persulfide prodrug through 1,6-elimination; (b) a ROS-sensitive benzyl persulfide prodrug through a 1,4-O,S-relay reaction; and (c) a ROS-sensitive persulfide prodrug through lactonization.

Along a similar line, Chakrapani et al. also used vinyl boronic esters as the H_2O_2-sensing moiety to develop a prodrug of benzyl persulfide **26** [61], which was designed to generate an unstable intermediate **27** by H_2O_2-mediated deprotection of the boronate ester group and subsequently undergo a 1,4-O,S-relay to release the persulfide species **28** (Figure 12.5b). The deprotection rate was measured to be to be 5.3×10^{-2} min^{-1} in the presence of H_2O_2 (10 equiv) in PBS buffer (pH = 7.4, 20% MeCN) at 37 °C. Both DNFB and mBB were used to trap the persulfide in the reaction mixture, and the corresponding products were detected by HPLC. However, the release yield was not reported. It should also be noted that, the byproduct, cinnamaldehyde, was found to react with GSH in PBS buffer (pH = 7.4). As such, cinnamaldehyde might also form an addition product with the persulfide. The protective effect of the prodrug **26** against an oxidative stress inducer, menadione, was tested in DLD-1 cells. It was found that 100 μM of the prodrug alleviated the cell death rate (caused by 50 μM menadione) from 70% to around 30%.

Lukesh et al. reported a ROS-sensitive persulfide donor based on a lactonization system [62]. As shown in Figure 12.5c, upon conversion of the aryl boronic ester group to a hydroxyl group by H_2O_2, the ensuing lactonization was designed to trigger the persulfide release. In ammonium bicarbonate buffer (pH = 7.4, 1 mM H_2O_2), the release yield of persulfide species (from 200 μM prodrug) was determined to be around 31% after 15 minutes of incubation using the DNFB trapping method. Pretreatment (1 h) of HeLa with this prodrug showed protective effects against H_2O_2 (200 μM). For example, compared to the vehicle group, cell viability was close to 100% in the presence of 200 μM of the prodrug RAH393 in contrast with 30% in the non-treated group. Further, treatment with the control compound C-RAH393 also offered some protection with the viability at about 50%. Again, this might be attributed to the ability for the boronate ester to consume peroxide. Further, the authors also noted that, the prodrug RAH393 might undergo thiol exchange reactions in cellular environments to produce a thiocarboxylic acid **31** and a disulfide compound **32**. These products cannot efficiently release the designated persulfide species, though **32** might still be able to afford persulfidation.

In general, the designs of the current ROS-sensitive prodrugs are based on installment of an aryl/vinyl boronic ester group as the sensing moiety. In the presence of H_2O_2, conversion of the boronic ester moiety to a hydroxyl group would trigger a cascade of reactions to release persulfide. ROS-sensitive prodrugs provide a new set of tools to deliver persulfide species to the desired sites with elevated ROS levels, which are normally associated with cancer or inflammation. Additionally, regarding the unique antioxidative activities of persulfide, these prodrugs might also be useful for developing persulfide-based antioxidant agents. However, the ability of boronate to consume ROS needs to be considered in evaluating these prodrugs in the context of attenuating oxidative stress. The responses of such prodrugs to endogenous ROS have not been explored yet. In the future, it would be interesting to examine the activities of these prodrugs in disease models.

12.2.4 pH-Sensitive Persulfide Prodrugs

To develop acid-sensitive persulfide prodrugs, Xian et al. designed an approach by installing a trimethylsilyl (TMS) or triethylsily (TES) group to cage an unstable "hydroxymethyl disulfide" intermediate [63]. Under acidic conditions, the deprotection of TMS or TES would reveal the "hydroxymethyl disulfide" intermediate, which would decompose to release the persulfide or H_2S_2 species (Figure 12.6). The persulfide release rates of the prodrugs were found to vary depending on pH. For 20 mM of prodrug **33**, at pH 7.4, the decomposition was not complete within 12 hours in DMF/PBS buffer (4 : 1). However, at pH 5.0, it only took 3.5 hours for the total decomposition of prodrug **33**. For the TES protecting group, it took a longer time to hydrolyze. At pH 5, the time needed for H_2S_2 prodrug **37** to decompose was 2.5 hours; however, it required 7 hours for H_2S_2 prodrug **38** to achieve full release under the same condition.

Galardon et al. utilized an intramolecular acyl transfer reaction to generate a persulfide species at physiological pH [64]. As shown in Figure 12.7a, prodrug **41** is designed to undergo an S-to-N intramolecular methoxycarbonyl transfer to generate the persulfide species **43**. The reaction kinetics were monitored by Ellman's reagent at different pH conditions. It was found that higher pH favors the transformation of the precursor to the persulfide as one would expect because the free amino form is the presumed nucleophilic species. When the pH increased from 6.5 to 8.5, the k_{obs} of the persulfide generation reaction also increased from 0.1 to 1.6 min^{-1}. To directly evaluate persulfide generation, *N*-ethylmaleimide (NEM) was used as a trapping agent in the reaction mixture, and its adduct with persulfide was characterized by LC-MS. Next, the persulfide prodrugs were used to examine its activities on 8-NO_2-cGMP and ferric cytochrome c (cyt c). It was found that, upon incubation with 1 mM prodrug **41** in Tris buffer (3 hours, at 37 °C), around

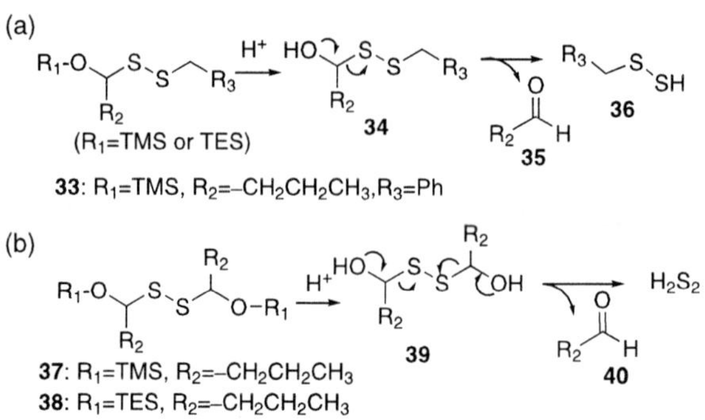

Figure 12.6 pH-sensitive persulfide species (a) and H_2S_2 (b) prodrugs via "hydroxymethyl disulfide" intermediate.

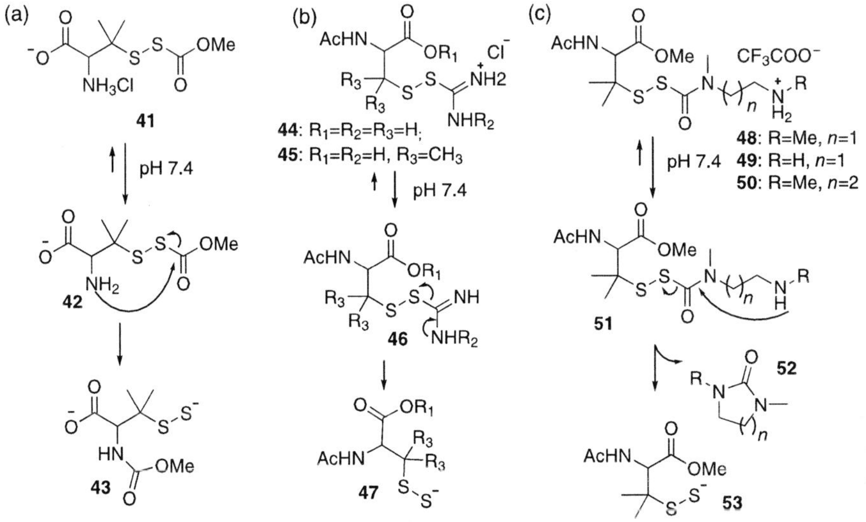

Figure 12.7 pH-sensitive persulfide prodrugs. (a) pH-sensitive persulfide prodrugs utilizing intramolecular acyl transfer to generate a persulfide species; (b) *S*-substituted-thioisothioureas as pH-sensitive persulfide prodrugs; and (c) pH-sensitive persulfide prodrugs based on an intramolecular cyclization reaction.

50% of 8-NO_2-cGMP was converted to the thiol derivative, 8-SH-cGMP. Compared to GSH, such a persulfide prodrug exhibited enhanced ability to reduce ferric cyt c to ferrous cyt c. At pH 7.4, the k_{obs} was around 0.3 min^{-1}.

Along this line, S-substituted-thioisothioureas were reported as persulfide precursors at physiological pH by Toscano et al. [65]. As illustrated in Figure 12.7b, the deprotonation of the urea moiety of prodrug **44** is designed to trigger the elimination reaction to generate the persulfide species. However, it was found that the released persulfide was able to react with **44** itself, leading to decreased yield. To avoid interfering reactions between the persulfide and the precursor, a *gem*-dimethyl moiety was installed next to the disulfide moiety to increase steric hindrance on prodrug **45**. By using NEM (20 mM) as the trapping agent, the persulfide release yield from prodrug **45** was determined to be 96% in PBS buffer at 37 °C (pH = 7.4, 10% D_2O). Toscano et al. also reported a series of persulfide prodrugs based on an intramolecular cyclization reaction [66]. In this design, the terminal amino group on prodrug **48** was kept in an ammonium salt form. Upon partial generation of the amine species at physiological pH (7.4), the intermediate **51** was designed to decompose to generate persulfide species and a cyclic urea byproduct **52** through an intramolecular cyclization reaction. The release yield of persulfide from prodrug **48** was studied by the reaction with a trapping reagent using an analog of iodoacetamide. The release yield was determined to be around 89% in PBS buffer (containing DTPA) at 37 °C. In order to tune the release rate of the prodrugs, the linker between the amine and perthiocarbamate and the substitution of the amine moiety were modified. It was found that increasing the length of the spacer (from 2 to 3 methylene units) or substituting the methyl group (R–) by –H increased the $t_{1/2}$ of prodrug. The half-lives of these persulfide prodrugs were tuned in the range of 1.4–484 minutes. For example, the $t_{1/2}$ of prodrugs **48** and **50** were determined to be 1.4 and 114 minutes (PBS buffer, at 37 °C), respectively, indicating the importance of the linker length. Similarly, the $t_{1/2}$ of **49** 16.7 minutes, which is much longer than that of **48**, which suggested a faster reaction rate for the secondary amine as a nucleophile. It is worth pointing out that these prodrugs tend to generate carbonyl sulfide (COS) in the presence of free thiols and thus serve as H_2S donors as well. This feature is covered in the H_2S donor chapters and is not discussed in detail here. Prodrug **49** was further tested for its cardioprotective effects on the Langendorff model of myocardial ischemia-reperfusion (*ex vivo*). 100 μM of **49** decreased the area of infarcted heart tissue from 42% to 16%, indicating the potential pharmacological usage of this prodrug in ameliorating myocardial ischemia-reperfusion.

In this section, pH-sensitive persulfide prodrugs are summarized. Their release rates vary in a pH-dependent fashion. For intramolecular ligand transfer-based prodrugs, steric hindrance needs to be put in place to avoid the interaction between persulfide and the prodrug molecule.

12.2.5 Photo-Sensitive Persulfide Prodrugs

Recently, a series photo-sensitive persulfide prodrugs were reported by Singh et al. [67]. In the first example, NAC persulfide was caged by an ortho-nitrobenzyl (ONB) group as a photo-sensitive trigger. The ONB moiety was designed to dissociate from the prodrug (**54** and **55**) to release the NAC persulfide **56** upon irradiation at ≥365 nm (Figure 12.8a). To quantify the persulfide release, persulfide prodrugs were irradiated for 55 (prodrug **54**) and 20 minutes (prodrug **55**), respectively, in the presence of 5 mM of DNFB in ACN/PBS (3 : 7) at 37 °C. The release yield was determined to be ~60% from 100 μM of the prodrugs (**54** and **55**). Additionally, the photochemical quantum yields of the prodrugs were also measured to be 0.07 (prodrug **54**) and 0.36 (prodrug **55**) by an HPLC method. It should be noted that the prodrugs exhibited fairly good stability toward thiol species. The extent of decomposition was found to be less than 20% upon incubating the prodrugs with 1 mM of thiols (NAC and GSH) for 10 days in the absence of light. The cytoprotective effects of the prodrugs against H_2O_2-induced toxicity in HeLa cells were assessed. It was found that 200 μM of the prodrug

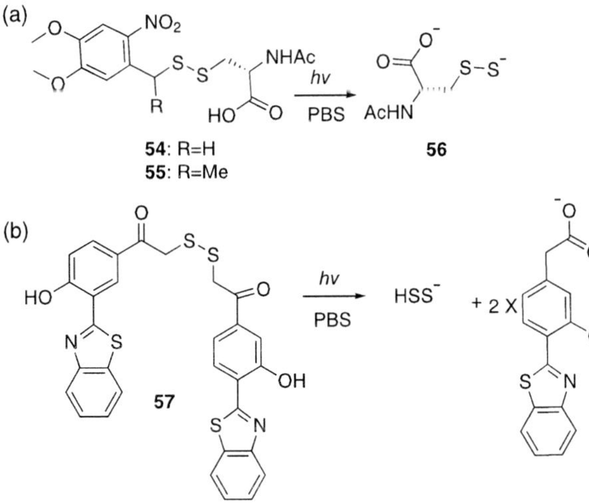

Figure 12.8 Light-sensitive prodrugs for (a) NAC persulfide and (b) H_2S_2.

(**54** and **55**) was able to rescue cells from 50% to 70% viability upon light irritation for 10 to 30 minutes.

Along the same line, a light-sensitive H_2S_2 prodrug (**57**) was also reported by Singh et al. [68]. As shown in Figure 12.8b, a disulfide moiety was symmetrically attached to a light-sensing moiety. This was designed to release of H_2S_2 through a proposed excited-state intramolecular proton transfer (ESIPT) process upon exposure to visible light (≥410 nm). The rate constant was determined to be $7.11\times 10^{-2}\ s^{-1}$ in PBS buffer (pH = 7.4, 5% DMSO) upon light irritation, and the reaction quantum yield was found to be 0.12 ± 0.05. For quantifying the H_2S_2, a fluorescent H_2S_n probe, DSP-3, was used as a trapping agent, and about 43 μM of the trapped product was detected from 100 μM of prodrug **57**. In protecting MDA-MB-468 cells from H_2O_2-induced damage, the prodrug was found to be more effective than H_2S and other thiol species (GSH and cysteine). Preincubation with 100 μM of prodrug **57** led to cell viability of around 70% as compared to 50% for the control experiments.

Currently, only a few light-sensitive persulfide prodrugs have been reported. Compared to other categories, these prodrugs offer the feature of spatiotemporal control. However, the release yields from these prodrugs are generally low and the wavelength is short, which might limit their application in the future.

12.2.6 H_2S Prodrugs That Release H_2S Via Persulfide Intermediate

Primary persulfide species are highly unstable and can generate H_2S by degradation or reaction with other free thiol species [20, 48, 69]. For this reason, persulfide

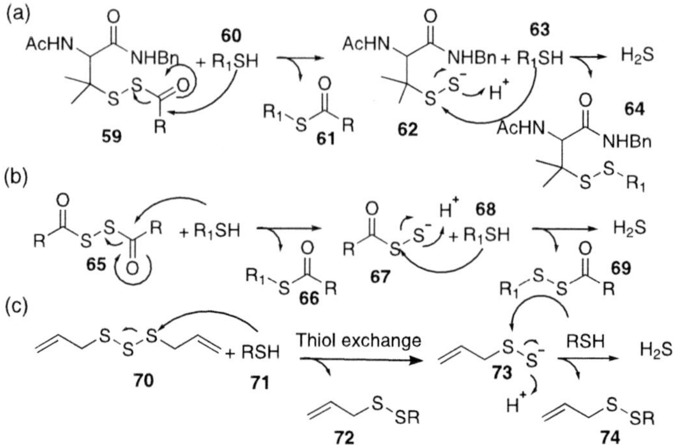

Figure 12.9 H_2S prodrugs that release H_2S via a persulfide intermediate. (a) *S*-Acyl disulfide derivatives; (b) dithioperoxyanhydrides; and (c) diallyl trisulfide as a H_2S prodrug via a persulfide intermediate in the presence of excess free thiols. Source: (a) Based on Zhao et al. [37, 69]. (b) Roger et al. [70]/John Wiley & Sons. (c) [71]/American Chemical Society.

species have been used as H_2S precursors and several H_2S prodrugs have utilized persulfide an intermediate for H_2S delivery. Xian et al. have prepared a series of S-acyl disulfides derivatives **59**. The acyl group is designed to be cleaved by excess free thiols, revealing a persulfide intermediate **62**, which would quickly react with free thiols to generate H_2S (Figure 12.9a) [37, 69]. A series of dithioperoxyanhydrides **65** have been developed by Galardon's lab to take advantage of the ability for persulfide to generate H_2S (Figure 12.9b) [70]. The thiol ester bond can be cleaved by free thiol species, leading to persulfide intermediate **67**, which would react with free thiols to release H_2S. Diallyl trisulfide (DATS, **70**) can release H_2S via S-allyl persulfide intermediate (**73**) in the presence of excess free thiols (Figure 12.9c) [71]. The release of a persulfide from these donors mentioned above relies on the presence of free thiols in excess. Although these thiol-activated prodrugs can generate persulfide, the released persulfide species would quickly exchange with the free thiols present in the system to generate H_2S. Thus, they are often viewed as prodrugs of H_2S instead of persulfide, although these prodrugs probably can function as both type of prodrugs.

12.3 Challenges in Persulfide Prodrug Design and Potential Therapeutic Applications

Research in developing and using persulfide prodrugs is still at its early stage and faces challenges both from the aspects of chemical design and therapeutic

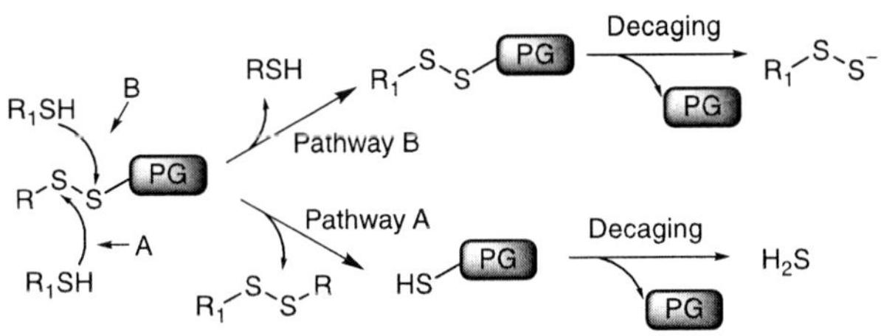

Figure 12.10 Degradation of persulfide prodrug in the presence of free thiol species.

applications. Firstly, the current persulfide prodrugs exclusively use disulfide bond or sulfur-thiol acid linkage as precursors of the persulfide species. These linkages contain electrophilic sulfur atoms and are vulnerable to other free thiol species, which could cause stability issues of the persulfide prodrug upon exposure to biological milieu. Further, the released persulfide can also react with the unreleased prodrug or the byproduct. One thing worth mentioning is that persulfide prodrugs with a disulfide bond or thiocarboxylic linkage are subject to free thiol attack; this process would end up releasing H_2S (Pathway A, Figure 12.10) instead of persulfide or releasing a different kind of persulfide species (Pathway B, Figure 12.10). To some extent, the persulfide activity would not be totally abolished because of the preservation of the collective oxidation state. It is highly possible that in the complicated biological system, it is a mixed process of releasing persulfide and H_2S, and persulfide release yield depends on several factors including the intrinsic stability of disulfide bond or thiocarboxylic linkage, the release rates, and the local free thiol concentration.

Secondly, the term "persulfide," defined as RSS_xH ($x \geq 1$), describes a diverse set of compounds. Whether the various species would lead to different activities/potency is subject to further experimental studies. For example, current persulfide prodrugs can release species such as GSSH, *N*-acetyl cysteine persulfide, benzyl persulfide, and ethyl persulfide. One can argue that the less polar species such as benzyl persulfide and ethyl persulfide would be more permeable across cellular membrane, while polar persulfide species may rely on specific transporters [72]. Thus, persulfides with different "R" substituents may have very different overall activities due to differences in reactivity, stability, and permeability. Whether the "R" substituent would affect the stability and reactivity of persulfide species is another interesting and important research topic. In this sense, the ability to develop prodrugs of H_2S and persulfide species is very different. For H_2S prodrugs, the cargo/active molecule is fixed, which is H_2S, leaving no room for structure and activity (SAR) analysis and modification [38]. However, in the case of persulfide prodrugs, the "R" substituents might play a role as discussed above, leaving space for activities or PK/PD improvement through SAR studies.

Thirdly, the release rate of persulfide species from their parent prodrugs is another critical factor that would directly affect the therapeutic output. One issue is the dependence of the "steady state" concentration of the active drug on the release rate [38]. For this reason, special attention is needed in considering the release rate during the design stage. It is known that H_2S has a "bell-shaped" dosage response curve with a narrow therapeutic window. Based on the current limited experimental evidence, persulfide species might have similar dosage response curve like H_2S [48], which means that factors that determine the local persulfide concentration such as release rate, persulfide species intrinsic stability, and free thiol species concentration will all need to be considered.

Besides the challenges discussed above, another critical issue is assessing for the suitability of a particular prodrug for a given clinical applications. Up-to-date, there are only a few reported animal model studies of persulfide prodrugs. However, current sulfur signaling mechanistic studies and *in vitro* and *in vivo* experiments suggest that H_2S and persulfide share the same or similar therapeutic indications, as one would expect [12, 16–19]. So, theoretically, in disease models where H_2S has been proven to be effective, persulfide prodrugs should have qualitatively similar effectiveness. H_2S prodrugs have been used to reduce inflammation, attenuate gastropathy induced by nonsteroidal anti-inflammatory drugs (NSAIDs) or ethanol, and protect heart tissue against myocardial ischemia-reperfusion (MI/R) [38, 50, 73–79]. Several H_2S prodrugs have reached clinical trials for the treatment of heart failure, arthritis, and idiopathic oligoasthenozoospermia [80–82]. These disease models could be the starting points to identify clinical applications for persulfide prodrugs. It is worth pointing out that in these potential therapeutic applications, these persulfide prodrugs will need to show improved properties in order to justify the utilization of the more challenging chemistry in designing persulfide prodrugs.

Despite all the challenges discussed above, the potential for further development of persulfide species is still very promising. Besides the elevated activity of persulfide species in various signaling pathways and thus the potential improvement in therapeutic effectiveness compared with H_2S, there are many reasons to be optimistic. First of all, many of the persulfide species such as GSSH, cysteine persulfide, and H_2S_2 are endogenous signaling molecules and can be metabolized quickly. Therefore, their potential toxicity is not a serious concern compared with other traditional small-molecule drugs as long as the persulfide species are not suddenly given in a bolus administration. Second, because of the shared characteristics between H_2S and persulfide species, the lessons/knowledge accumulated in the clinical trials of various H_2S prodrugs can serve as guidelines for developing persulfide species prodrugs. SG1002, a H_2S prodrug, have reached clinical trials for the treatment of congestive heart failure and are currently sold as a prescription medical food under the trade name "Sulfzix" [80, 82]. Although SG1002 is broadly considered as a H_2S prodrug, the major component of SG1002 is S_8, which is sulfane sulfur with oxidation state of −1 or 0 [10, 83, 84] and can release H_2S under reducing conditions such as in the presence of free thiols [85, 86]. H_2S does not contain sulfane sulfur while persulfide species do. Thus, one can reason that S_8 share more similarities with persulfide species than with H_2S. The success of SG1002 so far in clinical trials has been an inspiration for the development of persulfide prodrugs. Recently, using a series of esterase-sensitive sulfur species prodrugs, Ke et al demonstrated that a H_2S_2 prodrug exhibited elevated analgesic effects and decreased toxicity *in vivo* compared with a H_2S prodrug, which further highlights the therapeutic potentials of persulfide prodrug [87].

References

1 Rouault, T.A. and Tong, W.H. (2008). Iron–sulfur cluster biogenesis and human disease. *Trends Genet.* 24 (8): 398–407.

2 Fukuto, J.M., Vega, V.S., Works, C., and Lin, J. (2020). The chemical biology of hydrogen sulfide and related hydropersulfides: interactions with biologically relevant metals and metalloproteins. *Curr. Opin. Chem. Biol.* 55: 52–58.

3 Corsello, T., Komaravelli, N., and Casola, A. (2018). Role of hydrogen sulfide in Nrf2- and sirtuin-dependent maintenance of cellular redox balance. *Antioxidants* 7 (10): 129.

4 Tyagi, N., Moshal, K.S., Sen, U. et al. (2008). H_2S protects against methionine-induced oxidative stress in brain endothelial cells. *Antioxid. Redox. Signal.* 11 (1): 25–33.

5 Vitvitsky, V., Miljkovic, J.L., Bostelaar, T. et al. (2018). Cytochrome c reduction by H_2S potentiates sulfide signaling. *ACS Chem. Biol.* 13 (8): 2300–2307.

6 Mostofa, M.G., Rahman, A., Ansary, M.M.U. et al. (2015). Hydrogen sulfide modulates cadmium-induced physiological and biochemical responses to alleviate cadmium toxicity in rice. *Sci. Rep.* 5 (1): 14078.

7 Terzić, V., Padovani, D., Balland, V. et al. (2014). Electrophilic sulfhydration of 8-nitro-cGMP involves sulfane sulfur. *Org. Biomol. Chem.* 12 (29): 5360–5364.

8 Paul, B.D. and Snyder, S.H. (2012). H_2S signalling through protein sulfhydration and beyond. *Nat. Rev. Mol. Cell Biol.* 13 (8): 499–507.

9 Nishida, M., Nishimura, A., Matsunaga, T. et al. (2017). Redox regulation of electrophilic signaling by reactive persulfides in cardiac cells. *Free Radic. Biol. Med.* 109: 132–140.

10 Filipovic, M.R., Zivanovic, J., Alvarez, B., and Banerjee, R. (2018). Chemical biology of H_2S signaling through persulfidation. *Chem. Rev.* 118 (3): 1253–1337.

11 Yu, B., Yuan, Z., Yang, X., and Wang, B. (2020). Prodrugs of persulfides, sulfur dioxide, and carbon disulfide: important tools for studying sulfur signaling at various oxidation states. *Antioxid. Redox. Signal.* 33 (14): 1046–1059.

12 Yang, C., Devarie-Baez, N., Hamsath, A. et al. (2019). S-persulfidation: chemistry, chemical biology, and significance in health and disease. *Antioxid. Redox. Signal.* 33 (15): 1092–1114.

13 Galardon, E., Huguet, F., Herrero, C. et al. (2017). Reactions of persulfides with the heme cofactor of oxidized myoglobin and microperoxidase 11: reduction or coordination. *Dalton Trans.* 46 (24): 7939–7946.

14 Akiyama, M., Shinkai, Y., Unoki, T. et al. (2017). The capture of cadmium by reactive polysulfides attenuates cadmium-induced adaptive responses and hepatotoxicity. *Chem. Res. Toxicol.* 30 (12): 2209–2217.

15 Abiko, Y., Yoshida, E., Ishii, I. et al. (2015). Involvement of reactive persulfides in biological bismethylmercury sulfide formation. *Chem. Res. Toxicol.* 28 (6): 1301–1306.

16 Francoleon, N.E., Carrington, S.J., and Fukuto, J.M. (2011). The reaction of H_2S with oxidized thiols: generation of persulfides and implications to H_2S biology. *Arch. Biochem. Biophys.* 516 (2): 146–153.

17 Cuevasanta, E., Lange, M., Bonanata, J. et al. (2015). Reaction of hydrogen sulfide with disulfide and sulfenic acid to form the strongly nucleophilic persulfide. *J. Biol. Chem.* 290 (45): 26866–26880.

18 Xu, S., Wang, Y., Parent, Z., and Xian, M. (2020). Diacyl disulfides as the precursors for hydrogen persulfide (H_2S_2). *Bioorg. Med. Chem. Lett.* 30 (4): 126903.

19 Ida, T., Sawa, T., Ihara, H. et al. (2014). Reactive cysteine persulfides and S-polythiolation regulate oxidative stress and redox signaling. *Proc. Natl. Acad. Sci. U.S.A.* 111 (21): 7606–7611.

20 Yuan, Z., Zheng, Y., Yu, B. et al. (2018). Esterase-sensitive glutathione persulfide donor. *Org. Lett.* 20 (20): 6364–6367.

21 Powell, C.R., Dillon, K.M., Wang, Y. et al. (2018). A persulfide donor responsive to reactive oxygen species: insights into reactivity and therapeutic potential. *Angew. Chem. Int. Ed.* 57 (21): 6324–6328.

22 Nishida, M., Sawa, T., Kitajima, N. et al. (2012). Hydrogen sulfide anion regulates redox signaling via electrophile sulfhydration. *Nat. Chem. Biol.* 8 (8): 714–724.

23 Akaike, T., Ihara, H., and Sawa, T. (2012). 8-SH-cGMP endogenously formed from 8-nitro-cGMP as a second messenger of hydrogen sulfide. *Nitric Oxide* 27: S14–S15.

24 Kasamatsu, S., Nishimura, A., Morita, M. et al. (2016). Redox signaling regulated by cysteine persulfide and protein polysulfidation. *Molecules* 21 (12): 1721.

25 Marutani, E., Yamada, M., Ida, T. et al. (2015). Thiosulfate mediates cytoprotective effects of hydrogen sulfide against neuronal ischemia. *J. Am. Heart. Assoc.* 4 (11): e002125.

26 Du, C., Lin, X., Xu, W. et al. (2018). Sulfhydrated sirtuin-1 increasing iIts deacetylation activity is an essential epigenetics mechanism of anti-atherogenesis by hydrogen sulfide. *Antioxid. Redox. Signal.* 30 (2): 184–197.

27 Hourihan, J.M., Kenna, J.G., and Hayes, J.D. (2012). The gasotransmitter hydrogen sulfide induces Nrf2-target genes by inactivating the Keap1 ubiquitin ligase substrate adaptor through formation of a disulfide bond between Cys-226 and Cys-613. *Antioxid. Redox. Signal.* 19 (5): 465–481.

28 Zivanovic, J., Kouroussis, E., Kohl, J.B. et al. (2019). Selective persulfide detection reveals evolutionarily conserved antiaging effects of S-sulfhydration. *Cell Metab.* 30 (6): 1152–1170.

29 Fukuto, J.M., Ignarro, L.J., Nagy, P. et al. (2018). Biological hydropersulfides and related polysulfides – a new concept and perspective in redox biology. *FEBS Lett.* 592 (12): 2140–2152.

30 Yu, B., Zheng, Y., Yuan, Z. et al. (2018). Toward direct protein S-persulfidation: a prodrug approach that directly delivers hydrogen persulfide. *J. Am. Chem. Soc.* 140 (1): 30–33.

31 Dóka, É., Pader, I., Bíró, A. et al. (2016). A novel persulfide detection method reveals protein persulfide- and polysulfide-reducing functions of thioredoxin and glutathione systems. *Sci. Adv.* 2 (1): e1500968.

32 Roos, G. and Messens, J. (2011). Protein sulfenic acid formation: from cellular damage to redox regulation. *Free Radical Biol. Med.* 51 (2): 314–326.

33 Zhang, D., Macinkovic, I., Devarie-Baez, N.O. et al. (2014). Detection of protein S-sulfhydration by a tag-switch technique. *Angew. Chem. Int. Ed.* 53 (2): 575–581.

34 Kimura, Y., Mikami, Y., Osumi, K. et al. (2013). Polysulfides are possible H2S-derived signaling molecules in rat brain. *FASEB J.* 27 (6): 2451–2457.

35 Yuan, S., Pardue, S., Shen, X. et al. (2016). Hydrogen sulfide metabolism regulates endothelial solute barrier function. *Redox Biol.* 9: 157–166.

36 Wedmann, R., Onderka, C., Wei, S. et al. (2016). Improved tag-switch method reveals that thioredoxin acts as depersulfidase and controls the intracellular levels of protein persulfidation. *Chem. Sci.* 7 (5): 3414–3426.

37 Zhao, Y., Wang, H., and Xian, M. (2011). Cysteine-activated hydrogen sulfide (H2S) donors. *J. Am. Chem. Soc.* 133 (1): 15–17.

38 Zheng, Y., Yu, B., De La Cruz, L.K. et al. (2018). Toward hydrogen sulfide based therapeutics: critical drug delivery and developability issues. *Med. Res. Rev.* 38 (1): 57–100.

39 Park, C.-M., Johnson, B.A., Duan, J. et al. (2016). 9-Fluorenylmethyl (Fm) disulfides: biomimetic precursors for persulfides. *Org. Lett.* 18 (5): 904–907.

40 Pan, J. and Carroll, K.S. (2013). Persulfide reactivity in the detection of protein S-sulfhydration. *ACS Chem. Biol.* 8 (6): 1110–1116.

41 Böhme, H. and Zinner, G. (1954). Über Darstellung und Eigenschaften von Alkyl-hydro-polysulfiden. *Justus Liebigs Ann. Chem.* 585 (1): 142–149.

42 Heimer, N.E., Field, L., and Waites, J.A. (1985). Organic disulfides and related substances. 44. Preparation and characterization of 1-adamantyl hydrodisulfide as a stable prototype of the series. *J. Org. Chem.* 50 (21): 4164–4166.

43 Bailey, T.S., Zakharov, L.N., and Pluth, M.D. (2014). Understanding hydrogen sulfide storage: probing conditions for sulfide release from hydrodisulfides. *J. Am. Chem. Soc.* 136 (30): 10573–10576.

44 Bailey, T.S. and Pluth, M.D. (2015). Reactions of isolated persulfides provide insights into the interplay between H_2S and persulfide reactivity. *Free Radic. Biol. Med.* 89: 662–667.

45 Kang, J., Ferrell, A.J., Chen, W. et al. (2018). Cyclic acyl disulfides and acyl selenylsulfides as the precursors for persulfides (RSSH), selenylsulfides (RSeSH), and hydrogen sulfide (H_2S). *Org. Lett.* 20 (3): 852–855.

46 Yang, Y.-h., Aloysius, H., Inoyama, D., Chen, Y., Hu, L.-q. (2011). Enzyme-mediated hydrolytic activation of prodrugs. *Acta Pharm. Sin. B* 1 (3): 143-159.

47 Bernt, K.M., Steinwaerder, D.S., Ni, S. et al. (2002). Enzyme-activated prodrug therapy enhances tumor-specific replication of adenovirus vectors. *Cancer Res.* 62 (21): 6089.

48 Zheng, Y., Yu, B., Li, Z. et al. (2017). An esterase-sensitive prodrug approach for controllable delivery of persulfide species. *Angew. Chem. Int. Ed.* 56 (39): 11749–11753.

49 Sawahata, T. and Neal, R.A. (1982). Use of 1-fluoro-2,4-dinitrobenzene as a probe for the presence of hydrodisulfide groups in proteins. *Anal. Biochem.* 126 (2): 360–364.

50 Elrod, J.W., Calvert, J.W., Morrison, J. et al. (2007). Hydrogen sulfide attenuates myocardial ischemia-reperfusion injury by preservation of mitochondrial function. *Proc. Natl. Acad. Sci. U.S.A.* 104 (39): 15560–15565.

51 Szabo, C. (2016). Gasotransmitters in cancer: from pathophysiology to experimental therapy. *Nat. Rev. Drug Discov.* 15 (3): 185–203.

52 Dillon, K.M., Carrazzone, R.J., Wang, Y. et al. (2020). Polymeric persulfide prodrugs: mitigating oxidative stress through controlled delivery of reactive sulfur species. *ACS Macro Lett.* 9 (4): 606–612.

53 Zheng, Y., Yu, B., Ji, K. et al. (2016). Esterase-sensitive prodrugs with tunable release rates and direct generation of hydrogen sulfide. *Angew. Chem. Int. Ed.* 55 (14): 4514–4518.

54 Levine, M.N. and Raines, R.T. (2012). Trimethyl lock: a trigger for molecular release in chemistry, biology, and pharmacology. *Chem. Sci.* 3 (8): 2412–2420.

55 Checconi, P., Limongi, D., Baldelli, S. et al. (2019). Role of glutathionylation in infection and inflammation. *Nutrients* 11 (8): 1952.

56 Lermant, A. and Murdoch, C.E. (2019). Cysteine glutathionylation acts as a redox switch in endothelial cells. *Antioxidants* 8 (8): 315.

57 Mailloux, R.J. and Treberg, J.R. (2016). Protein S-glutathionlyation links energy metabolism to redox signaling in mitochondria. *Redox Biol.* 8: 110–118.

58 Jarosz, A.P., Wei, W., Gauld, J.W. et al. (2015). Glyceraldehyde 3-phosphate dehydrogenase (GAPDH) is inactivated by S-sulfuration in vitro. *Free Radical Biol. Med.* 89: 512–521.

59 Mittal, M., Siddiqui, M.R., Tran, K. et al. (2014). Reactive oxygen species in inflammation and tissue injury. *Antioxid. Redox. Signal.* 20 (7): 1126–1167.

60 Liou, G.-Y. and Storz, P. (2010). Reactive oxygen species in cancer. *Free Radical Res.* 44 (5): 479–496.

61 Bora, P., Chauhan, P., Manna, S., and Chakrapani, H. (2018). A vinyl-boronate ester-based persulfide donor controllable by hydrogen peroxide, a reactive oxygen species (ROS). *Org. Lett.* 20 (24): 7916–7920.

62 Hankins, R.A., Suarez, S.I., Kalk, M.A. et al. (2020). An innovative hydrogen peroxide-sensing scaffold and insight towards its potential as an ROS-activated persulfide donor. *Angew. Chem. Int. Ed.* 59 (49): 22238–22245.

63 Kang, J., Xu, S., Radford, M.N. et al. (2018). O→S relay deprotection: a general approach to controllable donors of reactive sulfur species. *Angew. Chem. Int. Ed.* 57 (20): 5893–5897.

64 Artaud, I. and Galardon, E. (2014). A persulfide analogue of the nitrosothiol SNAP: formation, characterization and reactivity. *ChemBioChem* 15 (16): 2361–2364.

65 Khodade, V.S. and Toscano, J.P. (2018). Development of S-substituted thioisothioureas as efficient hydropersulfide precursors. *J. Am. Chem. Soc.* 140 (50): 17333–17337.

66 Khodade, V.S., Pharoah, B.M., Paolocci, N., and Toscano, J.P. (2020). Alkylamine-substituted perthiocarbamates: dual precursors to hydropersulfide and carbonyl sulfide with cardioprotective actions. *J. Am. Chem. Soc.* 142 (9): 4309–4316.

67 Chaudhuri, A., Venkatesh, Y., Das, J. et al. (2019). One- and two-photon-activated cysteine persulfide donors for biological targeting. *J. Org. Chem.* 84 (18): 11441–11449.

68 Chaudhuri, A., Venkatesh, Y., Jena, B.C. et al. (2019). Real-time monitoring of a photoactivated hydrogen persulfide donor for biological entities. *Org. Biomol. Chem.* 17 (39): 8800–8805.

69 Zhao, Y., Bhushan, S., Yang, C. et al. (2013). Controllable hydrogen sulfide donors and their activity against myocardial ischemia-reperfusion injury. *ACS Chem. Biol.* 8 (6): 1283–1290.

70 Roger, T., Raynaud, F., Bouillaud, F. et al. (2013). New biologically active hydrogen sulfide donors. *ChemBioChem* 14 (17): 2268–2271.

71 Liang, D., Wu, H., Wong, M.W., and Huang, D. (2015). Diallyl trisulfide is a fast H2S donor, but diallyl disulfide is a slow one: the reaction pathways and intermediates of glutathione with polysulfides. *Org. Lett.* 17 (17): 4196–4199.

72 Lin, J., Akiyama, M., Bica, I. et al. (2019). The uptake and release of polysulfur cysteine species by cells: physiological and toxicological implications. *Chem. Res. Toxicol.* 32 (3): 447–455.

73 Wang, X., Huang, X., Ochs, T. et al. (2011). Effect of sulfur dioxide preconditioning on rat myocardial ischemia/reperfusion injury by inducing endoplasmic reticulum stress. *Basic. Res. Cardiol.* 106 (5): 865–878.

74 Liang, W., Chen, J., Li, L. et al. (2019). Conductive hydrogen sulfide-releasing hydrogel encapsulating ADSCs for myocardial infarction treatment. *ACS Appl. Mater. Interfaces* 11 (16): 14619–14629.

75 Distrutti, E., Sediari, L., Mencarelli, A. et al. (2006). 5-Amino-2-hydroxybenzoic acid 4-(5-thioxo-5-H-[1,2]dithiol-3yl)-phenyl ester (ATB-429), a hydrogen sulfide-releasing derivative of mesalamine, exerts antinociceptive effects in a model of postinflammatory hypersensitivity. *J. Pharmacol. Exp. Ther.* 319 (1): 447–458.

76 Whiteman, M., Li, L., Rose, P. et al. (2009). The effect of hydrogen sulfide donors on lipopolysaccharide-induced formation of inflammatory mediators in macrophages. *Antioxid. Redox. Signal.* 12 (10): 1147–1154.

77 Wallace, J.L. (2007). Hydrogen sulfide-releasing anti-inflammatory drugs. *Trends Pharmacol. Sci.* 28 (10): 501–505.

78 Hirata, I., Naito, Y., Takagi, T. et al. (2011). Endogenous hydrogen sulfide is an anti-inflammatory molecule in dextran sodium sulfate-induced colitis in mice. *Dig. Dis. Sci.* 56 (5): 1379–1386.

79 Szabo, C. and Papapetropoulos, A. (2017). International Union of Basic and Clinical Pharmacology. CII: Pharmacological modulation of H_2S levels: H_2S donors and H_2S biosynthesis inhibitors. *Pharmacol. Rev.* 69 (4): 497–564.

80 Wallace, J.L., Vaughan, D., Dicay, M. et al. (2017). Hydrogen sulfide-releasing therapeutics: translation to the clinic. *Antioxid. Redox. Signal.* 28 (16): 1533–1540.

81 Martinez, A.M., Sordia-Hernández, L.H., Morales, J.A. et al. (2015). A randomized clinical study assessing the effects of the antioxidants, resveratrol or SG1002, a hydrogen sulfide prodrug, on idiopathic oligoasthenozoospermia. *Asian Pac. J. Reprod.* 4 (2): 106–111.

82 Polhemus, D.J., Li, Z., Pattillo, C.B. et al. (2015). A novel hydrogen sulfide prodrug, SG1002, promotes hydrogen sulfide and nitric oxide bioavailability in heart failure patients. *Cardiovasc. Ther.* 33 (4): 216–226.

83 Ono, K., Akaike, T., Sawa, T. et al. (2014). Redox chemistry and chemical biology of H_2S, hydropersulfides, and derived species: Implications of their possible biological activity and utility. *Free Radic. Biol. Med.* 77: 82–94.
84 Wang, X.-B., Jin, H.-F., Tang, C.-S., and Du, J.-B. (2011). The biological effect of endogenous sulfur dioxide in the cardiovascular system. *Eur. J. Biochem.* 670 (1): 1–6.
85 Wood, J.L. (1987). Sulfane sulfur. In: *Methods in Enzymology*, vol. 143 (ed. W.B. Jakoby and O.W. Griffith), 25–29. Cambridge, Massachusetts: Academic Press.
86 Greiner, R. and Dick, T.P. (2015). Real-time assays for monitoring the influence of sulfide and sulfane sulfur species on protein thiol redox states. *Methods Enzymol.* 555: 57–77.
87 Yu, B., Kang, T., Xu, Y., Liu, Y., Ma, Y., and Ke, B. (2022). Prodrugs of persulfide and sulfide: is there a pharmacological difference between the two in the context of rapid exchanges among various sulfur species in vivo? *Angew. Chem. Int. Ed.* 61 (20): e202201668.

13

COS-Based H_2S Donors

Annie K. Gilbert and Michael D. Pluth

Department of Chemistry and Biochemistry, Institute of Molecular Biology, Knight Campus for Accelerating Scientific Impact, Materials Science Institute, University of Oregon, Eugene, Oregon 97403, USA

13.1 Introduction

Small molecule hydrogen sulfide (H_2S) donors have been developed to provide controlled-releasing sources of H_2S that contrast the large bolus release of inorganic sulfide salts. With slower sustained release, these H_2S donors can better mimic endogenous H_2S production. Moreover, the slower release profiles can allow for diffusion into areas of therapeutic need (i.e. oxidative environments and areas of inflammation). Despite this progress, scaffolds that produce H_2S directly often have limited tunability and generally lack widely used control compounds to delineate effects of H_2S from resulting byproducts. As a strategy to overcome these limitations, our group as well as other researchers have developed modular and tunable H_2S donors that function by the intermediate release of precursors that are quickly converted to H_2S. Inspired by the enzymatic conversion of carbonyl sulfide (COS) to H_2S, we reasoned that COS could serve as a vehicle for H_2S delivery [1]. No longer tied to releasing H_2S directly, COS-based H_2S donors provide access to established prodrug platforms and chemistry that has been known for years but not recognized as a potential platform for H_2S delivery. This general approach has provided COS/H_2S donors that can be activated by biological stimuli, deliver fluorophores and prodrugs alongside COS/H_2S, and can incorporate targeting groups to further localize H_2S delivery [2, 3]. In this chapter, we highlight the basic properties of COS and summarize current COS-based H_2S delivery strategies.

13.2 Properties of COS

COS was initially misidentified as a mixture of CO_2 and H_2S, both of which are hydrolysis byproducts of COS and thus common impurities found in COS gas [4]. On its own, COS is a colorless and odorless gas that was discovered by Than in 1867 when mixing carbon monoxide with excess sulfur in a glowing porcelain tube [5]. This initial method to prepare COS contained significant carbon monoxide (CO) impurities. A more reliable synthesis of COS relies on the acid-mediated hydrolysis of thiocyanate salts and subsequent purification through a series of washing steps [6].

COS is the most abundant sulfur-containing gas in the atmosphere (~0.5 ppb) and is emitted by a variety of abiotic, biotic, and anthropogenic sources including volcanoes, oceans, microorganisms, and coal [6]. Beyond its identity as a major atmospheric gas in the global sulfur cycle, prior work suggests that COS could be an important biological molecule and a potential gasotransmitter [1]. For example, exposure of α-amino acids to COS under prebiotic conditions generates peptides, implicating the possible role of COS in origin of life peptide ligation (Figure 13.1) [7, 8]. COS has also been detected in various biological tissues and in the exhaled breath of humans, further supporting its significance in biology, although enzymatic pathways for mammalian COS production are yet to be identified [9–11].

COS is converted to H_2S by a variety of enzymes with varying efficiency. These enzymes include CS_2 hydrolase, nitrogenase, CO dehydrogenase, ribulose-

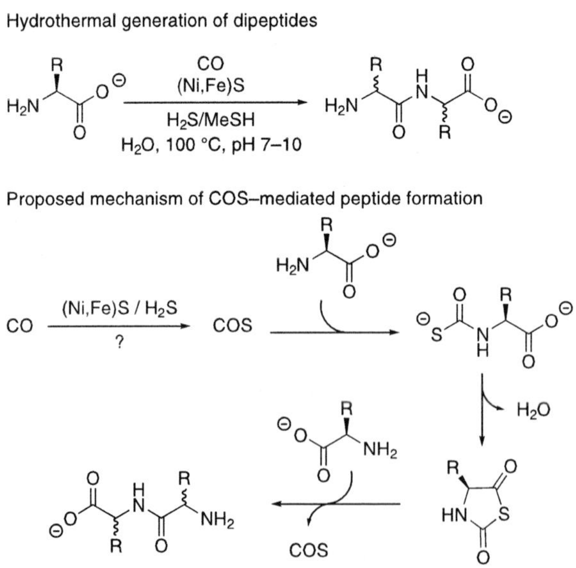

Figure 13.1 Hydrothermal generation of dipeptides in the presence of CO, (Ni,Fe)S, and H_2S/MeSH. COS generation serves as a key intermediate. Source: Based on Lemon et al. [7] and Huber and Wächtershäuser [8].

1,5-bisphosphate carboxylase (Rubisco), and carbonic anhydrase (CA) [12–16]. Although CS_2 hydrolase has the greatest catalytic efficiency for COS hydrolysis (8.2×10^7 M^{-1} s^{-1}), this enzyme is isolated from acidophilic and thermophilic archaea extremophiles and is not likely to contribute significantly to global COS hydrolysis nor COS conversion to H_2S in common terrestrial organisms [12]. CA, on the other hand, has a relatively high catalytic efficiency of COS hydrolysis (2.2×10^4 M^{-1} s^{-1}) and is widely distributed throughout cells [13]. The primary function of CA is to regulate biological pH by hydrolyzing CO_2 (k_{cat}/K_M = ~8×10^7 M^{-1} s^{-1}) [17]; however, the widespread localization of CA paired with its relatively high catalytic efficiency suggests it could be a significant contributor to endogenous COS consumption. Although these two factors are convenient for COS delivery in biological milieu to produce H_2S, they complicate direct biological investigations into the role of COS independent to that of H_2S. For example, in toxicological investigations of COS on rats, Chengelis and Neal reported pretreatment of rats with acetazolamide (AAA), a known CA inhibitor, resulted in reduced blood levels of H_2S and subsequently reduced the toxicity of COS, suggesting at least some of the toxicity observed from COS comes from CA-mediated H_2S generation [18].

13.3 COS-Based H_2S Delivery

The widespread localization of CA paired with its relatively high catalytic efficiently toward COS hydrolysis motivated the use of COS in H_2S donor development. As an alternative to producing H_2S directly, COS delivery can leverage already known biocompatible decarboxylation reactions, many of which can be activated by or tuned to specific analytes or environments. For example, self-immolative carbamates rely on cascade decarboxylation reactions and have been used previously as highly tunable methods for the controlled release of small molecules. These prodrugs are activated in the presence of specific analytes or environments to release CO_2, a quinone methide, and the desired payload. By simply replacing one oxygen atom with a sulfur allows for these decarboxylation reactions to be tuned to release COS instead of CO_2 and serve as COS-based H_2S donors (Figure 13.2).

The first application of the self-immolative thiocarbamate system described in Figure 13.2 was used to address the challenge of analyte consumption in activity based H_2S detection strategies. A common approach to develop activity-based H_2S-selective fluorescent probes is the use of H_2S-mediated azide reduction to produce a fluorescent amine. This reaction, however, consumes H_2S and could perturb cellular homeostasis or impact biological outcomes. To overcome these limitations, we developed an analyte replacement probe for H_2S that

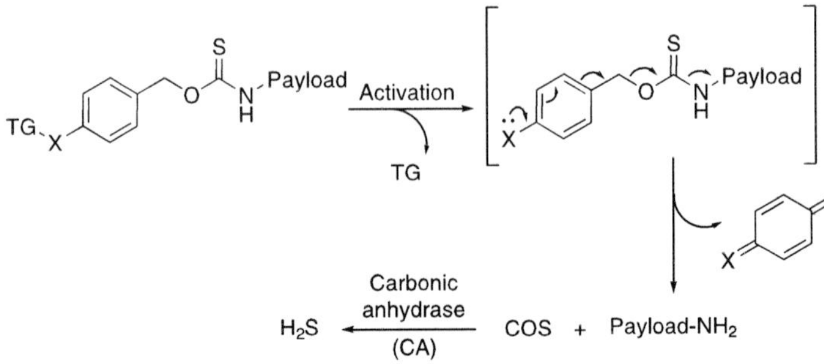

Figure 13.2 Self-immolation cascade reaction of caged thiocarbamates. Deprotection of a triggering group (TG) initiates decomposition to release COS which is rapidly hydrolyzed in the presence of carbonic anhydrase (CA) to produce H_2S.

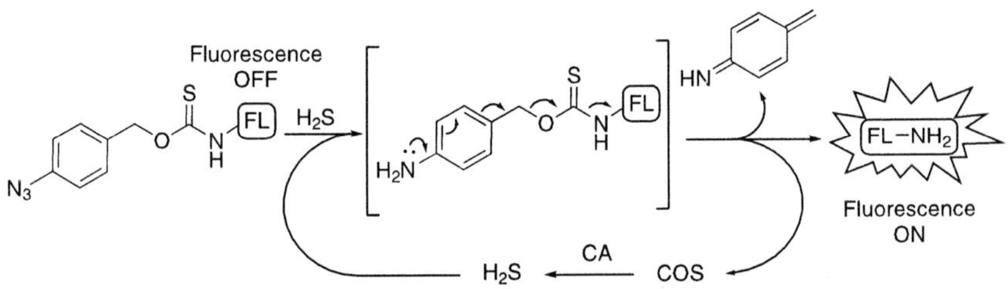

Figure 13.3 Analyte-replacement H_2S probe mechanism of COS/H_2S release with concomitant turn-on fluorescence. Source: Based on Steiger et al. [19].

yields both a fluorescent turn-on response and COS/H_2S release to replace the consumed H_2S for probe activation (Figure 13.3) [19]. A model thiocarbamate containing an azide triggering group was used to confirm biocompatibility by demonstrating H_2S production in whole mouse blood upon azide reduction using *tris*(2-carboxyethyl)phosphine (TCEP). Addition of AAA to inhibit CA resulted in decreased H_2S production, which confirmed that CA is required to convert the released COS to H_2S. Having confirmed the H_2S donating capabilities of the thiocarbamate scaffold, the H_2S-selective analyte replacement probe (MeRho-TCA) was prepared by functionalizing a quenched methyl rhodol fluorophore with a *p*-azidobenzylthiocarbamate. Reduction of the azide triggering group by H_2S initiates decomposition to release COS/H_2S and the methyl rhodol payload to generate a 65-fold fluorescent turn-on response with excellent selectivity over other reactive sulfur, oxygen, and nitrogen species. One limitation to this approach is that the fluorescence response is not directly proportional to the initial H_2S concentration because the probe releases COS/H_2S upon reaction with H_2S. Furthermore, complete azide reduction requires two equivalents of H_2S, which suggests that only one-half of the consumed H_2S is replaced. Despite these drawbacks, this work provided an important foundation for further expansion of COS-based H_2S donors. More recently, this azide-reduction triggering approach has been expanded to include polymeric donors [20]. Since this initial report of COS-based H_2S delivery, the available library of COS-based H_2S donors has expanded significantly to include donors activated by a wide variety of stimuli and conditions.

13.3.1 Stimuli Responsive COS/H_2S Donors

The first application of caged thiocarbamates as stimulus-responsive COS/H_2S donors used a boronate ester trigger (Peroxy-TCM-1) to develop donors that were responsive to reactive oxygen species ROS [21]. In the presence of ROS, the boronate ester triggering group is converted to a phenol, which initiates self-immolative decomposition to produce COS/H_2S (Figure 13.4). Using ROS as a stimulus for H_2S delivery not only localizes COS/H_2S where it can protect against ROS but also combines the ROS scavenging abilities of boronate esters with cytoprotective effects of H_2S. Using the ROS-activated Peroxy-TCM-1, H_2O_2-dependent release of COS/H_2S was demonstrated in the presence of CA using an H_2S selective electrode. No COS/H_2S release was observed with the

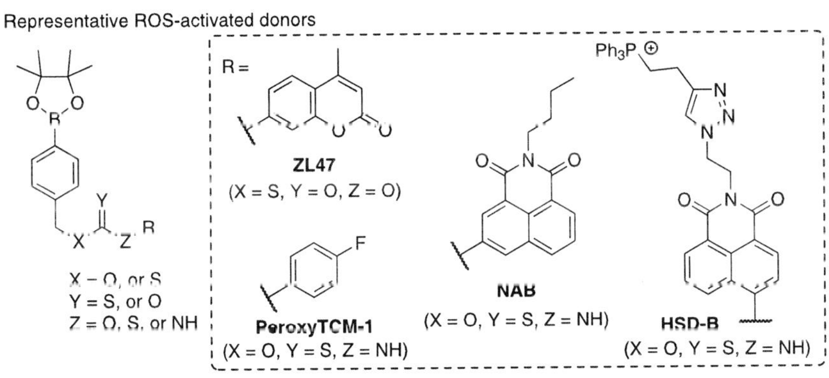

Figure 13.4 Mechanism of ROS-activation of aryl boronates and representative ROS-activated COS/H_2S donors.

carbamate and triggerless control compounds confirming that the thiocarbamate and boronate trigger are critical for COS/H_2S release. Furthermore, COS/H_2S release was confirmed in live cell environments. Incubation of HeLa cells with a fluorescent H_2S probe (HSN2) and Peroxy-TCM-1 resulted in a turn-on response when treated with exogenous H_2O_2. An increase in fluorescence was also observed when Raw 264.7 cells were treated with phorbol myristate acetate (PMA) to increase endogenous ROS levels in the presence of Peroxy-TCM-1 and the H_2S-responsive fluorescent probe HSN2. Furthermore, the pretreatment of HeLa cells with the ROS-activated donor provided significant cytoprotective effects against exogenous H_2O_2.

The ROS-activated donor scaffold has served as useful platform for studying the self-immolative COS/H_2S donor system. Since the initial report, a variety of isomeric cores (*O*-alkyl and *S*-alkyl thiocarbamates/thiocarbonates) and a variety of payloads (electron donating and electron withdrawing) have been incorporated with the ROS-activated scaffold to study release kinetics [22, 23]. Overall, the second order rate constant (k_2) for COS/H_2S delivery from *O*-alkyl thiocarbamates is over an order of magnitude greater than *S*-alkyl thiocarbamates ($k_2 = 1.16\,M^{-1}\,s^{-1}$ compared to $0.16\,M^{-1}\,s^{-1}$). This difference was attributed to the greater ground state stability of the *S*-alkyl isomer due to enhanced resonance stabilization in the amide moiety of the thiocarbamate. Incorporating electron withdrawing groups (EWG) were hypothesized to accelerate COS/H_2S release and electron donating groups (EDG) were predicted to slow COS/H_2S release. This trend, however, was not observed uniformly across *O*-alkyl donors suggesting the acidification of the thiocarbamate NH proton could be interfering with COS release. For *S*-alkyl thiocarbamates, Chakrapani and coworkers reported the rate of COS/H_2S release could be altered predictably by modulating the pK_a of the thiocarbamate NH. Aniline-based derivatives with lower pK_a values (4.75–5.43) resulted in faster H_2S release compared to derivatives with higher pK_a values (8.36–10.53). This work also contributed a novel ROS-activated H_2S-releasing nonsteroidal anti-inflammatory drug (NSAID) based on mesalamine, which could have additional potential applications for H_2S delivery. The general approach of using an ROS-trigger to deliver COS/H_2S has also been used for the development of fluorescent COS/H_2S donors [24–26], which are covered in more detail in Section 13.3.6.

13.3.2 Bio-orthogonal Donor Activation

Bio-orthogonal activation strategies have also been used to develop COS/H_2S donors. For example, the click-and-release COS/H_2S donor (TCO-1) was developed that uses an inverse-electron demand Diels-Alder (IEDDA) reaction between a thiocarbamate-functionalized *trans*-cyclooctene and tetrazine [27]. The IEDDA

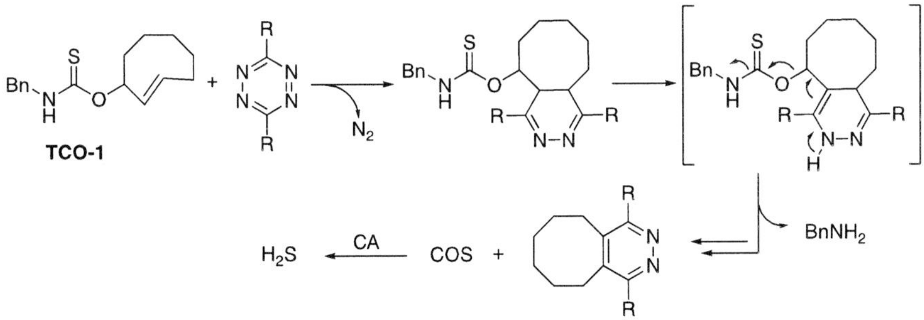

Figure 13.5 Click-and-release mechanism of TCO-1 to produce COS/H_2S.

reaction between these coupling partners produces a dihydropyridazine that spontaneously tautomerizes then deprotonates and rearomatizes to release COS/H_2S (Figure 13.5). This bio-orthogonal approach remained viable in both diluted sheep and bovine blood without the need for exogenous CA addition, which confirms that endogenous CA levels are sufficient for COS conversion to H_2S. Although this work demonstrated preliminary biological compatibility, further investigations are needed to improve upon the efficiency and rate of COS release in more complex environments.

Expanding the COS/H_2S donors that are activated by bio-orthogonal stimuli, a number of groups have reported photoactivated COS/H_2S donors. In the initial report, our group demonstrated the compatibility of using a photocleavable *o*-nitrobenzyl group to develop COS/H_2S donors [28]. Irradiation with UV light ($\lambda = 365$ nm) results in *o*-nitrobenzyl group cleavage by a Norrish type II mechanism to produce COS and an aniline byproduct (Figure 13.6). Consistent with previous findings, the photocleavage and rate of COS/H_2S release increases when electron-donating substituents are added to the *o*-nitrobenzyl group. Gou and coworkers provided an additional photolabile COS/H_2S donor using the *o*-nitrobenzyl photolabile group and a napthalimide payload (Nap-Sul-ONB) to enable real-time tracking of COS release, which is covered in more detail in Section 13.3.6 [29]. Although these *o*-nitrobenzyl containing photoactivatable COS/H_2S donors demonstrate an on-demand approach to H_2S release, the requirement of UV light for activation significantly reduces the biocompatibility.

Expanding the palette of light-activated COS/H_2S donors, Chakrapani and coworkers developed the visible light activated BODIPY COS/H_2S donor BODIPY-TCM (Figure 13.7) [30]. The BODIPY-based photolabile group is cleaved by 470 nm light via a photoinduced electron transfer process, which then initiates COS release through a self-immolative decomposition. To evaluate the cell viability during photocleavage of BODIPY-TCM, HeLa cells were incubated with BODIPY-TCM and exposed to 470 nm light for 5 minutes. No cytotoxicity

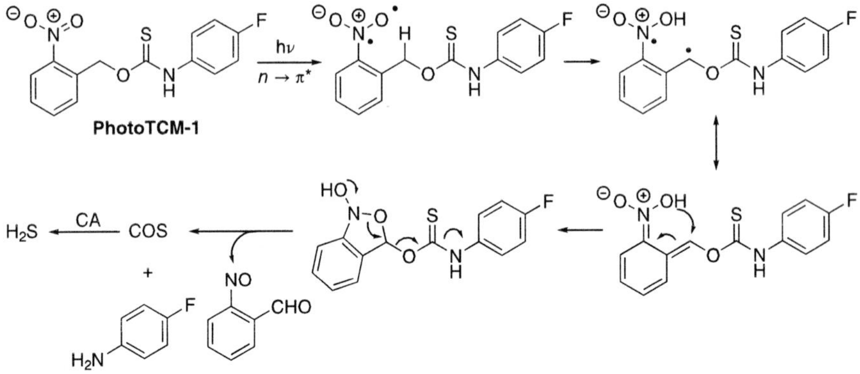

Figure 13.6 Light-activated COS/H_2S release of PhotoTCM-1.

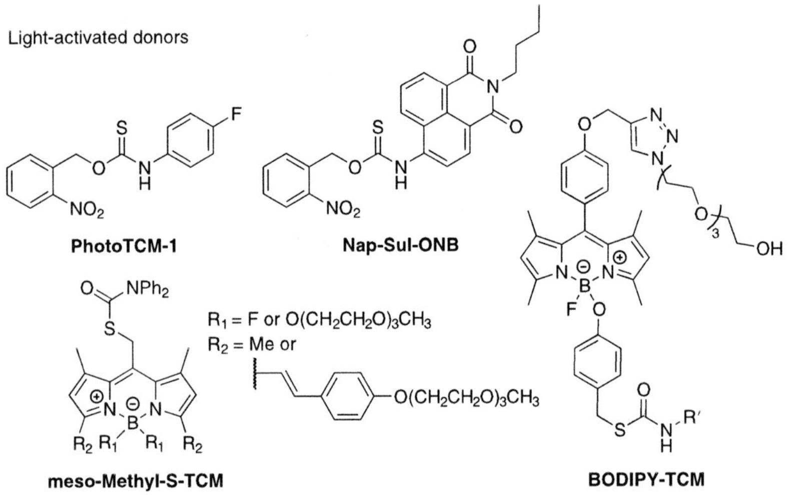

Figure 13.7 Representative light-activated COS-based H_2S donors.

was observed demonstrating this compound and the light exposure is not cytotoxic and can be applied to cell-based applications. To expand photoactivatable COS/H_2S donors to longer-wavelength platforms, Klán and coworkers reported a NIR-activated donor compatible with the tissue-transparent region (600–900 nm) with altering the substitution of the thiocarbamate in the BODIPY system to the *meso*-position (meso-Methyl-S-TCM) [31]. To increase solubility, the fluorine atoms of the BF_2 motif were substituted with polyethylene glycol (PEG) chains. Further installment of a PEG-substituted styryl units to the BODIPY core shifted the absorption to the NIR region due to the extended π-conjugation. Irradiation of the resulting NIR-activated donor with 700 nm light in HepG2 cells treated with an H_2S-selective fluorescent probe results in a significant increase in fluorescence intensity compared to the control. Collectively, this donor demonstrates potential for on-demand delivery of H_2S to deep-tissue.

13.3.3 Donors Activated by Nucleophiles

N-Thiocarboxyanhydrides (NTAs) are one of the first classes of COS-based H_2S donors and the first COS donor platform to avoid generation of electrophilic byproducts [32]. Matson and coworkers demonstrated that this class of donors is ring-opened upon nucleophilic attack of biological amines to release COS (Figure 13.8). H_2S production from NTAs (10 μM) was confirmed in PBS buffer in the presence of glycine (1 mM) and CA (300 nM) using an H_2S-selective electrode. In the absence of CA, no H_2S generation was observed, confirming the necessity of CA in generating H_2S from COS. In subsequent work aimed at providing potential COS/H_2S releasing materials for localized H_2S delivery, the NTA donor platform was incorporated in a norbornene-based monomer and was successfully polymerized via ring-opening metathesis. The water soluble NTA functionalized co-polymer (polyNTA1, Figure 13.8) showed a rapid initial release followed by a plateau after 2 hours with sustained release for 20 hours. This extended release could serve beneficial for therapeutic delivery of H_2S in biological applications. Expanding the available COS/H_2S donating materials, the NTA scaffold was coupled with a series of functional groups used in bioconjugation and polymer chemistry including azides, alkynes, alkenes, and norbornenes (Figure 13.8) [33]. The functionalized NTAs were successfully conjugated to azides and alkynes, alkenes, and thiols using copper-catalyzed azide-alkyne

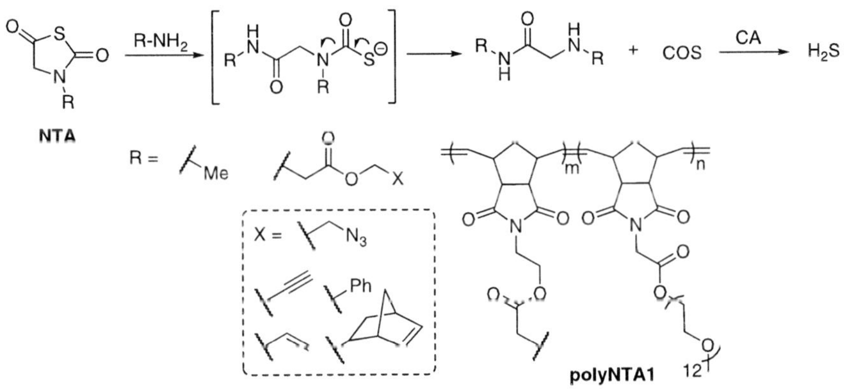

Figure 13.8 Example NTA-based COS/H_2S donors and the associated mechanism of release. Source: Based on Powell et al. [33].

cycloaddition (CuAAC), olefin cross metathesis, and radical thiol-ene reactions. Polymer-NTA conjugates showed some variation of H_2S release rates with the NTA functionalized poly(acryloyl morpholine) polymer peaking in H_2S release at 20 minutes, whereas the NTA PEG polymer peaks in H_2S concentration around 2 hours. Despite this variation, both polymeric donors show a gradual return to baseline H_2S concentrations, suggesting the polymer conjugates provide sustained release of H_2S. Collectively, this work highlights the versatility of the NTA scaffold to be incorporated in a variety of materials that provide sustained and localized H_2S release.

Adding to the amine-based nucleophile activation of NTAs, a number of different strategies have been developed that rely on thiol activation for COS/H_2S release. For example, our lab reported the cysteine-selective self-immolative COS/H_2S donor (OA-CysTCM) that uses the previously established cyclization between cysteine and acrylates commonly used to develop cysteine-selective probes [34, 35]. With an acrylate group in the trigger position, nucleophilic attack by cysteine followed by a cyclization with the pendant amine uncages COS from OA-CysTCM (Figure 13.9), which also imparts selectivity for cysteine over other biological thiols. The COS/H_2S release was further confirmed in biological environments with bEnd.3 cells and an H_2S-selective fluorescent probe.

Cyclic sulfenyl thiocarbamates (sulfenylTCMs) have also been demonstrated to release COS/H_2S upon reaction with thiols [36]. We demonstrated that sulfenylTCMs are stable in aqueous solution until activated by biological thiols to produce COS and an iminodisulfide intermediate (Figure 13.10). Further reaction of the iminodisulfide with thiols produces either a thioamide and H_2S or a thioamide and a persulfide which can further react with thiols to produce H_2S. Thioamides are one of the simplest direct-releasing H_2S donor motifs and have been coupled with pharmaceuticals like NSAIDs to help reduce GI damage [37]. Despite their accessibility, the mechanism of H_2S release from thioamides is not fully established and generally results in low H_2S releasing efficiencies

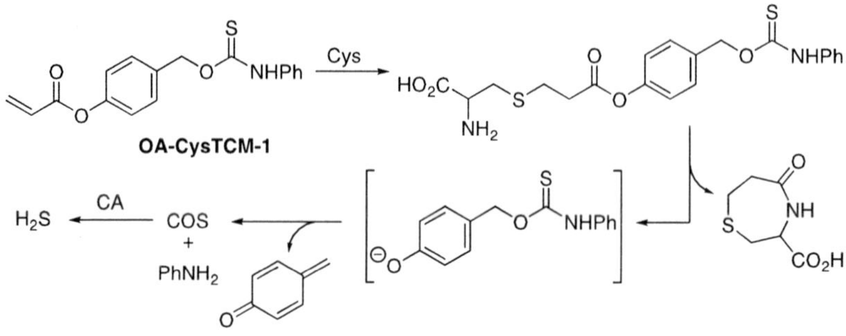

Figure 13.9 Cysteine-selective activation of OA-CysTCM-1.

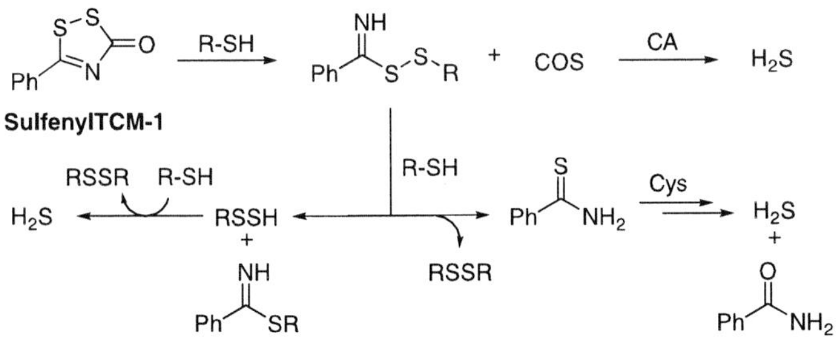

Figure 13.10 Proposed COS/H_2S release pathways from SulfenylTCM donors.

(1–2%). Because sulfenylTCMs generate a thioamide upon reaction with thiols, we developed the donor YZ-597 that release both COS/H_2S and the previously developed H_2S-releasing thiobenzamide naproxen derivative (ATB-346) upon reaction with thiols. Compared to ATB-346, YZ-597 demonstrated significantly more efficient H_2S release (90%), whereas ATB-346 only produced baseline levels. This enhanced efficiency was also demonstrated in live cell imaging with a strong fluorescent signal resulting from treatment of HeLa cells with YZ-597 and the H_2S-selective probe (SF7-AM).

Similar to the sulfenylTCM and NTA scaffolds, the cyclic dithiasuccinoyl (DTS) group can also react with nucleophiles, in this case thiols, to release COS/H_2S without producing electrophilic byproducts [38]. DTS groups have been previously used as protecting groups for amines in peptide synthesis [39] and its application to COS/H_2S delivery was reported independently by our group and Ji and coworkers [40]. This class of donors was hypothesized to have increased H_2S-releasing efficiency compared to other cyclic donor motifs because it reacts with thiols to uncage two equivalents of COS with an amine byproduct (Figure 13.11). As expected, the aryl DTS donors containing EWGs released 2 equivalents of H_2S in the presence of cysteine (20 equiv) and CA (25 μg/mL). Conversely, alkyl DTS donors resulted in significantly slower rates of H_2S release. We postulated that inductive donation of the alkyl amines likely stabilizes the thiocarbamic intermediate to decrease COS/H_2S-releasing kinetics.

Expanded mechanistic work by Ji and coworkers provided further insights into the COS and H_2S release from DTS-containing donors [40]. Notably, it was found that phenyl DTS, as well as other electron-withdrawing donors, produced a considerable amount of H_2S in the absence of CA, which indicated an alternative pathway for direct H_2S release (Figure 13.11). To investigate the role of electronic contributions to COS/H_2S or direct H_2S release, a series of differently substituted DTS based donors were developed which led to mechanistic bifurcation shown in Path A and Path B in Figure 13.11. For DTS motifs with EWGs, the carbonyl carbon of the perthiocarbamate intermediate becomes sufficiently

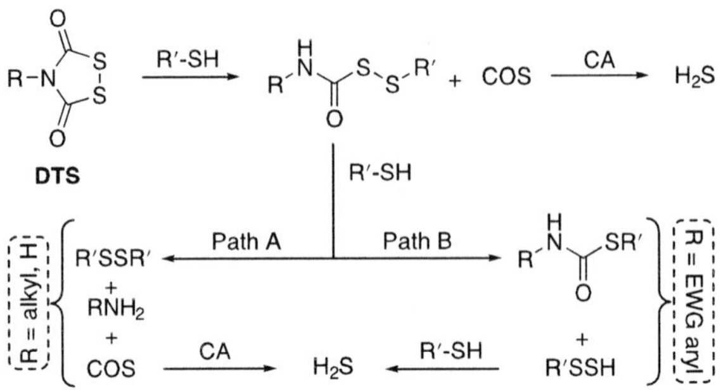

Figure 13.11 DTS mechanism of COS/H_2S release with alternative direct H_2S release pathway for alkyl groups.

electrophilic such that it can be attacked by a thiol to generate a subsequent persulfide intermediate, which can further react with thiols to produce H_2S (Path B in Figure 13.11). An additional byproduct is observed by LC-MS for electron withdrawing DTS donors which matches the thiocarbamate formed in Path B, supporting the direct-releasing H_2S mechanism. For alkyl or electron donating donors, attack by thiol occurs primarily at the disulfide motif in the perthiocarbamate intermediate, rather than at the carbonyl carbon, which results in COS release exclusively (Path A in Figure 13.11). This exclusive COS release from alkyl or electron donating donors is confirmed from H_2S-release experiments in the absence of CA resulting in no H_2S production. In addition to providing mechanistic insights to the DTS scaffold, Ji and coworkers also investigated the biocompatibility of the DTS motif. BV2 microglia cells treated with the H_2S-responsive probe HSip-1 and either phenyl or methyl DTS resulted in a significant fluorescent enhancement, even at low 2.5 μM concentrations of the donors. Furthermore, 1 and 2.5 μM levels of phenyl DTS provided dose-dependent anti-inflammatory effects, as indicated by the suppression of LPS-induced TNF-α production. These results not only confirm the biocompatibility of the DTS motif, but additionally reveal significant efficiency in delivering H_2S in biological milieu.

Although 1,2,4-thiadiazolidin-3,5-diones have been previously reported as cysteine-activated direct H_2S releasing donors, the reported production of H_2S (<6.5% efficiency) compared to the observed biological activity (vasodilatory effects observed as low as 10 nM) suggests that other reactive sulfur species or release mechanisms could be at play [41]. Initially, the proposed mechanism of H_2S production was thiol-initiated ring opening of the thiadiazolidindione followed by a subsequent thiol addition into the resulting disulfide and hydrolysis of the thiocarbamate (Figure 13.12). An alternative mechanism is that the thiocarbamate intermediate could dethiocarboxylate to produce COS, which would not be detected by amperometric measurements in the absence of CA but could result in H_2S production within biological environments. This proposed pathway

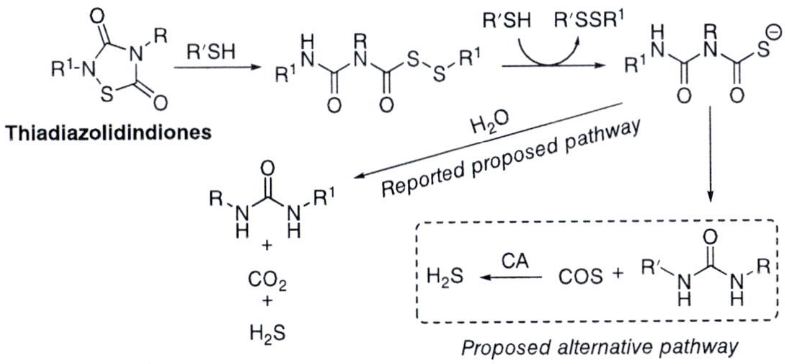

Figure 13.12 Proposed alternative pathway for COS-based H_2S release from 1,2,4-thiadiazolidin-3,5-diones.

would also be consistent with previously established chemistry of thiocarbamate intermediates in COS-based donors and could explain the low H_2S levels initially reported.

Acyclic perthiocarbamate derivatives are an additional class of thiol-activated COS/H_2S donors. Toscano and coworkers demonstrated *N*-alkyl perthiocarbamates function as tunable platforms for persulfide and/or COS release [42]. Deprotonation of the ammonium produces an amine that cyclizes to release a persulfide and a urea byproduct in the absence of thiols (Figure 13.13). Varying the substituents on the perthiocarbamates altered the half-lives of persulfide release (1.4–484 minutes) which were measured using an electrophilic trap to quantify the persulfide. Longer-lived persulfide precursors also resulted in COS

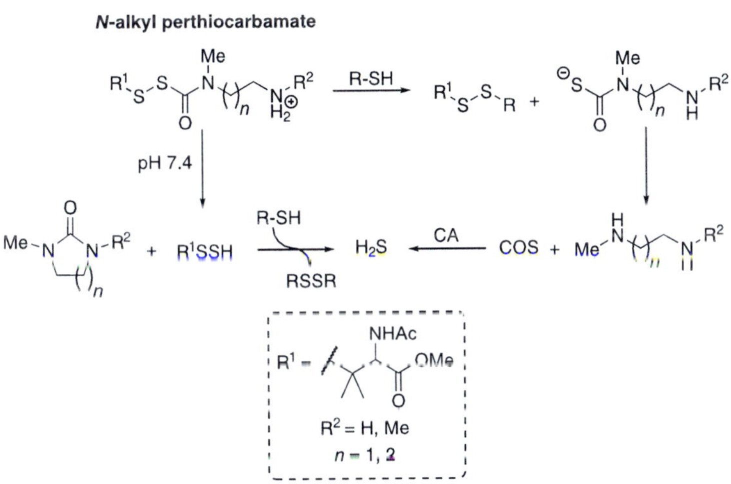

Figure 13.13 Persulfide and COS releasing pathways of *N*-alkyl perthiocarbamates.

release generated through thiol attack of the perthiocarbamate to generate a thiocarbamate intermediate followed by dethiocarboxylation. This COS release pathway, however, is minor with persulfide releasing efficiency ranging from 82–94%. The donor platform demonstrated biocompatibility and cytoprotective effects upon pretreatment of H9c2 cells with the perthiocarbamates prior to H_2O_2 exposure. Additionally, the donors provided cardioprotective effects in myocardial ischemic/reperfusion injury model which highlights the beneficial effects of combined persulfide and COS/H_2S delivery.

13.3.4 Enzyme-Activated Donors

Enzyme-mediated bond cleavage is a widely used approach in prodrug activation. For example, enzyme-activated prodrug therapy has been used as an effective cancer treatment to localize non-toxic prodrugs to tumors where enzymes within the tumor microenvironment can subsequently convert the prodrugs into highly toxic drugs [43]. In addition to being a useful delivery strategy, this approach also uses physiologically relevant stimuli while avoiding the consumption of biological nucleophiles that could perturb cellular homeostasis. Enzyme-activated H_2S delivery was first reported by Wang and coworkers in 2016 by developing H_2S donors that were initiated by esterase-mediated cleavage [44]. The following year, both Chakrapini and coworkers [45] and our group [46] independently reported esterase-activated COS-based H_2S donors using the self-immolative thiocarbamate platform. Esterase is expressed in most tissue types allowing for activation throughout various cells and organs. Another benefit of esterase-activated donors is that the rate of COS/H_2S release can be tuned by modifying the ester trigger group.

Chakrapani and coworkers developed a series of *S*-alkyl thiocarbamates and *S*-alkyl thiocarbonates with a *t*-butyl ester linked to the scaffold via an extended methylenedioxy linker (Figure 13.14a) [45]. In this work, they provide a library of aryl and benzyl payloads with various substituents and measure H_2S production using danzyl azide, an H_2S-sensitive fluorogenic dye. The rate of COS/H_2S release in the presence of porcine liver esterase (PLE, 1 U/mL) and CA was similar in magnitude across the different payloads (R = 4-NO_2Ph, Ph or CH_2Ph) for thiocarbonate cores ($k_{obs} = 6.5 - 7.0 \times 10^{-4}$ s^{-1}). The thiocarbamate cores with aryl payloads (R = 4-NO_2Ph and Ph) had comparable rates to the thiocarbonates, however, the thiocarbamate core with a benzylic payload demonstrated a slower rate of H_2S release compared to the other compounds. HPLC analysis revealed a longer lived intermediate, potentially the thiocarbamate ion, suggesting the *S*-alkyl thiocarbamate scaffold could modulate H_2S release through stereoelectronic effects on the NH. Moreover, the esterase-activated donors showed H_2S production in MCF-7 human cancer cells using an H_2S-selective fluorogenic probe (NBD-Fluorescein)

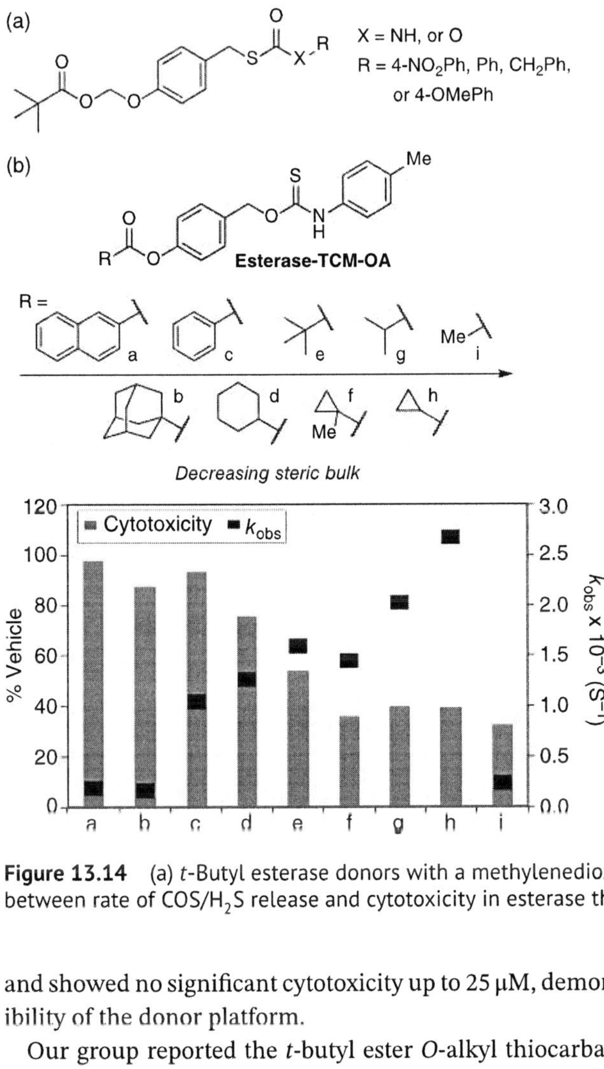

Figure 13.14 (a) *t*-Butyl esterase donors with a methylenedioxy linker. (b) Correlation between rate of COS/H_2S release and cytotoxicity in esterase thiocarbamates.

and showed no significant cytotoxicity up to 25 μM, demonstrating the biocompatibility of the donor platform.

Our group reported the *t*-butyl ester *O*-alkyl thiocarbamate Esterase-TCM-OA which released COS/H_2S in the presence of PLE and CA ($k_{obs} = 1.6 \times 10^{-3}\ s^{-1}$) [46]. Interestingly, treatment of human lung epithelial (BEAS2B) cells with low concentrations of this donor (10 μM) resulted in nearly complete cell death. Furthermore, the same concentrations of the carbamate and triggerless control compounds or Na_2S showed no cytotoxicity. We hypothesized this unexpected toxicity could be from a buildup of COS caused by the rapid release of COS from the donor. Because small ester hydrolysis is likely faster than COS hydrolysis (up to $5.8 \times 10^5\ M^{-1}\ s^{-1}$ compared to $2.2 \times 10^4\ M^{-1}\ s^{-1}$ for bovine CA II), a buildup of COS could be causing direct toxic effects. In a follow-up study to investigate this hypothesis, we prepared a library of donors with varying sizes of ester groups to alter the rate of COS release [47]. We found that the rate of release directly correlated with cytotoxicity in HeLa cells (Figure 13.14b). Although it is still unclear whether the cytotoxicity of these compounds was from direct COS effects or localization of COS/H_2S, this work demonstrated slower releasing esterase-activated donors were nontoxic and could be used to deliver COS/H_2S to live cells.

Expanding away from ester cleavage, Chakrapani and coworkers also developed a COS/H_2S donor activated by NAD(P)H quinoneoxidoreductase 1 (NQO1) [48]. NQO1 is a two-electron reducing cytosolic enzyme that reduces quinones to hydroquinones and is predominantly expressed under oxidative stress. The increased expression of NQO1 in oxidative environments makes it an ideal target for COS/H_2S delivery because H_2S should provide protective effects in such environments. The NQO1-activated donors contain a quinone triggering group linked to a thiocarbamate scaffold via an ethyl ester. Reduction of the quinone produces a hydroxyquinone that cyclizes and deprotects the caged thiocarbamate (Figure 13.15). The thiocarbamate intermediate subsequently undergoes self-immolative cascade to release COS which is converted to H_2S in the presence of CA. The persulfidation activity of the NQO1-activated donors was investigated in human colon adenocarcinoma (DLD-1) cells where NQO1 is overexpressed. DLD-1 cells were treated with the NQO1 donors (10 μM) for 1 hour followed by treatment of methyl sulfonylbenzathiazole and CN-BOT to trap and fluorescently tag persulfides, respectively. Protein persulfidation from the NQO1 donors was confirmed with a significant enhancement in fluorescence intensity compared to the untreated control. Furthermore, this platform was coupled with mesalamine (a drug used to treat colitis) which demonstrated cytoprotective effects against oxidative stress in DLD-1 cells, whereas mesalamine alone was not found to be as effective.

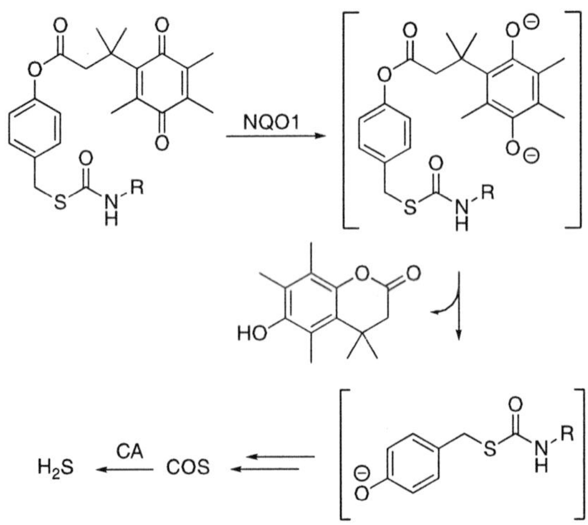

Figure 13.15 NQO1-mediated COS/H_2S release of NQO1 donors.

13.3.5 pH-Activated Donors

The use of acidic conditions as a stimulus to initiate H_2S release has recently gained attention to treat myocardial ischemia reperfusion injury and suppress tumor growth, both of which typically occur in mildly acidic environments. Xian and coworkers developed an acid-activated direct H_2S releasing donor (JK-1) that demonstrated significant therapeutic effects toward treatment of ischemic injury [49]. To further establish control over H_2S release within specific mildly acidic environments, we developed a COS-based acid-labile donor (*S*-pHTCM) that releases within a specific pH window [50]. An acid-labile imine triggering group was appended to the caged thiocarbamate where under acidic conditions, imine hydrolysis is catalyzed to initiate self-immolative decomposition and COS/H_2S release. Although acidic conditions increase the rate and efficiency of COS/H_2S release compared to neutral and basic pH, strongly acidic conditions protonate the aniline intermediate and decrease COS/H_2S production (Figure 13.16). Collectively, COS/H_2S release peaked within a window of pH based on the pK_a of the iminium (pK_a ~ 5–7) and anilinium (pK_a ~ 4.6) ions shown in Figure 13.16, confirming the preference of COS/H_2S release in mildly acidic environments. This donor further establishes control in localizing H_2S release in mildly acidic environments and could serve as a useful strategy in the design of orally administered hydrolysis-based H_2S donor motifs.

Another class of pH-dependent COS/H_2S donors include γ-ketothiocarbamates (γ-KetoTCM) [51], which are activated under basic conditions. This release mechanism proceeds through an enol intermediate, which subsequently undergoes β-elimination to release COS/H_2S without any electrophilic byproducts (Figure 13.17). Furthermore, the donor contains a *p*-nitroaniline (PNA) payload to produce a turn-on colorimetric response (λ_{max} = 381 nm) for real-time monitoring of COS release via UV/Vis spectroscopy. The rate constant obtained from fitting the absorbance of PNA ($k_{obs} = 8.6 \pm 0.9 \times 10^{-4}$ s^{-1}) matched the rate observed when measuring H_2S production ($k_{obs} = 8.1 \pm 0.3 \times 10^{-4}$ s^{-1}), which confirmed that the colorimetric response directly correlates with H_2S production. The rate of COS/H_2S release demonstrated pH-dependence with the fastest rate of COS/H_2S release observed at pH 8.0 ($k_{rel.} = 2.79$) and the slowest rate of COS/H_2S release observed at pH 6.0 ($k_{rel.} = 0.084$). Consistent with previous reports of bovine serum albumin (BSA)-catalyzed β-elimination reactions, the rate of PNA formation was significantly faster in the presence BSA (5 mg/mL) and suggests that COS/H_2S release may be faster in biological environments. This highlights the ability of γ-KetoTCM as a colorimetric donor to evaluate donor release in complex biological systems and observe differences between *in vitro* and *in vivo* donor behavior. The γ-KetoTCM donor additionally demonstrated H_2S delivery in HeLa cells as visualized with an H_2S-selective fluorescent probe (SF7-AM) and exhibited anti-inflammatory effects by decreasing lipopolysaccharide (LPS)-induced NO_2^- formation in Raw 264.7 cells.

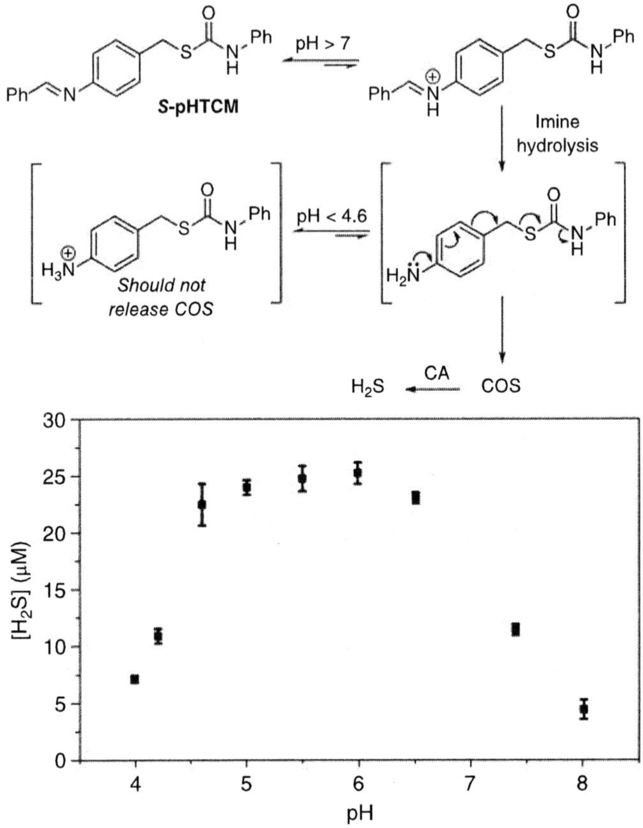

Figure 13.16 Proposed mechanism of pH-dependent release of *S*-pHTCM and pH curve of H_2S concentration at 5 h.

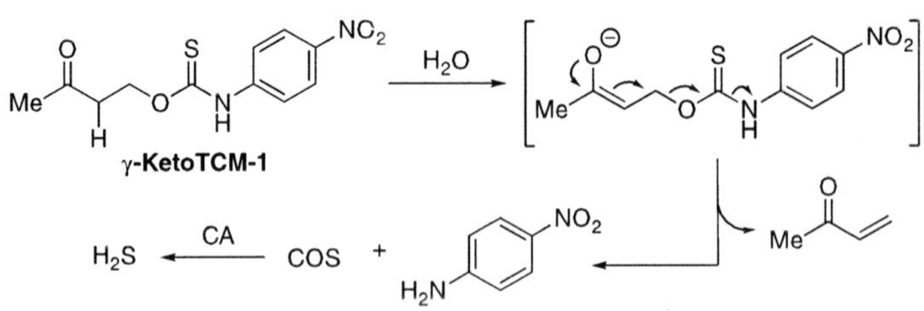

Figure 13.17 Mechanism of COS/H_2S release from γ-KetoTCM-1.

13.3.6 Fluorescent Donors

One limitation of many H_2S donors is the inability to monitor release dynamics in complex environments without measuring, and subsequently consuming, the delivered H_2S. One approach to overcome these limitations is to develop H_2S donors that provide a fluorescent response upon H_2S delivery, and therefore allow for direct monitoring of H_2S release in complex environments. As an early example of this approach, we developed a turn-on fluorescent COS/H_2S based on a thiol-activated sulfenyl thiocarbonate motif [52]. The resulting donor (FLD-1) is not fluorescent until exposed to thiols, which subsequently cleave each sulfenyl thiocarbonate motifs to release COS and fluorescein (Figure 13.18). Furthermore, this system does not produce electrophilic byproducts, which is a limitation in several self-immolative thiocarbamate donors. The COS/H_2S production from FLD-1 and concomitant fluorescent turn-on response was evaluated in the presence of excess cysteine (10 equiv) in PBS (pH 7.4, 10 mM) with CA (25 μg/mL). Cysteine activation of FLD-1 resulted in a 500-fold turn-on response that correlated directly to H_2S concentrations observed by the MBA. The donor platform was evaluated in HeLa cells where a fluorescence signal of the uncaged fluorescein was observed and H_2S production was further confirmed with the H_2S-selective fluorescent probe C7-Az. This donor also demonstrated significant biological activity where pretreatment of RAW 264.7 cells with FLD-1 protected against lipopolysaccharide-induced inflammation.

Expanding on fluorescent donor strategies, Ma and coworkers developed an ROS-activated COS/H_2S donor with a naphthalimide payload to produce a

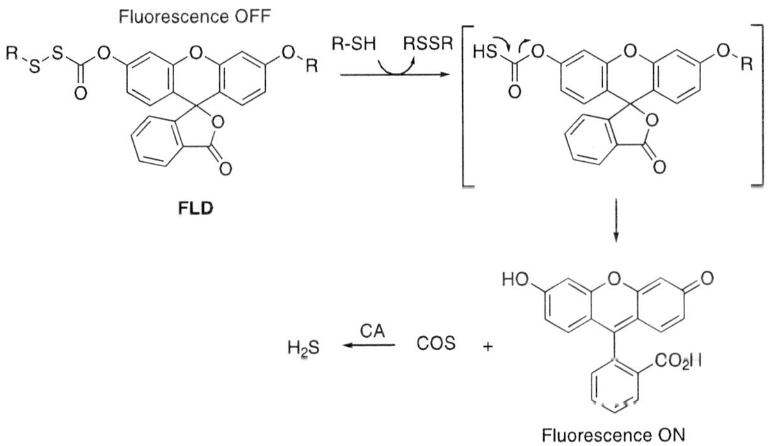

Figure 13.18 COS/H_2S release and simultaneous turn on fluorescence of FLD in the presence of thiols.

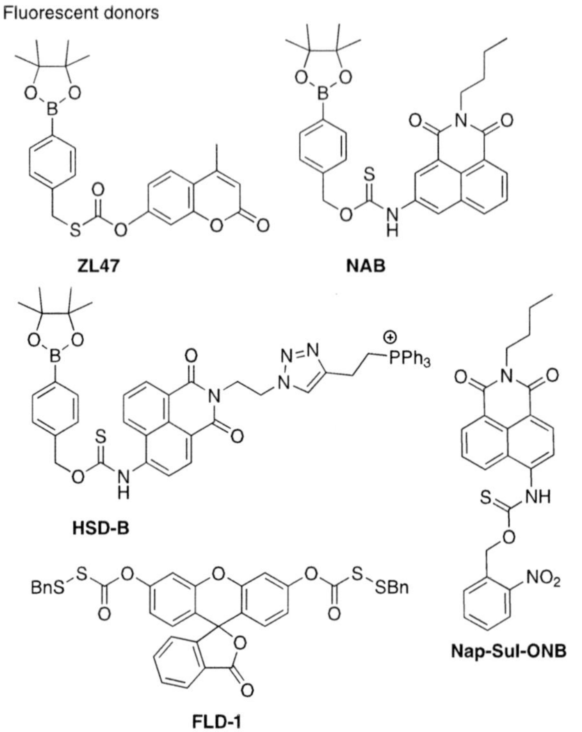

Figure 13.19 Fluorescent COS-based H_2S donors.

stimulus responsive donor with real-time tracking (NAB, Figure 13.19) [26]. The ROS-activated fluorescent donor exhibits a weak fluorescence when excited at 405 nm. Upon exposure to ROS, self-immolative decomposition of NAB releases COS with a 3-amino-*N*-butyl-1,8-naphthalimide that exhibits a strong fluorescence emission peak at 577 nm (λ_{ex} = 405 nm). To confirm H_2S delivery in cells and evaluate protective effects of NAB toward inflammation, Raw 264.7 cells were pretreated with an ROS-inducing agent (PMA) followed by treatment with NAB (10 μM). The fluorescence intensity increased in cells treated with both PMA and NAB compared to the control. NAB additionally recovered decreased cell viability of PMA treated cells and showed ROS scavenging capacity observed via monitoring intracellular ROS change with an ROS-sensitive dye (CRDR).

He and coworkers expanded from this approach by tethering a mitochondrially targeted triphenylphosphonium group to the naphthalimide fluorescent payload to localize COS/H_2S release to the mitochondria, which is a common site of ROS production [25]. The resulting donor (HSD-B, Figure 13.19) is inherently fluorescent with a fluorescence maximum at 475 nm (λ_{ex} = 380 nm), but in the presence of H_2O_2, the emission peak of HSD-B decreases with concomitant growth of an emission peak at 540 nm corresponding to the uncaged naphthalimide (HSD-G). In H9c2 myocardial cells, the fluorescence of unreacted donor overlapped with the mitochondrially tracking dye, Mito-Tracker Red, demonstrating localization in the mitochondria. Treatment of H9c2 myocardial cells with Rosup (15/50 μg/mL) to increase endogenous ROS levels resulted in an increase in the uncaged HSD-G fluorescence intensity and a decrease in HSD-B intensity. An H_2S-selective probe ($Cy\text{-}NO_2$) further confirmed H_2S production only in the presence of Rosup. Notably, HSD-B demonstrated protective effects against myocardial ischemia reperfusion injury, elevating cell viability in a hypoxia/reoxygenation-induced injury model by nearly 40% with only 2 μM of the donor.

As an additional example of a fluorescent ROS-activated donor, Lukesh and coworkers appended a 7-hydroxy-4-methylcoumarin to a boronate *S*-alkyl thiocarbonate scaffold (ZL47, Figure 13.19) [24]. Upon exposure to H_2O_2, a greater than 70-fold increase in fluorescence intensity was observed, which correlated linearly with increasing H_2O_2 concentrations. Furthermore, this donor showed an increase in fluorescence intensity when imaged in live HeLa cells, suggesting its potential to image and alleviate oxidative stress in live cells. In addition to the ROS-activated fluorescent donor scaffolds, Gou and coworkers developed a photolabile *o*-nitrobenzyl thiocarbamate donor with a naphthalamide payload (Nap-Sul-ONB, Figure 13.19) to provide signal feedback upon photoinitiated uncaging of COS [29]. Although cytotoxic effects are typically observed with UV light exposure, a turn-on response of Nap-Sul-ONB was successfully imaged in live MCF-7 cancer cells upon irradiation with 365 nm light for 2 minutes. Moreover, Nap-Sul-ONB provided antioxidative and cytoprotective effects against endogenous ROS.

13.4 Conclusions and Outlook

Since the discovery that COS could act as an indirect source of H_2S, the palette of available H_2S donors has grown substantially. The donors highlighted in this chapter increase the efficiency of previous direct H_2S-releasing donors, provide synthetically accessible donors that can be integrated readily with drugs and fluorophores, and afford a great deal of tunability to develop donors that are activated in specific biological environments. One limitation with some of these scaffolds is the generation of electrophilic byproducts. Specifically, the electrophilic quinone/iminoquinone methide byproduct from the self-immolative thiocarbamate scaffold is likely scavenged by water but could result in unwanted side reactions in biological settings. Despite this drawback, several COS/H_2S donors lack electrophilic byproducts while still retaining a great deal of tunability and have recently been developed.

Another limitation of the COS/H_2S donor platform is the difficulty to delineate COS versus H_2S effects. The abundance of CA in combination with its rapid hydrolysis activity suggests that most released COS is converted to H_2S, but in certain examples where the rate of COS release is competitive with the rate of CA hydrolysis, this may not be the case. The direct study of COS cellular effects is quite challenging due to the widespread expression of CA, its roles in regulating pH, and its various isoforms. CA knock-out and inhibition studies alter cellular CO_2/HCO_3^- equilibria and buffering capacity, which further complicates attribution of observed outcomes with COS effects directly. Furthermore, the CA isoforms expressed commonly in mammalian cells have different sensitivities toward common CA inhibitors, which further complicates broad CA inhibition investigations. Lastly, there are few biological regions without CA because of its critical role in regulating pH.

To complement the available tools for COS/H_2S delivery, additional methods for COS measurement in complex environments are needed. Such systems could allow for determination of whether COS buildup was occurring from different classes of COS/H_2S donors. As an early effort toward developing chemical tools for COS detection, we recently demonstrated that a benzobisimidazolium salt (TBBI) could function as a responsive probe for COS that provided different responses for COS, CO_2, and CS_2, albeit still requiring organic solution [53].

In summary, the strategy of indirect delivery of H_2S via COS has expanded the library of chemical tools to probe the chemical biology of H_2S in various biological settings. Ongoing work to provide tools to understand COS versus H_2S effects, including the development of activity-based COS probes, can continue to bridge the gap between outcomes of different classes of H_2S-releasing donors.

Acknowledgments

We thank prior researchers in our group and throughout the field who have contributed to the chemistry described in this chapter. Work on COS/H_2S delivery in our lab is supported by the NIH (R01GM113030).

References

1 Steiger, A.K., Zhao, Y., and Pluth, M.D. (2018). Emerging roles of carbonyl sulfide in chemical biology: sulfide transporter or gasotransmitter? *Antioxid. Redox Signaling* 28 (16): 1516–1532.

2 Levinn, C.M., Cerda, M.M., and Pluth, M.D. (2020). Activatable small-molecule hydrogen sulfide donors. *Antioxid. Redox Signaling* 32 (2): 96–109.

3 Levinn, C.M., Cerda, M.M., and Pluth, M.D. (2019). Development and application of carbonyl sulfide-based donors for H_2S delivery. *Acc. Chem. Res.* 52 (9): 2723–2731.

4 Ferm, R.J. (1957). The chemistry of carbonyl sulfide. *Chem. Rev.* 57 (4): 621–640.

5 Beck, M.T. and Kauffman, G.B. (1985). COS and C_3S_2: the discovery and chemistry of two important inorganic sulfur compounds. *Polyhedron* 4 (5): 775–781.

6 Svoronos, P.D.N. and Bruno, T.J. (2002). Carbonyl sulfide: a review of its chemistry and properties. *Ind. Eng. Chem. Res.* 41 (22): 5321–5336.

7 Leman, L., Orgel, L., and Ghadiri, M.R. (2004). Carbonyl sulfide-mediated prebiotic formation of peptides. *Science* 306 (5694): 283–286.

8 Huber, C. and Wächtershäuser, G. (1998). Peptides by activation of amino acids with CO on (Ni,Fe)S surfaces: implications for the origin of life. *Science* 281 (5377): 670.

9 Balazy, M., Abu-Yousef, I.A., Harpp, D.N., and Park, J. (2003). Identification of carbonyl sulfide and sulfur dioxide in porcine coronary artery by gas chromatography/mass spectrometry, possible relevance to EDHF. *Biochem. Biophys. Res. Commun.* 311 (3): 728–734.

10 Kamboures, M.A., Blake, D.R., Cooper, D.M. et al. (2005). Breath sulfides and pulmonary function in cystic fibrosis. *Proc. Natl. Acad. Sci. U.S.A.* 102 (44): 15762–15767.

11 Sehnert, S.S., Jiang, L., Burdick, J.F., and Risby, T.H. (2002). Breath biomarkers for detection of human liver diseases: preliminary study. *Biomarkers* 7 (2): 174–187.

12 Smeulders, M.J., Pol, A., Venselaar, H. et al. (2013). Bacterial CS_2 hydrolases from Acidithiobacillus thiooxidans strains are homologous to the archaeal catenane CS_2 hydrolase. *J. Bacteriol.* 195 (18): 4046–4056.

13 Haritos, V.S. and Dojchinov, G. (2005). Carbonic anhydrase metabolism is a key factor in the toxicity of CO_2 and COS but not CS_2 toward the flour beetle *Tribolium castaneum* [Coleoptera: Tenebrionidae]. *Comp. Biochem. Physiol. C Toxicol.* 140 (1): 139–147.

14 Lorimer, G. and Pierce, J. (1989). Carbonyl sulfide: an alternate substrate for but not an activator of ribulose-1, 5-bisphosphate carboxylase. *J. Biol. Chem.* 264 (5): 2764–2772.

15 Ensign, S.A. (1995). Reactivity of carbon monoxide dehydrogenase from rhodospirillum rubrum with carbon dioxide, carbonyl sulfide, and carbon disulfide. *Biochemistry* 34 (16): 5372–5381.

16 Seefeldt, L.C., Rasche, M.E., and Ensign, S.A. (1995). Carbonyl sulfide and carbon dioxide as new substrates, and carbon disulfide as a new inhibitor, of nitrogenase. *Biochemistry* 34 (16): 5382–5389.

17 Kernohan, J.C. (1964). The activity of bovine carbonic anhydrase in imidazole buffers. *Biochim. Biophys. Acta* 81 (2): 346–356.

18 Chengelis, C.P. and Neal, R.A. (1980). Studies of carbonyl sulfide toxicity: metabolism by carbonic anhydrase. *Toxicol. Appl. Pharmacol.* 55 (1): 198–202.

19 Steiger, A.K., Pardue, S., Kevil, C.G., and Pluth, M.D. (2016). Self-immolative thiocarbamates provide access to triggered H_2S donors and analyte replacement fluorescent probes. *J. Am. Chem. Soc.* 138 (23): 7256–7259.

20 Powell, C.R., Foster, J.C., Swilley, S.N. et al. (2019). Self-amplified depolymerization of oligo (thiourethanes) for the release of COS/H 2 S. *Polym. Chem.* 10 (23): 2991–2995.

21 Zhao, Y. and Pluth, M.D. (2016). Hydrogen sulfide donors activated by reactive oxygen species. *Angew. Chem.* 128 (47): 14858–14862.

22 Zhao, Y., Henthorn, H.A., and Pluth, M.D. (2017). Kinetic insights into hydrogen sulfide delivery from caged-carbonyl sulfide isomeric donor platforms. *J. Am. Chem. Soc.* 139 (45): 16365–16376.

23 Chauhan, P., Jos, S., and Chakrapani, H. (2018). Reactive oxygen species-triggered tunable hydrogen sulfide release. *Org. Lett.* 20 (13): 3766–3770.

24 Zhu, C., Suarez, S.I., and Lukesh, J.C. (2021). Illuminating and alleviating cellular oxidative stress with an ROS-activated, H_2S-donating theranostic. *Tetrahedron Lett.* 152944.

25 Zhang, N., Hu, P., Wang, Y. et al. (2020). A reactive oxygen species (ROS) activated hydrogen sulfide (H_2S) donor with self-reporting fluorescence. *ACS Sensors* 5 (2): 319–326.

26 Hu, Y., Li, X., Fang, Y. et al. (2019). Reactive oxygen species-triggered off-on fluorescence donor for imaging hydrogen sulfide delivery in living cells. *Chem. Sci.* 10 (33): 7690–7694.

27 Steiger, A.K., Yang, Y., Royzen, M., and Pluth, M.D. (2017). Bio-orthogonal "click-and-release" donation of caged carbonyl sulfide (COS) and hydrogen sulfide (H_2S). *Chem. Commun.* 53 (8): 1378–1380.

28 Zhao, Y., Bolton, S.G., and Pluth, M.D. (2017). Light-activated COS/H_2S donation from photocaged thiocarbamates. *Org. Lett.* 19 (9): 2278–2281.

29 Hua, W., Zhao, J., and Gou, S. (2020). A naphthalimide derivative can release COS and form H_2S in a light-controlled manner and protect cells against ROS with real-time monitoring ability. *Analyst* 145 (11): 3878–3884.

30 Sharma, A.K., Nair, M., Chauhan, P. et al. (2017). Visible-light-triggered uncaging of carbonyl sulfide for hydrogen sulfide (H_2S) release. *Org. Lett.* 19 (18): 4822–4825.

31 Štacko, P., Muchová, L., Vitek, L., and Klán, P. (2018). Visible to NIR light photoactivation of hydrogen sulfide for biological targeting. *Org. Lett.* 20 (16): 4907–4911.

32 Powell, C.R., Foster, J.C., Okyere, B. et al. (2016). Therapeutic delivery of H_2S via COS: small molecule and polymeric donors with benign byproducts. *J. Am. Chem. Soc.* 138 (41): 13477–13480.

33 Powell, C.R., Kaur, K., Dillon, K.M. et al. (2019). Functional N-substituted N-thiocarboxyanhydrides as modular tools for constructing H_2S donor conjugates. *ACS Chem. Biol.* 14 (6): 1129–1134.

34 Zhao, Y., Steiger, A.K., and Pluth, M.D. (2018). Cysteine-activated hydrogen sulfide (H_2S) delivery through caged carbonyl sulfide (COS) donor motifs. *Chem. Commun.* 54 (39): 4951–4954.

35 Yang, X., Guo, Y., and Strongin, R.M. (2011). Conjugate addition/cyclization sequence enables selective and simultaneous fluorescence detection of cysteine and homocysteine. *Angew. Chem. Int. Ed.* 50 (45): 10690–10693.

36 Zhao, Y., Steiger, A.K., and Pluth, M.D. (2019). Cyclic sulfenyl thiocarbamates release carbonyl sulfide and hydrogen sulfide independently in thiol-promoted pathways. *J. Am. Chem. Soc.* 141 (34): 13610–13618.

37 Wallace, J.L., Caliendo, G., Santagada, V., and Cirino, G. (2010). Markedly reduced toxicity of a hydrogen sulphide-releasing derivative of naproxen (ATB-346). *Br. J. Pharmacol.* 159 (6): 1236–1246.

38 Cerda, M.M., Mancuso, J.L., Mullen, E.J. et al. (2020). Use of dithiasuccinoyl-caged amines enables COS/H_2S release lacking electrophilic byproducts. *Chem. Eur. J.* 26 (24): 5374–5380.

39 Barany, G. and Merrifield, R.B. (1977). A new amino protecting group removable by reduction. Chemistry of the dithiasuccinoyl (Dts) function. *J. Am. Chem. Soc.* 99 (22): 7363–7365.

40 Zhou, S., Mou, Y., Liu, M. et al. (2020). Insights into the mechanism of thiol-triggered COS/H_2S release from *N*-dithiasuccinoyl amines. *J. Org. Chem.* 85 (13): 8352–8359.

41 Severino, B., Corvino, A., Fiorino, F. et al. (2018). 1,2,4-Thiadiazolidin-3, 5-diones as novel hydrogen sulfide donors. *Eur. J. Med. Chem.* 1431677–1431686.

42 Khodade, V.S., Pharoah, B.M., Paolocci, N., and Toscano, J.P. (2020). Alkylamine-substituted perthiocarbamates: dual precursors to hydropersulfide and carbonyl sulfide with cardioprotective actions. *J. Am. Chem. Soc.* 142 (9): 4309–4316.

43 Shahriari, M., Zahiri, M., Abnous, K. et al. (2019). Enzyme responsive drug delivery systems in cancer treatment. *J. Controlled Release* 308172–308189.

44 Zheng, Y., Yu, B., Ji, K. et al. (2016). Esterase-sensitive prodrugs with tunable release rates and direct generation of hydrogen sulfide. *Angew. Chem. Int. Ed.* 55 (11): 1511–1518.

45 Chauhan, P., Bora, P., Ravikumar, G. et al. (2017). Esterase activated carbonyl sulfide/hydrogen sulfide (H_2S) donors. *Org. Lett.* 19 (1): 62–65.

46 Steiger, A.K., Marcatti, M., Szabo, C. et al. (2017). Inhibition of mitochondrial bioenergetics by esterase-triggered COS/H_2S donors. *ACS Chem. Biol.* 12 (8): 2117–2123.

47 Levinn, C.M., Steiger, A.K., and Pluth, M.D. (2019). Esterase-triggered self-immolative thiocarbamates provide insights into COS cytotoxicity. *ACS Chem. Biol.* 14 (2): 170–175.

48 Chauhan, P., Gupta, K., Ravikumar, G. et al. (2019). Carbonyl sulfide (COS) donor induced protein persulfidation protects against oxidative stress. *Chem. Asian J.* 14 (24): 4717–4724.

49 Kang, J., Li, Z., Organ, C.L. et al. (2016). pH-controlled hydrogen sulfide release for myocardial ischemia-reperfusion injury. *J. Am. Chem. Soc.* 138 (20): 6336–6339.

50 Gilbert, A.K., Zhao, Y., Otteson, C.E., and Pluth, M.D. (2019). Development of acid-mediated H_2S/COS donors that respond to a specific pH window. *J. Organomet. Chem.* 84 (22): 14469–14475.

51 Zhao, Y., Steiger, A.K., and Pluth, M.D. (2018). Colorimetric carbonyl sulfide (COS)/hydrogen sulfide (H_2S) donation from γ-ketothiocarbamate donor motifs. *Angew. Chem. Int. Ed.* 57 (40): 13101–13105.

52 Zhao, Y., Cerda, M.M., and Pluth, M.D. (2019). Fluorogenic hydrogen sulfide (H_2S) donors based on sulfenyl thiocarbonates enable H_2S tracking and quantification. *Chem. Sci.* 10 (6): 1873–1878.

53 Cerda, M.M., Fehr, J.M., Sherbow, T.J., and Pluth, M.D. (2020). Progress toward colorimetric and fluorescent detection of carbonyl sulfide. *Chem. Commun.* 56 (67): 9644–9647.

14

Light-Activatable H_2S Donors

Petr Klán[1,2], Tomáš Slanina[3], and Peter Šťacko[4]

[1] *Department of Chemistry, Faculty of Science, Masaryk University, Brno, Czech Republic*
[2] *RECETOX, Faculty of Science, Masaryk University, Brno, Czech Republic*
[3] *Institute of Organic Chemistry and Biochemistry of the Czech Academy of Sciences, Prague, Czech Republic*
[4] *Department of Chemistry, University of Zurich, Zurich, Switzerland*

14.1 Introduction

Compared to other control stimuli, light offers unparalleled advantages in terms of availability, flexibility, high spatial and temporal precision, orthogonality toward chemical reactions, and often even a reduction in the number of waste products. The potential of light as a controllable trigger is widely recognized, as evidenced by the development of photoremovable (or photoactivatable) protecting groups (PPGs) [1, 2] and, more recently, also in the field of photopharmacology [3–5]. However, its application as a trigger for the release of hydrogen sulfide (H_2S) can be complicated by potential secondary photoinduced oxidation of H_2S due to rich chemistry and the availability of several higher oxidation states of sulfur (see Chapter 1). Many light-activatable H_2S photodonors have been developed in recent years, and their chemistry and applications are described in this chapter.

14.2 Photophysical and Photochemical Concepts

Here we introduce some basic concepts of photophysics and photochemistry. Photophysical/chemical processes in an organic molecule can be depicted in a *Jabłonski diagram* [6] (Figure 14.1). Electronic states are represented by thick horizontal lines (—) that are arranged to indicate relative energies and are labeled successively by increasing energy, starting with the *singlet ground state* S_0 and followed by *excited singlet states* S_1, S_2, etc. The *triplet states* (T_1 [the lowest state],

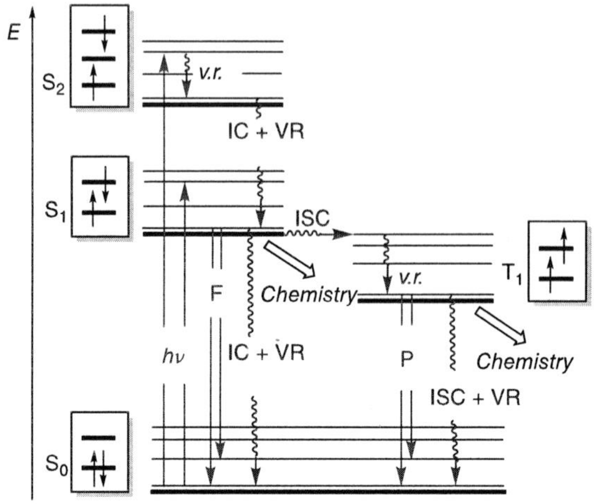

Figure 14.1 Jabłonski diagram. Source: Klan and Wirz [6]/with permission of John Wiley & Sons.

T_2, etc.) are placed in a separate column in the same way. The energy levels represent minima on the potential energy surfaces.

We distinguish *radiative transitions* (straight vertical lines: $h\nu$ = *absorption*; F = *fluorescence*; P = *phosphorescence*) or *radiationless transitions* (wavy arrows: VR = *vibrational relaxation*; IC = *internal conversion* as a transition between two electronic states of the same multiplicity; ISC = *intersystem crossing* as a transition between two electronic states of different multiplicity); in principle, no chemical changes take place during those transitions (Figure 14.1). IC and ISC are isoenergetic and principally irreversible processes because they are associated with an entropy increase. According to *Kasha's rule*, molecules generally emit light only from the lowest excited state of a given multiplicity, although there are many exceptions to this rule [6]. In a more general sense, the lowest S_1 and T_1 states are usually responsible for emission and primary chemical processes (depicted as "chemistry," Figure 14.1).

The term *quantum yield* $\Phi_x(\lambda)$ is equal to the amount n_x of photophysical/chemical events x that occurred, divided by the amount n_p of photons of the given irradiation wavelength λ absorbed by reactant:$\Phi_x(\lambda) = n_x/n_p$. In general, the Φ_x values are found in the range of $0 \leq \Phi_x \leq 1$ and are a measure of the probability that a compound undergoes the photophysical/photochemical process after absorption of a photon.

Photochemical processes are typically activated by one-photon excitation. The absorption spectrum of the *chromophore* (the molecule that absorbs light) is the first information that needs to be obtained before a photochemical application is considered. The multiplicity of the excited state, its *lifetime*, and ability (quantum yield or efficiency) to undergo a chemical reaction are the main factors in the design of new photoactive moieties. When a visible- or near-infrared-light-absorbing photoactive molecule is designed, specific chromophore modifications, such as extending a π-system, can be made.

14.3 Phototherapeutic Window

Irradiation wavelengths applicable in the human body are restricted by the toxicity of UV radiation and the presence of exogenous and endogenous chromophores or tissue inhomogeneity responsible for light scattering. The optical properties of tissue can be described using absorption coefficients (μ_a) and the Rayleigh and Mie light scattering (μ_s') contributions [7]. The absorption term is dominated by hemoglobin below 600 nm and by water above 900 nm. This region in the near-infrared region (NIR) of the light spectrum with the best tissue light penetration is called a phototherapeutic window (650–900 nm) [8]. Other chromophores such as melanin, bilirubin, β-carotene, and fat also influence μ_a. Therefore, the optical properties are dependent on the type of tissue and the content of water, fat, blood, and even oxygen due to variable hemoglobin/oxy-hemoglobin concentration ratio [9] according to Eqs. (14.1) and (14.2):

$$\mu_a = BS\mu_{a.oxyHGb} + B(1-S)\mu_{a.deoxyHGb} + W\mu_{a.water} + M\mu_{a.melan} + F\mu_{a.fat} + \dots \tag{14.1}$$

$$\mu_s'(\lambda) = a\left(f_{Ray}\left(\frac{\lambda}{500(nm)}\right)^{-4} + (1-f_{Ray})\left(\frac{\lambda}{500(nm)}\right)^{-b_{Mie}}\right) \tag{14.2}$$

where B is blood volume fraction, S is oxygen saturation of hemoglobin, W is water volume fraction, M is melanosome volume fraction, F is fat volume fraction, $\mu_{a\cdot i}$ are absorption coefficients of the corresponding chromophores, a is the value μ_s' (500 nm), f_{Ray} is the fraction of scattering at 500 nm due to Rayleigh scattering, b_{Mie} is the scattering power for Mie scattering, and λ is the wavelength.

The typical values vary in the ranges of μ_a = 0.3–1.6 and μ_s'= 1.2–40 cm^{-1}, depending on the type of tissue [10–12]. To put this into perspective, light can, in general, penetrate human tissue up to 2–3 mm at 600 nm and 4–5 mm at 800 nm.

14.4 Light Sources

In terms of light delivery to human tissue, light-activatable H_2S donors can significantly benefit from the technology and instrumentation developed for photodynamic therapy (PDT) purposes and facilitated by its translation into clinical praxis. PDT is a medical protocol for the treatment of cancer and skin diseases, which lends credibility to the use of light in the NIR region as the control stimulus in clinical solutions [13–15]. In the past, large lasers demanding extensive technical support were required for PDT, which limited its use to specialized facilities. However, the development of light-emitting diodes (LEDs) increases exponentially in terms of their efficiency and cost, as forecasted by Haitz's law [16, 17], providing access to cheap, reliable, portable, and easy-to-use light sources. The commonly used light sources can be divided into three categories:

- *Lasers* generate coherent monochromatic light with a very narrow bandwidth and high optical power of up to 1 W/cm^2. These include argon, metal vapor, or solid-state lasers. A pumped dye laser setup where organic dye is excited and used to tune the emission wavelength of the laser is also often used. They can be easily coupled to optical fibers or beam-expanding lenses. However, they are large, expensive, and require high maintenance. Therefore, most clinical PDT applications nowadays utilize high-power diode lasers, which are more compact, stable, cheaper but fixed in wavelengths. Standard air-cooled clinical units can generate up to 8 W of power and irradiance up to 1 W/cm^2.
- *Lamps* are cheap, portable, easy to operate, and include fluorescent, incandescent, xenon arc, metal halide, or sodium arc lamps. Their broadband emission spectrum (300–1200 nm) requires optical filters to filter out the desired irradiation wavelengths and remove the harmful UV and IR light. The large beam divergence makes coupling to optical guides, such as small optical fibers, very difficult and, in combination with filters, results in high losses of light intensity. Consequently, lamps are more suited for irradiation of large areas or cavities with typical irradiance of 20–250 mW/cm^2.
- *LEDs* are semiconductor devices that generate light with a somewhat broad spectral width and angular spread, with irradiance up to 200 mW/cm^2. These small (up to a fraction of mm) and extremely affordable devices can be assembled into arrays to increase the power output further. Moreover, they can be used directly for endoscopic or interstitial applications or be easily coupled to optical fibers. LED-based devices are often utilized in PDT.

Thus, besides direct irradiation, light delivery may also be realized endoscopically or interstitially using an ad hoc probe such as optical fibers (often sub-mm in diameter) made of silicate-based glasses, various plastics, or recently even biodegradable materials [18, 19]. To maintain uniform irradiation and based on the desired application, they can be terminated by microlenses, spherical or cylindrical diffusers, and balloon diffusers to irradiate large cavities. Dedicated applications have been developed for larynx [20], esophagus [21], pleural [22] or anal [23] cavities, bladder [24], cervix [25], or nasopharynx [26]. The total light doses currently employed in PDT treatments are typically 50–100 J/cm^2 [19]. Therefore, the development of light-activatable H_2S donors that operate in this range can significantly accelerate their potential translation into medical applications.

14.5 (Photo)Physical Properties of H_2S

The concentration of hydrogen sulfide (H_2S) in an aqueous solution under biological conditions ($pK_a = 6.76$ at 37 °C) [27] is determined by an equilibrium of the weak diacid (H_2S) and hydrosulfide anion (HS^-). At physiological pH and 37 °C, only 20% of sulfide exists as H_2S, a lipophilic species that readily crosses cell membranes [28]. The half-life of H_2S is shorter in cells compared to aqueous solutions because it reacts with intracellular species, such as peroxide, peroxynitrite, superoxide, and hypochlorite [27, 29–32]. In aqueous solutions at pH >6, the HS^- concentration is sufficiently high, and it can undergo oxidation by air oxygen via a chain mechanism, probably catalyzed by trace impurities of transition metal ions [27, 33].

The sensitivity of H_2S to oxidation makes photochemical applications difficult because they may be accompanied by the formation of various oxidants as side products. For example, Wang and coworkers demonstrated that H_2S as an antioxidant [34, 35] scavenges the triplet state of riboflavin as well as tryptophan and tyrosine radical species [36]. The molar absorption coefficient of HS^- above 254 nm is negligible [37], thus its direct photoactivation in biological experiments is not feasible. However, a possible photocatalytic degradation of H_2S in the presence of oxygen via photoinduced electron transfer [38–42] needs to be considered when a specific photoactivatable H_2S donor is designed.

14.6 Mechanisms and Examples of H_2S Photorelease

PPGs or caged compounds are used for spatiotemporal control over the release of molecules of interest [43, 44]. They consist of a light-absorbing moiety that undergoes chemical changes upon irradiation. We can identify several distinct basic mechanisms of the photochemically induced H_2S release (Scheme 14.1). This gaseous molecule can be released from (a) an excited state of a PPG, (b) a reactive intermediate formed from an excited PPG, or (c) a chemical species (HL) released from an excited PPG that further undergoes a chemical transformation resulting in the H_2S formation. Because H_2S is a dibasic acid, it can be protected with one or two PPGs, thus two consecutive (photo)processes are, in principle, needed to release a free molecule. Another strategy for activating the release with visible/NIR-light is to use a sensitizer that absorbs at longer wavelengths and can transfer energy to (d) a protected H_2S donor that becomes excited

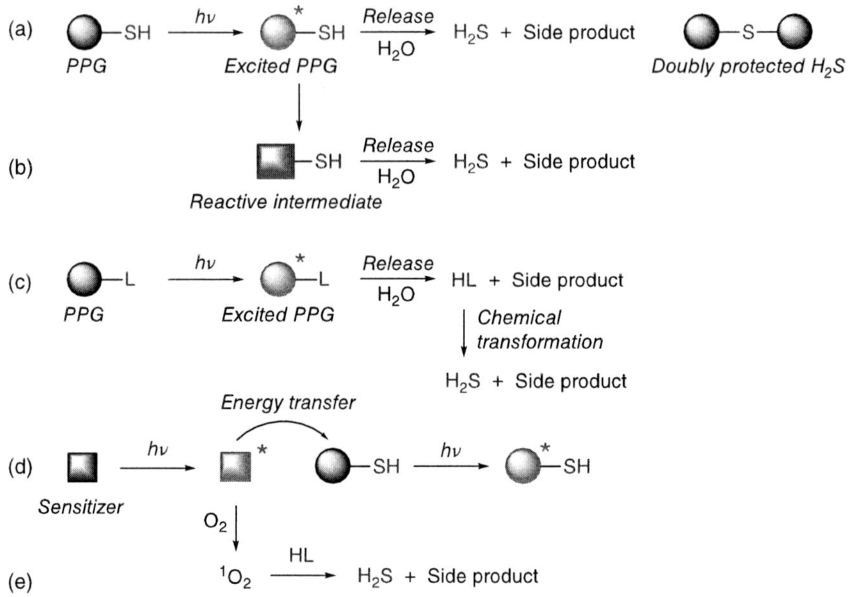

Scheme 14.1 Mechanisms of the photoinduced release of H_2S.

(the remaining pathway is identical to that of Scheme 14.1a), or (e) an auxiliary reagent produced by photosensitization, such as singlet oxygen ($^1O_2$), reacts with a sulfur-containing species to generate H_2S.

14.6.1 Photorelease of H_2S from Excited State

Analogous to the photochemistry of a hydroxylated malachite green derivative (triphenylmethane leucohydroxide) **1**, used as a photobase (HO^-) generator [45], HS^- can be liberated from thiolated malachite green **2** directly upon irradiation in the wavelength range of 254–365 nm with $\Phi_{254nm} = 0.22$ and $\Phi_{365nm} = 0.01$ in PBS aqueous buffer to give the malachite green carbocation **3** as a side-product (Scheme 14.2) [46]. The release is reversible because H_2S as a predominant form under the given conditions can slowly regenerate the starting material **2**. Diblock copolymer polymersomes were used in this case to stabilize compound **2** against dark hydrolysis in an aqueous buffer.

The photorelease of doubly protected H_2S is demonstrated on two different types of PPGs. The first system utilizes a *p*-hydroxyphenacyl (pHP) group **4**, which upon excitation and subsequent ISC generates a leaving group (LG^-) and a triplet biradical intermediate **5** that gives a Favorskii-rearrangement cyclopropanone moiety and subsequently *p*-hydroxyphenylacetic acid as the final side-product [43, 44, 47] (Scheme 14.3).

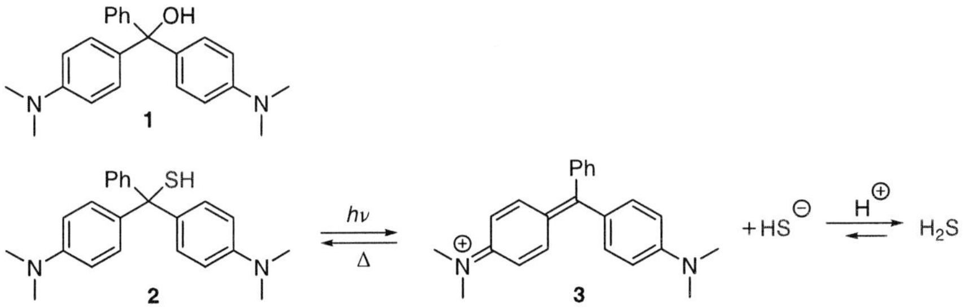

Scheme 14.2 Photoactivatable thiolated malachite green as a monoprotected H_2S donor. Source: Tron et al. [46]/Royal Society of Chemistry.

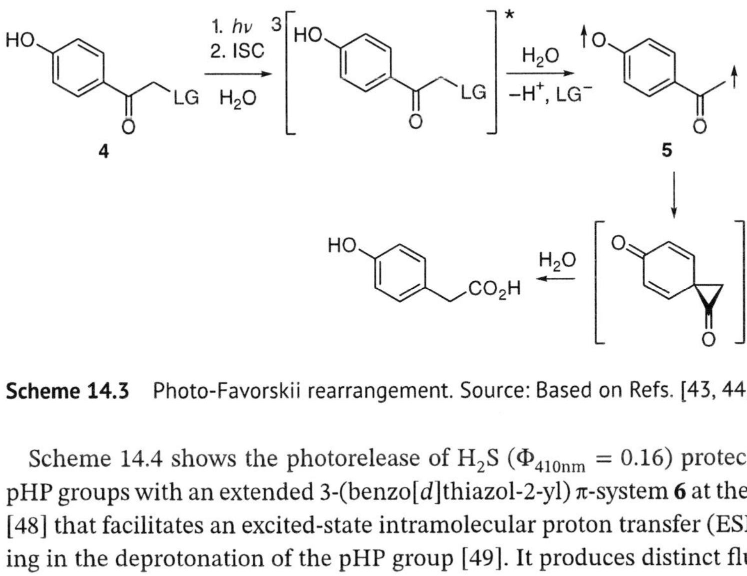

Scheme 14.3 Photo-Favorskii rearrangement. Source: Based on Refs. [43, 44, 47].

Scheme 14.4 shows the photorelease of H_2S ($\Phi_{410nm} = 0.16$) protected by two pHP groups with an extended 3-(benzo[*d*]thiazol-2-yl) π-system **6** at the 3-position [48] that facilitates an excited-state intramolecular proton transfer (ESIPT) assisting in the deprotonation of the pHP group [49]. It produces distinct fluorescence color change (green to blue) upon irradiation that can be used for self-monitoring of the uncaging process. While the derivative **6** absorbs at $\lambda_{max} = 350$ nm in non-hydrogen bonding solvents, its absorption is shifted bathochromically to $\lambda_{max} = 380$ nm in polar protic solvents due to keto-enol tautomerization. The tautomerization process was confirmed by the dual fluorescence of **6** in polar protic solvents, indicating that the tautomeric equilibrium between the enol form and the keto form is shifted in the excited state by ESIPT as compared to the ground state. Although Pradeep Singh and coworkers suggested that the release of H_2S occurs simultaneously from an excited triplet **7**, the authors of this review are instead convinced that the protecting group should be photoreleased in two independent steps upon the sequential absorption of two photons. The photorelease quantum yield of H_2S was determined to be 0.16, and the maximum

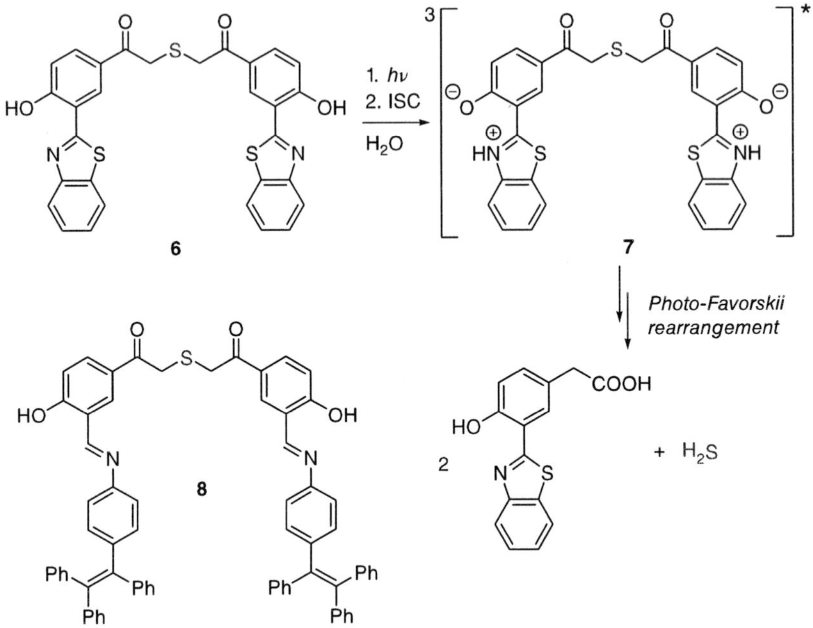

Scheme 14.4 Doubly protected H_2S with two pHP groups.

chemical yield of H_2S was ~40%. The compound efficiently penetrated through the cell membrane (HeLa cervical cancer cells) and was monitored by bright green fluorescence. The 3-(4,5-dimethylthiazol-2-yl)-2,5-diphenyltetrazolium bromide (MTT) assay of **6** and its photoproducts did not reveal any cytotoxicity in concentrations up to 20 μM.

Molecules **8** (Scheme 14.4), a tetraphenylethylene-substituted analog of **6**, aggregate in water forming organic nanoparticles of ~35 nm size with distinct aggregation-induced emission (AIE) [50].

It released H_2S in an aqueous solution upon irradiation with visible light ($\lambda_{irr} > 410$ nm) with good quantum yield ($\Phi_r = 0.18$), which can easily be monitored by the change of λ^{em} from 549 to 486 nm. The nanoparticle donor was efficiently internalized into HeLa cells and was selectively localized in lysosomes. No toxicity of **8** or its photoproducts was found in concentrations up to 20 μM by MTT assay.

The second example of H_2S protected with two *m*-aminobenzyl groups (**9**) [51] is shown in Scheme 14.5. The cleavage ($\Phi_{365nm} = 0.14$) probably occurs from the singlet excited state of individual chromophores and is based on the excited-state *meta* effect resulting from an increase in the electron density in the *meta* (benzyl) position bearing the leaving group (LG) [52]. A simultaneous photorelease of

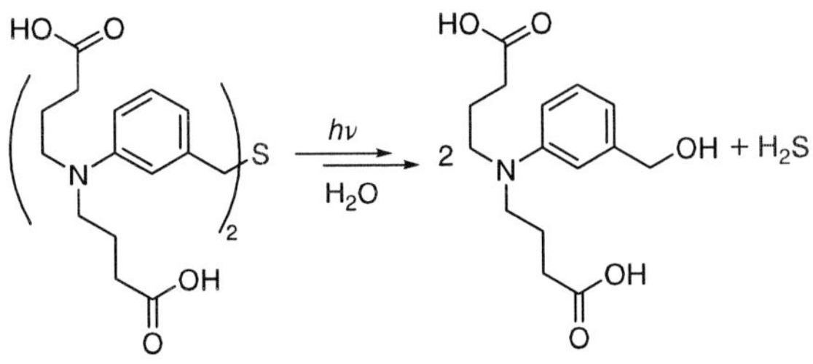

9

Scheme 14.5 Doubly protected H_2S with two *m*-aminobenzyl groups.

H_2S has again been proposed [51]. The released benzyl cation reacts with solvent molecules to form the corresponding benzyl alcohol. It was shown that **9** releases hydrogen sulfide in HeLa cells and exhibits low cytotoxicity.

14.6.2 Release of H_2S from a Reactive Intermediate

The LG liberation from an excited state is, in principle, the fastest process, which relies on the excited-state lifetime. However, the LG release from a reactive intermediate formed upon irradiation of a PPG (Scheme 14.1b) can also be very fast. The first strategy is based on an *o*-nitrobenzyl (oNB) group **10** (Scheme 14.6), which is a general-purpose and widely used PPG. Upon its excitation, intramolecular hydrogen abstraction by the nitro group, an *aci*-nitro intermediate **11** is formed and transformed in several steps to a hemiacetal intermediate **12**, which hydrolyzes to liberate an LG [43, 44, 54, 55]. The arylnitroso side-product **13** is a (photo)reactive and strongly absorbing species, which can be detrimental for some applications.

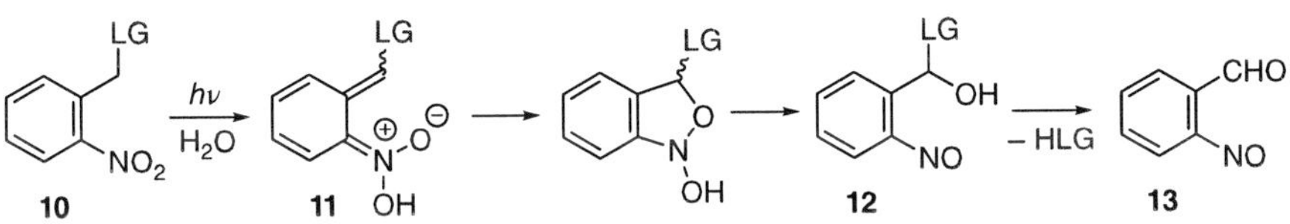

Scheme 14.6 Photochemistry of oNB group. Source: Based on Mitsuhashi et al. [30] and Devaric-Baez et al. [53].

Scheme 14.7 shows the photorelease of H_2S from a doubly protected system **14** developed Nakagawa and coworkers [56], which most probably liberates H_2S in two consecutive photochemical steps. The photochemical release of hydrogen sulfide using 300–350 nm light was successful, but its chemical yield, found by quantitative HSip-1 fluorescent assay [57], was very low (~0.5%) probably due to the reaction of the released H_2S with 2-nitrosobenzaldehyde **13**. Thus, the reaction was not studied further.

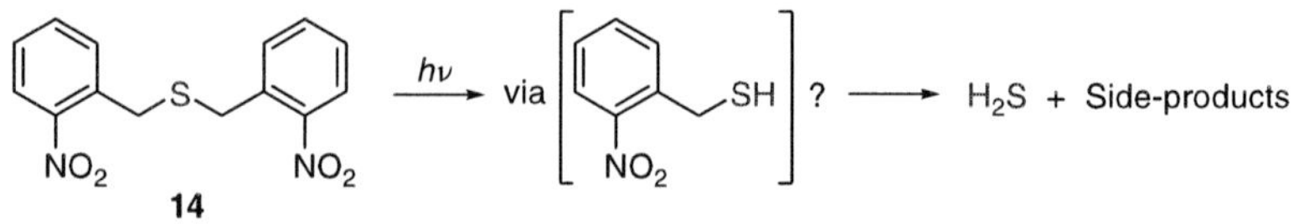

Scheme 14.7 Doubly protected H_2S with two oNB groups.

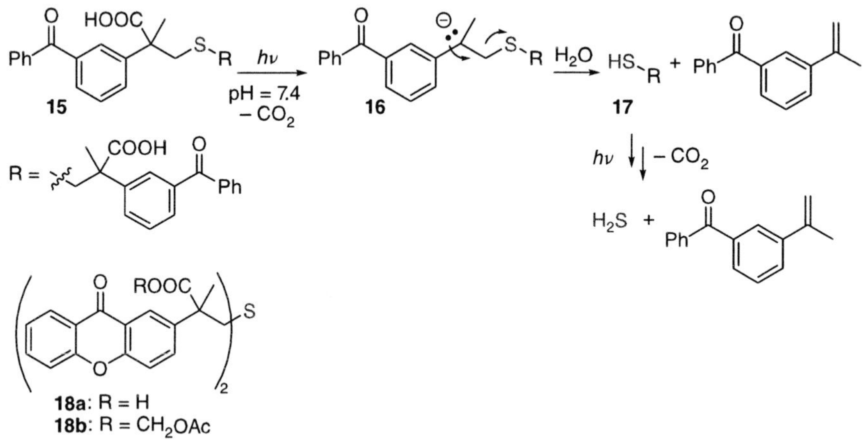

Scheme 14.8 Doubly protected H_2S with two ketoprofenate groups. Source: Based on Fukushima et al. [56, 58].

Nakagawa and coworkers have demonstrated a different strategy of H_2S liberation utilizing two ketoprofenate groups in 15 (Scheme 14.8) [56, 58]. The molecule **15** activatable upon irradiation at 300–350 nm was developed as a water-soluble alternative to oNB protected H_2S donors. The ketoprofenate PPG was designed by Scaiano and coworker [59], and its photochemistry is based on efficient photodecarboxylation to give a benzyl anion intermediate **16** that eliminates an LG [59], in this case H_2S. The monosubstituted thiol intermediate **17** was anticipated but never detected [60]. The release of H_2S can be accomplished in neutral aqueous-based buffers with an excellent photorelease quantum yield and uncaging cross-section ($\Phi_{313nm} = 0.42 \pm 0.08$; $\Phi_r\varepsilon(\lambda_{irr}) \sim 1000\,M^{-1}\,cm^{-1}$). The poor absorption properties of **15** in the ultraviolet A (UVA) region were further improved by the development of xanthone-based H_2S donor **18a** (Scheme 14.8), showing similar photochemical behavior [58]. Thanks to its more extended conjugation, the molecule can be activated by UVA irradiation (325–385 nm). While the carboxylate group increases water solubility, it reduces its cell membrane permeability. This was overcome by introducing an acetoxymethylated derivative **18b** that penetrates through cell membranes and is intracellularly hydrolyzed by cytosolic esterases to **18a**.

14.6.3 Photorelease of Potential H_2S Donors

Photochemical transformations can also be used to generate species that serve as H_2S donors under specific conditions (Scheme 14.1c). The photochemical step is essential for their release, however, the efficiency and rate of the release are dependent on the subsequent dark processes described above in detail.

Carbonyl sulfide (COS) can be converted in the presence of carbonic anhydrase, an omnipresent enzyme that catalyzes the hydration of carbon dioxide and the dehydration of bicarbonate [61], to H_2S [61, 62] (see Chapter 13). It has been shown that COS is formed from thiocarbamates by their hydrolysis, which can be photochemically liberated from a variety of PPGs. Scheme 14.9 shows the photorelease of thiocarbamate **19** from an oNB cage **20a** that is subsequently converted to COS to give H_2S (Scheme 14.6) finally [63]. The methoxy substituted derivative **20b** generated H_2S more efficiently (by the factor of ~2), whereas the relative chemical yield of released hydrogen sulfide was approximately 2–3 times lower as compared to that of the parent compound **20a**. The role of carbonic anhydrase was confirmed by the observed reduction in H_2S formation in the presence of acetazolamide, a well-known carbonic anhydrase inhibitor [64].

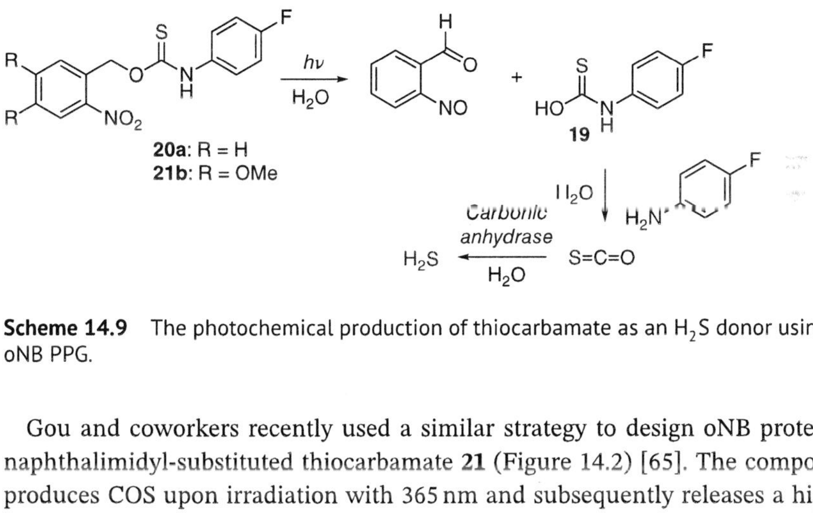

Scheme 14.9 The photochemical production of thiocarbamate as an H_2S donor using an oNB PPG.

Gou and coworkers recently used a similar strategy to design oNB protected naphthalimidyl-substituted thiocarbamate **21** (Figure 14.2) [65]. The compound produces COS upon irradiation with 365 nm and subsequently releases a highly fluorescent naphthimide derivative with an emission maximum at 544 nm. These self-reporting properties were used for cell imaging of H_2S release in MCF-7 cancer cells. It was also shown that **21** can reduce cellular reactive oxygen species (ROS) level by scavenging endogenous (phorbol-12-myristate-13-acetate, PMA) and exogenous (hydrogen peroxide) ROS to increase the cell survival rate.

A 4,4-difluoro-4-bora-3a,4a-diaza-*s*-indacene (BODIPY)-based PPG was used by Chakrapani and coworkers for the synthesis of the first H_2S donor activatable

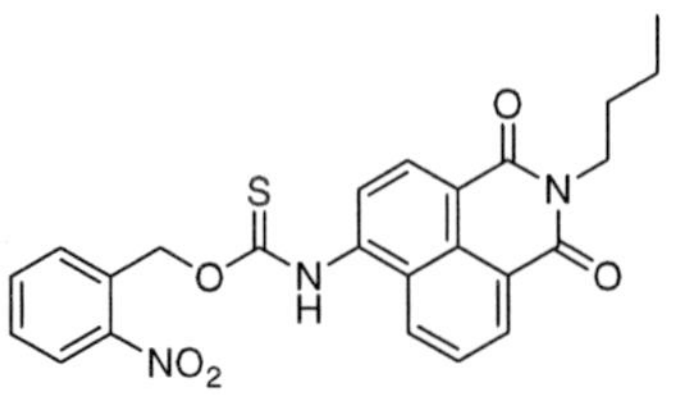

21

Figure 14.2 Naphthalimide-substituted H_2S donor **21**. Source: Hua et al. [65]/Royal Society of Chemistry.

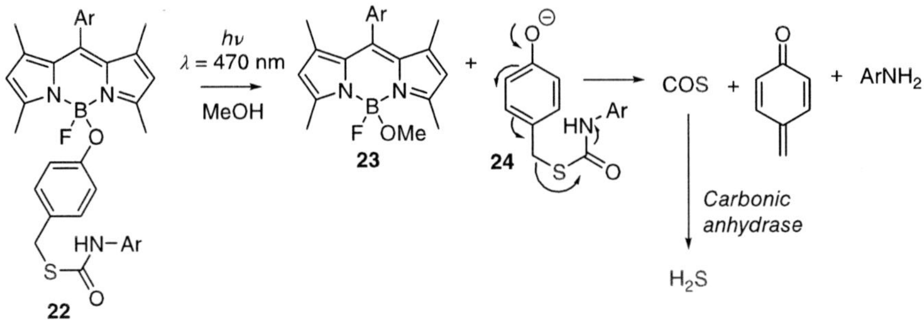

Scheme 14.10 BODIPY-based H_2S donor and the mechanism of its photoactivation.

by direct excitation with visible light [66]. The derivative **22** (Scheme 14.10) was substituted with a short oligoethylene glycol chain that increased the aqueous solubility. It undergoes photoinduced B—O bond cleavage by irradiation at 470 nm to release a methoxy-substituted BODIPY byproduct **23** and a thiocarbamate-substituted phenolate intermediate **24** that has previously been shown to thermally generate COS [67]. The rate of the photochemical step ($k_{release}$ = 0.16 min^{-1}), dependent on the light source intensity and experimental setup, was found to be higher than the subsequent thermal self-immolation of **24** (k_{immol} = 0.02 min^{-1}), which accumulated in the reaction mixture. The final transformation of COS to H_2S in the presence of carbonic anhydrase was rapid (k_{hydrol} = 1.82 s^{-1}). The H_2S yield was in the range of 30–40%, and its formation in HeLa cells was monitored by the fluorescence enhancement thanks to the highly emissive BODIPY-based photoproduct **23**. No cytotoxicity was found for **22** and its photoproducts in cell lines HeLa and A549 up to the concentration of 50 μM.

BODIPY-based H_2S donors **25–27** (Scheme 14.11), developed by Štacko, Klán, et al. [68], can also liberate thiocarbamate upon irradiation, which hydrolyses into COS. The absorption maxima of the derivatives were tuned by π-extending substituents in the range of 519–779 nm. While **25** is poorly soluble in water, the polyethylene glycol-substituted π-extended analog **26** is water-soluble and efficiently releases COS and Ph_2NH (λ_{max}(abs) = 513 nm in degassed aq. PBS, Φ_r = 15.1 × 10^{-4} at 365 nm) in ~86% yield. The NIR-absorbing derivative **27**

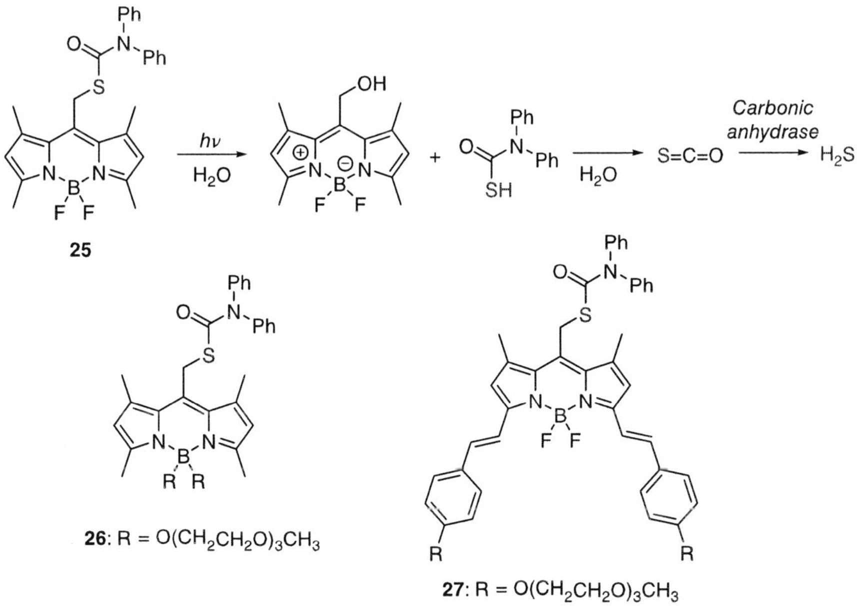

Scheme 14.11 The *meso*-methyl-substituted BODIPY-based H_2S-releasing molecules.

photoreleases thiocarbamate upon irradiation at 700 nm (λ_{max}(abs) = 688 nm in degassed aq. PBS, Φ_r = 9.7×10^{-4} at 365 nm, yield 69%). While the green light-absorbing derivatives **25** and **26** form BODIPY-based photoproducts of the nucleophilic substitution at the *meso*-methyl position, the π-extended analog **27** generates side product(s) with a hypsochromically shifted absorption (λ_{max}(abs) = 590 nm), formed presumably by the oxidation of one of the styryl moieties with singlet oxygen generated by the triplet state-sensitization of O_2. The formation of hydrogen sulfide from COS was demonstrated in crude mice liver homogenate containing natural concentrations of carbonic anhydrase and in HepG2 cells. The compound **26** exhibited mild toxicity above 25 μM, whereas **27** was not toxic up to the concentration of 150 μM. All photoproducts were found nontoxic (MTT assay, SHSY5Y cells).

gem-Dithiols (**28**, Scheme 14.12) represent another moiety that is efficiently hydrolyzed to H_2S in aqueous solutions in the dark [69, 70]. They are stable in the protected form and readily synthetically available via the $TiCl_4$-catalyzed condensation of substituted thiols with the corresponding ketone. *gem*-Dithiols can be released from the corresponding doubly protected thioethers **29** upon irradiation (Scheme 14.12), where a PPG is an oNB derivative (see also Scheme 14.6) [33, 53].

oNB-caged geminal *gem*-dithiol derivatives **30a–f** (Scheme 14.13) were introduced as the first known light-activatable H_2S donors by Xian and coworkers

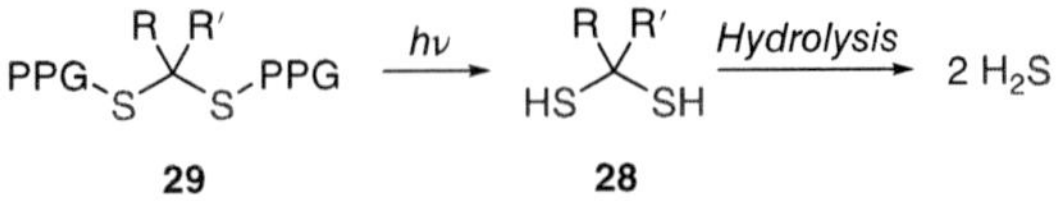

Scheme 14.12 Photorelease of *gem*-dithiols as H_2S donors.

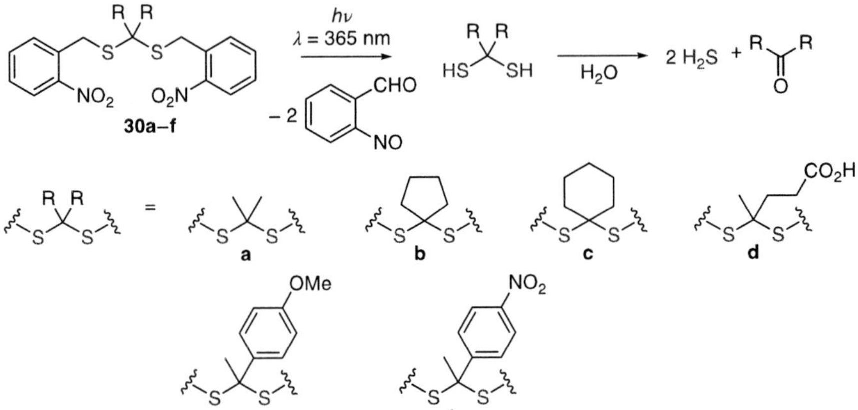

Scheme 14.13 *o*-Nitrobenzyl caged *gem*-dithiols and the mechanism of their deprotection.

in 2013 [53]. Upon irradiation of these compounds at 365 nm in the presence of water (phosphate buffer/acetonitrile 1 : 1 (v/v)), a free *gem*-dithiol is released and then hydrolyzed to liberate H_2S. Alkyl-substituted derivatives **30a–d** release H_2S in higher chemical yields (~10–17%) than aryl-substituted analogs **30e** and **30f** (<2%). The pH of the solution was found to affect the hydrolysis efficiency; the highest H_2S yield was obtained at a moderately acidic pH of 5.5. The release of hydrogen sulfide was followed by the standard methylene blue method [71] and was monitored in HeLa cells with an H_2S specific fluorescent probe [72]. This strategy was also used for water-soluble polymeric H_2S photodonors [73] and H_2S-releasing block copolymer nanoparticles [74]. Li and coworkers utilized oNB-protected thioketal units **30d** conjugated on the side chains of methacrylate-based [poly(AEMA-*co*-EGMA)] copolymer [73]. While the conjugate and its degradation product show no cytotoxic effects on NIH3T3 cells, the 2-nitrosobenzaldehyde photoproduct showed detrimental toxic effects on the cells. The molecule **30d** was also used to synthesize poly[tri(ethylene glycol) methyl ether acrylate] (PTEGMA)-*co*-PTK (polythioketal) block copolymer that assembles in an aqueous solution to give nanoparticles of a ~80-nm size, which can release H_2S *in vitro* by irradiation with 365 nm light [74].

Similarly, Chen and coworkers deposited oNB-protected thioketal on LiYF4: Yb/Tm upconversion nanoparticles (UCPs) coated with polyethylene glycol-octadecylamine through hydrophobic interactions [75]. Upon irradiation at 980 nm (E = 6.5–20 W/cm^2), the nanoparticles convert NIR light into UV photoemission (λ^{em} ~ 365 nm), leading to the deprotection of oNB groups and subsequent release of H_2S. Furthermore, the UCPs were shown to be monitored *in vivo* thanks to a secondary emission band at ~450–500 nm. The applicability of this system was demonstrated in *in vitro* experiments with L929 cells and in *ex vivo* experiments using pork skin as a light filter. However, the authors also acknowledged that the overlap of the irradiation wavelengths with the water absorption band (λ_{max} = 970 nm) and the use of relatively long irradiation times led to solution overheating.

Thioaldehydes, *S*-analogs of aldehydes, are reactive intermediates that can be converted to H_2S and the corresponding imines in the presence of amines [76]. Thiobenzaldehyde **31** was shown to be generated from the phenacyl derivative **32** upon irradiation at 355 nm (Scheme 14.14) [77] via C–S photofragmentation following Norrish type II hydrogen-atom abstraction [78]. A series of molecular and macromolecular moieties have been developed, including small-molecule crystalline compounds, water-soluble polymers, polystyrene films, and hydrogels, demonstrating the scalability and versatility of these H_2S donors. Upon irradiation with UV light, the liberated thiobenzaldehydes underwent nucleophilic substitution with auxiliary amines, such as glycine, with the simultaneous release of H_2S. A hydrogel and polystyrene polymer was used as a scaffold for growing 3T3 fibroblast cells which, upon irradiation with UV light, released and delivered H_2S into the cells. On the other hand, a water-soluble polymer led to inhibition of platelet activation upon irradiation with light and subsequently aggregation via suppressing the P-selectin expression, offering a platform for thrombogenesis regulation.

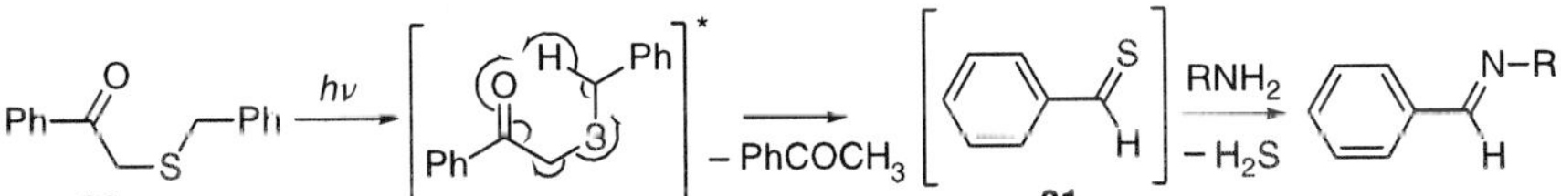

Scheme 14.14 Photorelease of thiobenzaldehyde as an H_2S donor. Source: Xiao et al. [77]/John Wiley & Sons.

Phosphinodithioate derivative **33** (Scheme 14.15) represents a reactive intermediate that is spontaneously hydrolyzed in aqueous solutions to give, in principle, two equivalents of H_2S [79, 80]. The release of **33** from ruthenium(II) complexes **34a–b** absorbing red light (λ_{max}(abs) = 581 nm for **34a** and 570 nm for **34b**) is an efficient (Φ_{626nm} = 0.85 for **34a**, Φ_{626nm} = 1.02 for **34b**) ligand exchange that

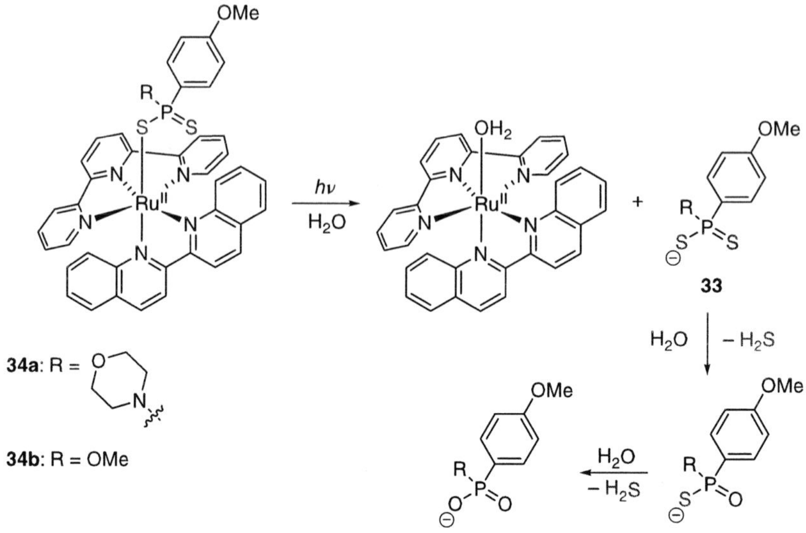

Scheme 14.15 Photorelease of phosphinodithioate as an H_2S donor.

proceeds via excitation to the corresponding metal to ligand charge transfer (^{1}MLCT) absorption band of the complex, followed by efficient intersystem crossing to give a dissociative triplet ligand field ^{3}LF state [81]. The released H_2S donor **33** further undergoes hydrolytic decomposition to give two equivalents of hydrogen sulfide, which was monitored by UV–vis absorption spectroscopy, $^{31}P\{^1H\}$ nuclear magnetic resonance (NMR) spectroscopy, and a fluorescence assay using 1,5-dansyl azide. The complexes were nontoxic at concentrations up to 300 μM at short incubation times (4 hours), whereas the photoproducts did not show any toxicity. Complexes **34a** and **34b** were successfully used to release H_2S in living cells (A549 lung cancer cells) and protect H9c2 cardiomyoblasts in an *in vitro* model of ischemia–reperfusion injury.

14.6.4 Photosensitized H_2S Release

An example of the photosensitized production of H_2S (Scheme 14.1e) is demonstrated in Scheme 14.16. Singlet oxygen ($^1O_2$), generated by visible-light sensitization of ground-state oxygen with an excited platinum(II) porphyrin (λ_{irr} = 500–550 nm; Φ_Δ = 0.30) or iridium-based (λ_{irr} = 380–500 nm; Φ_Δ = 0.41) photosensitizer (PS), is trapped by 1,3-diphenylisobenzothiophene **35** to give an endoperoxide intermediate **36**, which is thermally decomposed to H_2S [82].

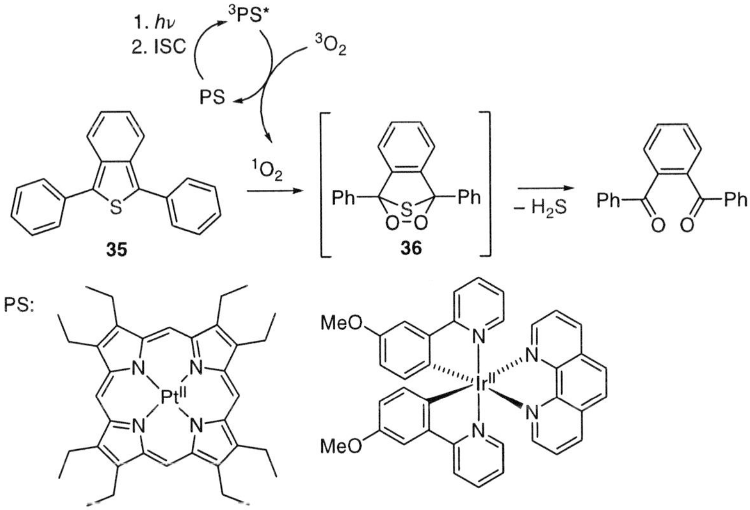

Scheme 14.16 Photosensitized H_2S release.

Encapsulation of these components in biocompatible Pluronic F-127 polymer resulted in spherical nanoparticles of a 10-nm diameter. The photorelease of H_2S upon irradiation with visible light was detected by both a fluorogenic probe SF4 and an H_2S-selective electrode. The oxidation was shown to be mediated selectively by $^1O_2$, whereas other ROS were ineffective. Finally, the biological utility was demonstrated on HeLa cells by intracellular delivery and cytoprotective action of H_2S against oxidative stress. The potential drawbacks of the photooxygenation approach include a decreased release efficiency in hypoxic conditions, deactivation of $^1O_2$ by endogenous quenchers, and oxidation of H_2S by photogenerated ROS.

Singh and coworkers reported another example of the sensitized H_2S release from a releasing system that uses fluorescent carbon dots (CDs) decorated with quinoline-based [83, 84] photoremovable protecting groups **37** (Scheme 14.17) [85]. Irradiation of CDs with visible light ($\lambda_{irr} > 410$ nm) in PBS (pH = 7.4) led to sensitization of quinoline-protected H_2S moieties and the subsequent release of H_2S in ~60% chemical yield with $\Phi_r = 0.09$. The system is also fluorescent in the visible region ($\lambda^{em} = 425$ nm, $\Phi_F = 0.078$). Its biological applicability was demonstrated on HeLa cells, where the H_2S release was monitored by an increase of the fluorescence intensity in the cells at 425 nm or using an H_2S-specific probe. The conjugate and the corresponding photoproducts showed no significant toxicity on a HeLa cell line at the studied concentrations.

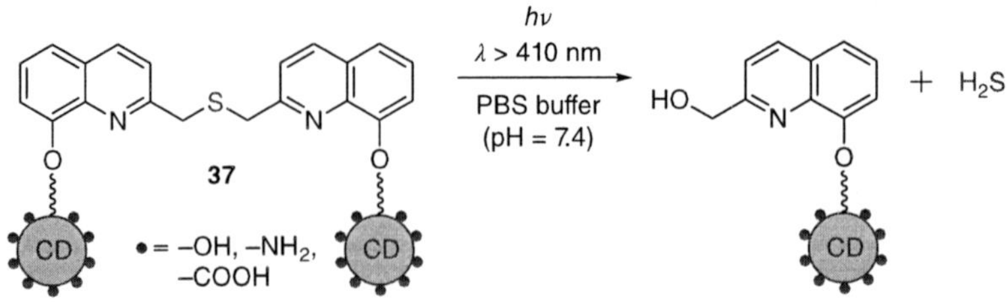

Scheme 14.17 Photorelease of H_2S from carbon dots decorated with quinoline-based PPGs. Source: Bera et al. [85]/Royal Society of Chemistry.

14.6.5 Photothermal Effect

Liu and coworkers utilized a combination of Au@Ag/Au core@shell plasmonic nanoparticles (pNPs) encapsulated in a thermo-labile polymer (PEI–DTC) containing dithiocarbamate moieties as H_2S donors and agarose hydrogel (Scheme 14.18) [86]. Irradiation at 780 nm resulted in a photothermal effect and temperature increase of the hydrogel up to 45 °C. This initialized the thermal conversion of dithiocarbamate units to isothiocyanate with the concomitant release of H_2S. The rate of the release (up to 2.4 μmol/(L min)) was shown to depend on both loading of the nanocomposite in agarose hydrogel and the irradiation intensity ($E = 0.06–3.6\ W/cm^2$). The system showed no cytotoxicity in cell viability tests, and it was successfully applied for anti-inflammatory treatment in rat toe to demonstrate the spatiotemporal release of H_2S *in vivo*.

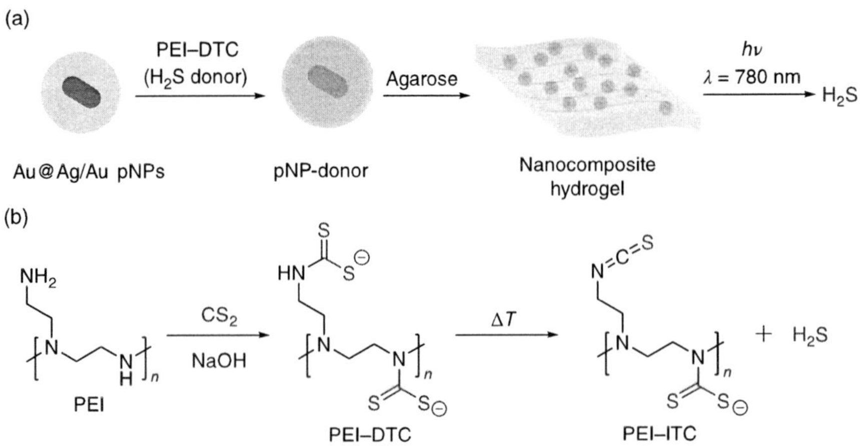

Scheme 14.18 (a) The release of H_2S from nanocomposite hydrogel using photothermal effect. (b) The formation of PEI–DTC and its utilization in (photo)thermal release of H_2S. Source: Huang et al. [86]/Elseiver.

The same concept was used utilized by Liu and coworkers to produce a nanocomposite of the PEI–DTC polymer adsorbed on nanosheets of reduced graphene oxide (rGO) [87]. In this case, the conversion of NIR light (λ_{irr} > 780 nm; E = 1.5–3.0 W/cm^2) into thermal energy was facilitated by rGO, with the temperature reaching up to 55 °C upon irradiation. The system displayed low intrinsic cytotoxicity in the dark. Simultaneous photothermal and H_2S treatment of a human breast cell line (MCF-7) by the nanocomposite demonstrated the inhibition of proliferation exceeding that of naked rGO.

14.7 Outlook

Despite some significant developments over the last decade, the field of light-activatable H_2S donors is still in its infancy, and several important challenges remain to be addressed. The major issue that complicates a direct photochemical release is the *sensitivity of H_2S to (photo)oxidation* in both caged and released forms. The presence of excited redox-active species (e. g. riboflavin) can lead to efficient photocatalytic degradation of H_2S in the presence of oxygen via photoinduced electron transfer [38–42]. Therefore, new photochemical pathways to H_2S release should be carefully designed, and additional advanced techniques for H_2S (photo)protection and maintaining of redox homeostasis (e.g. with sacrificial reductants) to prevent H_2S scavenging need to be developed.

Electron-rich functional groups attached to a chromophore, such as amines and sulfides, often undergo photoinduced electron transfer by one-electron oxidation of the central atom (N, S). This leads either to non-radiative quenching of the excited state or unwanted side-reactivity. Oxidative quenching can be avoided by the protection of *H_2S precursors* instead of hydrogen sulfide itself. However, the scope of available donors, such as COS, *gem*-dithiols, thioaldehydes, phosphinodithioates, is quite limited. Therefore, alternative types of precursors with tunable reactivity and high yields of H_2S liberation, forming biocompatible side-products, could still to be developed.

For bioapplications, the protecting/sensitizing chromophoric group should be *activated by the light within the phototherapeutic window* (650–900 nm), and the photorelease should be efficient and the photoproducts transparent at the irradiation wavelengths. Such photoactivatable compounds have been developed only recently [43, 44] and can serve as inspiration for designing the next generations of H_2S donors. In addition, the development of light-activatable H_2S donors that can release biologically relevant amounts of hydrogen sulfide upon irradiation with light doses currently employed in PDT treatment (50–100 J/cm^2) [19] can significantly accelerate their potential translation into medical applications.

Another desired property of the developed systems is the *ability to monitor the release of* H_2S. A self-reporting system that undergoes a photoinduced change of its absorption and/or emission properties (preferably a fluorescence turn-on response upon H_2S release) represents a very successful strategy. Alternatively, a combination of a photodonor with a H_2S-sensing molecule that does not consume the analyte can also be utilized.

Photoinduced release of sufficient amounts of H_2S can be accomplished either by the application of photodonors possessing high photorelease cross sections at the irradiation wavelengths, or by irradiation of *hybrid systems* where multiple equivalents of H_2S are released upon the absorption of one photon. Such high-molecular systems can combine multiple H_2S donors coupled with support (nano)particles and can be activated by direct irradiation, sensitization, or photothermal trigger.

The development of light-activated H_2S donors must always reflect the type of application, ranging from precise and self-reporting donors for cellular and tissue experiments, through nontoxic biocompatible formulations for *in vivo* studies, to H_2S-releasing biomaterials and drugs for medicinal research.

Acknowledgment

Support for this work was provided by the Czech Science Foundation (GA21-01799S (P.K.) and GJ19-20467Y (T.S.)), and the Czech Ministry of Education, Youth and Sports (LTAIN19166 (T.S.)). P.K. also thanks the RECETOX Research Infrastructure (No. LM2018121) financed by the Ministry of Education, Youth and Sports, and the Operational Programme Research, Development and Education (the CETOCOEN EXCELLENCE project No.CZ.02.1.01/0.0/0.0/17_043/0009632) for supportive background. This project was supported by the European Union's Horizon 2020 Research and Innovation Programme under grant agreement No.857560 (P.K.). This publication reflects only the author's view, and the European Commission is not responsible for any use that may be made of the information it contains. The Department of Chemistry of University of Zurich and the Swiss National Science Foundation (SNSF, PZ00P2_193425) are gratefully acknowledged for their financial support (P.Š).

References

1 Shrestha, P., Dissanayake, K.C., Gehrmann, E.J. et al. (2020). Efficient far-red/near-IR absorbing BODIPY photocages by blocking unproductive conical intersections. *J. Am. Chem. Soc.* 142 (36): 15505–15512.

2 Peterson, J.A., Wijesooriya, C., Gehrmann, E.J. et al. (2018). Family of BODIPY photocages cleaved by single photons of visible/near-infrared light. *J. Am. Chem. Soc.* 140 (23): 7343–7346.

3 Broichhagen, J., Frank, J.A., and Trauner, D. (2015). A roadmap to success in photopharmacology. *Acc. Chem. Res.* 48 (7): 1947–1960.

4 Hüll, K., Morstein, J., and Trauner, D. (2018). In vivo photopharmacology. *Chem. Rev.* 118 (21): 10710–10747.

5 Lerch, M.M., Hansen, M.J., van Dam, G.M. et al. (2016). Emerging targets in photopharmacology. *Angew. Chem. Int. Ed.* 55 (37): 10978–10999.

6 Klan, P. and Wirz, J. (2009). *Photochemistry of Organic Compounds: From Concepts to Practice*, 1e, 584. Chichester: Wiley.

7 Jacques, S.L. (2013). Optical properties of biological tissues: a review. *Phys. Med. Biol.* 58: R37.

8 Phan, T.G. and Bullen, A. (2010). Practical intravital two-photon microscopy for immunological research: faster, brighter, deeper. *Immun. Cell Biol.* 88 (4): 438–444.

9 Jacques, S.L. (2013). Optical properties of biological tissues: a review. *Phys. Med. Biol.* 58 (11): R37–R61.

10 Sandell, J.L. and Zhu, T.C. (2011). A review of in-vivo optical properties of human tissues and its impact on PDT. *J. Biophotonics* 4 (11, 12): 773–787.

11 Bashkatov, A., Berezin, K., Dvoretskiy, K. et al. (2018). Measurement of tissue optical properties in the context of tissue optical clearing. *J. Biomed. Opt.* 23 (9): 091416.

12 Wisotzky, E., Uecker, F., Dommerich, S. et al. (2019). Determination of optical properties of human tissues obtained from parotidectomy in the spectral range of 250 to 800 nm. *J. Biomed. Opt.* 24 (12): 125001.

13 Allison, R.R. and Sibata, C.H. (2010). Oncologic photodynamic therapy photosensitizers: a clinical review. *Photodiagn. Photodyn. Ther.* 7 (2): 61–75.

14 Agostinis, P., Berg, K., Cengel, K.A. et al. (2011). Photodynamic therapy of cancer: an update. *CA Cancer J. Clin.* 61 (4): 250–281.

15 Huang, Z. (2005). A review of progress in clinical photodynamic therapy. *Technol. Cancer Res. Treat.* 4 (3): 283–293.

16 Haitz's law. *Nat. Photonics* 1 (1): 23. (2007). https://www.nature.com/articles/nphoton.2006.78.

17 Steele, R.V. (2007). The story of a new light source. *Nat. Photonics* 1 (1): 25–26.

18 Brancaleon, L. and Moseley, H. (2002). Laser and non-laser light sources for photodynamic therapy. *Lasers Med. Sci.* 17 (3): 173–186.

19 Kim, M.M. and Darafsheh, A. (2020). Light sources and dosimetry techniques for photodynamic therapy. *Photochem. Photobiol.* 96 (2): 280–294.

20 Grossman, C., Zhu, T., Finlay, J. et al. (2010). Targeted laryngeal photodynamic therapy with a balloon diffusing light source. *Photodiagn. Photodyn. Ther.* 7 (3): 158–161.

21 Qumseya, B.J., David, W., and Wolfsen, H.C. (2013). Photodynamic therapy for Barrett's esophagus and esophageal carcinoma. *Clin. Endosc.* 46 (1): 30–37.

22 Zhu, T., Liang, X., Chang, C. et al. (2011). An IR navigation system for real-time treatment guidance of pleural PDT. *SPIE* 7886: 78860L.

23 Kruijt, B., van der Snoek, E.M., Sterenborg, H.J.C.M. et al. (2010). A dedicated applicator for light delivery and monitoring of PDT of intra-anal intraepithelial neoplasia. *Photodiagn. Photodyn. Ther.* 7 (1): 3–9.

24 Miyazaki, K., Morimoto, Y., Nishiyama, N. et al. (2012). A novel homogeneous irradiation fiber probe for whole bladder wall photodynamic therapy. *Lasers Surg. Med.* 44 (5): 413–420.

25 Soergel, P., Wang, X., Stepp, H. et al. (2008). Photodynamic therapy of cervical intraepithelial neoplasia with hexaminolevulinate. *Lasers Surg. Med.* 40 (9): 611–615.

26 Nyst, H.J., van Veen, R.L.P., Tan, I.B. et al. (2007). Performance of a dedicated light delivery and dosimetry device for photodynamic therapy of nasopharyngeal carcinoma: phantom and volunteer experiments. *Lasers Surg. Med.* 39 (8): 647–653.

27 Hughes, M.N., Centelles, M.N., and Moore, K.P. (2009). Making and working with hydrogen sulfide: the chemistry and generation of hydrogen sulfide in vitro and its measurement in vivo: a review. *Free Radic. Biol. Med.* 47 (10): 1346–1353.

28 Cuevasanta, E., Denicola, A., Alvarez, B., and Möller, M.N. (2012). Solubility and permeation of hydrogen sulfide in lipid membranes. *PLoS One* 7 (4): e34562.

29 Zhao, Y. and Pluth, M.D. (2016). Hydrogen sulfide donors activated by reactive oxygen species. *Angew. Chem. Int. Ed.* 55 (47): 14638–14642.

30 Mitsuhashi, H., Yamashita, S., Ikeuchi, H. et al. (2005). Oxidative stress-dependent conversion of hydrogen sulfide to sulfite by activated neutrophils. *Shock* 24 (6): 529–534.

31 Li, L. and Moore, P.K. (2008). Putative biological roles of hydrogen sulfide in health and disease: a breath of not so fresh air? *Trends Pharmacol. Sci.* 29 (2): 84–90.

32 Benchoam, D., Cuevasanta, E., Möller, M.N., and Alvarez, B. (2019). Hydrogen sulfide and persulfides oxidation by biologically relevant oxidizing species. *Antioxidants* 8 (2): 48.

33 Chen, K.Y. and Morris, J.C. (1972). Kinetics of oxidation of aqueous sulfide by oxygen. *Environ. Sci. Technol.* 6 (6): 529–537.

34 Misak, A., Grman, M., Bacova, Z. et al. (2018). Polysulfides and products of H_2S/*S*-nitrosoglutathione in comparison to H_2S, glutathione and antioxidant Trolox are potent scavengers of superoxide anion radical and produce hydroxyl radical by decomposition of H_2O_2. *Nitric Oxide* 76: 136–151.

35 Predmore, B.L., Lefer, D.J., and Gojon, G. (2012). Hydrogen sulfide in biochemistry and medicine. *Antioxid. Redox Signal.* 17 (1): 119–140.
36 Wang, M., Li, K., Zhu, R. et al. (2011). The protective function of hydrogen sulfide for lysozyme against riboflavin-sensitized photo-oxidation. *J. Photochem. Photobiol. B* 103 (2): 186–191.
37 Tzvi, Y. and Paz, Y. (2019). Highly efficient method for oxidation of dissolved hydrogen sulfide in water, utilizing a combination of UVC light and dissolved oxygen. *J. Photochem. Photobiol. A* 372: 63–70.
38 Dan, M., Yu, S., Li, Y. et al. (2020). Hydrogen sulfide conversion: how to capture hydrogen and sulfur by photocatalysis. *J. Photochem. Photobiol. C* 42: 100339.
39 Raevskaya, A.E., Stroyuk, A.L., and Kuchmii, S.Y. (2004). Photocatalytic oxidation of hydrosulfide ions by molecular oxygen over cadmium sulfide nanoparticles. *J. Nanopart. Res.* 6 (2): 149–158.
40 Iliev, V., Prahov, L., Bilyarska, L. et al. (2000). Oxidation and photooxidation of sulfide and thiosulfate ions catalyzed by transition metal chalcogenides and phthalocyanine complexes. *J. Mol. Catal.* 151 (1): 161–169.
41 Portela, R., Suárez, S., Rasmussen, S.B. et al. (2010). Photocatalytic-based strategies for H_2S elimination. *Catal. Today* 151 (1): 64–70.
42 Zhang, X., Tang, Y., Qu, S. et al. (2015). H_2S-selective catalytic oxidation: catalysts and processes. *ACS Catal.* 5 (2): 1053–1067.
43 Klán, P., Šolomek, T., Bochet, C.G. et al. (2013). Photoremovable protecting groups in chemistry and biology: reaction mechanisms and efficacy. *Chem. Rev.* 113 (1): 119–191.
44 Weinstain, R., Slanina, T., Kand, D., and Klán, P. (2020). Visible-to-NIR-light activated release: from small molecules to nanomaterials. *Chem. Rev.* 120 (24): 13135–13272.
45 Irie, M. (1983). Light-induced reversible pH change. *J. Am. Chem. Soc.* 105 (7): 2078–2079.
46 Tron, A., Peyret, A., Thevenot, J. et al. (2016). A prototype reversible polymersome-stabilized H_2S photoejector operating under pseudophysiological conditions. *Org. Biomol. Chem.* 14 (27): 6394–6397.
47 Givens, R.S., Heger, D., Hellrung, B. et al. (2008). The photo-Favorskii reaction of *p*-hydroxyphenacyl compounds is initiated by water-assisted, adiabatic extrusion of a triplet biradical. *J. Am. Chem. Soc.* 130 (11): 3307–3309.
48 Venkatesh, Y., Das, J., Chaudhuri, A. et al. (2018). Light triggered uncaging of hydrogen sulfide (H_2S) with real-time monitoring. *Chem. Commun.* 54 (25): 3106–3109.
49 Barman, S., Mukhopadhyay, S.K., Biswas, S. et al. (2016). A *p*-hydroxyphenacyl–benzothiazole–chlorambucil conjugate as a real-time-

monitoring drug-delivery system assisted by excited-state intramolecular proton transfer. *Angew. Chem. Int. Ed.* 55 (13): 4194–4198.

50 Parthiban, C., Pavithra, M., Reddy, L.V.K. et al. (2018). Tetraphenylethylene conjugated *p*-hydroxyphenacyl: fluorescent organic nanoparticles for the release of hydrogen sulfide under visible light with real-time cellular imaging. *Org. Biomol. Chem.* 16 (42): 7903–7909.

51 Bera, M., Maji, S., Paul, A. et al. (2019). A water soluble light activated hydrogen sulfide donor induced by an excited state *meta* effect. *Org. Biomol. Chem.* 17 (40): 9059–9064.

52 Wang, P. (2017). Developing photolabile protecting groups based on the excited state meta effect. *J. Photochem. Photobiol. A.* 335: 300–310.

53 Devarie-Baez, N.O., Bagdon, P.E., Peng, B. et al. (2013). Light-induced hydrogen sulfide release from "caged" *gem*-dithiols. *Org. Lett.* 15 (11): 2786–2789.

54 Il'ichev, Y.V., Schwörer, M.A., and Wirz, J. (2004). Photochemical reaction mechanisms of 2-nitrobenzyl compounds: methyl ethers and caged ATP. *J. Am. Chem. Soc.* 126 (14): 4581–4595.

55 Il'ichev, Y.V. and Wirz, J. (2000). Rearrangements of 2-nitrobenzyl compounds. 1. Potential energy surface of 2-nitrotoluene and its isomers explored with ab initio and density functional theory methods. *J. Phys. Chem. A* 104 (33): 7856–7870.

56 Fukushima, N., Ieda, N., Sasakura, K. et al. (2014). Synthesis of a photo-controllable hydrogen sulfide donor using ketoprofenate photocages. *Chem. Commun.* 50 (5): 587–589.

57 Sasakura, K., Hanaoka, K., Shibuya, N. et al. (2011). Development of a highly selective fluorescence probe for hydrogen sulfide. *J. Am. Chem. Soc.* 133 (45): 18003–18005.

58 Fukushima, N., Ieda, N., Kawaguchi, M. et al. (2015). Development of photo-controllable hydrogen sulfide donor applicable in live cells. *Bioorg. Med. Chem. Lett.* 25 (2): 175–178.

59 Lukeman, M. and Scaiano, J.C. (2005). Carbanion-mediated photocages: rapid and efficient photorelease with aqueous compatibility. *J. Am. Chem. Soc.* 127 (21): 7698–7699.

60 Nakagawa, H. (2018). Photo-controlled release of small signaling molecules to induce biological responses. *Chem. Rec.* 18 (12): 1708–1716.

61 Supuran, C.T. and Scozzafava, A. (2007). Carbonic anhydrases as targets for medicinal chemistry. *Bioorg. Med. Chem.* 15 (13): 4336–4350.

62 Chengelis, C.P. and Neal, R.A. (1980). Studies of carbonyl sulfide toxicity: metabolism by carbonic anhydrase. *Toxicol. Appl. Pharmacol.* 55 (1): 198–202.

63 Zhao, Y., Bolton, S.G., and Pluth, M.D. (2017). Light-activated COS/H_2S donation from photocaged thiocarbamates. *Org. Lett.* 19 (9): 2278–2281.

64 Hamidi, S. and Avoli, M. (2015). Carbonic anhydrase inhibition by acetazolamide reduces in vitro epileptiform synchronization. *Neuropharmacology* 95: 377–387.

65 Hua, W., Zhao, J., and Gou, S. (2020). A naphthalimide derivative can release COS and form H_2S in a light-controlled manner and protect cells against ROS with real-time monitoring ability. *Analyst* 145 (11): 3878–3884.

66 Sharma, A.K., Nair, M., Chauhan, P. et al. (2017). Visible-light-triggered uncaging of carbonyl sulfide for hydrogen sulfide (H_2S) release. *Org. Lett.* 19 (18): 4822–4825.

67 Chauhan, P., Bora, P., Ravikumar, G. et al. (2017). Esterase activated carbonyl sulfide/hydrogen sulfide (H_2S) donors. *Org. Lett.* 19 (1): 62–65.

68 Štacko, P., Muchová, L., Vítek, L., and Klán, P. (2018). Visible to NIR light photoactivation of hydrogen sulfide for biological targeting. *Org. Lett.* 20 (16): 4907–4911.

69 Cairns, T.L., Evans, G.L., Larchar, A.W., and McKusick, B.C. (1952). *gem*-Dithiols. *J. Am. Chem. Soc.* 74 (16): 3982–3989.

70 Berchtold, G.A., Edwards, B.E., Campaigne, E., and Carmack, M. (1959). The preparation of crystalline *gem*-dithiol under mild conditions. *J. Am. Chem. Soc.* 81 (12): 3148–3148.

71 Siegel, L.M. (1965). A direct microdetermination for sulfide. *Anal. Biochem.* 11 (1): 126–132.

72 Peng, B., Chen, W., Liu, C. et al. (2014). Fluorescent probes based on nucleophilic substitution–cyclization for hydrogen sulfide detection and bioimaging. *Chem. Eur. J.* 20 (4): 1010–1016.

73 Li, Z., Li, D., Wang, L. et al. (2019). Photocontrollable water-soluble polymeric hydrogen sulfide (H_2S) donor. *Polymer* 168: 16–20.

74 Wang, L., Ma, F., Zou, X., and Li, Z. (2015). Block copolymer nanoparticles as light-induced hydrogen sulfide (H_2S) donors. *J. Control. Release* 213: e71–e72.

75 Chen, W., Chen, M., Zang, Q. et al. (2015). NIR light controlled release of caged hydrogen sulfide based on upconversion nanoparticles. *Chem. Commun.* 51 (44): 9193–9196.

76 Pauloehrl, T., Welle, A., Oehlenschlaeger, K.K., and Barner-Kowollik, C. (2013). Spatially controlled surface immobilization of nucleophiles via trapping of photo-generated thioaldehydes. *Chem. Sci.* 4 (9): 3503–3507.

77 Xiao, Z., Bonnard, T., Shakouri-Motlagh, A. et al. (2017). Triggered and tunable hydrogen sulfide release from photogenerated thiobenzaldehydes. *Chem. Eur. J.* 23 (47): 11294–11300.

78 Caserio, M.C., Lauer, W., and Novinson, T. (1970). Ketonization of enols in aprotic solvents. Photoelimination of β-keto sulfides. *J. Am. Chem. Soc.* 92 (20): 6082–6084.

79 Li, L., Whiteman, M., Guan, Y.Y. et al. (2008). Characterization of a novel, water-soluble hydrogen sulfide-releasing molecule (GYY4137). *Circulation* 117 (18): 2351–2360.

80 Alexander, B.E., Coles, S.J., Fox, B.C. et al. (2015). Investigating the generation of hydrogen sulfide from the phosphonamidodithioate slow-release donor GYY4137. *MedChemComm* 6 (9): 1649–1655.

81 Woods, J.J., Cao, J., Lippert, A.R., and Wilson, J.J. (2018). Characterization and biological activity of a hydrogen sulfide-releasing red light-activated ruthenium(II) complex. *J. Am. Chem. Soc.* 140 (39): 12383–12387.

82 Yi, S.Y., Moon, Y.K., Kim, S. et al. (2017). Visible light-driven photogeneration of hydrogen sulfide. *Chem. Commun.* 53 (86): 11830–11833.

83 Zhu, Y., Pavlos, C.M., Toscano, J.P., and Dore, T.M. (2006). 8-Bromo-7-hydroxyquinoline as a photoremovable protecting group for physiological use: mechanism and scope. *J. Am. Chem. Soc.* 128 (13): 4267–4276.

84 Karthik, S., Saha, B., Ghosh, S.K., and Pradeep Singh, N.D. (2013). Photoresponsive quinoline tethered fluorescent carbon dots for regulated anticancer drug delivery. *Chem. Commun.* 49 (89): 10471–10473.

85 Bera, M., Maji, S., Paul, A. et al. (2020). Quinoline H_2S donor decorated fluorescent carbon dots: visible light responsive H_2S nanocarriers. *J. Mater. Chem. B* 8 (5): 1026–1032.

86 Huang, Y., Li, H., He, X. et al. (2020). Near-infrared photothermal release of hydrogen sulfide from nanocomposite hydrogels for anti-inflammation applications. *Chin. Chem. Lett.* 31 (3): 787–791.

87 Li, H., Yao, Y., Shi, H. et al. (2019). A near-infrared light-responsive nanocomposite for photothermal release of H_2S and suppression of cell viability. *J. Mater. Chem. B* 7 (39): 5992–5997.

15

Macromolecular and Supramolecular Approaches for H_2S Delivery

Sarah N. Swilley-Sanchez[1,2], *Zhao Li*[1,2], *and John B. Matson*[1,2]

[1] *Department of Chemistry, Virginia Polytechnic Institute and State University, Blacksburg, VA, USA*
[2] *Macromolecules Innovation Institute, Virginia Polytechnic Institute and State University, Blacksburg, VA, USA*

List of Abbreviations

4HTB	4-hydroxythiobenzamide
A	alanine
ACAT-1	acetyl-CoA acetyltransferase 1
Ad	adamantyl group
ADT	anethole dithiolethione
ADT-OH	5-(4-hydroxyphenyl)-3*H*-1,2-dithiole-3-thione
AEMA	2-(4-oxopentanoate) ethyl methacrylate
AIE	aggregation induced emission
BSA	bovine serum albumin
CD36	cluster of differentiation 36
CMC	critical micelle concentration
CM-cellulose	carboxymethyl cellulose
COS	carbonyl sulfide
CPDB	benzodithioate dithioester
CTA	chain transfer agent
Cys	cysteine
DADS	diallyl disulfide
DLS	dynamic light scattering
DPTU	depolymerizable poly(thiourethane)
dSATO	di-SATO-functionalized peptide

DSC	differential scanning calorimetry
E	glutamic acid
EGMA	ethylene glycol methyl ether methacrylate
F	phenylalanine
FBA	4-formylbenzamide
FBEMA	2-(4-formylbenzoyloxy)ethyl methacrylate
FeS@BSA	ferrous sulfide-embedded bovine serum albumin
GaOS	garlic oil-soluble extract
GMA	glycidyl methacrylate
GSH	glutathione
GYY4137	morpholin-4-ium 4-methoxyphenyl(morpholino) phosphinodithioate
H_2S	hydrogen sulfide
HA	hyaluronic acid
HNE	human neutrophil elastase
HO-1	heme oxygenase-1
HO·	hydroxyl radical
I	isoleucine
K	lysine
K_{ATP}	ATP-sensitive potassium
K_S	*S*-aroylthiooxime functionalized lysine residue
MAPK	mitogen-activated protein kinase
MEO_2MA	2-(2-methoxyethoxy)ethyl methacrylate
MRI	magnetic resonance imaging
NF-κB	nuclear factor kappa light chain enhancer of activated B cells
NSHD1	*N*-(benzoylthio)benzamide
OEGMA	oligo(ethylene glycol) methyl ether methacrylate
PACMO	poly(*N*-acryloyl morpholine)
PADT	poly(anethole trithione)
PAEMA	poly(2-(4-oxopentanoate) ethyl methacrylate)
PAEMATK	poly(2-(4-oxopentanoate) ethyl methacrylate thioketal)
PAM	poly(*N*-acryloyl morpholine)
PCL	poly(ε-caprolactone)
PDTE	(bis)phenyl dithioesters
PEG	poly(ethylene glycol)
PFHMA	poly(3-formyl-4-hydroxybenzyl methacrylate)
PFM	poly(D,L-lactic acid) fibrous membranes
PI-3K/Akt	phosphatidylinositol-3-kinase/protein kinase B
PISA	polymer-induced self-assembly
PLA	poly(D,L-lactic acid)
PLGA	poly(D,L-lactic-*co*-glycolic acid)

PMA	phorbol 12-myristate-13-acetate
PMMA	poly(methyl methacrylate)
POEGA	poly[oligo-*n*(ethylene glycol) methyl ether acrylate]
POEGMA	poly[oligo(ethylene glycol) methyl ether methacrylate]
PTBAG	poly(*tert*-butyl acryloylglycinate)
PVA	poly(vinyl alcohol)
qSATO	quad-SATO-functionalized peptide
RAFT	reversible addition–fragmentation chain transfer polymerization
RDRP	reversible-deactivation radical polymerization
ROS	reactive oxygen species
SATHA	*S*-aroylthiohydroxylamine
SATO	*S*-aroylthiooxime
SBTHA	*S*-benzoylthiohydroxylamine
sGC/cGMP	soluble guanylyl cyclase/cyclic guanosine monophosphate
SR-A	macrophage scavenger receptor
STF	simulated tear fluid
TBAG	*tert*-butyl acryloylglycinate
TEM	transmission electron microscopy
T_g	glass transition temperature
tSATO	tri-SATO-functionalized peptide
V	valine

15.1 Introduction

First coined by Wang in 2002, the term gasotransmitter refers to endogenous small molecule gases that are permeable to membranes and have specific physiological functions [1]. The three established gasotransmitters are nitric oxide, carbon monoxide, and hydrogen sulfide (H_2S). H_2S was the last to be discovered and was thought of as a toxic gas most known for its revolting smell. However, in 1996, Abe and Kimura showed that H_2S is present at high concentrations in the brain, and concluded that it likely serves a role in synaptic activity [1]. Since this breakthrough, it has been demonstrated over the past quarter century that H_2S therapy decreases left ventricular dilation, cardiac hypertrophy, and improves cardiac function in animal models [2]. Additionally, H_2S delivery reduces oxidative stress typically associated with heart failure, providing long-term protection against myocardial injury [3]. Thus, these properties, along with preservation of mitochondrial function, inhibition of cardiomyocyte apoptosis, and anti-inflammatory and anti-oxidative effects, gives H_2S the designation of a cardioprotective agent [4].

In addition to acting as a cardioprotective agent, H_2S relaxes vascular endothelial cells and vascular smooth muscle cells by activating K_{ATP} channels [5]. Downregulation of NF-κB or upregulation of HO-1 expression gives rise to the anti-inflammatory properties of H_2S [6]. Though the exact pathways are complex, H_2S likely promotes vascular endothelial cell proliferation via a combination of stimulation of the PI-3K/A_{kt} pathway, opening of K_{ATP} channels, and inhibition of the sGC/cGMP pathway [7]. H_2S also promotes angiogenesis, the formation of new blood vessels from existing ones, via the MAPK pathway [8]. Related to angiogenesis, downregulation of CD36, SR-A, and ACAT-1 expression prevents foam cells (M2 macrophages) from forming, therefore limiting atherosclerosis [9]. From these studies and many more, it has been extensively shown that H_2S plays a multitude of vital roles that impact human health.

As a result of its critical roles in a number of disease states, exogenous administration of H_2S holds therapeutic potential, but its efficacy and clinical utility will depend on effective delivery. Unfortunately, methods for direct administration of H_2S are limited due to the gaseous and reactive nature of H_2S. For example, delivery via inhalation presents safety concerns due to the toxicity and flammability of H_2S, and intravenous injections of sulfide salts (Na_2S and NaHS) generate a rapid spike in H_2S bloodstream concentration followed by a quick decline [10, 11]. We note that in this chapter when we refer to H_2S in aqueous solution, this includes both H_2S and HS^-, which are both present in appreciable quantities in neutral water due to the pK_{a1} of H_2S near 7. In addition, the above delivery methods would require repetitive administration to maintain therapeutic levels and require elevated doses, amplifying the risk of negative side effects [12]. In order to overcome the inherent challenges of delivering H_2S, including its gaseous nature, poor thermal and aqueous stability, short *in vivo* half-life, and toxicity at high concentrations [13], small molecules that decompose under specific conditions (termed H_2S donors) have been developed that provide control over the location, duration, and rate of H_2S release [14–16]. Beyond small molecules, macromolecular and supramolecular systems offer additional potential advantages, including improved drug solubility and stability, prolonged circulation time, reduced side effects, improved pharmacokinetics, and the potential for intracellular trafficking [17, 18].

Three common drug release profiles are shown below in Figure 15.1. The solid black line indicates a burst release/injection profile, while the dotted gray line indicates repeated dosing, representative of an ideal profile generated from taking medications daily. The third profile, shown as a dashed line, depicts a controlled release profile. Polymeric drug delivery systems are often designed to generate this type of steady bloodstream concentration profile. They may also offer better patient compliance as well as reduced dosing frequency compared to injections or daily medications [19]. Various materials have been studied for the delivery of

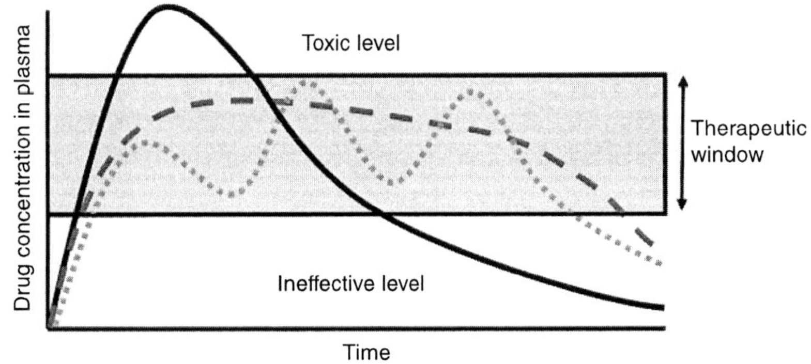

Figure 15.1 Release profile from different types of drug administration. The solid black line indicates one dose, the dotted gray line shows repeated doses, and the dashed dark gray line represents an ideal controlled release profile. Source: Modified from Liechty et al. [18].

drugs, biologics, and gasotransmitters such as H_2S. Polymers are perhaps the most widely studied drug delivery materials, and they have been utilized clinically over the last few decades to control release profiles for various drugs and to target drugs to a desired site of action [20]. In many cases, polymers greatly decrease the rate at which a drug is administered into a biological system, enabling a controlled release profile (Figure 15.1, dashed gray curve) [18]. Controlled release is helpful for many small molecule drugs, but it is critical for delivery of H_2S due to its short *in vivo* half-life (on the order of seconds to minutes) [1, 21]. Although the field of H_2S donors and H_2S delivery systems is young, it has evolved quickly. In this chapter, we discuss macromolecular and supramolecular systems that have been studied for the delivery of H_2S. This chapter is organized by macromolecular topology and type of bonding (covalent or supramolecular), and we highlight the advantages, disadvantages, and potential uses of these materials with a view toward a potential clinical future for H_2S.

15.2 H_2S-Donating Linear Polymers

Linear polymers are the simplest type of polymer topology, and many major industrial polymers are linear. There exist a wide variety of synthetic methods to make functional linear polymers with the ability to release H_2S. Linear polymers may be functionalized on the repeating units to install pendant H_2S donors (Figure 15.2a) or on one or both chain ends (Figure 15.2b), and several examples of H_2S-releasing polymers have been reported using both functionalization strategies. More recently, linear polymers have been reported that contain H_2S-releasing units in the backbone, where triggered depolymerization generates H_2S (Figure 15.2c). This section covers each type of linear H_2S-releasing polymer.

Figure 15.2 Functionalization methods for linear polymeric H_2S donors. (a) Pendant, (b) chain end, and (c) backbone H_2S donors. Black lines indicate polymers and gray spheres indicate H_2S-donating groups.

15.2.1 Pendant H_2S Donors

Polymers with pendant functionalities for donating drugs have been extensively studied [22], and early drug-releasing polymers for cancer treatment were originally of this type [23, 24]. Copolymers, polymers that contain two or more monomers, are typically utilized for applications such as these, where crystallinity, glass transition temperature, critical solution temperatures, drug release profiles, drug loading, and other properties can be tuned by varying the ratio of the monomers. In addition to the ratio of the monomers, the arrangement of the monomers also affects polymer properties, most notably whether the copolymer is fully soluble in water or forms aggregates such as micelles. As shown in Figure 15.3, monomer units in copolymers can be arranged to form either block, alternating, or statistical copolymers. Amphiphilic block copolymers, where one polymer block is water soluble and one is not, can form micelles and other types of aggregates in water. Alternating and statistical copolymers for drug delivery are also usually designed with a water-soluble monomer and a hydrophobic, drug-carrying monomer; in this case, distributing both types of monomer units along the entire backbone promotes solubility of the whole copolymer. Throughout the rest of this section, and in many cases in Sections 2.1,2.2, 3.1, 4.1, and 6.3, examples are discussed in which the ratio of two types of repeating units and the specific type of copolymer determines the solubility of the copolymer, total H_2S loading and release kinetics, and physicochemical properties from macromolecular and supramolecular H_2S donors [12, 22, 25–31].

Linear polymers with pendant H_2S donors are synthesized either by forming a pre-polymer and then installing H_2S donors onto the side chains, or by directly polymerizing a monomer that contains an H_2S donor that is stable to the polymerization conditions. An example of the post-polymerization modification approach

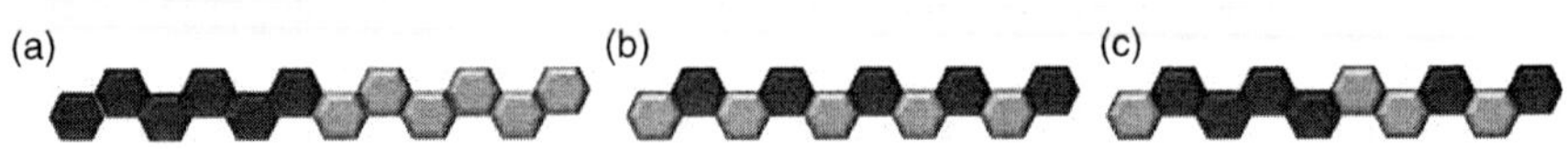

Figure 15.3 Types of copolymers: (a) block copolymer, (b) alternating copolymer, (c) statistical copolymer where light gray hexagons indicate one monomer and dark gray hexagons indicate a second type of monomer.

is Li's synthesis of a statistical copolymer of 2-(4-oxopentanoate) ethyl methacrylate (AEMA) and ethylene glycol methyl ether methacrylate (EGMA) (Figure 15.4) [25]. After isolation of the PAEMA-*co*-EGMA copolymer, the ketone functionality on the AEMA block was used to conjugate 2-nitrobenzenemethanethiol via a thioketal bond to generate PAEMATK-*co*-EGMA. The AEMA block contains caged *gem*-dithiols, which have previously been shown to release H_2S upon exposure to light [32]. In related work from Xian, small molecule caged gem dithiols have been reported to degrade at physiological pH and upon exposure to UV irradiation at 365 nm, resulting in peak H_2S release at 6 minutes [32]. The polymeric donors developed by Li achieved peak H_2S release at 3–6 minutes depending on light intensity. Although the small molecule donor and polymeric donors showed nearly identical release rates, the oligo(ethylene glycol) methyl ether methacrylate (OEGMA) units in the copolymer greatly increased the water solubility of the *gem*-dithiol H_2S donor [25].

Linear polymers with pendant H_2S-releasing groups are relatively easy to prepare, but many are based on polymers with entirely carbon-containing backbones, such as polyacrylates and polymethacrylates [26, 29, 33], polynorbornenes [34], and styrenics [35]. This is likely due to straightforward synthetic methods for making these types of polymers. Polymers derived from methacrylate, acrylate, or other vinyl or alkenyl monomers, however, leave behind aliphatic polymer backbones that are not readily biodegradable. Alternatively, polymers with aliphatic polyesters, polycarbonates, polysaccharides, and polyamide backbones may be degradable and more favorable for biological applications. Some examples of biodegradable polymers used in H_2S delivery vehicles are further discussed in Section 15.6.

15.2.2 H_2S Donors on Chain Ends

Reversible addition–fragmentation chain transfer (RAFT) polymerization is a common reversible-deactivation radical polymerization (RDRP) technique used to prepare polymers with control over molecular weight, molecular weight distribution (i.e. dispersity, Đ), and chain end functionality. This polymerization technique also offers the additional advantage of a wide range of monomers available for polymerization [36]. Additionally, such polymers are typically well-tolerated in biological systems at low concentrations [37]. Chain transfer agents (CTAs) derived from thiocarbonylthio species provide control in RAFT polymerizations, and the thiocarbonylthio species can be modified to install functional groups on one or both sides (termed the R and Z groups). In such systems, the R group ends up on the α chain end of the polymer, and the Z group is installed on the ω chain end [38, 39].

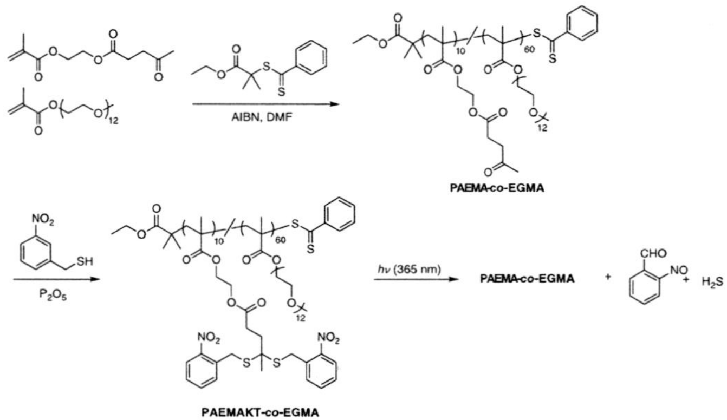

Figure 15.4 Chemical structure of $PAEMA_{10}$-*co*-$EGMA_{60}$, $PAEMATK_{10}$-*co*-$EGMA_{60}$, and subsequent H_2S release. Source: Li et al. [25]/Elsevier.

Dithiobenzoate moieties are used widely as CTAs in RAFT polymerizations [40], and more recently they were found to be thiol-triggered H_2S donors [39]. Release studies of the small molecule (bis)phenyl dithioester (PDTE) by Pluth and coworkers showed a plateau in H_2S release at ~30 minutes and a release efficiency of 68% after 60 minutes exposure to cysteine (Cys). Further functionalization of PDTE with various alkyl groups yielded a plateau in H_2S release at 15 minutes, and the addition of electron donating groups shifted the plateau in H_2S release closer to 5 minutes [39].

Drawing upon this work, Quinn and coworkers reported the synthesis of poly[oligo(ethylene glycol) methyl ether methacrylate] (POEGMA) and poly[oligo-*n*(ethylene glycol) methyl ether acrylate] (POEGA) synthesized using a benzodithioate dithioester, abbreviated CPDB, as the CTA (Figure 15.5a) [33]. The use of CPDB provided a dithioester unit on the ω chain end. Both POEGMA ($R = CH_3$) and POEGA ($R = H$) polymers were subjected to addition of 10 equiv of Cys, homocysteine, N-acetylcysteine, and glutathione (GSH). Each polymer had similar drug releasing kinetics upon exposure to Cys, with both showing peak H_2S concentration at 30 minutes. In contrast, after exposure to homocysteine, only a small amount of H_2S was released from either polymer, and neither *N*-acetylcysteine nor GSH produced significant release of H_2S from the polymer chain ends [33]. In comparing the ability of the polymer and small molecule donors to release H_2S in response to Cys, the polymer released only 25% of its payload within 50 minutes, while the small molecule released 68% of its H_2S within 60 minutes. Despite the low H_2S delivery capacity of chain-end-functionalized linear polymers, the evident decrease in release rate, along with the simplicity of installing a dithiobenzoate moiety onto a chain end using RAFT polymerization, highlights the promising nature of these types of donors for promoting water solubility and sustained H_2S release.

Similarly, polymer systems with acyl-protected perthiol chain termini have been developed (Figure 15.5b) [41]. POEGMA was synthesized as previously described with a benzodithioate (S(C=S)Ph) end group and subsequently modified to have a pyridyl disulfide (S-S-Py) endgroup (POEGMA-S-S-Py), which was converted to an acyl persulfide (POEGMA-S-S(C=O)Ph). Two other related acyl persulfide-terminated polymers were also synthesized utilizing butyl methacrylate (BMA), 2-(dimethylamino)ethyl methacrylate (DMAEMA), and 2-(diisopropylamino)ethyl methacrylate (DIPMA). Amphiphilic block copolymer POEGMA-*block*-BMA-S-S-(C=O)Ph was chosen in order to form micelles (further discussed in Section 15.4) where POEGMA is hydrophilic and poly(butyl methacrylate) (PBMA) is hydrophobic. Block copolymer (POEGMA-*co*-DMAEMA)-*block*-DIPMA-S-S-(C=O)Ph was also prepared where the DIPMA block was chosen to impart either hydrophobic or hydrophilic properties based on the pH of its environment.

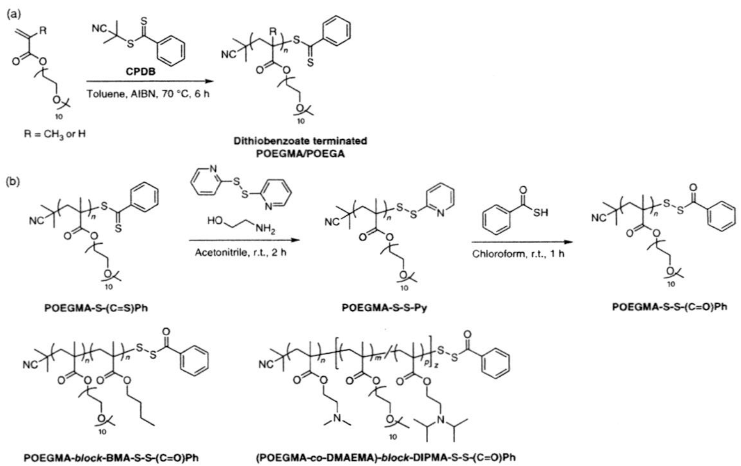

Figure 15.5 (a) Chemical structure of POEGMA (R = CH_3) and POEGA (R = H) with a dithiobenzoate endgroup. Source: Urquhart et al. [33]/American Chemical Society. (b) Chemical structures of POEGMA-S-(C=S)Ph, POEGMA-S-S-Py, POEGMA-S-S-(C=O)Ph, POEGMA-*block*-BMA-S-S-(C=O)Ph, and (POEGMA-*co*-DMAEMA)-*block*-DIPMA-S-S-(C=O)Ph. Source: Yu et al. [41]/Royal Society of Chemistry.

The H_2S releasing properties of these polymers were then investigated. In response to Cys, all three polymers exhibited peaking times in the range of 38–51 minutes at pH 7.4, but with widely varying H_2S peaking concentrations of 17, 1, and 0.2 μM for homopolymer POEGMA-S-S-(C=O)Ph, block copolymer POEGMA-*block*-BMA-S-S-(C=O)Ph, and pH-responsive block copolymer (POEGMA-*co*-DMAEMA)-*block*-DIPMA-S-S-(C=O)Ph, respectively. The authors attributed the dramatic decrease in H_2S released from the two block copolymers to their micellar structure, which sequesters the H_2S donors within the micelle core, thus limiting their exposure to Cys. In a related experiment, pH-responsive block polymer (POEGMA-*co*-DMAEMA)-*block*-DIPMA-S-S-(C=O)Ph exhibited a peaking time of 42 minutes when the pH was decreased to 5, and the concentration of H_2S released increased 20-fold to 4.4 μM, which is attributed to micelle disassembly when the DIPMA groups are protonated. Lastly, the group utilized HEK cells and a fluorescent H_2S chemosensor, SF4, to demonstrate the ability of these polymers to successfully deliver H_2S in a cellular environment. This chain end modification procedure provides a facile way to modify dithiobenzoate groups to acyl persulfide groups, which allows for H_2S release in response to a variety of thiols. Further, the ability to change polymer topology in response to pH may provide a way to selectively deliver H_2S to localized areas *in vivo*.

15.2.3 Depolymerizable Polymers for the Release of H_2S via COS

Linear self-immolative (depolymerizable) polymers were first introduced in 2008 by Shabat [42]. In this paper, the authors designed a poly(urethane) that underwent a single triggering event on the polymer chain end, which subsequently led to "unzipping" of the polymer via cascading decarboxylation and 1,6-elimation reactions along the polymer backbone. Every repeat unit that was eliminated from the polymer backbone released one molecule of a fluorophore [42]. Building on this work, one can envision a polymeric H_2S donor that depolymerizes to release multiple equivalents of H_2S in response to a single triggering event.

In 2016, Pluth and coworkers showed that small molecule thiocarbamates could decompose via decarboxylation and 1,6-elimination in the presence of reducing agents, generating carbonyl sulfide (COS) [43]. COS is quickly hydrolyzed *in vivo* by carbonic anhydrase into H_2S [44], and several types of small molecule H_2S donors that invoke a COS intermediate have been developed in the past few years [34, 45–51]. Small molecule donors, however, release a single molecule of COS/H_2S per triggering event [43]. A 2019 report from our group drew upon this idea of releasing COS from thiocarbamates by incorporating such linkages into a polymer backbone [52]. In this work, we used a condensation polymerization technique to generate a depolymerizable poly(thiourethane) (DPTU), ensuring that a thiocarbamate linkage was present in each repeat unit of the

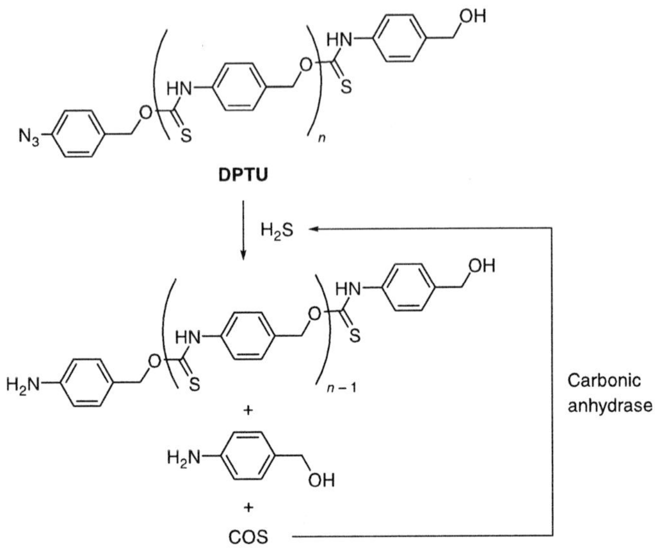

Figure 15.6 Chemical structure of depolymerizable poly(thiourethane) to generate COS/H_2S. Source: Powell et al. [52]/Royal Society of Chemistry.

polymer backbone (Figure 15.6). The polymer was then end-capped with an azide to provide a functional group that could be reduced by H_2S to initiate the depolymerization cascade. This polymer underwent alternating loss of COS and 1,6-elimation in a depolymerization reaction to release one molecule of COS per repeat unit. Furthermore, in the presence of carbonic anhydrase, the COS was converted into H_2S, which led to further depolymerization events, creating a self-amplified depolymerization reaction. Upon exposure to 0.1 equiv of Na_2S, the polymer produced a peak H_2S concentration at ~200 minutes, much longer than an analogous small molecule. Additionally, the ability to release multiple equivalents of H_2S from a single triggering event may be useful in generating H_2S-releasing polymers that can produce significant amounts of H_2S from a low level of trigger, such as an enzyme.

15.3 H_2S Delivery from Branched and Graft Polymer Topologies

Graft polymers contain a backbone polymer with pendant side chain polymers, (Figure 15.7a) [53], and they have been utilized as stimuli-responsive drug donors with triggering stimuli such as pH, redox, temperature, and UV irradiation [54]. Sometimes called brush polymers, comb polymers, or bottlebrush polymers,

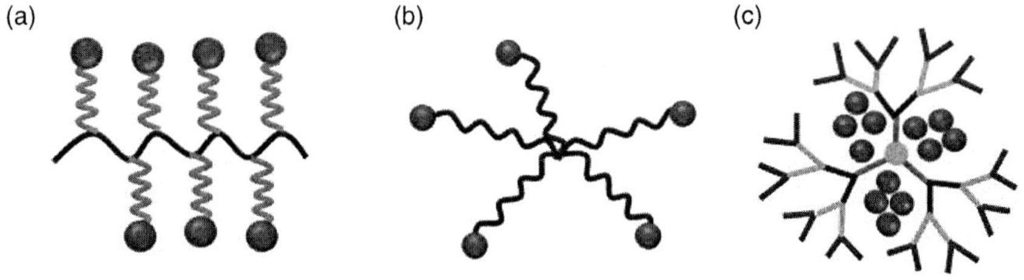

Figure 15.7 Types of branched and graft polymers. (a) Graft polymer topology, where black lines indicate the backbone polymer, gray lines indicates the grafted polymer, and spheres indicate H_2S-donating groups. Source: Based on Jenkins et al. [53]. (b) Star polymer topology, where black lines indicate polymers and green circles indicate H_2S-donating groups. (c) Dendrimer topology with physically sequestered donors, here the center circle indicates the dendrimer core (generation 0, G0), each set of lines spanning from the interior indicates one generation from G1 to G4; exterior circles indicate H_2S-donating groups. In all cases, drug donating molecules may be installed at the chain termini as shown, or placed pendant to the polymer chain (panels a and b) or within the dendrimer core (panel c).

graft polymers have been used as MRI contrast agents [55], chemotherapeutics [56], and therapeutic delivery systems, specifically for delivery of proteins and enzymes [57], and nucleic acids [58]; as such, they are promising candidates for stimuli-responsive drug delivery systems. Star polymers and dendrimers are also nonlinear polymer topologies, and while no example of H_2S-donating star or dendritic polymers exist, they may be worth considering as the field of polymeric H_2S donors advances. For example, star polymers (Figure 15.7b) offer a multitude of advantages when it comes to delivering small molecule drugs in a biological setting [59]. Due to the synthetic methods used to synthesize star polymers, they possess highly regular structures, and as such they have precise sites for drug conjugation. Arm lengths in star polymers can be easily tuned, enabling control over the hydrodynamic diameter. Additionally, there is a lot of control over chain end functionalities, and because of this they have high drug loading capabilities as well as the ability to introduce targeting ligands onto their surfaces [59]. Dendrimers also may be useful for delivering H_2S. Dendrimers are theoretically monodisperse macromolecules with an exponentially branching structure (Figure 15.7c) that were first introduced in 1978 by Vögtle [60]. Dendrimers are useful in the field of drug delivery as some readily penetrate biological barriers and cell membranes while offering high drug loading capacity via both internal cavities and multiple exposed chain ends that can be decorated with drug donor molecules [61]. As such, a variety of triggers have been employed such as temperature, hydrolysis, and pH [62]. Dendritic drug donors have been utilized in antineoplastic, anti-inflammatory, antibacterial, antiviral, and cardiovascular therapies [62].

15.3.1 Graft Polymers for the Delivery of H_2S

In 2014, our group reported on a copolymer comprised of 2-(4-formylbenzoyloxy) ethyl methacrylate (FBEMA) and 2-(2-methoxyethoxy)ethyl methacrylate (MEO_2MA). This polymethacrylate that contained arylaldehyde pendant groups was modified with *S*-aroylthiohydroxylamines (SATHAs) to install pendant *S*-aroylthiooxime (SATO) groups along the polymer backbone (Figure 15.8) [63]. The proposed mechanism of H_2S release from SATOs begins with a reversible thiol exchange between the SATO and Cys, thus producing an arylidenethiooxime (R–C=N–SH) and *N*-benzoylated Cys after an S → N acyl transfer step [64]. Next, the arylidenethiooxime reacts with a second equivalent of Cys to form thiocysteine (Cys–SSH), ammonia, and an aldehyde. Finally, a reaction between thiocysteine and Cys generates cystine and H_2S. We found that Cys-triggered H_2S release from the SATO groups could be tuned by varying the electronics at the para position of the SATO unit – by changing the R group from H, to CH_3, or CF_3, the H_2S release peaking time was changed from 36, to 53, to 25 minutes, respectively, and these results directly mimicked those of previous small molecule studies [64]. While the H_2S release peaking times for these polymeric H_2S donors were similar to those of analogous small molecules, the incorporation of MEO_2MA enhanced the water solubility of the macromolecular donor. Additionally, this study gives insight as to a facile way to modify polymers with SATO groups at high conversions and under mild reaction conditions.

In a 2020 report from Quinn and coworkers [26], trisulfide-containing polymethacrylate-*graft*-poly(ethylene glycol) graft polymers were synthesized via RAFT. Two types of trisulfide-containing monomers were prepared – one with the trisulfide unit between the methacrylate and the PEG side chain (M_1), and one with the trisulfide unit near the end of the PEG side chain (M_2). Each of these monomers was co-polymerized with OEGMA (Figure 15.9) at molar ratios of 10%, 20%, and 30% to produce a library of six polymers with trisulfide H_2S-donating moieties.

The authors utilized GSH and Cys as triggers to analyze the H_2S-donating ability of this polymer system. They found that the copolymers containing higher concentrations of M_1 and M_2 released more H_2S than their lower concentration counterparts, although all six polymers reached a plateau in H_2S release at around 30 minutes. Interestingly, the M_2 copolymers released less H_2S than M_1 copolymers despite having equivalent amounts of trisulfide monomers. Dynamic light scattering (DLS) analysis of the $M_1 - 30$ and $M_2 - 30$ polymers revealed that the block polymers containing M_1 formed smaller aggregates, suggesting a more tightly packed nanostructure that decreased the ability of the trigger to reach the trisulfide moiety [26]. A related study from a different group on the small molecule diallyl trisulfide showed that it reached a plateau in H_2S release

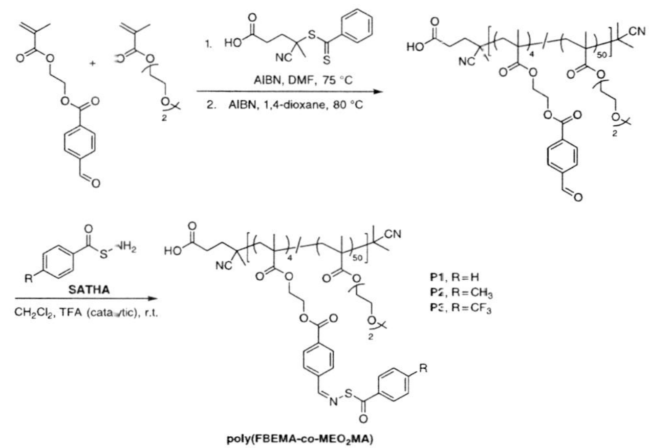

Figure 15.8 Synthesis of poly(FBEMA-*co*-MEO_2MA) polymers. Source: Foster and Matson [63]/American Chemical Society.

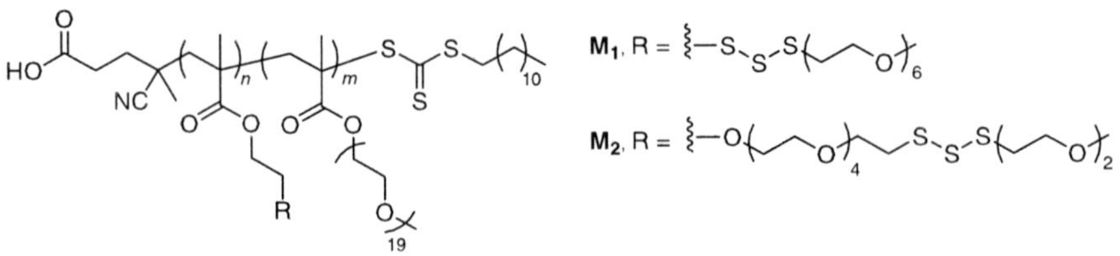

Figure 15.9 Chemical structures of OEGMA-*block*-M_1 and OEGMA-*block*-M_2 copolymers. Source: Dao et al. [26]/American Chemical Society.

in <15 minutes upon exposure to GSH [65]. Although the conditions were not exactly the same in the small molecule study and the polymer study, it appears that incorporating a multitude of trisulfide linkages into the side chains of a brush copolymer extended the time to peak H_2S release.

Because graft polymers have been utilized in a multitude of applications with various triggers to release drug payloads, there is fertile ground to employ them further as stimuli-responsive H_2S delivery vehicles with tunable stimuli and release kinetics. Additionally, we believe there is room for this field to advance by creating novel dendritic and star polymer H_2S donors because they possess features that are useful in the field of drug delivery, providing highly regular structures with high drug loading capabilities. Based on the limited studies available, it appears that these nonlinear topologies slow down the rate of release of H_2S compared with small molecule analogs; whether they release more slowly than similar linear polymers is not yet known. Regardless, branched and graft polymer topologies offer higher H_2S donor loading capacity on the chain ends than linear polymers, and they may also enable the possibility for installation of other surface or interior functional groups to provide targeting and/or imaging capacity.

15.4 Polymer Micelles for H_2S Delivery

Polymer micelles are usually made from a block copolymer with a hydrophobic and hydrophilic block. Upon dissolution in aqueous solution, a selective solvent for only the hydrophilic block, they can self-assemble to form spherical structures (Figure 15.10) referred to as micelles. To assess the stability of these structures, a critical micelle concentration value (CMC) can be determined, above which micelles form and create a hydrophobic pocket in which hydrophobic drugs can be sequestered [68]. Micelles exhibit lower CMCs than small molecule surfactants, and they are also typically large enough (20–100 nm) to prevent rapid renal clearance [68–70], making them in many cases more favorable in drug delivery applications than surfactants, linear polymers, and small molecules.

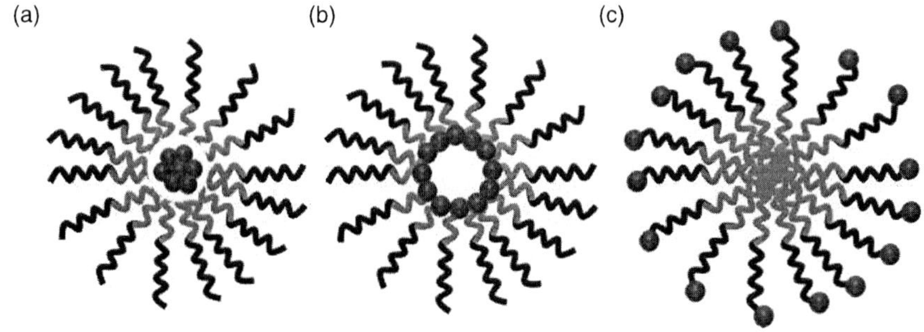

Figure 15.10 Various types of drug loading either within or on polymer micelles. (a) Physical encapsulation of drug or prodrug within micelle. Source: Based on Kabanov et al. [66]. (b) Covalent attachment of drug or prodrug to the micelle core block. (c) Covalent attachment of drug or prodrug to micelle shell block. Black lines indicate hydrophilic shell polymers, gray lines indicate hydrophobic core, and spheres indicate H_2S-donating groups. Source: Based on Bader et al. [67].

In 1989, Kabanov first proposed physically entrapping drugs or other molecules within a hydrophobic micelle core (Figure 15.10a) [66]. This approach is ideal especially when either the hydrophobic block utilized to form the micelle core and/or the drug to be delivered lacks a chemical handle in which to conjugate the two together [71, 72]. A separate approach, introduced 5 years prior by Ringsdorf, is that of chemical conjugation of a drug to either the core forming block (Figure 15.10b) or the corona forming block (Figure 15.10c) [67]. The chemical linkage that connects the two can then be altered to provide various stimuli-responsive handles that cleave the drug and release it from the micelle core [73–75]. Of these two methods, the chemical conjugation method has been used most widely thus far for delivering H_2S [27–29, 76–79].

15.4.1 H_2S Donors Covalently Attached to Polymer Amphiphiles

Several types of H_2S donors have been covalently attached to either the hydrophobic or hydrophilic block as a pendant group or on chain ends in block copolymer micelles, including SATOs [76, 78, 79], thioamides [29], trisulfides [80], and dithiolthiones [27, 28, 77, 81, 82]. In other manuscripts, we have discussed the relative strengths and weaknesses of these different types of small molecule H_2S donors, and questioned whether some even release H_2S under physiological conditions [14]; however, here we focus on how the micellar structure can be used to tune release rates or other aspects of H_2S delivery. Because most H_2S donors are hydrophobic, the majority have been conjugated to the hydrophobic block.

In an example of a covalently attached H_2S donor system, Hasegawa synthesized poly(*N*-acryloyl morpholine) (PACMO) as the water-soluble block and

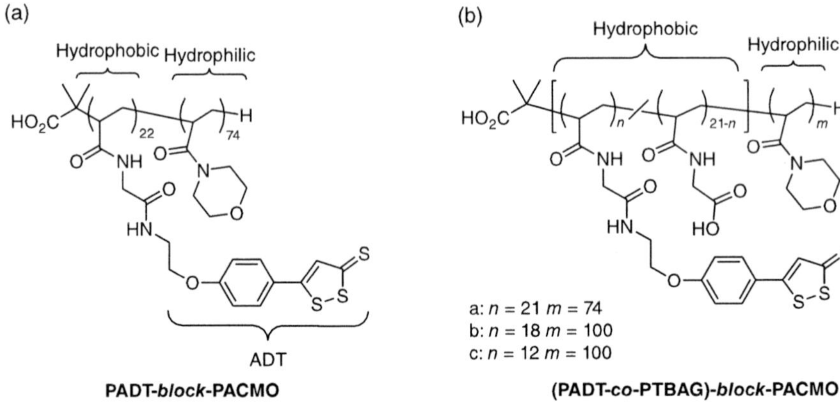

Figure 15.11 (a) PADT-*block*-PACMO. Source: Takatani-Nakase et al. [27]/Royal Society of Chemistry. (b) (PADT-*co*-PTBAG)-*block*-PACMO copolymers. Source: Chen et al. [28]/Royal Society of Chemistry.

then subsequently polymerized *tert*-butyl acryloylglycinate (TBAG) to create a block copolymer, PACMO-*block*-PTBAG [28]. The *tert*-butyl group on the PTBAG block was removed, and this block was further functionalized with anethole dithiolethione (ADT) via a post-polymerization modification to create the block copolymer PAM-*block*-PADT (PAM = poly(*N*-acryloyl morpholine), PADT = poly(anethole trithione)) (Figure 15.11a). H_2S release studies were done in order to compare the release rate from both the micelles and a small molecule analog, 5-(4-hydroxyphenyl)-3*H*-1,2-dithiole-3-thione (ADT-OH). The authors found that neither donor released H_2S in cell culture medium, but some H_2S was observed in the presence of cell lysate. In response to the cell lysate, H_2S release from the micelles was much slower as compared to ADT-OH, potentially due to sterics preventing a putative lysate-derived enzyme from entering the micelle core to trigger drug release [27].

More recently, a similar micelle system was proposed in which the PACMO-*block*-PTBAG block copolymer was modified by changing the degree of polymerization of the PACMO block as well as the degree of functionalization with ADT-OH in the PADT block (Figure 15.11b) [28]. Compared with the original polymer micelles, a change in H_2S release was observed that correlated with micelle stability (i.e. the CMC value), where polymer (a) formed the most stable micelles (lowest CMC value), and polymer (c) formed a soluble polymer (no aggregation). As more ADT groups were added to the PTBAG block, the rate of H_2S release decreased; the authors suggest that micelles with more hydrophobic ADT content have lower CMC values, leading to a more thermodynamically stable micellar structure and therefore slower H_2S release, though they do not remark on the reason behind this observation. Following this trend, the H_2S release from the micelles was fastest for dissolved polymer (c) and slowest for micelles from polymer (a). Together, these two studies indicate that sequestering ADT-OH within a micelle core as well as increasing the thermodynamic stability of micelles decreases the rate of release of H_2S from the micelle core.

In an example of polymer micelles with thioamide-based H_2S donors, Davis and coworkers reported a study in which linear polymers were compared to micelles that had either H_2S donors sequestered within the micelle core or on the outer hydrophilic shell (Figure 15.12) [29]. To create the library of polymers, the group POEGMA, a polymeric block containing an aryl thioamide (PTHA), and poly(methyl methacrylate) (PMMA) were copolymerized at various ratios to yield three unique polymer structures. Polymer (a) had H_2S donors installed in the micelle corona (Figure 15.12a), while polymer (b) had donors installed within the micelle core (Figure 15.12b). As a direct comparison, polymer (c) was synthesized (Figure 15.12c) to produce an analog of the hydrophilic segment of polymer (a), but without the hydrophobic PMMA block. This afforded a hydrophilic, molecularly dissolved polymer rather than micelles in solution.

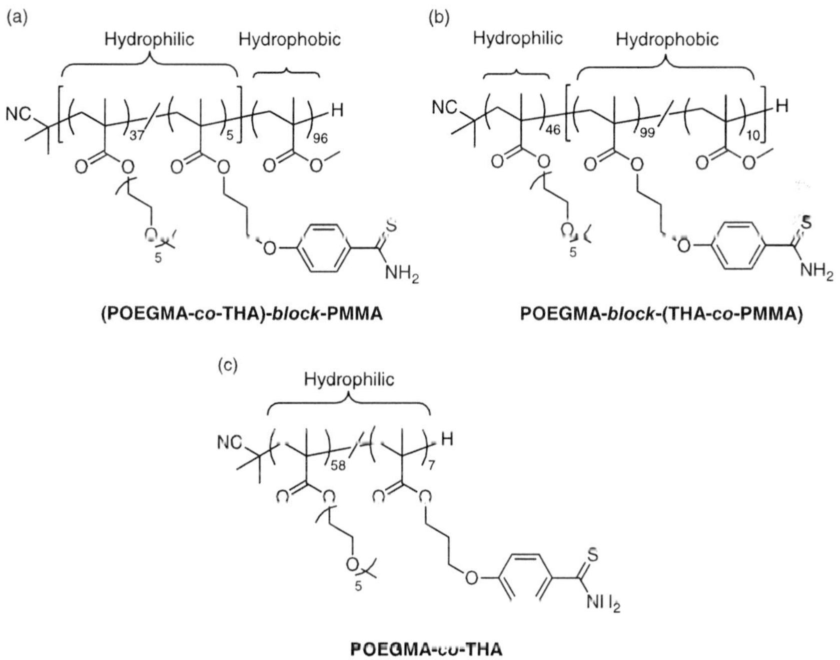

Figure 15.12 Chemical structure of micelle forming polymers (a) (POEGMA-*co*-THA)-*block*-PMMA, (b) POEGMA-*block*-(THA-*co*-PMMA), and (c) POEGMA-*co*-THA. Source: Ercole et al. [29]/American Chemical Society.

Thioamides release H_2S upon exposure to Cys or via hydrolysis, and all three polymers were studied under both types of triggering conditions. H_2S release rates under hydrolysis conditions (no added thiol) showed that polymer (a) had a peaking time of 60 minutes, polymer (b) had a longer peaking time at >120 minutes, and the linear analog (c) had a peaking time of 41 minutes [29]. These results revealed that the linear polymer (c) undergoes hydrolysis most rapidly, followed by polymers (b) and (a). This suggests that the donors sequestered within the hydrophobic micelle core hydrolyze more slowly because the hydrophobic micelle core provides a dehydrated environment. Upon exposure to Cys, however, the peaking times for all three polymers were nearly identical (12–14 minutes). Further analysis was done by comparing the percentage of thioamides that were converted into H_2S, and the authors showed that unimers released only 20% of their possible H_2S loading while polymers (a) and (b) released 88% and 58%, respectively. These data suggest that sequestering the thioamides within the micelle core restricts their accessibility to the Cys trigger, which decreases the amount of H_2S released in response to Cys. Further, the authors note that peaking time concentrations of H_2S were 4, 3, and 23 μmol/L for (a)–(c) due to hydrolysis, respectively, and this changed to 58, 67, and 19 μmol/L in when using Cys as a trigger. Taken together, these results suggest that the rate of H_2S release is much faster in response to Cys, but unimers are much more susceptible to hydrolysis than their micelle counterparts. Lastly, the authors note that there is a possible mechanism in which the thioamides hydrolyze but do not produce any H_2S, and this could also contribute to the release rates mentioned above [29].

In related work, in 2019 our group demonstrated the ability to tune the rate of H_2S release from micelle cores [79]. To begin, FBEMA was polymerized via RAFT using a PEGylated CTA (Figure 15.13). The resulting polymer, PFBEMA, consisted of a hydrophilic PEG block and a hydrophobic PFBEMA block. After the removal of the trithiocarbonate chain end, the aldehyde groups on PFBEMA were functionalized either with SATHAs to provide SATO H_2S donors, or an alkyl hydroxylamine to add a glass transition temperature (T_g)-lowering alkyl functional group (Figure 15.13) Homopolymer analogs mimicking just the core block were analyzed by differential scanning calorimetry (DSC), revealing that polymers containing solely SATO pendant groups possessed a T_g of 96 °C, while 100% alkyl-functionalized polymers had a T_g of 18 °C. From a set of five block copolymers with core block T_g values ranging from 18 to 96 °C, a library of micelles with variable micelle core mobility was generated.

The different block copolymers with variable core blocks were then analyzed to evaluate how micelle core T_g influenced the release rate of H_2S. It was found that micelles formed from polymers (a)–(d) had H_2S release half-lives of 4.2, 1.7, 0.39, and 0.18 hours, respectively, in response to 1 mM Cys (polymer (e) had no SATO groups so it did not release H_2S). As such, this demonstrated that as the amount of

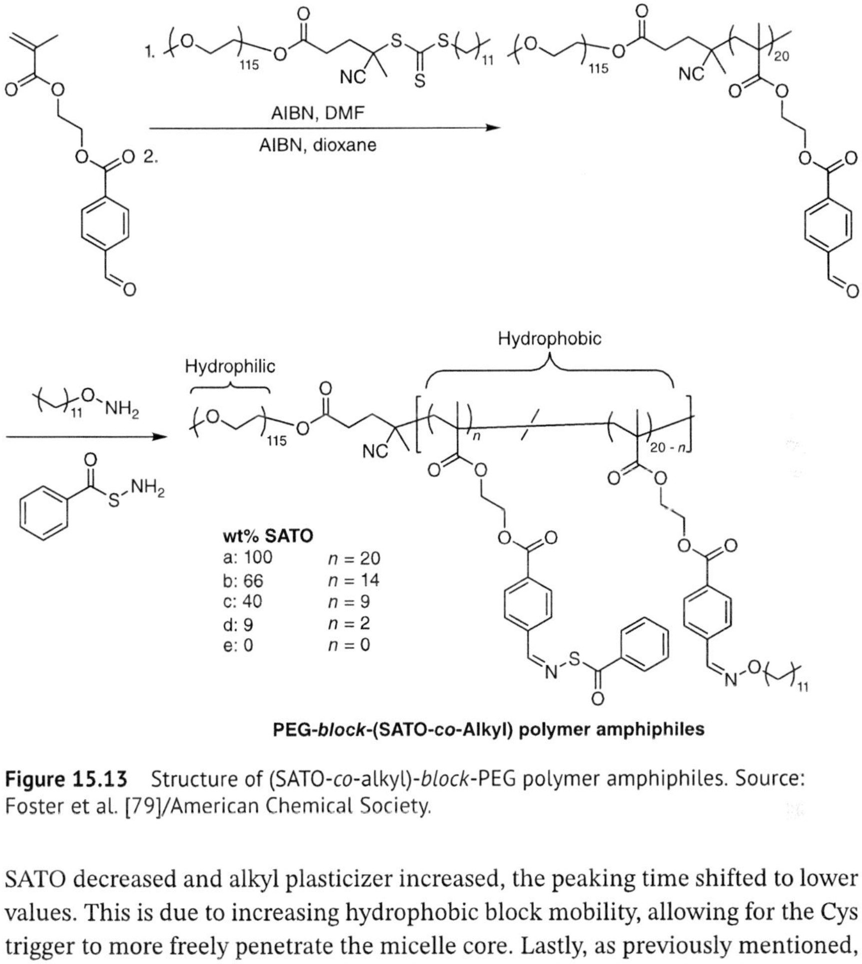

Figure 15.13 Structure of (SATO-*co*-alkyl)-*block*-PEG polymer amphiphiles. Source: Foster et al. [79]/American Chemical Society.

SATO decreased and alkyl plasticizer increased, the peaking time shifted to lower values. This is due to increasing hydrophobic block mobility, allowing for the Cys trigger to more freely penetrate the micelle core. Lastly, as previously mentioned, SATOs were previously reported to release H_2S with peaking times between 15 and 34 minutes depending on the substituents used [64]. A small molecule SATO utilized in this study displayed a peaking time of 0.4 ± 0.03 hours under similar conditions; therefore, by tuning the mobility of the micelle core, the rate of H_2S could be extended over 10 times that of the small molecule.

Each paper highlighted here demonstrates the ability to covalently attach H_2S donors to micelle-forming copolymers, not only increasing the solubility of the donor, but also drastically lengthening the duration of H_2S release through this type of synthetic strategy [27–29, 79]. Further, these works establish that the location of the H_2S donor in the micelle core or micelle shell affects H_2S release rate. Beyond donor location, micelle stability (i.e. CMC) and micelle core mobility are additional potential handles from which H_2S release can be tuned.

15.5 Polymer Networks for Localized H_2S Delivery

Polymer networks (Figure 15.14), consisting of polymers either blended together or chemically crosslinked to form a 3-D network, are of great interest to polymer chemists working to develop materials that controllably release drugs [83]. Such networks are typically composed of naturally occurring and biodegradable polymers, and applications of these include medical implants or scaffolds. We refer the reader to excellent recent reviews discussing the large scope of polymer networks used for drug delivery [84–86]. Hydrogels, which will be discussed later in this section, are one specific type of polymer network that contain hydrophilic polymers and therefore possess the ability to absorb large amounts of water within their porous structure [85]. These materials have gained interest as a way to deliver drugs because of their 3-D structures as well as the ability of oxygen and nutrients to readily diffuse throughout the network due to the porous structure [87, 88].

15.5.1 Physical Encapsulation of H_2S Donors Within Networks

Networks created from polymers can either contain physical or chemical crosslinks, and the resulting structures provide ample empty volume in which drugs can be sequestered, improving their solubility [84–86, 89–91]. Small molecule H_2S donors such as NaHS and Na_2S exhibit uncontrolled burst release profiles of H_2S once dissolved in solution [92]. While this offers the advantage of a rapid increase in the concentration of H_2S, the potential for acute side effects is possible, and this is far from the body's natural regulation of H_2S concentrations. Correspondingly, such donors only provide short term effects in tissues [93]. Sequestration of NaHS and other H_2S donors into polymer networks greatly increases the release time of H_2S, providing a controlled and elongated rate of release [12, 31, 94].

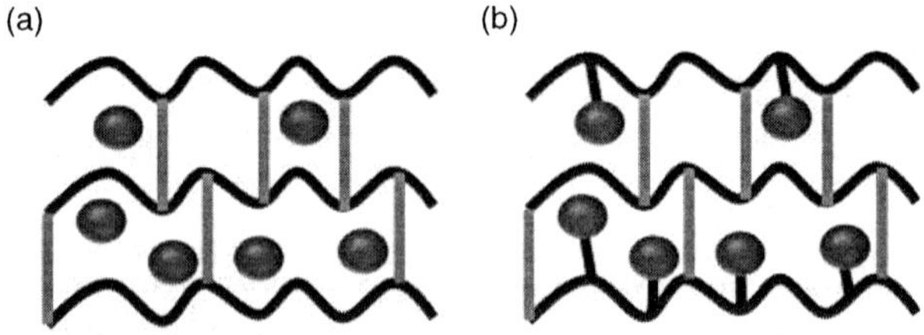

Figure 15.14 Methods of drug encapsulation within polymer networks. (a) Physical encapsulation and (b) covalent attachment. Black lines indicate polymers, gray lines indicate crosslinks within the network, and spheres indicate H_2S-donating groups. These networks are either swollen with water to make hydrogels if hydrophilic polymers are used, or slowly degrade in the body due to erosion by water if hydrophobic polymers are used.

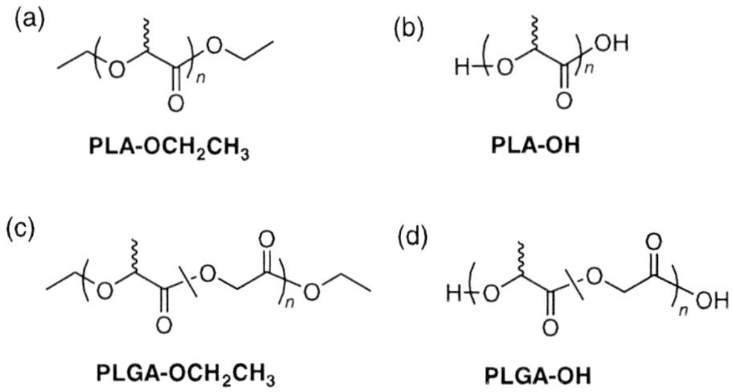

Figure 15.15 Library of PLA and PLGA polymers with various end groups. Source: Ali et al. [12]/Springer Nature.

One example of a polymeric network for the encapsulation of an H_2S donor is a study by Singh in which poly(D,L-lactic acid) (PLA) and poly(D,L-lactic-*co*-glycolic acid) (PLGA) copolymers were utilized to design a controlled release delivery system [12]. PLA and PLGA, with either carboxylic acid or ethyl ester end groups (Figure 15.15), were utilized to probe how polymer end groups change H_2S release. The four polymers were individually dissolved in a mixture of benzyl benzoate and benzyl alcohol at 10%, 15%, and 20% w/v and shaken until the mixture was homogenous, thus creating 12 separate formulations. The mixtures were then doped with NaHS (0.4% w/v) and subsequently injected into simulated tear fluid (STF).

As a positive control, NaHS solutions were prepared and were observed to release 100% of H_2S within a few minutes. A trend for all polymers was observed in that the cumulative release of H_2S from the formulations decreased as the concentration of polymer in solution increased. All four polymers released ~80% of the loaded H_2S after 72 hours at 10% w/v; however, at higher concentrations (20% w/v) the release from the gels differed. From (a) to (d) 60%, 75%, 50%, and 75% of H_2S was released after 72 hours, respectively. The authors attribute the difference in H_2S release to the relative hydrophilicity of (b) and (d) and the relative hydrophobicity of (a) and (c), where more hydrophobic polymer formulations aggregate more quickly and therefore sequester NaHS more efficiently. Additionally, they noted that the pore size of the resulting formulations decreased as the concentration of polymer in solution increased, which further enhanced the rate of H_2S release.

Injectable hydrogels that can form gels *in situ* after administration are of special interest as they can be administered more easily than 3-D structures implanted surgically and can take on the shape of a cavity [95, 96]. Xian's group has previously developed a class of small molecule phosphonamidothiolate-based H_2S donors termed "JK donors" [97]. In a recent study, they demonstrated the ability to sequester an H_2S donor named JK-1 into a hyaluronic acid (HA) hydrogel. HA was modified with methacrylic anhydride, and the resulting polymer was crosslinked with dithiothreitol (Figure 15.16) in a solution containing the JK-1 donor, resulting in a JK-1 doped HA hydrogel [30]. At pH 6, both the hydrogel and JK-1 released 70% of the available H_2S within 100 and 200 minutes, respectively. However, at physiological pH, JK-1 reached a plateau in H_2S release at 1200 minutes with ~50% of H_2S released. In comparison, the HA-JK-1 hydrogel plateaued at 500 minutes with a release capacity of ~25%. Thus, the results indicated that introduction of JK-1 into the hyaluronic acid hydrogel decreased the rate of H_2S release by more than half as compared to the small molecule alone.

15.5.2 Covalent Attachment of H_2S Donors Within Hydrogels

Covalently attaching drug donating molecules gives researchers the ability to tune drug release in two different ways: (i) controlling the degradability of the polymer matrix or (ii) controlling the stability of the covalent linkage between the polymer and the drug [98]. In 2020 our group demonstrated the ability to covalently attach SATO H_2S donors within a carboxymethyl cellulose (CM-cellulose) hydrogel to produce an H_2S-releasing hydrogel with release triggered by human neutrophil elastase (HNE) (Figure 15.17) [94]. To synthesize the SATO-hydrogel, a peptide composed of 4-formylbenzamide (FBA), valine (V), and lysine (K) was first synthesized with a final structure of FBA-VKVKVK; this peptide sequence was chosen because valine can be specifically cleaved via HNE on its C-terminal side. Next, CM-cellulose and carboxylic acid terminated PEG (HOOC–PEG–COOH) were added to the peptide along with a coupling agent to create a crosslinked hydrogel bearing pendant arylaldehydes. Different hydrogels were made at 35 : 65, 55 : 45, and 75 : 25 mol% CM-cellulose: PEG. Each hydrogel was subsequently reacted with *S*-benzoylthiohydroxylamine (SBTHA) to produce three SATO-hydrogels with varying degrees of stiffness.

H_2S release studies of the 35% CM-cellulose hydrogels, performed in the presence of Cys, showed that without the HNE trigger these hydrogels exhibited a peaking time of 240 ± 20 minutes, and upon exposure to one and two equivalents of HNE the peaking time increased to 354 ± 4 and 580 ± 50 minutes, respectively. Similarly, the peak concentration of H_2S changed from 0.6 ± 0.2 μM without HNE to 0.8 ± 0.1 and 1.4 ± 0.3 μM with 1 and 2 equiv of HNE. From this, along with mass spectrometry results, we concluded that there were SATO moieties both on the surface of the hydrogel and buried inside the hydrogel matrix. Degradation in response to HNE was relatively slow, and therefore, as more HNE was added to solution, both the peaking time and degradation rate increased as more HNE was added. These observations demonstrate the need for two triggers: HNE to degrade the polymer matrix, and Cys to trigger H_2S release from the SATO groups. Additionally, from the analysis of the 55% and 75% CM-cellulose hydrogels, we found that as the % CM-cellulose increased, the stiffness of the hydrogel also increased.

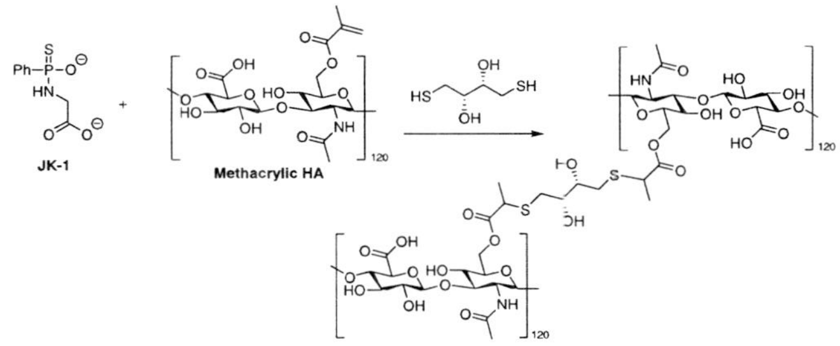

Figure 15.16 Chemical structure of the crosslinked HA network doped with JK-1. Source: Wu et al. [30]/Elsevier.

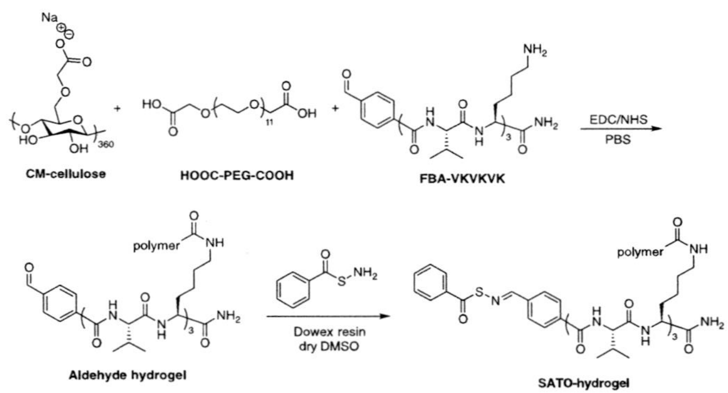

Figure 15.17 Peptide/SATO/PEG/CM-cellulose hydrogel synthesis. Source: Zhou et al. [94]/Royal Society of Chemistry.

Increased stiffness led to limited conversion when installing SATO moieties, and as a result, both peaking times and concentrations decreased compared with the less stiff 35% CM-cellulose hydrogel. This work showed how enzyme-triggered release could be used to create hydrogels with slow H_2S release in response to a specific biological signal [64].

15.6 Other Polymeric Systems for the Encapsulation of H_2S Donors

Micro scale drug delivery systems have been developed to address specific needs for controlled and localized drug delivery [99]. Examples include microfibers, polymer membranes, and microparticles, all of which have been evaluated for the delivery of H_2S (Figure 15.18) [100]. Microfibers are promising candidates for wound dressings, surgical implants, and scaffold materials. Additionally, during the manufacturing process to produce microfibers, they can be doped with drugs, for example using the Absorv® system by Zeus Inc. [101]. Similarly, polymer membranes have been used to deliver peptides, proteins, antibodies, and other drugs both topically and internally [102, 103]. Membranes can protect drugs from degrading too rapidly and also improve their solubility. Additionally, membranes can prevent burst release of drugs and decrease their release rate, which can be tuned by varying the pore size within the membrane [102]. Lastly, microparticles have also been introduced into the commercial setting, for example the Lupron Depot® used to deliver leuprolide [104]. Microparticle drug delivery systems have been developed in order to administer drugs via nasal sprays, pulmonary inhalation, oral delivery, and implant delivery [104, 105].

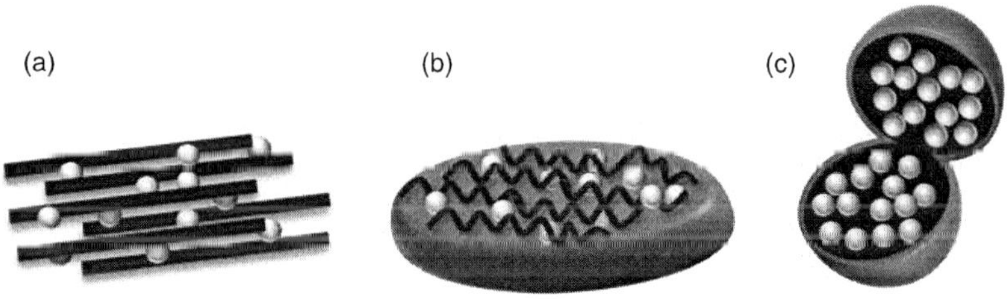

Figure 15.18 Micro scale polymeric drug delivery systems. (a) Polymeric microfibers, (b) polymer membranes, and (c) polymer microparticles. Black lines/filling indicates polymer chains, spheres are H_2S-donating groups, and gray or shell filling indicates a potential second polymer. Source: Based on Lawn et al. [100].

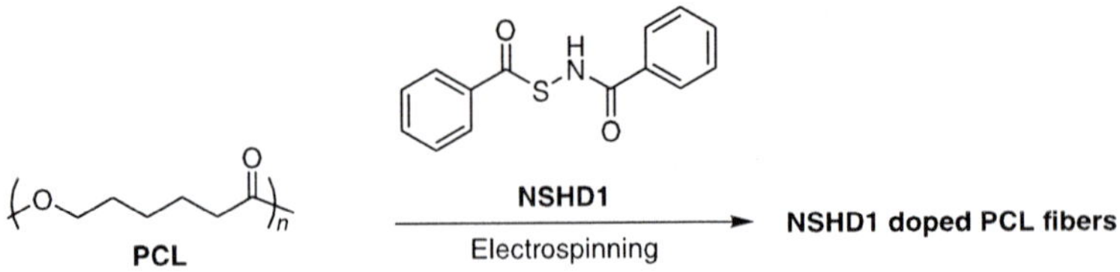

Figure 15.19 Structure of electrospun NSHD1-doped PCL microfibers. Source: Feng et al. [109]/Elsevier.

15.6.1 Microfibers as H_2S Donors

Microfibers (Figure 15.18a) are highly flexible, porous, and possess high surface area to volume ratios, and these types of systems have been studied widely for use in drug delivery [106–108]. Electrospinning is a common technique used for the facile production of microfibers from which 3-D polymer membranes can be assembled. This technique is cost effective as well as versatile – the scope of polymers that can be electrospun into fibers (ranging from nanometers to micrometers in size) is vast [106–108]. Further, such polymers can be mixed with small molecules in solution prior to electrospinning to create microfibers that are doped with drugs [109, 110].

Wang's group demonstrated a microfiber system composed of biodegradable polymers and the H_2S donor *N*-(benzoylthio)benzamide (NSHD1). Poly(ε-caprolactone) (PCL) was dissolved in 1,1,1,3,3,3-hexafluoro-2-propanol (HFIP) followed by the addition of NSHD1, and the resulting solution was electrospun to generate NSHD1-doped PCL microfibers (Figure 15.19) [109]. The small molecule NSHD1 alone had an H_2S-release peaking time of 40 minutes; however, when sequestered in PCL microfibers at 6%, 9%, and 12% w/v solutions, the peaking time increased to 120, 160, and 210 minutes, respectively; thus, indicating a three to fivefold increase in H_2S release rate. Moreover, these microfibers were cytocompatible and protected both H9c2 and NIH 3T3 cells from reactive oxygen species (ROS).

15.6.2 Membranes as H_2S Donors

Polymeric membranes (Figure 15.18b) possess two key capabilities: they can donate drugs at a controlled rate, and they provide a 3-D scaffold that mimics the extracellular matrix [110]. Materials designed to allow cells to grow in 3-D networks may provide more accurate *in vitro* drug models than traditional 2-D *in vitro* assays [111]. As mentioned above in Section 15.6.1, electrospinning has become an effective way to synthesize microfibers that can be used to assemble membranes to provide a 3-D scaffold that can be used in tissue engineering

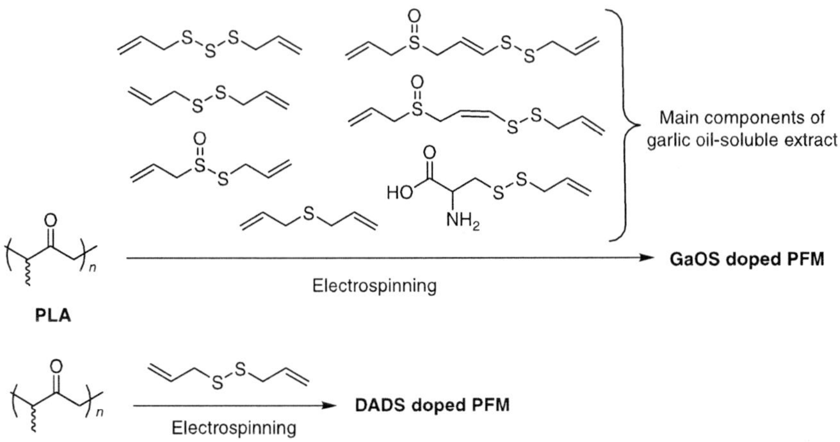

Figure 15.20 Assembly of GaOS-doped PFMs and DADS-doped PFMs via electrospinning. Source: Cacciotti et al. [110]/MDPI.

[106–108]. Figure 15.20 depicts a recent report that uses electrospinning to generate PLA polymer membranes (termed poly(D,L-lactic acid) fibrous membranes, PFMs) doped with either garlic oil-soluble extract GaOS (a mixture of several sulfur-containing compounds in garlic [112]) or one specific garlic-derived compound, diallyl disulfide (DADS) [110].

GaOS-PFM released 6 μM of H_2S after 1 hour incubation time in buffer solution at 37 °C as compared to 35 μM of H_2S from the DADS-PFM. The authors attribute the low amount of H_2S released from GaOS-PFM to reactions of some components of GaOS with solvents used during synthesis. Further studies with the DADS-PFM indicated that it released 55% of its potential H_2S after incubation in buffer solution for 7 days. In cell assays utilizing cardiac mesenchymal stem cells, neither of the doped PFMs showed toxicity, and DADS-PFM stimulated cell growth [110]. We note that computational work indicates that DADS is a slower and less effective H_2S donor that diallyl trisulfide, another component in GaOS [65, 113]. Because diallyl trisulfide likely decomposed during the electrospinning process, the full capabilities of this system may not have been reached.

15.6.3 Microparticles and Nanoparticles as H_2S Donors

Polymeric microparticles (Figure 15.18c) ranging between ~0.5 and 1000 μm in diameter are of significant interest as a drug delivery platform because of their potential for targeted delivery, tunable release kinetics, fixed and/or individualized doses, and better patient compliance compared with pills that must be taken daily [114]. These attributes stem in large part from the inability of microparticles to

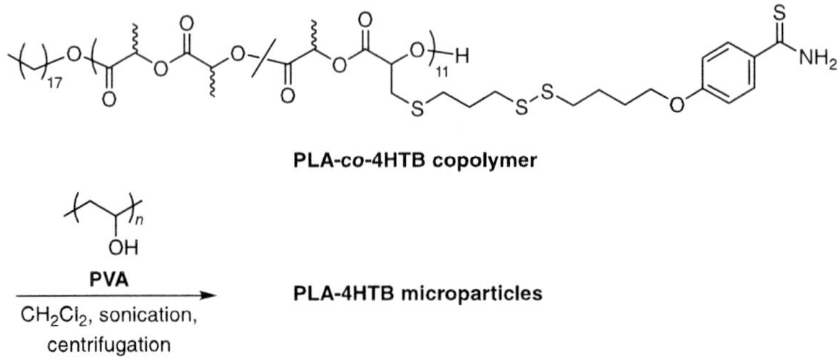

Figure 15.21 Poly(lactic acid) 4-hydroxythiobenzamide block copolymer (PLA-4HTB) and PLA-4HBT microparticles stabilized by PVA. Source: Based on Du et al. [31] and Long et al. [115].

transverse membranes, thus ensuring local activity. Polymeric nanoparticles, in contrast, are typically on the order of 10–100 nm, and like micelles, they are small enough to travel through the bloodstream, providing potential therapeutic benefits by targeting specific tissues. Both micro- and nanoparticles have been examined as H_2S delivery systems.

As discussed in Section 15.4.1, aryl thioamides are capable of releasing H_2S upon exposure to Cys or via hydrolysis [29]. In 2019, a polymeric microparticle system comprised of a poly(lactic acid) 4-hydroxythiobenzamide block copolymer (PLA-4HTB) encapsulated within poly(vinyl alcohol) (PVA) (Figure 15.21) was reported [31]. Extending from their previous paper [115], Bowden and Salem reported long-term H_2S release as well as the ability for their microparticles to protect kidney 293 cells from damage due to ROS [31].

These microparticles, assembled via a single solvent evaporation method, possessed an average size of 500 ± 200 nm. To monitor H_2S release, the microparticles were incubated in tissue culture media at 37 °C, and aliquots were taken over the span of 1 month. Within the first 50 hours of incubation, 45% of H_2S was released; the authors attribute this burst release profile to smaller particles potentially degrading more rapidly than larger ones, and thus releasing most of their H_2S in the first few days. However, after this burst release, a slow release was observed, and after 4 weeks 80% of total H_2S was released. The long release profile of H_2S is either due to the slow degradation of PLA, which can take up to years to degrade, or the slow hydrolysis of 4-HTB followed by its diffusion out of the microparticle core. Furthermore, the particles were not cytotoxic up to concentrations of 1 mg/mL and were able to protect HEK 293 cells from ROS damage [31].

In an example of nanoparticle H_2S delivery systems, Lu and coworkers reported on the synthesis of a novel, polymeric H_2S donor with the ability to "self-track" [116]. Poly(3-formyl-4-hydroxybenzyl methacrylate) (PFHMA) was synthesized

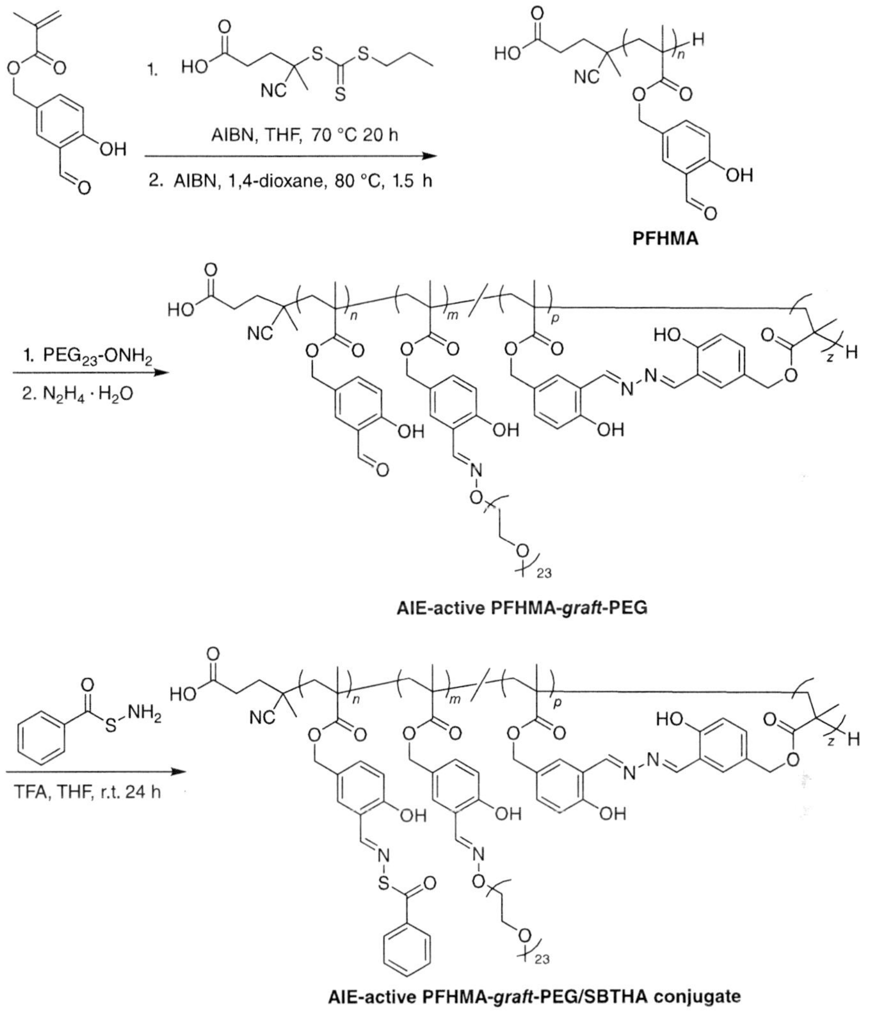

Figure 15.22 Chemical structures of AIE-active PFHMA-*graft*-PEG and AIE-active PFHMA-*graft*-PEG/SBTHA conjugate. Source: Lin et al. [116]/Royal Society of Chemistry.

via RAFT polymerization and then reacted with aminooxy terminated PEG to form a PFHMA-*graft*-PEG graft copolymer (Figure 15.22). Some of the residual pendant salicylaldehyde groups were subsequently reacted with hydrazine to link together PFHMA blocks in different polymer chains, thus forming a crosslinked PFHMA-*graft*-PEG nanoparticle that exhibited aggregation-induced emission (AIE). Lastly, a portion of the remaining salicylaldehydes groups were conjugated with SBTHA to form a SATO group, yielding the final AIE-active polymer nanoparticles PFHMA-*graft*-PEG/SBTHA. The dispersible nanoparticles aggregated together when introduced into water, enabling AIE fluorescence, and subsequent fluorescent imaging indicated the "self-tracking" ability of the donor as it fluoresced within cells. In response to 1 mM of Cys or GSH, these nanoparticles demonstrated a peaking time in H_2S release around 50–60 minutes, and cell studies revealed this system to be non-toxic up to concentrations of 100 μg/mL in L929 cells. Small molecule SATOs can have a range of peaking times, and a small molecule developed by our group similar in electronic structure to the SATO donor used in Lu's work displayed a peaking time of 24 minutes under similar conditions [64]. Therefore, conjugating the SATO H_2S donor to the PFHMA-*graft*-PEG copolymer roughly doubled the peaking time of H_2S release while also providing a fluorescence method to potentially follow these polymers *in vivo* [116].

In summary, several micro and nano-scale drug delivery systems for the delivery of H_2S have been recently developed, including microfibers, membranes, and nano- and microparticles. These types of nano- to macroscale materials hold promise in the potential translation for H_2S delivery into medical practice but with different applications depending on their size. The larger materials are too big to enter the bloodstream but could offer long-term, localized H_2S delivery in an area of interest. The ability to drastically increase the duration of drug release, combined with the ability to rescue cells from ROS damage highlighted in some of these examples, together give promising results for the advancement of localized H_2S-releasing materials. Additional materials, including hydrogels discussed in Section 15.5 and some of the supramolecular H_2S delivery systems noted below in Section 15.7, are also potential candidates for localized H_2S delivery.

15.7 H_2S Release via Supramolecular Systems

Supramolecular chemistry, characterized by the use of specific, directional, tunable, and reversible molecular recognition motifs, has been demonstrated as a useful strategy to prepare controlled and predictable materials organized across nanometer scales [117–122]. This field has progressed significantly as it has shown promise in both medical devices and therapeutics [123, 124]. A significant benefit in utilizing supramolecular chemistry to design biomaterials is that properties often follow directly from small molecule building blocks that lack the polydispersity and potential batch-to-batch variability of traditional (covalent) polymers, allowing for predictable, reversible, and highly tunable properties [118]. These properties give supramolecular materials significant potential to realize simple or complex H_2S releasing behaviors that can be applied in different medical or biological circumstances. Peptides [125], polymer nanoparticles [126], and proteins [127] have been used as supramolecular carriers to load and deliver H_2S in biomedical and biological studies. This section discusses several examples developed recently in the field of H_2S delivery systems based on supramolecular nanomaterials.

15.7.1 Self-Assembled, Peptide-Based Materials for H_2S Delivery

Peptides are useful materials to build self-assembled nanostructures for biomedical applications due to their inherent biodegradability and biocompatibility under many circumstances [128]. The straightforward synthesis of peptides allows for derivation of structure–property relationships via small changes in the chemical structure. By varying the position of a specific residue or functional group, or through changes to the peptide sequence, one can change nanostructures and ultimately the material properties of H_2S-releasing peptides. For example, we have described peptides that form nanoribbons, nanocoils, and crescent-shaped structures all from simple tetrapeptides (Figure 15.23); interestingly, such nanoassemblies exhibit different rates of H_2S release [129]. Additionally, the stepwise synthetic methods used to make peptides allow for the installation of a specific amount of H_2S-donating groups within the sequence [130], which helps develop different H_2S release behaviors. Herein we describe examples of supramolecular systems based on peptides for the controlled release of H_2S.

In 2015, we reported the first H_2S-releasing hydrogel using a self-assembling peptide functionalized with a SATO group [131]. The aromatic amphiphilic peptide sequence SATO-IAVEEE (SATO = H_2S-releasing SATO group, I = isoleucine, A = alanine, E = glutamic acid) (Figure 15.24a) has a β-sheet-rich IAV region (where peptide chains are arranged parallel/antiparallel to each other and are stabilized by a network of intermolecular H-bonds), a hydrophilic region containing three glutamic acid residues, and a SATO group on the N-terminus. Driven by hydrophobic interactions of SATO groups and stabilization via hydrogen bonding in the β-sheet-rich region, the H_2S releasing peptides self-assembled into nanofibers in aqueous solutions with an average diameter of 10 nm (Figure 15.24b). At concentrations as low as 1 wt%, a robust hydrogel formed upon charge screening through the addition of calcium chloride. Extended

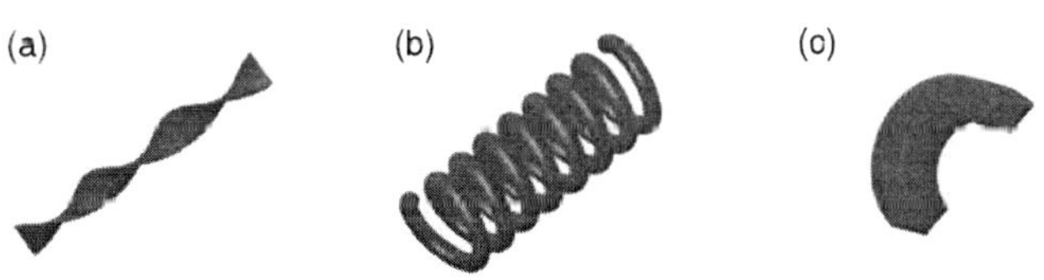

Figure 15.23 Supramolecular nanostructures of H_2S-releasing peptides. (a) Twisted ribbon, (b) nanocoil, and (c) nanocrescent.

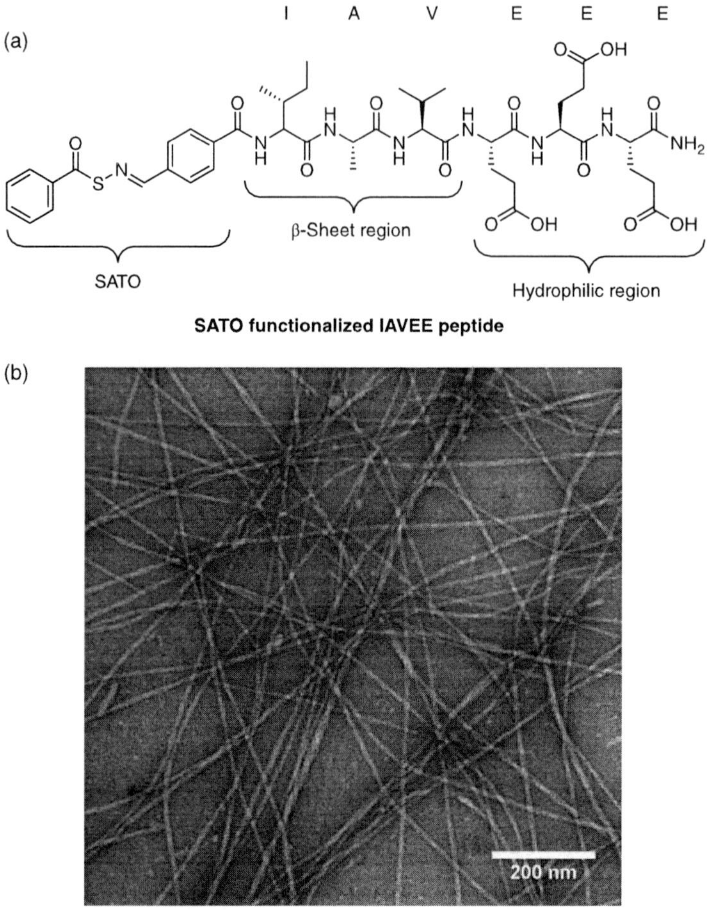

Figure 15.24 (a) Structure of SATO-functionalized peptide (IAVEEE). (b) Conventional TEM image of SATO-IAVEEE cast from water (stained with 2% uranyl acetate). Source: Carter et al. [131]/Royal Society of Chemistry. Image reproduced under a Creative Commons CC BY license from Wiley.

H_2S release was observed from these hydrogels in the presence of Cys, with measurable H_2S levels over 15 hours.

More recently, we used a similar peptide design to vary the rate of H_2S release by changing the substituents on the SATO unit or adding various linker segments between the SATO unit and the amphiphilic hexapeptide (Figure 15.25) [132, 133]. With substituent groups varying from electron-withdrawing groups to electron-donation groups, SATO-functionalized peptides with tightly packed self-assembled nanostructures displayed an increased duration of H_2S release in

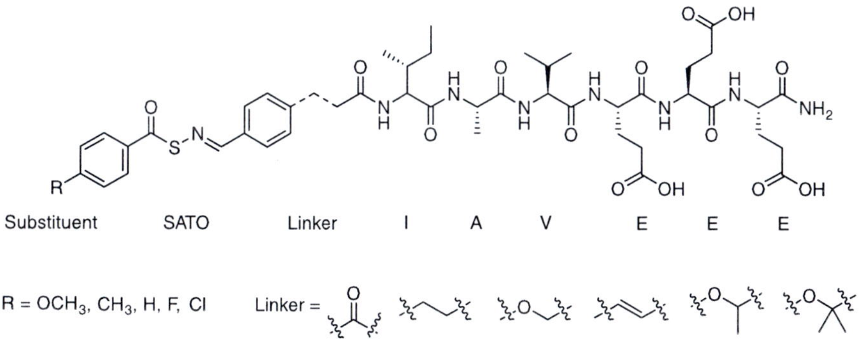

Figure 15.25 Chemical structure of SATO-functionalized peptides with different substituents and linker segments. Source: Based on Qian et al. [132] and Kaur et al. [133].

solution [132]. This is because the tight packing limited access of Cys to SATOs in the core of these nanostructures. In contrast, in the gel state SATO-peptides with flexible linkers, which packed more compactly to form a stiffer gel than peptides with rigid linkers, showed a shorter peaking time than that of peptides with rigid linkers [133]. We speculate that tight molecular packing did actually slow H_2S release, but that over the relatively longtime scales of the experiment, released H_2S was lost because of volatilization and oxidation before being detected. In general, a slow and sustained delivery of H_2S can be realized by tightly packed nanostructures resulting from molecular-level modifications. These types of gels could be beneficial to a number of disease states, and we have thus far examined them in the context of vascular repair and wound healing [134, 135].

H_2S releasing behaviors could also be controlled by varying the level of H_2S donor loading within the peptide sequence, and this has been achieved by controlling the number of H_2S releasing SATO functional groups near the N-terminus of three related peptides (Figure 15.26) [130]. The H_2S-release peaking time as measured on an H_2S-sensitive electrochemical probe depended on the morphology of the self-assembled nanostructures, where the peptides with two SATO units (dSATO) and three SATO units (tSATO) formed twisted nanoribbons, and the peptide with four SATO (qSATO) formed short, bundled nanoribbons. As the amount of SATO units that were conjugated to the peptides increased from 24%, to 28%, to 31%, the H_2S release behavior exhibited a longer peaking time from 127, to 136, to 181 minutes. Additionally, we found that qSATO, the peptide with the highest SATO loading percentage and the longest H_2S release peaking time, mitigated Dox-induced toxicity toward H9c2 cardiomyocytes most effectively.

In 2018, we reported on a different strategy to regulate H_2S release based on the supramolecular nanostructures of constitutionally isomeric peptides [129]. A group of tetrapeptides (K_SEK_SE, K_SK_SEE, K_SEEK_S) (Figure 15.27a–c) was

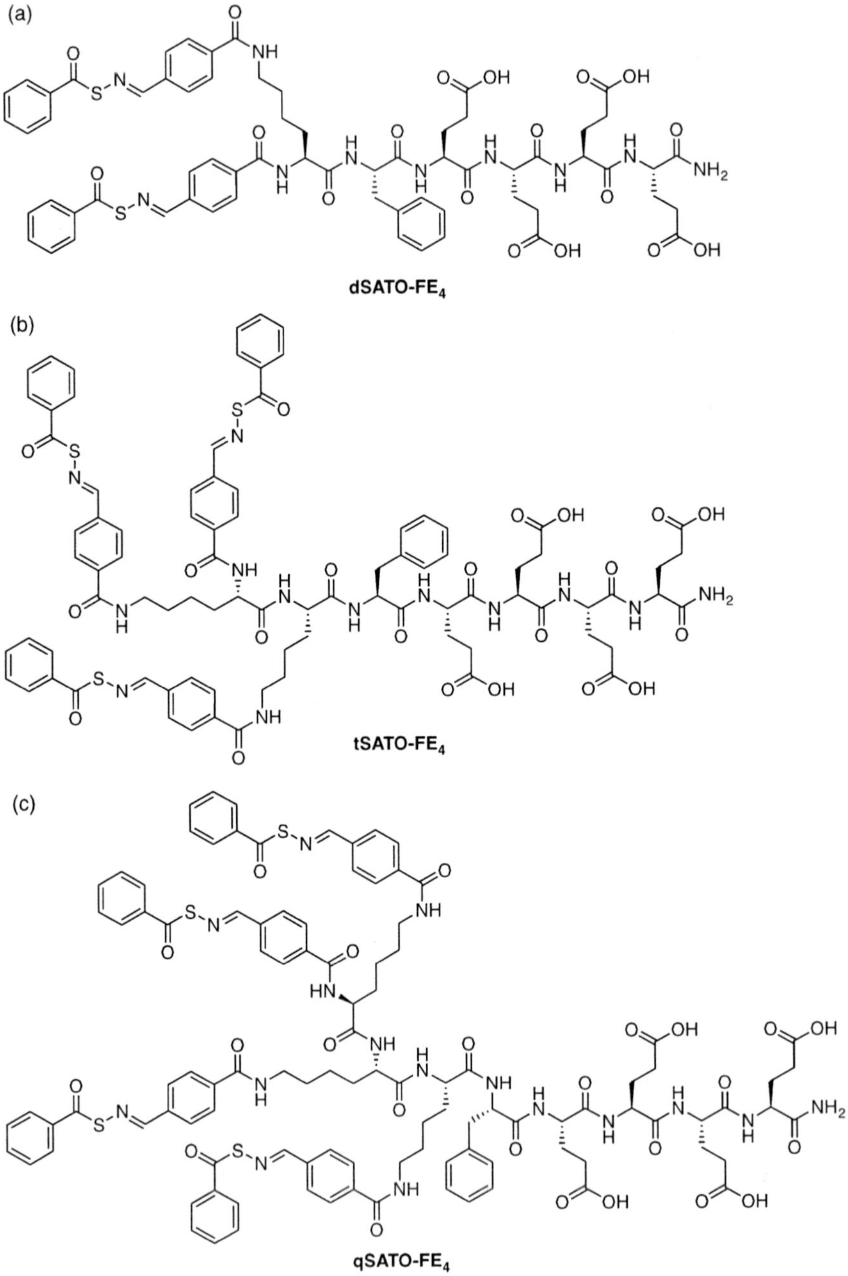

Figure 15.26 (a–c) Molecular structures of dSATO-FE_4, tSATO-FE_4, qSATO-FE_4 (F = phenylalanine). Source: Wang and Matson [130]/American Chemical Society.

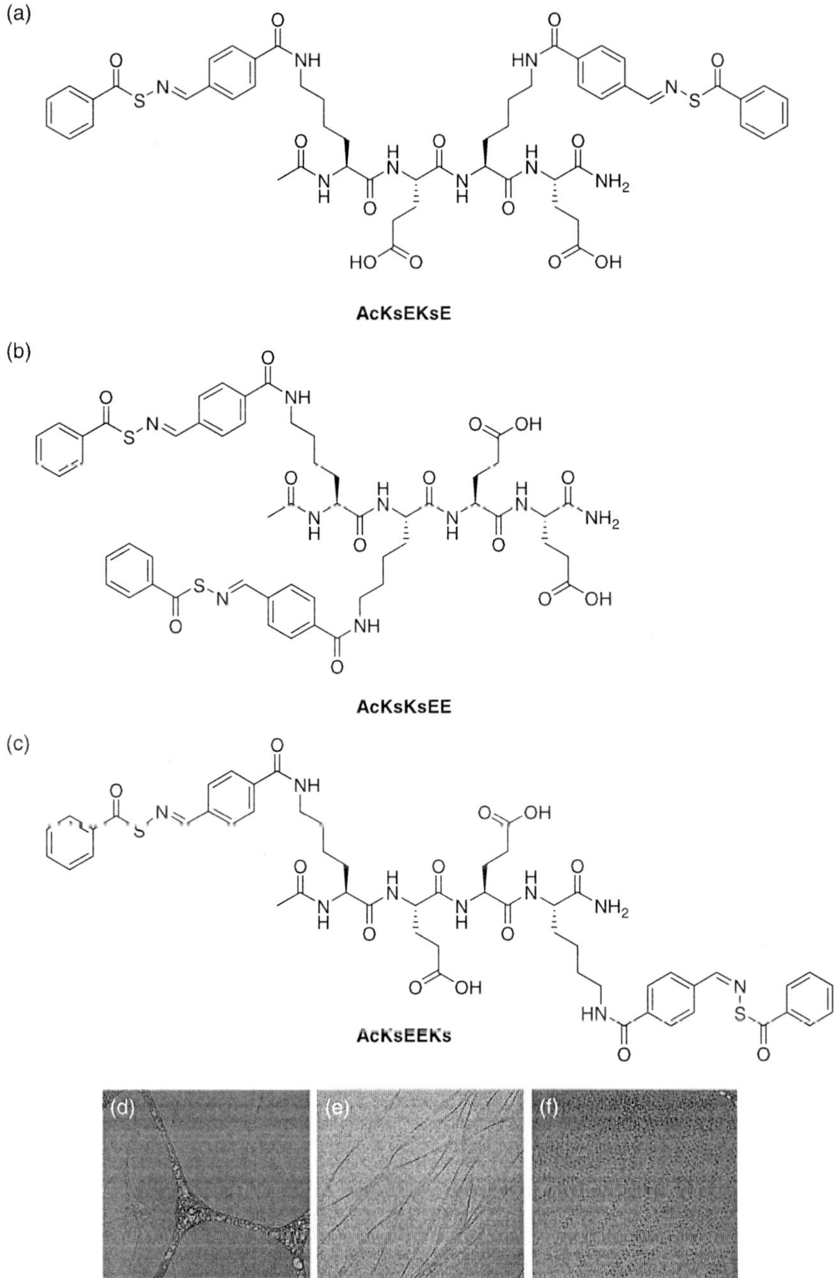

Figure 15.27 (a–c) Molecular structures of K_SEK_SE, K_SK_SEE, K_SEEK_S. (d–f) Cryo-TEM images of twisted ribbons formed by (d) K_SEK_SE and (e) K_SK_SEE, and nanocoils formed by K_SEEK_S (f) in aqueous solution (scale bar = 200 nm). Source: Wang et al. [129]/American Chemical Society. Image reproduced under a Creative Commons CC BY license from Wiley.

synthesized with two SATO-functionalized lysine residue (K_S) and two glutamic acid residues (E) in each amino acid sequence. Interestingly, the three peptides self-assembled into different nanostructures including twisted nanoribbons with varying dimensions and rigid nanocoils (Figure 15.27d–f), as was determined by conventional and cryogenic transmission electron microscopy (TEM). H_2S release studies demonstrated that nanocoil-forming peptides exhibited an initial period of 60 minutes where H_2S was slowly liberated. This initial period was not present in the profile of either of the twisted ribbons. This unique incubation period was attributed to a slower diffusion of Cys into the nanocoils than into the twisted ribbons. Furthermore, the nanocoils were more effective than either small-molecule H_2S donors (GYY4137 and Na_2S) or twisted nanoribbons in rescuing H9c2 cardiomyocytes from doxorubicin toxicity. This is likely due to the significant differences in H_2S release rates among the donors, where the slower donors enabled more of the released H_2S to activate cytoprotective pathways (versus non-productive consumption of H_2S via processes such as oxidation).

We also very recently discovered a unique crescent-shaped nanostructure based on adamantane-functionalized tetrapeptides with the general structure AdK_SK_SEX (Ad = adamantyl group, X = variable amino acid residue) (Figure 15.28a) [136]. This crescent-shaped nanostructure (Figure 15.28b) exhibited a robust amino acid tolerance on the C-terminal amino acid, but removal of the X residue led to the formation of short nanoworms (Figure 15.28c) on the size scale of the nanocrescents but without the persistent crescent shape. *In vitro* cell studies demonstrated that H_2S release from crescent nanostructures reduced ROS levels and mitigated toxicity induced by phorbol 12-myristate-13-acetate (PMA) in RAW 264.7 macrophage cells better than common H_2S donors (GYY4137 and Na_2S) and the self-assembled peptide nanoworms.

In these peptide-based supramolecular H_2S donors, H_2S release was regulated by the self-assembled nanostructures, which included nanofibers, twisted nanoribbons, curved nanoribbon bundles, nanocoils, and nanocrescents. Generally, nanostructures with a tighter molecular packing exhibited a longer and more sustained release profile by limiting the diffusion of Cys into SATO units in the nanostructure cores. This controllable release profile, combined with the monodisperse nature of the peptides and their inherent biodegradability, makes self-assembling peptides promising candidates for further development as H_2S delivery systems. Integration with bioactive peptide components may also be an avenue to create synergistic effects between peptide therapeutics and H_2S.

15.7.2 Self-Assembled Nanoparticles and Proteins for H_2S Delivery

In addition to peptide-based supramolecular H_2S delivery systems, nanoparticle-based H_2S delivery systems have also been developed, in this case by using

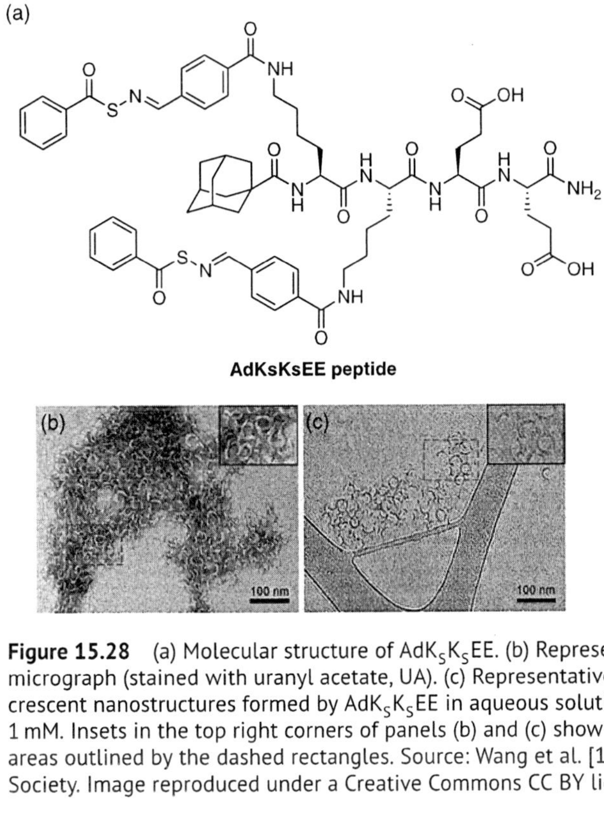

Figure 15.28 (a) Molecular structure of AdK_SK_SEE. (b) Representative conventional TEM micrograph (stained with uranyl acetate, UA). (c) Representative cryo-TEM micrograph of crescent nanostructures formed by AdK_SK_SEE in aqueous solution (PB, 10 mM, pH 7.4) at 1 mM. Insets in the top right corners of panels (b) and (c) show zoomed-in images of the areas outlined by the dashed rectangles. Source: Wang et al. [136]/American Chemical Society. Image reproduced under a Creative Commons CC BY license from Wiley.

polymer-induced self-assembly (PISA) and post-polymerization functionalization techniques. These nanoparticle systems, which have variable nanostructures, have demonstrated improved stabilities as compared to comparable small molecule donors, and therefore give rise to more controlled release profiles. In addition, the high solids content and synthetic ease of PISA methodologies can provide advantages over polymer micelles assembled from pre-formed block copolymers and supramolecular peptides, which tend to require a lot of solvent in their preparation and cannot be made at high concentrations in water.

Quinn et al. developed a shape-controlled H_2S delivery system by modification of polymer-based nanoparticles with spherical and worm-like morphologies [126]. The nanoparticles were prepared via PISA by using poly[glycidyl methacrylate-*block*-oligo(ethylene glycol) methyl ether methacrylate] (P[GMA-*b*-OEGMA]) as a macromolecular RAFT agent that was used to polymerize styrene, which acted as both the monomer and co-solvent in methanol (Figure 15.29). The GMA units were subsequently functionalized with benzoyl protected perthiol groups, generating polymer nanostructures with morphologies

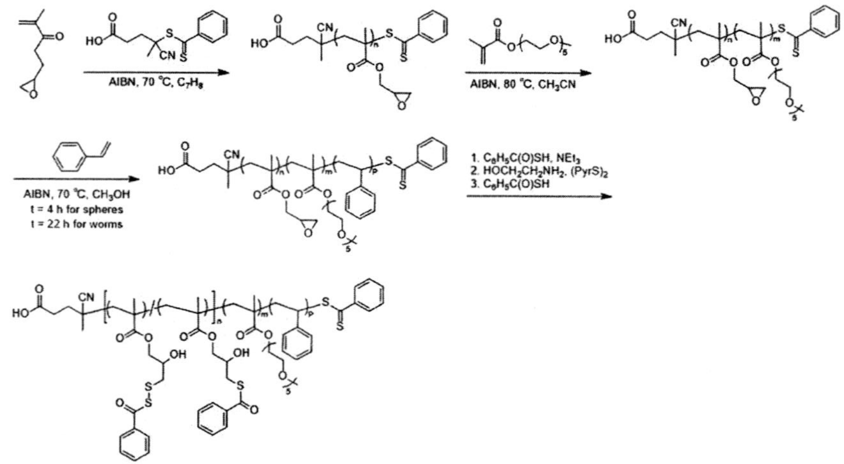

Figure 15.29 Synthesis of H_2S-donating spherical and worm-like nanoparticles via a PISA approach. Source: Yu et al. [126]/John Wiley & Sons.

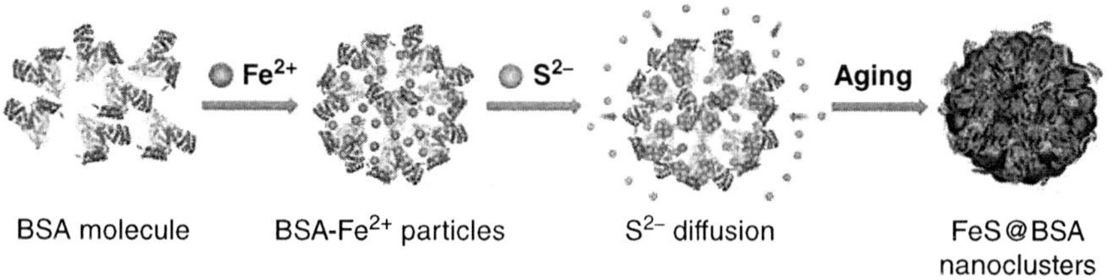

Figure 15.30 Schematic illustration of the formation of FeS@BSA nanoclusters during synthesis. Source: Xie et al. [127]/John Wiley & Sons/CC BY 4.0.

that depended on the length of the polystyrene block. After transferring into water, these thiol-triggerable H_2S polymeric donor systems released H_2S in the presence of Cys in solution and inside U2OS human bone osteosarcoma cells.

In an example of self-assembled, protein-based H_2S donors, Cui et al. designed and synthesized amorphous nanoclusters based on bovine serum albumin (BSA) nanoparticles [127]. They diffused Fe^{2+} and S^{2-} ions (note that S^{2-} does not exist in appreciable amounts in neutral water, so the S^{2-} ions, added as Na_2S, would actually be a mixture of H_2S and HS^-) into BSA nanoparticles to form ferrous sulfide-embedded BSA (FeS@BSA) nanoclusters through a self-assembly approach (Figure 15.30). When the surrounding micro-environment was mildly acidic and contained H_2O_2, mimicking the environment in many cancer tumors, Fe^{2+} induced ROS generation. The FeS@BSA nanoclusters triggered a rapid Fenton reaction to induce formation of toxic hydroxyl radicals (HO·) and released H_2S effectively within Huh7 cancer cells. The H_2S, released intracellularly, inhibited the activity of catalase, an essential antioxidant enzyme in cells that plays a crucial role in maintaining the balance of intracellular ROS. This process increased intracellular H_2O_2 levels and accelerated production of HO significantly. Interestingly, in this work H_2S did not act as a cell protecting gasotransmitter, but rather presented synergistic effects with the Fenton reaction to enhance the generation of toxic ROS species. Inspired by this application, supramolecular H_2S donors may have broader applications in cancer therapy.

In summary, different materials have been developed via supramolecular chemistry to build modular H_2S delivery systems, including peptides, polymers, and proteins. By introducing supramolecular interactions like hydrogen bonding, hydrophobic interactions, or metal coordination, these materials self-assemble into different nanostructures that improve the stability of H_2S donors and extend the release lifetime when compared with small molecular H_2S donors. More significantly, these supramolecular approaches afford nanostructures that may make targeted delivery possible, especially in the case of cancer.

15.8 Conclusions and Future Perspectives

In this chapter, we have highlighted various macro and supramolecular strategies that have been developed over the last few years to function as H_2S releasing systems. Many have longer release profiles compared to small molecules, which may require repetitive administration to maintain therapeutic levels and elevated doses, amplifying the risk of negative side effects. Macromolecular and supramolecular systems offer additional potential advantages, including improved H_2S donor solubility and stability, prolonged circulation time, reduced side effects, improved pharmacokinetics, and the potential for targeting and intracellular trafficking. These properties could overcome some of the inherent challenges of delivering H_2S, including its gaseous nature, poor thermal and aqueous stability, short *in vivo* half-life, and toxicity at high concentrations. However, with these advantages come some downsides, including potential side effects from the carriers and the more arduous regulatory hurdles associated with polydisperse materials versus well defined small molecule therapeutics. However, on the whole, we believe that polymeric and supramolecular materials are a promising method to push H_2S into clinical use as they not only elongate H_2S release, but also have the ability to provide localized, targeted delivery.

As has been shown throughout the entirety of this chapter, it is evident that many widely varying macro- and supramolecular H_2S delivery systems have been studied, and it appears to be a Sophie's choice as to which system might be most efficacious for a specific application or indication. One critical element to consider is the biodegradability and/or potential toxicity of the polymer carriers or excipients in any delivery system. Many materials discussed here are inherently biodegradable and show little to no toxicity, and as such are likely better suited for potential clinical applications than nondegradable carriers. Additionally, while there are currently no examples of stars, dendrimers, or physically encapsulated H_2S donors within micelles, we believe that there is potential to expand into these research areas as we have discussed some of the advantages in Sections 15.3 and 15.4.

While many polymeric systems mentioned here are biodegradable, such as those mentioned in Sections 15.5 and 15.7, many other systems (highlighted throughout all of Sections 15.3 and 15.4) are made of non-degradable hydrocarbon backbones. Synthetic polymers such as polyesters, polyurethanes (when the polyol segment consists of an ester), and polyanhydrides are commonly used biodegradable polymers because they degrade via hydrolysis or esterases [137]. Most such polymers degrade into benign products such as glycine, lactic acid, carbon dioxide, and water [138]. Additionally, some protein and polysaccharide-based delivery systems offer an alternative route to synthetic degradable polymers by utilizing naturally occurring polymer sources that are inherently biodegradable. Conversely, polymers with hydrocarbon backbones such as polyacrylates are not susceptible to hydrolysis, and an oxidation process is required in order for them to degrade; therefore, they are widely considered to be non-biodegradable [137]. For extensive discussion on polymer degradability, we refer the reader to reviews by Tighzert and Cato [137, 138]. In addition to polymeric carriers, many H_2S donors also have small molecule byproducts that remain after H_2S generation and must be considered. For example, gem-dithiols release H_2S in response to light (Figure 15.4) along with 2-nitrobenzaldehyde [25], which can block the generation of singlet oxygen, suppressing fatty acid photooxidation and decreasing cells' ability to break down long chain fatty acids into usable metabolites [139]. Another example is SATOs, as mentioned in Section 15.3.1, which along with H_2S release cystine, an aldehyde, and ammonia, all of which may be bioactive [64]. Thus, both the polymer and the H_2S donor byproducts must be taken into consideration if these systems are to be used safely in a clinical setting. Future development of macromolecular, supramolecular, and small molecule donors must be done consciously to develop H_2S donors with inert and biodegradable byproducts.

As for the future of H_2S releasing systems, we pose a few questions that need to be answered as this field progresses further. Will systems such as these allow for both localized and targeted release, and therefore require less H_2S overall compared with small molecule H_2S donors to impart biological effects? This seems likely, but more head-to-head comparisons with proper control studies are needed. Many macromolecular and supramolecular H_2S delivery systems slowly release H_2S over an extended period, and most results thus far indicate these types of release profiles at intermediate concentrations of H_2S are more efficacious than other donors, at least in cell studies. Finally, along these lines, slow and targeted release from macromolecular and supramolecular donor systems is more consistent with the endogenous production of H_2S than a burst release. Slow-releasing systems with few or no bioactive or toxic byproducts may allow biologists and pharmacologists to elucidate the effects of extended H_2S therapy, with the potential for delivery over the course of weeks to months. Comparative studies with daily administration of fast-releasing H_2S donors may reveal unexpected biological and pharmaceutical effects that inform how reactive sulfur species fit into the greater cell signaling web (recently referred to as the reactive species interactome [140]) and dictate how H_2S therapy could be best applied clinically.

References

1 Wang, R. (2002). Two's company, three's a crowd: can H_2S be the third endogenous gaseous transmitter? *FASEB J.* 16 (13): 1792–1798.

2 Calvert, J.W., Elston, M., Nicholson, C.K. et al. (2010). Genetic and pharmacologic hydrogen sulfide therapy attenuates ischemia-induced heart failure in mice. *Circulation* 122 (1): 11–19.

3 Calvert, J.W., Jha, S., Gundewar, S. et al. (2009). Hydrogen sulfide mediates cardioprotection through Nrf2 signaling. *Circ. Res.* 105 (4): 365–374.

4 Polhemus, D.J., Calvert, J.W., Butler, J. et al. (2014). The cardioprotective actions of hydrogen sulfide in acute myocardial infarction and heart failure. *Scientifica* 2014: 8.

5 Zhao, W., Zhang, J., Lu, Y. et al. (2001). The vasorelaxant effect of H_2S as a novel endogenous gaseous K_{ATP} channel opener. *EMBO J.* 20 (21): 6008–6016.

6 Szabó, C. (2007). Hydrogen sulphide and its therapeutic potential. *Nat. Rev. Drug Discov.* 6: 917.

7 Wang, R. (2012). Physiological implications of hydrogen sulfide: a Whiff exploration that blossomed. *Physiol. Rev.* 92 (2): 791–896.

8 Papapetropoulos, A., Pyriochou, A., Altaany, Z. et al. (2009). Hydrogen sulfide is an endogenous stimulator of angiogenesis. *Proc. Natl. Acad. Sci. U.S.A.* 106 (51): 21972–21977.

9 Laggner, H., Muellner, M.K., Schreier, S. et al. (2007). Hydrogen sulphide: a novel physiological inhibitor of LDL atherogenic modification by HOCl. *Free Radic. Res.* 41 (7): 741–747.

10 Wang, R. (2003). The gasotransmitter role of hydrogen sulfide. *Antioxid. Redox Signal.* 5 (4): 493–501.

11 Yang, G., Wu, L., Jiang, B. et al. (2008). H_2S as a physiologic vasorelaxant: hypertension in mice with deletion of cystathionine γ-lyase. *Science* 322 (5901): 587–590.

12 Ali, H., Opere, C., and Singh, S. (2014). In vitro-controlled release delivery system for hydrogen sulfide donor. *AAPS PharmSciTech* 15 (4): 910–919.

13 Kasparek, M.S., Linden, D.R., Kreis, M.E. et al. (2008). Gasotransmitters in the gastrointestinal tract. *Surgery* 143 (4): 455–459.

14 Powell, C.R., Dillon, K.M., and Matson, J.B. (2018). A review of hydrogen sulfide (H_2S) donors: chemistry and potential therapeutic applications. *Biochem. Pharmacol.* 149: 110–123.

15 Levinn, C.M., Cerda, M.M., and Pluth, M.D. (2019). Activatable small-molecule hydrogen sulfide donors. *Antioxid. Redox Signal.* 32 (2): 96–109.

16 Yang, C.-T., Chen, L., Xu, S. et al. (2017). Recent development of hydrogen sulfide releasing/stimulating reagents and their potential applications in cancer and glycometabolic disorders. *Front. Pharmacol.* 8: 664.

17 Grund, S., Bauer, M., and Fischer, D. (2011). Polymers in drug delivery—state of the art and future trends. *Adv. Eng. Mater.* 13 (3): B61–B87.

18 Liechty, W.B., Kryscio, D.R., Slaughter, B.V. et al. (2010). Polymers for drug delivery systems. *Annu. Rev. Chem. Biomol. Eng.* 1: 149–173.

19 Saikia, C. and Gogoi, P. (2015). Chitosan: a promising biopolymer in drug delivery applications. *J. Mol. Genet. Med.* S4: 006.

20 Misra, A. and Shahiwala, A. (2020). *Applications of Polymers in Drug Delivery*, 2e. Amsterdam, Netherlands: Elsevier.

21 Insko, M.A., Deckwerth, T.L., Hill, P. et al. (2009). Detection of exhaled hydrogen sulphide gas in rats exposed to intravenous sodium sulphide. *Br. J. Pharmacol.* 157 (6): 944–951.

22 Kamaly, N., Yameen, B., Wu, J. et al. (2016). Degradable controlled-release polymers and polymeric nanoparticles: mechanisms of controlling drug release. *Chem. Rev.* 116 (4): 2602–2663.

23 Duncan, R., Kopecková-Rejmanová, P., Strohalm, J. et al. (1987). Anticancer agents coupled to *N*-(2-hydroxypropyl)methacrylamide copolymers. I. Evaluation of daunomycin and puromycin conjugates in vitro. *Br. J. Cancer* 55 (2): 165–174.

24 Kopecek, J., Sprincl, L., and Lïm, D. (1973). New types of synthetic infusion solutions. I. Investigation of the effect of solutions of some hydrophilic polymers on blood. *J. Biomed. Mater. Res. Part A* 7 (2): 179–191.

25 Li, Z., Li, D., Wang, L. et al. (2019). Photocontrollable water-soluble polymeric hydrogen sulfide (H_2S) donor. *Polymer* 168: 16–20.

26 Dao, N.V., Ercole, F., Kaminskas, L.M. et al. (2020). Trisulfide-bearing PEG brush polymers donate hydrogen sulfide and ameliorate cellular oxidative stress. *Biomacromolecules* 21 (12): 5292–5305.

27 Takatani-Nakase, T., Katayama, M., Matsui, C. et al. (2017). Hydrogen sulfide donor micelles protect cardiomyocytes from ischemic cell death. *Mol. Biosyst.* 13 (9): 1705–1708.

28 Chen, J.J.Y., van der Vlies, A.J., and Hasegawa, U. (2020). Hydrogen sulfide-releasing micelles for promoting angiogenesis. *Polym. Chem.* 11 (27): 4454–4463.

29 Ercole, F., Mansfeld, F.M., Kavallaris, M. et al. (2016). Macromolecular hydrogen sulfide donors trigger spatiotemporally confined changes in cell signaling. *Biomacromolecules* 17 (1): 371–383.

30 Wu, J., Chen, A., Zhou, Y. et al. (2019). Novel H_2S-releasing hydrogel for wound repair via in situ polarization of M2 macrophages. *Biomaterials* 222: 119398.

31 Do, A.V., Smith, R., Tobias, P. et al. (2019). Sustained release of hydrogen sulfide (H_2S) from poly(lactic acid) functionalized 4-hydroxythiobenzamide microparticles to protect against oxidative damage. *Ann. Biomed. Eng.* 47 (8): 1691–1700.

32 Devarie-Baez, N.O., Bagdon, P.E., Peng, B. et al. (2013). Light-induced hydrogen sulfide release from "caged" *gem*-dithiols. *Org. Lett.* 15 (11): 2786–2789.

33 Urquhart, M.C., Dao, N.V., Ercole, F. et al. (2020). Polymers with dithiobenzoate end groups constitutively release hydrogen sulfide upon exposure to cysteine and homocysteine. *ACS Macro Lett.* 9 (4): 553–557.

34 Powell, C.R., Foster, J.C., Okyere, B. et al. (2016). Therapeutic delivery of H_2S via COS: small molecule and polymeric donors with benign byproducts. *J. Am. Chem. Soc.* 138 (41): 13477–13480.

35 Connal, L.A. (2018). The benefits of macromolecular hydrogen sulfide prodrugs. *J. Mater. Chem. B* 6 (44): 7122–7128.

36 Fairbanks, B.D., Gunatillake, P.A., and Meagher, L. (2015). Biomedical applications of polymers derived by reversible addition–fragmentation chain-transfer (RAFT). *Adv. Drug Deliv. Rev.* 91: 141–152.

37 Pissuwan, D., Boyer, C., Gunasekaran, K. et al. (2010). In vitro cytotoxicity of RAFT polymers. *Biomacromolecules* 11 (2): 412–420.

38 Perrier, S. (2017). 50th anniversary perspective: RAFT polymerization—a user guide. *Macromolecules* 50 (19): 7433–7447.

39 Cerda, M.M., Newton, T.D., Zhao, Y. et al. (2018). Dithioesters: simple, tunable, cysteine-selective H_2S donors. *Chem. Sci.* 10 (6): 1773–1779.

40 Keddie, D.J., Moad, G., Rizzardo, E. et al. (2012). RAFT agent design and synthesis. *Macromolecules* 45 (13): 5321–5342.

41 Yu, S.H., Ercole, F., Veldhuis, N.A. et al. (2017). Polymers with acyl-protected perthiol chain termini as convenient building blocks for doubly responsive H_2S-donating nanoparticles. *Polym. Chem.* 8 (41): 6362–6367.

42 Sagi, A., Weinstain, R., Karton, N. et al. (2008). Self-immolative polymers. *J. Am. Chem. Soc.* 130 (16): 5434–5435.

43 Steiger, A.K., Pardue, S., Kevil, C.G. et al. (2016). Self-immolative thiocarbamates provide access to triggered H_2S donors and analyte replacement fluorescent probes. *J. Am. Chem. Soc.* 138 (23): 7256–7259.

44 Steiger, A.K., Zhao, Y., and Pluth, M.D. (2017). Emerging roles of carbonyl sulfide in chemical biology: sulfide transporter or gasotransmitter? *Antioxid. Redox Signal.* 28 (16): 1516–1532.

45 Chauhan, P., Bora, P., Ravikumar, G. et al. (2017). Esterase activated carbonyl sulfide/hydrogen sulfide (H_2S) donors. *Org. Lett.* 19 (1): 62–65.

46 Sharma, A.K., Nair, M., Chauhan, P. et al. (2017). Visible-light-triggered uncaging of carbonyl sulfide for hydrogen sulfide (H_2S) release. *Org. Lett.* 19 (18): 4822–4825.

47 Cerda, M.M., Mancuso, J.L., Mullen, E.J. et al. (2020). Use of dithiasuccinoyl-caged amines enables COS/H_2S release lacking electrophilic byproducts. *Chem. Eur. J.* 26 (24): 5374–5380.

48 Gilbert, A.K., Zhao, Y., Otteson, C.E. et al. (2019). Development of acid-mediated H_2S/COS donors that respond to a specific pH window. *J. Org. Chem.* 84 (22): 14469–14475.

49 Zhao, Y., Steiger, A.K., and Pluth, M.D. (2019). Cyclic sulfenyl thiocarbamates release carbonyl sulfide and hydrogen sulfide independently in thiol-promoted pathways. *J. Am. Chem. Soc.* 141 (34): 13610–13618.

50 Zhao, Y., Steiger, A.K., and Pluth, M.D. (2018). Colorimetric carbonyl sulfide (COS)/hydrogen sulfide (H_2S) donation from γ-ketothiocarbamate donor motifs. *Angew. Chem. Int. Ed.* 57 (40): 13101–13105.

51 Zhao, Y., Steiger, A.K., and Pluth, M.D. (2018). Cysteine-activated hydrogen sulfide (H_2S) delivery through caged carbonyl sulfide (COS) donor motifs. *Chem. Commun.* 54 (39): 4951–4954.

52 Powell, C.R., Foster, J.C., Swilley, S.N. et al. (2019). Self-amplified depolymerization of oligo(thiourethanes) for the release of COS/H_2S. *Polym. Chem.* 10 (23): 2991–2995.

53 Jenkins, A.D., Kratochvil, P., Stepto, R.F.T. et al. (1996). Glossary of basic terms in polymer science. *Pure Appl. Chem.* 68: 2287–2311.

54 Das, S.S., Bharadwaj, P., Bilal, M. et al. (2020). Stimuli-responsive polymeric nanocarriers for drug delivery, imaging, and theragnosis. *Polymers* 12 (6): 1397.

55 Yar, Y., Khodadust, R., Akkoc, Y. et al. (2018). Development of tailored SPION-PNIPAM nanoparticles by ATRP for dually responsive doxorubicin delivery and MR imaging. *J. Mater. Chem. B* 6 (2): 289–300.

56 Zhu, R., Su, L., Dai, J. et al. (2020). Biologically responsive plasmonic assemblies for second near-infrared window photoacoustic imaging-guided concurrent chemo-immunotherapy. *ACS Nano* 14 (4): 3991–4006.

57 Yu, E., Galiana, I., Martínez-Máñez, R. et al. (2015). Poly(*N*-isopropylacrylamide)-gated Fe_3O_4/SiO_2 core shell nanoparticles with expanded mesoporous structures for the temperature triggered release of lysozyme. *Colloids Surf. B* 135: 652–660.

58 Krishnamoorthy, M., Li, D., Sharili, A.S. et al. (2017). Solution conformation of polymer brushes determines their interactions with DNA and transfection efficiency. *Biomacromolecules* 18 (12): 4121–4132.

59 England, R.M., Moss, J.I., Gunnarsson, A. et al. (2020). Synthesis and characterization of dendrimer-based polysarcosine star polymers: well-defined, versatile platforms designed for drug-delivery applications. *Biomacromolecules* 21 (8): 3332–3341.

60 Augustus, E.N., Allen, E.T., Nimibofa, A. et al. (2017). A review of synthesis, characterization and applications of functionalized dendrimers. *Am. J. Polym. Sci.* 7 (1): 8–14.

61 Cheng, Y., Xu, Z., Ma, M. et al. (2008). Dendrimers as drug carriers: applications in different routes of drug administration. *J. Pharm. Sci.* 97 (1): 123–143.

62 Chis, A.A., Dobrea, C., Morgovan, C. et al. (2020). Applications and limitations of dendrimers in biomedicine. *Molecules* 25 (17): 3982.

63 Foster, J.C. and Matson, J.B. (2014). Functionalization of methacrylate polymers with thiooximes: a robust postpolymerization modification reaction and a method for the preparation of H_2S-releasing polymers. *Macromolecules* 47 (15): 5089–5095.

64 Foster, J.C., Powell, C.R., Radzinski, S.C. et al. (2014). *S*-Aroylthiooximes: a facile route to hydrogen sulfide releasing compounds with structure-dependent release kinetics. *Org. Lett.* 16 (6): 1558–1561.

65 Liang, D., Wu, H., Wong, M.W. et al. (2015). Diallyl trisulfide is a fast H_2S donor, but diallyl disulfide is a slow one: the reaction pathways and intermediates of glutathione with polysulfides. *Org. Lett.* 17 (17): 4196–4199.

66 Kabanov, A.V., Chekhonin, V.P., Alakhov, V. et al. (1989). The neuroleptic activity of haloperidol increases after its solubilization in surfactant micelles. Micelles as microcontainers for drug targeting. *FEBS Lett.* 258 (2): 343–345.

67 Bader, H., Ringsdorf, H., and Schmidt, B. (1984). Watersoluble polymers in medicine. *Angew. Makromolek. Chem.* 123 (1): 457–485.

68 Cabral, H., Miyata, K., Osada, K. et al. (2018). Block copolymer micelles in nanomedicine applications. *Chem. Rev.* 118 (14): 6844–6892.

69 Nguyen, M.M., Carlini, A.S., Chien, M.-P. et al. (2015). Enzyme-responsive nanoparticles for targeted accumulation and prolonged retention in heart tissue after myocardial infarction. *Adv. Mater.* 27 (37): 5547–5552.

70 Ambade, A.V., Savariar, E.N., and Thayumanavan, S. (2005). Dendrimeric micelles for controlled drug release and targeted delivery. *Mol. Pharm.* 2 (4): 264–272.

71 Kabanov, A.V. and Alakhov, V.Y. (2002). Pluronic block copolymers in drug delivery: from micellar nanocontainers to biological response modifiers. *Crit. Rev. Ther. Drug Carrier Syst.* 19 (1): 1–72.

72 Allen, C., Maysinger, D., and Eisenberg, A. (1999). Nano-engineering block copolymer aggregates for drug delivery. *Colloids Surf. B.* 16 (1): 3–27.

73 Veronese, F.M. and Morpurgo, M. (1999). Bioconjugation in pharmaceutical chemistry. *Il Farmaco* 54 (8): 497–516.

74 D'Souza, A.J. and Topp, E.M. (2004). Release from polymeric prodrugs: linkages and their degradation. *J. Pharm. Sci.* 93 (8): 1962–1979.

75 Bae, Y., Fukushima, S., Harada, A. et al. (2003). Design of environment-sensitive supramolecular assemblies for intracellular drug delivery: polymeric micelles that are responsive to intracellular pH change. *Angew. Chem. Int. Ed.* 42 (38): 4640–4643.

76 Carrazzone, R.J., Foster, J.C., Li, Z. et al. (2020). Tuning small molecule release from polymer micelles: varying H_2S release through crosslinking in the micelle core. *Eur. Polym. J.* 141: 110077.

77 Hasegawa, U. and van der Vlies, A.J. (2015). Polymeric micelles for hydrogen sulfide delivery. *Med. Chem. Commun.* 6 (2): 273–276.

78 Foster, J.C., Radzinski, S.C., Zou, X. et al. (2017). H_2S-releasing polymer micelles for studying selective cell toxicity. *Mol. Pharm.* 14 (4): 1300–1306.

79 Foster, J.C., Carrazzone, R.J., Spear, N.J. et al. (2019). Tuning H_2S release by controlling mobility in a micelle core. *Macromolecules* 52 (3): 1104–1111.

80 Ercole, F., Whittaker, M.R., Halls, M.L. et al. (2017). Garlic-inspired trisulfide linkers for thiol-stimulated H_2S release. *Chem. Commun.* 53 (57): 8030–8033.

81 Hasegawa, U. and van der Vlies, A.J. (2014). Design and synthesis of polymeric hydrogen sulfide donors. *Bioconjugate Chem.* 25 (7): 1290–1300.

82 Hasegawa, U., Tateishi, N., Uyama, H. et al. (2015). Hydrolysis-sensitive dithiolethione prodrug micelles. *Macromol. Biosci.* 15 (11): 1512–1522.

83 Raina, N., Rani, R., Khan, A. et al. (2020). Interpenetrating polymer network as a pioneer drug delivery system: a review. *Polym. Bull.* 77 (9): 5027–5050.

84 Hoffman, A.S. (2002). Hydrogels for biomedical applications. *Adv. Drug Deliv. Rev.* 54 (1): 3–12.

85 Peppas, N.A., Bures, P., Leobandung, W. et al. (2000). Hydrogels in pharmaceutical formulations. *Eur. J. Pharm. Biopharm.* 50 (1): 27–46.

86 Bajpai, A.K., Shukla, S.K., Bhanu, S. et al. (2008). Responsive polymers in controlled drug delivery. *Prog. Polym. Sci.* 33 (11): 1088–1118.

87 Balakrishnan, B., Mohanty, M., Umashankar, P.R. et al. (2005). Evaluation of an in situ forming hydrogel wound dressing based on oxidized alginate and gelatin. *Biomaterials* 26 (32): 6335–6342.

88 Kirker, K.R., Luo, Y., Nielson, J.H. et al. (2002). Glycosaminoglycan hydrogel films as bio-interactive dressings for wound healing. *Biomaterials* 23 (17): 3661–3671.

89 Wischke, C., Neffe, A.T., Steuer, S. et al. (2009). Evaluation of a degradable shape-memory polymer network as matrix for controlled drug release. *J. Control. Release* 138 (3): 243–250.

90 Neffe, A.T., Hanh, B.D., Steuer, S. et al. (2009). Polymer networks combining controlled drug release, biodegradation, and shape memory capability. *Adv. Mater.* 21 (32, 33): 3394–3398.

91 Ma, W.-F., Wu, K.-Y., Tang, J. et al. (2012). Magnetic drug carrier with a smart pH-responsive polymer network shell for controlled delivery of doxorubicin. *J. Mater. Chem.* 22 (30): 15206–15214.

92 DeLeon, E.R., Stoy, G.F., and Olson, K.R. (2012). Passive loss of hydrogen sulfide in biological experiments. *Anal. Biochem.* 421 (1): 203–207.

93 Zhao, Y., Biggs, T.D., and Xian, M. (2014). Hydrogen sulfide (H_2S) releasing agents: chemistry and biological applications. *Chem. Commun.* 50 (80): 11788–11805.

94 Zhou, M., Qian, Y., Zhu, Y. et al. (2020). Elastase-triggered H_2S delivery from polymer hydrogels. *Chem. Commun.* 56 (7): 1085–1088.

95 Maturavongsadit, P., Luckanagul, J.A., Metavarayuth, K. et al. (2016). Promotion of in vitro chondrogenesis of mesenchymal stem cells using in situ hyaluronic hydrogel functionalized with rod-like viral nanoparticles. *Biomacromolecules* 17 (6): 1930–1938.

96 Sun, Y., Nan, D., Jin, H. et al. (2020). Recent advances of injectable hydrogels for drug delivery and tissue engineering applications. *Polym. Test.* 81: 106283.

97 Kang, J., Li, Z., Organ, C.L. et al. (2016). pH-controlled hydrogen sulfide release for myocardial ischemia-reperfusion injury. *J. Am. Chem. Soc.* 138 (20): 6336–6339.

98 Li, J. and Mooney, D.J. (2016). Designing hydrogels for controlled drug delivery. *Nat. Rev. Mater.* 1 (12): 16071.

99 Davoodi, P., Lee, L.Y., Xu, Q. et al. (2018). Drug delivery systems for programmed and on-demand release. *Adv. Drug Deliv. Rev.* 132: 104–138.

100 Lavin, D.M., Stefani, R.M., Zhang, L. et al. (2012). Multifunctional polymeric microfibers with prolonged drug delivery and structural support capabilities. *Acta Biomater.* 8 (5): 1891–1900.

101 https://www.zeusinc.com/products/biomaterials/absorv-bioabsorbable-extrusions (accessed 25 January 2020).

102 Jeon, G., Yang, S.Y., and Kim, J.K. (2012). Functional nanoporous membranes for drug delivery. *J. Mater. Chem.* 22 (30): 14814–14834.

103 McHugh, A.J. (2005). The role of polymer membrane formation in sustained release drug delivery systems. *J. Control. Release* 109 (1): 211–221.

104 Anselmo, A.C. and Mitragotri, S. (2014). An overview of clinical and commercial impact of drug delivery systems. *J. Control. Release* 190: 15–28.

105 Ong, B.Y.S., Ranganath, S.H., Lee, L.Y. et al. (2009). Paclitaxel delivery from PLGA foams for controlled release in post-surgical chemotherapy against glioblastoma multiforme. *Biomaterials* 30 (18): 3189–3196.

106 Venugopal, J., Low, S., Choon, A.T. et al. (2008). Interaction of cells and nanofiber scaffolds in tissue engineering. *J. Biomed. Mater. Res. B Appl. Biomater.* 84 (1): 34–48.

107 Liu, W., Thomopoulos, S., and Xia, Y. (2012). Electrospun nanofibers for regenerative medicine. *Adv. Healthcare Mater.* 1 (1): 10–25.

108 Chen, M., Li, Y.F., and Besenbacher, F. (2014). Electrospun nanofibers-mediated on-demand drug release. *Adv. Healthcare Mater.* 3 (11): 1721–1732.

109 Feng, S., Zhao, Y., Xian, M. et al. (2015). Biological thiols-triggered hydrogen sulfide releasing microfibers for tissue engineering applications. *Acta Biomater.* 27: 205–213.

110 Cacciotti, I., Ciocci, M., Di Giovanni, E. et al. (2018). Hydrogen sulfide-releasing fibrous membranes: potential patches for stimulating human stem cells proliferation and viability under oxidative stress. *Int. J. Mol. Sci.* 19 (8).

111 Geckil, H., Xu, F., Zhang, X. et al. (2010). Engineering hydrogels as extracellular matrix mimics. *Nanomedicine* 5 (3): 469–484.

112 Miroddi, M., Calapai, F., and Calapai, G. (2011). Potential beneficial effects of garlic in oncohematology. *Mini Rev. Med. Chem.* 11: 461–472.

113 Cai, Y.R. and Hu, C.H. (2017). Computational study of H_2S release in reactions of diallyl polysulfides with thiols. *J. Phys. Chem. B* 121 (26): 6359–6366.

114 Lengyel, M., Kállai, N., Antal, V. et al. (2019). Microparticles, microspheres, and microcapsules for advanced drug delivery. *Sci. Pharm.* 87: 20.

115 Long, T., Wongrakpanich, A., Do, A.-V. et al. (2015). Long-term release of a thiobenzamide from a backbone functionalized poly(lactic acid). *Polym. Chem.* 6: 7188–7195.

116 Lin, L., Qin, H., Huang, J. et al. (2018). Design and synthesis of an AIE-active polymeric H_2S-donor with capacity for self-tracking. *Polym. Chem.* 9 (21): 2942–2950.

117 Lehn, J.M. (1988). Supramolecular chemistry—scope and perspectives molecules, supermolecules, and molecular devices. *Angew. Chem. Int. Ed.* 27 (1): 89–112.

118 Lehn, J.M. (2002). Toward complex matter: supramolecular chemistry and self-organization. *Proc. Natl. Acad. Sci. U.S.A.* 99 (8): 4763–4768.

119 de Greef, T.F.A. and Meijer, E.W. (2008). Supramolecular polymers. *Nature* 453 (7192): 171–173.

120 De Greef, T.F.A., Smulders, M.M.J., Wolffs, M. et al. (2009). Supramolecular polymerization. *Chem. Rev.* 109 (11): 5687–5754.

121 Aida, T., Meijer, E.W., and Stupp, S.I. (2012). Functional supramolecular polymers. *Science* 335 (6070): 813–817.

122 Dong, R.J., Zhou, Y.F., Huang, X. et al. (2015). Functional supramolecular polymers for biomedical applications. *Adv. Mater.* 27 (3): 498–526.

123 Freeman, R., Boekhoven, J., Dickerson, M.B. et al. (2015). Biopolymers and supramolecular polymers as biomaterials for biomedical applications. *MRS Bull.* 40 (12): 1089–1101.

124 Webber, M.J., Appel, E.A., Meijer, E.W. et al. (2016). Supramolecular biomaterials. *Nat. Mater.* 15 (1): 13–26.

125 Carter, J.M., Qian, Y., Foster, J.C. et al. (2015). Peptide-based hydrogen sulphide-releasing gels. *Chem. Commun.* 51 (66): 13131–13134.

126 Yu, S.H., Esser, L., Khor, S.Y. et al. (2019). Development of a shape-controlled H_2S delivery system using epoxide-functional nanoparticles. *J. Polym. Sci., Part A: Polym. Chem.* 57 (18): 1982–1993.

127 Xie, C.K., Cen, D., Ren, Z.H. et al. (2020). FeS@BSA nanoclusters to enable H_2S-amplified ROS-based therapy with MRI guidance. *Adv. Sci.* 7 (7): 1903512.

128 Matson, J.B. and Stupp, S.I. (2012). Self-assembling peptide scaffolds for regenerative medicine. *Chem. Commun.* 48 (1): 26–33.

129 Wang, Y., Kaur, K., Scannelli, S.J. et al. (2018). Self-assembled nanostructures regulate H_2S release from constitutionally isomeric peptides. *J. Am. Chem. Soc.* 140 (44): 14945–14951.

130 Wang, Y. and Matson, J.B. (2019). Supramolecular nanostructures with tunable donor loading for controlled H_2S release. *ACS Appl. Bio Mater.* 2 (11): 5093–5098.

131 Carter, J.M., Qian, Y., Foster, J.C. et al. (2015). Peptide-based hydrogen sulphide-releasing gels. *Chem. Commun.* 51 (66): 13131–13134.

132 Qian, Y., Kaur, K., Foster, J.C. et al. (2019). Supramolecular tuning of H_2S release from aromatic peptide amphiphile gels: effect of core unit substituents. *Biomacromolecules* 20 (2): 1077–1086.

133 Kaur, K., Wang, Y., and Matson, J.B. (2020). Linker-regulated H_2S release from aromatic peptide amphiphile hydrogels. *Biomacromolecules* 21 (3): 1171–1178.

134 Longchamp, A., Kaur, K., Macabrey, D. et al. (2019). Hydrogen sulfide-releasing peptide hydrogel limits the development of intimal hyperplasia in human vein segments. *Acta Biomater.* 97: 374–384.

135 Qian, Y., Altamimi, A., Yates, S.A. et al. (2020). H_2S-releasing amphiphilic dipeptide hydrogels are potent *S. aureus* biofilm disruptors. *Biomater. Sci.* 8 (9): 2564–2576.

136 Wang, Y., Li, Z., Shmidov, Y. et al. (2020). Crescent-shaped supramolecular tetrapeptide nanostructures. *J. Am. Chem. Soc.* 142 (47): 20058–20065.

137 Vroman, I. and Tighzert, L. (2009). Biodegradable polymers. *Materials* 2 (2): 307–344.

138 Nair, L.S. and Laurencin, C.T. (2007). Biodegradable polymers as biomaterials. *Prog. Polym. Sci.* 32 (8): 762–798.

139 Hajimohammadi, M., Vaziri Sereshk, A., Schwarzinger, C. et al. (2018). Suppressing effect of 2-nitrobenzaldehyde on singlet oxygen generation, fatty acid photooxidation, and dye-sensitizer degradation. *Antioxidants* 7 (12): 194.

140 Cortese-Krott, M.M., Koning, A., Kuhnle, G.G.C. et al. (2017). The reactive species interactome: evolutionary emergence, biological significance, and opportunities for redox metabolomics and personalized medicine. *Antioxid. Redox Signal.* 27 (10): 684–712.

16

H_2S and Hypertension

Vincenzo Brancaleone[1], Mariarosaria Bucci[2], and Giuseppe Cirino[2]

[1] *Department of Science, University of Basilicata, Potenza, Italy*
[2] *Department of Pharmacy, School of Medicine, University of Naples Federico II, Naples, Italy*

List of Abbreviations

3-MP	3-mercaptopyruvate
3-MST	3-mercaptosulfutransferase
3-MST-SSH	3-mercaptosulfurtransferase persulfide
AC	adenylate cyclase
ACE	angiotensin converting enzyme
ACEi	angiotensin converting enzyme inhibitor
BP	blood pressure
CAT	cysteine aminotransferase
CBS	cystathionine-β-synthase
CDO	cysteine dioxygenase
CSAD	cysteine sulfinate decarboxylase
CSE	cystathionine-γ-lyase
DAO	D-aminoacid oxidase
EDHF	endothelium dependent hyperpolarizing factors
eNOS	endothelial NO synthase
ETC	electronic transport chain
GSSG	oxidized glutathione
H_2S	hydrogen sulphide
HPLC	high performance liquid chromatography
L-NAME	L-nitroarginine methyl-ester
NLRP3	(NOD)-like receptor protein-3
NO	nitric oxide
NOD	non obese diabetic mice

PAG	propargylglycine
PDE	phosphodiesterases
PLP	pyridoxal-5′-phosphate
RAAS	renin–angiotensin–aldosterone system
sGC	soluble guanylate cyclase
SHR	spontaneously hypertensive rats
SO	sulphite oxidase
SR	sulfurtransferase
TMT	thiol-*S*-methyltransferase
TSR	thiosulfate reductase
TSST	thiosulfate sulfurtransferase
VEGFR2	vascular endothelial growth factor receptor-2
WKY	Wistar–Kyoto rat

16.1 Hypertension, Vascular Homeostasis and Mediators Controlling Blood Pressure

Hypertension is a multifactorial pathology often resulting in fatal events associated with long-term arterial remodeling and dysfunction [1]. In the world's adult population, almost 30% suffers hypertension, and it has been estimated that this percentage will increase [2]. Blood pressure (BP) is the result of multiple factors acting within the body and controlling cardiovascular homeostasis. In early studies focusing on elevated BP, referred to as hypertension, molecules able to modify arterial pressure were identified and, initially, these seemed to derive mainly from the kidney. The discovery of angiotensin (AT) opened a new field that started to enlarge the relevance of the knowledge on hypertension. The ability of AT to increase BP was first described as a solely vasocontractile effect. However, following this finding, Page's research group demonstrated that AT could increase BP only after 24 hours following low-rate infusion [3]. This evidence expanded the knowledge to very close to what it is now established. Indeed, vasoconstriction is not the only mechanism involved in the onset of hypertension, as other organs and tissues are part of the whole mechanistic path, including the kidney (renin–angiotensin–aldosterone system, RAAS), vasculature (mediators controlling reactivity and caliber), and central nervous system (adrenergic innervation) [4].

Within the cardiovascular system, both contractile as well as relaxant factors are produced. It is now established that diverse mediators are produced and generate vasodilatory effects including the gaseous mediators namely nitric oxide (NO) and hydrogen sulfide (H_2S). A harmonized balance between the biosynthesis and the release of molecules with these opposing properties is responsible for the BP homeostasis, while disruption of this dynamic equilibrium leads to the development of hypertension. The reasons underlying these changes are to be found within two main vascular perturbations: (i) the enhancement of vasoconstrictor agents effects; and (ii) the suppression of vasodilating molecules pathways. In addition, the first type of perturbation is often coupled with alterations in vascular function promoting vasoconstriction as well as suppress vasodilation.

Several studies on isolated vessels from hypertensive animals have shown that both conduit vessels and arterioles display an increased contractile response to a variety of agents. Interestingly, this is often associated with the reduction of NO biosynthesis that frequently paralleled a drop in H_2S levels [5–7]. Of note, one mechanism by which NO production is suppressed relies on the inhibition of NO synthase (eNOS) activity operated by Rho-kinase through phosphorylation of Ser495 that results in increased vasoconstriction [8]. Interestingly, this mechanism can be modulated by H_2S that by inhibiting the Rho kinase activation, and the subsequent contraction, which leads to a vasodilatory effect [9].

Thus, a crucial event in hypertension is the impairment of vasodilation associated with a reduced NO bioavailability. However, this event is more evident in large conduit vessels; in smaller arteries and arterioles, instead, there is a major involvement of vasodilatory prostaglandins (PGI_2) and endothelium-dependent hyperpolarizing factors (EDHFs) [10–12]. NO exerts its vasodilatory effect by activating soluble guanylate cyclase (sGC) thus increasing cGMP, while PGI_2 relaxes vascular smooth muscle cells through the enhancement of adenylate cyclase (AC) activity resulting in augmentation of cAMP levels. Differently, EDHF, whose chemical nature is still under discussion, generates the hyperpolarization of smooth muscle cells that allows the dilation of the vascular bed. Interestingly, H_2S has been proposed as one of the EDHFs for its vasodilatory properties and its capacity to modulate ion channels opening [13–15].

It is now well accepted that changes in both vascular structure and function occurring through different mechanisms (e.g. oxidative stress, inflammation, endothelial dysfunction) strongly affect the homeostatic control of BP paving the way to the pathogenesis of hypertension [16]. This chapter will be focused on the role played by the H_2S pathway in the regulation of BP.

16.2 Generation of H_2S in the Cardiovascular System

16.2.1 Biosynthetic Pathways

H_2S is ubiquitously generated in the human body by enzymatic and nonenzymatic sources. The predominant mechanism to generate H_2S is the transsulfuration pathway. The majority of endogenous H_2S is synthesized by three enzymes, cystathionine-β-synthase (CBS), cystathionine-γ-lyase (CSE), and 3-mercaptosulfutransferase (3-MST) [17–19]. Cardiovascular tissues do express all three enzymes, although the relative abundance of each type is diverse within the different cell types. Although identified in vascular tissue, CBS is the predominant H_2S-synthesizing enzyme in the brain and CNS, while CSE does account for the majority of H_2S generated within the cardiovascular system under physiological conditions. In addition, 3-MST also provides H_2S biosynthesis, in cooperation with cysteine aminotransferase (CAT), in functional vascular endothelium.

Besides the different distribution and the relative abundance of H_2S-synthesizing enzymes in the cardiovascular environment, their different localization has also been documented within cells. Indeed, CSE and CBS are only cytosolic, while 3-MST displays both cytosolic and mitochondrial localization, with the latter being most prominent [20, 21]. However, CSE remains the major source of endogenous H_2S biosynthesis in the vasculature. Indeed, CSE is expressed in both vascular endothelium and smooth muscle cells, while 3-MST, together with CAT, is only found in the endothelium. This expression pattern is confirmed by higher levels of H_2S detected in vessels compared to other tissues [22, 23].

Different enzymes, then, show different biochemical reactions and substrates leading to the generation of H_2S. Nevertheless, CBS, CSE, and CAT all require L-cysteine as substrate and pyridoxal-5′-phosphate (PLP) as a cofactor. CBS and CSE can generate H_2S from cysteine or cysteine in presence of homocysteine [24]. 3-MST produces endogenous H_2S and pyruvate starting from 3-MP derived by CAT or from thioredoxin, dihydrolipoic acid, and D-cysteine, along with D-aminoacid oxidase (DAO) [25, 26]. These enzymes are all part of a cycle where homocysteine also enters as substrate and is converted into cystathionine and cysteine following a sulfur transfer under CBS and CSE catalysis. CBS catalyzes cysteine (and all thiols) through a β-substitution reaction. CSE catalyzes, differently from CBS, three different types of reactions: α,β-cleavage (cysteine), α,γ-cleavage (homocysteine), and γ-substitution (two molecules of homocysteine). Instead, being a sulfurtransferase (SR), 3-MST can transfer the sulfur group from mercaptopyruvate to an active cysteine site, thus forming the persulfide intermediate 3-mercaptosulfurtransferase persulfide (3MST-SSH). Except for thioredoxin, a variety of small molecules such as dihydrolipoic acid, homocysteine, cysteine and glutathione (GSH) release H_2S by receiving the persulfide group in the presence of a reductant.

16.2.2 Catabolic Pathway for H_2S

The amount of H_2S is correlated to its biosynthesis, but catabolic pathways involved in its degradation are similarly relevant. H_2S is normally underlying a pH-dependent balance where hydrosulfide (HS^-) is the primary species in solution, the undissociated form of H_2S makes up about one third or less, and sulfide anions (S^{2-}) are negligible in concentration. These anions can be catabolized to generate several sulfur-containing molecules, including sulfite and thiosulfate. The nonenzymatic route is one of the preferential pathways, however, some enzymes are involved in the metabolism of H_2S. In particular, sulfite derives from thiosulfate, which is metabolized by thiosulfate reductase (TSR) in the liver, brain, or kidney. Alternatively, thiosulfate can be also substrate for thiosulfate sulfurtransferase (TSST) in the liver and it could be sequentially oxidized to sulfate via sulfite oxidase (SO), in GSH-dependent manner. Additional means of elimination for H_2S are represented by the enzymatic intervention of rhodanese, the methylation to methanethiol (CH_3SH), the formation of methemoglobin (sulfo-hemoglobin), or the scavenging reactions involving superoxide, NO, metallo- or disulfide-containing molecules (e.g. GSSG). Nonenzymatic degradation of H_2S also leads to the formation of elemental sulfur, dithionate, polysulfides, and polythionates.

Entering into details, the first step of the H_2S catabolic pathway is associated with the activity of sulfide quinone oxidoreductase (SQR) that performs an oxidative reaction within the mitochondrial matrix, thus generating a persulfide (R-SSH) [27]. These derivatives are then oxidized by the persulfide dioxygenase (ETHE1) that produces sulfite (SO_3^{2-}) as an intermediate metabolite. Following sulfite formation, additional oxidative steps are required to obtain sulfates (SO_4^{2-}), which are carried out by SO. Alternatively, rhodanese can also metabolize sulfites to thiosulfates ($S_2O_3^{2-}$) by TSR that releases H_2S as a byproduct in the presence of GSH [28]. Sulfates represent the end product of H_2S catabolism that is excreted in the urine. H_2S can also be eliminated as it is through expiration; however, its levels could be detected only when exhaled H_2S is highly concentrated, e.g. following sodium sulfide administration [29]. Furthermore, persulfide intermediates can be converted into thiosulfates by the sulfurtransferase (SR) that represents an alternative enzymatic branch in the H_2S catabolic pathway [28].

Thus, control of H_2S metabolism needs to be very efficient since it can generate detrimental effects, often observed at the mitochondrial level [30]. Indeed, an excessive concentration of H_2S inhibits oxygen binding to mitochondrial complex IV, blocking the electron transport chain (ETC), and suppressing the generation of mitochondrial ATP [31]. This action is mainly because H_2S is largely processed within mitochondria, which represents the primordial origin as a bacterium with sulfide oxidizing activity. Together with the oxidative pathway, methylation of H_2S is an alternative metabolic reaction leading to H_2S elimination. Methylation reactions are slower than oxidative ones and occur in the cytoplasm, where H_2S is converted to CH_3SH and CH_3SCH_3 (dimethylsulfide) by Thiol-*S*-Methyltransferase (TMT).

In physiology, the above mentioned catabolic activity keeps H_2S half-life within the range of 2–10 minutes for different tissues (liver, kidney, and brain) [32]. Nevertheless, when oxygen levels are low, oxidative reactions are reduced and H_2S is poorly metabolized, thus increasing its levels. This cross-talk suggests the possibility to think of H_2S as a reliable sensor for oxygen levels within the cells [32].

Interestingly, the control of H_2S levels is not only operated by its metabolization but is also regulated by modulation of intracellular levels of its precursor L-cysteine via oxidative pathways. Indeed, when the L-cysteine level is above the physiological threshold is oxidized to sulfinate through the enzymatic activity of cysteine dioxygenase (CDO) [33]. The sulfinate intermediate is therefore further converted to sulfites following the enzymatic action of both cytosolic CAT and mitochondrial SO, or to taurine by cysteine sulfinate decarboxylase (CSAD).

16.3 Relevance of H_2S in Hypertension

H_2S represents one of the key elements in the regulation of vascular tone that reflects the control of BP. It is now well established that the impairment in H_2S levels, associated with reduced production or deficiency of H_2S-synthesizing enzymes, results in hypertension. These conditions are relevant in preclinical models of elevated BP as well as in human patients suffering hypertension, all indicating a key role for H_2S in the control of vascular homeostasis.

16.3.1 Preclinical Evidence

The role of H_2S in the mechanisms that regulate BP has been massively investigated, and the data obtained led to consistent observations. Zhao et al. have proposed the first evidence somehow associated with the control of BP operated by H_2S in 2001, when they discovered that H_2S can activate $K^{+}{}_{ATP}$ channels, thus generating hyperpolarization and subsequent vasorelaxation [18]. In addition, in 2003 Du et al. studied the role of H_2S in an animal model of hypertension by using spontaneously hypertensive rats (SHRs) (Figure 16.1) [34, 35]. In this study, it was found that when SHR rats develop hypertension, (9 weeks old) plasma and aortic levels of H_2S are significantly decreased in comparison to control rats (Wistar–Kyoto rats, WKY). A proof of concept study of the role played by H_2S was obtained by restoring H_2S levels by administration of NaSH, an exogenous source of H_2S. Indeed, following this oral treatment, BP was significantly lowered in SHR rats. Conversely, inhibition of endogenous hydrogen sulfide synthesis following the oral administration of D,L-propargylglycine (PAG), a CSE inhibitor, leads to the development of hypertension in WKY rats. This study represents the first report where endogenous H_2S deficiency was associated with the development

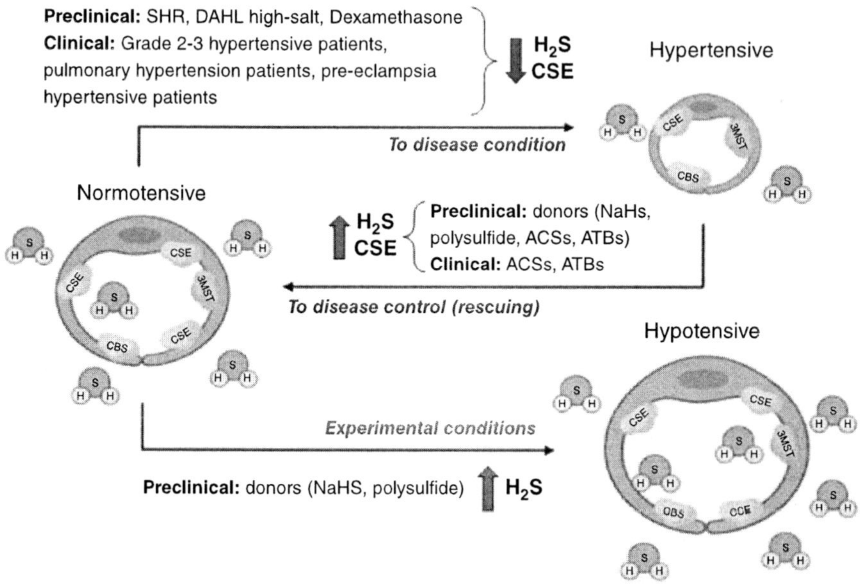

Figure 16.1 Schematic view of the role of H_2S in the control of blood pressure. SHR, Spontaneous hypertensive rats; ACSs, H_2S donors from CTG Pharma; ATBs, H_2S donors from Antibe Therapeutics.

of hypertension, and this evidence has been consistently confirmed in multiple studies [36]. Following this study, several pieces of evidence have been achieved and further corroborated the key role of endogenous H_2S in the control of BP. The biological function of H_2S within the vasculature does not end in the regulation of BP but it is also involved, through other mechanisms, with the vascular function control. In particular, reduced levels of H_2S, as reported in SHR animals, have been associated with medial thickening of intramyocardial coronary arterioles, interstitial fibrosis, and ROS production, effects that are all ameliorated by administration of exogenous NaSH. Interestingly, these beneficial effects have been associated with the activation of $K^+{}_{ATP}$ channels operated by H_2S [37].

However, this is not the only mechanism used by H_2S to exert vasodilatory and antihypertensive effects. Indeed, phosphodiesterases (PDEs) have been demonstrated to be inhibited by H_2S leading to an increase of cGMP in the vasculature, which in turn activates a downstream pathway inducing vasorelaxation [38, 39]. The possible role played by H_2S donation in the regulation of BP has been also demonstrated. Angiotensin-converting enzyme inhibitors (ACEis) are widely used as therapeutic agents in the treatment of hypertension, also because they are effective in associated conditions, including ischemic heart disease and left ventricular dysfunction in experimental studies and clinical trials [40, 41]. The main mechanism of ACEi is consistent with the dual inhibition of angiotensin II generation coupled to bradykinin degradation, and this leads to additional cardioprotective actions in hypertensive patients, including the reduction of preload and afterload, the suppression of adrenergic firing and the improvement of vascular function [42]. Interestingly, sulfhydryl-group containing ACEi, e.g. zofenopril, shows enhanced clinical efficacy in comparison with other ACEi when used in patients with hypertension, myocardial infarction, or vascular impairment following diabetes [43–45]. In this regard, it has been demonstrated that zofenopril displays its beneficial effects on vascular recovery in the SHR model through the modulation of the H_2S pathway. This effect has been shown to be beyond its ACE inhibition activity since this effect is also shared by the *R* isomer of zofenopril, which does not inhibit ACE [46]. In addition, it has been also demonstrated that H_2S can directly modulate ACE activity through its binding to Zn^{2+} within the active site, thus further enhancing ACE blocking activity [47]. Therefore, the use of drugs with the additional feature to modulate the H_2S pathway may represent a novel perspective in the scenario of the therapeutic approach to hypertension. All evidence so far discussed consolidates the role of H_2S in the control of BP in health as well as in disease, e.g. hypertension and atherosclerosis.

In cardiovascular diseases, there is an underlying endothelial dysfunction that drives the impairment of vascular function and that is paralleled by localized inflammation. In this regard, it has been recently demonstrated that H_2S, beyond the regulation of BP, also modulates the activation of (NOD)-like receptor protein 3 (NLRP3) inflammasome and oxidative stress burst by acting as a physiological brake. In fact, exogenous administration of H_2S ameliorated both endothelial-dependent reactivity paralleled by the suppression of NLRP3 inflammasome activity as well as ROS formation in SHR [48]. In addition, another study reported that H_2S counteract endothelial dysfunction and hypertension by increasing the expression of vascular endothelial growth factor-2 (VEGFR2) [49, 50]. This latter finding is relevant concerning the preservation of endothelial function since VEGFR2 can trigger different pathways, such as PI3K/Akt/eNOS signaling, a crucial pathway in vascular homeostasis. It is therefore clear that H_2S unbalance might result in BP elevation but the reasons why this occur might be multiple. Indeed, administration of some drugs such as corticosteroids (e.g. dexamethasone) induces an increase in BP, and this effect has been associated with a reduction in H_2S biosynthesis paired with a reduction of CSE and CBS expression [51]. The beneficial role of H_2S is again underlined in high-salt diet Dahl rats. In this context, hypertension is associated with suppression of H_2S production and CBS expression and increased activation of the RAAS. Again, replenishment of hydrogen sulfide by using H_2S donors ameliorated all cardiovascular parameters, thus, preserving vascular function and reducing the risks following hypertension [52, 53].

The relevance of the endogenous H_2S pathway in the onset of hypertension depends on several mechanisms that are all associated with a reduction of H_2S bioavailability. Administration of exogenous H_2S as an inorganic salt (NaSH or Na_2S) is the most used method to study *in vivo* effects on high BP. Alternatively, a similar approach is represented by the use of the substrate L-cysteine, which leads to an enzyme-mediated increase of H_2S, thus mimicking the effect of endogenously generated hydrogen sulfide. A study by Hsu et al. showed that early dosage of cysteine prevented both hypertension and kidney damage in the SHR model [54]. Therefore, the re-establishment of H_2S physiological levels either using an exogenous source or stimulating the endogenous synthesis leads to beneficial effects in the control of vascular function and BP. It is worthy to note that there is a cross-talk between the H_2S and NO pathway, and this aspect needs to be always acknowledged when we consider the vascular role of H_2S by itself. Indeed, the activation of common signaling pathways with NO (e.g. PI3K, Akt) and the presence of this cross-talk has been demonstrated by several studies (reviewed in [55]). For example, eNOS activation by PI3K/Akt is triggered by H_2S, and the inhibition of eNOS by L-nitroarginine methyl-ester (L -NAME) induces an increase in BP markedly improved by H_2S administration [56].

Despite the big body of evidence from preclinical models about the role of H_2S in hypertension, there is still a bit of debate concerning the role of H_2S-synthesizing enzymes in the cardiovascular system. The complete genetic deficiency of CBS (as in a homozygote knockout mouse) exhibits a neonatally lethal phenotype because of liver dysfunction and could not be used as a study model [57], while 3-MST knockout is not lethal to mice. In this regard, there is very little information about any possible effect on cardiovascular disease [58, 59]. Nonetheless, it has been recently shown that genetic ablation of 3-MST in mice impacts the cardiovascular system in an age-dependent manner. In particular, the absence of 3-MST seems to be cardioprotective in young adult mice, while predisposing to hypertension and cardiac hypertrophy in older animals [60].

CSE is the major enzyme-producing H_2S in the vasculature, and many studies are carried out by using animals where the CSE gene has been knocked out ($CSE^{-/-}$). In particular, BP of $CSE^{-/-}$ mice is higher than that measured in control mice, suggesting that a deficiency of CSE-derived H_2S contributes to hypertension [61]. However, this observation is not univocal as other studies controversially reported that CSE deficiency exhibits normal BP [62]. This discrepancy could be due to linked differences in the genetic background of knockout mice. An additional limitation could be represented by the fact that CSE deficiency is displayed within the whole body and not just in the cardiovascular system (conditional knockout mice). Interestingly, Lucock et al. found two genetic variants associated with CSE (G1364T) and CBS (844ins68) that are biologically crucial for H_2S biosynthesis and its regulation of BP [63]. Furthermore, confirming the crucial role of CSE in vascular tissue, the reduction of H_2S levels, associated with CSE inactivation, leads to an increased vascular inflammation based on a low level of RNA sulfhydration. In particular, a low amount of H_2S allows for stabilization of CD62E mRNA, which in turn accelerate endothelial dysfunction and atherosclerosis, effects reversed by administration of polysulfide donor [64].

16.3.2 Clinical Evidence

Several studies suggest that H_2S is critical in the pathogenesis and development of hypertension in humans. For instance, a clinical study reported that patients with severe hypertension display lower plasma levels of H_2S [65]. In addition, in patients with mild hypertension, plasma H_2S levels have been found lower than in subjects with normal BP [66, 67]. Similarly, the endogenous concentration of H_2S in patients with portal hypertension (PH) has been reported to be lower than that measured in healthy controls. Interestingly, PH patients display an inverse correlation between the severity of disease and low plasma H_2S levels and this also relates to portal vein diameter [68]. In addition, patients with pulmonary hypertension (PHT) also show a suppression of both H_2S levels and CSE expression [69].

Another form of hypertension is that driven by preeclampsia, which affects about 5% of all pregnancies and mainly contribute to maternal and fetal morbidity and mortality. Plasma H_2S and CSE mRNA levels are reduced in preeclamptic patients compared with the normal placenta [70]. Conversely, some reports describe an unchanged level of CSE in severe preeclamptic patients, while CBS mRNA expression in the placenta is low in early preeclampsia [71]. These findings seem to be contradictory to the role of H_2S; however, in some cases, this is due to the small patient sample size considered in the studies [71, 72]. In addition, it has been suggested that the early onset (<34 weeks) versus late (>34 weeks) onset of preeclampsia represent different phenotypes of this syndrome. Thus, the onset timing can critically influence the clinical outcomes and the relative biochemical quantification, including H_2S levels in plasma [73].

The overall data in the literature about vascular homeostasis and H_2S seem to be therefore univocal toward a beneficial role of this gaseous molecule in controlling BP, independently from the typology of hypertension considered. Thus, it is widely accepted that hypertension involves impairment in H_2S levels, due to enzymes (CSE, CBS, 3MST) downregulation or reduced activation. Nevertheless, it cannot be excluded that H_2S deficiency relates to additional mechanisms, including transcriptional regulation or post-translational modification of the H_2S-synthesizing enzymes. In fact, little is still known about biochemical reactions such as sulfhydration, cysteinylation, or nitrosylation, and these aspects require a more consistent experimental approach.

16.4 Conclusions

The amount of findings reported in the literature of H_2S biology provides robust evidence that this gasotransmitter is critical for cardiovascular physiology. The protective effect exerted by endogenous H_2S has highlighted the chance to use a pharmacological or genetic approach to understand and therefore modulate its pathway in disease. Therefore, regarding the cardiovascular system, one of the most challenging goals in the field of drug discovery is to efficiently modulate the levels of H_2S.

Thioacetamide is a water-soluble organosulfur compound used as a source of H_2S and its derivatives. However, it is a carcinogen agent classified in class 2B and also shows hepatotoxicity. Thioacetamide was widely used in the classical qualitative inorganic analysis as an in situ source for sulfide ions. It is not used as a direct H_2S-releasing molecule, rather it is employed as a moiety in the chemical synthesis of H_2S donors. Another historical H_2S releaser is represented by Lawesson's reagent, a chemical compound used in organic synthesis as a thiation agent, also being a H_2S releaser. Lawesson's reagent has a four-membered ring of alternating sulfur and phosphorus atoms. The rings can open to form reactive intermediates that decompose and release H_2S. Based on Lawesson's compound, several compounds have been synthesized and one of them is represented by GYY4137, which slowly releases H_2S both *in vitro* and *in vivo*. GYY4137 exerts vasodilator and antihypertensive effects and seems to be a useful H_2S-releasing molecule [74].

Nevertheless, the majority of the studies report the effect of exogenous hydrogen sulfide by using NaSH as a readily available source of H_2S [75]. However, cells produce endogenous levels of H_2S far lower than those achieved following exogenous NaSH administration. Thus, it is extremely difficult to infer that data obtained with this donor reflect the biological effects displayed by endogenous H_2S [76]. Additionally, there are some unsolved issues associated with the limited understanding of the actual physiological concentrations of H_2S in blood as well as in tissues, therefore limiting the knowledge about the effective/toxic concentrations of H_2S. This issue is mainly due to the variability of the experimental approach utilized in the sample processing and H_2S determination, ranging from photometrical to fluorescence-based or high performance liquid chromatography (HPLC) methods. Besides the quantification issue, an ideal H_2S-releasing molecule designed for therapeutic purposes should be able to release H_2S at the same rate and concentration as the endogenous generation [75].

Many efforts have been directed toward the synthesis or identification of donors displaying a controlled release near the physiological amounts of H_2S within the cardiovascular system. Considering the pharmacological and adverse effects of H_2S, pharmaceutical companies directed part of their research toward this field, aiming to formulate specific donors starting to form well-established parent compounds. For instance, CTG Pharma developed a series of H_2S-releasing compounds, referred to as ACS, to check possible employment in hypertension, metabolic syndrome, thrombosis, and arthritis (www.ctgpharma.com). Interestingly, these compounds have been shown effective in different experimental settings. For example, ACS6, H_2S-releasing sildenafil can suppress superoxide formation and has beneficial effects in preclinical models of acute respiratory distress syndrome (ARDS) [77, 78].

Another company, Antibe Therapeutics, has developed several derivatives able to release H_2S, known as ATBs, for the treatments of inflammatory bowel disease (ATB-429), joint pain (ATB-346), and irritable bowel syndrome (ATB-294) (http://www.antibe-therapeutics.com). Although these compounds have not been used for cardiovascular diseases, it would be intriguing to test whether they could exert any beneficial effect at the vascular level.

Furthermore, the compound IK1001, from the company Ikaria, is an injectable form of Na_2S as calcium-cross-linked alginate polymer, which is pure, pH neutral, and stable. IK-1001 has been used in several basic studies and processed into clinical trials. One is a phase I safety trial designed to assess the pharmacokinetics of intravenous injection of IK-1001 (http://ClinicalTrials.gov ID: NCT00879645). Another is a phase II efficacy trial where IK-1001 was administered in patients undergoing surgery for coronary artery bypass (http://ClinicalTrials.gov ID: NCT00858936).

In addition to industry-derived compounds, many other organic molecules generate H_2S. Some polysulfide is of natural origin and derive from garlic. These compounds release H_2S through a relatively slow mechanism, thus recreating endogenous-like conditions. Several reports show beneficial effects of garlic-derived compound diallyldisulfide (DADS) and diallyltrisulfide (DATS) [79].

Overall, all donors used in the experimental preclinical approach have been found effective in lowering BP, including NaSH [80–82]. However, at the present stage, an ideal H_2S-releasing molecule designed for therapeutic purposes has not been as yet found, although organic polysulfide derivatives may represent useful templates to generate H_2S-releasing compounds as innovative therapeutic agents for cardiovascular diseases.

References

1 Pinheiro, L.C., Tanus-Santos, J.E., and Castro, M.M. (2017). The potential of stimulating nitric oxide formation in the treatment of hypertension. *Expert Opin. Ther. Targets* 21 (5): 543–556.

2 Kearney, P.M., Whelton, M., Reynolds, K. et al. (2005). Global burden of hypertension: analysis of worldwide data. *Lancet* 365 (9455): 217–223.

3 McCubbin, J.W., DeMoura, R.S., Page, I.H., and Olmsted, F. (1965). Arterial hypertension elicited by subpressor amounts of angiotensin. *Science* 149 (3690): 1394–1395.

4 Harrison, D.G., Coffman, T.M., and Wilcox, C.S. (2021). Pathophysiology of hypertension: the mosaic theory and beyond. *Circ. Res.* 128 (7): 847–863.

5 Li, Q., Youn, J.Y., and Cai, H. (2015). Mechanisms and consequences of endothelial nitric oxide synthase dysfunction in hypertension. *J. Hypertens.* 33 (6): 1128–1136.

6 Touyz, R.M. (2000). Oxidative stress and vascular damage in hypertension. *Curr. Hypertens. Rep.* 2 (1): 98–105.

7 Lv, B., Chen, S., Tang, C. et al. (2021). Hydrogen sulfide and vascular regulation – an update. *J. Adv. Res.* 27: 85–97.

8 Sugimoto, M., Nakayama, M., Goto, T.M. et al. (2007). Rho-kinase phosphorylates eNOS at threonine 495 in endothelial cells. *Biochem. Biophys. Res. Commun.* 361 (2): 462–467.

9 Aydinoglu, F., Adibelli, E.O., Yilmaz-Oral, D., and Ogulener, N. (2019). Involvement of RhoA/Rho-kinase in L-cysteine/H_2S pathway-induced inhibition of agonist-mediated corpus cavernosal smooth muscle contraction. *Nitric Oxide* 85: 54–60.

10 Sandoo, A., van Zanten, J.J., Metsios, G.S. et al. (2010). The endothelium and its role in regulating vascular tone. *Open Cardiovasc. Med. J.* 4: 302–312.

11 Bryan, R.M. Jr.,, You, J., Golding, E.M., and Marrelli, S.P. (2005). Endothelium-derived hyperpolarizing factor: a cousin to nitric oxide and prostacyclin. *Anesthesiology* 102 (6). 1261–1277.

12 Gao, X., Martinez-Lemus, L.A., and Zhang, C. (2011). Endothelium-derived hyperpolarizing factor and diabetes. *World J. Cardiol.* 3 (1): 25–31.

13 Mustafa, A.K., Sikka, G., Gazi, S.K. et al. (2011). Hydrogen sulfide as endothelium-derived hyperpolarizing factor sulfhydrates potassium channels. *Circ. Res.* 109 (11): 1259–1268.

14 Wang, M., Hu, Y., Fan, Y. et al. (2016). Involvement of hydrogen sulfide in endothelium-derived relaxing factor-mediated responses in rat cerebral arteries. *J. Vasc. Res.* 53 (3, 4): 172–185.

15 Wen, J.Y., Zhang, J., Chen, S. et al. (2021). Endothelium-derived hydrogen sulfide acts as endothelium-derived hyperpolarizing factor and exerts neuroprotective effects via activation of large-conductance Ca^{2+}-activated K^+ channels. *Br. J. Pharmacol.* 178 (20): 4155–4175

16 Higashi, Y., Kihara, Y., and Noma, K. (2012). Endothelial dysfunction and hypertension in aging. *Hypertens. Res.* 35 (11): 1039–1047.

17 Szabo, C. (2007). Hydrogen sulphide and its therapeutic potential. *Nat. Rev. Drug Discov.* 6 (11): 917–935.

18 Zhao, W., Zhang, J., Lu, Y., and Wang, R. (2001). The vasorelaxant effect of H_2S as a novel endogenous gaseous K_{ATP} channel opener. *EMBO J.* 20 (21): 6008–6016.

19 Shibuya, N., Mikami, Y., Kimura, Y. et al. (2009). Vascular endothelium expresses 3-mercaptopyruvate sulfurtransferase and produces hydrogen sulfide. *J. Biochem.* 146 (5): 623–626.

20 Pan, L.L., Liu, X.H., Gong, Q.H. et al. (2012). Role of cystathionine γ-lyase/hydrogen sulfide pathway in cardiovascular disease: a novel therapeutic strategy? *Antioxid. Redox Signal.* 17 (1): 106–118.

21 Li, L., Hsu, A., and Moore, P.K. (2009). Actions and interactions of nitric oxide, carbon monoxide and hydrogen sulphide in the cardiovascular system and in inflammation – a tale of three gases! *Pharmacol. Ther.* 123 (3): 386–400.

22 Levitt, M.D., Abdel-Rehim, M.S., and Furne, J. (2011). Free and acid-labile hydrogen sulfide concentrations in mouse tissues: anomalously high free hydrogen sulfide in aortic tissue. *Antioxid. Redox Signal.* 15 (2): 373–378.

23 Yu, X.H., Cui, L.B., Wu, K. et al. (2014). Hydrogen sulfide as a potent cardiovascular protective agent. *Clin. Chim. Acta* 437: 78–87.

24 Kimura, H. (2015). Signaling of hydrogen sulfide and polysulfides. *Antioxid. Redox Signal.* 22 (5): 347–349.

25 Mikami, Y., Shibuya, N., Kimura, Y. et al. (2011). Thioredoxin and dihydrolipoic acid are required for 3-mercaptopyruvate sulfurtransferase to produce hydrogen sulfide. *Biochem. J.* 439 (3): 479–485.

26 Yadav, P.K., Yamada, K., Chiku, T. et al. (2013). Structure and kinetic analysis of H_2S production by human mercaptopyruvate sulfurtransferase. *J. Biol. Chem.* 288 (27): 20002–20013.

27 Kabil, O. and Banerjee, R. (2014). Enzymology of H_2S biogenesis, decay and signaling. *Antioxid. Redox Signal.* 20 (5): 770–782.

28 Roman, H.B., Hirschberger, L.L., Krijt, J. et al. (2013). The cysteine dioxgenase knockout mouse: altered cysteine metabolism in nonhepatic tissues leads to excess H_2S/HS^- production and evidence of pancreatic and lung toxicity. *Antioxid. Redox Signal.* 19 (12): 1321–1336.

29 Toombs, C.F., Insko, M.A., Wintner, E.A. et al. (2010). Detection of exhaled hydrogen sulphide gas in healthy human volunteers during intravenous administration of sodium sulphide. *Br. J. Clin. Pharmacol.* 69 (6): 626–636.

30 Murphy, B., Bhattacharya, R., and Mukherjee, P. (2019). Hydrogen sulfide signaling in mitochondria and disease. *FASEB J.* 33 (12): 13098–13125.

31 Abou-Hamdan, A., Guedouari-Bounihi, H., Lenoir, V. et al. (2015). Oxidation of H_2S in mammalian cells and mitochondria. *Methods Enzymol.* 554: 201–228.

32 Vitvitsky, V., Kabil, O., and Banerjee, R. (2012). High turnover rates for hydrogen sulfide allow for rapid regulation of its tissue concentrations. *Antioxid. Redox Signal.* 17 (1): 22–31.

33 Stipanuk, M.H. and Ueki, I. (2011). Dealing with methionine/homocysteine sulfur: cysteine metabolism to taurine and inorganic sulfur. *J. Inherit. Metab. Dis.* 34 (1): 17–32.

34 Yan, H., Du, J., and Tang, C. (2004). The possible role of hydrogen sulfide on the pathogenesis of spontaneous hypertension in rats. *Biochem. Biophys. Res. Commun.* 313 (1): 22–27.

35 Du, J., Yan, H., and Tang, C. (2003). Endogenous H_2S is involved in the development of spontaneous hypertension. *Beijing Da Xue Xue Bao Yi Xue Ban* 35 (1): 102.

36 Roy, A., Khan, A.H., Islam, M.T. et al. (2012). Interdependency of cystathione γ-lyase and cystathione β-synthase in hydrogen sulfide-induced blood pressure regulation in rats. *Am. J. Hypertens.* 25 (1): 74–81.

37 Shi, Y.X., Chen, Y., Zhu, Y.Z. et al. (2007). Chronic sodium hydrosulfide treatment decreases medial thickening of intramyocardial coronary arterioles, interstitial fibrosis, and ROS production in spontaneously hypertensive rats. *Am. J. Physiol. Heart Circ. Physiol.* 293 (4): H2093–H2100.

38 Bucci, M., Papapetropoulos, A., Vellecco, V. et al. (2010). Hydrogen sulfide is an endogenous inhibitor of phosphodiesterase activity. *Arterioscler. Thromb. Vasc. Biol.* 30 (10): 1998–2004.

39 Bucci, M., Papapetropoulos, A., Vellecco, V. et al. (2012). cGMP-dependent protein kinase contributes to hydrogen sulfide-stimulated vasorelaxation. *PLoS One* 7 (12): e53319.

40 Napoli, C., Cicala, C., D'Armiento, F.P. et al. (1999). Beneficial effects of ACE-inhibition with zofenopril on plaque formation and low-density lipoprotein oxidation in watanabe heritable hyperlipidemic rabbits. *Gen. Pharmacol.* 33 (6): 467–477.

41 Napoli, C., Sica, V., de Nigris, F. et al. (2004). Sulfhydryl angiotensin-converting enzyme inhibition induces sustained reduction of systemic oxidative stress and improves the nitric oxide pathway in patients with essential hypertension. *Am. Heart J.* 148 (1): e5.

42 Borghi, C., Bacchelli, S., Degli Esposti, D., and Ambrosioni, E. (2004). A review of the angiotensin-converting enzyme inhibitor, zofenopril, in the treatment of cardiovascular diseases. *Expert Opin. Pharmacother.* 5 (9): 1965–1977.

43 Borghi, C., Bacchelli, S., Esposti, D.D., and Ambrosioni, E. (2003). Effects of the early ACE inhibition in diabetic nonthrombolyzed patients with anterior acute myocardial infarction. *Diabetes Care* 26 (6): 1862–1868.

44 Borghi, C., Omboni, S., Reggiardo, G. et al. (2017). Efficacy of zofenopril compared with placebo and other angiotensin-converting enzyme inhibitors in patients with acute myocardial infarction and previous cardiovascular risk factors: a pooled individual data analysis of 4 randomized, double-blind, controlled, prospective studies. *J. Cardiovasc. Pharmacol.* 69 (1): 48–54.

45 Borghi, C., Omboni, S., Novo, S. et al. (2016). Zofenopril and ramipril in patients with left ventricular systolic dysfunction after acute myocardial infarction: a propensity analysis of the Survival of Myocardial Infarction Long-term Evaluation (SMILE) 4 study. *J. Renin Angiotensin Aldosterone Syst.* 17 (4): 1–9.

46 Bucci, M., Vellecco, V., Cantalupo, A. et al. (2014). Hydrogen sulfide accounts for the peripheral vascular effects of zofenopril independently of ACE inhibition. *Cardiovasc. Res.* 102 (1): 138–147.

47 Laggner, H., Hermann, M., Esterbauer, H. et al. (2007). The novel gaseous vasorelaxant hydrogen sulfide inhibits angiotensin-converting enzyme activity of endothelial cells. *J. Hypertens.* 25 (10): 2100–2104.

48 Li, J., Teng, X., Jin, S. et al. (2019). Hydrogen sulfide improves endothelial dysfunction by inhibiting the vicious cycle of NLRP3 inflammasome and oxidative stress in spontaneously hypertensive rats. *J. Hypertens.* 37 (8): 1633–1643.

49 Zhu, M.L., Zhao, F.R., Zhu, T.T. et al. (2021). The antihypertension effect of hydrogen sulfide (H_2S) is induced by activating VEGFR2 signaling pathway. *Life Sci.* 267: 118831.

50 Tao, B.B., Liu, S.Y., Zhang, C.C. et al. (2013). VEGFR2 functions as an H_2S-targeting receptor protein kinase with its novel Cys1045-Cys1024 disulfide bond serving as a specific molecular switch for hydrogen sulfide actions in vascular endothelial cells. *Antioxid. Redox Signal.* 19 (5): 448–464.

51 d'Emmanuele di Villa Bianca, R., Mitidieri, E., Donnarumma, E. et al. (2015). Hydrogen sulfide is involved in dexamethasone-induced hypertension in rat. *Nitric Oxide* 46: 80–86.

52 Huang, P., Chen, S., Wang, Y. et al. (2015). Down-regulated CBS/H_2S pathway is involved in high-salt-induced hypertension in Dahl rats. *Nitric Oxide* 46: 192–203.

53 Huang, S., Li, H., and Ge, J. (2015). A cardioprotective insight of the cystathionine γ-lyase/hydrogen sulfide pathway. *Int. J. Cardiol. Heart Vasc.* 7: 51–57.

54 Hsu, C.N., Lin, Y.J., Lu, P.C., and Tain, Y.L. (2018). Early supplementation of D-cysteine or L-cysteine prevents hypertension and kidney damage in spontaneously hypertensive rats exposed to high-salt intake. *Mol. Nutr. Food Res.* 62 (2), 1700596.

55 Cirino, G., Vellecco, V., and Bucci, M. (2017). Nitric oxide and hydrogen sulfide: the gasotransmitter paradigm of the vascular system. *Br. J. Pharmacol.* 174 (22): 4021–4031.

56 Ji, W., Liu, S., Dai, J. et al. (2014). Hydrogen sulfide defends against the cardiovascular risk of Nw-nitro-L-argininemethyl ester-induced hypertension in rats via the nitric oxide/endothelial nitric oxide synthase pathway. *Chin. Med. J.* 127 (21): 3751–3757.

57 Watanabe, M., Osada, J., Aratani, Y. et al. (1995). Mice deficient in cystathionine β-synthase: animal models for mild and severe homocyst(e)inemia. *Proc. Natl. Acad. Sci. U.S.A.* 92 (5): 1585–1589.

58 Nagahara, N. (2013). Regulation of mercaptopyruvate sulfurtransferase activity via intrasubunit and intersubunit redox-sensing switches. *Antioxid. Redox Signal.* 19 (15): 1792–1802.

59 Nagahara, N., Nagano, M., Ito, T. et al. (2013). Antioxidant enzyme, 3-mercaptopyruvate sulfurtransferase-knockout mice exhibit increased anxiety-like behaviors: a model for human mercaptolactate-cysteine disulfiduria. *Sci. Rep.* 3: 1986.

60 Peleli, M., Bibli, S.I., Li, Z. et al. (2020). Cardiovascular phenotype of mice lacking 3-mercaptopyruvate sulfurtransferase. *Biochem. Pharmacol.* 176: 113833.

61 Yang, G., Wu, L., Jiang, B. et al. (2008). H_2S as a physiologic vasorelaxant: hypertension in mice with deletion of cystathionine γ-lyase. *Science* 322 (5901): 587–590.

62 Ishii, I., Akahoshi, N., Yamada, H. et al. (2010). Cystathionine γ-lyase-deficient mice require dietary cysteine to protect against acute lethal myopathy and oxidative injury. *J. Biol. Chem.* 285 (34): 26358–26368.

63 Lucock, M., Yates, Z., Martin, C. et al. (2013). Hydrogen sulphide-related thiol metabolism and nutrigenetics in relation to hypertension in an elderly population. *Genes Nutr.* 8 (2): 221–229.

64 Bibli, S.I., Hu, J., Looso, M. et al. (2020). Mapping the endothelial cell *S*-sulfhydrome highlights the crucial role of integrin sulfhydration in vascular function. *Circulation* 143 (9): 935–948.

65 Meng, G., Ma, Y., Xie, L. et al. (2015). Emerging role of hydrogen sulfide in hypertension and related cardiovascular diseases. *Br. J. Pharmacol.* 172 (23): 5501–5511.

66 Sun, N.L., Xi, Y., Yang, S.N. et al. (2007). Plasma hydrogen sulfide and homocysteine levels in hypertensive patients with different blood pressure levels and complications. *Zhonghua Xin Xue Guan Bing Za Zhi* 35 (12): 1145–1148.

67 Sun, H.J., Wu, Z.Y., Nie, X.W., and Bian, J.S. (2020). Role of endothelial dysfunction in cardiovascular diseases: the link between inflammation and hydrogen sulfide. *Front. Pharmacol.* 10. 1568.

68 Wang, C., Han, J., Xiao, L. et al. (2014). Role of hydrogen sulfide in portal hypertension and esophagogastric junction vascular disease. *World J. Gastroenterol.* 20 (4): 1079–1087.

69 Sun, L., Sun, S., Li, Y. et al. (2014). Potential biomarkers predicting risk of pulmonary hypertension in congenital heart disease: the role of homocysteine and hydrogen sulfide. *Chin. Med. J.* 127 (5): 893–899.

70 Wang, K., Ahmad, S., Cai, M. et al. (2013). Dysregulation of hydrogen sulfide producing enzyme cystathionine γ-lyase contributes to maternal hypertension and placental abnormalities in preeclampsia. *Circulation* 127 (25): 2514–2522.

71 Holwerda, K.M., Bos, E.M., Rajakumar, A. et al. Hydrogen sulfide producing enzymes in pregnancy and preeclampsia. *Placenta* 33 (6): 518–521.

72 Cindrova-Davies, T., Herrera, E.A., Niu, Y. et al. (2013). Reduced cystathionine γ-lyase and increased miR-21 expression are associated with increased vascular resistance in growth-restricted pregnancies: hydrogen sulfide as a placental vasodilator. *Am. J. Pathol.* 182 (4): 1448–1458.

73 Raymond, D. and Peterson, E. (2011). A critical review of early-onset and late-onset preeclampsia. *Obstet. Gynecol. Surv.* 66 (8): 497–506.

74 Li, L., Salto-Tellez, M., Tan, C.H. et al. (2009). GYY4137, a novel hydrogen sulfide-releasing molecule, protects against endotoxic shock in the rat. *Free Radic. Biol. Med.* 47 (1): 103–113.

75 Martelli, A., Testai, L., Breschi, M.C. et al. (2010). Hydrogen sulphide: novel opportunity for drug discovery. *Med. Res. Rev.* 32 (6): 1093–1130.

76 Li, L., Whiteman, M., Guan, Y.Y. et al. (2008). Characterization of a novel, water-soluble hydrogen sulfide-releasing molecule (GYY4137): new insights into the biology of hydrogen sulfide. *Circulation* 117 (18): 2351–2360.

77 Muzaffar, S., Jeremy, J.Y., Sparatore, A. et al. (2008). H_2S-donating sildenafil (ACS6) inhibits superoxide formation and gp91phox expression in arterial endothelial cells: role of protein kinases A and G. *Br. J. Pharmacol.* 155 (7): 984–994.

78 Shukla, N., Rossoni, G., Hotston, M. et al. (2009). Effect of hydrogen sulphide-donating sildenafil (ACS6) on erectile function and oxidative stress in rabbit isolated corpus cavernosum and in hypertensive rats. *BJU Int.* 103 (11): 1522–1529.

79 Liang, D., Wu, H., Wong, M.W., and Huang, D. (2015). Diallyl trisulfide is a fast H_2S donor, but diallyl disulfide is a slow one: the reaction pathways and intermediates of glutathione with polysulfides. *Org. Lett.* 17 (17): 4196–4199.

80 Ni, X., Zhang, L., Peng, M. et al. (2018). Hydrogen sulfide attenuates hypertensive inflammation via regulating connexin expression in spontaneously hypertensive rats. *Med. Sci. Monit.* 24: 1205–1218.

81 Meng, Q.H., Yang, G., Yang, W. et al. (2007). Protective effect of hydrogen sulfide on balloon injury-induced neointima hyperplasia in rat carotid arteries. *Am. J. Pathol.* 170 (4): 1406–1414.

82 Yang, F., Liu, Z., Wang, Y. et al. Hydrogen sulfide endothelin-induced myocardial hypertrophy in rats and the mechanism involved. *Cell Biochem. Biophys.* 70 (3): 1683–1686.

17

H_2S Supplementation and Augmentation: Approaches for Healthy Aging

Christopher Hine[1], Jie Yang[1], Aili Zhang[1], Natalia Llarena[1,2], and Christopher Link[1]

[1]*Department of Cardiovascular and Metabolic Sciences, Cleveland Clinic Lerner Research Institute, Cleveland, OH, USA*
[2]*Department of Reproductive Endocrinology and Infertility, Cleveland Clinic Women's Health Institute, Cleveland, OH, USA*

List of Abbreviations

3-MST	3-mercaptopyruvate sulfur transferase
ARMs	aging-related morbidities
CR	caloric restriction
CBS	cystathionine β-synthase
CGL, CTH, CSE	cystathionine γ-lyase
DR	dietary restriction
H_2S	hydrogen sulfide
MetR	methionine restriction
ppb	parts per billion
ROS	Reactive oxygen species

17.1 Introduction and Background

17.1.1 Global Aging Populations

The maximum attainable life expectancy in humans, approximately 120 years, does not appear to be a malleable threshold limit [1]; however, the number of people living above 100 and near to this threshold has dramatically increased over the past century [2]. Dramatic increases in average life expectancy across the globe can be primarily attributed to improvements in public health policies,

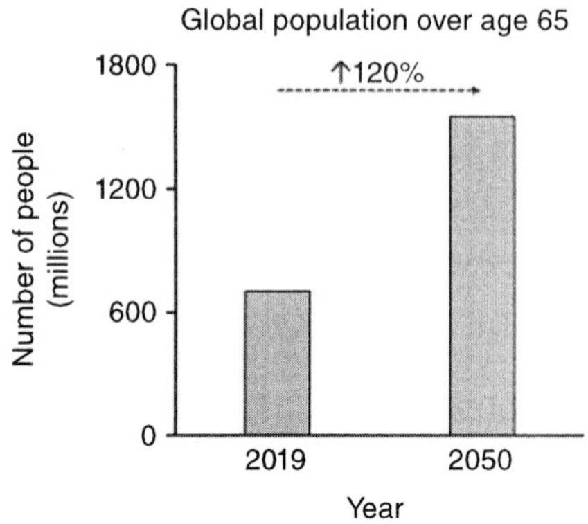

Figure 17.1 Growth of Global Population over age 65 by year 2050. Data used for generating the graph derived from United Nations Department of Economic and Social Affairs, Population Division. Source: Based on Ref. [3].

sanitation control, and the advent of both preventative and reactive medical innovations such as vaccines and antibiotics, respectively. It is also theorized that this trend will continue well into the twenty-first century and beyond barring anthropogenic and/or natural catastrophes. For example, the global population over age 65 in 2021 is around 700 million, which in just 30 years is estimated to grow over 120% to 1550 million [3] (Figure 17.1).

The dramatic numerical growth of this age bracket is not due solely to an ever-expanding total global population, but also a proportional increase of this specific age range. In 1990, those over age 65 made up approximately 6% of the global population, with that percentage increasing to near 10% in 2019, and expected to climb to over 16% by year 2050 [3] (Figure 17.2).

While an increase in the global population of those over age 65 has obvious idealistic personal and societal benefits such as the retention and teaching of institutional knowledge, support in familial and community growth, and extended enjoyment of life post retirement, it is not without its drawbacks. This is primarily due to the fact that extension of life does not inherently confer extension of health. As the increase in life expectancy has been dramatic in such a relatively

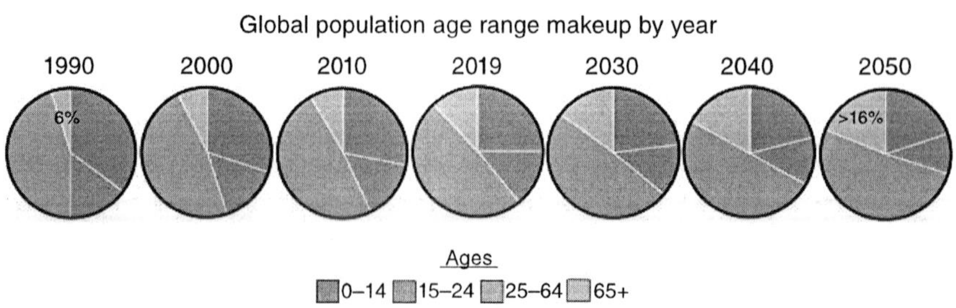

Figure 17.2 Expansion of the global population percentage of those aged 65 and over between 1990 and 2050. Data used for generating the graph derived from United Nations Department of Economic and Social Affairs, Population Division. Source: Based on Ref. [3].

short period of time on a human evolutionary timescale, so has the arrival of aging-related morbidities (ARMs) and diseases. Due to this, natural selection has not had time to "catch up" and fine-tune our inherent biological defenses against ARMs in later ages. Thus, there exists the need to address ARMs from medial, public health, and cultural perspectives.

While public health and cultural approaches to care and aid for an ever-increasing aging population, while immensely important, are out of scope for this chapter, in this chapter, we give an overview of preclinical and clinical medical strategies to improve healthy aging based on hydrogen sulfide and sulfur amino acid metabolism.

17.1.2 Pathophysiological Aspects of Aging

While chronological aging describes the passing of time in measurable increments, be it years for humans or days for yeast, biological aging is far more difficult to depict and quantify. In the context of humans, the biological aging process surprisingly begins in one's mid-30s. Whereas this is not a hard set point for accelerated deterioration of cellular and organ systems, it does represent the beginnings of gradual losses in many faculties, including lean mass [4], female reproductive health [5], and mental capacities to create and learn [6]. Thus, even during this seemingly innocuous period, there is precedence for the implantation of anti-aging interventions. As the years progress into the fifth, sixth, and seventh decades of life, ARMs become much more apparent and the rate of aging accelerates [7]. Hallmarks of the pathophysiological aspects of aging during these later decades include cardiovascular diseases, neurodegeneration, immune deficiencies, eye diseases, cancer, localized and systemic inflammation, metabolic diseases, kidney failure, and respiratory disease [8].

Causes of aging are numerous; however, a central dogma in the aging field is that aberrant and sustained production and exposure to reactive oxygen species (ROS) and free radicals drives the aging process, termed the free radical theory of aging by Denham Harman [9, 10]. While still widely accepted, the free radical theory of aging has been amended in recent years with the understanding that some free radical and ROS signaling are essential for normal physiology and may actually promote healthy aging by preconditioning and protecting against larger oxidative and toxicological insults [11–13]. Nonetheless, loss of redox homeostasis leads to a downward spiral of oxidative damage to macromolecules including DNA, lipids, and proteins resulting in the seemingly irreversible and self-perpetuating aspects of aging. Thus, investigating mechanisms to safely and effectively maintain redox homeostasis through (i) dampening the production of ROS and free radicals through modifying metabolic pathways, (ii) enhancing endogenous anti-oxidant processes, and/or (iii) applying these protective molecules exogenously, can ultimately provide therapeutic interventions against aging and ARMs.

17.1.3 Alterations in Sulfur Amino Acid Metabolism and Hydrogen Sulfide During Aging

The versatility of sulfur having oxidation states from −2 to +6 permits it to act as an acceptor and donator of electrons [14]. These chemical properties make it an excellent element to build upon for redox-based interventions and therapies. The sulfur on hydrogen sulfide (H_2S) has an oxidation state of −2, indicative of a reducing agent, and thus H_2S performs several non-mutually exclusive biochemical reactions *in vivo*. These include mitochondrial electron transfer, alterations of iron–sulfur and heme centers, antioxidant activity, and protein posttranslational modification via sulfhydration; aka persulfidation or S-sulfhydration, of reactive cysteine residues to form persulfide tails (R-SS_nH) [15]. Persulfidation functions as both a signaling event [14] as well as a protective modification preventing irreversible oxidation and damage of cysteine residues [16]. As H_2S and its associated hydrosulfide anion (HS^-) and sulfide anion (S^{2-}), herein collectively referred to as H_2S in this chapter, act as a redox modifying metabolite and a pharmacological agent, its study and applications in the realm of aging and ARMs have been and continues to be of great importance.

In addition to the increase of aberrant ROS and free radicals giving way to oxidative damage during aging, there are aging-related changes in endogenous H_2S production and signaling. In mammals, H_2S is enzymatically produced during the metabolism of sulfur containing amino acids, primarily cysteine and homocysteine, via the activities of cystathionine β-synthase (*aka* CBS), cystathionine γ-lyase (*aka* CGL, CTH, or CSE), and 3-mercaptopyruvate sulfurtransferase (*aka* 3-MST) [17] (Figure 17.3a,b). H_2S is also produced systemically to a limited extent non-enzymatically by coordinated catalysis of cysteine via iron and vitamin B_6 [18] (Figure 17.3c). Lack of CGL, CBS, and/or 3-MST expression and/or activity are attributed to ARMs such as hypertension [19], neurodegeneration [20], and the inability to properly respond to dietary and endocrine cues [21–25]. Conversely, increased CGL expression and H_2S production capacity are hallmarks of numerous dietary, genetic, hormonal, and pharmaceutical models of extended lifespan and may serve as molecular biomarkers associated with longevity [26, 27].

One of the first studies to recognize the decline in endogenous enzymatic H_2S levels and/or production in the realm of aging independent of a specific disease state was by Predmore et al. in 2010 [28]. They found in rats that aortic H_2S concentrations, and contractile responses to H_2S, decrease with advanced age. Similarly, in the worm *Caenorhabditis elegans* expression of CGL and the resultant H_2S production capacity naturally decrease with age [29]. More recently, the Filipovic group thoroughly displayed not only accelerated aging in *C. elegans* lacking 3-MST concomitant with decreased protein persulfidation with advancing age but also losses in protein persulfidation and expression of all three H_2S generating enzymes

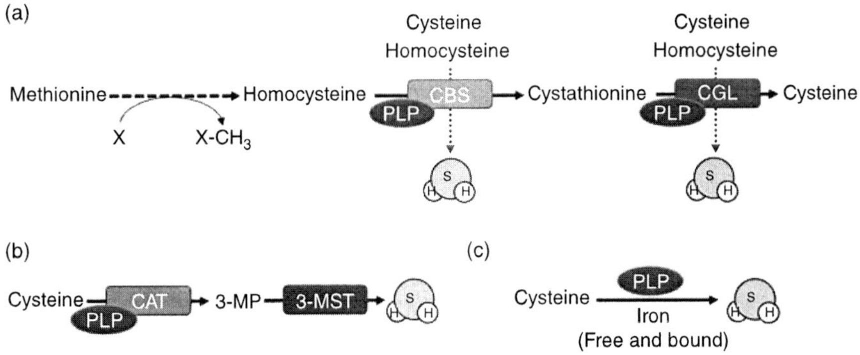

Figure 17.3 Endogenous H_2S production pathways in mammalian systems. (a) Generalized depiction of the trans-sulfuration pathway, in which methionine is converted to cysteine through multiple enzymatic steps, including those catalyzed by cystathionine β-synthase (CBS) and cystathionine γ-lyase (CGL). The vertical dotted arrows show alternative catalytic activities of CBS and CGL, in that they catabolize sulfur containing amino acids to produce hydrogen sulfide (H_2S) gas. CBS and CGL enzymatic activities depend on their cofactor pyridoxal phosphate (PLP), *aka* Vitamin B_6. (b) Stepwise deamination of cysteine to 3-mercaptopyruvate (3-MP) by cysteine/asparatate aminotransferase (CAT) with co-factor PLP and further release of H_2S via 3-mercaptopyruvate sulfurtransferase (3-MST). Source: (a, b) Based on Kabil and Banerjee [17]. (c) Nonenzymatic H_2S production catalyzed by the coordinated activities of PLP and iron on the sulfur containing amino acid cysteine. Source: Modified from Yang et al. [18].

in the brains, hearts, and livers of 24-month-old rats relative to 1-month-old rats [30]. Strikingly, this same group discovered declines in protein persulfidation and increases in oxidative damage in human fibroblasts taken from the same individual at different ages, being 31 and 48 years of age [30]. These latter results highlight that aging is not necessarily a process that only occurs in the sixth, seventh, and eight decades of life, but commences much earlier during middle age. Thus, interventions to slow the aging process by targeting H_2S-related biology could be implemented as early as one's 30s.

When factoring in specific aging-related diseases and ARMs, it becomes more apparent that loss of H_2S production and/or expression of its generating enzymes play pathologic roles underlying the mechanisms of these diseases. As cardio/cerebro-vascular diseases, cancer, and neurodegenerative disorders are three of the major aging-related causes of death, we detail below what is known in regards to changes in H_2S-biology for these three. However, it should be mentioned here that in the studies described below and throughout this chapter different methods and approaches to measure H_2S, sulfide, and persulfides were used that are specific to that study or research model. This can result in different absolute concentration levels from study to study. Thus, while H_2S or other sulfide detection methods are useful for generating values for internal comparison and relative production capacities and levels within the same study, the values they generate may be difficult to compare between independent studies. For more on this topic, please see the thorough reviews and papers by Filipovic et al. [14], Bogdandi et al. [31], and Hartle and Pluth [32].

Aging-related cardiovascular/cerebrovascular diseases include hypertension, stroke, heart failure, atherosclerosis, and peripheral artery disease. One of the first studies to identify the physiological roles of CGL in the cardiovascular system identified mice with hypertension that were deficient in CGL expression and/or activity [19]. In humans with various forms of cardiovascular disease, circulating sulfide levels were decreased in a gender- and ethnic-dependent manner [33]. Similarly, individuals with diabetes-related vascular inflammation had decreased circulating sulfide levels [34]. Endogenous production capacity of H_2S and associated sulfides are also blunted in cardiovascular diseases, chiefly in ischemic–reperfusion injury and heart failure, with the attenuation in enzymatic production causative of disease progression [35, 36]. In cerebrovascular disease, *aka* stroke, buildup of cystathionine in the plasma serves as an independent risk factor for increased incidence of total and ischemic stroke [37], with increased CBS and/or H_2S protecting against neuroinflammation following stroke [38]. Due to the importance of H_2S in the cardiovascular system, much has been invested in terms of basic research and clinical translation of H_2S-based therapeutics.

The majority of cancers and oncologic diseases are aging-related or have their onset in later life. While juvenile and young adult cancers do exist, particularly those of hematologic origin, the incidence of most cancers dramatically increased during the fifth and sixth decades of life [39]. Although decreased H_2S is strongly associated with and well-studied in cardiovascular and neurodegenerative disorders, it is less understood and a somewhat paradoxical situation related to cancer. This is primarily due to cancer being a heterogeneous disease consisting of multi-tissue and multicellular origins [40] with H_2S having a multimodal cellular dose response that is shifted dependent on the tissue of origin [41, 42]. Of note, H_2S is pro-tumorigenic in the realm of colorectal [43, 44] and ovarian [45] cancer, primarily due to its pro-angiogenesis attributes [44, 46]. Conversely, H_2S, particularly when delivered exogenously, provides anti-tumorigenic capabilities to numerous other cancer cell types *in vitro* and *in vivo* [47], due to its alterations in cancer cell metabolism and/or sensitization of cells to apoptosis [48, 49]. Moreover, there is emerging evidence that aging-related loss of H_2S can alter tumor immunity due to its inability to suppress indoleamine 2,3-dioxygenase 1 (IDO1), which enhances immune suppression in the tumor microenvironment [50]. In glioblastoma (GBM), one of the most lethal forms of cancer, endogenous enzymatic H_2S synthesis inhibits proliferation of cultured GBM tumor cells [51]. In parallel, the enzymatic activity of both CGL and 3-MST decreases with increasing grade/metastatic spread and is essentially nonfunctional in the context of advanced GBM [52]. Thus, there appears a great deal to learn in terms of how aging-related changes impact specific forms of both primary and metastatic cancers.

The field of H_2S in neuro-cognitive function may be the best and longest studied with regard to H_2S-related biomedical research. The first report of the significance of endogenous enzymatic production of H_2S having a physiological role was published by Kazuho Abe and Hideo Kimura in 1996 in their seminal work "The possible role of hydrogen sulfide as an endogenous neuromodulator" [53]. Since then, the field of H_2S in neuromodulation and neurodegeneration has expanded from the most basic molecular and biochemical [54] all the way to the translation and development of sulfide and trans-sulfuration based therapies and diagnostics for Alzheimer's [55], Huntington's [20], and Parkinson's diseases [56, 57]. H_2S, sulfide levels, persulfidation, and/or expression of H_2S generating enzymes have all been found to be decreased in aging-related neurodegenerative diseases [56], with supplementation of H_2S-based donor molecules or sulfur containing amino acids providing therapeutic improvements in preclinical cellular and animal models.

While many cases of H_2S decline and disruption of trans-sulfuration in ARMs can be classified as correlative, the resounding evidence from cellular, animal, and early human trials addressing causative losses in H_2S related to ARMs has been very promising. But just how to safely and effectively enhance endogenous H_2S production and/or apply it exogenously for prevention and/or treatment of ARMs continues to be a major question in the field. Nonetheless, applying geroscience approaches to slow the aging process, possibly involving H_2S-based therapeutics, may eventually lessen the severity of aging related diseases, delay their onset, and/or eradicate them completely.

17.1.4 Geroscience Approaches to Address Longevity and Improved Healthspan, and Their Connection to Hydrogen Sulfide

Extension of lifespan and the delay of aging have been of great interest for medical and scientific pursuits for centuries. However, it was not until the latter part of the twentieth and early twenty-first centuries that the molecular targets and geroscience approaches needed to make this a reality were revealed through rigorous aging research in model organisms from yeast, worms, flies, rodents, and even nonhuman primates [58]. The geroscience approach, largely defined and championed by Felipe Sierra, attempts to identify the biological underpinnings for the age-dependency of most chronic diseases. The hypothesis is that if aging is malleable, as well as the main risk factor for chronic diseases and morbidities, then addressing the basic biology of aging is likely to provide a better payoff than addressing diseases one at a time [59]. This approach is displayed in Figure 17.4, with current medical treatments providing scenario A, future advancements

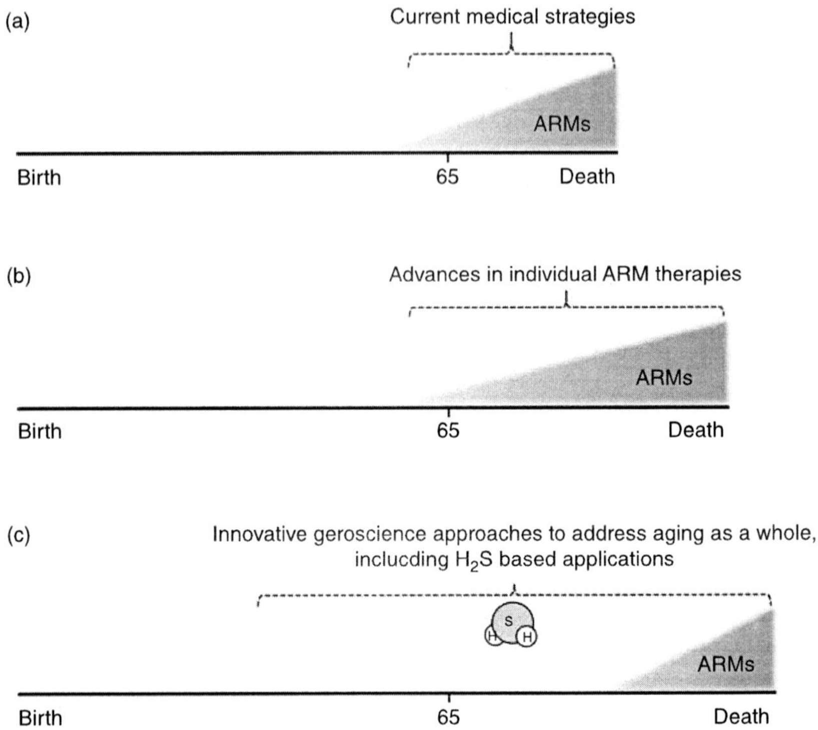

Figure 17.4 Current and future methods to treat and/or prevent aging-related morbidities (ARMs). (a–c) Lifespan and healthspan (*aka* period of life without ARMs) predictions from current medical practices (a), future improvements in treating individual ARMs (b), and incorporation of geroscience approaches starting in middle age in addition to the standard medical advancements (c). H_2S-based therapies, including the use of exogenous donor molecules and/or enhancements of endogenous H_2S production, would fit into the geroscience approaches to slow aging as well as provide therapeutic treatment against ARMs.

in individual ARM treatments providing scenario B, and applied geroscience approaches, including the use of H_2S-based therapies, to slow the aging process to delay the onset and/or treat multiple ARMs simultaneously providing scenario C.

What are these geroscience approaches and what are their links to H_2S, if any? Established means to slow the aging process and extend lifespan in model laboratory organisms across evolutionary boundaries include dietary, genetic, and pharmacological approaches. Dietary restriction (DR) without malnutrition is a term applied to a broad set of dietary interventions which include, but are not limited to, conventional caloric restriction (CR), restriction of specific macronutrients such as sulfur amino acids (*aka* methionine restriction, MetR), and intermittent fasting (IF) [58, 60]. In mammals, these interventions can be applied short term; 1 day to 3 months, or longer term; 3 months to adult lifespan, depending on the target endpoint. In yeast and other non-mammalian organisms, nutrient restriction is often applied, such as altering the ratios of proteins and carbohydrates or simply diluting nutrients in liquid medium. Implementation of DR has been reported to extend lifespan and healthspan starting in the 1930's [61], and since then many insights into the nutritional triggers [62] and molecular mechanisms [21, 63] have been revealed. Evidence for altered trans-sulfuration and enhanced H_2S production and/or persulfide signaling during the various forms of DR are evident in yeast [21], worms [21], flies [21, 24], mice [21, 22, 30, 64–66], and rats [28]. Mammalian genetic models of longevity have primarily focused on reduction of growth hormone (GH) and insulin like growth factor (IGF)-1 signaling as evident in Ames and Snell Dwarf mice and growth hormone receptor knockout mice [67]. Like DR, alterations in trans-sulfuration and enhanced H_2S production have been detected in these genetic and hormonal-induced longevity extensions [25, 68–70]. Pharmacological means to extend lifespan have been a great focus of the Interventions Testing Program (ITP) of the National Institute on Aging (NIA) [71–73]. In this program, small molecules and drugs that have shown promise in other model organisms or in short-term murine studies are put through rigorous multi-center longevity tests in heterogeneous mouse models. As of 2017, six compounds showed significant extension of lifespan, and these include aspirin, rapamycin, 17α-Estradiol, acarbose, nordihydroguaiaretic acid, and Protandim [73]. While these drugs have been verified by the ITP, other possible candidates, such as metformin [74] and resveratrol [75], have also been shown to extend lifespan and/or slow aging under specific conditions. Just like dietary and genetic models of longevity, pharmacological interventions have also shown to stimulate endogenous H_2S production and/or impact trans-sulfuration [76]. Remarkably, the lab of Vadim Gladyshev has recently shown that increased CGL expression in liver is a hallmark of numerous dietary, genetic, hormonal, and pharmaceutical models of extended lifespan, and they propose that increased CGL expression may serve as a molecular biomarker associated with longevity [26].

Here, we provide select examples in which H_2S itself, donor molecules, or the stimulation of its endogenous production by way of the geroscience interventions mentioned above were used to extend lifespan, promote healthspan, and counter aging-related diseases. Additionally, we also provide a few examples in which enhanced endogenous H_2S production and/or altered trans-sulfuration were found in models of exceptional longevity. While other chapters in this book go into more detail in regards to some specific aging-related diseases and how H_2S can act as a therapeutic, such as in neurodegeneration and cardiovascular disease, here we only provide a general overview in each of these topics, while including the many organisms across evolutionary boundaries in which H_2S impacts their specific aging trajectories.

17.2 Hydrogen Sulfide Metabolism and Applications in Non-mammalian Aging

17.2.1 Plants

While plant aging is usually not a heavily studied subject in the academic sense, over the past 10 years interest has grown in the economic value of H_2S in agriculture and plant-based food industry uses. Overly rapid plant ripening and spoiling greatly limit the time between harvest to consumer end use, resulting in waste. Exogenous H_2S, usually with the use of NaHS, effectively extends postharvest shelf life, defers ripening and senescence, and maintains a high quality consumer product in a variety of fruits and vegetables, including strawberries (*Fragaria × ananassa*) [77], broccoli (*Brassica oleracea*) [78], grapes (*Vitis vinifera* L. × *Vitis labrusca* L.) [79], banana (*Musa* spp., AAA group) [80], tomato (*Solanum lycopersicum* L.) [81, 82], avocado (*Persea americana* Mill) [83], kiwifruit (*Actinidia chinensis*) [84, 85], and daylilies (*Hemerocallis fulva*) [86]. Most of these studies attribute the protective effects of H_2S to an increase in antioxidant systems, a decrease in ROS production and accumulation, and its antagonizing effect in ethylene production. The implications of H_2S as a signaling molecule and resultant benefits in fresh produce of fruits/vegetables are far less explored, although it has been shown that H_2S can regulate plant catalase, a key protein during fruit ripening, via persulfidation [87], and that persulfidation in plants has wide ranging biological implications for regulation of pathways and systems biology [88].

17.2.2 Bacteria

While aging and senescence in bacteria are documented [89, 90], they are often understudied fields. Bacterial aging involves the sequestration and partitioning of metabolic toxins and damaged macromolecules away from "daughter cells" and retention in the "mother cells" [91]. However, what is often better studied is the role of bacteria, and more specifically, the microbiota that reside within the gut of the host, in diseases and ARMs. Gut microbiota-derived H_2S may be the greatest source of H_2S and/or sulfide synthesis in the body. Free H_2S levels, bound sulfane sulfur, and CGL activity are reduced in germ-free mice compared to control animals [92]. Indeed, microbiota, together with the mucosa where they reside, form an important barrier against external damage/injury. Disruption of the barrier, such as by antibiotics or lifestyle factors (high meat intake, high fat diet, stress), can cause overgrowth of specific/undesired species of bacteria and ultimately increase the susceptibility of mucosa to damage and injury. Numerous studies have focused on the difference of the gut microbiota in healthy subjects versus those with a particular intestinal disorder, including inflammatory bowel disease, irritable bowel syndrome, and colorectal cancer. Associations have also been reported between gut microbiota and distant organ dysfunction such as chronic kidney disease (CKD) and cardiovascular disease [93, 94]. Not surprisingly, it has been postulated that gut microbiota-derived H_2S may contribute to disease progression [95]. The studies described below aim to exploit H_2S in mucosal defense and microbiota function, via pharmaceutical drug developments and dietary interventions, which as a result restore gastrointestinal health and mitigate inflammation.

Much progress has been made in utilizing H_2S to reduce nonsteroidal anti-inflammatory drugs (NSAIDs)-induced gastrointestinal inflammation/toxicity (Table 17.1). Administration of conventional NSAIDs produces significant gastrointestinal toxicity, such as ulceration and bleeding. Therapeutic delivery of H_2S, via oral administration of ATB-346, a H_2S-releasing naproxen derivative as a new class of NSAID, significantly protected rats from conventional naproxen-induced ulceration and bleeding [96]. Meanwhile, ATB-346 also attenuates stress-induced gastric lesions [98], exhibits more efficient protection against abdominal surgery induced postoperative ileus [97], and more effective chemopreventive effects in both drug-induced colon cancer mouse model [99] and hereditary intestinal cancer APC^{Min+} mouse models [100]. A phase-II-B study further supports the anti-inflammatory effect of this H_2S-releasing NSAID drug, where subjects with ATB-346 (250 mg once daily) showed fewer ulcers and a remarkable reduction of gastrointestinal complication compared to conventional NSAID naproxen (550 mg twice daily) [101]. Mechanistically, ATB-346 administration produces an equivalent inhibition of cyclooxygenase (COX) to naproxen [101], modifies intestinal microbiota, as well as reduces the cytotoxicity of bile caused by traditional NSAIDs [96].

The gut microbiome is also associated with distant host organ functions, including but not limited to the heart, kidney, and brain. The ability of H_2S in modifying gut microbiota has been exploited in treating CKD. Therapeutic delivery of H_2S, via intraperitoneal injection of GYY4137 (133 μM/kg/d), a slow-releasing H_2S donor, improves glomerular filtration in drug-induced hypertensive kidney disease in mice. GYY4137 supplementation reduces the Erysipelotrichia class of bacteria, which is enriched in hypertensive animals, suggesting the normalization of hypertensive kidney function through modulation of the gut microbiome [103].

17.2.3 Yeast

While yeast appear to be simple single-celled eukaryotic organisms, much of the groundbreaking and impactful molecular discoveries in the aging field were made, and continue to be made, in yeast, and more specifically in the species *Saccharomyces cerevisiae*. These discoveries include nutrient sensing pathways, the anti-aging benefits of caloric restriction and methionine restriction, and aberrations in pro-growth/anabolic signal transduction [104–107]. Aging is

Table 17.1 H_2S-derivatized drugs and their protection from GI toxicity.

H_2S donor	Study subject	Route	Effect on gut	Reference
ATB-346, Antibe Therapeutics	Wistar rats	Oral administration	Protects from conventional naproxen-induced ulceration and bleeding, most likely by modifying intestinal microbiota as well as reducing the cytotoxicity of bile, compared with naproxen treated animals	Blackler et al. (2015) [96]
ATB-346, Antibe Therapeutics GYY4137, National University of Singapore	C57Bl6J mice	ATB-346 16 mg/kg, intragastrically; GYY4137: 50 mg/kg, intraperitoneally	Prevented abdominal surgery induced postoperative ileus (POI), compared to the conventional drug naproxen. Results with GYY4137 show that H_2S *per se* can prevent POI	Van Dingenen et al. (2019) [97]
ATB-346, Antibe Therapeutics	Wistar rats	Intragastrically equimolar dose to naproxen (20 mg/kg)	Attenuates stress-induced gastric lesions	Magierowski et al. (2017) [98]
ATB-346, Antibe Therapeutics	Drug-induced colon cancer mice		Exerts more effective chemopreventive effects than its parent drug, naproxen, and did not induce GI injury as observed in naproxen treatment	Elsheikh et al. (2014) [99]
ATB-346, Antibe Therapeutics	Hereditary intestinal cancer APCMin+ mouse model	Orally treated with equimolar dose to naproxen (20 mg/kg)	Prevents and reverses the development of pre-cancerous lesions. These effects may be related to the combined effects of suppression of cyclooxygenase (COX) and release of H_2S, and correction of most of the APC mutation-associated alterations in the transcriptome	Paul-Clark et al. (2016) [100]
ATB-346, Antibe Therapeutics	Human	250 mg once daily	A proof-of-concept, Phase 2 clinical trial of the gastro-intestinal safety of a H_2S-releasing anti-inflammatory drug In humans, ATB-346 (250 mg once daily) was found to inhibit COX to the same extent as naproxen (550 mg twice daily)	Wallace et al. (2020) [101]
GIC-1001, GIcare Pharma	C57Bl6 mice	30 and 60 mg/kg	Reduces nociceptive response to increasing pressures of colorectal distension, more efficiently than its parent drug, trimebutine	Cenac et al. (2016) [102]

assessed in multiple ways in *S. cerevisiae,* with the most common means being replicative lifespan via measuring the number of replications a cell can undergo, or by measuring the actual chronological time a cell spends in a non-dividing state before losing viability [108]. Complicating the matter of how H_2S and sulfur metabolism play roles in yeast aging is the increased complexity of sulfur and sulfide metabolism in yeast compared to metazoans. In the yeast *S. cerevisiae,* there are several mechanisms by which H_2S is produced, including (i) catabolism of sulfur-containing amino acids via the yeast trans-sulfuration CYS3 and CYS4 genes encoding CGL/STR1 and CBS/STR4, respectively, (ii) reduction of elemental sulfur, or (iii) via the assimilatory pathway for the reduction of sulfites and sulfates [109, 110]. Thus, modulating endogenous H_2S production in yeast is possible from many different targets and pathways.

Evidence for alterations in sulfide and/or sulfur amino acid metabolism impacting aging in yeast can be found in studies identifying that MetR extends chronological and replicative lifespan as well as enhancing stress resistance [105, 107, 111]. Likewise, supplementation of methionine on top of glucose restriction abrogates the age-delaying benefits of glucose restriction [112]. Genetic manipulation of yeast to decrease methionine synthesis/uptake via single deletion assays on seven of the eight genes encoding enzymes responsible for methionine/cysteine biosynthesis also extends lifespan [112]. One such gene/enzyme identified is Met15 (*aka* Met17/Met25) [105, 112]. Deletion of Met15 results in the inability to assimilate exogenous sulfides (such as H_2S) for the production of homocysteine and methionine, thus resulting in a buildup of H_2S as well as a decrease in methionine synthesis. Due to the buildup of H_2S in Met15 mutants, these strains are often used for colony selection when grown on lead-containing media for determining loss/gain of external artificial plasmids/chromosomes containing a gene of interest tied into an extra copy of functional Met15 [113]. Thus, when using the Met15 deletion strains, such as the common BY4743 strain, for selection purposes in aging studies one must be cognizant to perform proper controls to address their already extended lifespan and the inherent augmentation of H_2S.

Separating out the contributions of increased endogenous production and/or incorporation of H_2S or reduced methionine biosynthesis providing the mechanism for extended longevity in these nutrient restriction and/or methionine/cysteine biosynthesis genetic models is difficult and complicated [112, 114]. However, the sufficiency for exogenous addition of H_2S and/or H_2S donor molecules to extend yeast lifespan has been established [21]. The addition of the H_2S salt NaHS and the H_2S donor GYY4137 in culture media started during early logarithmic growth leads to extended chronological lifespan in yeast [21]. Additionally, NaHS treatment at the late growth stage of *S. cerevisiae* also extends chronological lifespan [115]. Thus, despite difficulties of discerning correlation versus causation in endogenous H_2S production being the driver for increased lifespan in yeast, the sufficiency of exogenous H_2S to deliver this benefit champions the cause for enhanced endogenous production contributing at least some mechanistic function for longevity.

17.2.4 Worms

Despite their relatively small size, many of the genetic and molecular mechanisms behind delayed aging and lifespan extension have been uncovered in the nematode worm *C. elegans* [29, 116, 117]. On average, the median lifespan of *C. elegans* is approximately 17 days, with extensions or shortening by at least 10–15% deemed biologically significant [117]. Extension of lifespan and improved activity at later ages in these organisms is possibly via geroscience approaches such as in the pharynx-disrupting *eat* mutations that functionally result in DR [116], disruption of insulin/IGF-1 and germline signaling [68, 118], and treatment with the drug metformin [119]. In some of these models, increased enzymatic production capacity of H_2S and/or persulfide signaling has been correlated, and in some cases, necessary for lifespan extension [21, 30, 68]. Under dietary or nutrient restriction, which is modeled by the *eat-2* mutation in *C. elegans*, median lifespan is increased by almost 15 days and correlates with increased whole body H_2S production. However, deletion of CBS abrogates this lifespan increase [21]. Likewise, overexpression of CBS alone increased median lifespan by approximately 10 days, indicating the necessity and sufficiency of increased CBS expression/activity to lengthen lifespan in these worms [21]. In another independent study, examining the connections between endogenous H_2S production and signaling with food restriction-induced longevity, the *eat-2* mutant worms were paired with mutations in CGL or 3-MST [30]. While the *eat-2* mutants had a similar ~15 day increase in median lifespan concurrent with enhanced protein persulfidation, those with CGL or 3-MST deletions lost the longevity benefit and the boost in protein persulfidation [30]. Paired with findings that deletion of 3-MST in *C. elegans* shortens lifespan but can be rescued with exogenous addition of the H_2S donor GYY4137 [120], these data highlight the central importance for H_2S and sulfur amino acid metabolism in slowing the aging process via DR. Increased H_2S production is also required for extended longevity following the removal of germ cells [68], which is a distinct model of longevity extension in worms independent of nutrient intake.

While many studies have examined endogenous H_2S production and/or the presence of H_2S-generating enzymes as necessity for lifespan extension, the first *C. elegans* aging studies related to H_2S involved exogenous sources. Longevity extension and stress resistance driven by H_2S was first shown by Miller and Roth in 2007 by exposing worms to 50-ppm H_2S in air during early development and throughout life. The authors showed an extension in mean lifespan of 70%

and the attainment of thermotolerance [121]. The same group showed that hypoxia-inducible factor-1 (HIF-1) and transcription factor skinhead-1 (SKN-1) are required to coordinate the transcriptional response to H_2S in worms [122]. Furthermore, treatment early in adulthood with the H_2S generating garlic constituent diallyl trisulfide also extends worm longevity in a SKN-1-dependent manner [123]. The slow-releasing H_2S donor GYY4137 also extends lifespan in worms [120, 124]. Thus, the presence of H_2S alone is sufficient to provide lifespan and healthspan benefits in *C. elegans*. However, recent data proposes that the lifespan and healthspan benefits of exogenous H_2S are independent from the effects downstream of the *eat-2* mutation in worms, suggesting that exogenous H_2S does not mimic DR completely [125]. Hence, there are both shared as well as independent biological responses to enhanced endogenous H_2S production compared to exogenous H_2S, and the level of overlap may depend on dosage, vehicle, timing, developmental stage of the organism, and interacting factors impacting redox balance.

17.2.5 Flies

The fruit fly *Drosophila melanogaster*, also known as the vinegar fly, has long been used in dietary, genetic, and environmental aging research studies [126]. Due to a relatively short median lifespan of 70 days and a maximum of 90 days, along with the ability to monitor aging-related physiological changes in metabolism, behavior, stress resistance, and reproductive capacity, these flies are a great tool for multifaceted aging studies [126]. Much like in other organisms, methionine restriction extends fly lifespan [127]. Additionally, methionine restriction mimics the effects of DR in flies and requires the target of rapamycin (TOR) suppression activity of Tsc2 [128].

Strong evidence for the role of sulfur amino acid metabolism and trans-sulfuration in fly aging and longevity emerged in the work by Kabil et al. in 2011 [24]. In this landmark paper, it was discovered that increased trans-sulfuration mediates longevity conferred by DR in *Drosophila*. Specifically, the authors revealed both CBS and CGL as positive regulators for lifespan extension and improved physiology under DR [24]. Importantly, they showed that neuron-specific transgenic overexpression of the fly version of CBS enhanced longevity independent of diet, indicating the necessity and sufficiency of enhanced H_2S production capacity to drive lifespan extension [24]. The finding of enhanced CBS expression and/or activity driving longevity was confirmed in an independent study from the Moskalev group, in which overexpression of CBS increased median lifespan by 12.5% and maximum lifespan by 6.9%, as well as improved locomotor activity [129]. The Moskalev group has also recently shown in *Drosophila* that overexpression of CBS and CGL concurrent with supplementation of the precursors for H_2S synthesis cysteine and *N*-acetyl-L-cysteine provides maximum geroprotective potential [130]. Additionally, maximal endogenous H_2S production capacity from whole fly lysates was detected in flies fed the optimal methionine to essential amino acid ratio diet that directly correlated to extended longevity [21, 128].

Unlike in worms and yeast, there have been no definitive studies published showing exogenous addition of H_2S or H_2S donors extends lifespan in flies. However, similar to worms the exogenous addition of H_2S does offer stress resistance in the fly. One of the earliest studies to indicate the stress resistance capabilities of exogenous H_2S comes from 1969, in which *Drosophila* eggs irradiated in the presence of H_2S had reduced embryonic mortality compared to eggs not provided H_2S, providing a clue for the radioprotective effect of H_2S [131]. Additional stress resistance afforded due to H_2S exposure in *Drosophila* comes in the form of desiccation tolerance. Flies exposed to H_2S gas with a K_2S donor had increased survival under arid and food-free conditions, with a concomitant increase in the desiccation tolerance-related genes Frost and Desat2 [132]. Thus, exogenous use of H_2S provides stress resistance and hints at its mechanistic importance during periods of limited food intake in flies.

17.3 Hydrogen Sulfide Metabolism and Applications in Nonhuman Mammalian Aging

17.3.1 Standard Laboratory Rodents (Mice and Rats)

While the time, cost, and potential risk expenses of performing aging research and lifespan studies in mammals compared to the non-mammals described above is high, its translational opportunities and human relevancies are vital. This is exemplified with the National Institutes of Health performing their Interventions Testing program to determine lifespan and healthspan extending interventions on mice [73]. Surprisingly, initial studies into DR benefits in rodent models began in the early 1900s with Peyton Rouse showing its prevention of spontaneous and induced tumors [133] and later in the 1930s with Clive McCay with extension of lifespan [61]. Starting in the early 1980s and into present day, investigation into the nutritional, genetic, and molecular triggers of delayed aging, particularly changes in sulfur amino acid metabolism, brought upon via geroscience approaches became much more rigorous and adapted by laboratories with varying interests and methodologies [65, 134–137].

Evidence for the importance of sulfur amino acid metabolism playing a role in mammalian lifespan extension was first presented by the Orentreich group in 1993 and 1994 [138, 139]. In this set of complimentary studies, they showed the restriction of methionine and cysteine in the diet, whereas the amount of methionine in the diet was reduced from 0.86% down to 0.17% and cysteine completely removed, resulted in a 30% increase in lifespan of male Fischer 344 rats [138] concurrent with increased blood glutathione [139]. These findings were followed up a decade later by the Miller group showing sulfur amino acid restriction, even when accounting for inherent aversion to the diet, increases maximal lifespan, immune homeostasis, and stress resistance in mice, concurrent with lowering levels of serum IGF-1, insulin, glucose, and thyroid hormone [136]. While these diets have been historically termed methionine restriction (MetR), they do require combined cysteine restriction for many of their benefits [140, 141]. Likewise, the longevity, delayed aging, and stress resistance outcomes in mice and rats undergoing standard caloric restriction have been linked to the decrease of sulfur amino acid intake, where the addback of sulfur amino acids dampens the benefits of the diet restriction including longevity reversal [21, 66, 142].

However, what is ultimately changed in sulfur amino acid metabolism under DR or other forms of exceptional longevity in rodents that functionally lends to beneficial endpoints? A common molecular phenomenon among several dietary and endocrine models of rodent longevity is the altered metabolism of sulfur containing amino acids methionine and cysteine, which entails flux through trans-sulfuration [69, 70, 143] and increased production capacity and/or bioavailability of hydrogen sulfide (H_2S) gas [27]. In regard to dietary geroscience approaches, comparative analysis of trans-sulfuration gene and protein expression and metabolomics profiling in liver and kidney of mice suggests increased H_2S production upon short-term DR and long-term DR [66], as a key molecular event leading to stress resistance and longevity [21]. Our group recently performed H_2S and persulfidation studies in CGL wildtype (WT) and knockout (KO) mice, and determined short-term 50% CR in young adult mice enhances H_2S production capacity in the liver and kidney, concurrent with enhanced protein persulfidation in liver, kidney, skeletal muscle, and brain [65]. These changes in production and persulfidation were largely dependent on CGL. Additionally, we showed longer-term every other day fasting in aged mice enhanced kidney H_2S production and protein persulfidation in liver, kidney, and brain, and these increases directly correlated with improved metabolic fitness and cognitive performance in the aged animals [65, 144]. These results are in line with data from the Filipovic group showing caloric restriction enhances tissue-specific protein persulfidation in both young adult and aged mice [30]. As a function of the % caloric restriction, the jump from 20% to 40% results in a dose-dependent increases of liver H_2S production in male and female mice [145]. Independent of diet, long-lived mouse models of hypopituitary dwarfism and/or aberrations in growth hormone and thyroid hormone signaling display increased CGL expression and H_2S

production in their liver [25]. This boost in production may serve causative roles in pro-longevity endocrine homeostasis in regards to thyroid hormone, IGF-1, and estrogen [25, 27, 146–148]. Also independent of diet, pharmaceuticals and small molecule drugs shown to extend mouse lifespan and healthspan such as metformin [74, 149] enhance H_2S production/bioavailability in murine tissues [76, 149]. Thus, aiming for increased endogenous H_2S production for delaying the onset of ARMs can be accomplished via dietary, hormonal, and pharmacologic means.

However, the strain of mouse is an important factor when planning aging and longevity experiments, as there are positive, neutral, and negative responders to DR [150]. Interestingly, when examining hepatic H_2S production in positive, neutral, and negative responders to DR for longevity endpoints in females from three ILSXISS recombinant inbred mouse strains, it was found that after 10 months of 40% DR, only in the positive responders (strain TejJ89) H_2S production was increased, while in the neutral (strain TejJ48) and in the negative (strain TejJ114), H_2S production was not increased [151]. This was met with increased hepatic 3-MST expression in the positive responding strain [151]. Work by Yoshida et al. comprehensively highlights dietary scenarios that directly contribute to lifespan and healthspan extension as a function of not only caloric intake but also amino acid intake, and specifically methionine, with links to renal H_2S production [66]. In their work, aged mice were fed ad libitum, simple DR, or DR with recovering nonessential amino acids (NEAAs) or essential amino acids (EAAs). They found simple DR extended lifespan and ameliorated age-related kidney injury. However, these benefits were canceled and/or completely reversed by supplementing with dietary EAAs but not NEAAs. In the ad libitum and DR plus EAA groups, the renal content of methionine was high, and this was accompanied by lower H_2S production in the kidney. Importantly, removal of methionine from the EAA supplementation reversed the detrimental effects of dietary EAA on lifespan and kidney injury as well as led to the recovery in renal H_2S production capacity. Ultimately, their data imply renal methionine metabolism, the trans-sulfuration pathway, and H_2S production/signaling could be promising nutritional and/or pharmacologic targets for preventing aging-related kidney function decline and for promoting healthy aging [66]. In rats, CGL and CBS expression and H_2S production decline in multiple tissues as a function of increasing age, however, these declines are dampened by long-term 10–40% CR from 8 to 38 months of age [28]. Likewise, 30% CR for 6 months increases renal CGL and CBS expression, and H_2S production [152]. Independent of total calories, sulfur amino acid restriction in rats increased hepatic CGL expression [153].

While these studies show the correlation for increased H_2S production with longevity extending diets and genetic models, unlike in worms and flies, they do not show the necessity of enhanced endogenous H_2S production capacity and/or

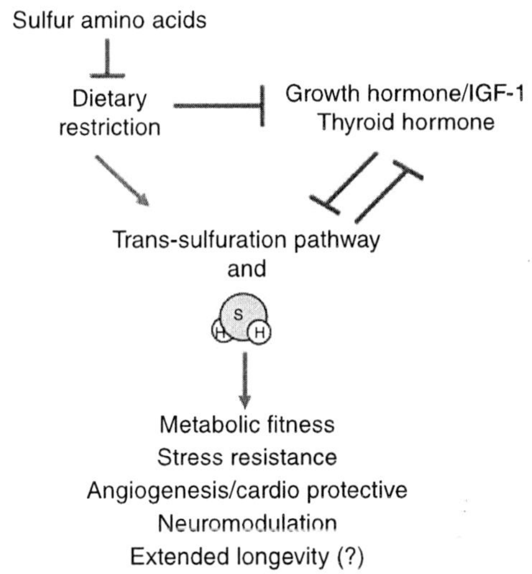

Figure 17.5 Trans-sulfuration and H_2S as central mediators for downstream benefits in standard rodent geroscience models of diet restriction and hypopituitary dwarfism.

CGL to confer the longevity benefit of DR. Preliminary data we have generated in our lab during a multiyear lifespan study in male and female CGL WT and KO mice suggests CGL is required for the lifespan and healthspan extending benefits of young adult-initiated caloric restriction [144]. While the study is currently being completed and finalized, it suggests, along with studies investigating stress resistance [21], angiogenesis [22], and metabolism [25], that CGL and endogenous H_2S production are required for anti-aging benefits of DR in mammals. This hypothesis is highlighted in Figure 17.5.

Unlike in yeast and worms, the impact of exogenous H_2S on extending rodent longevity are not yet known. While there are numerous examples of exogenous H_2S and H_2S donors used to treat aging related diseases in rodents such as atherosclerosis [154], hypertension [19, 155], cancer [47], neuro-cognitive decline [55], inflammation [156], and metabolic syndrome [157], its use to prolong median and maximum lifespan has not been reported. However, aging-dependent declines in tissue H_2S in rats and mice [28, 158] are attenuated by exogenous H_2S supplementation [158]. On a cellular level, NaHS treatment to CGL deficient mouse embryonic fibroblasts prevents premature cellular aging and replicative senescence [159]. It should be noted that currently, there are three separate trials listed on the NIA Interventions Testing Program website (https://www.nia.nih.gov/research/dab/interventions-testing-program-itp/compounds-testing) examining the use of the H_2S donor SG1002 to increase mouse lifespan. The first trial listed in the 2016 cohort will use 240 ppm SG1002 in food starting at 19 months of age, the second trial listed in the 2018 cohort will use 240 ppm SG1002 starting at 18 months of age, and the third trial listed in the 2019 cohort will use 240 ppm

SG1002 starting at 6 months of age. As mouse longevity experiments can take at least 4–6 years from start to finish in addition to final publication timetables, ideally we can expect outcomes from these interventional studies to be released between the years 2022 and 2025.

17.3.2 Naked Mole-Rats

While not considered a "standard" laboratory rodent commonly used for biomedical research, the naked mole-rat, *Heterocephalus glaber*, is an important, curious, and indispensable organism for aging and cancer studies [160, 161]. This eusocial, euthermic, and subterranean rodent displays an extraordinary lifespan of up to 35 years [162] with mortality rates defying Gompertzian laws by not increasing with age [163]. Additionally, they are highly resistant, but not completely immune [164], to the development of spontaneous and induced tumors and neoplasms [160, 165]. Thus, there is great scientific and translational interests into uncovering the molecular mechanisms these organisms use to reach such feats.

Investigations into the role sulfur amino acid metabolism and H_2S play in naked mole-rat biology have thus far provided valuable insights into their remarkable physiology. Plasma metabolite profiles taken from an unbiased metabolomics screen revealed low circulating levels of specific amino acids, and in particular methionine and methionine-derived metabolites, and remarkably these profiles resembled those found in methionine restricted rats, long-lived hypopituitary dwarf mice, and caloric restricted mice [161]. Interestingly, it was reported that naked mole-rats have low circulating sulfide levels in addition to low CBS activity under non-stressed conditions [166, 167]. However, naked mole-rat CBS has an unusually high degree of activation by *S*-adenosylmethionine (SAM), which is linked to regions on naked mole-rat specific CBS outside of the putative regulatory domain [166]. It is proposed that this high activation level of naked mole-rat CBS contributes to the extraordinary long lifespan via altering flux into trans-sulfuration [168]. Thus, like mice and rats, alterations in sulfur amino acid metabolism and its associated metabolites, such as H_2S, may play a role in extended longevity [167].

17.4 Hydrogen Sulfide Metabolism and Applications in Human Aging and Aging-Related Disorders

17.4.1 Human Exposure to H_2S and Advances in Clinical Biomarker and Interventional H_2S Approaches

Epidemiologic studies from regions of high ambient H_2S provide insight into the effects of chronic, low-dose exposure in humans. Rotorua, New Zealand is an active geothermal area, where mean levels of H_2S are approximately

20 ppb and range from 0 to 64 ppb. Levels of H_2S in Rotorua are among the highest in the world, and are significantly elevated compared to non-geothermal regions such as London (0.1 ppb) [169]. Elevated H_2S levels can also be found in regions with industrial sources of H_2S, including Reykjavik, Iceland and Dakota City and South Sioux, Nebraska. High levels of ambient H_2S are toxic; exposures at 1,000,000–2,000,000 ppb result in respiratory paralysis, levels of 320,000–530,000 ppb can cause pulmonary edema, and levels between 10,000 and 50,000 ppb can cause eye and respiratory irritation [170]. The effects of chronic, low-level exposure to H_2S, however, are less clear.

Early studies in Rotorua using area-level estimates of H_2S exposure suggested an association between H_2S and increased rates of nervous system and respiratory diseases; however, subsequent studies, in which H_2S exposure was estimated in individual participants, have shown no association [171–175]. A study of over 1200 participants in Rotorua showed no increase in the risk of chronic obstructive pulmonary disease (COPD) or asthma, and a possible protective effect of H_2S against lung damage in smokers [174]. Conversely, a study in Reykjavik, Iceland, demonstrated an increase in prescriptions for anti-asthmatic or adrenergic drugs in the days after spikes in H_2S [176]. Data from Dakota City and South Sioux, Nebraska showed an association between H_2S levels and hospital visits for asthma on the following day [177].

With regard to cognitive function, a study of 1637 patients in Rotorua found no association between H_2S levels and episodic memory, attention, fine motor skills, or mood [171]. A trend toward faster reaction times, as measured by motor response to a visual cue, was noted in those with higher H_2S exposure; however, the difference was not statistically significant [171]. Because animal models of Parkinson's disease have demonstrated a benefit of H_2S for the protection of dopaminergic neurons, it has been hypothesized that the improvements in reaction time may be due to increased concentrations of H_2S in dopaminergic neurons [169]. However, these experimental animal models evaluate the impact of acute rather than chronic exposure.

The variability in outcomes across studies of chronic H_2S exposure is likely due to differences in study methodology [178]. Whereas some studies used hospital discharge data, others used functional outcomes. The methods used to measure H_2S exposure also differed, with some using area-level estimates, and others using estimates individualized for each participant. Additionally, when the source of H_2S is industrial rather than geothermal, it may be difficult to control for confounding industrial exposures. Thus far, the preponderance of data suggests that chronic, low-level H_2S exposure does not increase the risks of respiratory or neurologic disease, and may be protective in some circumstances. The knowledge that inhaled H_2S is safe allows for the possibility of controlled experiments to further evaluate the impact of long-term H_2S exposure.

A role for H_2S-based therapies in the treatment of aging-related disorders such as hypertension, cardiovascular disease, and cancer has been established in animal models; however, the translation of these findings to clinical use is in its infancy. Although few clinical trials of H_2S therapies have been published, a number of pharmacologic approaches to increasing H_2S levels have been evaluated in humans either *in vitro* or *in vivo*, including sulfide salts, synthetic H_2S donors/generating molecules, cysteine analogs, and sulfide-releasing NSAIDs [179].

In 2010, a company called Ikaria Therapeutics published a phase I safety trial of IK-1001, a parenteral formulation of Na_2S [180]. The trial, which included 52 healthy participants, demonstrated elevated plasma levels of H_2S and thiosulfate minutes after intravenous administration and no adverse events [180]. The company subsequently registered several clinical trials to evaluate the impact of IK-1001 in the setting of myocardial infarction and coronary artery bypass graft surgery [181]. Although the results from this phase I trial appeared promising, the clinical utility of sulfide salts is limited by their instantaneous release of H_2S in solution and resulting time-limited presence in tissue [179].

In 2013, Sulfagenix patented a long-acting H_2S prodrug called SG1002. The compound was shown to improve left ventricular function in an animal model of heart failure, and in 2015, a small phase I clinical trial was performed in seven healthy patients and eight with heart failure [182]. Results showed increases in plasma H_2S and nitrite levels in participants who received SG1002. Among healthy patients who received the experimental drug, 3/6 (50%) developed mild gastrointestinal side effects [182]. No further clinical trials were undertaken, but SG1002 was later sold as a dietary supplement under the brand name Sulfzi and is currently being tested in mouse lifespan studies as part of the Interventions Testing Program at the NIH/NIA as mentioned previously.

Several H_2S-releasing NSAIDs have been developed, including *S*-diclofenac, *S*-naproxen, and a nitric oxide-hydrogen-sulfide-releasing aspirin [179]. These compounds consist of a traditional NSAID covalently attached to an H_2S-releasing moiety. They appear to have equivalent analgesic properties but a substantially lower risk of gastrointestinal injury compared to their parent NSAIDs [179]. The H_2S-releasing NSAID otenproxesul (previously known as ATB-346), developed by Antibe Therapeutics, has been tested in phase 2B clinical trials among participants with osteoarthritis and demonstrated a substantial reduction in gastrointestinal (GI) toxicity and ulcer incidence with equivalent suppression of COX compared to its parent NSAID naproxen [101].

GIC-1001 (trimebutine 3-thiocarbamoylbenzenesulfonate) was designed to increase the analgesic effects of trimebutine with H_2S release. Trimebutine is an anti-spasmodic agent and opioid agonist, and H_2S is thought to have anti-nociceptive properties [102]. GIC-1001 has been tested in phase 2 clinical trials as a treatment for visceral pain in patients undergoing sedation-free colonoscopy. Although full results have not yet been published, initial reports showed that 60–70% of patients felt that the drug provided adequate analgesia during colonoscopy (http://clinicaltrials.gov study number NCT01926444).

Finally, a group at the National University of Singapore developed the H_2S donor GYY4137. The release of H_2S from morpholin-4-ium 4 methoxyphenyl (morpholino) phosphainodithioate (GYY4137) was first described in 2008 [183]. The advantage of GYY4137 over sulfide salts is that the release of H_2S is slower and yields longer-lasting effects, more closely mimicking physiologic conditions [183]. When administered to rats, GYY4137 results in aortic relaxation and reduced levels of proinflammatory cytokines [183, 184]; however, there have not yet been any clinical trials evaluating GYY4137 in humans.

Opposite of environmental exposure to H_2S or through pharmacological donors, studies on foods naturally high in H_2S-releasing molecules (or precursors), such as alliums like garlic or cruciferous vegetables, have been completed for aging-related diseases [185, 186]. Of note, human red blood cells convert garlic-derived organic polysulfides, which include diallyl disulfide and diallyl trisulfide, into H_2S in a chemical mechanism involving glucose-maintained cytosolic glutathione levels and reduced thiols in or on the red blood cell membrane resulting in aortic vasorelaxation [187]. Not only improvements to blood flow are possible with garlic but aging-related neurodegeneration and heart-disease risks are also lessened [188]. And while long-term longevity studies with garlic in humans may not be feasible, in worms the garlic constituent diallyl trisulfide increases lifespan in a SKN-1 dependent manner [123]. Sulforaphane is an aliphatic isothiocyanate sulfur-containing organic compound found in the cruciferous vegetables broccoli and cabbage, and releases H_2S in the presence of mammalian tissues [189]. Akin to garlic extracts, sulforaphane's clinical potential has been shown in metabolic disorders and inflammation [190], however human aging studies have not been attempted. Nonetheless, sulforaphane extends the lifespan of worms, this time via regulation of DAF-16, DAF-2, and insulin/IGF-1 signaling [191]. Thus, the old adage of eating one's fruits and vegetables now has mechanistic rationale and offers an opportunity to introduce one's children to the wonders of H_2S biology at the dinner table.

The discussion that follows will focus on *in vitro* and *in vivo* data in humans regarding H_2S therapeutics for the aging-related diseases hypertension and cancer.

17.4.2 Cardiovascular Diseases

Hypertension is a well-established aging-related risk factor for cardiovascular and cerebrovascular disease. The condition affects 46% of adults in the United States, and is the most common reason for outpatient visits and the chronic use of prescription medication [192, 193]. Approximately one third of those diagnosed with hypertension have inadequately controlled blood pressure [193]. Emerging evidence suggests a role for hydrogen sulfide (H_2S) in the pathogenesis of hypertension, and serum H_2S levels are decreased in both animals and humans with hypertension [194]. H_2S promotes vasodilation through multiple mechanisms, including the activation of adenosine triphosphate (ATP)-sensitive and calcium-activated potassium channels in vascular smooth muscle cells, and the release of vasodilators such as nitric oxide (NO) and endothelium-derived relaxing factor. Additionally, Bucci et al. demonstrated that the anti-hypertensive effects of some angiotensin converting enzyme (ACE) inhibitors may be related to H_2S release that occurs independently of ACE inhibition [195]. Given the contribution of H_2S to the pathogenesis of hypertensive disorders, the gaseous molecule has been proposed as a therapeutic.

The endothelium regulates vascular tone, in part through the synthesis of NO, and endothelial dysfunction plays a major role in the pathophysiology of hypertension [196–198]. *In vitro* data suggest that H_2S administration may help mitigate endothelial dysfunction in hypertension by suppressing the renin–angiotensin system (RAS) with angiotonin II (AngII)/angiotensin II type 1 receptor (AT1) and upregulating peroxisome proliferator-activated receptor delta (PPARδ). PPARδ is a transcription factor that is thought to protect the endothelium by reducing oxidative stress and increasing endothelial NO synthase (eNOS) phosphorylation [198, 199]. PPARδ has been proposed as a novel target for anti-hypertensive therapies [198]. In a study of renal arteries obtained from 37 patients with and without hypertension undergoing partial nephrectomy, levels of H_2S and CGL expression were significantly lower in those with hypertension [198]. Culturing renal arteries of hypertensive patients with NaHS attenuated the expression of AT1, which is elevated in the setting of hypertension. Additionally, the expression of PPARδ and the phosphorylation of Akt, AMP-activated protein kinase (AMPK), and eNOS are all decreased in hypertensive patients; renal arteries cultured with NaHS showed increases in all four proteins. Furthermore, culture of human umbilical vein endothelial cells (HUVECs) with NaHS increased levels of NO, which are reduced by angiotonin II in hypertension [198]. These data, in combination with similar findings from *in vivo* animal models [198] and the hypertensive phenotype of CGL KO mice [19], suggest a therapeutic role for H_2S in mitigating endothelial dysfunction in hypertensive patients.

Female reproductive function and health are one of the first major systems to dramatically age in humans, with age 35 considered near the older spectrum of maternal age [5]. One aging-related complication during pregnancy is preeclampsia, which is characterized by high blood pressure with potential serious danger to both mother and child. Decreased H_2S activity has been implicated in the pathophysiology of preeclampsia, and the H_2S-generating compound GYY4137 may have therapeutic potential for treating women with pregnancies complicated by hypertensive disorders [200]. Plasma levels of H_2S and CGL are decreased in women with preeclampsia compared to age-matched controls. Additionally, the inhibition of CGL by DL-propargylglycine (PAG) in first-trimester placental explants resulted in poor trophoblast invasion *in vitro* and decreased production of placental growth factor (PlGF), suggesting that enzymatic H_2S production has an important role in normal placental vascular function [200]. Furthermore, inhibiting CGL expression *in vitro* by small-interfering RNA or adenovirus in HUVECs increased the release of both soluble fms-like tyrosine kinase-1 (sFlt-1) and soluble endoglin (sENG), both of which are placental angiogenic factors that are elevated in the setting of preeclampsia [200]. Conversely, overexpression of CGL inhibited sFlt-1 and sEng, providing further evidence that diminished CGL activity contributes to the development of hypertension in pregnancy. The addition of H_2S to cultured HUVECs reversed the anti-angiogenic effects of elevated sFlt-1 levels *in vitro* [200]. Although the H_2S generator GYY4317 has not been tested in pregnant women, administration of this compound in a mouse model of preeclampsia ameliorated fetal growth restriction and restored normal placental vasculature [200]. This suggests that increasing H_2S in pregnant women with preeclampsia may reduce help restore healthy placental vasculature.

17.4.3 Oncological Diseases

The role of hydrogen sulfide in cancer biology is complex as H_2S has both anti-tumor and pro-tumor activities [44, 201]. Endogenous H_2S has the capacity to induce angiogenesis, accelerate cell cycle, and inhibit apoptosis, which functions to promote tumorigenesis [202, 203]. However, treating cancers with H_2S donors at high concentrations or for long durations suppresses tumor growth by inducing cellular acidification and regulating signaling pathways involved in cell cycle function, proliferation, apoptosis, and metastasis [202]. Generally, H_2S functions vary in a cell and cancer specific fashion. To explain these disparate findings, a "bell-shaped model" has been suggested, in which endogenous H_2S or low levels of exogenous H_2S promote tumor growth and higher levels of exogenous H_2S inhibit cancer cell survival [201].

H_2S plays critical roles in cancer biology. H_2S increases glucose uptake by stimulating the activity of glucose transporter (GluT), which enhances the rate of glycolysis. At the same time, sulfide quinone reductase (SQR) uses H_2S to generate electrons, which increases the production of ATP and supports cancer cell proliferation [201]. H_2S also functions as a pro-angiogenic mediator by promoting the persulfidation of K_{ATP} channel and the expression of hypoxia inducible factor (HIF-1α) [204–210]. Moreover, H_2S is able to activate multiple anti-apoptotic pathways by persulfidation of the proteins NF-κB p65, Keap1, and

MEK1 [159, 211, 212]. H_2S is also reported to induce the activation of extracellular signal-regulated kinase (ERK) in colon cancer [213] and hepatoma cells [214]. Further, H_2S downregulates the expression of replication protein A70 (RpA-70) and retinoblastoma protein (RB1), but upregulates the expression of proliferation cell nuclear antigen (PCNA) and cyclin-dependent kinase 4 (CDK4), resulting in enhanced cell proliferation [215]. Furthermore, H_2S has been reported to inhibit tumor necrosis factor (TNFα) and lipopolysaccharides-stimulated NF-κB activation and thus prevents cancer cell apoptosis [184, 216]. Despite the above-mentioned pro-tumor functions, several studies indicate that H_2S has antitumor properties. H_2S enhances glycolysis by increasing the uptake of glucose, which in turn disrupts the export of intracellular acid through inhibiting the activity of anion exchanger (AE) and sodium/proton exchanger (NHE), leading to cancer cell death [217]. H_2S promotes oral cancer cell apoptosis by decreasing the expression of the apoptotic suppressor pleckstrin homology-like domain-A1 (PHLDA1) [218]. In addition, H_2S causes cell cycle arrest via upregulation of cyclin-dependent kinase inhibitor $p21^{Cip1}$ in colon cancer cell lines [219].

Paradoxically, CBS inhibitors and H_2S donors have both demonstrated anti-cancer properties *in vitro* and *in vivo*. Human colon cancer biopsies and colon cancer cell lines both show increased expression of CBS and elevated levels of H_2S production. *In vitro*, the general PLP inhibitor aminooxyacetic acid (AOAA), once thought to be a specific CBS inhibitor, diminished the proliferation, migration, and invasion of the human colon cancer cell line HCT116 [44]. *In vivo*, treatment of mice with AOAA mitigated the growth of xenografts derived from human colon cancer cell lines [44]. Similar *in vitro* results have been observed with the inhibition of CBS in breast cancer and ovarian cancer cell lines [45].

In direct contrast to the anticancer effects seen following the inhibition of CBS, H_2S donors/generating molecules also show anti-tumor properties. *In vitro* data regarding the impact of GYY4137 on cancer cell survival are available from human cell lines. One study compared the impact of GYY4137 and NaHS on the survival of a variety of human cancer and non-cancer cells *in vitro*, and correlated the effects with changes in the concentration of H_2S [47]. Whereas release of H_2S from NaHS in culture was fast, peaking at 20 minutes and falling to undetectable levels by 90 minutes, administration of GYY4137 resulted in lower peak levels of H_2S that were sustained above baseline for 7 days. NaHS and GYY4137 were each cultured with seven human cancer cell lines: MCF-7 (breast adenocarcinoma), MV4–11 (acute promyelocytic leukemia), HL-60 (myelomonocytic leukemia), cervical carcinoma (HeLa), colorectal carcinoma (HCT-116), hepatocellular carcinoma (Hep G2), and osteosarcoma (U2OS). NaHS resulted in a modest reduction (15–30%) in cell survival among the HCT-116, Hep G2, and MCF-7 cell lines, but failed to reduce cell survival in the other four cancer cell lines. In hepatocellular carcinoma cell lines and the HepG2 xenograft model, GYY4137 suppressed phosphorylation of STAT3 to inhibit STAT3 activation and the downstream target gene transcription, thereby inhibiting tumor growth [220]. GYY4137, however, significantly reduced cell survival by 75–95% in all seven cancer cell lines. Notably, neither NaHS nor GYY4137 altered survival of human fibroblast cells in culture. These data are supported by xenograft studies in severe combined immunodeficient (SCID) mice, in which tumors were established by injection of HL-60 or MVF-11 cells [47]. Daily administration of GYY4137 to the xenografted SCID mice resulted a 50% reduction in tumor volume. These results indicate a potential role for H_2S donors as anti-cancer agents, and suggest that the sustained elevations in H_2S achieved by GYY4137 may be important for obtaining therapeutic benefit.

S-Propargyl cysteine (SPRC), a stimulator of endogenous H_2S and a structural analogue of *S*-allycysteine, a derivative of garlic, has also been evaluated as an anticancer agent. Treatment of gastric cancer cells with SPRC *in vitro* resulted in increased CGL expression, reduced cell survival, increased apoptosis, and cell cycle arrest at the G1/S phase [221]. Additionally, treatment of nude mice with SPRC reduced tumor size by 40–75% in gastric cancer implants. Notably, when CGL was inhibited by PAG, the impact of SPRC on tumor growth was also inhibited, suggesting that the effects of SPRC on tumor growth are mediated in part through H_2S production [221].

H_2S-releasing NSAIDs may also have a role in tumor suppression. Chattopadhyay et al. evaluated the impact of 4 different HS-NSAIDs on the growth of 11 human cancer cell lines in 6 different tissues (colon, breast, pancreatic, prostate, lung, and leukemia) [222]. HS-NSAIDs attenuated the growth of all 11 cancer cell lines by inhibiting cell proliferation, increasing apoptosis, and inducing a G0/G1 cell cycle block. The inhibition conferred by NS-NSAIDs was 28 to >3000-fold greater than the parent NSAID alone [222]. In another study, the HS-NSAID ATB-346 inhibited the proliferation of human melanoma cells *in vitro* by preventing the activation of pro-survival pathways involving NF-κB and Akt [223]. Oral treatment with the compound also prevented melanoma progression in mice *in vivo* [223].

In summary, H_2S as a gaseous signaling molecule plays important roles in multiple biological processes in cancers. The three key H_2S-producing enzymes CGL, CBS, and 3-MST have altered expression in a variety of cancers in a cancer-type specific manner. Therefore, CGL, CBS, and 3-MST may be potential biomarkers and therapeutic targets for cancer diagnosis and treatment. Further data are needed to clarify the role of H_2S in the pathogenesis of human cancers and the potential role of H_2S based therapeutics. Given the paradoxical findings regarding the impact of H_2S inhibition and exogenous administration, additional preclinical data are needed before testing these compounds in clinical trials.

17.5 Conclusions and Summary

While in its relative infancy compared to other geroscience approaches, the utilization of H_2S in aging-related diseases has great potential as a promising treatment option. Through preclinical and early stage clinical trials we are beginning to see a clearer picture of what can be possible with H_2S and the steps needed to advance its use to the next stage. However, much still needs to be learned, including the upstream and downstream mechanisms regulating endogenous production and signaling, efficacy and utility of endogenous production enhancement versus exogenous delivery, and the obvious issues of bioavailability, safety, timing, and desired endpoints. Additionally, when focusing on the development of exogenous H_2S donors, there are crucial components to keep in mind. These include release times – should it be acute and fast-releasing or slow-releasing? While there are still cautions that must be taken to ensure they are put to optimal and safe use [224], enthusiasm in the field is undeniable for the possibilities of incorporating H_2S-based approaches as treatment options to address human's "forever problem" of aging.

Acknowledgments

This work was supported by grants from the National Institutes of Health to CH (NIA R00AG050777 and NHLBI R01HL148352).

References

1 Dong, X., Milholland, B., and Vijg, J. (2016). Evidence for a limit to human lifespan. *Nature* 538 (7624): 257–259.
2 Robine, J.M. and Cubaynes, S. (2017). Worldwide demography of centenarians. *Mech. Ageing Dev.* 165 (Pt B): 59–67.
3 (2019). *World Population Ageing 2019: Highlights*. United Nations Department of Economic and Social Affairs, Population Division.
4 Volpi, E., Nazemi, R., and Fujita, S. (2004). Muscle tissue changes with aging. *Curr. Opin. Clin. Nutr. Metab. Care* 7 (4): 405–410.
5 Llarena, N., Hine, C. (2021). Reproductive longevity and aging: geroscience approaches to maintain long-term ovarian fitness. *J. Gerontol. A Biol. Sci. Med. Sci.*, 76(9):1551–1560.
6 Strittmatter, A., Sunde, U., and Zegners, D. (2020). Life cycle patterns of cognitive performance over the long run. *Proc. Natl. Acad. Sci. U.S.A.* 117 (44): 27255–27261.

7 Li, T., Yang, Y.C., and Anderson, J.J. (2013). Mortality increase in late-middle and early-old age: heterogeneity in death processes as a new explanation. *Demography* 50 (5): 1563–1591.

8 Hohn, A., Weber, D., Jung, T. et al. (2017). Happily (n)ever after: aging in the context of oxidative stress, proteostasis loss and cellular senescence. *Redox Biol.* 11: 482–501.

9 Harman, D. (2009). Origin and evolution of the free radical theory of aging: a brief personal history, 1954–2009. *Biogerontology* 10 (6): 773–781.

10 Barja, G. (2013). Updating the mitochondrial free radical theory of aging: an integrated view, key aspects, and confounding concepts. *Antioxid. Redox Signal.* 19 (12): 1420–1445.

11 Ristow, M. and Zarse, K. (2010). How increased oxidative stress promotes longevity and metabolic health: the concept of mitochondrial hormesis (mitohormesis). *Exp. Gerontol.* 45 (6): 410–418.

12 Hine, C.M. and Mitchell, J.R. (2012). NRF2 and the phase II response in acute stress resistance induced by dietary restriction. *J. Clin. Exp. Pathol.* S4 (4): 7329.

13 Gladyshev, V.N. (2014). The free radical theory of aging is dead. Long live the damage theory! *Antioxid. Redox Signal.* 20 (4): 727–731.

14 Filipovic, M.R., Zivanovic, J., Alvarez, B. et al. (2018). Chemical biology of H_2S signaling through persulfidation. *Chem. Rev.* 118 (3): 1253–1337.

15 Filipovic, M.R. (2015). Persulfidation (S-sulfhydration) and H_2S. *Handb. Exp. Pharmacol.* 230: 29–59.

16 Doka, E., Ida, T., Dagnell, M. et al. (2020). Control of protein function through oxidation and reduction of persulfidated states. *Sci. Adv.* 6 (1): eaax8358.

17 Kabil, O. and Banerjee, R. (2014). Enzymology of H_2S biogenesis, decay and signaling. *Antioxid. Redox Signal.* 20 (5): 770–782.

18 Yang, J., Minkler, P., Grove, D. et al. (2019). Non-enzymatic hydrogen sulfide production from cysteine in blood is catalyzed by iron and vitamin B6. *Commun. Biol.* 2 (1): 194.

19 Yang, G., Wu, L., Jiang, B. et al. (2008). H_2S as a physiologic vasorelaxant: hypertension in mice with deletion of cystathionine γ-lyase. *Science* 322 (5901): 587–590.

20 Paul, B.D., Sbodio, J.I., Xu, R. et al. (2014). Cystathionine γ-lyase deficiency mediates neurodegeneration in Huntington's disease. *Nature* 509 (7498): 96–100.

21 Hine, C., Harputlugil, E., Zhang, Y. et al. (2015). Endogenous hydrogen sulfide production is essential for dietary restriction benefits. *Cell* 160 (1, 2): 132–144.

22 Longchamp, A., Mirabella, T., Arduini, A. et al. (2018). Amino acid restriction triggers angiogenesis via GCN2/ATF4 regulation of VEGF and H_2S production. *Cell* 173 (1): 117–129.e114.

23 Nakano, S., Ishii, I., Shinmura, K. et al. (2015). Hyperhomocysteinemia abrogates fasting-induced cardioprotection against ischemia/reperfusion by limiting bioavailability of hydrogen sulfide anions. *J. Mol. Med.* 93 (8): 879–889.

24 Kabil, H., Kabil, O., Banerjee, R. et al. (2011). Increased transsulfuration mediates longevity and dietary restriction in *Drosophila*. *Proc. Natl. Acad. Sci. U.S.A.* 108 (40): 16831–16836.

25 Hine, C., Kim, H.J., Zhu, Y. et al. (2017). Hypothalamic–pituitary axis regulates hydrogen sulfide production. *Cell Metab.* 25 (6): 1320–1333.e1325.

26 Tyshkovskiy, A., Bozaykut, P., Borodinova, A.A. et al. (2019). Identification and application of gene expression signatures associated with lifespan extension. *Cell Metab.* 30 (3): 573–593.e578.

27 Hine, C., Zhu, Y., Hollenberg, A.N. et al. (2018). Dietary and endocrine regulation of endogenous hydrogen sulfide production: implications for longevity. *Antioxid. Redox Signal.* 28 (16): 1483–1502.

28 Predmore, B.L., Alendy, M.J., Ahmed, K.I. et al. (2010). The hydrogen sulfide signaling system: changes during aging and the benefits of caloric restriction. *Age* 32 (4): 467–481.

29 Narayan, V., Ly, T., Pourkarimi, E. et al. (2016). Deep proteome analysis identifies age-related processes in *C. elegans*. *Cell Syst.* 3 (2): 144–159.

30 Zivanovic, J., Kouroussis, E., Kohl, J.B. et al. (2020). Selective persulfide detection reveals evolutionarily conserved antiaging effects of S-sulfhydration. *Cell Metab.* 31 (1): 207.

31 Bogdandi, V., Ida, T., Sutton, T.R. et al. (2019). Speciation of reactive sulfur species and their reactions with alkylating agents: do we have any clue about what is present inside the cell? *Br. J. Pharmacol.* 176 (4): 646–670.

32 Hartle, M.D. and Pluth, M.D. (2016). A practical guide to working with H_2S at the interface of chemistry and biology. *Chem. Soc. Rev.* 45 (22): 6108–6117.

33 Rajpal, S., Katikaneni, P., Deshotels, M. et al. (2018). Total sulfane sulfur bioavailability reflects ethnic and gender disparities in cardiovascular disease. *Redox Biol.* 15: 480–489.

34 Jain, S.K., Bull, R., Rains, J.L. et al. (2010). Low levels of hydrogen sulfide in the blood of diabetes patients and streptozotocin-treated rats causes vascular inflammation? *Antioxid. Redox Signal.* 12 (11): 1333–1337.

35 Li, Z., Polhemus, D.J., and Lefer, D.J. (2018). Evolution of hydrogen sulfide therapeutics to treat cardiovascular disease. *Circ. Res.* 123 (5): 590–600.

36 Elrod, J.W., Calvert, J.W., Morrison, J. et al. (2007). Hydrogen sulfide attenuates myocardial ischemia-reperfusion injury by preservation of mitochondrial function. *Proc. Natl. Acad. Sci. U.S.A.* 104 (39): 15560–15565.
37 Dhar, I., Svingen, G.F.T., Ueland, P.M. et al. (2018). Plasma cystathionine and risk of incident stroke in patients with suspected stable angina pectoris. *J. Am. Heart Assoc.* 7 (17): e008824.
38 Zhang, M., Wu, X., Xu, Y. et al. (2017). The cystathionine β-synthase/hydrogen sulfide pathway contributes to microglia-mediated neuroinflammation following cerebral ischemia. *Brain Behav. Immun.* 66: 332–346.
39 White, M.C., Holman, D.M., Boehm, J.E. et al. (2014). Age and cancer risk: a potentially modifiable relationship. *Am. J. Prev. Med.* 46 (3 Suppl 1): S7–S15.
40 Fisher, R., Pusztai, L., and Swanton, C. (2013). Cancer heterogeneity: implications for targeted therapeutics. *Br. J. Cancer* 108 (3): 479–485.
41 Kabil, O., Vitvitsky, V., Xie, P. et al. (2011). The quantitative significance of the transsulfuration enzymes for H_2S production in murine tissues. *Antioxid. Redox Signal.* 15 (2): 363–372.
42 Lagoutte, E., Mimoun, S., Andriamihaja, M. et al. (2010). Oxidation of hydrogen sulfide remains a priority in mammalian cells and causes reverse electron transfer in colonocytes. *Biochim. Biophys. Acta* 1797 (8): 1500–1511.
43 Hellmich, M.R., Coletta, C., Chao, C. et al. (2015). The therapeutic potential of cystathionine beta-synthetase/hydrogen sulfide inhibition in cancer. *Antioxid. Redox Signal.* 22 (5): 424–448.
44 Szabo, C., Coletta, C., Chao, C. et al. (2013). Tumor-derived hydrogen sulfide, produced by cystathionine-β-synthase, stimulates bioenergetics, cell proliferation, and angiogenesis in colon cancer. *Proc. Natl. Acad. Sci. U.S.A.* 110 (30): 12474–12479.
45 Bhattacharyya, S., Saha, S., Giri, K. et al. (2013). Cystathionine β-synthase (CBS) contributes to advanced ovarian cancer progression and drug resistance. *PLoS One* 8 (11): e79167.
46 Papapetropoulos, A., Pyriochou, A., Altaany, Z. et al. (2009). Hydrogen sulfide is an endogenous stimulator of angiogenesis. *Proc. Natl. Acad. Sci. U.S.A.* 106 (51): 21972–21977.
47 Lee, Z.W., Zhou, J., Chen, C.S. et al. (2011). The slow-releasing hydrogen sulfide donor, GYY4137, exhibits novel anti-cancer effects in vitro and in vivo. *PLoS One* 6 (6): e21077.
48 Wu, D., Si, W., Wang, M. et al. (2015). Hydrogen sulfide in cancer: friend or foe? *Nitric Oxide* 50: 38–45.
49 Xu, Q., Li, Y., Gao, X. et al. (2020). HNF4α regulates sulfur amino acid metabolism and confers sensitivity to methionine restriction in liver cancer. *Nat. Commun.* 11 (1): 3978.

50 Yang, D., Li, T., Li, Y. et al. (2019). H_2S suppresses indoleamine 2,3-dioxygenase 1 and exhibits immunotherapeutic efficacy in murine hepatocellular carcinoma. *J. Exp. Clin. Cancer Res.* 38 (1): 88.

51 Takano, N., Sarfraz, Y., Gilkes, D.M. et al. (2014). Decreased expression of cystathionine β-synthase promotes glioma tumorigenesis. *Mol. Cancer Res.* 12 (10): 1398–1406.

52 Wrobel, M., Czubak, J., Bronowicka-Adamska, P. et al. (2014). Is development of high-grade gliomas sulfur-dependent? *Molecules* 19 (12): 21350–21362.

53 Abe, K. and Kimura, H. (1996). The possible role of hydrogen sulfide as an endogenous neuromodulator. *J. Neurosci.* 16 (3): 1066–1071.

54 Vandiver, M.S., Paul, B.D., Xu, R. et al. (2013). Sulfhydration mediates neuroprotective actions of parkin. *Nat. Commun.* 4: 1626.

55 Giovinazzo, D., Bursac, B., Sbodio, J.I. et al. (2021). Hydrogen sulfide is neuroprotective in Alzheimer's disease by sulfhydrating GSK3β and inhibiting Tau hyperphosphorylation. *Proc. Natl. Acad. Sci. U.S.A.* 118 (4): e2017225118.

56 Paul, B.D. and Snyder, S.H. (2018). Gasotransmitter hydrogen sulfide signaling in neuronal health and disease. *Biochem. Pharmacol.* 149: 101–109.

57 Kida, K., Yamada, M., Tokuda, K. et al. (2011). Inhaled hydrogen sulfide prevents neurodegeneration and movement disorder in a mouse model of Parkinson's disease. *Antioxid. Redox Signal.* 15 (2): 343–352.

58 Fontana, L. and Partridge, L. (2015). Promoting health and longevity through diet: from model organisms to humans. *Cell* 161 (1): 106–118.

59 Sierra, F. (2016). The emergence of geroscience as an interdisciplinary approach to the enhancement of health span and life span. *Cold Spring Harb. Perspect. Med.* 6 (4): a025163.

60 Redman, L.M., Smith, S.R., Burton, J.H. et al. (2018). Metabolic slowing and reduced oxidative damage with sustained caloric restriction support the rate of living and oxidative damage theories of aging. *Cell Metab.* 27 (4): 805–815.e804.

61 McCay, C., Crowell, M., and Maynard, L. (1935). The effect of retarded growth upon length of lifespan and upon ultimate body size. *J. Nutr.* 10: 63–79.

62 Solon-Biet, S.M., McMahon, A.C., Ballard, J.W. et al. (2014). The ratio of macronutrients, not caloric intake, dictates cardiometabolic health, aging, and longevity in ad libitum-fed mice. *Cell Metab.* 19 (3): 418–430.

63 Bonkowski, M.S., Dominici, F.P., Arum, O. et al. (2009). Disruption of growth hormone receptor prevents calorie restriction from improving insulin action and longevity. *PLoS One* 4 (2): e4567.

64 Trocha, K.M., Kip, P., Tao, M. et al. (2019). Short-term preoperative protein restriction attenuates vein graft disease via induction of cystathionine γ-lyase. *Cardiovasc. Res.* 116 (2): 416–428.

65 Bithi, N., Link, C., Henderson, Y.O. et al. (2021). Dietary restriction transforms the mammalian protein persulfidome in a tissue-specific and cystathionine γ-lyase-dependent manner. *Nat. Commun.* 12 (1): 1745.

66 Yoshida, S., Yamahara, K., Kume, S. et al. (2018). Role of dietary amino acid balance in diet restriction-mediated lifespan extension, renoprotection, and muscle weakness in aged mice. *Aging Cell* 17 (4): e12796.

67 Selman, C. and Withers, D.J. (2011). Mammalian models of extended healthy lifespan. *Philos. Trans. R Soc. Lond. B Biol. Sci.* 366 (1561): 99–107.

68 Wei, Y. and Kenyon, C. (2016). Roles for ROS and hydrogen sulfide in the longevity response to germline loss in *Caenorhabditis elegans*. *Proc. Natl. Acad. Sci. U.S.A.* 113 (20): E2832–E2841.

69 Uthus, E.O. and Brown-Borg, H.M. (2006). Methionine flux to transsulfuration is enhanced in the long living Ames dwarf mouse. *Mech. Ageing Dev.* 127 (5): 444–450.

70 Brown-Borg, H.M., Rakoczy, S.G., and Uthus, E.O. (2005). Growth hormone alters methionine and glutathione metabolism in Ames dwarf mice. *Mech. Ageing Dev.* 126 (3): 389–398.

71 Harrison, D.E., Strong, R., Allison, D.B. et al. (2014). Acarbose, 17-α-estradiol, and nordihydroguaiaretic acid extend mouse lifespan preferentially in males. *Aging Cell* 13 (2): 273–282.

72 Miller, R.A., Harrison, D.E., Astle, C.M. et al. (2011). Rapamycin, but not resveratrol or simvastatin, extends life span of genetically heterogeneous mice. *J. Gerontol. A Biol. Sci. Med. Sci.* 66 (2): 191–201.

73 Nadon, N.L., Strong, R., Miller, R.A. et al. (2017). NIA interventions testing program: investigating putative aging intervention agents in a genetically heterogeneous mouse model. *EBioMedicine* 21: 3–4.

74 Martin-Montalvo, A., Mercken, E.M., Mitchell, S.J. et al. (2013). Metformin improves healthspan and lifespan in mice. *Nat. Commun.* 4: 2192.

75 Baur, J.A., Pearson, K.J., Price, N.L. et al. (2006). Resveratrol improves health and survival of mice on a high caloric diet. *Nature* 444 (7117): 337–342.

76 Wiliński, B., Wiliński, J., Somogyi, E. et al. (2013). Metformin raises hydrogen sulfide tissue concentrations in various mouse organs. *Pharmacol. Rep.* 65 (3): 737–742.

77 Hu, L.Y., Hu, S.L., Wu, J. et al. (2012). Hydrogen sulfide prolongs postharvest shelf life of strawberry and plays an antioxidative role in fruits. *J. Agric. Food Chem.* 60 (35): 8684–8693.

78 Li, S.P., Hu, K.D., Hu, L.Y. et al. (2014). Hydrogen sulfide alleviates postharvest senescence of broccoli by modulating antioxidant defense and senescence-related gene expression. *J. Agric. Food. Chem.* 62 (5): 1119–1129.

79 Ni, Z.J., Hu, K.D., Song, C.B. et al. (2016). Hydrogen sulfide alleviates postharvest senescence of grape by modulating the antioxidant defenses. *Oxid. Med. Cell Longev.* 2016: 4715651.

80 Ge, Y., Hu, K.D., Wang, S.S. et al. (2017). Hydrogen sulfide alleviates postharvest ripening and senescence of banana by antagonizing the effect of ethylene. *PLoS One* 12 (6): e0180113.

81 Yao, G.F., Wei, Z.Z., Li, T.T. et al. (2018). Modulation of enhanced antioxidant activity by hydrogen sulfide antagonization of ethylene in tomato fruit ripening. *J. Agric. Food Chem.* 66 (40): 10380–10387.

82 Liu, D., Li, J., Li, Z. et al. (2020). Hydrogen sulfide inhibits ethylene-induced petiole abscission in tomato (*Solanum lycopersicum* L.). *Hortic. Res.* 7: 14.

83 Joshi, N.C., Yadav, D., Ratner, K. et al. (2020). Sodium hydrosulfide priming improves the response of photosynthesis to overnight frost and day high light in avocado (*Persea americana* Mill, cv. 'Hass'). *Physiol. Plant* 168 (2): 394–405.

84 Lin, X., Yang, R., Dou, Y. et al. (2020). Transcriptome analysis reveals delaying of the ripening and cell-wall degradation of kiwifruit by hydrogen sulfide. *J. Sci. Food Agric.* 100 (5): 2280–2287.

85 Zhu, L., Wang, W., Shi, J. et al. (2014). Hydrogen sulfide extends the postharvest life and enhances antioxidant activity of kiwifruit during storage. *J. Sci. Food Agric.* 94 (13): 2699–2704.

86 Liu, D., Xu, S., Hu, H. et al. (2017). Endogenous hydrogen sulfide homeostasis is responsible for the alleviation of senescence of postharvest daylily flower via increasing antioxidant capacity and maintained energy status. *J. Agric. Food Chem.* 65 (4): 718–726.

87 Palma, J.M., Mateos, R.M., Lopez-Jaramillo, J. et al. (2020). Plant catalases as NO and H_2S targets. *Redox Biol.* 34: 101525.

88 Aroca, A., Benito, J.M., Gotor, C. et al. (2017). Persulfidation proteome reveals the regulation of protein function by hydrogen sulfide in diverse biological processes in Arabidopsis. *J. Exp. Bot.* 68 (17): 4915–4927.

89 Lapinska, U., Glover, G., Capilla-Lasheras, P. et al. (2019). Bacterial ageing in the absence of external stressors. *Philos. Trans. R Soc. Lond. B Biol. Sci.* 374 (1786): 20180442.

90 Stephens, C. (2005). Senescence: even bacteria get old. *Curr. Biol.* 15 (8): R308–R310.

91 Ackermann, M., Stearns, S.C., and Jenal, U. (2003). Senescence in a bacterium with asymmetric division. *Science* 300 (5627): 1920.

92 Shen, X., Carlstrom, M., Borniquel, S. et al. (2013). Microbial regulation of host hydrogen sulfide bioavailability and metabolism. *Free Radic. Biol. Med.* 60: 195–200.

93 Tang, W.H.W., Li, D.Y., and Hazen, S.L. (2019). Dietary metabolism, the gut microbiome, and heart failure. *Nat. Rev. Cardiol.* 16 (3): 137–154.

94 Tang, W.H. and Hazen, S.L. (2017). The gut microbiome and its role in cardiovascular diseases. *Circulation* 135 (11): 1008–1010.

95 Singh, S.B. and Lin, H.C. (2015). Hydrogen sulfide in physiology and diseases of the digestive tract. *Microorganisms* 3 (4): 866–889.

96 Blackler, R.W., De Palma, G., Manko, A. et al. (2015). Deciphering the pathogenesis of NSAID enteropathy using proton pump inhibitors and a hydrogen sulfide-releasing NSAID. *Am. J. Physiol. Gastrointest. Liver Physiol.* 308 (12): G994–G1003.

97 Van Dingenen, J., Pieters, L., Vral, A. et al. (2019). The H_2S-releasing naproxen derivative ATB-346 and the slow-release H_2S donor GYY4137 reduce intestinal inflammation and restore transit in postoperative ileus. *Front. Pharmacol.* 10: 116.

98 Magierowski, M., Magierowska, K., Surmiak, M. et al. (2017). The effect of hydrogen sulfide-releasing naproxen (ATB-346) versus naproxen on formation of stress-induced gastric lesions, the regulation of systemic inflammation, hypoxia and alterations in gastric microcirculation. *J. Physiol. Pharmacol.* 68 (5): 749–756.

99 Elsheikh, W., Blackler, R.W., Flannigan, K.L. et al. (2014). Enhanced chemopreventive effects of a hydrogen sulfide-releasing anti-inflammatory drug (ATB-346) in experimental colorectal cancer. *Nitric Oxide* 41: 131–137.

100 Paul-Clark, M., Elsheikh, W., Kirkby, N. et al. (2016). Profound chemopreventative effects of a hydrogen sulfide-releasing NSAID in the APCMin/+ mouse model of intestinal tumorigenesis. *PLoS One* 11 (2): e0147289.

101 Wallace, J.L., Nagy, P., Feener, T.D. et al. (2020). A proof-of-concept, phase 2 clinical trial of the gastrointestinal safety of a hydrogen sulfide-releasing anti-inflammatory drug. *Br. J. Pharmacol.* 177 (4): 769–777.

102 Cenac, N., Castro, M., Desormeaux, C. et al. (2016). A novel orally administered trimebutine compound (GIC-1001) is anti-nociceptive and features peripheral opioid agonistic activity and hydrogen sulphide-releasing capacity in mice. *Eur. J. Pain* 20 (5): 723–730.

103 Weber, G.J., Foster, J., Pushpakumar, S.B. et al. (2018). Altered microRNA regulation of short chain fatty acid receptors in the hypertensive kidney is normalized with hydrogen sulfide supplementation. *Pharmacol. Res.* 134: 157–165.

104 Mirzaei, H., Suarez, J.A., and Longo, V.D. (2014). Protein and amino acid restriction, aging and disease: from yeast to humans. *Trends Endocrinol. Metab.* 25 (11): 558–566.

105 Johnson, J.E. and Johnson, F.B. (2014). Methionine restriction activates the retrograde response and confers both stress tolerance and lifespan extension to yeast, mouse and human cells. *PLoS One* 9 (5): e97729.

106 Longo, V.D. (1999). Mutations in signal transduction proteins increase stress resistance and longevity in yeast, nematodes, fruit flies, and mammalian neuronal cells. *Neurobiol. Aging* 20 (5): 479–486.

107 Ruckenstuhl, C., Netzberger, C., Entfellner, I. et al. (2014). Lifespan extension by methionine restriction requires autophagy-dependent vacuolar acidification. *PLoS Genet.* 10 (5): e1004347.

108 Janssens, G.E. and Veenhoff, L.M. (2016). Evidence for the hallmarks of human aging in replicatively aging yeast. *Microb. Cell* 3 (7): 263–274.

109 Linderholm, A.L., Findleton, C.L., Kumar, G. et al. (2008). Identification of genes affecting hydrogen sulfide formation in *Saccharomyces cerevisiae*. *Appl. Environ. Microbiol.* 74 (5): 1418–1427.

110 Hine, C. and Mitchell, J.R. (2015). Calorie restriction and methionine restriction in control of endogenous hydrogen sulfide production by the transsulfuration pathway. *Exp. Gerontol.* 68: 26–32.

111 Wu, Z., Song, L., Liu, S.Q. et al. (2013). Independent and additive effects of glutamic acid and methionine on yeast longevity. *PLoS One* 8 (11): e79319.

112 Zou, K., Rouskin, S., Dervishi, K. et al. (2020). Life span extension by glucose restriction is abrogated by methionine supplementation: cross-talk between glucose and methionine and implication of methionine as a key regulator of life span. *Sci. Adv.* 6 (32): eaba1306.

113 Cost, G.J. and Boeke, J.D. (1996). A useful colony colour phenotype associated with the yeast selectable/counter-selectable marker MET15. *Yeast* 12 (10): 939–941.

114 Choi, K.M., Kim, S., Kim, S. et al. (2019). Sulfate assimilation regulates hydrogen sulfide production independent of lifespan and reactive oxygen species under methionine restriction condition in yeast. *Aging* 11 (12): 4254–4273.

115 Shah, A.A., Liu, B., Tang, Z. et al. (2021). Hydrogen sulfide treatment at the late growth stage of *Saccharomyces cerevisiae* extends chronological lifespan. *Aging* 13.

116 Lakowski, B. and Hekimi, S. (1998). The genetics of caloric restriction in *Caenorhabditis elegans*. *Proc. Natl. Acad. Sci. U.S.A.* 95 (22): 13091–13096.

117 Tissenbaum, H.A. (2015). Using *C. elegans* for aging research. *Invertebr. Reprod. Dev.* 59 (sup1): 59–63.

118 Lin, K., Hsin, H., Libina, N. et al. (2001). Regulation of the *Caenorhabditis elegans* longevity protein DAF-16 by insulin/IGF-1 and germline signaling. *Nat. Genet.* 28 (2): 139–145.

119 Cabreiro, F., Au, C., Leung, K.Y. et al. (2013). Metformin retards aging in *C. elegans* by altering microbial folate and methionine metabolism. *Cell* 153 (1): 228–239.

120 Qabazard, B., Li, L., Gruber, J. et al. (2014). Hydrogen sulfide is an endogenous regulator of aging in *Caenorhabditis elegans*. *Antioxid. Redox Signal.* 20 (16): 2621–2630.

121 Miller, D.L. and Roth, M.B. (2007). Hydrogen sulfide increases thermotolerance and lifespan in *Caenorhabditis elegans*. *Proc. Natl. Acad. Sci. U.S.A.* 104 (51): 20618–20622.

122 Miller, D.L., Budde, M.W., and Roth, M.B. (2011). HIF-1 and SKN-1 coordinate the transcriptional response to hydrogen sulfide in *Caenorhabditis elegans*. *PLoS One* 6 (9): e25476.

123 Powolny, A.A., Singh, S.V., Melov, S. et al. (2011). The garlic constituent diallyl trisulfide increases the lifespan of *C. elegans* via skn-1 activation. *Exp. Gerontol.* 46 (6): 441–452.

124 Qabazard, B., Ahmed, S., Li, L. et al. (2013). *C. elegans* aging is modulated by hydrogen sulfide and the sulfhydrylase/cysteine synthase cysl-2. *PLoS One* 8 (11): e80135.

125 Ng, L.T., Ng, L.F., Tang, R.M.Y. et al. (2020). Lifespan and healthspan benefits of exogenous H_2S in *C. elegans* are independent from effects downstream of *eat-2* mutation. *NPJ Aging Mech. Dis.* 6: 6.

126 Piper, M.D.W. and Partridge, L. (2018). Drosophila as a model for ageing. *Biochim. Biophys. Acta, Mol. Basis Dis.* 1864 (9 Pt A): 2707–2717.

127 Troen, A.M., French, E.E., Roberts, J.F. et al. (2007). Lifespan modification by glucose and methionine in *Drosophila melanogaster* fed a chemically defined diet. *Age* 29 (1): 29–39.

128 Lee, B.C., Kaya, A., Ma, S. et al. (2014). Methionine restriction extends lifespan of *Drosophila melanogaster* under conditions of low amino-acid status. *Nat. Commun.* 5: 3592.

129 Shaposhnikov, M., Proshkina, E., Koval, L. et al. (2018). Overexpression of CBS and CSE genes affects lifespan, stress resistance and locomotor activity in *Drosophila melanogaster*. *Aging* 10 (11): 3260–3272.

130 Shaposhnikov, M.V., Zemskaya, N.V., Koval, L.A. et al. (2021). Geroprotective potential of genetic and pharmacological interventions to endogenous hydrogen sulfide synthesis in *Drosophila melanogaster*. *Biogerontology* 22 (2): 197–214.

131 Matter, B.E., Wurgler, F.E., and Ulrich, H. (1969). On the radioprotective effect of hydrogen sulphide in *Drosophila melanogaster*. *Int. J. Radiat. Biol. Relat. Stud. Phys. Chem. Med.* 15 (6): 557–562.

132 Zhong, J.F., Wang, S.P., Shi, X.Q. et al. (2010). Hydrogen sulfide exposure increases desiccation tolerance in *Drosophila melanogaster*. *J. Insect Physiol.* 56 (12): 1777–1782.

133 Rous, P. (1914). The influence of diet on transplanted and spontaneous mouse tumors. *J. Exp. Med.* 20 (5): 433–451.

134 Weindruch, R. and Walford, R.L. (1982). Dietary restriction in mice beginning at 1 year of age: effect on life-span and spontaneous cancer incidence. *Science* 215 (4538): 1415–1418.

135 Weindruch, R., Walford, R.L., Fligiel, S. et al. (1986). The retardation of aging in mice by dietary restriction: longevity, cancer, immunity and lifetime energy intake. *J. Nutr.* 116 (4): 641–654.

136 Miller, R.A., Buehner, G., Chang, Y. et al. (2005). Methionine-deficient diet extends mouse lifespan, slows immune and lens aging, alters glucose, T4, IGF-I and insulin levels, and increases hepatocyte MIF levels and stress resistance. *Aging Cell* 4 (3): 119–125.

137 Brown-Borg, H.M. (2015). Reduced growth hormone signaling and methionine restriction: interventions that improve metabolic health and extend life span. *Ann. N.Y. Acad. Sci.*

138 Orentreich, N., Matias, J.R., DeFelice, A. et al. (1993). Low methionine ingestion by rats extends life span. *J. Nutr.* 123 (2): 269–274.

139 Richie, J.P., Leutzinger, Y., Parthasarathy, S. et al. (1994). Methionine restriction increases blood glutathione and longevity in F344 rats. *FASEB J.* 8 (15): 1302–1307.

140 Elshorbagy, A.K., Valdivia-Garcia, M., Mattocks, D.A. et al. (2011). Cysteine supplementation reverses methionine restriction effects on rat adiposity: significance of stearoyl-coenzyme A desaturase. *J. Lipid Res.* 52 (1): 104–112.

141 Elshorbagy, A.K., Valdivia-Garcia, M., Mattocks, D.A. et al. (2013). Effect of taurine and *N*-acetylcysteine on methionine restriction-mediated adiposity resistance. *Metabolism* 62 (4): 509–517.

142 López-Torres, M. and Barja, G. (2008). Lowered methionine ingestion as responsible for the decrease in rodent mitochondrial oxidative stress in protein and dietary restriction possible implications for humans. *Biochim. Biophys. Acta* 1780 (11): 1337–1347.

143 Uthus, E.O. and Brown-Borg, H.M. (2003). Altered methionine metabolism in long living Ames dwarf mice. *Exp. Gerontol.* 38 (5): 491–498.

144 Henderson, Y.O., Bithi, N., Link, C. et al. (2021). Late-life intermittent fasting decreases aging-related frailty and increases renal hydrogen sulfide production in a sexually dimorphic manner. *Geroscience* 43 (4): 1527–1554.

145 Mitchell, S.J., Madrigal-Matute, J., Scheibye-Knudsen, M. et al. (2016). Effects of sex, strain, and energy intake on hallmarks of aging in mice. *Cell Metab.* 23 (6): 1093–1112.

146 Shuang, T., Fu, M., Yang, G. et al. (2021). Interaction among estrogen, IGF-1, and H_2S on smooth muscle cell proliferation. *J. Endocrinol.* 248 (1): 17–30.

147 Shuang, T., Fu, M., Yang, G. et al. (2018). The interaction of IGF-1/IGF-1R and hydrogen sulfide on the proliferation of mouse primary vascular smooth muscle cells. *Biochem. Pharmacol.* 149: 143–152.

148 Li, H., Mani, S., Wu, L. et al. (2017). The interaction of estrogen and CSE/H_2S pathway in the development of atherosclerosis. *Am. J. Physiol. Heart Circ. Physiol.* 312 (3): H406–H414.

149 Alfaras, I., Mitchell, S.J., Mora, H. et al. (2017). Health benefits of late-onset metformin treatment every other week in mice. *NPJ Aging Mech. Dis.* 3: 16.

150 Liao, C.Y., Rikke, B.A., Johnson, T.E. et al. (2010). Genetic variation in the murine lifespan response to dietary restriction: from life extension to life shortening. *Aging Cell* 9 (1): 92–95.

151 Wilkie, S.E., Mulvey, L., Sands, W.A. et al. (2020). Strain-specificity in the hydrogen sulphide signalling network following dietary restriction in recombinant inbred mice. *Geroscience* 42 (2): 801–812.

152 Wang, W.J., Cai, G.Y., Ning, Y.C. et al. (2016). Hydrogen sulfide mediates the protection of dietary restriction against renal senescence in aged F344 rats. *Sci. Rep.* 6: 30292.

153 Sikalidis, A.K. and Stipanuk, M.H. (2010). Growing rats respond to a sulfur amino acid-deficient diet by phosphorylation of the alpha subunit of eukaryotic initiation factor 2 heterotrimeric complex and induction of adaptive components of the integrated stress response. *J. Nutr.* 140 (6): 1080–1085.

154 Du, C., Lin, X., Xu, W. et al. (2019). Sulfhydrated sirtuin-1 increasing its deacetylation activity is an essential epigenetics mechanism of anti-atherogenesis by hydrogen sulfide. *Antioxid. Redox Signal.* 30 (2): 184–197.

155 Huang, P., Chen, S., Wang, Y. et al. (2015). Down-regulated CBS/H_2S pathway is involved in high-salt-induced hypertension in Dahl rats. *Nitric Oxide* 46: 192–203.

156 Faller, S., Hausler, F., Goeft, A. et al. (2018). Hydrogen sulfide limits neutrophil transmigration, inflammation, and oxidative burst in lipopolysaccharide-induced acute lung injury. *Sci. Rep.* 8 (1): 14676.

157 Cheng, Z., Garikipati, V.N., Nickoloff, E. et al. (2016). Restoration of hydrogen sulfide production in diabetic mice improves reparative function of bone marrow cells. *Circulation* 134 (19): 1467–1483.

158 Hou, C.L., Wang, M.J., Sun, C. et al. (2016). Protective effects of hydrogen sulfide in the ageing kidney. *Oxid. Med. Cell Longev.* 2016: 7570489.

159 Yang, G., Zhao, K., Ju, Y. et al. (2013). Hydrogen sulfide protects against cellular senescence via S-sulfhydration of Keap1 and activation of Nrf2. *Antioxid. Redox Signal.* 18 (15): 1906–1919.

160 Tian, X., Azpurua, J., Hine, C. et al. (2013). High-molecular-mass hyaluronan mediates the cancer resistance of the naked mole rat. *Nature* 499 (7458): 346–349.

161 Lewis, K.N., Rubinstein, N.D., and Buffenstein, R. (2018). A window into extreme longevity; the circulating metabolomic signature of the naked mole-rat, a mammal that shows negligible senescence. *Geroscience* 40 (2): 105–121.

162 Buffenstein, R. (2005). The naked mole-rat: a new long-living model for human aging research. *J. Gerontol. A Biol. Sci. Med. Sci.* 60 (11): 1369–1377.

163 Ruby, J.G., Smith, M., and Buffenstein, R. (2018). Naked mole-rat mortality rates defy Gompertzian laws by not increasing with age. *Elife* 7: e31157.

164 Taylor, K.R., Milone, N.A., and Rodriguez, C.E. (2017). Four cases of spontaneous neoplasia in the naked mole-rat (*Heterocephalus glaber*), a putative cancer-resistant species. *J. Gerontol. A Biol. Sci. Med. Sci.* 72 (1): 38–43.

165 Buffenstein, R. (2008). Negligible senescence in the longest living rodent, the naked mole-rat: insights from a successfully aging species. *J. Comp. Physiol. B* 178 (4): 439–445.

166 Dziegelewska, M., Holtze, S., Vole, C. et al. (2016). Low sulfide levels and a high degree of cystathionine beta-synthase (CBS) activation by *S*-adenosylmethionine (SAM) in the long-lived naked mole-rat. *Redox Biol.* 8: 192–198.

167 Brown-Borg, H.M. and Buffenstein, R. (2017). Cutting back on the essentials: can manipulating intake of specific amino acids modulate health and lifespan? *Ageing Res. Rev.* 39: 87–95.

168 Olecka, M., Huse, K., and Platzer, M. (2018). The high degree of cystathionine β-synthase (CBS) activation by *S*-adenosylmethionine (SAM) may explain naked mole-rat's distinct methionine metabolite profile compared to mouse. *Geroscience* 40 (4): 359–360.

169 Cakmak, Y.O. (2017). Rotorua, hydrogen sulphide and Parkinson's disease – a possible beneficial link? *N. Z. Med. J.* 130 (1455): 123–125.

170 Chou, C.H.S.J. (2003). *Hydrogen Sulfide: Human Health Aspects*, W.H.O.I.P.o.C. Safety. World Health Organization.

171 Reed, B.R., Crane, J., Garrett, N. et al. (2014). Chronic ambient hydrogen sulfide exposure and cognitive function. *Neurotoxicol. Teratol.* 42: 68–76.

172 Bates, M.N., Garrett, N., and Shoemack, P. (2002). Investigation of health effects of hydrogen sulfide from a geothermal source. *Arch. Environ. Health* 57 (5): 405–411.

173 Bates, M.N., Garrett, N., Graham, B. et al. (1997). Air pollution and mortality in the Rotorua geothermal area. *Aust. N. Z. J. Public Health* 21 (6): 581–586.

174 Bates, M.N., Crane, J., Balmes, J.R. et al. (2015). Investigation of hydrogen sulfide exposure and lung function, asthma and chronic obstructive pulmonary disease in a geothermal area of New Zealand. *PLoS One* 10 (3): e0122062.

175 Durand, M. and Wilson, J.G. (2006). Spatial analysis of respiratory disease on an urbanized geothermal field. *Environ. Res.* 101 (2): 238–245.

176 Carlsen, H.K., Zoega, H., Valdimarsdottir, U. et al. (2012). Hydrogen sulfide and particle matter levels associated with increased dispensing of anti-asthma drugs in Iceland's capital. *Environ. Res.* 113: 33–39.

177 Campagna, D., Kathman, S.J., Pierson, R. et al. (2004). Ambient hydrogen sulfide, total reduced sulfur, and hospital visits for respiratory diseases in northeast Nebraska, 1998–2000. *J. Expo. Anal. Environ. Epidemiol.* 14 (2): 180–187.

178 Lim, E., Mbowe, O., Lee, A.S. et al. (2016). Effect of environmental exposure to hydrogen sulfide on central nervous system and respiratory function: a systematic review of human studies. *Int. J. Occup. Environ. Health* 22 (1): 80–90.

179 Kashfi, K. and Olson, K.R. (2013). Biology and therapeutic potential of hydrogen sulfide and hydrogen sulfide-releasing chimeras. *Biochem. Pharmacol.* 85 (5): 689–703.

180 Toombs, C.F., Insko, M.A., Wintner, E.A. et al. (2010). Detection of exhaled hydrogen sulphide gas in healthy human volunteers during intravenous administration of sodium sulphide. *Br. J. Clin. Pharmacol.* 69 (6): 626–636.

181 Wallace, J.L. and Wang, R. (2015). Hydrogen sulfide-based therapeutics: exploiting a unique but ubiquitous gasotransmitter. *Nat. Rev. Drug Discov.* 14 (5): 329–345.

182 Polhemus, D.J., Li, Z., Pattillo, C.B. et al. (2015). A novel hydrogen sulfide prodrug, SG1002, promotes hydrogen sulfide and nitric oxide bioavailability in heart failure patients. *Cardiovasc. Ther.* 33 (4): 216–226.

183 Csiszar, A., Labinskyy, N., Podlutsky, A. et al. (2008). Vasoprotective effects of resveratrol and SIRT1: attenuation of cigarette smoke-induced oxidative stress and proinflammatory phenotypic alterations. *Am. J. Physiol. Heart Circ. Physiol.* 294 (6): H2721–H2735.

184 Whiteman, M., Li, L., Rose, P. et al. (2010). The effect of hydrogen sulfide donors on lipopolysaccharide-induced formation of inflammatory mediators in macrophages. *Antioxid. Redox Signal.* 12 (10): 1147–1154.

185 Santin-Marquez, R., Alarcon-Aguilar, A., Lopez-Diazguerrero, N.E. et al. (2019). Sulforaphane – role in aging and neurodegeneration. *Geroscience* 41 (5): 655–670.

186 Wlosinska, M., Nilsson, A.C., Hlebowicz, J. et al. (2020). The effect of aged garlic extract on the atherosclerotic process – a randomized double-blind placebo-controlled trial. *BMC Complement. Med. Ther.* 20 (1): 132.

187 Benavides, G.A., Squadrito, G.L., Mills, R.W. et al. (2007). Hydrogen sulfide mediates the vasoactivity of garlic. *Proc. Natl. Acad. Sci. U.S.A.* 104 (46): 17977–17982.

188 Borek, C. (2006). Garlic reduces dementia and heart-disease risk. *J. Nutr.* 136 (Suppl 3): 810S–812S.

189 Pei, Y., Wu, B., Cao, Q. et al. (2011). Hydrogen sulfide mediates the anti-survival effect of sulforaphane on human prostate cancer cells. *Toxicol. Appl. Pharmacol.* 257 (3): 420–428.

190 Houghton, C.A. (2019). Sulforaphane: its "coming of age" as a clinically relevant nutraceutical in the prevention and treatment of chronic disease. *Oxid. Med. Cell Longev.* 2019: 2716870.

191 Qi, Z., Ji, H., Le, M. et al. (2021). Sulforaphane promotes *C. elegans* longevity and healthspan via DAF-16/DAF-2 insulin/IGF-1 signaling. *Aging* 13 (2): 1649–1670.

192 Muntner, P., Carey, R.M., Gidding, S. et al. (2018). Potential US population impact of the 2017 ACC/AHA high blood pressure guideline. *Circulation* 137 (2): 109–118.

193 Yoon, S.S., Gu, Q., Nwankwo, T. et al. (2015). Trends in blood pressure among adults with hypertension: United States, 2003 to 2012. *Hypertension* 65 (1): 54–61.

194 van Goor, H., van den Born, J.C., Hillebrands, J.L. et al. (2016). Hydrogen sulfide in hypertension. *Curr. Opin. Nephrol. Hypertens.* 25 (2): 107–113.

195 Bucci, M., Vellecco, V., Cantalupo, A. et al. (2014). Hydrogen sulfide accounts for the peripheral vascular effects of zofenopril independently of ACE inhibition. *Cardiovasc. Res.* 102 (1): 138–147.

196 Vanhoutte, P.M., Shimokawa, H., Tang, E.H. et al. (2009). Endothelial dysfunction and vascular disease. *Acta Physiol.* 196 (2): 193–222.

197 Gkaliagkousi, E., Gavriilaki, E., Triantafyllou, A. et al. (2015). Clinical significance of endothelial dysfunction in essential hypertension. *Curr. Hypertens. Rep.* 17 (11): 85.

198 Xiao, L., Dong, J.H., Teng, X. et al. (2018). Hydrogen sulfide improves endothelial dysfunction in hypertension by activating peroxisome proliferator-activated receptor delta/endothelial nitric oxide synthase signaling. *J. Hypertens.* 36 (3): 651–665.

199 Brown, J.D. and Plutzky, J. (2007). Peroxisome proliferator-activated receptors as transcriptional nodal points and therapeutic targets. *Circulation* 115 (4): 518–533.

200 Wang, K., Ahmad, S., Cai, M. et al. (2013). Dysregulation of hydrogen sulfide producing enzyme cystathionine γ-lyase contributes to maternal hypertension and placental abnormalities in preeclampsia. *Circulation* 127 (25): 2514–2522.

201 Cao, X., Ding, L., Xie, Z.Z. et al. (2019). A review of hydrogen sulfide synthesis, metabolism, and measurement: is modulation of hydrogen sulfide a novel therapeutic for cancer? *Antioxid. Redox Signal.* 31 (1): 1–38.

202 Li, H., Xu, F., Gao, G. et al. (2020). Hydrogen sulfide and its donors: novel antitumor and antimetastatic therapies for triple-negative breast cancer. *Redox Biol.* 34: 101564.

203 Szabo, C. (2021). Hydrogen sulfide, an endogenous stimulator of mitochondrial function in cancer cells. *Cells* 10 (2): 220.

204 Bir, S.C., Kolluru, G.K., McCarthy, P. et al. (2012). Hydrogen sulfide stimulates ischemic vascular remodeling through nitric oxide synthase and nitrite reduction activity regulating hypoxia-inducible factor-1α and vascular endothelial growth factor-dependent angiogenesis. *J. Am. Heart Assoc.* 1 (5): e004093.

205 Johansen, D., Ytrehus, K., and Baxter, G.F. (2006). Exogenous hydrogen sulfide (H_2S) protects against regional myocardial ischemia-reperfusion injury – evidence for a role of K ATP channels. *Basic Res. Cardiol.* 101 (1): 53–60.

206 Liu, X., Pan, L., Zhuo, Y. et al. (2010). Hypoxia-inducible factor-1α is involved in the pro-angiogenic effect of hydrogen sulfide under hypoxic stress. *Biol. Pharm. Bull.* 33 (9): 1550–1554.

207 Szabó, C. and Papapetropoulos, A. (2011). Hydrogen sulphide and angiogenesis: mechanisms and applications. *Br. J. Pharmacol.* 164 (3): 853–865.

208 Tang, G., Wu, L., Liang, W. et al. (2005). Direct stimulation of K_{ATP} channels by exogenous and endogenous hydrogen sulfide in vascular smooth muscle cells. *Mol. Pharmacol.* 68 (6): 1757–1764.

209 Zhang, Z., Huang, H., Liu, P. et al. (2007). Hydrogen sulfide contributes to cardioprotection during ischemia-reperfusion injury by opening K_{ATP} channels. *Can. J. Physiol. Pharmacol.* 85 (12): 1248–1253.

210 Zhao, W., Zhang, J., Lu, Y. et al. (2001). The vasorelaxant effect of H_2S as a novel endogenous gaseous K_{ATP} channel opener. *EMBO J.* 20 (21): 6008–6016.

211 Sen, N., Paul, B.D., Gadalla, M.M. et al. (2012). Hydrogen sulfide-linked sulfhydration of NF-κB mediates its antiapoptotic actions. *Mol. Cell* 45 (1): 13–24.

212 Zhao, K., Ju, Y., Li, S. et al. (2014). S-Sulfhydration of MEK1 leads to PARP-1 activation and DNA damage repair. *EMBO Rep.* 15 (7): 792–800.

213 Cai, W.J., Wang, M.J., Ju, L.H. et al. (2010). Hydrogen sulfide induces human colon cancer cell proliferation: role of Akt, ERK and p21. *Cell Biol. Int.* 34 (6): 565–572.

214 Pan, Y., Ye, S., Yuan, D. et al. (2014). Hydrogen sulfide (H_2S)/cystathionine γ lyase (CSE) pathway contributes to the proliferation of hepatoma cells. *Mutat. Res.* 763, 764: 10–18.

215 Ma, Z., Bi, Q., and Wang, Y. (2015). Hydrogen sulfide accelerates cell cycle progression in oral squamous cell carcinoma cell lines. *Oral Dis.* 21 (2): 156–162.

216 Gong, Q.H., Wang, Q., Pan, L.L. et al. (2011). *S*-Propargyl-cysteine, a novel hydrogen sulfide-modulated agent, attenuates lipopolysaccharide-induced spatial learning and memory impairment: involvement of TNF signaling and NF-κB pathway in rats. *Brain Behav. Immun.* 25 (1): 110–119.

217 Lee, Z.W., Teo, X.Y., Tay, E.Y. et al. (2014). Utilizing hydrogen sulfide as a novel anti-cancer agent by targeting cancer glycolysis and pH imbalance. *Br. J. Pharmacol.* 171 (18): 4322–4336.

218 Murata, T., Sato, T., Kamoda, T. et al. (2014). Differential susceptibility to hydrogen sulfide-induced apoptosis between PHLDA1-overexpressing oral cancer cell lines and oral keratinocytes: role of PHLDA1 as an apoptosis suppressor. *Exp. Cell Res.* 320 (2): 247–257.

219 Wu, Y.C., Wang, X.J., Yu, L. et al. (2012). Hydrogen sulfide lowers proliferation and induces protective autophagy in colon epithelial cells. *PLoS One* 7 (5): e37572.

220 Lu, S., Gao, Y., Huang, X. et al. (2014). GYY4137, a hydrogen sulfide (H_2S) donor, shows potent anti-hepatocellular carcinoma activity through blocking the STAT3 pathway. *Int. J. Oncol.* 44 (4): 1259–1267.

221 Ma, K., Liu, Y., Zhu, Q. et al. (2011). H_2S donor, *S*-propargyl-cysteine, increases CSE in SGC-7901 and cancer-induced mice: evidence for a novel anti-cancer effect of endogenous H_2S? *PLoS One* 6 (6): e20525.

222 Chattopadhyay, M., Kodela, R., Nath, N. et al. (2012). Hydrogen sulfide-releasing NSAIDs inhibit the growth of human cancer cells: a general property and evidence of a tissue type-independent effect. *Biochem. Pharmacol.* 83 (6): 715–722.

223 De Cicco, P., Panza, E., Ercolano, G. et al. (2016). ATB-346, a novel hydrogen sulfide-releasing anti-inflammatory drug, induces apoptosis of human melanoma cells and inhibits melanoma development in vivo. *Pharmacol. Res.* 114: 67–73.

224 Sokolov, A.S., Nekrasov, P.V., Shaposhnikov, M.V. et al. (2021). Hydrogen sulfide in longevity and pathologies: inconsistency is malodorous. *Ageing Res. Rev.* 67: 101262.

18

Aberrant Hydrogen Sulfide Signaling in Alzheimer's Disease

Bindu D. Paul[1,2,3]

[1]*Department of Pharmacology and Molecular Sciences, Johns Hopkins University School of Medicine, Baltimore, MD, USA*
[2]*Department of Psychiatry and Behavioral Sciences, Johns Hopkins University School of Medicine, Baltimore, MD, USA*
[3]*The Solomon H. Snyder Department of Neuroscience, Johns Hopkins University School of Medicine, Baltimore, MD, USA*

List of Abbreviations

3-MST	3-mercaptopyruvate sulfur transferase
AD	Alzheimer's disease
APP	amyloid precursor protein
CBS	cystathionine β-synthase
CSE	cystathionine γ-lyase
CO	carbon monoxide
CVD	cardiovascular disease
DADS	diallyl disulfide
DATS	diallyl trisulfide
EOAD	early onset Alzheimer's disease
ERN	erucin
GAPDH	glyceraldehyde 3-phosphate dehydrogenase
GSK3β	glycogen synthase kinase 3β
H_2S	hydrogen sulfide
Hcy	homocysteine
HHcy	hyperhomocysteinemia
LOAD	late onset Alzheimer's disease
LTP	long-term potentiation
NAC	*N*-acetyl cysteine
NaSH	sodium hydrogen sulfide

NF-κB	nuclear factor κB
NFT	neurofibrillary tangles
NMDA	*N*-methyl D-aspartate
NO	nitric oxide
pK_a	acid dissociation constant
PHF	paired helical fragments
ROS	reactive oxygen species
RPS3	ribosomal protein S3
SAC	*S*-allyl cysteine
SAM	*S*-adenosyl methionine
SFN	sulforaphane
SOH	sulfenic acid
SO_2H	sulfinic acid
SO_3H	sulfonic acid
SPRC	*S*-propargyl-cysteine
SSO_2H	perthiosulfinic acid
SSO_3H	perthiosulfonic acid

18.1 Introduction

18.1.1 Hydrogen Sulfide

Hydrogen sulfide, a gaseous signaling molecule, was a major component of the ancient Earth's atmosphere before oxygen took over. H_2S was considered to be a byproduct of metabolic processes and a noxious gas until the discovery that it regulates vascular tone and blood pressure [1]. Similar to the gaseous molecules, nitric oxide (NO) and carbon monoxide (CO), H_2S has cytoprotective roles at low concentrations [2]. All three gasotransmitters are derived from amino acids – NO and H_2S directly from arginine and cysteine, respectively, and CO from glycine via heme [3, 4]. H_2S is generated by three enzymes in mammals: cystathionine γ-lyase (CSE), cystathionine β-synthase (CBS), and 3-mercaptopyruvate sulfurtransferase (3-MST), which synthesize the gas via the reverse trans-sulfuration pathway (Figure 18.1). CSE and CBS generate H_2S from L-cysteine as a substrate and pyridoxal 5-phosphate as a cofactor. 3-MST utilizes 3-mercaptopyruvate to produce H_2S [4, 5]. Additionally, CSE and CBS may use homocysteine (Hcy) as well, to generate the gaseous signaling molecule. All three enzymes are present in the peripheral systems as well as the brain; however, CSE is the major peripheral enzyme and CBS, the main H_2S generating enzyme present in the brain. 3-MST, which is also present in the brain, also generates H_2S using D-cysteine as a substrate [6].

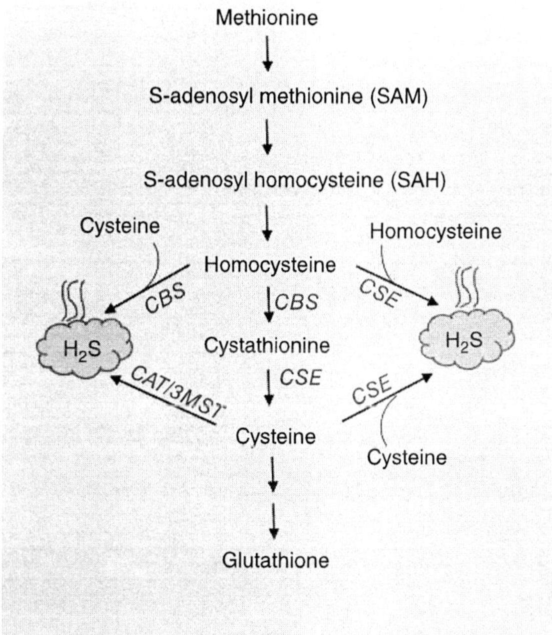

Figure 18.1 Biosynthesis of hydrogen sulfide. H_2S is generated via the reverse trans sulfuration pathway. Trans sulfuration refers to the transfer of sulfur from homocysteine to cysteine. Methionine, an essential amino acid for mammals, is converted to homocysteine via S-adenosyl methionine (SAM) and S-adenosyl homocysteine (SAH). Next, homocysteine is condensed with serine to generate cystathionine by cystathionine β-synthase (CBS). Cystathionine is cleaved by cystathionine γ-lyase (CSE) to produce the semi-essential amino acid cysteine. Cysteine can either be utilized to synthesize GSH and other sulfur-containing molecules or used as a substrate to generate hydrogen sulfide (H_2S). Both homocysteine and cysteine may be utilized to produce H_2S. While CSE generates H_2S from either cysteine or homocysteine, CBS prefers a combination of cysteine and homocysteine to generate H_2S. The 3-mercaptopyruvate sulfur transferase (3MST), in conjunction with cysteine aminotransferase (CAT), a third enzyme, also produces H_2S from cysteine.

The presence of H_2S in the brain was first detected in the brain by Warenycia and associates in 1989, who reported that inhalation of H_2S increased brain sulfide levels that was proportional to the dose of the gas inhaled. The study also reported the presence of sulfides in the untreated brain [7]. Increased levels of sulfide were also detected in the human brain in two cases of fatal H_2S poisoning, which spurred studies on production of H_2S in the brain [8]. Although all three biosynthetic enzymes for H_2S are present in the brain, there appears to be a spatial compartmentalization, in that CSE is expressed in neurons and CBS, in astrocytes [9–11]. The presence of 3-MST has been reported in both neurons and astrocytes [12, 13].

18.1.2 Protein Sulfhydration/Persulfidation

One of the modes by which H_2S regulates physiological processes is through persulfidation/sulfhydration, where –SH groups of reactive cysteine residues is converted to persulfide or –SSH groups [4, 5, 14, 15]. H_2S cannot directly modify reduced cysteine residues, and persulfidation may occur in the following ways: (i) nucleophilic attack of sulfide on oxidized cysteine residues such as cysteine sulfenic acids (Cys-SOH) or cysteine disulfides (–S–S–); (ii) by reaction between oxidized sulfide species such as polysulfides and cysteine thiols; and/or (iii) by a reaction of H_2S_2 with a cysteine thiol [5]. H_2S and sulfhydration modulate a myriad of physiological processes, which include response to inflammation and other forms of stress such as DNA damage, protein degradation, autophagy, vascular functions, mitochondrial bioenergetics, and organellar function, all of which contribute to modulation of healthspan and lifespan [4, 16–21] (Figure 18.2). Sulfhydration predominantly occurs at cysteine residues with a low pK_a, which are more susceptible to sulfhydration, because they exist as thiolate anions (RS^-) at physiological conditions. These residues are also more prone to oxidation to sulfenic acid (SOH), sulfinic acid (SO_2H), or sulfonic acid (SO_3H) derivatives. However, if these cysteines are sulfhydrated, although these same oxidation events occur, the cysteine can be restored to its native thiol form by cellular reductants. The $CySSO_2H$ (perthiosulfinic) and $CySSO_3H$ (perthiosulfonic) oxidation products of persulfides can be recycled back by the reduction of their S–S moieties, which does not occur in the case of the $CySO_3H$ oxidation product of unmodified (unsulfhydrated) cysteine residues on proteins. Several studies have demonstrated that sulfhydration prevents proteins from irreversible oxidation [22, 23].

18.1.3 Reciprocity of Protein Sulfhydration and Nitrosylation

Sulfhydration is analogous to nitrosylation, a modification mediated by the gaseous signaling molecule, nitric oxide (NO), where –SH groups of reactive cysteine residues are converted to –SNO groups [24]. Very often sulfhydration and nitrosylation occur on the same cysteine residue, exerting opposite functions [4, 5]. Sulfhydration is substantially more prevalent than nitrosylation, with about 25% of proteins being basally sulfhydrated in the liver [14]. Sulfhydration increases the reactivity of the cysteine residue being modified, whereas nitrosylation decreases its reactivity. In the case of the mouse glycolytic enzyme, glyceraldehyde phosphate dehydrogenase (GAPDH), sulfhydration of Cys150

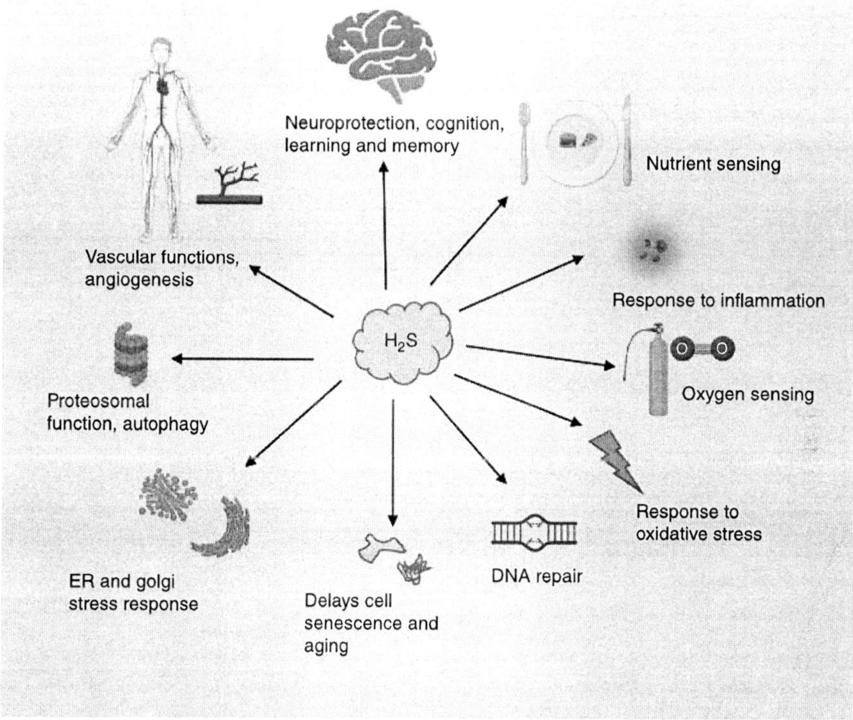

Figure 18.2 Physiological functions of hydrogen sulfide. H_2S modulates almost all aspects of cellular physiology, which are also linked to aging. H_2S regulates antioxidant defense, oxygen sensing, vascular functions, nutrient sensing, response to inflammation, mitochondrial function, endoplasmic reticulum (ER) and Golgi stress responses, aging, memory and long term potentiation and proteostasis. Dysregulation of these processes are also hallmarks of aging. Source: Based on Refs. [4, 16–21].

enhances its catalytic activity, while nitrosylation decreases it [14, 25]. A separate study involving in vitro assays using purified human GAPDH did not report sulfhydration at this residue, likely reflecting differences in *in vitro* versus *in vivo* conditions and/or species-specific differences [26]. Another example involves reciprocal sulfhydration of parkin, an E3 ubiquitin ligase, which is involved in clearance of misfolded proteins and mitophagy. Nitrosylation of parkin decreases its E3-ubiquitin ligase activity, while sulfhydration enhances it [16, 27]. Similarly, waves of sulfhydration and nitrosylation were observed during response to inflammation by the transcription factor nuclear factor κB (NF-κB). The p65 subunit of NF κB is first sulfhydrated at Cys38, which enhances its binding to the coactivator ribosomal protein S3 (RPS3) to stimulate the transcription of anti-apoptotic genes [4]. Sulfhydration of p65 predominates early after TNF-α treatment, then declines and is followed by a reciprocal enhancement of p65 nitrosylation. Nitrosylation of p65 decreases its association with the promoter of anti-apoptotic genes.

We have shown previously that dysregulation of H_2S, cysteine metabolism, and sulfhydration occurs in neurodegenerative diseases such as Parkinson's disease and Huntington's disease [16, 21, 28, 29]. We now discuss the neuroprotective roles of H_2S and its dysregulation in Alzheimer's disease (AD).

18.2 Alzheimer's Disease

AD is a complex, progressive neurodegenerative disease, which is the most common cause of dementia worldwide [30, 31]. The cause of the disease can be multifactorial with both familial and sporadic causes. AD can either be early (EOAD) or late onset (LOAD), with a vast majority of cases occurring after 65. Less than 5% of AD occurs earlier than 65 years of age. It is estimated that in the United States alone, about 6.2 million Americans over 65 suffer from AD and this figure could grow to 13.8 million by 2060 [32]. Early clinical symptoms of AD include difficulty remembering conversations and events, which progress to symptoms which range from disorientation, impaired executive functions, confusion, behavioral changes, and finally difficulty swallowing, speaking and deficits in motor functions. Presently, there is no cure or effective therapeutics for the disease despite decades of intense research. The entorhinal cortex region is the first region of the brain to affected in AD. The disease mainly affects the cerebral cortex and the hippocampus of the brain.

18.2.1 Neuropathology of AD

The neuropathological hallmarks of AD are deposition of extracellular neuritic plaques, constituted by the amyloid β-peptide and neurofibrillary tangles (NFTs) and paired helical fragments (PHFs) composed of the protein tau [31, 33–35]. The Aβ peptide, was first sequenced from the meningeal blood vessels of AD patients and individuals with Down syndrome (DS) and later identified as the component of the plaques in AD [36, 37]. The Aβ peptides are derived from amyloid precursor protein (APP), a transmembrane protein enriched in neuronal synapses [38]. In AD, aberrant processing of APP via the amyloidogenic pathway by action of β-secretase and γ-secretases leads to the generation of Aβ peptides, a major component of amyloid plaques and has been the subject of several reviews [33].

Tau, the component of NFTs, is a microtubule-binding protein, normally localized to the axonal cytoplasm, but may also be present in the presynaptic and postsynaptic compartments and associated with the nuclear membrane

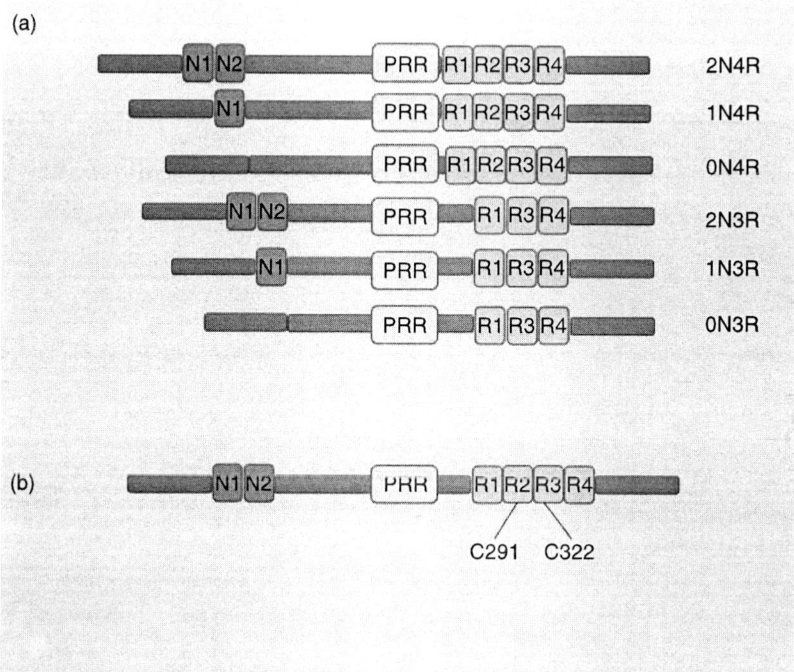

Figure 18.3 Isoforms of tau. Schematic representation of the six isoforms of tau generated by alternative splicing of the *MAPT* gene. N1 and N2 denote N-terminal inserts; R1 to R4 indicate microtubule-binding repeat domains, whose differential inclusion results in the formation of either 3R or 4R tau. PRR refers to the proline-rich region.

[39–43]. There are six isoforms of tau and the full-length tau protein has four repeat domains, which in conjunction with flanking sequences constitute the microtubule-binding domain and referred to as 4R tau. The isoforms are generated due to alternate splicing exons 2, 3, and 10, resulting in 0, 1, or 2 inserts in the N-terminal projection domain (N1 and N2 isoforms), and 3 or 4 repeats (3R and 4R isoforms) (Figure 18.3). Tau undergoes several post-translational modifications such as phosphorylation, acetylation, and ubiquitination. Tau harbors more than 80 predicted phosphorylation sites at Ser, Thr, Tyr residues, of which >50 are found to be modified [44]. Hyperphosphorylation of Tau occurs in AD, leading to its decreased binding to microtubules, aggregation, and mislocalization, eliciting neurotoxicity via multiple mechanisms, including changes in axonal transport, cytoskeletal organization, nucleocytoplasmic transport, and mitochondrial respiration [34, 42, 45–48].

18.2.2 H_2S Signaling in Alzheimer's Disease

Physiological concentrations of H_2S enhance the *N*-methyl D-aspartate (NMDA) receptors, which is involved in modulation of long-term potentiation (LTP), a process important for learning and memory [49]. The NMDA receptor is an ion channel which is activated by glutamate and its coagonists (glycine, D-serine or D-cysteine) [50]. This study constituted the first evidence that H_2S may function as a neuromodulator in the brain. Later a decrease in H_2S levels in the plasma of AD patients, which correlated with the degree of cognitive decline, was reported [51]. In this study, plasma from 31 AD patients, 28 vascular dementia patients, 20 cerebrovascular disease (CVD) patients, and 23 normal controls were evaluated for not only H_2S but also Hcy levels. Hcy levels in the three disease states were significantly lower than those from normal individuals. A point to be noted is that both CSE and CBS utilize Hcy as a substrate to produce H_2S (Figure 18.1). Additionally, hyperhomocysteinemia (HHcy) or elevated levels of Hcy is an independent risk factor for both vascular disease as well as dementia [52]. Several studies have correlated levels of Hcy with the severity of AD [53, 54]. S-Adenosylhomocysteine, which is the precursor of Hcy, reportedly causes aberrant processing of the APP and elicits neurotoxicity [55]. Elevated Hcy levels occur in traumatic brain injury which is a greater risk factor for AD as compared to stroke [56]. HHcy can occur by deficits in several pathways including depletion of cofactors of enzymes of the reverse trans-sulfuration pathway [57]. S-Adenosyl methionine (SAM), an allosteric activator of CBS is depleted in the human AD brain as well as in the cerebrospinal fluid, which could contribute to the dysregulated H_2S signaling observed in AD [58, 59]. Enriching the diet of the 3xTg-AD mouse model of AD with SAM delayed amyloid β and tau pathology [60]. Injection of $A\beta_{1-42}$ into rat hippocampi elicited cognitive impairment and expression of endogenous CBS and 3-MST and decrease in H_2S levels [61]. Intraperitoneal injection of the H_2S donor, NaSH, ameliorated cognitive deficits as well as neuroinflammation. Several studies have confirmed beneficial effects of administering H_2S donors to mouse models of AD; however, molecular mechanisms underlying these effects have not been investigated in detail [62–64]. NaSH ameliorated lipopolysaccharide (LPS)-induced cognitive deficits while decreasing inflammation and lipid peroxidation [65].

18.2.3 Sulfhydration in Aging and AD

Aging is the biggest risk factor for developing age-related neurodegenerative diseases such as AD. In a landmark study involving more than ten laboratories across the globe, we reported a decrease in levels of the H_2S producing enzymes, CSE and 3-MST during aging in the brain of Wistar rats [66]. Sulfhydration was also diminished during aging with a concomitant increase in cysteine sulfinylation. In addition, by sulfhydrating cysteine residues on target proteins, H_2S prevented irreversible oxidation of cysteine residues as demonstrated previously. Another important study demonstrated that dietary restriction, which is known to improve healthspan and lifespan, operates through H_2S [67]. Sulfhydration mediated by H_2S regulates all the reported hallmarks of aging [68, 69]. More recently, we analyzed the role of sulfhydration in both the 3xTg-AD mouse model of AD as well as postmortem human AD samples. Using the dimedone switch assay, we showed that the overall level of sulfhydration is diminished in both a mouse model as well as human AD samples. We utilized the 3xTg-AD mouse model of AD, which harbors the mutations PS1M146V, APPSwe, and Tau P301L and develops both NFTs and amyloid plaques in this study [70]. We further showed that CSE, which is localized to neurons, binds wild type but not mutant P301L tau. Wild type stimulates activity of CSE, which may have physiological functions, which are yet to be explored in the context of tau. We also showed that tau is sulfhydrated at Cys322, present in the repeat domain R3 by H_2S donors and that the H_2S donor, NaSH or H_2S generated by the action of CSE on L-cysteine, inhibited phosphorylation of tau by glycogen synthase 3β (GSK3β), the major kinases modifying it *in vivo*. Mutation of cysteine residues on tau did not prevent inhibition of GSK3β by NaSH, indicating that cysteine residues on tau did not play a role in the phosphorylation status of tau (Figure 18.4). Using the dimedone switch assay, we further showed that NaSH caused sulfhydration of GSK3β itself at Cys218. Analysis and 3-dimensional modeling of the amino acid sequence of GSK3β revealed that Cys218 lies close to Asp181 in the active site, which is involved in hydrogen bond formation for catalysis. Sulfhydration of Cys218 could disrupt the active site conformation and thereby inhibit its catalytic activity. These studies also revealed that sulfhydration of GSK3β is diminished in human AD. Administering the slow releasing H_2S donor, NaGYY4137, rescued cognitive and motor dysfunction in the 3xTg-AD mouse model and rescued overall levels of sulfhydration [57, 71].

18.3 Therapeutic Avenues

H_2S levels are tightly regulated in cells. Excess H_2S interferes with mitochondrial functions and bioenergetics and is implicated in a state of suspended animation caused by inhibition of complex IV of the electron transport chain [20, 72]. Interestingly, DS, caused by trisomy of chromosome 21 (on which CBS resides), leads to excess H_2S production and inhibition of mitochondrial function. APP is also present on chromosome 21, and these DS patients ultimately develop amyloid pathology in the brain. Thus, the combination of excess H_2S and other

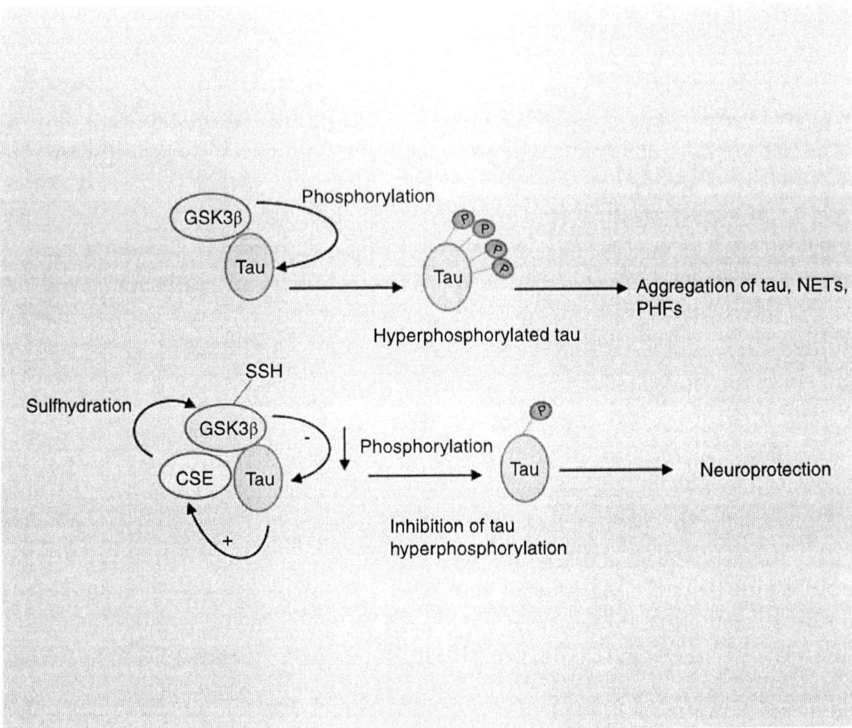

Figure 18.4 Model depicting neuroprotective role of hydrogen sulfide in Alzheimer's disease. A possible mode of neuroprotection mediated by H_2S. Glycogen synthase kinase 3 β (GSK3β) binds tau and phosphorylates it (labeled as "P"), which causes it to aggregate and form neurofibrillary tangles (NFTs), paired helical filaments (PHFs) and elicit AD pathology in the 3xTg-AD mice. H_2S produced by CSE sulfhydrates GSK3β (denoted as "SSH") and inhibits phosphorylation of tau to confer neuroprotection.

factors such as excess APP processing is deleterious. In AD, as paucity of H_2S and diminished sulfhydration has been reported by us and others. Agents that deliver H_2S or boost its production may be beneficial. Studies involving delivery of cysteine derivatives such as *N*-acetyl cysteine (NAC) have proven to be beneficial in several mouse models exhibiting neurodegeneration and neuropsychiatric symptoms and also in patients with these conditions [28, 73, 74]. In addition to its role as an antioxidant (being a stable cysteine donor) NAC has also been shown to boost sulfane sulfur production by boosting H_2S levels [75]. Natural products present in vegetables such as onion, garlic, broccoli, watercress, and cabbage that act as H_2S donors have also been considered as therapeutics. Among compounds present in garlic, allicin has been well studied. In aqueous media, allicin decomposes into the H_2S-releasing compounds: diallyl sulfide (DAS), diallyl disulfide (DADS), and diallyl trisulfide (DATS), among others [76–78]. The allicin compound DATS has been reported to be the major H_2S releasing compound as the others release the gas at very slow rates [79, 80]. Another organosulfur molecule derived from garlic includes S-allyl cysteine (SAC) and its structural analogue *S*-propargyl-cysteine (SPRC) which protects adult rat hearts and neonatal cardiomyocytes from ischemia/hypoxia injury [81, 82].

Hybrid molecules that not only release H_2S but have other protective functions are also being developed. A hybrid molecule derived from memantine, the NMDA receptor antagonist, memit, releases H_2S by a cysteine-mediated mechanism and prevented neuronal inflammation and production of reactive oxygen species (ROS). In addition, memit decreased $A\beta_{1-42}$ aggregation and prevented Aβ oligomers-induced damage in human neurons and rat microglia cells. Memit also enhanced autophagy [83]. Other hybrid molecules derived from the natural products, sulforaphane (SFN) and erucin (ERN), have been proposed as potential drugs for AD [84]. Developing mitochondria-targeted donors, such as AP39 would also afford therapeutic benefits. Drugs which cross the blood-barrier would be especially critical to ensure delivery to the brain.

To be considered is the fact that supplementation of H_2S donors may not restore the complete spectrum of abnormalities caused by dysregulation of H_2S production. Besides, caution must be exercised while characterizing H_2S donors as some of these compounds may release H_2S in an uncontrolled, either too fast or too slow (such as sulfide salts or GYY4137, respectively). Additionally, the byproducts formed after H_2S release may not have been identified. Compounds that stimulate the trans-sulfuration pathway as a whole may provide improved neuroprotection. Thus, small molecules which stimulate CSE or CBS may be harnessed for therapies targeting not only AD but also other diseases involving dysregulated H_2S signaling.

Acknowledgments

This work was supported by the American Heart Association (AHA)/Paul Allen Frontiers Group; grant number 19PABH134580006 from the AHA/Allen Initiative in Brain Health and Cognitive Impairment team to B.D.P. and associates and the grants DA044123 (NIDA/NIH), 1R21AG073684-01 (NIA/NIH) and R01AG071512-01A1 (NIA/NIH) to B.D.P.

References

1 Yang, G., Wu, L., Jiang, B. et al. (2008). H_2S as a physiologic vasorelaxant: hypertension in mice with deletion of cystathionine gamma-lyase. *Science* 322 (5901): 587–590.

2 Mustafa, A.K., Gadalla, M.M., and Snyder, S.H. (2009). Signaling by gasotransmitters. *Sci. Signal.* 2 (68): re2.

3 Paul, B.D. and Snyder, S.H. (2018). Production of NO-The L-arginine/NOS/NO system. In: *Gasotransmitters* (ed. R. Wang), 29–43. Cambridge, UK: Royal Society of Chemistry.

4 Sen, N., Paul, B.D., Gadalla, M.M. et al. (2012). Hydrogen sulfide-linked sulfhydration of NF-kappaB mediates its antiapoptotic actions. *Mol. Cell* 45 (1): 13–24.

5 Paul, B.D. and Snyder, S.H. (2015). H_2S: a novel gasotransmitter that signals by sulfhydration. *Trends Biochem. Sci* 40 (11): 687–700.

6 Shibuya, N. and Kimura, H. (2013). Production of hydrogen sulfide from d-cysteine and its therapeutic potential. *Front. Endocrinol. (Lausanne)* 4: 87.

7 Warenycia, M.W., Goodwin, L.R., Benishin, C.G. et al. (1989). Acute hydrogen sulfide poisoning. Demonstration of selective uptake of sulfide by the brainstem by measurement of brain sulfide levels. *Biochem. Pharmacol.* 38 (6): 973–981.

8 Goodwin, L.R., Francom, D., Dieken, F.P. et al. (1989). Determination of sulfide in brain tissue by gas dialysis/ion chromatography: postmortem studies and two case reports. *J. Anal. Toxicol.* 13 (2): 105–109.

9 Morikawa, T., Kajimura, M., Nakamura, T. et al. (2012). Hypoxic regulation of the cerebral microcirculation is mediated by a carbon monoxide-sensitive hydrogen sulfide pathway. *Proc. Natl. Acad. Sci. U.S.A.* 109 (4): 1293–1298.

10 Linden, D.R., Sha, L., Mazzone, A. et al. (2008). Production of the gaseous signal molecule hydrogen sulfide in mouse tissues. *J. Neurochem.* 106 (4): 1577–1585.

11 Enokido, Y., Suzuki, E., Iwasawa, K. et al. (2005). Cystathionine beta-synthase, a key enzyme for homocysteine metabolism, is preferentially expressed in the radial glia/astrocyte lineage of developing mouse CNS. *FASEB J.* 19 (13): 1854–1856.

12 Shibuya, N., Tanaka, M., Yoshida, M. et al. (2009). 3-Mercaptopyruvate sulfurtransferase produces hydrogen sulfide and bound sulfane sulfur in the brain. *Antioxid. Redox Signaling* 11 (4): 703–714.

13 Zhao, H., Chan, S.J., Ng, Y.K., and Wong, P.T. (2013). Brain 3-mercaptopyruvate sulfurtransferase (3MST): cellular localization and downregulation after acute stroke. *PLoS One* 8 (6): e67322.

14 Mustafa, A.K., Gadalla, M.M., Sen, N. et al. (2009). H_2S signals through protein S-sulfhydration. *Sci. Signal.* 2 (96): ra72.

15 Paul, B.D. and Snyder, S.H. (2015). Protein sulfhydration. *Methods Enzymol.* 555: 79–90.

16 Vandiver, M.S., Paul, B.D., Xu, R. et al. (2013). Sulfhydration mediates neuroprotective actions of parkin. *Nat. Commun.* 4: 1626.

17 Jiang, X., MacArthur, M.R., Trevino-Villarreal, J.H. et al. (2021). Intracellular H_2S production is an autophagy-dependent adaptive response to DNA damage. *Cell. Chem. Biol.*

18 Aroca, A., Yruela, I., Gotor, C., and Bassham, D.C. (2021). Persulfidation of ATG18a regulates autophagy under ER stress in Arabidopsis. *Proc. Natl. Acad. Sci. U.S.A.* 118 (20).

19 Mustafa, A.K., Sikka, G., Gazi, S.K. et al. (2011). Hydrogen sulfide as endothelium-derived hyperpolarizing factor sulfhydrates potassium channels. *Circ. Res.* 109 (11): 1259–1268.

20 Paul, B.D., Snyder, S.H., and Kashfi, K. (2021). Effects of hydrogen sulfide on mitochondrial function and cellular bioenergetics. *Redox Biol.* 38: 101772.

21 Sbodio, J.I., Snyder, S.H., and Paul, B.D. (2018). Golgi stress response reprograms cysteine metabolism to confer cytoprotection in Huntington's disease. *Proc. Natl. Acad. Sci. U.S.A.* 115 (4): 780–785.

22 Doka, E., Ida, T., Dagnell, M. et al. (2020). Control of protein function through oxidation and reduction of persulfidated states. *Sci. Adv.* 6 (1): eaax8358.

23 Shirasaki, D.I., Greiner, E.R., Al-Ramahi, I. et al. (2012). Network organization of the huntingtin proteomic interactome in mammalian brain. *Neuron* 75 (1): 41–57.

24 Stamler, J.S., Simon, D.I., Osborne, J.A. et al. (1992). S-nitrosylation of proteins with nitric oxide: synthesis and characterization of biologically active compounds. *Proc. Natl. Acad. Sci. U.S.A.* 89 (1): 444–448.

25 Hara, M.R., Agrawal, N., Kim, S.F. et al. (2005). S-nitrosylated GAPDH initiates apoptotic cell death by nuclear translocation following Siah1 binding. *Nat. Cell Biol.* 7 (7): 665–674.

26 Jarosz, A.P., Wei, W., Gauld, J.W. et al. (2015). Glyceraldehyde 3-phosphate dehydrogenase (GAPDH) is inactivated by S-sulfuration in vitro. *Free Radical Biol. Med.* 89: 512–521.

27 Chung, K.K., Thomas, B., Li, X. et al. (2004). S-nitrosylation of parkin regulates ubiquitination and compromises Parkin's protective function. *Science* 304 (5675): 1328–1331.

28 Paul, B.D., Sbodio, J.I., Xu, R. et al. (2014). Cystathionine gamma-lyase deficiency mediates neurodegeneration in Huntington's disease. *Nature* 509 (7498): 96–100.

29 Sbodio, J.I., Snyder, S.H., and Paul, B.D. (2016). Transcriptional control of amino acid homeostasis is disrupted in Huntington's disease. *Proc. Natl. Acad. Sci. U.S.A.* 113 (31): 8843–8848.

30 Lane, C.A., Hardy, J., and Schott, J.M. (2018). Alzheimer's disease. *Eur. J. Neurol.* 25 (1): 59–70.

31 Masters, C.L., Bateman, R., Blennow, K. et al. (2015). Alzheimer's disease. *Nat Rev Dis Primers* 1: 15056.

32 (2021). 2021 Alzheimer's disease facts and figures. *Alzheimers Dement.* 17 (3): 327–406.

33 Knopman, D.S., Amieva, H., Petersen, R.C. et al. (2021). Alzheimer disease. *Nat. Rev. Dis. Primers* 7 (1): 33.

34 Grundke-Iqbal, I., Iqbal, K., Tung, Y.C. et al. (1986). Abnormal phosphorylation of the microtubule-associated protein tau (tau) in Alzheimer cytoskeletal pathology. *Proc. Natl. Acad. Sci. U.S.A.* 83 (13): 4913–4917.

35 Iqbal, K., Grundke-Iqbal, I., Smith, A.J. et al. (1989). Identification and localization of a tau peptide to paired helical filaments of Alzheimer disease. *Proc. Natl. Acad. Sci. U.S.A.* 86 (14): 5646–5650.

36 Glenner, G.G. and Wong, C.W. (2012). Alzheimer's disease: initial report of the purification and characterization of a novel cerebrovascular amyloid protein. 1984. *Biochem. Biophys. Res. Commun.* 425 (3): 534–539.

37 Glenner, G.G. and Wong, C.W. (1984). Alzheimer's disease and Down's syndrome: sharing of a unique cerebrovascular amyloid fibril protein. *Biochem. Biophys. Res. Commun.* 122 (3): 1131–1135.

38 Haass, C. and Selkoe, D.J. (2007). Soluble protein oligomers in neurodegeneration: lessons from the Alzheimer's amyloid beta-peptide. *Nat. Rev. Mol. Cell Biol.* 8 (2): 101–112.

39 Witman, G.B., Cleveland, D.W., Weingarten, M.D., and Kirschner, M.W. (1976). Tubulin requires tau for growth onto microtubule initiating sites. *Proc. Natl. Acad. Sci. U.S.A.* 73 (11): 4070–4074.

40 Gallardo, G. and Holtzman, D.M. (2019). Amyloid-beta and Tau at the crossroads of Alzheimer's disease. *Adv. Exp. Med. Biol.* 1184: 187–203.

41 Pooler, A.M., Noble, W., and Hanger, D.P. (2014). A role for tau at the synapse in Alzheimer's disease pathogenesis. *Neuropharmacology* 76 (Pt A): 1–8.

42 Eftekharzadeh, B., Daigle, J.G., Kapinos, L.E. et al. (2018). Tau protein disrupts nucleocytoplasmic transport in Alzheimer's disease. *Neuron* 99 (5): 925–940 e7.

43 Weingarten, M.D., Lockwood, A.H., Hwo, S.Y., and Kirschner, M.W. (1975). A protein factor essential for microtubule assembly. *Proc. Natl. Acad. Sci. U.S.A.* 72 (5): 1858–1862.

44 Drepper, F., Biernat, J., Kaniyappan, S. et al. (2020). A combinatorial native MS and LC-MS/MS approach reveals high intrinsic phosphorylation of human

Tau but minimal levels of other key modifications. *J. Biol. Chem.* 295 (52): 18213–18225.

45 Jouanne, M., Rault, S., and Voisin-Chiret, A.S. (2017). Tau protein aggregation in Alzheimer's disease: an attractive target for the development of novel therapeutic agents. *Eur. J. Med. Chem.* 139: 153–167.

46 Medina, M., Hernandez, F., and Avila, J. (2016). New Features about Tau Function and Dysfunction. *Biomolecules* 6 (2).

47 Wegmann, S., Biernat, J., and Mandelkow, E. (2021). A current view on Tau protein phosphorylation in Alzheimer's disease. *Curr. Opin. Neurobiol.* 69: 131–138.

48 Lindwall, G. and Cole, R.D. (1984). Phosphorylation affects the ability of tau protein to promote microtubule assembly. *J. Biol. Chem.* 259 (8): 5301–5305.

49 Abe, K. and Kimura, H. (1996). The possible role of hydrogen sulfide as an endogenous neuromodulator. *J. Neurosci.* 16 (3): 1066–1071.

50 Lee, C.H., Lu, W., Michel, J.C. et al. (2014). NMDA receptor structures reveal subunit arrangement and pore architecture. *Nature* 511 (7508): 191–197.

51 Liu, X.Q., Liu, X.Q., Jiang, P. et al. (2008). Plasma levels of endogenous hydrogen sulfide and homocysteine in patients with Alzheimer's disease and vascular dementia and the significance thereof. *Zhonghua Yi Xue Za Zhi* 88 (32): 2246–2249.

52 Seshadri, S., Beiser, A., Selhub, J. et al. (2002). Plasma homocysteine as a risk factor for dementia and Alzheimer's disease. *N. Engl. J. Med.* 346 (7): 476–483.

53 Kitzlerova, E., Fisar, Z., Jirak, R. et al. (2014). Plasma homocysteine in Alzheimer's disease with or without co morbid depressive symptoms. *Neuro. Endocrinol. Lett.* 35 (1): 42–49.

54 Farina, N., Jerneren, F., Turner, C. et al. (2017). Homocysteine concentrations in the cognitive progression of Alzheimer's disease. *Exp. Gerontol.* 99: 146–150.

55 Lin, H.C., Hsieh, H.M., Chen, Y.H., and Hu, M.L. (2009). S-adenosylhomocysteine increases beta-amyloid formation in BV-2 microglial cells by increased expressions of beta-amyloid precursor protein and presenilin 1 and by hypomethylation of these gene promoters. *Neurotoxicology* 30 (4): 622–627.

56 Rahmani, A., Hatefi, M., Dastjerdi, M.M. et al. (2016). Correlation between serum homocysteine levels and outcome of patients with severe traumatic brain injury. *World Neurosurg.* 87: 507–515.

57 Paul, B.D. (2021). Neuroprotective roles of the reverse transsulfuration pathway in Alzheimer's disease. *Front. Aging Neurosci.* 13: 659402.

58 Morrison, L.D., Smith, D.D., and Kish, S.J. (1996). Brain S-adenosylmethionine levels are severely decreased in Alzheimer's disease. *J. Neurochem.* 67 (3): 1328–1331.

59 Linnebank, M., Popp, J., Smulders, Y. et al. (2010). S-adenosylmethionine is decreased in the cerebrospinal fluid of patients with Alzheimer's disease. *Neurodegener. Dis.* 7 (6): 373–378.

60 Lee, S., Lemere, C.A., Frost, J.L., and Shea, T.B. (2012). Dietary supplementation with S-adenosyl methionine delayed amyloid-beta and tau pathology in 3xTg-AD mice. *J. Alzheimers Dis.* 28 (2): 423–431.

61 Liu, H., Deng, Y., Gao, J. et al. (2015). Sodium hydrosulfide attenuates beta-amyloid-induced cognitive deficits and neuroinflammation via modulation of MAPK/NF-kappaB pathway in rats. *Curr. Alzheimer Res.* 12 (7): 673–683.

62 Giuliani, D., Ottani, A., Zaffe, D. et al. (2013). Hydrogen sulfide slows down progression of experimental Alzheimer's disease by targeting multiple pathophysiological mechanisms. *Neurobiol. Learn. Mem.* 104: 82–91.

63 Zhao, F.L., Fang, F., Qiao, P.F. et al. (2016). AP39, a mitochondria-targeted hydrogen sulfide donor, supports cellular bioenergetics and protects against Alzheimer's disease by preserving mitochondrial function in APP/PS1 mice and neurons. *Oxid. Med. Cell Longev.* 2016: 8360738.

64 Vandini, E., Ottani, A., Zaffe, D. et al. (2019). Mechanisms of hydrogen sulfide against the progression of severe Alzheimer's disease in transgenic mice at different ages. *Pharmacology* 103 (1-2): 50–60.

65 Aboulhoda, B.E., Rashed, L.A., Ahmed, H. et al. (2021). Hydrogen sulfide and mesenchymal stem cells-extracted microvesicles attenuate LPS-induced Alzheimer's disease. *J. Cell. Physiol.* 236 (8): 5994–6010.

66 Zivanovic, J., Kouroussis, E., Kohl, J.B. et al. (2019). Selective persulfide detection reveals evolutionarily conserved antiaging effects of S-sulfhydration. *Cell Metab.* 30 (6): 1152–1170 e13.

67 Hine, C., Harputlugil, E., Zhang, Y. et al. (2015). Endogenous hydrogen sulfide production is essential for dietary restriction benefits. *Cell* 160 (1–2): 132–144.

68 Petrovic, D., Kouroussis, E., Vignane, T., and Filipovic, M.R. (2021). The role of protein persulfidation in brain aging and neurodegeneration. *Front. Aging Neurosci.* 13: 674135.

69 Lopez-Otin, C., Blasco, M.A., Partridge, L. et al. (2013). The hallmarks of aging. *Cell* 153 (6): 1194–1217.

70 Oddo, S., Caccamo, A., Shepherd, J.D. et al. (2003). Triple-transgenic model of Alzheimer's disease with plaques and tangles: intracellular Abeta and synaptic dysfunction. *Neuron* 39 (3): 409–421.

71 Giovinazzo, D., Bursac, B., Sbodio, J.I. et al. (2021). Hydrogen sulfide is neuroprotective in Alzheimer's disease by sulfhydrating GSK3beta and inhibiting Tau hyperphosphorylation. *Proc. Natl. Acad. Sci. U.S.A.* 118 (4).

72 Blackstone, E., Morrison, M., and Roth, M.B. (2005). H_2S induces a suspended animation-like state in mice. *Science* 308 (5721): 518.

73 Wright, D.J., Gray, L.J., Finkelstein, D.I. et al. (2016). N-acetylcysteine modulates glutamatergic dysfunction and depressive behavior in Huntington's disease. *Hum. Mol. Genet.* 25 (14): 2923–2933.

74 Deepmala, Slattery, J., Kumar, N. et al. (2015). Clinical trials of N-acetylcysteine in psychiatry and neurology: A systematic review. *Neurosci. Biobehav. Rev.* 55: 294–321.

75 Ezerina, D., Takano, Y., Hanaoka, K. et al. (2018). N-acetyl cysteine functions as a fast-acting antioxidant by triggering intracellular H_2S and sulfane sulfur production. *Cell. Chem. Biol.* 25 (4): 447–459 e4.

76 Corvino, A., Frecentese, F., Magli, E. et al. (2021). Trends in H_2S-donors chemistry and their effects in cardiovascular diseases. *Antioxidants (Basel)* 10 (3).

77 Amagase, H. (2006). Clarifying the real bioactive constituents of garlic. *J. Nutr.* 136 (3 Suppl): 716S–725S.

78 Zhao, Y., Pacheco, A., and Xian, M. (2015). Medicinal chemistry: insights into the development of novel H_2S donors. *Handb. Exp. Pharmacol.* 230: 365–388.

79 Liang, D., Wu, H., Wong, M.W., and Huang, D. (2015). Diallyl trisulfide is a fast H_2S donor, but diallyl disulfide is a slow one: the reaction pathways and intermediates of glutathione with polysulfides. *Org. Lett.* 17 (17): 4196–4199.

80 Bolton, S.G., Cerda, M.M., Gilbert, A.K., and Pluth, M.D. (2019). Effects of sulfane sulfur content in benzyl polysulfides on thiol-triggered H_2S release and cell proliferation. *Free Radical Biol. Med.* 131: 393–398.

81 Kim, J.M., Lee, J.C., Chang, N. et al. (2006). S-allyl-L-cysteine attenuates cerebral ischemic injury by scavenging peroxynitrite and inhibiting the activity of extracellular signal-regulated kinase. *Free Radical Res.* 40 (8): 827–835.

82 Wang, Q., Liu, H.R., Mu, Q. et al. (2009). S-propargyl-cysteine protects both adult rat hearts and neonatal cardiomyocytes from ischemia/hypoxia injury: the contribution of the hydrogen sulfide-mediated pathway. *J. Cardiovasc. Pharmacol.* 54 (2): 139–146.

83 Sestito, S., Daniele, S., Pietrobono, D. et al. (2019). Memantine prodrug as a new agent for Alzheimer's disease. *Sci. Rep.* 9 (1): 4612.

84 Sestito, S., Pruccoli, L., Runfola, M. et al. (2019). Design and synthesis of H_2S-donor hybrids: a new treatment for Alzheimer's disease? *Eur. J. Med. Chem.* 184: 111745.

19

Multifaceted Actions of Hydrogen Sulfide in the Kidney

Balakuntalam S. Kasinath[1,2,3] *and Hak Joo Lee*[1]

[1]*Department of Medicine, Center for Renal Precision Medicine, University of Texas Health, San Antonio, TX, USA*
[2]*Barshop Institute for Longevity and Aging Studies, University of Texas Health, San Antonio, TX, USA*
[3]*Geriatric Research, Education & Clinical Center, South Texas Veterans Health Care System, San Antonio, TX, USA*

List of Abbreviations

3-MST	3-mercaptopyruvate sulfurtransferase
AGE	advanced glycation end products
AKI	acute kidney injury
AMPK	AMP (adenosine monophosphate)-activated protein kinase
AQP-2	aquaporin-2
ATP	adenosine triphosphate
ATPase	adenosine triphosphatase
CBS	cystathionine β-synthase
CLCA1	chloride channel accessory 1
ccRCC	clear cell renal cell carcinoma
CHF	congestive heart failure
CKD	chronic kidney disease
CO	carbon monoxide
CSE	cystathionine γ-lyase
DAO	D-amino acid oxidase
DKD	diabetic kidney disease
DR	dietary restriction
EGFR	epidermal growth factor receptor
eGFR	estimated GFR
EMT	epithelium to mesenchymal transition
ENaC	epithelial Na^+ channel

ESKD	end stage kidney disease
FGF23	fibroblast growth factor 23
GFR	glomerular filtration rate
H_2S	hydrogen sulfide
HFD	high fat diet
Hhcy	hyperhomocysteinemia
HIF	hypoxia inducible factor
HS^-	hydrosulfide
IGF1	insulin-like growth factor 1
IR	insulin receptor
IRI	ischemia reperfusion injury
KPTIRKO	kidney proximal tubule insulin receptor knock out
miR	microRNA
MMP	matrix metalloprotease
mTORC1	mechanistic target of rapamycin complex 1
NaHS	sodium hydrosulfide
NFκB	nuclear factor κB
NO	nitric oxide
NOX	NADPH (reduced nicotinamide adenine dinucleotide phosphate) oxidase
Nrf-2	nuclear factor erythroid-2 related factor-2
PARP-1	poly(ADP ribose) polymerase-1
PI 3 kinase	phosphoinositide 3 kinase
PTEN	phosphatase and tensin homolog
RCC	renal cell carcinoma
ROS	reactive oxygen species
SASP	senescence associated secretory phenotype
sFlt1	soluble fms-like tyrosine kinase
SIRT	sirtuin
TET	ten eleven translocation
TGFβ	transforming growth factor β
TMEM16A	transmembrane member 16A
TNFα	tumor necrosis factor α
VEGF	vascular endothelial growth factor

19.1 Introduction

In addition to waste elimination, the kidneys perform many other functions including reabsorption of most of the water and electrolytes that have been filtered thereby affecting contents of the body fluid compartments, blood pressure, and functions of several organ systems [1]. Kidneys synthesize erythropoietin that regulates maturation of erythrocytes, and vitamin 1 : 25 D_3 essential for bone metabolism. They are also involved in the metabolism of many peptide hormones such as insulin. Given the spectrum of their functions, it should not be surprising that kidney disease commonly results in wide spread disturbance in functions of other organs. Emerging evidence indicates a key role for hydrogen sulfide (H_2S) in the regulation of many of the aforementioned functions of the kidney. The topic of H_2S and the kidney has been extensively reviewed recently [2, 3].

19.2 H_2S Synthesis in the Kidney

In the 1980s, Stipanuk and Beck reported that the kidney generated H_2S by the regulation of L-cysteine metabolism [4]. There are several pathways of H_2S synthesis in the kidney: the trans-sulfuration pathway in the cytosol involving L-cysteine, D-cysteine metabolism in the peroxisome, and the mitochondrial pathway (Figure 19.1) [1, 2, 5]. In the trans-sulfuration pathway, H_2S synthesis is catalyzed by cystathionine β-synthase (CBS) and cystathionine γ-lyase (CSE), whereas 3-mercaptopyruvate sulfurtransferase (3-MST) regulates H_2S generation in the mitochondria; peroxisomal H_2S generation involves D-amino oxidase (DAO). CSE is reported to be more abundant than CBS in the kidney [6]. Although they exist in the cytosol, nuclear translocation of these enzymes has been reported [7]. Quantitatively, the D-cysteine pathway generates more H_2S than other pathways [8]. Glucose, glutathione, inorganic and organic polysulfides, and elemental sulfur can generate H_2S by nonenzymatic mechanisms [9]; whether this occurs in the kidney requires further study. With a pK_a of 6.8, most intracellular H_2S exists as HS^- [8, 10]. Transport of H_2S into and out of cells is not well understood. It may involve anion exchanger −1 as shown in erythrocytes [11]. A HS^- channel has been described in *Clostridium difficile* [10]. Methemoglobin has been suggested as a H_2S carrier in the blood [12].

The end products of H_2S metabolism include thiosulfate, sulfite, and sulfate. Urinary sulfate and thiosulfate levels are associated with dietary intake of sulfur containing amino acids and systemic H_2S [13, 14]. Caution should be taken in interpreting urinary sulfate measurement because (i) sulfate filtered by the glomerulus undergoes tubular reabsorption by transporters such as sodium sulfate 1 and Sat-1 [15], and (ii) bacterial flora in the intestines can generate H_2S and contribute to urinary sulfate [16].

Several cell types of the nephron, the structural and functional unit of the kidney, express CSE, CBS, and 3-MST [17, 18]. In the glomerulus, a vascular filter at the head of the nephron where the urine formation starts, the constituent mesangial cells and glomerular endothelial cells express CBS and CSE [19–21], whereas

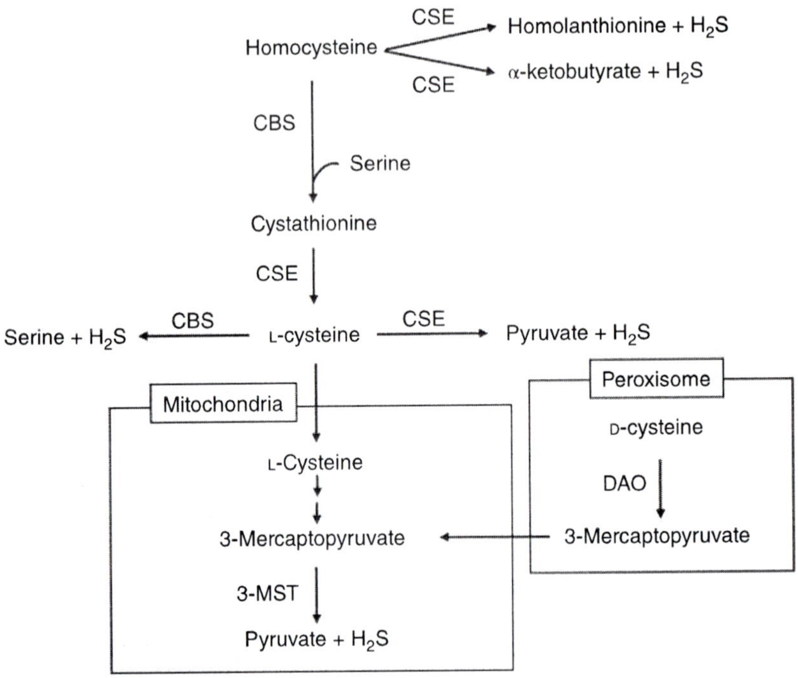

Figure 19.1 Hydrogen sulfide synthesis in the kidney. A simplified schematic of H_2S synthesis in the kidney; for clarity, many details have not been included. H_2S is generated during homocysteine metabolism in the trans-sulfuration pathway in the cytosol catalyzed by cystathionine β-synthase (CBS) and cystathionine γ-lyase (CSE). Cysteine undergoes complex metabolism in the mitochondria during which H_2S is generated catalyzed by 3-mercaptopyruvate sulfurtransferase (3-MST). Peroxisome is a site of H_2S generation from D-cysteine in the kidney under the regulation of D-amino acid oxidase (DAO).

the podocytes express CSE [18]. Kidney proximal tubular epithelial cells of the nephron express CBS [22–24]. Recent studies have shown that H_2S affects several major processes in the kidney by regulating redox biology, protein synthesis, and inflammation to name a few [1]. Additionally, post-translational modification of proteins by persulfidation is another major mechanism by which H_2S regulates their functions [25]. Reductant labile sulfur (sulfane sulfur) can exist as persulfide, polysulfide, thiosulfate, polythionates, thiosulfonate, elemental sulfur, and protein-associated sulfur [26]. Bound sulfane sulfur is characterized by the presence of two sulfur atoms bound by a covalent linkage, e.g. outer sulfur atoms of the persulfides and the inner chain atoms of polysulfides; reducing conditions release sulfane sulfur [27]. Polysulfides have extensive biologic effects and may account for many effects of H_2S [28]. Furthermore, H_2S inhibits adenosine triphosphate (ATP)- sensitive K^+ channel, e.g. pancreatic β cells [29]; however, how the regulation of K^+ATP channels figures in H_2S regulation of kidney cell biology has not been well studied.

19.3 H_2S and Kidney Physiology

A more detailed account of regulation of kidney physiology can be found in recent publications [1, 30]. The basic unit of the kidney is the nephron. Aortic blood enters the renal artery and its succeeding branches, and through the afferent arteriole enters the glomerulus, a capillary filter, to undergo filtration [31]. Whether plasma components enter the glomerular filtrate depends on their size and charge. Glomerular filtration rate (GFR) of circulating creatinine, an endogenous molecule, is commonly measured as an index of waste elimination function of the kidney. Elevation of serum creatinine implies a reduced GFR. Cells in blood circulation such as the red blood cells are prevented from entering the filtrate. Very small amounts of proteins such as albumin are allowed to escape which are further reabsorbed by the proximal tubules downstream of the glomerulus. Thus, increase in urinary albumin excretion can imply glomerular or tubular injury or both. The glomerular capillary branches reunite to form the efferent arteriole which carries blood away from the glomerulus at the end of filtration. Thus, glomerular capillary is unique in being interposed between two arterioles. Glomerular capillary blood flow undergoes complex hemodynamic regulation subject to the status of constriction or dilatation of the arterioles. Renal vascular infusion of sodium hydrosulfide (NaHS), which releases H_2S, increases GFR by increasing blood flow to the glomerulus [32]; increased renal blood flow and GFR has also been shown in mice following administration of GYY4137, an H_2S donor [33]. Renin is an enzyme that catalyzes conversion of angiotensinogen to angiotensin I; it is secreted by the interstitial cells around the afferent arteriole [34, 35]. Angiotensin converting enzyme modifies angiotensin I to angiotensin II, a powerful vasoconstrictor. H_2S reduces renin activity [36], which inhibits angiotensin II downstream, and favors more blood to flow through the glomerulus.

After glomerular filtration, the filtrate enters a series of tubules: proximal tubule, descending limb, and ascending limb of loop of Henle, distal convoluted tubule, connecting duct, cortical and medullary collecting ducts. Although distinct epithelial cells line these tubular segments, recent studies have shown transitional cell types may give rise to new epithelial cells, e.g. collecting ducts [37]. Tubules reabsorb most of the water and electrolytes filtered by the glomerulus, and also secrete acid into the filtrate. Tubules also metabolize peptide hormones such as insulin. Distinct intricate pathways are employed in handling electrolytes and water. For example, sodium, the major cation in the extracellular fluid compartment of which blood is a part, is freely filtered at the glomerulus and bulk of it is reabsorbed at the proximal tubule. The apically located Na^+-H^+ anion exchanger 3 transports Na^+ from the filtrate into the cytoplasm of proximal epithelial cells; this is followed by exit of Na^+ from the cytoplasm to the peritubular capillaries under the control of Na^+-K^+ adenosine triphosphatase (ATPase) located in the basolateral membrane [38, 39]. From the peritubular capillaries, Na^+ is returned to the systemic circulation through renal venous drainage. Additional mechanisms of Na^+ reabsorption at the proximal tubule involve cotransport with sulfate, glucose, phosphate, and amino acids [40]. The remaining filtrate Na^+ is successively reabsorbed by the Na^+-K^+-$2Cl^-$ transporter in the ascending limb of loop of Henle, and Na^+-Cl^- cotransporter in the distal convoluted tubule, respectively. Intrarenal infusion of NaHS resulted in increase in the urinary excretion of Na^+ and K^+ [32] that was attributed to inhibition of Na^+-K^+-$2Cl^-$ cotransporter and Na^+-K^+ ATPase, although more comprehensive studies are needed. Cell culture studies suggest that H_2S can inhibit the Na^+-H^+ exchanger 3 [41].

The final site of regulation of Na^+ handling is the collecting duct where the epithelial Na^+ channel (ENaC) transports Na^+ from the filtrate into cytoplasm under the control of aldosterone; this effect is mediated by reactive oxygen species (ROS) [42, 43]. Hydrogen peroxide, a source of OH^- radicals, stimulates phosphoinositide 3 kinase (PI 3 kinase) by inhibiting its constitutive inhibitor phosphatase and tensin homolog, resulting in increase in ENaC activity [44]. H_2S prevents inhibition of phosphatase and tensin homolog, leading to reduction in PI 3 kinase activity and the function of ENaC; this can contribute to augmented Na^+ excretion in the urine. As mentioned above, transport of reabsorbed Na^+ out of tubular cell cytoplasm to peritubular capillaries depends on the activity of Na^+-K^+ ATPase. H_2S impairs the activity of Na^+-K^+ ATPase by promoting endocytosis of its subunits. An interesting new mechanism involves H_2S stimulating the epidermal growth factor receptor (EGFR), which activates PI 3 kinase. *In vivo* support for the involvement of EGFR in the H_2S effect was obtained in hypertensive rats treated with NaHS [41]. Furthermore, cysteine 797/798 in the EGFR seems to play an important role in its mediation of H_2S effect [41] although whether this involves persulfidation is not known.

Water reabsorption from the filtrate occurs in the proximal nephron by passive transport. However, in the collecting duct, aquaporin-2 (AQP-2) promotes water reabsorption by forming water channels on the apical membrane [45]. Lithium inhibits AQP-2 and augments volume of water excretion. The resulting frequent urine passage limits patient acceptance of lithium in the treatment of psychiatric disorders. Recent studies have shown that H_2S stimulates AQP-2 expression and its organization into apical water channels [46]. Furthermore, H_2S inhibits lithium effect on urine volume by increasing AQP-2 expression and its organization into apical water channels [46]. H_2S inhibition of lithium effect on water excretion may have an important clinical application in psychiatry to improve acceptance of lithium as a therapeutic agent. Another toxic effect of lithium is the fibrosis of the interstitial compartment of the kidney which can result in discontinuation of an otherwise effective therapeutic agent. As H_2S ameliorates fibrosis in chronic kidney disease of diverse etiology (reviewed later), research is needed to test whether it can limit lithium-induced kidney fibrosis.

Kidney interstitium is the area surrounding the glomerulus and the tubules. The interstitial perivascular cells surrounding the afferent arteriole synthesize renin which plays an important role in angiotensin metabolism and hemodynamics as described above [34, 35]. The stellate fibroblast-like cells surrounding the tubules synthesize erythropoietin, an erythrocyte maturation factor the deficiency of which causes anemia in chronic kidney disease [35]. Erythropoietin synthesis is stimulated by hypoxia in the kidney. Integrated actions of H_2S, CO, and NO regulate oxygen sensing in the carotid body leading to appropriate changes in ventilation, heart rate, and blood pressure [1, 47]. H_2S appears to regulate oxygen sensing in the heart and gastrointestinal tract [48, 49]. Therefore, it is likely that H_2S has a similar role in oxygen sensing by the kidney, and affect erythropoietin synthesis. This is an area that merits investigation.

Similar to the handling of Na^+, complex and elegant pathways exist to regulate excretion of other electrolytes and acid base regulation in the kidney. It is important to study how H_2S affects these vital processes in the kidney.

19.4 H_2S and the Aging Kidney

A comprehensive review of relevance of H_2S to the aging process has recently appeared [50]. The importance of understanding the aging process is not only to extend life span but also to improve quality of life termed health span. The latter requires maintenance of integrity of individual organ structure and function including the kidney [51]. Aging results in nephrosclerosis, a composite term that includes replacement of glomerular tissue with fibrous tissue (glomerulosclerosis), atrophy of the tubules, interstitial fibrosis, and thickening of the blood vessels resulting in narrowing of their lumens (arteriolosclerosis). Nephrosclerosis is accompanied by loss of GFR in humans although this may not be universal [52, 53]; in aged mice, reduced GFR is indicated by increase in serum cystatin C [54]. In a cross-sectional study, loss of eGFR (estimated GFR) only partly correlated with histologic changes of nephrosclerosis suggesting the existence of other yet to be identified factors; [55–57]. In aging mice and marmosets, albuminuria and proteinuria are reported [54, 58, 59].

Dietary restriction (DR) of calorie intake is an established method of delaying aging related organ dysfunction including the kidney. DR affords protection to the kidney against ischemic injury, and also prolongs lifespan in animals across evolutionary boundaries; both processes involve stimulation of H_2S generation [60]. DR promotes CSE expression and H_2S generation in the kidney [61]. Dietary methionine restriction protects the kidney against aging related injury in mice and rats [61, 62]. DR includes decreased intake of the essential amino acid methionine, which leads to increased generation of H_2S by the trans-sulfuration pathway. The importance of methionine restriction was elegantly shown by the reversal of DR-induced protection of aging related kidney injury by dietary supplementation of methionine [61]. Vascular changes are an important component of nephrosclerosis in the aged kidney. Vascular density is decreased in the aged kidney along with increased renal vascular resistance which contributes to reduced GFR; these changes are ameliorated by the administration of GYY4137 [33]. A recent report described that NAD-Sirtuin 1 (SIRT1) axis was impaired in aged rodents which was rectified by the administration of H_2S resulting in improved vascular integrity [63].

Multiple mechanisms involving hemodynamic changes, angiotensin II and other growth factors, oxidative stress, advanced glycation end (AGE) products, cell cycle events, and metabolic reactions lead to senescence in the kidney [64, 65]. All these mechanisms involve signaling cascades which have received increasing attention recently. Increased expression of phosphorylated mechanistic target of rapamycin (mTOR) was observed in the aged kidney [66]. mTOR is a versatile kinase which regulates cell senescence, autophagy, and gene expression at the levels of transcription and mRNA translation [67]. SIRT-1 dependent mitochondrial autophagy is impaired in the aged kidney and it is mitigated by DR [68]. Infusion of blood from young mice into aging mice by parabiosis ameliorates kidney injury [69]. These studies did not examine if H_2S had a role in the beneficiary effects of DR and parabiosis.

Recent advances in biotechnology have allowed unbiased discovery approaches to identify new mechanisms of kidney injury. We performed RNA Seq and compared changes in mRNA expression in the kidney cortexes of young versus aged mice to identify hitherto unidentified pathways of injury. We found that chloride channel accessory 1 (CLCA1) mRNA was significantly increased in the aged kidney [70]. CLCA1 is a secreted protein with a N-terminal domain that is self-cleaved from the molecule by its protease activity [71]. The latter associates with transmembrane member 16A (TMEM16A, anoctamin-1), a Ca^{2+}-dependent Cl^- channel, and facilitates its activity [72]. Immunohistochemistry showed that CLCA1 expression was robustly increased in the kidney tubules in aged mice. Increase in kidney cortical CLCA1 expression was also seen in both aging marmosets and aging human subjects indicating evolutionary conservation.

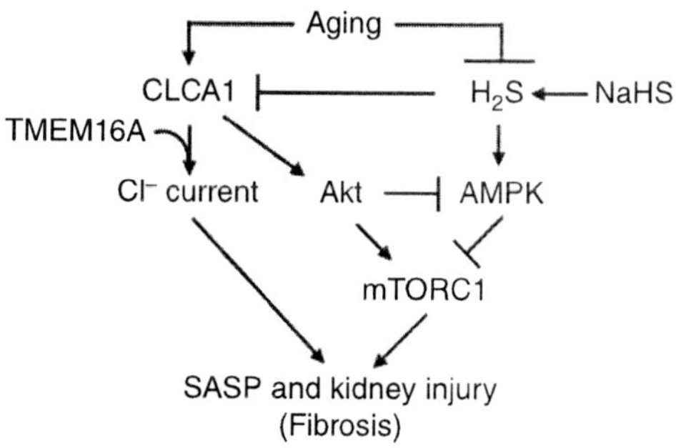

Figure 19.2 Hydrogen sulfide deficiency contributes to aging-related kidney injury. Aging is associated with increased kidney expression of chloride channel accessory 1 (CLCA1) protein and deficiency of H_2S generation, which may be upstream of CLCA1 increase. Deficiency of H_2S also results in inhibition of AMPK activity. CLCA1 associates with TMEM16A (anoctamin-1), a Ca^{++} dependent Cl^- channel, augments Cl^- current, and promotes senescence associated secretory phenotype (SASP). Increased CLCA1 expression also promotes Akt- mTORC1 axis. Activation of mTORC1 and H_2S deficiency promote synthesis of proteins including matrix proteins and albuminuria. NaHS, a source of H_2S, inhibits kidney CLCA1 increase in aged mice, and mitigates signaling reactions, matrix protein increase, and SASP induced by CLCA1 overexpression in proximal tubular epithelial cells.

We stably overexpressed CLCA1 in mouse kidney proximal tubule epithelial cells to explore its biological effects. This maneuver resulted in activation of Cl^- current measured by patch clamp, and increase in matrix protein fibronectin expression (Figure 19.2). Associated signaling changes consisted of inhibition of AMPK (AMP-activated protein kinase) activity and activation of Akt-mTORC1 pathway, similar to that seen in kidneys of aged mice [58]. Interestingly, overexpression of CLCA1 was sufficient to stimulate SASP (senescence associated secretory phenotype) in kidney epithelial cells. Changes in fibronectin, SASP, and signaling were inhibited by T16Ainh-A01, an inhibitor of TMEM16A, suggesting that these effects of CLCA1 are mediated by Cl^- channel activity of TMEM16A. Interestingly, CLCA1 overexpression decreased the expression of CSE without affecting H_2S generation. NaHS completely inhibited changes in Cl^- current, fibronectin expression, signaling changes and SASP induced by CLCA1 overexpression. Administration of NaHS to aging mice abolished aging related increased CLCA1 expression in the kidneys. Taken together these data show the following: (i) TMEM16A, a Ca^{2+}-dependent Cl^- channel, is activated due to increase in CLCA1 expression in the kidneys of aged mice (ii) A cell surface ionic channel sends cues to the tubular epithelial cell to augment matrix protein expression, activate Akt-mTORC1 signaling axis, and induce SASP. (iii) H_2S deficiency may be a contributing factor to stimulation of CLCA1-dependent TMEM16A activation in aging kidney.

Decreased expression of CSE and CBS and reduced H_2S generation in the aged kidney was shown to be associated with impaired anti-oxidant defense due to diminished expression of superoxide dismutase 2 and hemoxygenase 1. Expression of both these enzymes is under the control of nuclear factor erythroid-2 related factor-2 (Nrf-2), a transcription factor; these changes are reversed by the administration of H_2S in the form of NaHS resulting in reduced albuminuria and fibrosis [73]. MicroRNAs (miRs) have a role in aging related kidney injury. miRs are short forms of RNA that affect gene expression mostly by post-transcriptional regulation at the level of translation [74]. Studies in aged rats have shown that kidney expression of superoxide dismutase 2 and thioredoxin reductase 2, mitochondrial anti-oxidant enzymes, is reduced due to overexpression of miRs-335 and 34a; this contributes to increased production of ROS in the aged kidney [75]. In the galactoside-induced rapid aging model in mice, reduced CSE expression and H_2S generation in the kidney was rectified by NaHS [76].

Kidney fibrosis constitutes the final common pathway that leads to progressive injury including aging [77] It involves increased synthesis and decreased degradation of extracellular matrix proteins under the control of signaling reactions. We have reported that Akt-mTORC1 pathway is a major regulator of matrix protein increase in diabetic kidney disease [78–81]. We examined if H_2S has a role in the regulation of mTORC1 in the context of accumulation of matrix proteins in the kidneys of aged mice. In aged mice, the kidney cortical expression of CBS, CSE, and generation of H_2S were decreased; these changes were associated with increased insulin receptor (IR) phosphorylation indicating its activation, and stimulation of the canonical Akt-mTORC1 pathway [58]. AMPK is an energy sensor which constitutively inhibits mTORC1-induced mRNA translation and protein synthesis [18, 82]. AMPK activity was reduced in the aged kidney suggesting removal of a constitutive break on mTORC1.

SASP is a process in which a subset of cells in aging tissues undergoes cell cycle arrest and assumes a phenotype of secreting inflammatory cytokines promoting tissue injury [83]. In the aged kidney SASP manifested as increased expression of p21, and interleukins 1 and 6 [58]. Administration of NaHS to 18-/19-month old mice for 5 months resulted in prevention of the aforementioned signaling events, fibrotic changes, and SASP; functional improvements consisted of prevention of progressive increase in albuminuria and reduced clearance function of the aged kidney [58]. These observations establish a fundamental role for H_2S deficiency in aging related kidney injury.

Marmosets offer the advantages of evolutionary proximity to humans; they also have a short life span which permits interventional studies to be performed [84]. Studies in aged marmosets also showed association of nephrosclerosis and proteinuria with reduced H_2S generation, and stimulation of Akt-mTORC1 axis in the kidney; however, the upstream regulator appeared to be insulin-like growth factor 1 (IGF 1) receptor rather than IR [59]. Transforming growth factor β (TGF β) 1 is an important fibrogenic cytokine in the kidney; we observed increased TGF β1 in the kidneys of both aged mice and marmosets [54, 59]. These data indicate that H_2S deficiency and activation of signaling pathways in aging related kidney injury are evolutionarily conserved.

Kidney distal tubular epithelium synthesizes αKlotho, an antiaging hormone [85]. Mice lacking αKlotho age faster, while αKlotho transgenic mice resist aging [86]. Decreased αKlotho in aged kidneys correlates with reduced GFR [87]. Aging is associated with reduced Klotho in the kidney including the marmosets [59]. Interestingly, Klotho inhibits signaling by receptor tyrosine kinases such as the IR and IGF1 receptor [86] which could at least partly explain activation of these pathways in aging mice and marmosets [54, 59]. Klotho protects against angiotensin II-induced kidney injury [88] which is relevant to the mechanistic role played by that mediator in aging-induced kidney damage.

19.5 H_2S and Acute Kidney Injury (AKI)

Sudden reduction in kidney function with a rise in serum creatinine is called AKI, which can have serious consequences including dysfunction of other organs, progression to chronic kidney disease, and death. Injury to one kidney is compensated by the augmented function of the other, and does not result in AKI. Thus, AKI occurs when both kidneys are affected in people with two kidneys. However, in a person with a single kidney, e.g., kidney transplant recipient, injury to that kidney can lead to AKI. Interruption of adequate blood flow into the glomerulus (pre-renal), parenchymal kidney disease (intrinsic renal), and obstruction to the flow of urine out of the kidney (post-renal) are broad categories of AKI. Recent investigations have revealed a significant role for H_2S in AKI.

19.5.1 H_2S in AKI Due to Intrinsic Kidney Injury

Intrinsic causes of AKI include ischemia due to lack of blood supply, administration of agents that are toxic to the kidney, e.g. cis-platinum, and inflammatory injury to the kidney, e.g., glomerulonephritis.

19.5.1.1 Ischemia-Induced AKI

Ischemia reperfusion injury (IRI) occurs when blood flow to the nephron is suddenly interrupted followed by restoration of perfusion. It is characterized by oxidative stress, inflammation, and tubular cell death. In rats, IRI was associated with apoptosis, activation of stress mitogen-activated protein (MAP) kinases, and activation of NFκB (nuclear factor κB); these abnormalities were resolved by

NaHS [89]. Administration of NaHS by inhalation induced a hypometabolic state akin to hibernation and protected the kidney against IRI [90]. In mice, IRI decreased H_2S generation due to reduction in CSE expression; infusion of NaHS resolved the kidney injury [17]. IRI is of particular concern during kidney transplantation as the donor kidney is removed and implanted in the recipient. A positive correlation between CSE expression in the transplanted kidney and its subsequent function suggested that capacity for H_2S generation may impact function of the graft [17]. Urinary thiosulfate levels in kidney transplant recipients correlated with systemic H_2S levels and bore a positive correlation with future function of the kidney transplant [14, 90]. It is important to note that some of the H_2S in the body can be released by bacteria in the gut and affect urinary metabolites of H_2S. In one study, addition of H_2S to the preservative solution of the donor kidney appeared to improve its function [91]. Mitochondria are a site of H_2S generation by the actions of 3-MST as reviewed above. AP39, a H_2S donor targeted to mitochondria, has been shown to rescue kidney from IRI by inhibiting oxidative stress and inflammation [92].

Aging increases the risk of IRI [93]. Deficiency in kidney H_2S generation is seen in both aging and IRI, raising the possibility that together they may place the kidney at greater jeopardy than either condition alone. miR-21 is increased in IRI. Pushpakumar et al examined the roles of H_2S and miR-21 in IRI in aging mice [33]. Following IRI, renal expression of miR-21 was increased associated with reduced CBS expression, decreased GFR, increased matrix protein expression, macrophage polarization into injury phenotype in both young and old mice; however, the extent of increase in miR-21 and the intensity of injury occurred to a greater extent in old mice. These injury indices were significantly ameliorated by the administration of anti-miR 21, a miR-21 antagonist, and GYY4137 although functional recovery was less complete in old mice. In vitro studies in glomerular endothelial cells anti-miR-21 augmented the expression of CBS and 3-MST. These data suggest that miR-21 targets enzymes in the trans-sulfuration pathway to reduce H_2S generation in IRI [33]. These studies identify miR-21 and H_2S deficiency as mediators of increased injury in IRI in aged mice.

Whole body cooling followed by rewarming is another model of IRI in which the expression of CBS, CSE, and 3-MST is reduced; dopamine infusion stabilized H_2S generation and ameliorated ischemic injury [94]. During hibernation, animals manifest a hypometabolic state; following arousal, they do not incur IRI and tubular epithelium remains intact [95]. As described by Bos et al H_2S induces a torpor like state [90]. Understanding mechanisms underlying structural integrity of the kidneys in hibernation may help us prevent IRI. DR is another intervention that protects the kidney against IRI in a H_2S-dependent manner [60].

19.5.1.2 Rhabdomyolysis-Induced AKI

AKI is frequently associated with skeletal muscle injury (rhabdomyolysis) encountered in the battle field, vehicular accidents, and earth quakes. It may be due to toxic effects of injurious agents released from the injured muscles and ischemia due to hypotension. Administration of NaHS inhibited AKI in a rodent model of rhabdomyolysis [96].

19.5.1.3 Nephrotoxic AKI

AKI may be caused by agents used either in the investigation of disease, e.g. radiocontrast media, or in the management of disease, e.g. antibiotics, chemotherapeutic agents, e.g. gentamicin, cis-platinum.

Aminoglycoside-induced AKI. Gentamicin and other aminoglycosides are concentrated in the proximal tubule and frequently result in toxic cell injury. Several investigators have reported mitigation of gentamicin-induced AKI by administration of NaHS and diallyl sulfides present in garlic [97–99]. In this setting, there could be a complex interaction among the gasotransmitters. Gentamicin promoted inflammation and inducible NO synthase expression that were inhibited by NaHS. Additionally, zinc protoporphyrin, an inhibitor of hemoxygenase-1, when co-administered with NaHS, annulled the salutary effects of the latter suggesting possible involvement of CO [100].

cis-Platinum-induced AKI. Injury to the kidney is one of the limiting factors in employing chemotherapeutic agents in the treatment of cancer. *Cis*-platinum is commonly employed in the management of solid cancers but its use is limited as nearly one third of the patients develop kidney injury [101]. The role of H_2S in *cis*-platinum injury has been reviewed in depth recently [102, 103]. There is some controversy on the part played by H_2S in *cis*-platinum-induced AKI. Some studies suggest that H_2S mediates kidney injury [104, 105]; however, concerns are expressed on the experimental approaches employed in these studies including the unselective nature of propargylglycine, a chemical inhibitor of CSE, and the dose and solvent employed in the preparation of GYY4137 [103, 106, 107].

In contrast to these reports [104, 105], other studies demonstrate a therapeutic role for H_2S. Reduced kidney CSE expression and decreased H_2S generation occur in rats treated with cis-platinum [106]. H_2S reduced cis-platinum-induced ROS generation by inhibition of NADPH oxidase (NOX); the latter appears to involve persulfidation of p47 phox, a component of NOX [106]. In addition to oxidative stress, H_2S also inhibits inflammation and cell death [108].

In addition to NaHS and GYY4137 [106], polysulfides e.g., sodium tetrasulfide, also exert a mitigating effect on *cis*-platinum-induced AKI; this involves Nrf-2 induction of antioxidant genes, and inhibition of apoptosis [109]. Cis-platinum induces tumor necrosis factor α expression via NFκB [110], which is inhibited by NaHS [111]. Diallyl sulfides in garlic also reduce cis-platinum injury to the kidney by inhibiting NFκB and apoptosis [112, 113]. cis-platinum administration activation of signal transducer and activator of transcription 3 (STAT3) is involved in increased expression of tumor necrosis factor α and cytokines, and macrophage infiltration consistent with inflammatory injury to the kidney; these indices were ameliorated by NaHS, GYY4137, and sodium tetrasulfide [114] which involves persulfidation [114].

SIRT-3 is a NAD-dependent mitochondrial deacetylase which removes acetyl groups from substrates including superoxide dismutase, ATP synthase β, and optic atrophy protein-1, and enhances anti-oxidant defense [115]. Cis-platinum decreases SIRT-3 in the kidney which was rescued by mesenchymal stem cell transplantation [116]. Administration of NaHS restored SIRT-3 expression, reduced apoptosis and mitochondrial fragmentation, ATP production, and improved kidney function in mice receiving cis-platinum [117]. NaHS also promoted persulfidation in the zinc finger motif domain of SIRT-3, which appears to be required to inhibit cis-platinum-induced kidney injury [117]. Given the abundance of studies favoring a protective effect of H_2S in cis-platinum associated AKI, drugs that release H_2S should be explored for their clinical use.

19.5.1.4 Glomerulonephritis-Associated AKI

Inflammatory injury to the kidney glomerulus results in increased excretion of urinary protein, red blood cells in the urine (hematuria), and can result in progressive loss of kidney function, e.g. IgA glomerulonephritis, lupus nephritis. Glomerular mesangial cells are frequently the target of inflammatory injury in the kidney. Genetic deficiency of CBS reduces H_2S generation and results in accumulation of L-homocysteine (hyperhomocysteinemia) characterized by kidney injury. Incubation of mouse mesangial cells *in vitro* with L-homocysteine augments inflammatory cytokine production and reduces H_2S generation; these changes are reversed by restoration of H_2S production by re-constituting CBS or CSE gene expression [20]. Platelet derived growth factor (PDGF) is one of the mediators of inflammatory glomerular injury. In vitro studies in mesangial cells have shown that PDGF BB augments CSE expression possibly by stimulating JAK-STAT pathway, promoting ROS generation, and activating Nrf-2; glomerular expression of CSE and Nrf-2 correlated with mesangial cell injury in rats with anti-Thy-1 glomerulonephritis, a commonly employed model of mesangial proliferative glomerulonephritis [3, 19]. In view of the ameliorative effects of H_2S on mesangial cell injury in hyperhomocysteinemia and high glucose incubation [118, 119], the authors suggest that induction of CSE may be an effort at defense against PDGF-mediated mesangial cell injury [19].

19.5.2 H_2S in AKI Due to Obstruction of the Genitourinary Tract

A variety of diseases of the ureters, urinary bladder, prostate, and urethra can cause obstruction to the flow of urine out of the kidney and back pressure on the kidneys. AKI in the short term and chronic kidney injury in the long term can be seen when kidneys are obstructed. In a mouse model of obstruction, renal CBS expression was decreased leading to reduced H_2S generation, stimulation of TGFβ expression, and kidney fibrosis [22]. These pathological changes could be reversed by the administration of NaHS, establishing the importance of H_2S deficiency as a mechanism of kidney injury [22]. TGFβ-induced epithelial to mesenchyme transition, a contributing mechanism in renal fibrosis, appears to involve β-catenin and extracellular-signaling-regulated kinase (Erk) signaling; these processes were abolished by H_2S [120, 121]. CSE deficient mice have been reported to develop more severe obstruction-induced kidney injury suggesting a protective role for that enzyme [122]. Renal fibrosis in the ureteral obstruction model is associated with hypoxia which augments methylation of klotho promoter resulting in decreased klotho expression [123]. This involves inhibition of ten eleven translocation, a DNA demethylase, which maintains the demethylated status of the klotho promoter. Administration of NaHS improves hypoxia, restores ten eleven translocation activity and demethylation of the klotho promoter, which leads to increase in klotho expression, and mitigation of kidney fibrosis [123]. The role of H_2S in obstruction-induced kidney injury has been reviewed in depth recently [102].

19.5.3 Injurious Role of H_2S in AKI

There are reports that assign a pathologic role for H_2S in AKI, e.g. AKI in the setting of multiple organ injury in burns was less severe in mice lacking CSE [124]. Similar protection against AKI due to endotoxemia has been reported in CBS+/− mice [92]. H_2S has been suggested to mediate injury in IRI [125]. These divergences among investigations on the role of H_2S in AKI in different models need to be addressed.

19.6 H_2S in Chronic Kidney Disease (CKD)

CKD is defined as a gradual loss of kidney function of a minimum of 3-month duration whatever the etiology [126]. Approximately 850 million people worldwide are estimated to have CKD [127]. Diabetes is a major cause of CKD in the West. CKD causes serious morbidity and mortality, and the prognosis is similar to or worse than many forms of cancer [128]. CKD occurs due to ongoing or repeated injury to the kidney, and is characterized by replacement of functioning kidney parenchyma with fibrosis with progressive loss of renal function that can culminate in end stage kidney disease (ESKD) needing renal replacement therapy. CKD is associated with dysfunction of a wide spectrum of non-renal organ systems including the cardiovascular system. In addition, the costs of renal replacement therapy are huge, placing a financial burden on the society. The currently available treatment strategies may reduce the rate of progression of CKD but they do not arrest it. Clearly, new approaches are needed to control CKD.

Renal fibrosis is one of the most commonly studied aspects of CKD. It consists of fibrosis of the glomeruli (glomerulosclerosis) and the interstitium the latter being associated with tubular atrophy (tubulointerstitial fibrosis). Kidney fibrosis is the end result of several injury pathways including hypertension, ongoing proteinuria, hemodynamic changes in the glomerular circulation, and activation of growth factors such as angiotensin II and TGFβ. Of late tubulointerstitial fibrosis is acknowledged as equally or more important than glomerulosclerosis in contributing to progressive loss of kidney function in CKD [77]. Fibrosis in the tubulointerstitial compartment involves matrix deposition by myofibroblasts, tubular atrophy, and capillary rarefaction [77]. Proximal tubular epithelial cells are of major importance in this process as they are subject to direct injury by the underlying disease process, e.g. diabetes, they are metabolically transformed, and they synthesize mediators of fibrosis including angiotensin II and TGFβ [129, 130]. CKD predisposes to vascular injury due to atherosclerosis, hypertension, and direct effects of the underlying disease; the end result is narrowing of the vascular lumen, loss of capillaries, and ischemic injury to the nephron. Recently, parallels have been drawn between injury pathways operative in aging and CKD and the latter has been termed a premature aging condition [131]. SASP has been implicated in renal fibrosis seen in CKD.

Given the versatile nature of H_2S affecting metabolism, hemodynamics, and protein synthesis, it has gained increasing attention as a regulator of processes involved in progressive kidney injury. Dialysis patients manifest lower levels of H_2S, CBS, and CSE [132]. Similar changes have been reported in circulating mononuclear cells in CKD subjects not on dialysis, with a correlation between H_2S levels and eGFR [133]. Serum homocysteine levels were increased in CKD patients with a negative correlation with eGFR [133], perhaps suggesting its underutilization as a substrate for the generation of H_2S [1]. Serum levels of lanthionine and homolanthionine, intermediaries in the trans-sulfuration pathway of H_2S synthesis (Figure 19.1), are increased in dialysis patients [134]. Lanthionine inhibits CBS activity by reducing glutothionylation leading to impaired H_2S production in endothelial cells [135]. CKD affecting transplanted kidney can culminate in graft failure. As reviewed above, CSE expression in the kidney at the time of its implantation directly correlates with its function in transplant patients [17]. Additionally, urinary thiosulfate excretion, a surrogate marker of H_2S generation and bioavailability, bears an inverse relationship with failure of transplanted kidney [136]. These data suggest that higher the capacity for H_2S generation better the preservation of function of the transplanted kidney although increased urinary sulfate excretion may be related to high protein intake and sulfate conjugation of molecules.

Significant insights on the role of H_2S in CKD have come from animal studies. In rodents with removal of 5/6 of the total kidney functioning mass the expression of CBS, CSE, and 3-MST was reduced along with increased NOX4 activity and ROS generation suggesting an anti-oxidant role for H_2S [137]. Low methionine diet, which augments H_2S generation [60], ameliorated oxidant stress and inflammation resulting in decreased kidney fibrosis and urinary protein losses [138]. Mice with reduced CBS expression and activity develop hyperhomocysteinemia (Hhcy); studies in these mice have provided direct evidence of participation of H_2S deficiency in CKD [139, 140]. In a feedback loop, Hhcy inhibits the expression of CSE, CBS, and 3-MST leading to H_2S deficiency [141]. Hhcy inhibition of CSE expression appears to involve hypermethylation of its promoter [142]. Kidney injury in Hhcy mice is accompanied by oxidative stress, inflammation, and mitophagy with clinical manifestations of kidney fibrosis, hypertension, proteinuria, and reduced clearance function; these are mitigated by NaHS administration [118, 143–146]. Additional mechanism of kidney injury in Hhcy involves homocysteinylation of endothelial NO synthase (eNOS) resulting in inhibition of its activity; this is also resolved following NaHS treatment [147].

Myofibroblasts in the kidney interstitium deposit extracellular matrix proteins resulting in fibrosis. Angiotensin II and TGFβ are among the growth factors that are involved in stimulating myofibroblasts. In obstructive uropathy, TGFβ has been implicated [22]. The lineage of myofibroblasts is a hotly debated topic with some advancing the view that they arise from epithelium to mesenchymal transition (EMT) of the proximal tubule epithelial cells while others have suggested perivascular pericytes as the cells of origin. H_2S inhibits EMT induced by angiotensin II-TGFβ axis [148]. Lineage tracing studies favor origin of myofibroblasts in vascular pericytes [149]; recent studies have estimated that less than 3% of myofibroblasts arise from EMT [129]. Given the salutary effect of H_2S on tubulointerstitial fibrosis, studies on its regulation of transformation of perivascular pericytes into myofibroblasts are needed.

Anemia is a serious complication of CKD. It is multifactorial in origin; the major mechanism being lack of erythropoietin, a red cell maturation factor mainly synthesized by stellate fibroblast cells in the corticomedullary junction area of the kidney interstitium [34, 35]. Hypoxia is a strong stimulus for erythropoietin synthesis by the recruitment of hypoxia inducible factors (HIFs) HIF1α and HIF2α [150, 151]. CBS knock out mice display anemia and iron overload associated with reduced erythropoietin synthesis in the liver; these abnormalities were reversed by restoring CBS expression using an adenoviral vector [152, 153]. CSE knock out mice are unable to augment erythropoietin synthesis in the face of hypoxia; this was corrected by the administration of H_2S suggesting that H_2S promotes the expression of erythropoietin [154]. However, $CSE^{-/-}$ mice develop polycythemia suggesting an inhibitory role for CSE in erythropoiesis [155]. Clearly, more studies are needed to define the role of H_2S and its synthesizing enzymes in erythropoietin production with focus on the kidney in CKD.

Calcification of blood vessels and deposition of calcium deposits in the kidney and the heart are encountered in CKD. Sodium thiosulfate administration reduced these deposits in rats with CKD although bone strength was impaired [156]. Strategies that employ H_2S donors to remove calcium deposits from these organs but do not impair bone integrity are needed. H_2S promotes bone repair and may be of therapeutic significance in osteoporosis [157].

Bone pathology is commonly encountered in CKD in association with abnormalities in calcium and phosphate metabolism. It results in significant morbidity including fractures. Serum phosphate levels rise even as kidney function is impaired in CKD leading to stimulation of fibroblast growth factor 23 (FGF23) secretion [158]. FGF23 inhibits the type 2 Na-dependent phosphate transporters in the kidney to augment urinary phosphate excretion and restoration of its serum levels [158]. Klotho serves as a co-factor for the kidney receptor for FGF23 [159, 160]. With advancing CKD, serum klotho levels decrease leading to impairment in phosphate excretion governed by FGF23 [161]. H_2S maintains demethylation of the klotho promoter by stimulating TET, a demethylase, and this results in increase in klotho expression [123] as reviewed in the section on AKI. It will be of considerable interest to explore whether H_2S can restore klotho levels in CKD and prevent CKD-associated bone disease.

Nearly every organ system including cardiovascular and neurological systems is affected in CKD. Cardiovascular diseases are the leading cause of death in CKD [162]. Neuropathy in CKD can be debilitating. All these areas in CKD await exploration of the regulatory role of H_2S.

19.6.1 H_2S in Obesity-Related CKD

With the advances in food production and delivery, obesity is emerging as a burgeoning problem, particularly in the West. It is estimated that nearly 40% of adults in the United States are obese with a projection of 50% prevalence by the year 2030 [163]. In addition to predisposing to type 2 diabetes [164], obesity injures the kidney in its own right. Manifestations of obesity-induced kidney injury include enlargement of the glomerulus, scarring of the glomerulus (glomerulosclerosis including focal and segmental glomerulosclerosis), tubulointerstitial fibrosis, and proteinuria [165]. High-fat diet (HFD)–induced obesity in mice is associated with reduced kidney expression of CBS and reduced H_2S generation [166]. Obesity promotes kidney injury due to inflammation and matrix expansion that are mitigated by NaHS administration [167].

Multiple pathways involving angiotensin II, adiponectin deficiency, oxidative stress, hemodynamics, lipotoxicity, and immune injury have been implicated in CKD due to obesity [168–171]. Obesity is associated with resistance to insulin-driven glucose uptake in peripheral tissues resulting in elevation of circulating insulin levels (hyperinsulinemia). Whether hyperinsulinemia activates IR in the kidney and drives pathologic events including H_2S deficiency is not known. We tested the hypothesis that IR activation by insulin in the proximal tubule of the kidney mediates kidney injury [172]. Using a sodium glucose transporter 2 promoter, we generated mice with selective knock out of IR in the kidney proximal tubule (KPTIRKO) [172]. HFD augmented serum insulin levels in both genotypes. Increase in matrix protein expression, albuminuria, and hypertension were seen in HFD fed control mice, which were significantly reduced in KPTIRKO mice. Signaling studies showed Akt and mTORC1 activation and reduced AMPK activity in the kidneys of control mice fed HFD; these events were prevented in KPTIRKO mice. Kidney expression of CBS and H_2S generation were reduced by HFD in control mice although CSE expression was unaffected. Interestingly, although the kidney expression of CBS was also reduced in KPTIRKO mice fed HFD, yet, H_2S generation was maintained at normal levels indicating a CBS-independent mechanism. Taken together, these data show that in mice fed HFD, hyperinsulinemia activates IR in the proximal tubules of the kidney, leading to reduced generation of H_2S, activation of Akt-mTORC1 pathway, and reduced AMPK activity, which lead to kidney injury (Figure 19.3). We have reported that H_2S protection against kidney cell injury induced by high glucose is dependent on AMPK stimulation [173, 174]. Similarly, failure of HFD to reduce H_2S generation and preservation of AMPK activity in KPTIRKO mice may be the mechanism by which they are protected against kidney injury. Studies are underway to understand mechanisms by which insulin activation of IR leads to reduced H_2S generation in HFD-induced obesity.

19.6.2 H_2S in Diabetic Kidney Disease (DKD)

Among the causes of CKD and ESKD, diabetes reigns supreme in the West. Approximately one third of patients with type 1 (insulin deficient) and type 2 (insulin resistant) diabetes develop significant CKD. Interestingly, in early stages of DKD, GFR is increased; however, GFR progressively falls with the development of kidney fibrosis and proteinuria, culminating in ESKD. Fibrotic lesions in the glomerulus encroach on the filtering surface of capillaries resulting in reduction in GFR. Tubulointerstitial fibrosis correlates with loss of GFR in DKD [77].

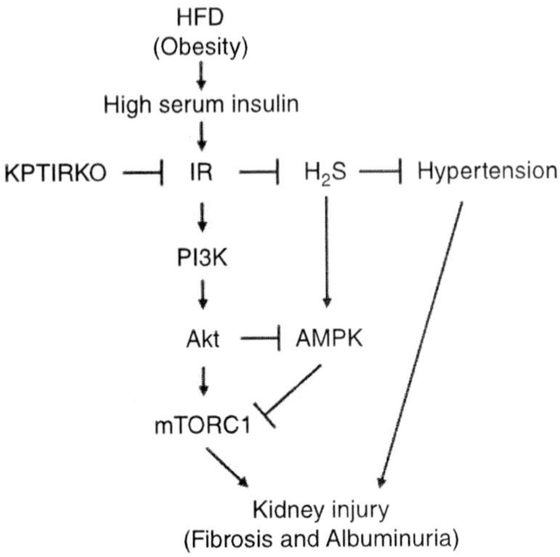

Figure 19.3 Hydrogen sulfide deficiency participates in obesity-associated kidney injury. High-fat diet (HFD)–induced obesity leads to high circulating insulin levels (hyperinsulinemia) and activation of insulin receptor (IR) in the proximal tubules of the kidney. This leads to deficiency of H_2S generation, activation of the canonical PI 3 kinase-Akt-mTOR pathway, and inhibition of AMPK activity. Together, these reactions contribute to H_2S deficiency, hypertension, and stimulation of synthesis of proteins including matrix proteins and albuminuria. Hypertension can contribute to kidney injury. These events are mitigated in kidney proximal tubule IR knock out (KPTIRKO) mice.

Detailed reviews of H_2S role in DKD have recently appeared [175–178]. H_2S has a complex role in diabetes. H_2S is thought to mediate injury to pancreatic β cells from streptozotocin, a commonly employed agent to induce experimental diabetes [179]. H_2S has also been reported to increase insulin-stimulated glucose uptake [180]. A majority of publications have posited that DKD is a state of H_2S deficiency and that administration of H_2S donors ameliorates kidney injury [119, 181–184]. Clinical studies have shown reduced circulating H_2S levels in diabetic subjects with CKD and in those on dialysis [185, 186]. Deficiency in kidney expression of CSE and CBS, under the regulation of matrix metalloproteinase 9 (MMP9), contributes to reduced H_2S generation in DKD in rodent models [18, 119, 181, 187, 188]. Recent investigations have suggested additional mechanisms for decreased expression of CBS in DKD including increased rate of degradation due to ubiquitination and reduced mRNA content [24]. In vitro studies in proximal tubular epithelial cells exposed to high glucose indicated that while proteasomal degradation contributed to reduced CBS expression in the short term, reduced mRNA content may be involved in the long term [24]. Further support for a pathologic role for H_2S deficiency in DKD comes from reports in which administration of H_2S donors has been shown to mitigate matrix accumulation and albuminuria in rodent models of DKD [24, 181, 184, 189].

Recent investigations have advanced our understanding of cellular mechanisms under regulation of H_2S in DKD:

Hemodynamic mechanisms underlying increased GFR in early stage of DKD involve increased renal plasma flow. Although H_2S has been reported to augment renal blood flow in normal rodents [32], its role in hemodynamic changes in early stage of DKD remains to be explored. As the disease progresses, vascular disease in afferent arterioles impairs autoregulation leading to increased blood pressure in the glomerulus; at this stage too, changes in H_2S generation need to be studied. Peritubular capillaries undergo destruction resulting in ischemic injury to the kidney in DKD, which is ameliorated by H_2S administration of [188, 190, 191]. The cellular mechanism in renal vascular injury in DKD appears to involve MMP9 activation of *N*-methyl D-aspartate receptor leading to Ca^{++} influx into the mitochondria, opening of the permeability transition pore, and ROS generation; these pathologic processes are suppressed by H_2S [187, 192]. Vascular remodeling involves increased expression of connexins 40 and 43 induced by *N*-methyl D-aspartate receptor activation [187]. Rectifying increased vascular resistance and decreased blood flow in the renal circulation by H_2S leads to improvement in fibrosis linking ischemia to the scarring process in the kidney [189].

Many *molecular mediators* participate in kidney hypertrophy and extracellular matrix expansion, the latter leading to fibrotic replacement of functioning kidney parenchyma. Angiotensin II is a major mediator of kidney cell injury in DKD [178], and the current clinical management of DKD depends on inhibition of its signaling pathway. Angiotensin II is produced locally in the nephron; for example, high glucose increases the expression of angiotensinogen and angiotensin converting enzyme and AT1, an angiotensin II receptor, in mesangial cells leading to augmented production of TGFβ and collagen [182]. As reviewed above, H_2S suppresses renin generation in the kidney [36], which can affect local production of angiotensin II in the nephron. TGFβ is another agent that mediates kidney injury in DKD [193, 194]. H_2S inhibits augmented production of TGFβ in the kidney in DKD [119, 181]. Other growth factors such as vascular endothelial growth factor [195–198], IGF1 [199], epidermal growth factor [200] also participate in kidney pathology in DKD; however, if they are under regulation by H_2S is not known.

H_2S regulates *signaling pathways* that control protein synthesis in kidney cells. Increased synthesis and decreased break down of matrix proteins such as type I and IV collagens, laminin, and fibronectin lead to matrix accumulation and fibrosis in the kidney in DKD. Signaling studies have shown that the canonical PI 3kinase-Akt-mTORC1 pathway is activated in the kidney in DKD [78, 81, 82, 201–203]. mTORC1 inhibition ameliorates kidney injury in diabetes in rodent models [81, 204]. mTORC1 regulates mRNA translation, a rate limiting step in protein synthesis [79]. Emerging studies have emphasized the role of cellular energy sensor AMPK which normally inhibits mTORC1 and keeps protein synthesis under check. In diabetes, AMPK activity is inhibited due to increase in ATP/AMP ratio [205]. Akt inhibits AMPK activity by phosphorylation of Ser485 linking PI 3 kinase-Akt axis to AMPK [206, 207]. Reduced AMPK activity leads to mTORC1 activation and augmented protein synthesis which contributes to kidney hypertrophy and matrix expansion; these changes are prevented by agents such as metformin and AICAR without affecting blood glucose levels [82, 208] AMPK is a crucial mediator of ameliorative effects of H_2S on kidney cell injury in obese mice [172], aging mice [58], and in the context of NOX4-mediated oxidative stress [174]. AMPK activation appears to mediate the ameliorative effects of H_2S in DKD also. In renal epithelial cells, high glucose-induced hypertrophy and matrix protein synthesis was abolished by H_2S by recruiting AMPK; mechanism appears to involve stimulation of calcium calmodulin-dependent kinase kinase β and possibly liver kinase B1, upstream enzymes that activate AMPK [18]. Metformin, a well-known activator of AMPK, is capable of stimulating H_2S generation [209]; whether this involves regulation at the level of expression of CSE, CBS, and/or 3-MST is not known and should be of great interest. The intolerable smell of H_2S makes it an unlikely therapeutic agent acceptable to humans. We examined whether tadalafil, a phosphodiesterase 5 inhibitor, whose effects are NO-dependent, could mimic H_2S effects on high glucose-induced kidney injury in podocytes [173]. High glucose-induced cell hypertrophy and increase in matrix protein expression were inhibited by tadalafil involving stimulation of AMPK and mTORC1 inhibition [173]. The salutary effects of tadalafil required H_2S generation by increased expression of CSE, the latter being regulated at the level of mRNA translation [173]. A clinical trial with a phosphodiesterase 5 inhibitor PF-00489791 has shown encouraging results with modest reduction in proteinuria in diabetic subjects [210].

Oxidative stress due to ROS is a widely recognized pathway of kidney injury in DKD [211, 212]. H_2S amelioration of kidney damage in DKD partly involves its antioxidant actions. H_2S facilitates the actions of Nrf-2, a transcription factor that augments antioxidant defense [181]. ROS-induced kidney injury involves many mediators. One such mediator is poly (ADP ribose) polymerase-1 (PARP1) which is involved in DNA strand repair. Kidney PARP1 expression is increased in DKD, and PARP1 knock out mice are protected from kidney injury in diabetes [213]. In diabetic Akita mice, GYY4137 reduced the expression of ROS, PARP-1, HIF1α, MMP-9, MMP-13, and mitigated vascular remodeling and matrix protein accumulation; interestingly, MMP-14 was reduced in diabetic mice that was rectified by H_2S administration [188]. These changes seem to be due to reduced expression of miR-194 that was corrected by GYY4137. In vitro data in glomerular endothelial cells indicated that high glucose-induced reduction in miR-194 was associated with increase in the expression of PARP1, MMPs-9, 13, and enzymes involved in remodeling of matrix collagen proteins [188]. It is important to determine the upstream mechanism that links H_2S deficiency to reduced miR-194 expression and whether the 3′UTRs in mRNA of these target proteins have binding sites for miR-194. Increased NOX4 in the kidneys of diabetic mice has been implicated in both fibrosis and proteinuria in diabetic rodents [214–217]. High glucose-induced increase in NOX4 expression in kidney proximal tubular epithelial cells was abolished by NaHS by stimulation of AMPK activity as well as by augmenting NO generation [174]. The source of ROS in DKD is controversial. Dugan et al showed that mitochondria were probably not the source of ROS in kidneys of diabetic mice [208] raising the possibility that ROS is generated at other sites in the kidney cells. Nitrosative stress also occurs in DKD and may be linked to reduced expression of CBS in the kidney [24].

Inflammatory injury with infiltration of macrophages into the kidney parenchyma is evident in DKD [218]. NFκB is an important signaling regulator of inflammation in the kidney, and it is inhibited by NaHS in diabetic rodents [181].

The translational significance of H_2S effect in preclinical models is becoming apparent from recent clinical studies. Higher eGFR and lower albuminuria showed an association with higher urinary sulfate level in a combined cohort of SUN-MACRO and PREVEND studies; multivariate analysis showed urinary sulfate level as an independent protective factor against kidney injury in DKD [219]. All-cause mortality rates in general population also appear to have inverse relationship with urinary sulfate level [220].

Novel therapeutic approaches are badly needed to manage DKD. We need deeper investigation to understand the role of H_2S in kidney injury in diabetes including its participation in hemodynamic changes, and potential effect on kidney cell injury caused by advanced glycation end products (AGE). High glucose-induced increase in methylglyoxal, a precursor of AGE, was abolished by a H_2S-releasing aspirin formulation resulting in amelioration of NOX4-mediated ROS generation in vascular smooth muscle cells [221]. AGE-induced increase in receptor for AGE expression and consequent oxidative stress was mitigated by NaHS in SH-SY5Y neuroblastoma cells [222]. H_2S effect on AGEs-induced kidney cell injury needs to be directly studied. H_2S has been reported to increase CO generation by stimulation of hemoxygenase 1 in mesangial cells [16]; greater understanding of H_2S interaction with NO and CO in the setting of DKD is another area ripe for exploration. From the studies to date H_2S appears to counteract nearly all major pathways of kidney injury in DKD. Properly designed prospective studies are needed to evaluate H_2S as a therapeutic target in patients with DKD by administering it in a form that is acceptable to human subjects.

19.6.3 H_2S in Congestive Heart Failure (CHF) Associated CKD

Congestive heart failure (CHF) is a state in which the left ventricular output into systemic circulation is impaired resulting in inadequate perfusion of tissues including the kidneys. Because kidneys receive nearly a quarter of the cardiac output, they are highly susceptible for dysfunction in acute or chronic CHF (type 1 and 2 cardiorenal syndrome, respectively) In type 3 and 4 cardiorenal syndromes, CHF occurs due to kidney diseases associated with AKI or CKD, respectively [223]. CHF results in maladaptive responses including activation of renin–angiotensin system and sympathetic nervous system; inhibition of these two pathways is a major therapeutic strategy in CHF. In an aortic constriction and pressure over load model of type 2 cardiorenal syndrome, JK1, a H_2S donor, inhibited activation of renin and sympathetic nervous system and mitigated cardiac fibrosis, resulting in improvement of cardiac and renal function [224]. This study provides a good rationale to further explore the therapeutic potential of H_2S in CHF.

19.7 H_2S and Preeclampsia

Preeclampsia consists of hypertension and proteinuria after 20th week of pregnancy without a previous history of CKD and hypertension; it affects about 2–5% of pregnant women [225, 226]. Placental ischemia due to remodeling of spiral arteries occurs and results in impaired fetal growth. Recent advances have identified contributing mechanisms as increased placental production of soluble fms-like tyrosine kinase (sFlt1), a vasoconstrictor and a splice variant of vascular endothelial growth factor receptor 1, and, reduced generation of placental growth factor, a vasodilator [227]. sFlt1 binds vascular endothelial growth factor and decreases its availability leading to endothelial dysfunction and vasoconstriction [226]. There is an increased predisposition for preeclampsia in subjects with single nucleotide polymorphism of CSE gene indicating a genetic susceptibility [228]. Reduced risk for preeclampsia has been reported to be associated with single nucleotide polymorphism of CBS gene [229]. These data support a possible association of genetic variations in H_2S generation and preeclampsia. In rodents, preeclampsia induced by adenoviral overproduction of sFlt1 is ameliorated by NaHS; inhibition of CSE promotes sFlt1 generation linking H_2S deficiency to a key event in preeclampsia [230, 231]. miR-133b expression is reduced in preeclamptic placentas; it is increased by H_2S to suppress the generation of sFlt1 in syncitiotrophoblasts [232]. Mitochondrial dysfunction contributes to preeclampsia; cytotrophoblasts from human preeclamptic placentae in culture respond to hypoxia by increasing mitochondrial ROS generation along with increased expression of sFlt1 and HIF1α, which are all mitigated by AP39, a mitochondrial targeted H_2S donor [233]. Kidney injury is not limited to short term in preeclampsia; in the long term, preeclamptic patients are at high risk for CKD [234]. Every effort should be made to explore the feasibility of H_2S donors as therapeutic agents in this disease.

19.8 H_2S and Genitourinary Cancers

Malignancies of the kidney, ureters, bladder, prostate, and urethra can remain undetected and present in advanced stages at the time of clinical diagnosis. For example, renal cell carcinoma (RCC) may present as a wide spread disease in nearly one third of the patients at the time of its diagnosis [235]. Among the histological types of RCC, the clear cell variety (ccRCC) is the most common. Persistent activation of HIF1α and HIF2α due to inactivation of von Hipple Lindau factor has been implicated in ccRCC. H_2S may have a complex role in renal cancer [236]. CBS expression in ccRCC has been correlated with invasion capability of the tumor [237]. CSE expression is low in RCC and has been linked to resistance to apoptosis which implies restoration of CSE may induce tumor cell death [238]. Additionally, the proangiogenic property of H_2S may facilitate tumor spread. More investigation is needed to define the precise role of H_2S in RCC and other genitourinary cancers.

19.9 Conclusion and Future Directions

It is highly likely that H_2S has been an integral component of animal physiology throughout organic evolution. Therefore, it is not surprising that it has such an extensive influence on basic cellular processes such as transcription, epigenetic control of transcription, mRNA translation, post-translational modification of proteins, affecting development, function, injury, repair, and malignant transformation of a wide variety of tissues including the kidney. As stated in different sections of the above review, there is much to be studied regarding the regulatory role of H_2S in kidney physiology and pathology. A number of pharmaceutical agents that release or inhibit H_2S have been developed and can aid in clinical evaluation [239, 240]. Even the scant data available on H_2S effects in the kidney have been so interesting; it is very likely that even more exciting discoveries will be made and these will aid in making intelligent therapeutic use of this primordial gas

Acknowledgments

We wish to express gratitude to students, physicians in training, post-doctoral fellows, and collaborators for their important contributions to the work done in our laboratory. We also acknowledge funding support from a variety of agencies such as the Veterans Administration, NIH, and private foundations over the years. We thank Prof. Goutam Ghosh Choudhury, University of Texas Health, San Antonio, for critical reading of the manuscript. The authors wish to thank their families for their forbearance and understanding over the years.

References

1 Kasinath, B.S., Feliers, D., and Lee, H.J. (2018). Hydrogen sulfide as a regulatory factor in kidney health and disease. *Biochem. Pharmacol.* 149: 29–41.
2 Ngowi, E.E., Sarfraz, M., Afzal, A. et al. (2020). Roles of hydrogen sulfide donors in common kidney diseases. *Front. Pharmacol.* 11: 564281.
3 Beck, K.F. and Pfeilschifter, J. (2021). Gasotransmitter synthesis and signalling in the renal glomerulus. Implications for glomerular diseases. *Cell. Signalling* 77: 109823.
4 Stipanuk, M.H. and Beck, P.W. (1982). Characterization of the enzymic capacity for cysteine desulphhydration in liver and kidney of the rat. *Biochem. J* 206 (2): 267–277.
5 Kimura, H. (2015). Signaling molecules: hydrogen sulfide and polysulfide. *Antioxid. Redox Signaling* 22 (5): 362–376.
6 Kabil, O., Vitvitsky, V., Xie, P., and Banerjee, R. (2011). The quantitative significance of the transsulfuration enzymes for H2S production in murine tissues. *Antioxid. Redox Signaling* 15 (2): 363–372.
7 Vicente, J.B., Malagrino, F., Arese, M. et al. (2016). Bioenergetic relevance of hydrogen sulfide and the interplay between gasotransmitters at human cystathionine beta-synthase. *Biochim. Biophys. Acta* 1857 (8): 1127–1138.
8 Shibuya, N., Koike, S., Tanaka, M. et al. (2013). A novel pathway for the production of hydrogen sulfide from D-cysteine in mammalian cells. *Nat. Commun.* 4: 1366.
9 Kolluru, G.K., Shen, X., Bir, S.C., and Kevil, C.G. (2013). Hydrogen sulfide chemical biology: pathophysiological roles and detection. *Nitric Oxide* 35: 5–20.
10 Czyzewski, B.K. and Wang, D.N. (2012). Identification and characterization of a bacterial hydrosulphide ion channel. *Nature* 483 (7390): 494–497.

11 Jennings, M.L. (2013). Transport of H2S and HS(-) across the human red blood cell membrane: rapid H2S diffusion and AE1-mediated Cl(-)/HS(-) exchange. *Am. J. Physiol. Cell Physiol.* 305 (9): C941–C950.

12 Jensen, B. and Fago, A. (2020). A novel possible role for met hemoglobin as carrier of hydrogen sulfide in the blood. *Antioxid. Redox Signaling* 32 (4): 258–265.

13 Levitt, M.D., Furne, J., Springfield, J. et al. (1999). Detoxification of hydrogen sulfide and methanethiol in the cecal mucosa. *J. Clin. Invest.* 104 (8): 1107–1114.

14 van den Berg, E., Pasch, A., Westendorp, W.H. et al. (2014). Urinary sulfur metabolites associate with a favorable cardiovascular risk profile and survival benefit in renal transplant recipients. *J. Am. Soc. Nephrol.* 25 (6): 1303–1312.

15 Beck, L. and Silve, C. (2001). Molecular aspects of renal tubular handling and regulation of inorganic sulfate. *Kidney Int.* 59 (3): 835–845.

16 van den Born, J.C., Frenay, A.S., Koning, A.M. et al. (2019). Urinary excretion of sulfur metabolites and risk of cardiovascular events and all-cause mortality in the general population. *Antioxid. Redox Signaling* 30 (17): 1999–2010.

17 Bos, E.M., Wang, R., Snijder, P.M. et al. (2013). Cystathionine gamma-lyase protects against renal ischemia/reperfusion by modulating oxidative stress. *J. Am. Soc. Nephrol.* 24 (5): 759–770.

18 Lee, H.J., Mariappan, M.M., Feliers, D. et al. (2012). Hydrogen sulfide inhibits high glucose-induced matrix protein synthesis by activating AMP-activated protein kinase in renal epithelial cells. *J. Biol. Chem.* 287 (7): 4451–4461.

19 Hassan, M.I., Boosen, M., Schaefer, L. et al. (2012). Platelet-derived growth factor-BB induces cystathionine gamma-lyase expression in rat mesangial cells via a redox-dependent mechanism. *Br. J. Pharmacol.* 166 (8): 2231–2242.

20 Sen, U., Givvimani, S., Abe, O.A. et al. (2011). Cystathionine beta-synthase and cystathionine gamma-lyase double gene transfer ameliorate homocysteine-mediated mesangial inflammation through hydrogen sulfide generation. *Am. J. Physiol. Cell Physiol.* 300 (1): C155–C163.

21 Kundu, S., Pushpakumar, S., Khundmiri, S.J., and Sen, U. (2014). Hydrogen sulfide mitigates hyperglycemic remodeling via liver kinase B1-adenosine monophosphate-activated protein kinase signaling. *Biochim. Biophys. Acta* 1843 (12): 2816–2826.

22 Song, K., Wang, F., Li, Q. et al. (2014). Hydrogen sulfide inhibits the renal fibrosis of obstructive nephropathy. *Kidney Int.* 85 (6): 1318–1329.

23 Yuan, X., Zhang, J., Xie, F. et al. (2017). Loss of the protein cystathionine beta-synthase during kidney injury promotes renal tubulointerstitial fibrosis. *Kidney Blood Press Res.* 42 (3): 428–443.

24 Yu, Y., Xiao, L., Ren, Z. et al. (2021). Glucose-induced decrease of cystathionine beta-synthase mediates renal injuries. *FASEB J.* 35 (5): e21576.

25 Filipovic, M.R., Zivanovic, J., Alvarez, B., and Banerjee, R. (2018). Chemical biology of H2S signaling through persulfidation. *Chem. Rev.* 118 (3): 1253–1337.

26 Shen, X., Peter, E.A., Bir, S. et al. (2012). Analytical measurement of discrete hydrogen sulfide pools in biological specimens. *Free Radical Biol. Med.* 52 (11-12): 2276–2283.

27 Predmore, B.L. and Lefer, D.J. (2011). Hydrogen sulfide-mediated myocardial pre- and post-conditioning. *Expert Rev. Clin. Pharmacol.* 4 (1): 83–96.

28 Yuan, S., Shen, X., and Kevil, C.G. (2017). Beyond a gasotransmitter: hydrogen sulfide and polysulfide in cardiovascular health and immune response. *Antioxid. Redox Signaling* 27 (10): 634–653.

29 Shoji, T., Hayashi, M., Sumi, C. et al. (2019). Pharmacological polysulfide suppresses glucose-stimulated insulin secretion in an ATP-sensitive potassium channel-dependent manner. *Sci. Rep.* 9 (1): 19377.

30 Cao, X. and Bian, J.S. (2016). The role of hydrogen sulfide in renal system. *Front. Pharmacol.* 7: 385.

31 Pollak, M.R., Quaggin, S.E., Hoenig, M.P., and Dworkin, L.D. (2014). The glomerulus: the sphere of influence. *Clin. J. Am. Soc. Nephrol.* 9 (8): 1461–1469.

32 Xia, M., Chen, L., Muh, R.W. et al. (2009). Production and actions of hydrogen sulfide, a novel gaseous bioactive substance, in the kidneys. *J. Pharmacol. Exp. Ther.* 329 (3): 1056–1062.

33 Pushpakumar, S., Kundu, S., Weber, G., and Sen, U. (2021). Exogenous hydrogen sulfide and miR-21 antagonism attenuates macrophage-mediated inflammation in ischemia reperfusion injury of the aged kidney. *Geroscience* 43 (3): 1349–1367.

34 Zeisberg, M. and Kalluri, R. (2015). Physiology of the renal interstitium. *Clin. J. Am. Soc. Nephrol.* 10 (10): 1831–1840.

35 Kurtz, A. (2017). Endocrine functions of the renal interstitium. *Pflugers Archiv.* 469 (7-8): 869–876.

36 Lu, M., Liu, Y.H., Goh, H.S. et al. (2010). Hydrogen sulfide inhibits plasma renin activity. *J. Am. Soc. Nephrol.* 21 (6): 993–1002.

37 Park, J., Shrestha, R., Qiu, C. et al. (2018). Single-cell transcriptomics of the mouse kidney reveals potential cellular targets of kidney disease. *Science* 360 (6390): 758–763.

38 Curthoys, N.P. and Moe, O.W. (2014). Proximal tubule function and response to acidosis. *Clin. J. Am. Soc. Nephrol.* 9 (9): 1627–1638.

39 Clausen, M.V., Hilbers, F., and Poulsen, H. (2017). The structure and function of the Na,K-ATPase isoforms in health and disease. *Front. Physiol.* 8: 371.

40 Hoenig, M.P. and Zeidel, M.L. (2014). Homeostasis, the milieu interieur, and the wisdom of the nephron. *Clin. J. Am. Soc. Nephrol.* 9 (7): 1272–1281.

41 Ge, S.N., Zhao, M.M., Wu, D.D. et al. (2014). Hydrogen sulfide targets EGFR Cys797/Cys798 residues to induce Na(+)/K(+)-ATPase endocytosis and inhibition in renal tubular epithelial cells and increase sodium excretion in chronic salt-loaded rats. *Antioxid. Redox Signaling* 21 (15): 2061–2082.

42 Pearce, D., Soundararajan, R., Trimpert, C. et al. (2015). Collecting duct principal cell transport processes and their regulation. *Clin. J. Am. Soc. Nephrol.* 10 (1): 135–146.

43 Yu, L., Bao, H.F., Self, J.L. et al. (2007). Aldosterone-induced increases in superoxide production counters nitric oxide inhibition of epithelial Na channel activity in A6 distal nephron cells. *Am. J. Physiol. Renal. Physiol.* 293 (5): F1666–F1677.

44 Zhang, J., Chen, S., Liu, H. et al. (2013). Hydrogen sulfide prevents hydrogen peroxide-induced activation of epithelial sodium channel through a PTEN/PI(3,4,5)P3 dependent pathway. *PLoS One* 8 (5): e64304.

45 Knepper, M.A., Kwon, T.H., and Nielsen, S. (2015). Molecular physiology of water balance. *N. Engl. J. Med.* 373 (2): 196.

46 Luo, R., Hu, S., Liu, Q. et al. (2019). Hydrogen sulfide upregulates renal AQP-2 protein expression and promotes urine concentration. *FASEB J.* 33 (1): 469–483.

47 Yuan, G., Vasavda, C., Peng, Y.J. et al. (2015). Protein kinase G-regulated production of H2S governs oxygen sensing. *Sci. Signal.* 8 (373): ra37.

48 Olson, K.R., Dombkowski, R.A., Russell, M.J. et al. (2006). Hydrogen sulfide as an oxygen sensor/transducer in vertebrate hypoxic vasoconstriction and hypoxic vasodilation. *J. Exp. Biol.* 209 (Pt 20): 4011–4023.

49 Dombkowski, R.A., Naylor, M.G., Shoemaker, E. et al. (2011). Hydrogen sulfide (H(2)S) and hypoxia inhibit salmonid gastrointestinal motility: evidence for H(2)S as an oxygen sensor. *J. Exp. Biol.* 214 (Pt 23): 4030–4040.

50 Sokolov, A.S., Nekrasov, P.V., Shaposhnikov, M.V., and Moskalev, A.A. (2021). Hydrogen sulfide in longevity and pathologies: inconsistency is malodorous. *Ageing Res. Rev.* 67: 101262.

51 Hodes, R.J., Sierra, F., Austad, S.N. et al. (2016). Disease drivers of aging. *Ann. N.Y. Acad. Sci.* 1386 (1): 45–68.

52 Lindeman, R.D., Tobin, J., and Shock, N.W. (1985). Longitudinal studies on the rate of decline in renal function with age. *J. Am. Geriatr. Soc.* 33 (4): 278–285.

53 Eriksen, B.O., Palsson, R., Ebert, N. et al. (2020). GFR in healthy aging: an individual participant data meta-analysis of iohexol clearance in european population-based cohorts. *J. Am. Soc. Nephrol.* 31 (7): 1602–1615.

54 Sataranatarajan, K., Feliers, D., Mariappan, M.M. et al. (2012). Molecular events in matrix protein metabolism in the aging kidney. *Aging Cell.* 11 (6): 1065–1073.

55 Li, P., Gupta, S., Mothi, S.S. et al. (2021). Histopathologic correlates of kidney function: insights from nephrectomy specimens. *Am. J. Kidney Dis.* 77 (3): 336–345.

56 Denic, A., Glassock, R.J., and Rule, A.D. (2021). Kidney histology, kidney function, and age. *Am. J. Kidney Dis.* 77 (3): 312–314.

57 Denic, A., Lieske, J.C., Chakkera, H.A. et al. (2017). The substantial loss of nephrons in healthy human kidneys with aging. *J. Am. Soc. Nephrol.* 28 (1): 313–320.

58 Lee, H.J., Feliers, D., Barnes, J.L. et al. (2018). Hydrogen sulfide ameliorates aging-associated changes in the kidney. *Geroscience* 40 (2): 163–176.

59 Lee, H.J., Gonzalez, O., Dick, E.J. et al. (2019). Marmoset as a model to study kidney changes associated with aging. *J. Gerontol. A Biol. Sci. Med. Sci.* 74 (3): 315–324.

60 Hine, C., Harputlugil, E., Zhang, Y. et al. (2015). Endogenous hydrogen sulfide production is essential for dietary restriction benefits. *Cell* 160 (1-2, 132): –44.

61 Yoshida, S., Yamahara, K., Kume, S. et al. (2018). Role of dietary amino acid balance in diet restriction-mediated lifespan extension, renoprotection, and muscle weakness in aged mice. *Aging Cell* 17 (4): e12796.

62 Komninou, D., Malloy, V.L., Zimmerman, J.A. et al. (2020). Methionine restriction delays aging-related urogenital diseases in male Fischer 344 rats. *Geroscience* 42 (1): 287–297.

63 Das, A., Huang, G.X., Bonkowski, M.S. et al. (2018). Impairment of an endothelial NAD(+)-H2S signaling network is a reversible cause of vascular aging. *Cell* 173 (1): 74–89 e20.

64 Choudhury, D. and Levi, M. (2011). Kidney aging--inevitable or preventable? *Nat. Rev. Nephrol.* 7 (12): 706–717.

65 Bolignano, D., Mattace-Raso, F., Sijbrands, E.J., and Zoccali, C. (2014). The aging kidney revisited: a systematic review. *Ageing Res. Rev.* 14: 65–80.

66 Zhuo, L., Cai, G., Liu, F. et al. (2009). Expression and mechanism of mammalian target of rapamycin in age-related renal cell senescence and organ aging. *Mech. Ageing Dev.* 130 (10): 700–708.

67 Zoncu, R., Efeyan, A., and Sabatini, D.M. (2011). mTOR: from growth signal integration to cancer, diabetes and ageing. *Nat. Rev. Mol. Cell Biol.* 12 (1): 21–35.

68 Kume, S., Uzu, T., Horiike, K. et al. (2010). Calorie restriction enhances cell adaptation to hypoxia through Sirt1-dependent mitochondrial autophagy in mouse aged kidney. *J. Clin. Invest.* 120 (4): 1043–1055.

69 Huang, Q., Ning, Y., Liu, D. et al. (2018). A young blood environment decreases aging of senile mice kidneys. *J. Gerontol. A Biol. Sci. Med. Sci.* 73 (4): 421–428.

70 Lee, H.J., Donati, A., Feliers, D. et al. (2021). Chloride channel accessory 1 integrates chloride channel activity and mTORC1 in aging-related kidney injury. *Aging Cell.* 20 (7): e13407.

71 Sala-Rabanal, M., Yurtsever, Z., Nichols, C.G., and Brett, T.J. (2015). Secreted CLCA1 modulates TMEM16A to activate Ca(2+)-dependent chloride currents in human cells. *eLife* 4: e05875.

72 Sala-Rabanal, M., Yurtsever, Z., Berry, K.N. et al. (2017). Modulation of TMEM16A channel activity by the von Willebrand factor type A (VWA) domain of the calcium-activated chloride channel regulator 1 (CLCA1). *J. Biol. Chem.* 292 (22): 9164–9174.

73 Hou, C.L., Wang, M.J., Sun, C. et al. (2016). Protective effects of hydrogen sulfide in the ageing kidney. *Oxid. Med. Cell Longev.* 2016: 7570489.

74 Kasinath, B.S. and Feliers, D. (2011). The complex world of kidney microRNAs. *Kidney Int.* 80 (4): 334–337.

75 Bai, X.Y., Ma, Y., Ding, R. et al. (2011). miR-335 and miR-34a promote renal senescence by suppressing mitochondrial antioxidative enzymes. *J. Am. Soc. Nephrol.* 22 (7): 1252–1261.

76 Wu, W., Hou, C.L., Mu, X.P. et al. (2017). H2S donor NaHS changes the production of endogenous H2S and NO in D-galactose-induced accelerated ageing. *Oxid. Med. Cell Longev.* 2017: 5707830.

77 Humphreys, B.D. (2018). Mechanisms of renal fibrosis. *Annu. Rev. Physiol.* 80: 309–326.

78 Feliers, D., Duraisamy, S., Faulkner, J.L. et al. (2001). Activation of renal signaling pathways in db/db mice with type 2 diabetes. *Kidney Int.* 60 (2): 495–504.

79 Kasinath, B.S., Feliers, D., Sataranatarajan, K. et al. (2009). Regulation of mRNA translation in renal physiology and disease. *Am. J. Physiol. Renal. Physiol.* 297 (5): F1153–F1165.

80 Mariappan, M.M., Feliers, D., Mummidi, S. et al. (2007). High glucose, high insulin, and their combination rapidly induce laminin-beta1 synthesis by regulation of mRNA translation in renal epithelial cells. *Diabetes* 56 (2): 476–485.

81 Sataranatarajan, K., Mariappan, M.M., Lee, M.J. et al. (2007). Regulation of elongation phase of mRNA translation in diabetic nephropathy: amelioration by rapamycin. *Am. J. Pathol.* 171 (6): 1733–1742.

82 Lee, M.J., Feliers, D., Mariappan, M.M. et al. (2007). A role for AMP-activated protein kinase in diabetes-induced renal hypertrophy. *Am. J. Physiol. Renal. Physiol.* 292 (2): F617–F627.

83 Acosta, J.C., Banito, A., Wuestefeld, T. et al. (2013). A complex secretory program orchestrated by the inflammasome controls paracrine senescence. *Nat. Cell Biol.* 15 (8): 978–990.

84 Ross, C.N., Adams, J., Gonzalez, O. et al. (2019). Cross-sectional comparison of health-span phenotypes in young versus geriatric marmosets. *Am. J. Primatol.* 81 (2): e22952.

85 Kuro-o, M. (2013). Klotho, phosphate and FGF-23 in ageing and disturbed mineral metabolism. *Nat. Rev. Nephrol.* 9 (11): 650–660.

86 Kurosu, H., Yamamoto, M., Clark, J.D. et al. (2005). Suppression of aging in mice by the hormone Klotho. *Science* 309 (5742): 1829–1833.

87 Drew, D.A., Katz, R., Kritchevsky, S. et al. (2017). Association between soluble klotho and change in kidney function: the health aging and body composition study. *J. Am. Soc. Nephrol.* 28 (6): 1859–1866.

88 Mitani, H., Ishizaka, N., Aizawa, T. et al. (2002). In vivo klotho gene transfer ameliorates angiotensin II-induced renal damage. *Hypertension* 39 (4): 838–843.

89 Tripatara, P., Patel, N.S., Collino, M. et al. (2008). Generation of endogenous hydrogen sulfide by cystathionine gamma-lyase limits renal ischemia/reperfusion injury and dysfunction. *Lab. Invest.* 88 (10): 1038–1048.

90 Bos, E.M., Leuvenink, H.G., Snijder, P.M. et al. (2009). Hydrogen sulfide-induced hypometabolism prevents renal ischemia/reperfusion injury. *J. Am. Soc. Nephrol.* 20 (9): 1901–1905.

91 Lobb, I., Davison, M., Carter, D. et al. (2015). Hydrogen sulfide treatment mitigates renal allograft ischemia-reperfusion injury during cold storage and improves early transplant kidney function and survival following allogeneic renal transplantation. *J. Urol.* 194 (6): 1806–1815.

92 Ahmad, A., Gero, D., Olah, G., and Szabo, C. (2016). Effect of endotoxemia in mice genetically deficient in cystathionine-gamma-lyase, cystathionine-beta-synthase or 3-mercaptopyruvate sulfurtransferase. *Int. J. Mol. Med.* 38 (6): 1683–1692.

93 Coca, S.G. (2010). Acute kidney injury in elderly persons. *Am. J. Kidney Dis.* 56 (1): 122–131.

94 Dugbartey, G.J., Talaei, F., Houwertjes, M.C. et al. (2015). Dopamine treatment attenuates acute kidney injury in a rat model of deep hypothermia and rewarming – The role of renal H2S-producing enzymes. *Eur. J. Pharmacol.* 769: 225–233.

95 Dugbartey, G.J., Hardenberg, M.C., Kok, W.F. et al. (2017). Renal mitochondrial response to low temperature in non-hibernating and hibernating species. *Antioxid. Redox Signaling* 27 (9): 599–617.

96 Teksen, Y., Kadioglu, E., Kocak, C., and Kocak, H. (2019). Effect of hydrogen sulfide on kidney injury in rat model of crush syndrome. *J. Surg. Res.* 235: 470–478.

97 Otunctemur, A., Ozbek, E., Dursun, M. et al. (2014). Protective effect of hydrogen sulfide on gentamicin-induced renal injury. *Ren. Fail.* 36 (6): 925–931.

98 Pedraza-Chaverri, J., Gonzalez-Orozco, A.E., Maldonado, P.D. et al. (2003). Diallyl disulfide ameliorates gentamicin-induced oxidative stress and nephropathy in rats. *Eur. J. Pharmacol.* 473 (1): 71–78.

99 Pedraza-Chaverri, J., Maldonado, P.D., Medina-Campos, O.N. et al. (2000). Garlic ameliorates gentamicin nephrotoxicity: relation to antioxidant enzymes. *Free Radical Biol. Med.* 29 (7): 602–611.

100 Aziz, N.M., Elbassuoni, E.A., Kamel, M.Y., and Ahmed, S.M. (2020). Hydrogen sulfide renal protective effects: possible link between hydrogen sulfide and endogenous carbon monoxide in a rat model of renal injury. *Cell Stress Chaperones* 25 (2): 211–221.

101 Manohar, S. and Leung, N. (2018). Cisplatin nephrotoxicity: a review of the literature. *J. Nephrol.* 31 (1): 15–25.

102 Lin, S., Juriasingani, S., and Sener, A. (2018). Is hydrogen sulfide a potential novel therapy to prevent renal damage during ureteral obstruction? *Nitric Oxide* 73: 15–21.

103 Cao, X., Zhang, W., Moore, P.K., and Bian, J. (2019). Protective smell of hydrogen sulfide and polysulfide in cisplatin-induced nephrotoxicity. *Int. J. Mol. Sci.* 20 (2): 313.

104 Della Coletta Francescato, H., Cunha, F.Q., Costa, R.S. et al. (2011). Inhibition of hydrogen sulphide formation reduces cisplatin-induced renal damage. *Nephrol. Dial. Transplant.* 26 (2): 479–488.

105 Liu, M., Jia, Z., Sun, Y. et al. (2016). A H_2S donor GYY4137 exacerbates cisplatin-induced nephrotoxicity in mice. *Mediators Inflamm.* 2016: 8145785.

106 Cao, X., Xiong, S., Zhou, Y. et al. (2018). Renal protective effect of hydrogen sulfide in cisplatin-induced nephrotoxicity. *Antioxid. Redox Signaling* 29 (5): 455–470.

107 Asimakopoulou, A., Panopoulos, P., Chasapis, C.T. et al. (2013). Selectivity of commonly used pharmacological inhibitors for cystathionine beta synthase (CBS) and cystathionine gamma lyase (CSE). *Br. J. Pharmacol.* 169 (4): 922–932.

108 Dugbartey, G.J., Bouma, H.R., Lobb, I., and Sener, A. (2016). Hydrogen sulfide: A novel nephroprotectant against cisplatin-induced renal toxicity. *Nitric Oxide* 57: 15–20.

109 Cao, X., Nie, X., Xiong, S. et al. (2018). Renal protective effect of polysulfide in cisplatin-induced nephrotoxicity. *Redox Biol.* 15: 513–521.

110 Ramesh, G. and Reeves, W.B. (2002). TNF alpha mediates chemokine and cytokine expression and renal injury in cisplatin nephrotoxicity. *J. Clin. Invest.* 110 (6): 835–842.

111 Abdollahzade Fard, A., Ahangarpour, A., Gharibnaseri, M.K. et al. (2013). Effects of hydrogen sulfide on oxidative stress, TNF-α level and kidney histological changes in cisplatin nephrotoxicity in rat. *J. Phys. Pharm. Adv.* 3 (3): 57–65.

112 Chiarandini Fiore, J.P., Fanelli, S.L., de Ferreyra, E.C., and Castro, J.A. (2008). Diallyl disulfide prevention of cis-diamminedichloroplatinum-induced nephrotoxicity and leukopenia in rats: potential adjuvant effects. *Nutr. Cancer* 60 (6): 784–791.

113 Elkhoely, A. and Kamel, R. (2018). Diallyl sulfide alleviates cisplatin-induced nephrotoxicity in rats via suppressing NF-kappaB downstream inflammatory proteins and p53/Puma signalling pathway. *Clin. Exp. Pharmacol. Physiol.* 45 (6): 591–601.

114 Sun, H.J., Leng, B., Wu, Z.Y., and Bian, J.S. (2020). Polysulfide and hydrogen sulfide ameliorate cisplatin-induced nephrotoxicity and renal inflammation through persulfidating STAT3 and IKKbeta. *Int. J. Mol. Sci.* 21 (20): 7805.

115 Samant, S.A., Zhang, H.J., Hong, Z. et al. (2014). SIRT3 deacetylates and activates OPA1 to regulate mitochondrial dynamics during stress. *Mol. Cell. Biol.* 34 (5): 807–819.

116 Perico, L., Morigi, M., Rota, C. et al. (2017). Human mesenchymal stromal cells transplanted into mice stimulate renal tubular cells and enhance mitochondrial function. *Nat. Commun.* 8 (1): 983.

117 Yuan, Y., Zhu, L., Li, L. et al. (2019). S-sulfhydration of SIRT3 by hydrogen sulfide attenuates mitochondrial dysfunction in cisplatin-induced acute kidney injury. *Antioxid. Redox Signaling* 31 (17): 1302–1319.

118 Sen, U., Sathnur, P.B., Kundu, S. et al. (2012). Increased endogenous H2S generation by CBS, CSE, and 3MST gene therapy improves ex vivo renovascular relaxation in hyperhomocysteinemia. *Am. J. Physiol. Cell Physiol.* 303 (1): C41–C51.

119 Yuan, P., Xue, H., Zhou, L. et al. (2011). Rescue of mesangial cells from high glucose-induced over-proliferation and extracellular matrix secretion by hydrogen sulfide. *Nephrol. Dial Transplant.* 26 (7): 2119–2126.

120 Lin, S., Visram, F., Liu, W. et al. (2016). GYY4137, a slow-releasing hydrogen sulfide donor, ameliorates renal damage associated with chronic obstructive uropathy. *J. Urol.* 196 (6): 1778–1787.

121 Guo, L., Peng, W., Tao, J. et al. (2016). Hydrogen sulfide inhibits transforming growth factor-beta1-induced EMT via Wnt/catenin pathway. *PLoS One* 11 (1): e0147018.

122 Han, S.J., Noh, M.R., Jung, J.M. et al. (2017). Hydrogen sulfide-producing cystathionine gamma-lyase is critical in the progression of kidney fibrosis. *Free Radical Biol. Med.* 112: 423–432.

123 Gu, Y., Chen, J., Zhang, H. et al. (2020). Hydrogen sulfide attenuates renal fibrosis by inducing TET-dependent DNA demethylation on Klotho promoter. *FASEB J.* 34 (9): 11474–11487.

124 Ahmad, A., Druzhyna, N., and Szabo, C. (2017). Cystathionine-gamma-lyase deficient mice are protected against the development of multiorgan failure and exhibit reduced inflammatory response during burn. *Burns* 43 (5): 1021–1033.

125 Marko, L., Szijarto, I.A., Filipovic, M.R. et al. (2016). Role of cystathionine gamma-lyase in immediate renal impairment and inflammatory response in acute ischemic kidney injury. *Sci. Rep.* 6: 27517.

126 Webster, A.C., Nagler, E.V., Morton, R.L., and Masson, P. (2017). Chronic kidney disease. *Lancet* 389 (10075): 1238–1252.

127 Levin, A., Tonelli, M., Bonventre, J. et al. (2017). Global kidney health 2017 and beyond: a roadmap for closing gaps in care, research, and policy. *Lancet* 390 (10105): 1888–1917.

128 Afkarian, M., Sachs, M.C., Kestenbaum, B. et al. (2013). Kidney disease and increased mortality risk in type 2 diabetes. *J. Am. Soc. Nephrol.* 24 (2): 302–308.

129 Wu, H., Lai, C.F., Chang-Panesso, M., and Humphreys, B.D. (2020). Proximal tubule translational profiling during kidney fibrosis reveals proinflammatory and long noncoding RNA expression patterns with sexual dimorphism. *J. Am. Soc. Nephrol.* 31 (1): 23–38.

130 Wu, H., Uchimura, K., Donnelly, E.L. et al. (2018). Comparative analysis and refinement of human PSC-derived kidney organoid differentiation with single-cell transcriptomics. *Cell Stem Cell* 23 (6): 869–881 e8.

131 Xu, J., Zhou, L., and Liu, Y. (2020). Cellular senescence in kidney fibrosis: pathologic significance and therapeutic strategies. *Front. Pharmacol.* 11: 601325.

132 Perna, A.F., Luciano, M.G., Ingrosso, D. et al. (2009). Hydrogen sulphide-generating pathways in haemodialysis patients: a study on relevant metabolites and transcriptional regulation of genes encoding for key enzymes. *Nephrol. Dial Transplant.* 24 (12): 3756–3763.

133 Kuang, Q., Xue, N., Chen, J. et al. (2018). Low plasma hydrogen sulfide is associated with impaired renal function and cardiac dysfunction. *Am. J. Nephrol.* 47 (5): 361–371.

134 Perna, A.F., Di Nunzio, A., Amoresano, A. et al. (2016). Divergent behavior of hydrogen sulfide pools and of the sulfur metabolite lanthionine, a novel uremic toxin, in dialysis patients. *Biochimie* 126: 97–107.

135 Vigorito, C., Anishchenko, E., Mele, L. et al. (2019). Uremic toxin lanthionine interferes with the transsulfuration pathway, angiogenetic signaling and increases intracellular calcium. *Int. J. Mol. Sci.* 20 (9): 2269.

136 Said, M.Y., Post, A., Minovic, I. et al. (2020). Urinary sulfate excretion and risk of late graft failure in renal transplant recipients – a prospective cohort study. *Transpl. Int.* 33 (7): 752–761.

137 Aminzadeh, M.A. and Vaziri, N.D. (2012). Downregulation of the renal and hepatic hydrogen sulfide (H2S)-producing enzymes and capacity in chronic kidney disease. *Nephrol. Dial. Transplant.* 27 (2): 498–504.

138 Cooke, D., Ouattara, A., and Ables, G.P. (2018). Dietary methionine restriction modulates renal response and attenuates kidney injury in mice. *FASEB J.* 32 (2): 693–702.

139 Miller, J.W., Nadeau, M.R., Smith, D., and Selhub, J. (1994). Vitamin B-6 deficiency vs folate deficiency: comparison of responses to methionine loading in rats. *Am. J. Clin. Nutr.* 59 (5): 1033–1039.

140 Jager, A., Kostense, P.J., Nijpels, G. et al. (2001). Serum homocysteine levels are associated with the development of (micro)albuminuria: the Hoorn study. *Arterioscler. Thromb. Vasc. Biol.* 21 (1): 74–81.

141 Wang, Y., Shi, S., Dong, S. et al. (2015). Sodium hydrosulfide attenuates hyperhomocysteinemia rat myocardial injury through cardiac mitochondrial protection. *Mol. Cell. Biochem.* 399 (1-2): 189–200.

142 Li, J.J., Li, Q., Du, H.P. et al. (2015). Homocysteine triggers inflammatory responses in macrophages through inhibiting CSE-H2S signaling via DNA hypermethylation of CSE promoter. *Int. J. Mol. Sci.* 16 (6): 12560–12577.

143 Kamat, P.K., Kalani, A., Givvimani, S. et al. (2013). Hydrogen sulfide attenuates neurodegeneration and neurovascular dysfunction induced by intracerebral-administered homocysteine in mice. *Neuroscience* 252: 302–319.

144 Pushpakumar, S., Kundu, S., and Sen, U. (2014). Endothelial dysfunction: the link between homocysteine and hydrogen sulfide. *Curr. Med. Chem.* 21 (32): 3662–3672.

145 Sen, U., Basu, P., Abe, O.A. et al. (2009). Hydrogen sulfide ameliorates hyperhomocysteinemia-associated chronic renal failure. *Am. J. Physiol. Renal. Physiol.* 297 (2): F410–F419.

146 Sen, U., Munjal, C., Qipshidze, N. et al. (2010). Hydrogen sulfide regulates homocysteine-mediated glomerulosclerosis. *Am. J. Nephrol.* 31 (5): 442–455.

147 Pushpakumar, S., Kundu, S., and Sen, U. (2019). Hydrogen sulfide protects hyperhomocysteinemia-induced renal damage by modulation of caveolin and eNOS interaction. *Sci. Rep.* 9 (1): 2223.

148 Huang, Y., Zhang, Z., Huang, Y. et al. (2018). Induction of inactive TGF-beta1 monomer formation by hydrogen sulfide contributes to its suppressive effects on Ang II- and TGF-beta1-induced EMT in renal tubular epithelial cells. *Biochem. Biophys. Res. Commun.* 501 (2): 534–540.

149 Humphreys, B.D., Lin, S.L., Kobayashi, A. et al. (2010). Fate tracing reveals the pericyte and not epithelial origin of myofibroblasts in kidney fibrosis. *Am. J. Pathol.* 176 (1): 85–97.

150 Leigh, J., Saha, M.N., Mok, A. et al. (2016). Hydrogen sulfide induced erythropoietin synthesis is regulated by HIF proteins. *J. Urol.* 196 (1): 251–260.

151 Warnecke, C., Zaborowska, Z., Kurreck, J. et al. (2004). Differentiating the functional role of hypoxia-inducible factor (HIF)-1alpha and HIF-2alpha (EPAS-1) by the use of RNA interference: erythropoietin is a HIF-2alpha target gene in Hep3B and Kelly cells. *FASEB J.* 18 (12): 1462–1464.

152 Zhou, Y.F., Wu, X.M., Zhou, G. et al. (2018). Cystathionine beta-synthase is required for body iron homeostasis. *Hepatology* 67 (1): 21–35.

153 Zhao, P., Qian, C., Chen, Y.J. et al. (2019). Cystathionine beta-synthase (CBS) deficiency suppresses erythropoiesis by disrupting expression of heme biosynthetic enzymes and transporter. *Cell Death Dis.* 10 (10): 708.

154 Leigh, J., Juriasingani, S., Akbari, M. et al. (2018). Endogenous H2S production deficiencies lead to impaired renal erythropoietin production. *Can. Urol. Assoc. J.* 13 (7): E210–E219.

155 Modis, K., Ramanujam, V.S., Govar, A.A. et al. (2019). Cystathionine-gamma-lyase (CSE) deficiency increases erythropoiesis and promotes mitochondrial electron transport via the upregulation of coproporphyrinogen III oxidase and consequent stimulation of heme biosynthesis. *Biochem. Pharmacol.* 169: 113604.

156 Pasch, A., Schaffner, T., Huynh-Do, U. et al. (2008). Sodium thiosulfate prevents vascular calcifications in uremic rats. *Kidney Int.* 74 (11): 1444–1453.

157 Gambari, L., Grigolo, B., and Grassi, F. (2019). Hydrogen sulfide in bone tissue regeneration and repair: state of the art and new perspectives. *Int. J. Mol. Sci.* 20 (20): 5231.

158 Vogt, I., Haffner, D., and Leifheit-Nestler, M. (2019). FGF23 and phosphate-cardiovascular toxins in CKD. *Toxins (Basel).* 11 (11): 647.

159 Shimada, T., Hasegawa, H., Yamazaki, Y. et al. (2004). FGF-23 is a potent regulator of vitamin D metabolism and phosphate homeostasis. *J. Bone Miner. Res.* 19 (3): 429–435.

160 Urakawa, I., Yamazaki, Y., Shimada, T. et al. (2006). Klotho converts canonical FGF receptor into a specific receptor for FGF23. *Nature* 444 (7120): 770–774.

161 Lu, X. and Hu, M.C. (2017). Klotho/FGF23 axis in chronic kidney disease and cardiovascular disease. *Kidney Dis. (Basel).* 3 (1): 15–23.

162 Go, A.S., Chertow, G.M., Fan, D. et al. (2004). Chronic kidney disease and the risks of death, cardiovascular events, and hospitalization. *N. Engl. J. Med.* 351 (13): 1296–1305.

163 Ward, Z.J., Bleich, S.N., Cradock, A.L. et al. (2019). Projected U.S. state-level prevalence of adult obesity and severe obesity. *N. Engl. J. Med.* 381 (25): 2440–2450.

164 Hashimoto, Y., Tanaka, M., Okada, H. et al. (2015). Metabolically healthy obesity and risk of incident CKD. *Clin. J. Am. Soc. Nephrol.* 10 (4): 578–583.

165 Choung, H.G., Bomback, A.S., Stokes, M.B. et al. (2019). The spectrum of kidney biopsy findings in patients with morbid obesity. *Kidney Int.* 95 (3): 647–654.

166 Liu, M., Deng, M., Su, J. et al. (2018). Specific downregulation of cystathionine beta-synthase expression in the kidney during obesity. *Phys. Rep.* 6 (13): e13630.

167 Wu, D., Gao, B., Li, M. et al. (2016). Hydrogen sulfide mitigates kidney injury in high fat diet-induced obese mice. *Oxid. Med. Cell Longev.* 2016: 2715718.

168 Decleves, A.E., Mathew, A.V., Cunard, R., and Sharma, K. (2011). AMPK mediates the initiation of kidney disease induced by a high-fat diet. *J. Am. Soc. Nephrol.* 22 (10): 1846–1855.

169 Sharma, K., Ramachandrarao, S., Qiu, G. et al. (2008). Adiponectin regulates albuminuria and podocyte function in mice. *J. Clin. Invest.* 118 (5): 1645–1656.

170 Ecelbarger, C.M., Rash, A., Sinha, R.K., and Tiwari, S. (2010). The effect of chronic candesartan therapy on the metabolic profile and renal tissue cytokine levels in the obese Zucker rat. *Mediators Inflamm.* 2010: 841343.

171 de Jong, P.E., Verhave, J.C., Pinto-Sietsma, S.J. et al. (2002). Obesity and target organ damage: the kidney. *Int. J. Obes. Relat. Metab. Disord.* 26 (Suppl 4): S21–S24.

172 Lee, H.J., Mariappan, M.M., Norton, L. et al. (2021). Proximal tubular epithelial insulin receptor mediates high-fat diet-induced kidney injury. *JCI Insight* 6 (3): e143619.

173 Lee, H.J., Feliers, D., Mariappan, M.M. et al. (2015). Tadalafil integrates nitric oxide-hydrogen sulfide signaling to inhibit high glucose-induced matrix protein synthesis in podocytes. *J. Biol. Chem.* 290 (19): 12014–12026.

174 Lee, H.J., Lee, D.Y., Mariappan, M.M. et al. (2017). Hydrogen sulfide inhibits high glucose-induced NADPH oxidase 4 expression and matrix increase by recruiting inducible nitric oxide synthase in kidney proximal tubular epithelial cells. *J. Biol. Chem.* 292 (14): 5665–5675.

175 van den Born, J.C., Hammes, H.P., Greffrath, W. et al. (2016). Gasotransmitters in vascular complications of diabetes. *Diabetes* 65 (2): 331–345.

176 Sen, U. and Pushpakumar, S. (2016). Mini-review: diabetic renal complications, a potential stinky remedy. *Am. J. Physiol. Renal. Physiol.* 310 (2): F119–F122.

177 Dugbartey, G.J. (2017). Diabetic nephropathy: a potential savior with 'rotten-egg' smell. *Pharmacol. Rep.* 69 (2): 331–339.

178 Sun, H.J., Wu, Z.Y., Cao, L. et al. (2019). Hydrogen sulfide: recent progression and perspectives for the treatment of diabetic nephropathy. *Molecules* 24 (15): 2857.

179 Yang, G., Tang, G., Zhang, L. et al. (2011). The pathogenic role of cystathionine gamma-lyase/hydrogen sulfide in streptozotocin-induced diabetes in mice. *Am. J. Pathol.* 179 (2): 869–879.

180 Xue, R., Hao, D.D., Sun, J.P. et al. (2013). Hydrogen sulfide treatment promotes glucose uptake by increasing insulin receptor sensitivity and ameliorates kidney lesions in type 2 diabetes. *Antioxid. Redox Signaling* 19 (1): 5–23.

181 Zhou, X., Feng, Y., Zhan, Z., and Chen, J. (2014). Hydrogen sulfide alleviates diabetic nephropathy in a streptozotocin-induced diabetic rat model. *J. Biol. Chem.* 289 (42): 28827–28834.

182 Xue, H., Yuan, P., Ni, J. et al. (2013). H(2)S inhibits hyperglycemia-induced intrarenal renin-angiotensin system activation via attenuation of reactive oxygen species generation. *PLoS One* 8 (9): e74366.

183 Qian, X., Li, X., Ma, F. et al. (2016). Novel hydrogen sulfide-releasing compound, S-propargyl-cysteine, prevents STZ-induced diabetic nephropathy. *Biochem. Biophys. Res. Commun.* 473 (4): 931–938.

184 Sun, H.J., Xiong, S.P., Cao, X. et al. (2021). Polysulfide-mediated sulfhydration of SIRT1 prevents diabetic nephropathy by suppressing phosphorylation and acetylation of p65 NF-kappaB and STAT3. *Redox Biol.* 38: 101813.

185 Jain, S.K., Bull, R., Rains, J.L. et al. (2010). Low levels of hydrogen sulfide in the blood of diabetes patients and streptozotocin-treated rats causes vascular inflammation? *Antioxid. Redox Signaling* 12 (11): 1333–1337.

186 Li, H., Feng, S.J., Zhang, G.Z., and Wang, S.X. (2014). Correlation of lower concentrations of hydrogen sulfide with atherosclerosis in chronic hemodialysis patients with diabetic nephropathy. *Blood Purif.* 38 (3-4): 188–194.

187 Kundu, S., Pushpakumar, S.B., Tyagi, A. et al. (2013). Hydrogen sulfide deficiency and diabetic renal remodeling: role of matrix metalloproteinase-9. *Am. J. Physiol. Endocrinol. Metab.* 304 (12): E1365–E1378.

188 John, A., Kundu, S., Pushpakumar, S. et al. (2017). GYY4137, a hydrogen sulfide donor modulates miR194-dependent collagen realignment in diabetic kidney. *Sci. Rep.* 7 (1): 10924.

189 Kundu, S., Pushpakumar, S., and Sen, U. (2015). MMP-9- and NMDA receptor-mediated mechanism of diabetic renovascular remodeling and kidney dysfunction: hydrogen sulfide is a key modulator. *Nitric Oxide* 46: 172–185.

190 Kimura, N., Kimura, H., Takahashi, N. et al. (2015). Renal resistive index correlates with peritubular capillary loss and arteriosclerosis in biopsy tissues from patients with chronic kidney disease. *Clin. Exp. Nephrol.* 19 (6): 1114–1119.

191 Yamamoto, J., Sato, W., Kosugi, T. et al. (2013). Distribution of hydrogen sulfide (H(2)S)-producing enzymes and the roles of the H(2)S donor sodium hydrosulfide in diabetic nephropathy. *Clin. Exp. Nephrol.* 17 (1): 32–40.

192 Papu John, A.S., Kundu, S., Pushpakumar, S. et al. (2019). Hydrogen sulfide inhibits Ca(2+)-induced mitochondrial permeability transition pore opening in type-1 diabetes. *Am. J. Physiol. Endocrinol. Metab.* 317 (2): E269–E283.

193 Kanwar, Y.S., Sun, L., Xie, P. et al. (2011). A glimpse of various pathogenetic mechanisms of diabetic nephropathy. *Annu. Rev. Pathol.* 6: 395–423.

194 Sharma, K. (2014). Obesity, oxidative stress, and fibrosis in chronic kidney disease. *Kidney Int. Suppl. (2011)* 4 (1): 113–117.

195 Senthil, D., Choudhury, G.G., McLaurin, C., and Kasinath, B.S. (2003). Vascular endothelial growth factor induces protein synthesis in renal epithelial cells: a potential role in diabetic nephropathy. *Kidney Int.* 64 (2): 468–479.

196 Feliers, D., Duraisamy, S., Barnes, J.L. et al. (2005). Translational regulation of vascular endothelial growth factor expression in renal epithelial cells by angiotensin II. *Am. J. Physiol. Renal. Physiol.* 288 (3): F521–F529.

197 Feliers, D., Gorin, Y., Ghosh-Choudhury, G. et al. (2006). Angiotensin II stimulation of VEGF mRNA translation requires production of reactive oxygen species. *Am. J. Physiol. Renal. Physiol.* 290 (4): F927–F936.

198 Feliers, D., Lee, M.J., Ghosh-Choudhury, G. et al. (2007). Heterogeneous nuclear ribonucleoprotein K contributes to angiotensin II stimulation of vascular endothelial growth factor mRNA translation. *Am. J. Physiol. Renal. Physiol.* 293 (2): F607–F615.

199 Senthil, D., Choudhury, G.G., Abboud, H.E. et al. (2002). Regulation of protein synthesis by IGF-I in proximal tubular epithelial cells. *Am. J. Physiol. Renal. Physiol.* 283 (6): F1226–F1236.

200 Chen, J., Chen, J.K., and Harris, R.C. (2015). EGF receptor deletion in podocytes attenuates diabetic nephropathy. *J. Am. Soc. Nephrol.* 26 (5): 1115–1125.

201 Feliers, D., Lee, H.J., and Kasinath, B.S. (2016). Hydrogen sulfide in renal physiology and disease. *Antioxid. Redox Signaling* 25 (13): 720–731.

202 Mariappan, M.M., D'Silva, K., Lee, M.J. et al. (2011). Ribosomal biogenesis induction by high glucose requires activation of upstream binding factor in kidney glomerular epithelial cells. *Am. J. Physiol. Renal. Physiol.* 300 (1): F219–F230.

203 Mariappan, M.M., Shetty, M., Sataranatarajan, K. et al. (2008). Glycogen synthase kinase 3beta is a novel regulator of high glucose- and high insulin-induced extracellular matrix protein synthesis in renal proximal tubular epithelial cells. *J Biol Chem.* 283 (45): 30566–30575.

204 Lloberas, N., Cruzado, J.M., Franquesa, M. et al. (2006). Mammalian target of rapamycin pathway blockade slows progression of diabetic kidney disease in rats. *J. Am. Soc. Nephrol.* 17 (5): 1395–1404.

205 Miyamoto, S., Hsu, C.C., Hamm, G. et al. (2016). Mass spectrometry imaging reveals elevated glomerular ATP/AMP in diabetes/obesity and identifies sphingomyelin as a possible mediator. *EBioMedicine* 7: 121–134.

206 Valentine, R.J., Coughlan, K.A., Ruderman, N.B., and Saha, A.K. (2014). Insulin inhibits AMPK activity and phosphorylates AMPK Ser(4)(8)(5)/(4)(9)(1) through Akt in hepatocytes, myotubes and incubated rat skeletal muscle. *Arch. Biochem. Biophys.* 562: 62–69.

207 Hawley, S.A., Ross, F.A., Gowans, G.J. et al. (2014). Phosphorylation by Akt within the ST loop of AMPK-alpha1 down-regulates its activation in tumour cells. *Biochem. J* 459 (2): 275–287.

208 Dugan, L.L., You, Y.H., Ali, S.S. et al. (2013). AMPK dysregulation promotes diabetes-related reduction of superoxide and mitochondrial function. *J. Clin. Invest.* 123 (11): 4888–4899.

209 Wilinski, B., Wilinski, J., Somogyi, E. et al. (2013). Metformin raises hydrogen sulfide tissue concentrations in various mouse organs. *Pharmacol. Rep.* 65 (3): 737–742.

210 Scheele, W., Diamond, S., Gale, J. et al. (2016). Phosphodiesterase type 5 inhibition reduces albuminuria in subjects with overt diabetic nephropathy. *J. Am. Soc. Nephrol.* 27 (11): 3459–3468.

211 Jha, J.C., Banal, C., Chow, B.S. et al. (2016). Diabetes and kidney disease: role of oxidative stress. *Antioxid. Redox Signaling* 25 (12): 657–684.

212 Jha, J.C., Ho, F., Dan, C., and Jandeleit-Dahm, K. (2018). A causal link between oxidative stress and inflammation in cardiovascular and renal complications of diabetes. *Clin. Sci. (Lond).* 132 (16): 1811–1836.

213 Shevalye, H., Maksimchyk, Y., Watcho, P., and Obrosova, I.G. (2010). Poly(ADP-ribose) polymerase-1 (PARP-1) gene deficiency alleviates diabetic kidney disease. *Biochim. Biophys. Acta* 1802 (11): 1020–1027.

214 Gorin, Y., Cavaglieri, R.C., Khazim, K. et al. (2015). Targeting NADPH oxidase with a novel dual Nox1/Nox4 inhibitor attenuates renal pathology in type 1 diabetes. *Am. J. Physiol. Renal. Physiol.* 308 (11): F1276–F1287.

215 Jha, J.C., Gray, S.P., Barit, D. et al. (2014). Genetic targeting or pharmacologic inhibition of NADPH oxidase nox4 provides renoprotection in long-term diabetic nephropathy. *J. Am. Soc. Nephrol.* 25 (6): 1237–1254.

216 Sedeek, M., Gutsol, A., Montezano, A.C. et al. (2013). Renoprotective effects of a novel Nox1/4 inhibitor in a mouse model of Type 2 diabetes. *Clin. Sci. (Lond).* 124 (3): 191–202.

217 Thallas-Bonke, V., Jha, J.C., Gray, S.P. et al. (2014). Nox-4 deletion reduces oxidative stress and injury by PKC-alpha-associated mechanisms in diabetic nephropathy. *Phys. Rep.* 2 (11): e12192.

218 Tesch, G.H. (2017). Diabetic nephropathy – is this an immune disorder? *Clin. Sci. (Lond).* 131 (16): 2183–2199.

219 van den Born, J.C., Frenay, A.R., Bakker, S.J. et al. (2016). High urinary sulfate concentration is associated with reduced risk of renal disease progression in type 2 diabetes. *Nitric Oxide* 55-56: 18–24.

220 Shen, X., Carlstrom, M., Borniquel, S. et al. (2013). Microbial regulation of host hydrogen sulfide bioavailability and metabolism. *Free Radical Biol. Med.* 60: 195–200.

221 Huang, Q., Sparatore, A., Del Soldato, P. et al. (2014). Hydrogen sulfide releasing aspirin, ACS14, attenuates high glucose-induced increased methylglyoxal and oxidative stress in cultured vascular smooth muscle cells. *PLoS One* 9 (6): e97315.

222 Zhou, H., Ding, L., Wu, Z. et al. (2017). Hydrogen sulfide reduces RAGE toxicity through inhibition of its dimer formation. *Free Radical Biol. Med.* 104: 262–271.

223 Rangaswami, J., Bhalla, V., Blair, J.E.A. et al. (2019). Cardiorenal syndrome: classification, pathophysiology, diagnosis, and treatment strategies: a scientific statement from the american heart association. *Circulation* 139 (16): e840–e878.

224 Li, Z., Organ, C.L., Kang, J. et al. (2018). Hydrogen sulfide attenuates renin angiotensin and aldosterone pathological signaling to preserve kidney function and improve exercise tolerance in heart failure. *JACC Basic Transl. Sci.* 3 (6): 796–809.

225 Hernandez-Diaz, S., Toh, S., and Cnattingius, S. (2009). Risk of pre-eclampsia in first and subsequent pregnancies: prospective cohort study. *BMJ* 338: b2255.

226 Maynard, S.E., Min, J.Y., Merchan, J. et al. (2003). Excess placental soluble fms-like tyrosine kinase 1 (sFlt1) may contribute to endothelial dysfunction, hypertension, and proteinuria in preeclampsia. *J. Clin. Invest.* 111 (5): 649–658.

227 Zeisler, H., Llurba, E., Chantraine, F. et al. (2016). Predictive value of the sFlt-1:PlGF ratio in women with suspected preeclampsia. *N. Engl. J. Med.* 374 (1): 13–22.

228 Mrozikiewicz, P.M., Bogacz, A., Omielanczyk, M. et al. (2015). The importance of rs1021737 and rs482843 polymorphisms of cystathionine gamma-lyase in the etiology of preeclampsia in the Caucasian population. *Ginekol. Pol.* 86 (2): 119–125.

229 Holwerda, K.M., Weedon-Fekjaer, M.S., Staff, A.C. et al. (2016). The association of single nucleotide polymorphisms of the maternal cystathionine-beta-synthase gene with early-onset preeclampsia. *Pregnancy Hypertens.* 6 (1): 60–65.

230 Holwerda, K.M., Burke, S.D., Faas, M.M. et al. (2014). Hydrogen sulfide attenuates sFlt1-induced hypertension and renal damage by upregulating vascular endothelial growth factor. *J. Am. Soc. Nephrol.* 25 (4): 717–725.

231 Wang, K., Ahmad, S., Cai, M. et al. (2013). Dysregulation of hydrogen sulfide producing enzyme cystathionine gamma-lyase contributes to maternal hypertension and placental abnormalities in preeclampsia. *Circulation* 127 (25): 2514–2522.

232 Hu, T.X., Guo, X., Wang, G. et al. (2017). MiR133b is involved in endogenous hydrogen sulfide suppression of sFlt-1 production in human placenta. *Placenta* 52: 33–40.

233 Covarrubias, A.E., Lecarpentier, E., Lo, A. et al. (2019). AP39, a modulator of mitochondrial bioenergetics, reduces antiangiogenic response and oxidative stress in hypoxia-exposed trophoblasts: relevance for preeclampsia pathogenesis. *Am. J. Pathol.* 189 (1): 104–114.

234 Turbeville, H.R. and Sasser, J.M. (2020). Preeclampsia beyond pregnancy: long-term consequences for mother and child. *Am. J. Physiol. Renal. Physiol.* 318 (6): F1315–F1326.

235 Rini, B.I., Campbell, S.C., and Escudier, B. (2009). Renal cell carcinoma. *Lancet* 373 (9669): 1119–1132.

236 Akbari, M., Sogutdelen, E., Juriasingani, S., and Sener, A. (2019). Hydrogen sulfide: emerging role in bladder, kidney, and prostate malignancies. *Oxid. Med. Cell Longev.* 2019: 2360945.

237 Shackelford, R.E., Abdulsattar, J., Wei, E.X. et al. (2017). Increased nicotinamide phosphoribosyltransferase and cystathionine-beta-synthase in renal oncocytomas, renal urothelial carcinoma, and renal clear cell carcinoma. *Anticancer Res.* 37 (7): 3423–3427.

238 Breza, J. Jr., Soltysova, A., Hudecova, S. et al. (2018). Endogenous H_2S producing enzymes are involved in apoptosis induction in clear cell renal cell carcinoma. *BMC Cancer* 18 (1): 591.

239 Li, H., Xu, F., Gao, G. et al. (2020). Hydrogen sulfide and its donors: novel antitumor and antimetastatic therapies for triple-negative breast cancer. *Redox Biol.* 34: 101564.

240 Szabo, C. (2016). Gasotransmitters in cancer: from pathophysiology to experimental therapy. *Nat. Rev. Drug Discovery* 15 (3): 185–203.

Index

C

d

q

r